A Guided Tour of the Living Cell

A GUIDED TOUR OF THE LIVING CELL

Christian de Duve

Illustrated by Neil O. Hardy

This book is published in collaboration with
The Rockefeller University Press

Scientific American Books, Inc.
New York

Library of Congress Cataloging in Publication Data

De Duve, Christian.
A guided tour of the living cell.

"This book is published in collaboration with the Rockefeller University Press."
Includes index.
1. Cytology. I. Hardy, Neil O. II. Title.
QH581.2.D43 1985 574.87 84-29818
ISBN 0-7167-6002-9 (pbk.)

Printed in the United States of America

Book design by Malcolm Grear Designers

Distributed by W. H. Freeman and Company,
41 Madison Avenue, New York, New York 10010.

1 2 3 4 5 6 7 8 9 KP 3 2 1 0 8 9 8 7 6 5

For Thierry, Anne, Françoise, and Alain

To their mother

What wonders would he discover, who could so fit his eyes to all sorts of objects, as to see, when he pleased, the figure and motion of the minute particles in the blood, and other juices of animals, as distinctly as he does, at other times, the shape and motion of the animals themselves!

JOHN LOCKE,

AN ESSAY CONCERNING HUMAN UNDERSTANDING

Contents

Preface

Every year at Christmas time, The Rockefeller University in New York invites some 550 selected high-school students to a series of four lectures by one of its professors. In 1976, my turn came to deliver the Alfred E. Mirsky Christmas Lectures on Science, as they are now called in memory of the distinguished biologist—an expert on the cell nucleus—who founded them in 1959.

To give such lectures is both a rewarding experience and a demanding challenge. One is faced with an audience that combines in a unique way the eager receptiveness of youth with a daunting degree of sophisticated knowledge in certain areas. Reflecting on how to give these youngsters, who probably knew all about DNA but little about other cell components, some sort of balanced view of cellular organization, I came upon the idea of taking them on a tour. We would shrink to the size of bacteria or, alternatively, blow up the cell a millionfold, which amounts to the same thing. We could then conveniently walk—or, rather, swim—around, look at the different cell parts to observe their structure, and watch them in action to understand their function. Once I got into the spirit of the game, I found it most enjoyable. And the audience was appreciative.

What started more than seven years ago as a four-hour fantasy has become an unwieldy nineteen-chapter opus. Writing it, like preparing the Christmas lectures, has been great fun. So much so, in fact, that the reader came to be somewhat forgotten in the process. The Tour is not a textbook organized so as to offer students an equitable coverage of a topic pitched at their required level of comprehension. It is even less the kind of critically documented scholarly volume that is written for the benefit of experts. But neither is it a popularization work of the sort that covers difficulty with a deceptive cloak of simplicity. It is just what the title says—a guided tour—with all that such an appellation implies in arbitrariness, including the exasperating privilege that tourist guides arrogate to themselves to linger in certain places in order to tell a story or propound some private opinion, only to rush through the next part of the tour to make up for lost time. It is a sharing of a very personal view, gained in decades of roaming through what Albert Claude has called the mansion of our birth and musing over its wonders, a view unavoidably colored, therefore, by preference, prejudice, and familiarity (or the lack of it). As it is written, it is not specifically directed at the student, the expert, or the layman. My hope is that each may find in it some parts or aspects that will make up for its shortcomings as a whole.

From the original "guided tour," I have retained the idea of reducing the observer or expanding the cell a millionfold, thereby making the cell components visible to the naked eye, aided solely, in certain cases, by "molecular eyeglasses." To readers who object to such a device in a serious scientific context and who resent being turned into minuscule "cytonauts," I offer my sympathy but no excuses. There is nothing in the nature of science that demands it to be solemn. There is, however, a danger in metaphor as a vehicle for concepts not easily defined in ordinary language, and even more so in an anthropomorphic description of processes such as those that occur in living cells, which we too readily are tempted to endow with our own brand of purposiveness. I hope the playful imagery adopted in parts of this tour will not prove misleading.

Cells contain a wealth of intricate structures. To try to make these structures visible by magnification is perfectly legitimate. We do not do otherwise with our optical instruments and electron microscopes. Even molecules have an anatomy, and a good part of modern biochemistry is concerned with their faithful three-dimensional representation. In the tour, a special effort has been made, thanks to the untiring and enthusiastic collaboration of Neil Hardy, to depict these structures as accurately as possible and at scale, within the limits imposed by the available information and by the demands for readability. In a number of cases, unfortunately, imagination had to be enlisted, and controversy had to be muted. In spite of the "biological revolution," knowledge of the cell remains fragmentary. It is also advancing rapidly, and the best one can do is to capture as late a stage as possible, but hardly the final one, in this progression. For factual illustration, I have tried to assemble a collection of pictures that have some historical interest, to the point of sometimes choosing an old master over a technically superior, more recent document. But I have also enlisted the collaboration of a few "young masters."

Anatomy, even animated by descriptions of moving parts, rarely reveals function and hardly ever elucidates mechanisms. In the living cell, this can be done only with the help of biochemistry. Not wishing to burden the reader with arrays of complex molecules and strings of intricate reactions, yet loath to resort to the kind of superficial and evasive language often used as a substitute, I have tried to focus on what I see as the main dynamic lineaments of life, stripped of the many individual shapes and forms that clothe and, to some extent, hide them. My guide in this attempt has been energetics, rather than organic chemistry. This may not be to the taste of those more interested in real mechanisms than in abstract concepts, especially since I have gone so far as to invent some rather outlandish terms, such as "oxphos unit" and "Janus intermediate," to avoid entering into specifics. I can only ask those who might be antagonized by this treatment to overcome, or at least to delay, their reaction and bear with me for a while. They may discover that this sort of analysis offers a certain global vision of the metabolic forest that does not demand a knowledge of any individual trees. However, I realize that I may have succumbed to a Gallic tendency that does not export well to other shores. To help the reader wade through the more biochemical parts of the book, I have collected the most important notions of descriptive biochemistry and bioenergetics in two appendices.

A work such as this does not see the light of day without much outside help and support. My first thanks go to the 550-odd high-school students whom I took on the original "guided tour of the living cell" on December 27–28, 1976. Their enthusiastic response and eager questions have been a major incentive to me. Yet I would probably not have started what turned out to be a major undertaking without the kind insistence of Bill Bayless and his colleagues of The Rockefeller University Press and without the encouraging interest of Jaime Etcheverry, of Buenos Aires, in the recordings of the lectures.

Two persons have proved invaluable collaborators, not only by the quality and importance of their contributions, but also because, with them, what is often drudgery, sometimes even agony, turned into genuine fun. I have already mentioned Neil Hardy, a gifted artist, whose professional conscience led him to learn more biochemistry than I could ever teach him and whose good nature I ruthlessly exploited in my groping search for clarity and intelligibility. The other is Helene Jordan Waddell, former director of The Rockefeller University Press, who tirelessly went through innumerable versions of the manuscript, weeding out faulty constructions, stylistic abominations, and other monstrosities with uncompromising firmness, yet always maintaining an unobtrusive and probably often mistaken respect of the author's idiosyncrasies. Both have become great friends. In later stages, I have received much helpful and sympathetic support from the publishers at Scientific American Books: Neil Patterson, Linda Chaput, and in particular Patty Mittelstadt, who has carried out with unfailing devotion, strict professional competence, and admirable forbearance the arduous task of giving the book its final polishing and putting it together for production under what at times proved rather stressful conditions.

Many friends and colleagues have responded generously to my sometimes intemperate requests for illustration material. They include Pierre Baudhuin, Wolfgang Beermann, Marcel Bessis, Robert Bloodgood, Daniel Branton, Ralph Brinster, John Cairns, Pierre Chambon, David Chase, Isabelle Coppens, Richard Dickerson, David Dressler, Marilyn Farquhar, Walter Fiers, Brian Ford, Werner Franke, Yukio Fujiki, Joseph Gall, Ian Gibbons, Jerome Gross, Pierre Guiot, Françoise Haguenau, Etienne de Harven, John Heuser, James Hirsch, Hans-Peter Hoffman, David Hogness, Hugh Huxley, James Jamieson, Morris Karnovsky, John Kendrew, Richard Kessel, Ulrich Laemmli, Emmanuel Margoliash, Arvid Maunsbach, Oscar Miller, Eldon Newcomb, Alex and Phyllis Novikoff, George Palade, Donald Parsons, Keith Porter, Evans Roth, Helen Shio, Samuel Silverstein, Sidney Tamm, Herman Van den Berghe, Marten Veenhuis, Leo Vernon, Eugene Vigil, Luc Waterkeyn, James Watson, Maurice Wilkins, and Heinz Günther Wittmann. I thank them all most heartily, as I do also the Arp Foundation in Paris, the late Buckminster Fuller, Dan Dixon, Karl Koopman, Henry Moore, Enrico di Rovasenda, the Metropolitan Museum of Art and the Museum of Modern Art of New York, and the Vatican Museum, for their artistic contributions.

Finally, I owe a special debt of gratitude to what, in retrospect, looks like an army of secretaries endlessly committing new drafts to the memories of their word processors. I cannot name them all, but I must mention Norma Musiek and Patricia Lahy in Brussels and especially Karrie Polowetzky, who, in New York, may be said to have seen it all and, thanks to a happy blend of shockproof resilience, unflappable equanimity, and ever-obliging dedication, survived.

My thanks also go to Miklós Müller, who read the whole manuscript and made many helpful suggestions, to Jacques Berthet, who looked over the appendix on bioenergetics, and to a number of anonymous reviewers who, often in no uncertain terms, pointed out mistakes and inaccuracies, as well as what they considered objectionable terms or modes of presentation. I have tried to correct all factual errors as much as possible. But I have not always followed my critics' advice on style and terminology. For all the many blemishes, minor and major, that remain, I must claim sole responsibility.

To this long list of creditors I must finally add my wife, my children, my friends, and my colleagues on both sides of the Atlantic, who have had to suffer with increasing patience the growing and seemingly never-ending demands that "the book" came to make on my time, my attention, and my temper. There is clearly no way for me to make up for all these cumulative derelictions. If, at least, I have succeeded in conveying some of the feeling of wonder and reverence that the living cell inspires in those who are fortunate enough to busy themselves with its exploration, and also some of the joy and excitement that the explorers derive from their expeditions, I will consider my own time and efforts not entirely wasted.

CHRISTIAN de DUVE
New York and Nethen
April, 1984

A Guided Tour of the Living Cell

1 Preparing for the Tour

We are about to enter a strange world, fascinating, mysterious, but very far removed from our everyday experience. It is a world that exists in each of us, multiplied more than 10,000 billion times, and in every other living being. All are made of one or more units of microscopic size called cells, each of which is capable of leading an independent life if given a suitable environment. In touring the cell, we will in fact be looking at life itself, in its most elementary and basic form. Before we start, some briefing is called for.

The World of Cells

To begin with, let it be clear that there is no such thing as *the* living cell. There are only living cells, innumerable varieties of them. Should we look simply at their more obvious features—such as size, shape, pattern of movement, and other external manifestations—we would find them so diverse that the fundamental kinship between all cells might well escape us, as it escaped the early microscopists for more than 150 years. But when we probe deeper, as we will in this tour, unity reveals itself. By the time we reach the world of submicroscopic structures and of molecular functions, the differences between cells are largely erased. When we speak of the living cell as the object of our tour, we mean some sort of common denominator of all living cells that shares with them the main attributes of life.

Much of our tour will be conducted through this composite entity, but we will make reference also to real cells.

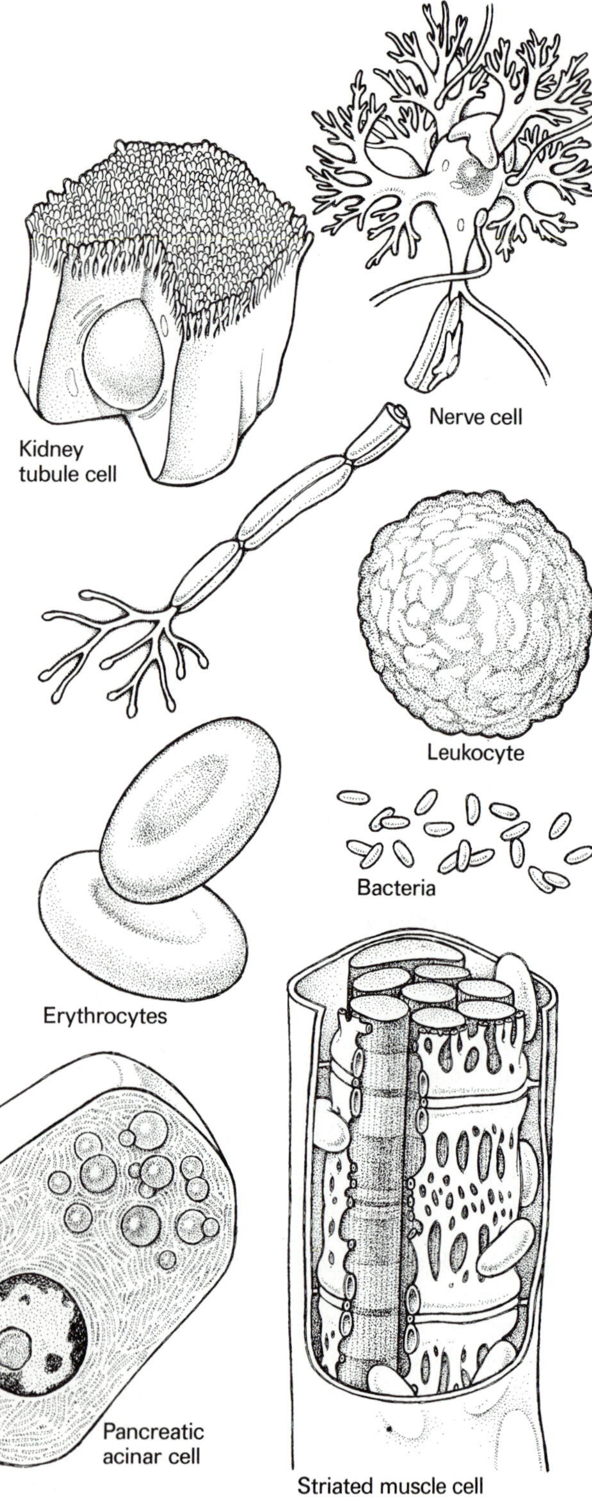

The diversity of living cells.

It is therefore necessary, before we start, to get some idea of the different kinds of cells that can be encountered.

Our own bodies are composed of some few hundred cell types, each represented by up to thousands of billions of individuals. They are the nerve cells, the muscle cells, the gland cells, the blood cells, and so on, of which in each case a number of different subtypes exist. As these names indicate, cells of a given type tend to be grouped into organs, or systems, to perform specific functions. The manner in which they are grouped often plays a decisive role in the expression of these functions. For instance, all striated muscle cells look alike and share the ability to contract. How they are associated makes the difference between the hundreds of different muscles found in the human body. Association patterns between cells reach their highest degree of complexity in the central nervous system, which is made up of tens of billions of cells, each of which may be connected with as many as ten thousand of its congeners. Such connections are established by cells that somehow seek and recognize each other, and then join together. Their associations are stabilized and supported by a variety of extracellular structural elements, which are mainly responsible for maintaining the characteristic architecture of each tissue.

Our close cousins, the other mammals, are constructed very much like us, with essentially the same types of cells. In fact, similar cells are encountered far down the animal scale. Typical muscle or nerve cells are found in fishes, in insects, in mollusks, in worms, but their arrangements become progressively simpler. Down at the level of the lower invertebrates, such as the sponges, the pattern itself begins to change—from that of a multicellular organism to that of a colony of semi-independent cells. At the bottom of the ladder are the fully independent protozoa, such as amoebae, which consist of single cells.

Plants also are made up of different cell types held together by structural elements. But the organization of plants is different from that of animals. They are built entirely around a solar economy conditioned by the occurrence in their cells of a special kind of light-powered factory, the green chloroplasts. Take these away, and you are left with something much like an animal cell. As with animals, there are different degrees of complexity in the organization of plants, from the highly complex flowering plants and trees, down to the lowly unicellular algae. Their nonphotosynthetic relatives, the fungi, display a similar range of complexity, descending from mushrooms to molds and yeasts.

All these cells, which make up the animal and the plant kingdoms, are constructed according to the same general blueprint. In particular, their bodies house a voluminous central structure of characteristic shape, called a nucleus, and are subdivided into many distinct compartments by membranous partitions. Such cells are called eukaryotes, which in Greek means that they have a good (*eu*) nucleus (*karyon*). They are the cells we plan to visit, remaining mostly in our own animal kingdom, with an occasional excursion into the plant world.

Next to the eukaryotes, there is a simpler form of living cells, called prokaryotes because they have only a primitive sort of nucleus. The prokaryotes are the bacteria. They are much smaller than eukaryotes, live singly or in very crude colonies, and show little internal organization. Yet, they occur in an enormous variety of species and have succeeded in invading the most inhospitable of environments, including the steaming ponds of hot springs and the almost solid brine of drying seas. There are bacteria everywhere and they carry out many of the essential mechanisms whereby the constituents of dead organisms are recycled back into forms that can again sustain life. Without bacteria, eukaryotic life would soon become extinct. Some bacteria are harmful, however, through their ability to invade higher organisms and cause diseases.

Finally, if we go one order of magnitude lower again, we encounter the viruses. These are no longer considered cells, as they do not have the capacity for independent life. Nevertheless, they possess one key property of life, which is the ability to supply the instructions for their own reproduction when provided with the required machinery. This they find by the simple device of entering an authentic cell, either eukaryote or prokaryote. Once inside, they appropriate the cell's copying devices and are thereby multiplied manifold, most often causing the invaded cell to degenerate and die. We will have more to say about the viruses when visiting the main information and duplication centers of the cell.

The World of Molecules

If we wish to understand how cells are constructed and how they function, we must use the language of chemistry. We must even use a highly sophisticated form of this language, in view of the exceptional degree of chemical wizardry displayed by living cells. Reflecting this complexity and the progress in our understanding of it, the science of biochemistry has developed enormously in recent years.

Not all tourists can be scholars, however. And it would be a great pity if the beauty and fascination of the world of cells should be reserved to that small minority of cognoscenti who are familiar with the world of biomolecules. Certainly, we would like to take along as many people as we possibly can, and every effort will be made to facilitate their participation. But some chemistry is going to be needed. Without it, much of the tour will be wasted.

It will be assumed, therefore, that everybody has some familiarity with the concepts and laws of chemistry. When possible, images and models will be used to convey key notions. Scientific rigor and accuracy will not, however, be sacrificed to the requirement for simplicity. A more systematic survey of the main constituents of living cells and of their chemical structures is given in Appendix 1. Essential physicochemical concepts are recalled in Appendix 2.

It is hoped that the tour will be accessible to many in this way. Even so, all of you are urged to delve deeper into the world of molecules, since it will heighten your enjoyment of the world of cells. You need not be an Egyptologist to enjoy the pyramids and the treasures of King Tut. But the more you know of the history of these famous objects, the more benefit and pleasure you derive from seeing them.

The Problem of Size

Cells are measured in micrometers (1 μm = one-millionth of a meter), molecules in nanometers (1 nm = one-billionth of a meter). Such small dimensions are very difficult to visualize. Take an average eukaryotic cell, for example. Roughly spherical in shape, it has a diameter of about 25 μm, or one-thousandth of an inch, which means that one billion cells fit snugly within one cubic inch. Bacteria are about 1 μm in diameter, which means that more than 10,000 can fit into a single eukaryotic cell. Many viruses are so small that thousands can be accommodated in a single bacterial cell, or tens of millions of billions within one cubic inch. Imagination refuses to follow.

In our tour, we will solve this difficulty by shrinking to the size of bacteria, say by one million times in each of the three dimensions. This is equivalent to saying that we remain as we are and blow up everything around us one million times. Magnified in this way, the earth would reach far beyond the present location of the sun, to the point where it would take more than 18 hours for a ray of light to travel from one pole to another, and the cell would grow to the convenient size of a large auditorium. We can now stop by any part that catches our fancy and discern every individual detail, down to single molecules.

The Fourth Dimension

Biology, like geology and cosmology, also concerns itself with historical events. Its objects cover a span of several billion years. This fourth dimension became evident some 200 years ago, when fossils were discovered and recognized for what they are: not, as some would have had it, the remains of victims of the Flood or the creation of a playful deity that planted a few dead species among the living ones, but the bones and shells of animals long extinct, the petrified imprints of plants that flourished eons ago. As geological dating progressed, a historical pattern began to reveal itself: the more ancient

The history of living forms (the fourth dimension).

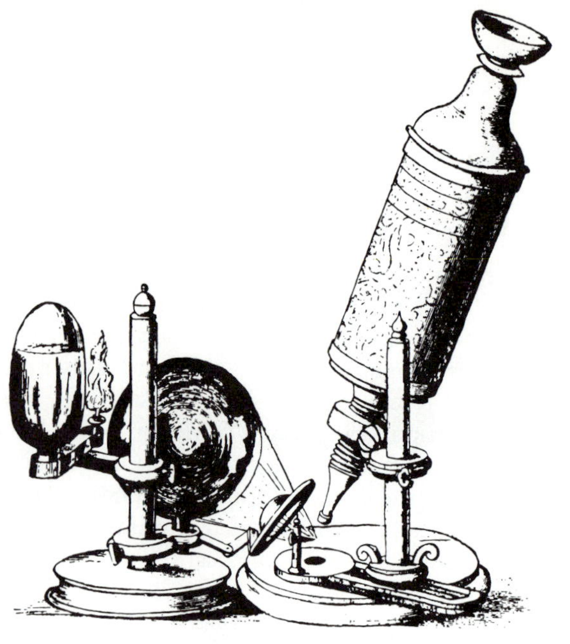

Robert Hooke's microscope, together with the oil lamp that he used for illumination, as drawn by the inventor in his *Micrographia.*

Historical drawing by Robert Hooke of the microscopic structure of a thin slice of cork, illustrating the small cavities that Hooke referred to as cells.

the fossil remnant, the more rudimentary its degree of organization. Mollusks went back farther than fishes, which themselves antedated reptiles. Birds and mammals came later, to be succeeded finally by the early humanoids. Out of this pattern, the concept of evolution emerged during the first half of the nineteenth century, culminating in the publication in 1858 of Darwin's seminal work, *On the Origin of Species by Means of Natural Selection*.

Although the fossil record holds few clues to the evolution of cells, recent advances in biochemistry and molecular biology have provided powerful new means of reconstructing the past by probing the present. The discoveries made in this way have generated considerable excitement, and the fourth dimension has invaded cell biology, permeating our concepts of the living cell and of its contents. We cannot ignore it on a tour like this. Occasionally, as we stop for a break, we will reflect on the origin and evolutionary history of what we see.

Tools and Their Development

Hardly more than 300 years have elapsed between the day when a living cell was first glimpsed and the present era of massive tourist invasion and media popularization. Every milestone on the way that led the early explorers deeper into the cell bears the name of a new tool or instrument. The main steps of this progression are worth recalling.

The Morphological Approach

The world of cells is invisible to the unaided eye. It remained entirely unknown and unexplored until the middle of the seventeenth century, when men of an inquisitive mind served by skilled hands started grinding lenses and using them to extend their power of vision. One of the first makers of microscopes was the Englishman Robert Hooke—physicist, meteorologist, biologist, engineer, architect—a most remarkable product of his

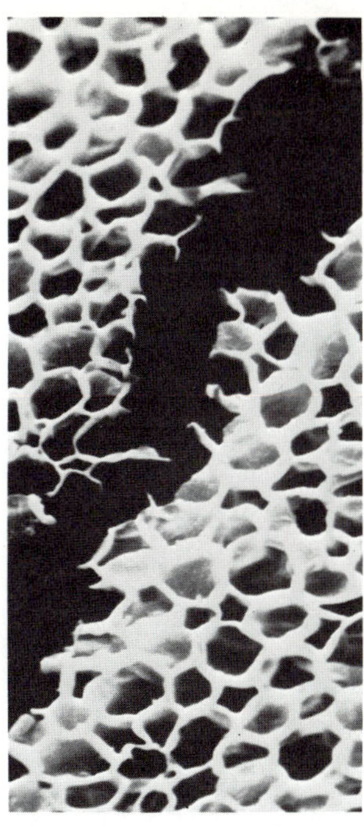

This picture was taken by the British biologist Brian J. Ford, at Cardiff University, by means of a modern instrument known as the scanning electron microscope, which emphasizes the relief of objects. It shows a section of cork similar to, and contemporary with, that drawn by Robert Hooke. The section was cut by Antonie van Leeuwenhoek in 1674 and discovered by Ford in 1981 in a file of Leeuwenhoek's letters to the Royal Society of London (of which Hooke was the secretary).

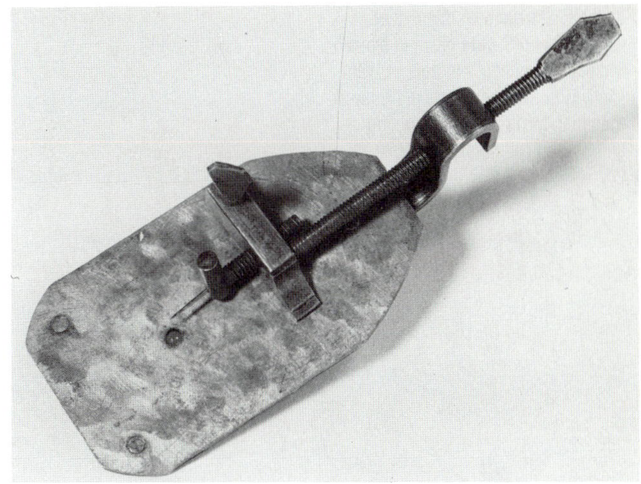

A replica of Leeuwenhoek's microscope. Objects were impaled on a needle and examined through a glass bead inserted in a copper plate.

time. In 1665, he published a beautiful collection of drawings called *Micrographia*, describing his microscopic observations; among them was that of a thin slice of cork showing a honeycomb structure, a regular array of "microscopic pores" or "cells." In this description Hooke used the word cell in its original meaning of small chamber, as in the cell of a prisoner or a monk. The word has remained, not to describe the little holes that Hooke saw in dead bark, but rather to designate the little blobs of matter that are the inmates of the holes in the living tree.

One of Hooke's most gifted contemporaries was the Dutchman Antonie van Leeuwenhoek, who made more than two hundred microscopes of a very special design. They consisted simply of a small bead of glass inserted in a copper plate. By holding this contraption close to his eye and peering through the glass bead at objects held on a needle that he could manipulate with a screw, Leeuwenhoek succeeded in obtaining up to 270-fold magnifications and made remarkable findings. He was able to see for the first time what he called "animalcules" in blood, sperm, and the water of marshes and ponds. Amazingly, he even saw bacteria, which he drew so accurately that specialists can identify them today.

Not all early users of microscopes were as perceptive. Especially when it came to objects as small as living cells, the images they were able to observe with their simple instruments were so blurred that most details had to be filled in by the imagination. Many showed admirable restraint in the use of this faculty. Others took full advantage of it, to the point of distinguishing, as did the Frenchman Gautier d'Agosty, an enthusiastic adept of the "preformist" theory, a fully formed baby within the head of a sperm cell.

And so, for a long time, microscopy did little more than hover around the world of cells until, in 1827, the Italian physicist Giovanni Battista Amici succeeded in correcting the major optical aberrations of lenses. The increase in the sharpness of the images was dramatic; so

This is one of the plates illustrating Theodor Schwann's *Mikroskopische Untersuchungen*, published in 1839. It is shown in support of the theory that plants and animals are made of similar cells. Figures 1, 2, 3, and 14 are from plants, the others from animals. Schwann mentions that Figures 2 and 3 were given to him by Mathias Schleiden, the botanist who jointly with Schwann is generally given the credit for first enunciating the generalized cell theory.

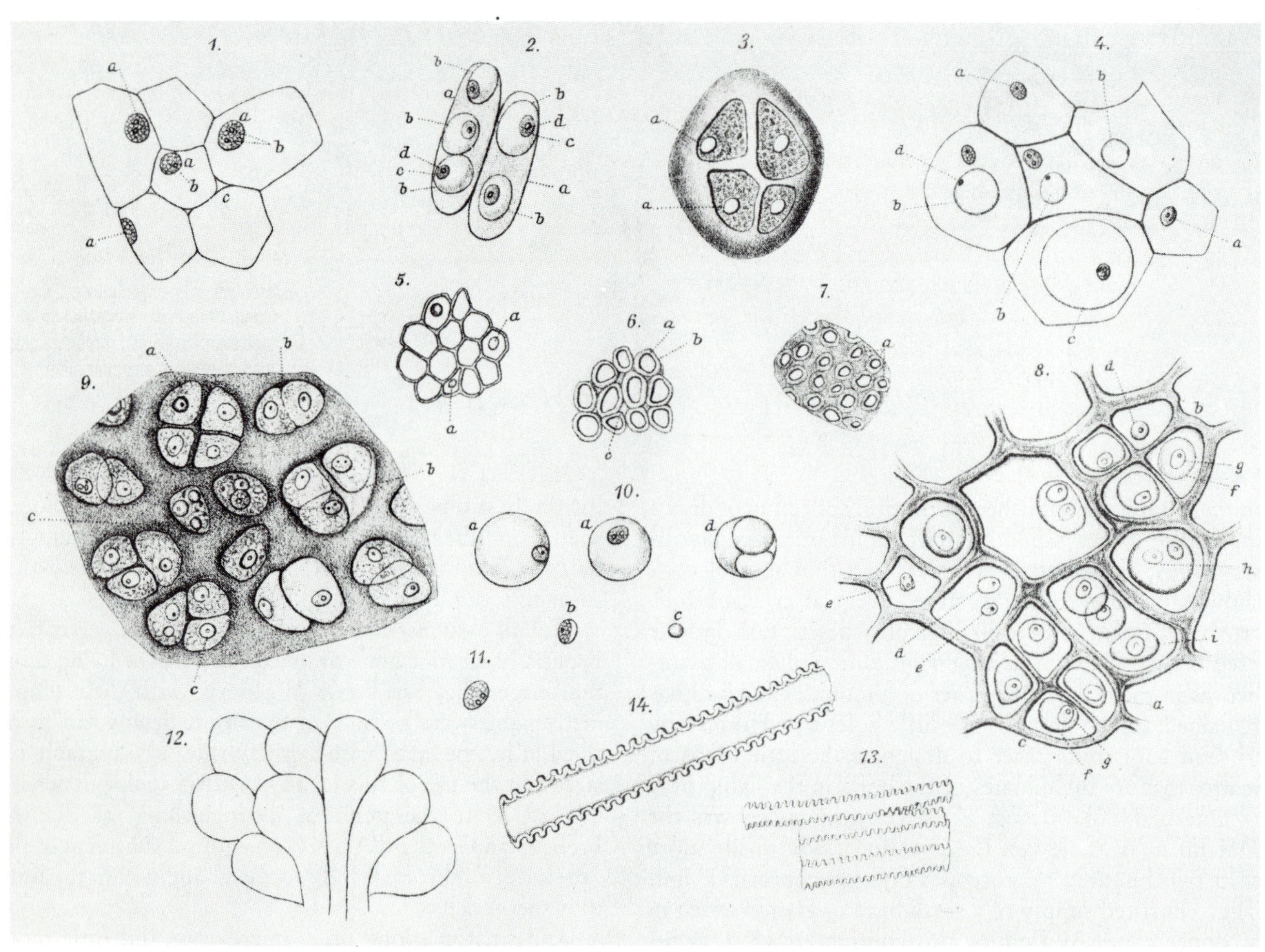

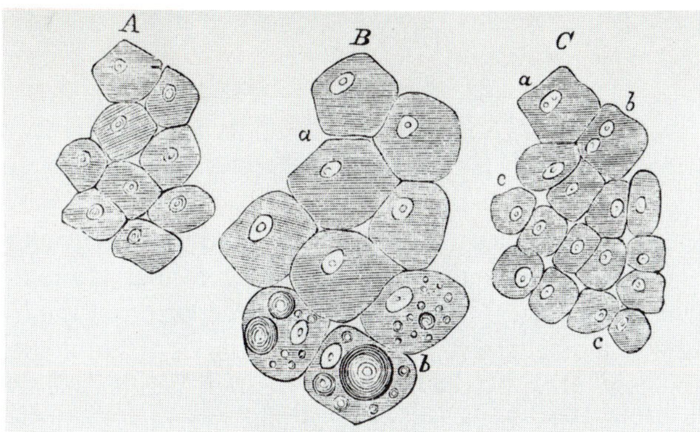

One of the 144 engravings illustrating *Cellularpathologie*, which was published in 1858 by Rudolf Virchow, who first recognized clearly that cells originate from cells, not by crystallization from some amorphous plasma, as was believed by Schwann.

much so that only a few years later the generalized theory could be formulated that all plants and animals are made of one or more similar units, the cells. This theory was proposed for plants in 1837 by the German botanist Mathias Schleiden, and was extended to animals by his friend, the physiologist Theodor Schwann. It was subsequently completed by the pathologist Rudolf Virchow, when he proclaimed in 1855: "*omnis cellula e cellula,*" every cell arises from a cell—actually a paraphrase of "*omne vivum ex ovo,*" every living being arises from an egg, an affirmation made by William Harvey, the English physician who discovered blood circulation and who died just a few years before Robert Hooke's discovery. Virchow also championed the extension of the cell theory to pathology, as witnessed by the title of his *Cellularpathologie*, published in 1858. By the middle of the nineteenth century, the cell theory was firmly established, and the science of cells, or cytology (Greek *kytos*, cavity), started to flourish. The first journal devoted exclusively to cell biology was started in 1884 by Jean-Baptiste Carnoy, at the Catholic University of Louvain, Belgium, under the name *La Cellule*. By the turn of the century, a number of important cell parts had been described and named.

Progressively, however, investigators ran into a new obstacle, apparently insurmountable, as it was set by the very laws of physics. Even with a perfect instrument, no detail smaller than about half the wavelength of the light used can be perceived, which puts the absolute limit of resolution of a microscope utilizing visible light at about 0.25 μm. In the world of cells, such a dimension is quite large, relatively speaking. Just think of what we would miss of our own world if no detail smaller than ten inches could be distinguished. That is all the classical microscop-

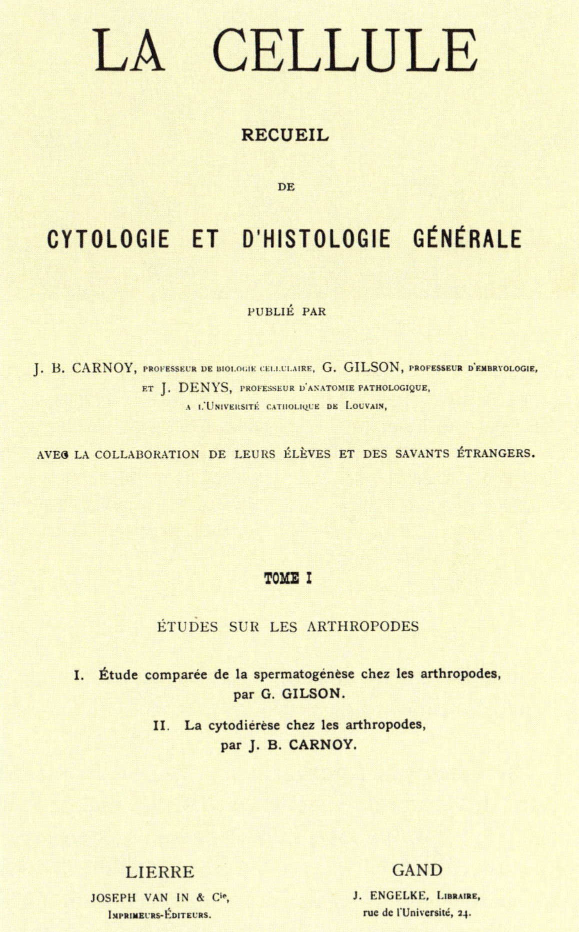

LA CELLULE

RECUEIL

DE

CYTOLOGIE ET D'HISTOLOGIE GÉNÉRALE

PUBLIÉ PAR

J. B. CARNOY, PROFESSEUR DE BIOLOGIE CELLULAIRE, G. GILSON, PROFESSEUR D'EMBRYOLOGIE, ET J. DENYS, PROFESSEUR D'ANATOMIE PATHOLOGIQUE, A L'UNIVERSITÉ CATHOLIQUE DE LOUVAIN,

AVEC LA COLLABORATION DE LEURS ÉLÈVES ET DES SAVANTS ÉTRANGERS.

TOME I

ÉTUDES SUR LES ARTHROPODES

I. Étude comparée de la spermatogénèse chez les arthropodes, par G. GILSON.

II. La cytodiérèse chez les arthropodes, par J. B. CARNOY.

LIERRE
JOSEPH VAN IN & Cie, IMPRIMEURS-ÉDITEURS.

GAND
J. ENGELKE, LIBRAIRE, rue de l'Université, 24.

Title page of the first issue of the first journal of cytology, published in 1884.

Portrait of Lavoisier and his wife, painted by Louis David in 1788, six years before the French physicist was condemned to death by guillotine by a judge who commented that "La République n'a pas besoin de savants" [the Republic has no need for scientists]. This historic painting hung for many years in the library of The Rockefeller Institute for Medical Research, now The Rockefeller University, where many of the discoveries in modern cell biology were made. It is now at The Metropolitan Museum of Art in New York.

ists would have seen if they had toured the living cell magnified a millionfold, as we are about to do.

The Chemical Approach

In the meantime, however, while microscopists struggled to improve their instruments, a second type of exploration of the cell was started, thanks to the discoveries by such scientists as the Frenchman Antoine de Lavoisier, the Englishman Joseph Priestley, and others, who, in the later part of the eighteenth century, founded the new science of chemistry. In contrast with morphology, which progresses from larger to smaller entities, chemistry moves from the smaller to the larger. It started by identifying the elements, the atoms, and then went on to characterize some of their simpler molecular combinations. A historical landmark in the penetration of the living world by chemistry is the first synthesis of a biological molecule, urea, by the German Friedrich Wöhler in 1828. The boundary between mineral and organic chemistry, believed by many to be passable only with the help of a special vital force, was crossed.

During the next hundred years, considerable advances were made in our understanding of the chemical composition of living cells. The amino acids, the sugars, the fats, the purines, the pyrimidines, and other relatively small natural molecules were recognized, purified, analyzed structurally, and reproduced synthetically. Some insight was gained into the metabolic transformations they undergo in the organism, as well as into their mode of association into the major biological macromolecules—the proteins, the polysaccharides, and the nucleic acids. But here again the obstacles to progress became increasingly formidable. When structural complexity reaches the degree found in these large molecules, the tools of classical chemistry become almost powerless.

The Experimental Approach

For a long time cells were studied mostly by observation. But, as the experimental method became progressively developed in the physical sciences, it began to be applied to living organisms. This move was aided powerfully by the great upsurge of biomedical exploration in the second half of the nineteenth century. Physiology, pharmacology, genetics, bacteriology, immunology, experimental embryology, comparative and evolutionary biology—all made important inroads into the world of cells. An especially significant development came in the early part of the present century, when the American Ross Harrison and the French-American Alexis Carrel showed that animal cells could be cultured in the test tube, like unicellular microorganisms. They thereby demonstrated the cells' capacity for independent life and set up a technique that is still rendering major services today.

Powerful as they were, all these advances utilized means that, perforce, had to remain indirect and circui-

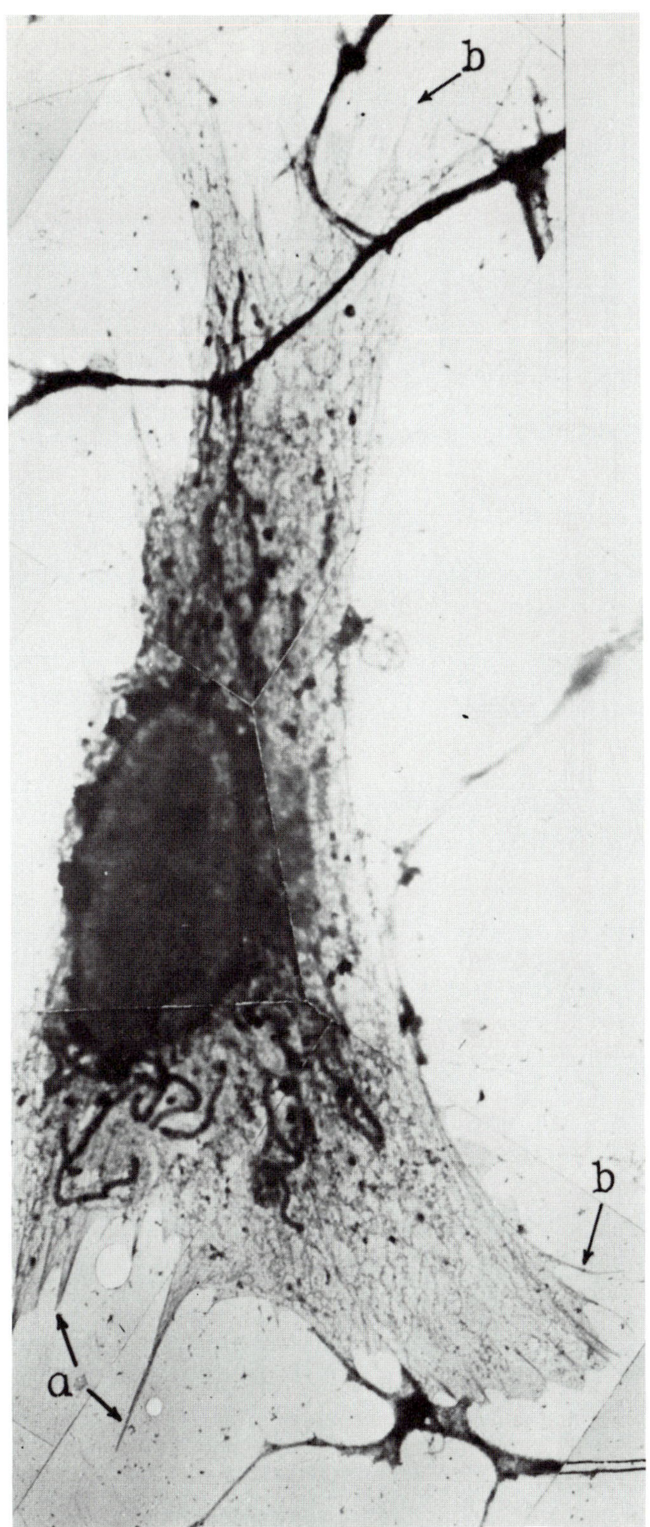

The first electron-microscope picture of a cell was taken by Keith R. Porter, Albert Claude, and Ernest F. Fullam and published in 1945 in the *Journal of Experimental Medicine.*

tous, leaving the cells themselves as virtually unopened "black boxes." And so, in spite of much progress on all fronts, there remained, between the smallest entity discernible in the light microscope and the largest molecular size accessible to chemistry, an unexplored no-man's-land extending over two orders of magnitude, a vast region that had to be labeled terra incognita on the map of the living cell. Scientists knew that this mysterious territory held major notions and concepts without which the life of cells would forever remain ununderstandable. But they could only stare in frustration at its seemingly impenetrable boundaries. Some did not give up, however. Like their predecessors, they put ingenuity to the service of inquisitiveness, and searched for the only solution: better tools.

The 1945 Breakthrough

This long, continuous effort came to sudden fruition at the end of World War II, when, through a remarkable combination of circumstances, a battery of powerful new instruments and techniques became available at more or less the same time. The morphologist's share of this boon was the electron microscope. This instrument, which was invented in the 1930s, has a sufficient resolution to traverse the whole unknown cellular territory, down to the nanometer level. But the weak penetrating power of the electron beam requires specimens to be extremely thin—a few millionths of an inch at most—and to be examined under high vacuum. These requirements posed great technical difficulties, which discouraged many. But a few held on, fired by the perspectives of progress opened by the new instrument. In a surprisingly short time, they succeeded in developing techniques for the preparation of specimens and constructed instruments that would slice the specimens into ultrathin films. Images of increasing quality were obtained, so that by the early 1960s much of the unknown territory had already been mapped out morphologically.

At the same time, biochemistry also acquired a number of incisive new tools. The most important ones among these were chromatography and isotopes. Of the two,

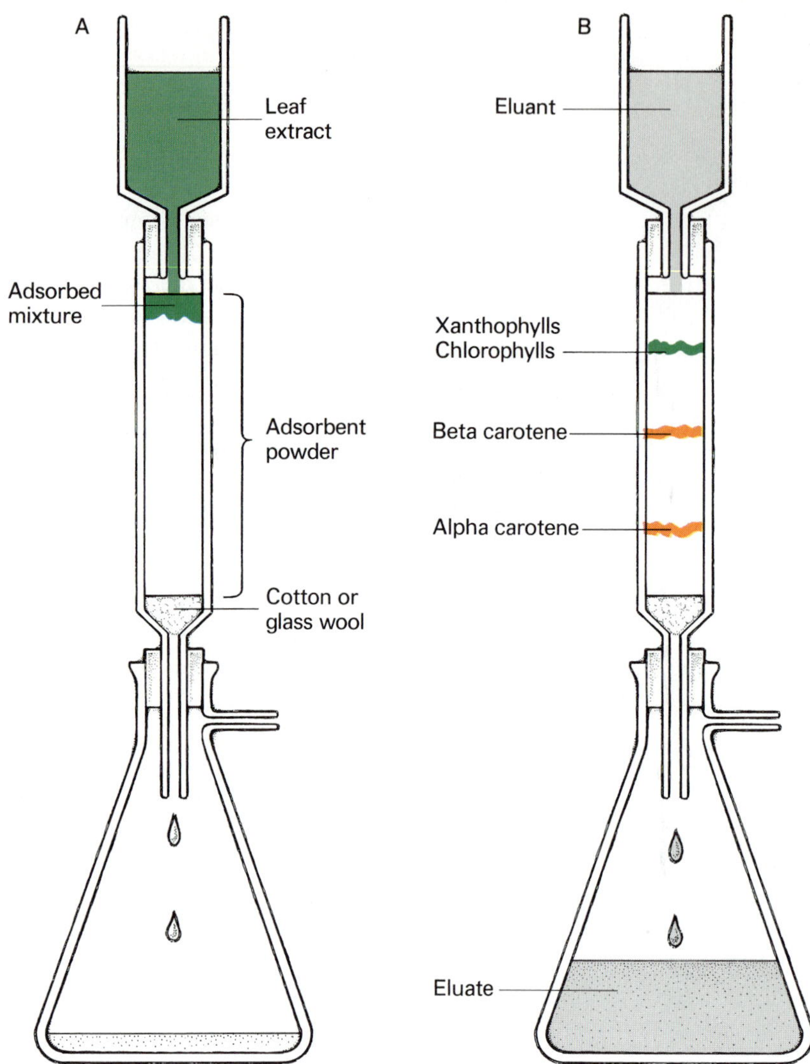

Chromatography. A simple device similar to that diagrammed here was used by Tswett to separate the main pigments in a leaf extract. It consists of a column of adsorbent powder, mounted so as to allow fluid to seep through.

A. A leaf extract is allowed to run through the column. All pigments remain bound on top of the adsorbent column.

B. After the column has been charged, a suitable fluid (eluant) is let through, displacing the different pigments at different rates, depending on the strength of their binding to the powder.

chromatography is especially remarkable, as it utilizes a very simple phenomenon, familiar to everyone who has ever seen a drop of ink spread on a piece of blotting paper or tried to remove a stain with cleaning fluid: the fringe, or halo, surrounding the spotted area. It is explained by the fact that different dyes do not move at the same rate with the spreading fluid. Some may move together with the solvent front, but many are retarded to a greater or lesser degree by binding to the paper or cloth fibers. They thus form concentric rings. In the beginning of this century, a half-Italian, half-Russian botanist, Mikhail Semenovich Tswett, became the first to make use of this phenomenon. By passing a leaf extract through a vertical tube packed with some adsorbent powder, he was able to separate the leaves' main green and orange pigments, which appeared as distinct colored bands or rings on the column. Hence the name "chromatography" that he gave to his technique (Greek *khrôma,* color; *graphein,* to write).

Tswett died relatively young, and the potentialities of his remarkable technique remained largely unexploited until it was revived in the early 1940s. There are now innumerable variants of chromatography, not restricted to pigmented molecules, of course, but applicable to all

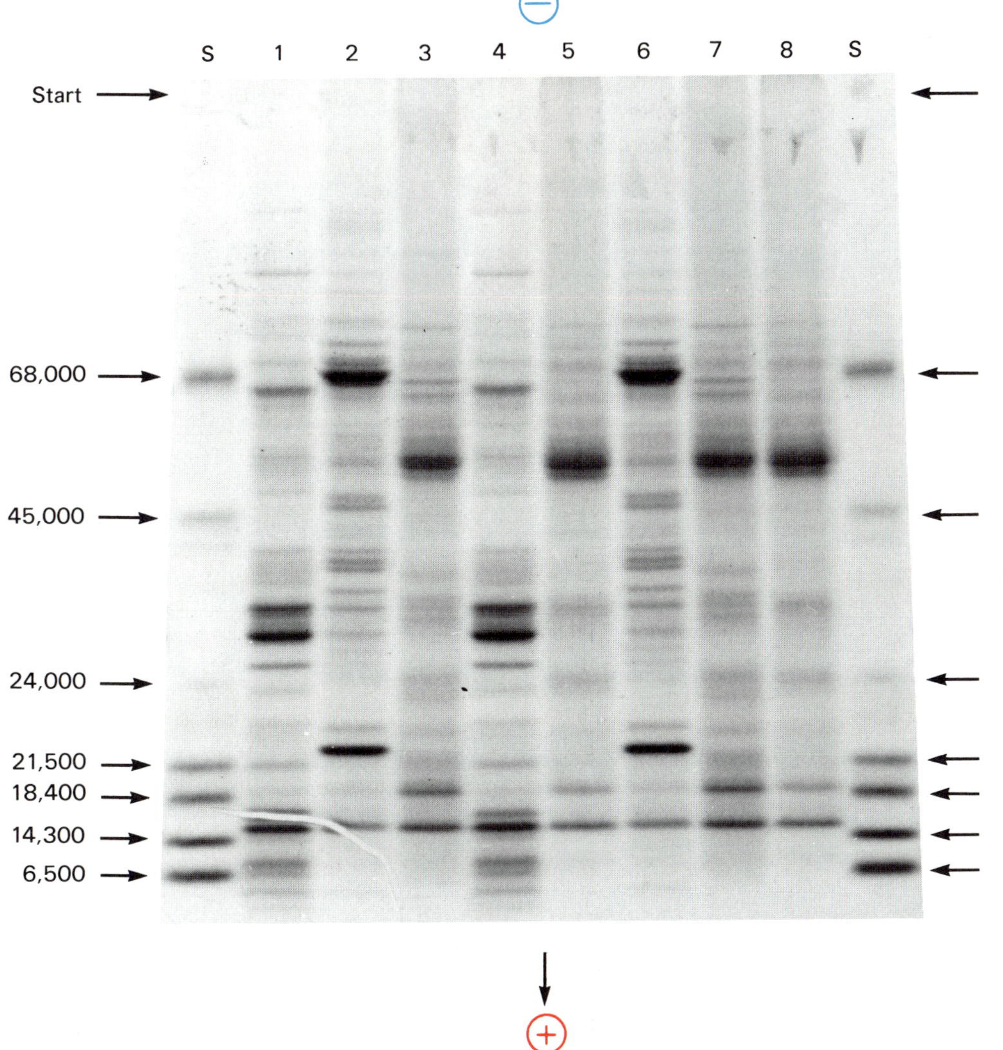

Separation of proteins by sodium dodecyl sulfate polyacrylamide gel electrophoresis (SDS-PAGE). Like most modern versions of chromatography, this technique uses a flat substrate—looking somewhat like a slab of jelly—instead of a cylindrical column. However, it relies on an electric field, not on a flow of fluid, to cause the different kinds of electrically charged protein molecules to move apart from each other. Mixtures are deposited as narrow bands at the start position. Electromotive force is applied and causes different protein species (all negatively charged by binding of SDS) to move toward the positive electrode, at rates inversely related to their molecular masses. At the end of the experiment, protein bands are revealed by staining with a dye. Lanes marked S contain standards of indicated molecular masses. Other lanes contain membrane proteins extracted from different organelles: mitochondria (1 and 4), peroxisomes (2 and 6), rough-surfaced endoplasmic reticulum (3 and 7), and smooth-surfaced endoplasmic reticulum (5 and 8). Note the differences between samples of different origin, enhanced by the reproducibility of duplicates.

substances that can be recognized by some analytical procedure. Related to chromatography is the technique of gel electrophoresis, in which an electromotive force, rather than a flow of solvent, moves the (electrically charged) components to be separated. These developments have, in one sweep, revolutionized the whole field of chemical separation and analysis. What in earlier days either could not be accomplished at all or required laboriously repeated extractions, precipitations, or crystallizations, to be performed on large amounts of starting material, can now be done without effort on trace quantities of almost any kind of mixture.

The second tool that has radically changed the chemical exploration of living cells is the isotopic-tracer method. Isotopes are species of the same element that have different atomic weights. Some exist naturally, and many more can be made artificially by nuclear reactions. For instance, in addition to the hydrogen atom ^{1}H, of atomic weight 1, by far the most abundant in nature, there is a heavy natural isotope, ^{2}H (deuterium), and an even heavier artificial one, ^{3}H (tritium). Chemically, these three forms of hydrogen have closely similar properties. They all combine with oxygen to form water, with carbon to make hydrocarbons, and so on. But they can be distinguished from

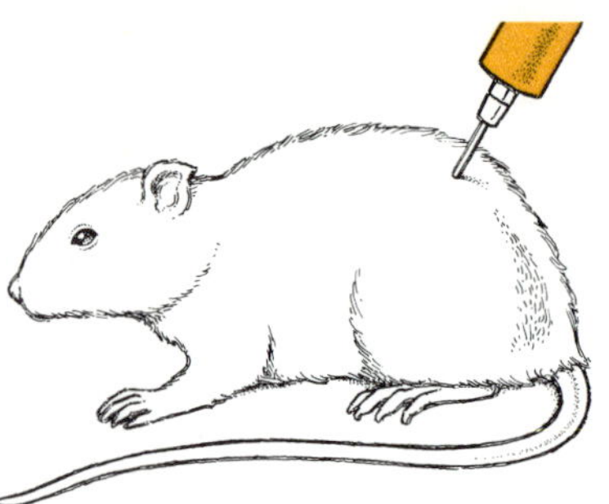

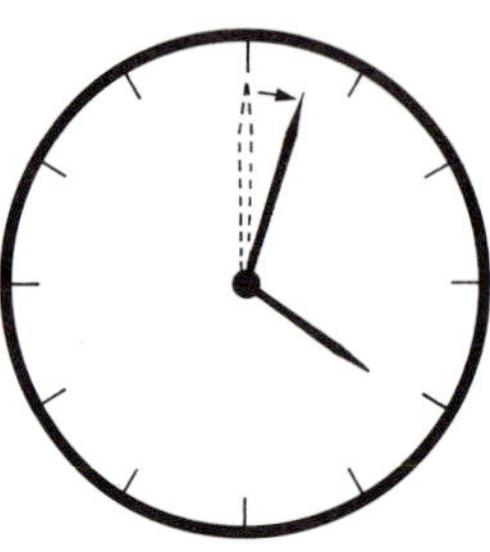

A

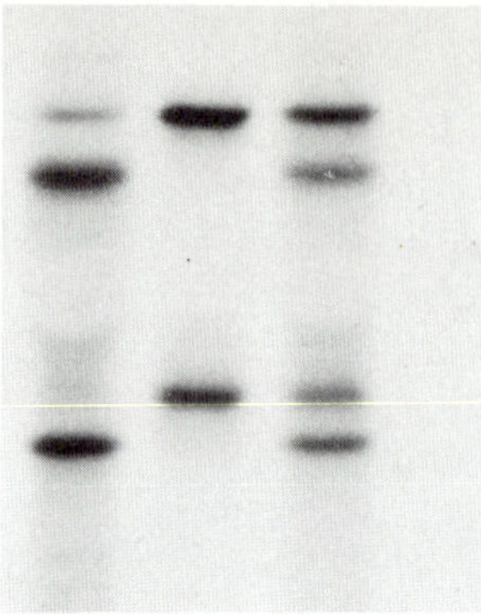

B

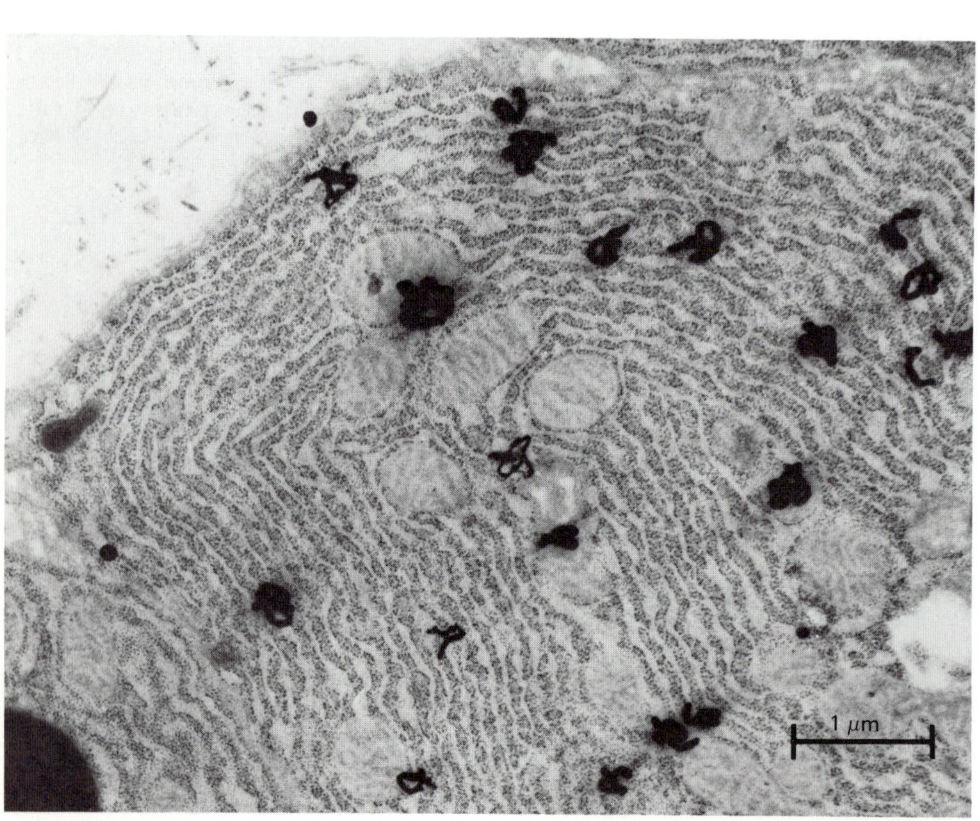

Three examples illustrating the use of radioactive isotopes.

A. Use of labeled amino acids as substrates allows the accurate measurement of the amount of protein synthesized from them, even though this amount is exceedingly small with respect to the total amount of protein present.

B. Impression of a radiographic film by emitted radiation reveals position of proteins synthesized from radioactive amino acids (as in part A), after SDS-PAGE separation (see the figure on p. 13).

C. High-resolution autoradiography combined with electron microscopy serves to localize newly made proteins in the part of the cell occupied by rough-surfaced endoplasmic reticulum. In this section through a cell from the pancreas of a guinea pig that received an injection of a tritium-labeled amino acid (^{3}H-leucine) 3 minutes before killing, each dense curlicue represents a track left in the photographic emulsion by a single high-energy electron emitted locally by a radioactive protein.

each other by a mass spectrograph, which, as the name indicates, separates atoms according to mass. Tritium is particularly easy to detect, because, in addition, it is radioactive, as are most of the isotopes used as tracers.

The advantage of isotopes is that they can be used to label specifically certain molecules, or parts of molecules, so that these can be distinguished and recognized from their congeners with almost no disturbance of the system. One of the many fruitful applications of this technique has been in the analysis of biosynthetic processes, most of which could not have been unraveled otherwise. For instance, as soon as labeled amino acids became available it was possible to study their assembly into proteins in the living animal or in the test tube, even though the amount of new protein manufactured was infinitesimally small relative to the amount of protein present. Although undetectable chemically, the newly made proteins could be recognized and even measured accurately, thanks to their radioactivity. In actual fact, the use of isotopes for studies of this sort had started before World War II, with the few natural (^{2}H, ^{15}N) or artificial (^{32}P) isotopes then available. But only after nuclear reactors were developed and a wide variety of radioisotopes could be manufactured on a large scale at a reasonable price did the technique truly come into its own. Without it, the revolutionary advances of the past decades in cellular and molecular biology would have been impossible. It represents one major peaceful benefit of nuclear power.

With morphology and biochemistry thus both becoming immeasurably strengthened, there still remained the need for a bridge between the two. This was provided by the development of procedures for separating the cell parts in such a way that they could be analyzed and their properties and functions recognized. For this to succeed, biochemists first had to learn to break cells open as gently as possible, so that the fragile cell constituents would be simply let loose but not significantly harmed. Once this is accomplished, one can take advantage of the differences in physical properties, mostly size and density, among the various cell constituents to separate them from each other. The procedures used for such fractionations depend mostly on centrifugation. In this respect, the development of centrifugation as an analytical tool and the construction of high-speed ultracentrifuges, which took place in the 1920s and 1930s, played an important role.

By some remarkable historical coincidence, all these new tools became available more or less simultaneously in the mid-1940s. Among the many names associated with these events, that of Albert Claude, who died in May 1983, deserves special mention. Born in Belgium at the turn of the century, Claude spent the years from 1929 to 1949 at The Rockefeller Institute for Medical Research in New York and there developed almost single-handedly the first applications of electron microscopy to the study of cells, as well as the main techniques for fractionating cells by centrifugation. He thus led both the morphological and the chemical armies into the cell's virgin territories and saw them effect their fateful junction. Today the ground that he opened is so familiar as to allow even the kind of excursion we are about to undertake. One part of the cell, however—actually its most centrally important part, the nucleus—would have remained largely impenetrable had not one additional development taken place.

The Molecular Biology Boom

This development started at the same time as the main invasion of the eukaryotic cell, but at the very fringe of the living world, with an attempt to analyze by means of genetic methods the properties of some simple viruses that infect bacteria and are called bacteriophages—literally bacterium eaters—for this reason. This humble beginning turned out to be the right approach to the problem of genetic organization, which even in these simplest of all organisms has proved to be of enormous complexity. For a long time, the new discipline, now known as molecular biology, remained confined to the study of viruses and of bacteria. But in the last few years, it has broken into the domain of eukaryotes and provided investigators with incisive new tools with which they are now exploring the most hidden recesses of the cell. It has also spawned the powerful new recombinant DNA technology, which may have profound influences on the future of humanity.

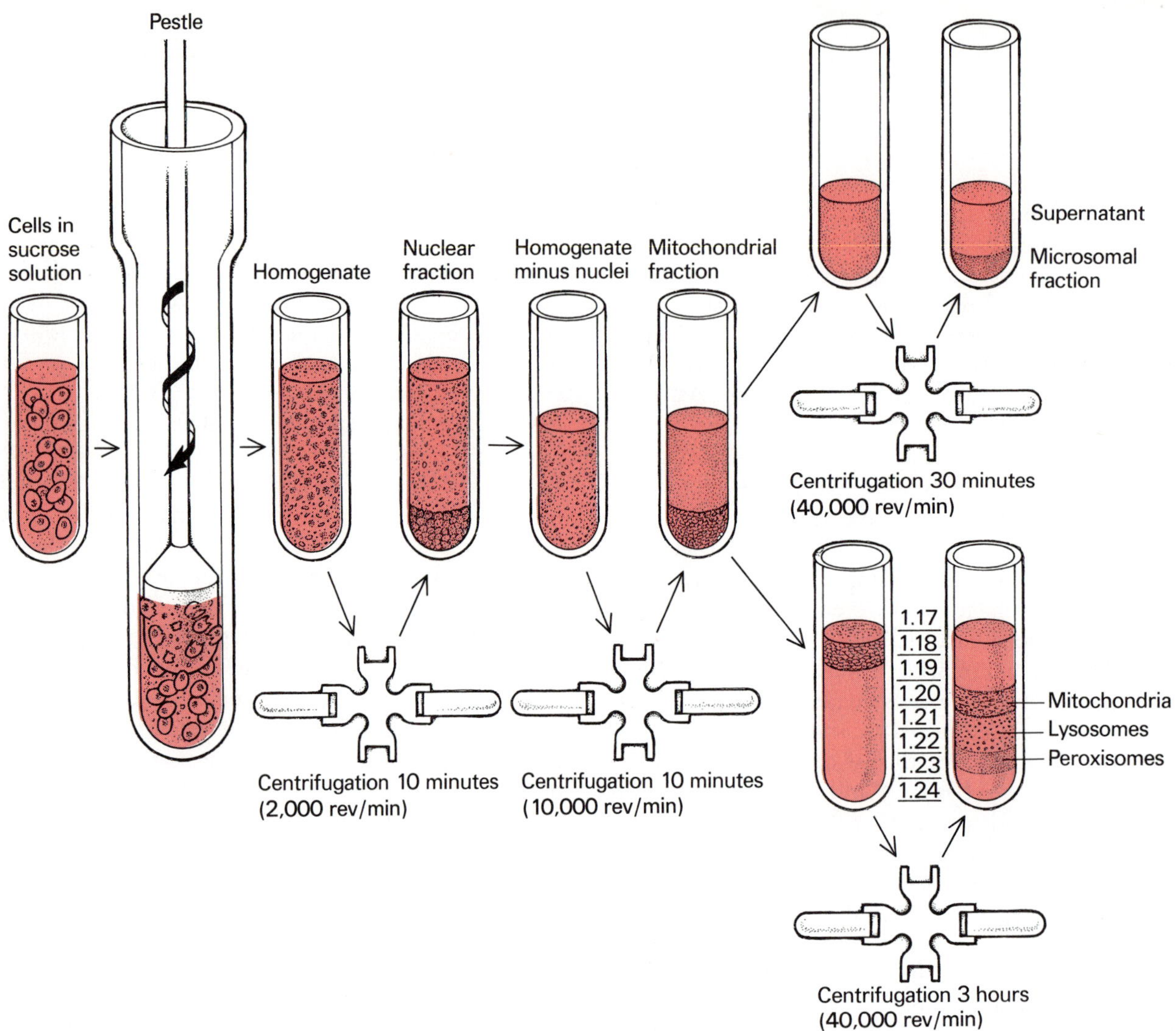

The fractionation of cells by differential centrifugation as first developed by Albert Claude and his pupils Walter Schneider, George Hogeboom, and George Palade. Cells are first ground in a solution of sucrose. (The grinder consists of a heavy-walled glass tube and of a tight-fitting pestle that is simultaneously rotated and moved up and down.) The resulting "homogenate" (an inappropriate term, as the main virtue of the preparation is its heterogeneity) is spun at low speed in a centrifuge, so that only the heavier components (mostly nuclei) are sedimented. The supernatant is then centrifuged at higher speed to separate the mitochondrial fraction. Very high centrifugal speeds (40,000 or more revolutions per minute) are used to bring down the small cell particles that make up the microsomal fraction. Soluble components and very small particles remain in the final supernatant. Altogether, four fractions are separated by this technique of differential centrifugation. Each fraction is grossly impure and heterogeneous and requires further subfractionation by more refined methods. One such method, called density-gradient centrifugation, is illustrated at the lower right. A crude mitochondrial fraction is layered above a density gradient. High centrifugal force is then applied to cause the particles to move down in the gradient until they reach a position at which their density equals that of the medium (equilibrium position). This technique therefore separates the cell constituents according to their density. In this experiment, particles present in the crude mitochondrial fraction have been separated into three distinct groups: mitochondria, lysosomes, and peroxisomes.

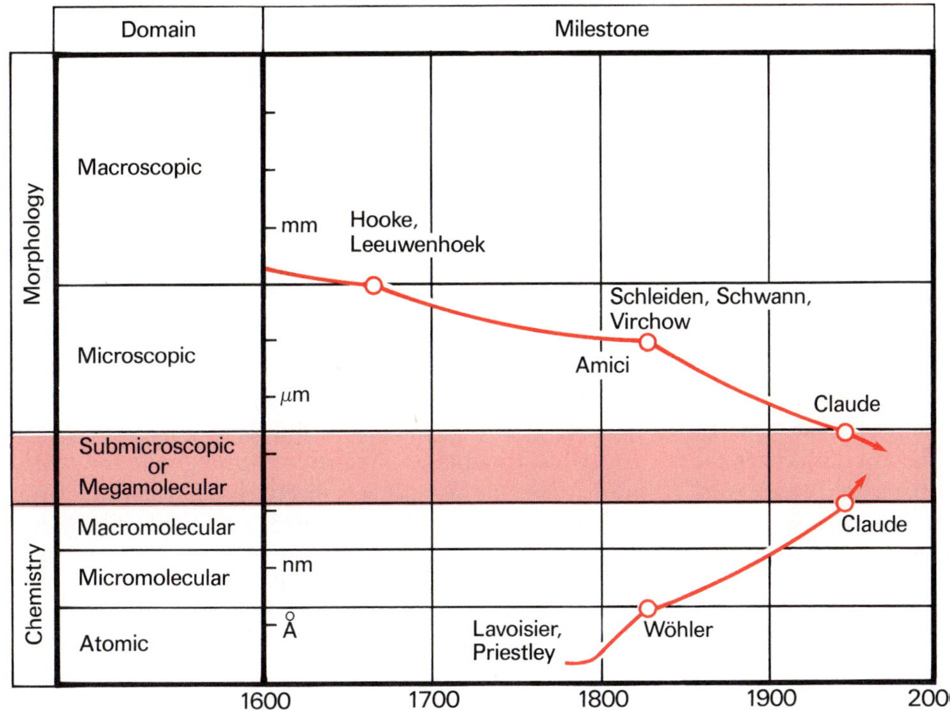

Some milestones in the invasion of the living cell. This diagram shows, as a function of time, the progress of morphology in the resolution of objects of decreasing size and that of chemistry in the characterization of molecules of increasing size. The no-man's-land between the microscopic and the macromolecular domains was invaded simultaneously from both sides immediately after the end of World War II.

An important lesson to remember from this brief historical survey is the crucial role of instruments and techniques in the progress of science. Not that the creative faculties of intellect, intuition, imagination, sometimes even genius, may not be of critical significance at some stage. But these faculties are powerless without the means of establishing contact with reality. As has been said by Claude, "In the history of cytology, it is repeatedly found that further advance had to await the accident of technical progress." Many such "accidents" have happened since Robert Hooke first turned his microscope on a slice of cork. Their consequences have been increasingly far reaching, culminating in the major discoveries of the last decades. Although it is always difficult to judge one's own time in historical perspective, one cannot help the feeling that the second half of this century will be remembered for one of the great breakthroughs of human knowledge—perhaps the greatest to date, as it concerns the basic mechanisms of life.

The Personal Factor

Scientists, it has been said, do not read the book of Nature. They write it. This does not mean, of course, that works of science are works of fiction. On the contrary. Scientists do the best they can to stick to the facts, and the collective process whereby the facts are recorded and put together aims at objectivity.

The point is that, however hard one may try, perfect objectivity is unattainable. There is no such thing as an isolated fact. There are only recordings and interpretations of facts by individuals. Even the simplest observations come to us through our senses, which act as highly selective, as well as subjective, filters that let through only certain limited aspects of reality. We have greatly extended the range of our senses by means of instruments, but only at the cost of interposing additional filters. A lens may reveal more than the eye does, but not without distortion. We can probe even deeper with an electron microscope, reach atoms by X-ray diffraction, detect subatomic particles with a high-energy accelerator. But the resulting picture increasingly depends on elaborate theories, complex machines, and delicate manipulations.

The choice of facts to be collected introduces an additional subjective element. This is true even in the observational sciences. An astronomer has no power over the stars he looks at, but he nevertheless decides in what direction to point his telescope and thereby influences the development of astronomy by his personal preferences. The individual's decision plays a much more important role in the experimental sciences, where facts are provoked. Every experiment is a question asked of Nature. As in all interrogations, the answer depends to some extent on how the question is asked. Scientists are well aware of this danger but, try as they may, cannot avoid it entirely.

Finally, and most importantly, we must remember that the collection of facts is only a small part of the scientific process. What counts is the way in which the facts are interpreted and organized into theories that advance our *understanding*. The construction of a theory is a highly creative process that bears the imprint of its author. That is why we speak of Einstein's relativity theory as we speak of Leonardo's *Mona Lisa*. We do so, not only to give Einstein his well-deserved credit, but also to indicate that the theory of relativity may, for all its universal significance, include lingering traces of the mind that conceived it. Science aims at perfect objectivity, but it progresses toward this goal by successive approximation. We may at any given stage have the impression that we have reached the goal, but subsequent events prove this to be an illusion.

This warning should be kept in mind throughout our tour. Every object, every site, every happening, every process, every mechanism that will be pointed out as though it were there to be seen is actually a product of individual human brains mulling and churning over collections of images and sets of figures, themselves the products of recordings made by intricate instruments on biological materials subjected to complex experimental manipulations. There is no other way of entering the cellular microcosm, and scientists are justifiably proud of having already reached as far as they have. But the risks of distortion are considerable. And so is the amount of guesswork needed to fill in all the gaps in our knowledge. As a result, even fairly well established facts may be presented very differently by different scientists, depending on their personal biases and particular areas of expertise.

Ideally, you should be given all the data and allowed to make your own judgment. But this would mean giving you what thousands of scientists throughout the world are hardly able to master collectively. Then, should not their names at least be mentioned, as is Einstein's, partly to give credit where credit is due, and especially to identify individual responsibilities? Even that, however, can be done only exceptionally, because of the way in which individual contributions have become interwoven into the complex fabric of modern science. For this reason, very few scientists will be mentioned by name in the course of our tour. But they are there anonymously—often, it must be added, overshadowed by the shortcomings and biases of the guide—behind every object that we will be given to contemplate, behind every process that we will try to understand. This must be kept in mind.

The Itineraries

A Look at the Map

Before we set off on our tour, a brief look at a cell map is in order, so we can get our bearings and have some idea of where we are going and how the visit is to be organized. In this perfunctory survey, we will do no more than identify by name the main parts of the cell and their functions. Clarification will come in due time, in the course of our tour.

The most obvious feature of any eukaryotic cell, perceived even by the early cytologists, is the distinction between the centrally located nucleus and the surrounding cytoplasm. These two parts of the cell are related to each other somewhat like the stone—nucleus is derived from the Latin word for nut—and the pulp in a cherry. Like a cherry, the cell is covered by a skin, or membrane, called the plasmalemma, or plasma membrane.

Map of the cell.

The nucleus is the repository of the cell's genetic library, which is filed in chemically coded form in distinct units, the chromosomes. As a rule, these are so intermingled as to form what appears to be a single mass of chromatin, irregularly subdivided into denser (heterochromatin) and less dense (euchromatin) parts, and including one or more specialized structures called nucleoli. A membranous envelope completely encloses this mass, which is impregnated by a fluid named nucleoplasm.

The main functions of the nucleus are directly related to information processing. They include conservation and, if need be, restoration of the genetic library and, especially, transcription, a complex and highly selective process by which certain specific instructions are read from the information store and sent out to the cytoplasm for expression. Genes exert their commanding influence over the cell by these mechanisms. When a cell prepares to divide, the nucleus has to perform an additional duty, which consists in the complete and accurate duplication of the genetic library. Subsequently, the nucleus goes through a complex reorganization called mitosis, in which the chromosomes become temporarily visible as discrete rods, and which results in the formation of two nuclei.

The cytoplasm consists of a formless jelly, the cytosol, bolstered by a cytoskeleton and containing a number of embedded organelles. These subserve a variety of functions, which can be roughly classified under "internal affairs" and "external affairs."

The internal affairs of a living cell are mainly concerned with biosynthesis and energy production. Biosynthesis is an ongoing activity, even in a nongrowing cell, because cells are not chemically static. They continuously break down and rebuild most of their constituents in a highly dynamic fashion. This activity consumes a great deal of energy. So do the other forms of work that cells accomplish in relation to motion, molecular transport, generation of electricity, information transfer, and sometimes light emission. Cells cover these requirements by breaking down energy-rich foodstuffs supplied either from the outside or from the cell's own reserves and, in green plants and photosynthetic bacteria, by absorbing and utilizing sunlight. The sum total of these reactions makes up what is generally referred to as metabolism, itself subdivided into anabolism (biosynthetic processes) and catabolism (breakdown reactions).

The systems that carry out these various activities are situated in the cytosol and in a number of organelles that maintain close relationships with it. Among these are the mitochondria, often designated as the cell's power plants, sites of the major oxidative reactions and of the mechanisms whereby the energy released by these reactions is retrieved and made available to the cell in utilizable form; the chloroplasts, which house the photosynthetic machinery of green plant cells; the microbodies, a heterogeneous family of metabolic organelles, of which peroxisomes are the most important members; a number of different motor units concerned with cell movement; the ribosomes, which are the centers of protein synthesis and, as such, the main executors of the genetic commands issued by the nucleus; and, finally, a variety of cytomembranes, which are primarily involved with the cell's external affairs (below), but which, in addition, house a number of important metabolic systems.

Under the heading of external affairs we group the various activities involved in communication and in exchanges of matter between the cell and the world around it. These activities are shared between the plasma membrane, which is the actual cell boundary, and an elaborate network of intracellular membranes related to the plasma membrane and organized into a large number of closed, pouchlike structures. Capable, directly or indirectly, of establishing transient connections with each other or with the plasma membrane, these structures serve in the storage, processing, and intracellular transport of materials that are either taken in from the outside and broken down intracellularly or made inside the cell for extracellular discharge. A key feature of these exchanges is that they occur without the membranes involved ever suffering the slightest gap or tear. There is thus always a membrane boundary between the contents of the pouches and the cytosol. For this reason, these membranes act like the plasma membrane to the extent that they also mediate exchanges between the cell and its surroundings by means of the segregated spaces that they delimit.

Designated vacuome by the early cytologists—the term is largely abandoned, but deserves to be resurrected—this cytomembrane system consists of two distinct, though closely interconnected, sections, dedicated respectively to import and to export, and themselves subdivided into functionally distinct subsections. The import department includes specialized areas of the plasma membrane concerned with uptake of extracellular materials by a mechanism called endocytosis; a storage compartment made up of endosomes, concerned mainly with the sorting and routing of the materials taken up; and a complex of digestive vacuoles or lysosomes, in which these materials are broken down. The export system starts with the rough-surfaced endoplasmic reticulum, which collects and processes newly made export proteins manufactured by ribosomes bound to its limiting membranes. This structure communicates, by means of the smooth-surfaced endoplasmic reticulum (without attached ribosomes), with a complex system known as the Golgi apparatus. Further processing, as well as sorting, of the export materials goes on in these two subdivisions. From the Golgi apparatus, the materials are then eventually directed, after storage and concentration in secretion granules, toward the cell periphery, where they are discharged by exocytosis. A special set of vesicles conveys materials from the Golgi apparatus to the lysosomes. Others are involved in recycling of the membranes that participate in these various processes.

Based on this blueprint, our tour will be divided into three distinct itineraries.

Itinerary 1: The Outskirts and Surface of the Cell, the Vacuome

We will start by approaching the cell progressively, coming from a blood vessel, so as to have an opportunity to see some of the extracellular structures. Then, after a good look at the cell surface, we will explore the vacuome, entering by endocytosis and exiting by exocytosis. In between, we will take advantage of the local mass-transportation system to visit all the different parts of the complex and to watch firsthand, and practically share, the intracellular travels and experiences of such materials as either are imported by the cell from the outside or are manufactured by it for export. This first part of our tour will also provide us with convenient opportunities for becoming acquainted with the main cell constituents—the proteins, polysaccharides, and lipids—and with some of the structures that they build together.

Itinerary 2: The Cytosol and Cytoplasmic Organelles

Our second itinerary will take us directly into the cytosol, from which we will be able to call successively at all the major organelles that are connected with it. In doing so, we will be able to learn something about metabolism and the major principles that govern energy transformations and biosynthetic mechanisms. By the time we reach the ribosomes, we will receive our first introduction to biological information transfer and to the molecules that mediate it, the nucleic acids.

Itinerary 3: The Nucleus

Finally, our third itinerary will cover the nucleus, where we will see the genes in action, as well as the chromosomes going through the complex transformations that are associated with mitosis and with a related process called meiosis. Briefed as we will be by then, we will be able to address some key biological problems, including the origin of life and the mechanism of evolution, as well as such important issues as the mechanism of cancer and the future of biotechnology. We will take advantage of the turmoil that accompanies cell division to effect our final exit.

Underwater Equipment Needed

One last point before we set out: cells live in a watery world, even when the organisms of which they are part do not. Take our own cells, for instance. Except for the outer layers of our skin, which consist of mummified dead cells, all cells are immersed in liquid, either blood or some fluid derived from blood. Similarly, plant cells are bathed by sap. Even the hardiest bacteria need some moisture around them. They may survive complete dryness, but only in a dormant state, with all their processes arrested until they are reawakened by water. The same may happen with more complex cells, such as those of molds and of seeds.

Thus, cell tourists really resemble those underwater explorers popularized by the films of Jacques Cousteau. They carry out their visits swimming. Indeed, much of what they see would shrivel away but for the surrounding water, much like the delicate creatures that cover the surface of submerged rocks. So strap on your aqualung and plunge. We are off.

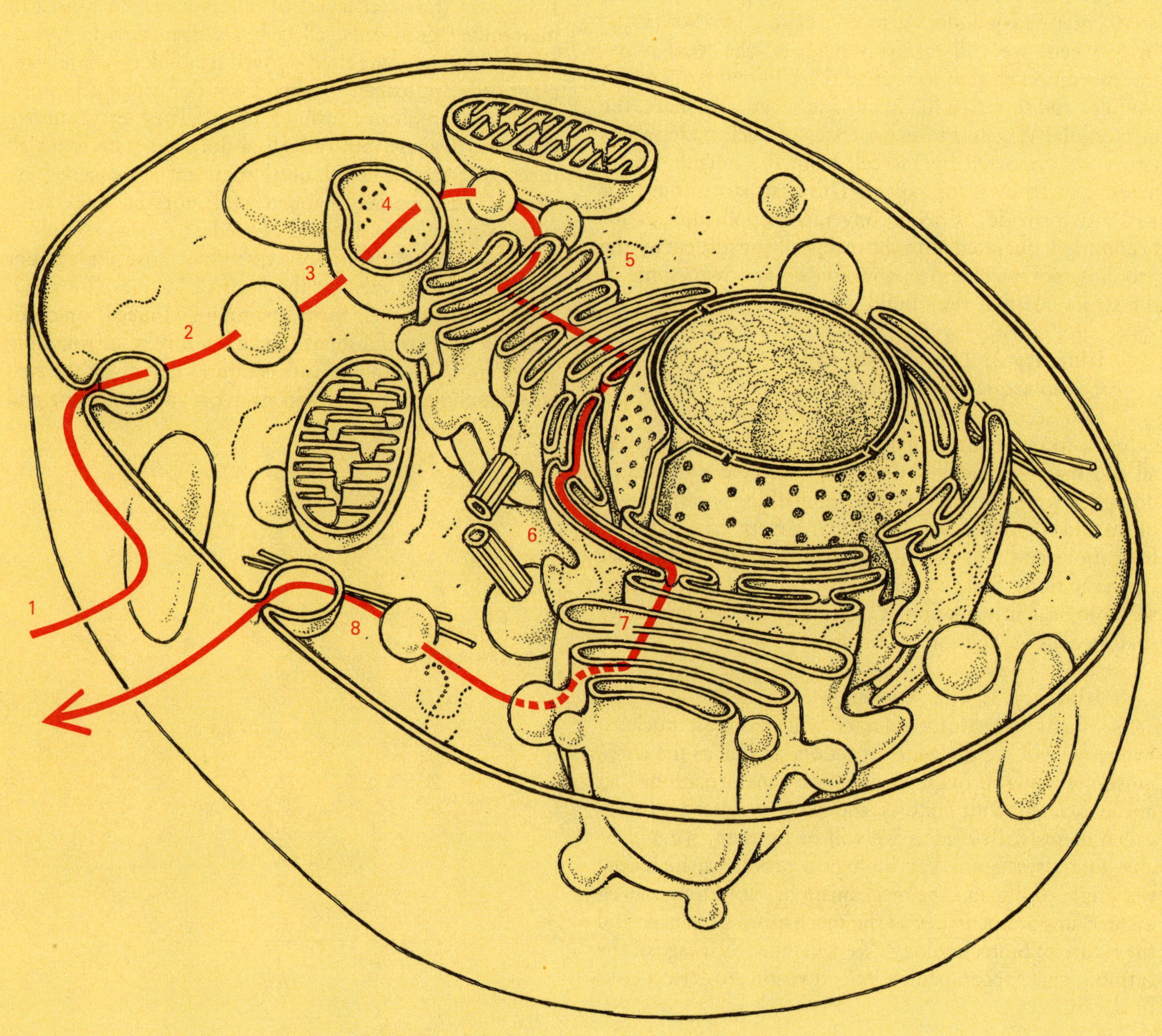

1
2
3
4
5
6
7
8

ITINERARY I

The Outskirts and Surface of the Cell, the Vacuome

1 Approach the cell through extracellular structures and explore its surface.

2 Enter by way of endocytosis and watch sorting process in endosome.

3 Reach lysosome by endosome-lysosome fusion.

4 Visit lysosomal space.

5 Board lysosome-Golgi shuttle and proceed immediately upstream into endoplasmic reticulum to reach a rough-surfaced recess.

6 Watch synthesis of secretory proteins and follow their processing down endoplasmic reticulum and back into Golgi apparatus.

7 Visit Golgi apparatus.

8 Board secretion granule and leave by exocytosis.

2 Extracellular Structures, With an Introduction to Polysaccharides and Proteins

Before we actually enter a living cell, we should take a look at its surface, which has many interesting features. But it is rarely a simple matter to find a cell, for most cells are surrounded by a more or less elaborate network of outer defenses and scaffoldings that often completely hide them from view and greatly hamper approach to them. Such structures are not parts of the cells but are built from precursor materials that are secreted by the cells and that subsequently join together into a variety of combinations of almost every possible shape and consistency, from the softest of jellies to the toughest of shells and woods. Sometimes they become impregnated with minerals, which may provide their organic matrix with the durability of stone or the hardness of enamel. These extracellular constructions act as props for the cells; they provide every sort of visible form that life creates on our planet. Without them, there would be no trees, no flowers, no animals; nothing but an amorphous covering of oozy slime made of a myriad of naked cells crawling over each other.

An Architect's Dream

Extracellular structures are particularly extensive in plants, where they form a continuous rigid scaffolding within which each individual cell is provided with a small chamber of its own—the original "cells" first seen by Robert Hooke as microscopic holes in a slice of cork. The main component of this framework is appropriately called cellulose. Most abundant of all organic substances on earth, cellulose is described by chemists as a

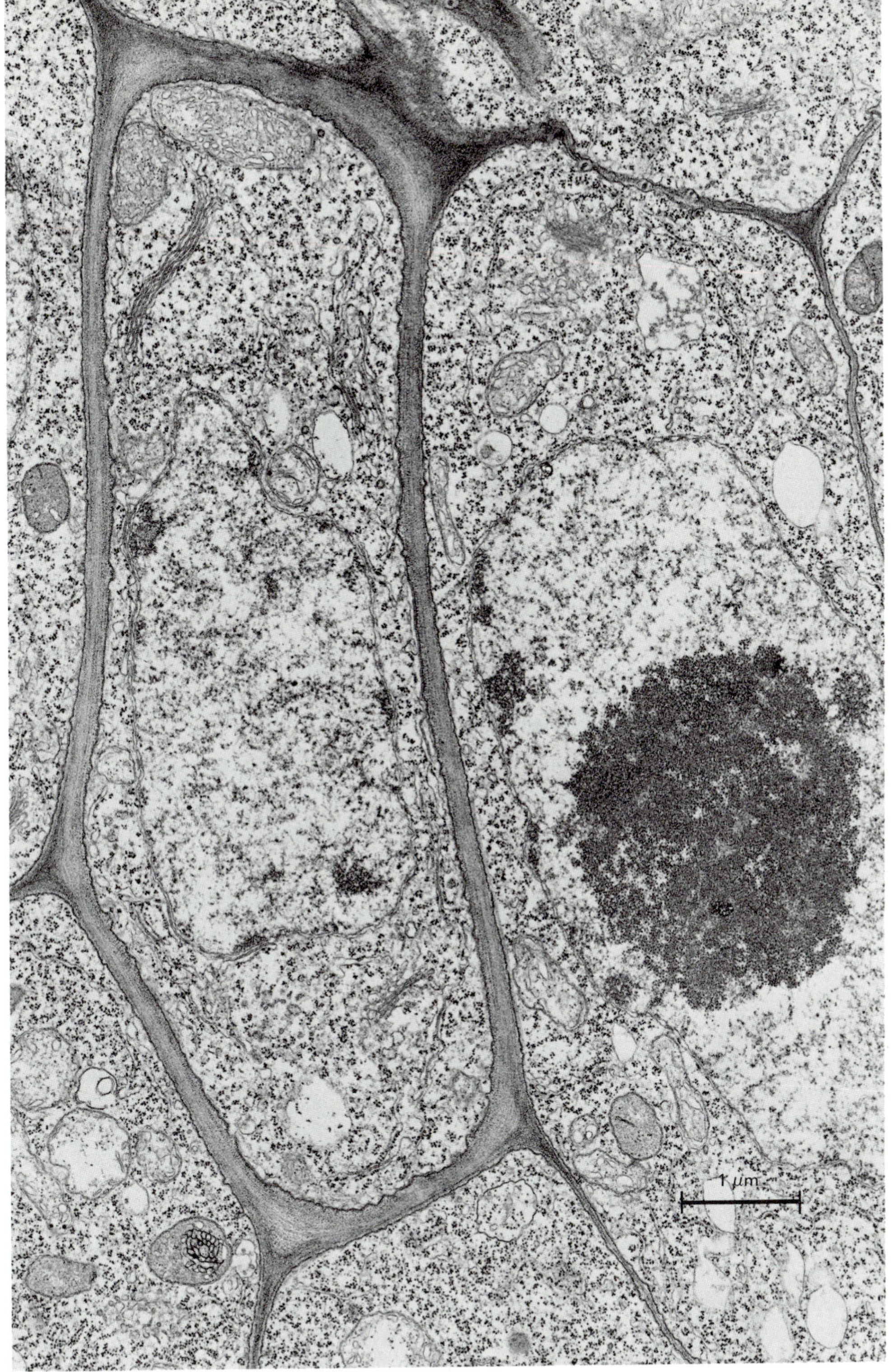

Electron micrograph of a transverse section of a root tip of the bean *Phaseolus vulgaris* showing how each cell occupies a closed chamber. Only the walls of the chambers were seen by Robert Hooke (see the illustration on p. 6).

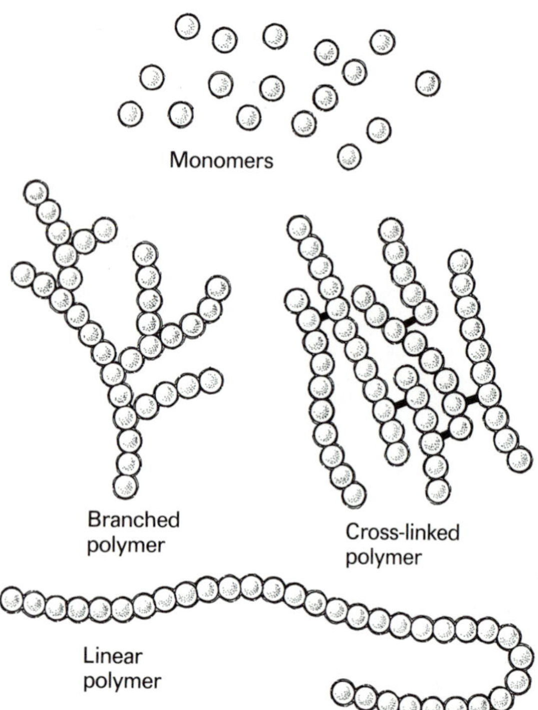

Biological polymers are made of monomers, which are variously shaped molecules ranging between about 0.3 and 1 nm in size. Magnified a millionfold, they would be just visible as tiny flecks.

polymer of the simple sugar glucose. Polymers are giant molecules made of many (Greek *polys*) parts (Greek *meros*) called monomers (Greek *monos*, single). All our plastics and artificial fibers are polymers. So are the main constituents of living organisms, including the polysaccharides (Greek *sakkharon*, sugar), to which cellulose belongs, and the proteins, which we will consider in a moment. Unlike most man-made polymers, which are usually of an indefinite size, limited essentially by the mold in which they are cast at the time of polymerization, natural polymers have definite molecular sizes. They consist of macromolecules (Greek *makros*, large), in which the building blocks are assembled in a specific fashion. Their molecular masses range mostly from a few thousands to a few hundreds of thousands of daltons. (The dalton is the mass of the hydrogen atom.) Exceptionally, they may reach or exceed one million daltons. The corresponding molecular sizes are of the order of a few nanometers, if the molecules are globular. Fibrous molecules may be as much as several hundred nanometers long. Most structural macromolecules are fibrous and are naturally endowed with the property of combining into characteristic multimacromolecular assemblages. Quite often these assemblages are reinforced by chemical cross-links.

Cellulose is a particularly simple polymer, made entirely of a single type of monomer, glucose. It is thus identical in gross composition with starch, the main carbohydrate reserve substance of plant cells, and with glycogen, its animal counterpart, which are also simple glucose polymers. But, through a quirk of its chemical structure, cellulose is extremely resistant to degradation, both biological and chemical. After purification, it serves as the almost indecomposable fibrous component of cotton, linen, and paper, and of such chemical derivatives as cellulose acetate or cellophane. Only certain microorganisms can degrade cellulose. Their favorite ecological niche is the digestive tract of herbivores, much to the advantage of the animals, who owe their ability to utilize cellulose to these friendly guests. Most other mammals—including ourselves—pass out cellulose essentially intact. It is the "fiber" in our food and forms much of the solid part of stools. Plants use a number of other structural molecules besides cellulose. Trees, in particular, depend largely for their exceptional resilience on a phenolic polymer called lignin (Latin *lignus*, wood).

Bacteria also are completely surrounded by a rigid wall. It includes a number of substances that are peculiar to the bacterial world. This wall acts as a protective shell that allows the bacterial cell to withstand very harsh environmental conditions. It is also important from the medical point of view. Infectious microbes depend largely on

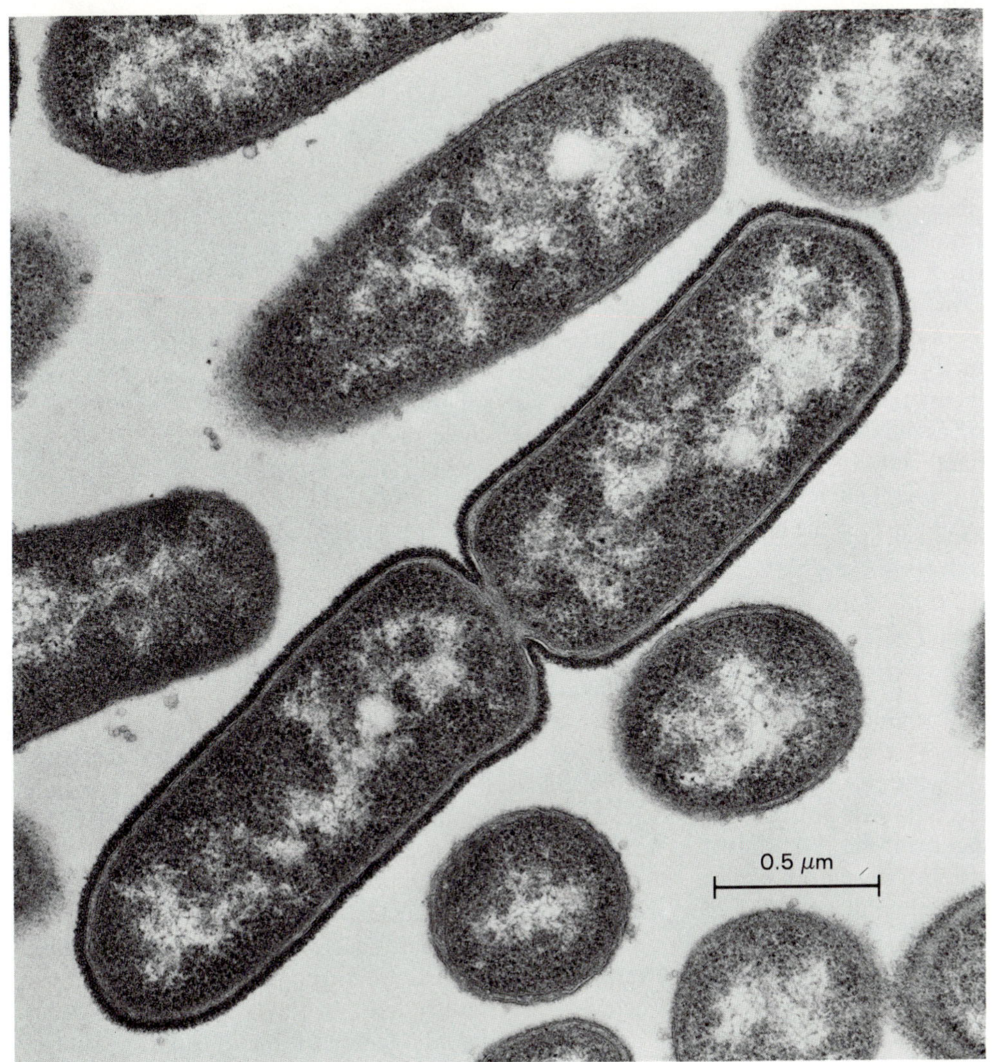

Electron micrograph of a section through a group of *Bacillus fragilis* illustrating the bacterial cell wall. The cells in the middle are in the process of separating.

their outer covering to resist or elude our defenses. Some even use it posthumously to launch a toxic attack. In turn, our own cells, especially the white blood cells, which form our main defense corps, "see" only the wall of the invader. They are alerted to it, and they ultimately kill the bacteria by a complex series of motions triggered off by that chemical recognition. (Some of the mechanisms are considered in Chapters 4 and 5.) Interestingly, penicillin and related antibiotics actually kill microbes by interfering with the construction of the cell wall, thus blocking bacterial growth. Lysozyme, a natural antibacterial agent occurring in white cells and in tears, kills bacteria by breaking down their cell walls.

In the animal world, cells are rarely encased individually. Most often they huddle into intimately clustered groups of characteristic shape, which themselves are then bolstered and buttressed by extracellular structures of various sorts. This kind of arrangement is largely responsible for the richness of organization and evolutionary capability of animals. For one thing, it has allowed the development of nervous systems. Unfortunately, time does not permit us to explore the admirable manner in which each particular tissue or organ is architecturally adapted to perform its function. But we should at least take a look at the main components that are used in this remarkable construction work. Several of these components belong to the all-important class of proteins, of which we will encounter many representatives all along our tour. Here is obviously a case where some chemical briefing is mandatory.

A Look at Proteins

In 1838, the Dutch chemist Gerardus Johannes Mulder, a pioneer in the analysis of "albuminoid" substances (Latin *albus*, white; *albumen*, egg white), adopted the name protein (Greek *prôteios*, primary) to designate what he thought was the basic constituent of heat-coagulable nitrogenous substances such as blood fibrin, milk casein, and egg albumin, which were beginning to be recognized at that time as belonging to a common class. The term protein was suggested to Mulder in a letter by the Swedish scientist Jöns Jacob Berzelius, one of the founders of chemistry and the father of the concept of "catalysis."

Mulder and Berzelius were guilty of vast oversimplification, inevitable in their day, but the proposed terminology was prophetic. Not only have the proteins turned out to be the primary agents of all living processes, but they also share with the Greek sea-god Proteus the ability to take on innumerable different shapes. Proteins are also protean.

We can readily appreciate this by adjusting our molecular eyeglasses to a high degree of resolution, so that we can examine in detail objects only a few nanometers in size. At our adopted millionfold magnification, such objects would still be only about the size of a gnat. Under an appropriate magnifying glass, however, gnats and other small creatures of similar size reveal a wealth of different shapes and structures. So do proteins if we increase the magnification further, except that their forms are more abstract, resembling in their sleek curvatures the biomorphic sculptures of Hans Arp, rather than the articulated angularities of insects. Some are globoid, almost spherical; others have more tormented shapes, bent, twisted, or bulging with tuberosities; yet others thin out to slender threads, often helically coiled. These shapes are not immutable. They swell, pulsate, elongate, contract, or uncoil, sometimes with dramatic suddenness.

Most proteins are situated inside cells, where their main function is concerned with catalysis, a fact which, had he but known it, would have delighted Berzelius. The Swedish chemist coined this word (Greek *kata*, down; *lysis*, loosening) to designate the property of substances (called catalysts) that facilitate the occurrence of chemical transformations without themselves being consumed in the reactions. Inorganic catalysts are known and are used in the chemical industry, but the real magicians of catalysis are to be found in living cells. Life, even that of a humble microbe, depends on the performance of thousands of chemical reactions, most of which either cannot be reproduced under artificial conditions or can be so only under conditions—such as high pressure, temperature, or acidity—that are incompatible with life. In living cells, these reactions take place with remarkable ease and rapidity, thanks to the mediation of specific catalysts called enzymes. This word, which is derived from the Greek *zymê*, yeast, recalls the historical role of the study of alcoholic fermentation in the characterization of the first intracellular catalysts (see Chapter 7). The term ferment has also been used, for the same reason, but is now abandoned.

Virtually every chemical reaction that occurs in living organisms is catalyzed by a specific enzyme, often assisted, as we shall see in the second part of our tour, by one or more cofactors, or coenzymes. All known enzymes are proteins. Thus the number of different protein varieties is considerable—several thousand, at least, in any given cell, where they generally account for some 20 per cent of the wet weight of the cell, or more than half its dry weight. In addition to their functions as catalysts,

Top: *Serpent,* by Hans Arp.

Bottom: Plasticine model of a molecule of myoglobin, the oxygen carrier of red muscle cells. The white wormlike structure represents a coiled polypeptide chain (see the illustration on p. 33), the dark disk an iron-containing heme group, which binds oxygen (see Chapter 9). Hemoglobin, the oxygen carrier of red blood cells, consists of four subunits shaped like the myoglobin molecule.

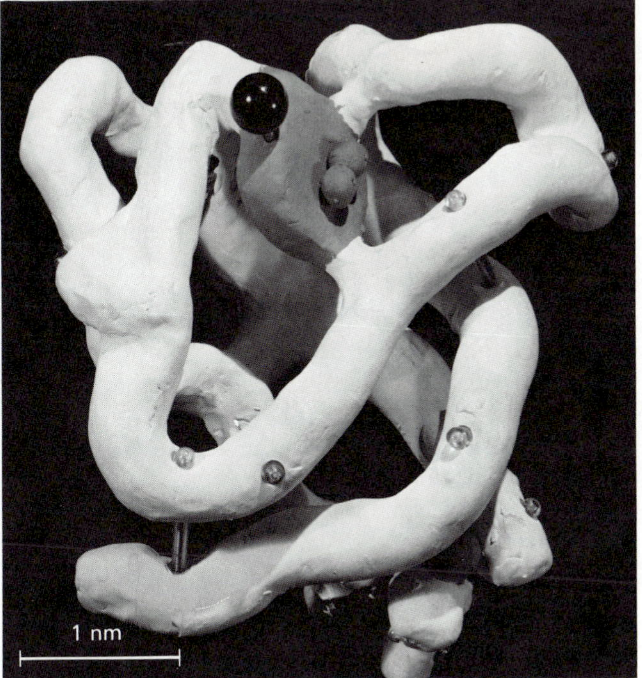

proteins also serve in regulation, transport, locomotion, and many other activities. They also play an important role as structural components, both intracellularly and extracellularly. The proteins are indeed the primary constituents of living cells; they deserve their name.

Proteins readily lose their unique shape, as well as their catalytic power, when subjected to heat or other physical or chemical agents. This process is called denaturation. Long believed to be irreversible, protein denaturation has often been used as an argument in support of the need for a special force or principle to guide the assembly of living structures. This view is no longer held, and denaturation has, in fact, been shown to be reversible in a number of cases. There is nothing exceptional about the shape of a protein molecule. It is none other than the most probable conformation, the one the molecule is most likely to assume if given the opportunity to do so.

It is instructive to look with our molecular eyeglasses at a mixture of denatured protein molecules. Except for differences in size, they now have the same general appearance. All we see, in lieu of the beautiful variety of shapes and structures, is a tangle of long, exceedingly thin threads. Now we understand better why denaturation often behaves as an irreversible phenomenon. Unless special conditions are provided, each thread will remain caught in the tangle, unable to wind back into its characteristic shape. Even more important, denaturation tells us something of major significance about the structure of protein molecules, which otherwise might well have escaped our scrutiny. All proteins are filamentous molecules. The innumerable shapes with which they present us are all generated by coils and twists of a threadlike arrangement of atoms.

As we know from chemical studies, the common structure behind all these different shapes is a backbone made of a simple, six-atom repetitive unit recurring as many as several hundred times:

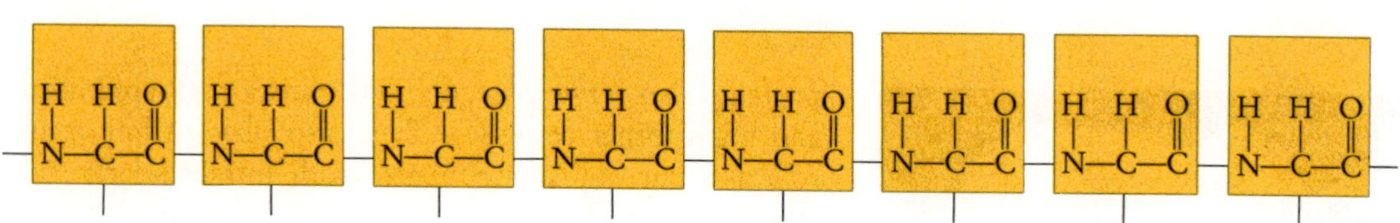

The central carbon of each unit has one valence that is not engaged in the backbone. It bears one of twenty distinct chemical groups, which are responsible for the specific properties that distinguish different protein molecules from each other. The link between the units is called the peptide bond, a name that goes back to the discovery that this bond is hydrolyzed by the digestive enzyme pepsin, a component of gastric juice (Greek *pepsis*, digestion):

$$-\overset{\overset{\mathrm{O}}{\|}}{\mathrm{C}}-\overset{\overset{\mathrm{H}}{|}}{\mathrm{N}}- + \mathrm{H_2O} \longrightarrow -\overset{\overset{\mathrm{O}}{\|}}{\mathrm{C}}-\mathrm{OH} + \mathrm{H}-\overset{\overset{\mathrm{H}}{|}}{\mathrm{N}}-$$

When all such bonds in a given backbone have been split, the resulting products all have the structure:

$$\mathrm{H}-\overset{\overset{\mathrm{H}}{|}}{\mathrm{N}}-\underset{\underset{\mathrm{R}}{|}}{\overset{\overset{\mathrm{H}}{|}}{\mathrm{C}}}-\overset{\overset{\mathrm{O}}{\|}}{\mathrm{C}}-\mathrm{OH}$$

They are designated amino acids, because they contain both the amino group ($-\mathrm{NH_2}$) and the carboxyl acid group ($-\mathrm{COOH}$); more specifically, α-amino acids, because the two groups are attached to the α-carbon; and, even more specifically, L-α-amino acids, because they all (except for the optically inactive glycine) exhibit the L type of configuration around the asymmetric α-carbon. Their stereoisomers, the D-amino acids, occur in certain bacterial constituents, including some antibiotics, but not in proteins.

Twenty different amino acids take part in the formation of proteins. Their structures (see Appendix 1) conform (with one minor variant, proline) to the structure near the top of this page, differing from each other by the nature of the side group, R. The names of the amino acids are listed in the table on the facing page, along with the abbreviations and symbols used to represent them. As we shall see later, some of the symbols also represent another important group of substances, the nucleosides. Within the appropriate context, there is no risk of confusion. If there is, the three-letter abbreviations are employed.

Amino acid	Abbreviation	Symbol
Alanine	Ala	A
Arginine	Arg	R
Asparagine	Asn	N
Aspartic acid	Asp	D
Cysteine	Cys	C
Glutamic acid	Glu	E
Glutamine	Gln	Q
Glycine	Gly	G
Histidine	His	H
Isoleucine	Ile	I
Leucine	Leu	L
Lysine	Lys	K
Methionine	Met	M
Phenylalanine	Phe	F
Proline	Pro	P
Serine	Ser	S
Threonine	Thr	T
Tryptophan	Trp	W
Tyrosine	Tyr	Y
Valine	Val	V

Amino acids attached to each other by peptide bonds are designated residues. The resulting chains are called peptides. The term peptide is often preceded by a Greek prefix indicating the number of amino-acid residues in the chain—for instance, *di*, two; *tri*, three; *tetra*, four; *penta*, five; *oligo*, a few; *poly*, many.

Peptide chains conform to the general structure:

$$\mathrm{H_2N{-}\underset{\displaystyle R_1}{\underset{|}{C}H}{-}CO{-}NH{-}\underset{\displaystyle R_2}{\underset{|}{C}H}{-}CO{-}\cdots\cdots{-}NH{-}\underset{\displaystyle R_{n-1}}{\underset{|}{C}H}{-}CO{-}NH{-}\underset{\displaystyle R_n}{\underset{|}{C}H}{-}COOH}$$

They have a free amino end-group (N-terminal at the left), a free carboxyl end-group (C-terminal at the right), and $n - 1$ peptide bonds.

Peptides differ from each other by the number (n), nature, and ordering, or sequence, of their amino-acid residues. They may be likened to words of variable length written with an alphabet of twenty letters. The simplified one-letter symbols make this particularly clear. For instance, the tetrapeptide Cys-Glu-Leu-Leu becomes CELL, and the nonapeptide Ala-Arg-Cys-His-Glu-Thr-Tyr-Pro-Glu becomes ARCHETYPE. In fact, this whole book could be written in peptide language, were it not for the absence of B, J, O, U, X, and Z in that alphabet.

Peptides are flexible structures, owing to the possibility of rotation around the N—C and C—C axes in the backbone, though not around the peptide bond itself. As a result, they adopt more or less contorted shapes, depending on the attractions and repulsions existing between their parts and on the ability of these parts to either bind or exclude water molecules. In addition, the folding of a peptide chain is often influenced by the presence of substances with which it is able to associate. Many of these interactions are physical and do not involve the formation of true (covalent) chemical bonds. They depend on two types of forces, which are often designated by the names of the scientists who discovered them: the French physicist Coulomb and the Dutch chemist van der Waals.

Coulomb forces are electrostatic. They govern the attraction between electric charges of opposite sign and the repulsion between charges of the same sign. Some amino acids have R groups that are charged, either positively or negatively, under physiological conditions. Others, without being charged, are polarized—that is, they show a local charge displacement that creates a positive and a negative pole. All such groups can interact electrostatically.

Polar groups bearing a hydrogen atom can bind electrostatically to negative or negatively polarized groups by means of a special bond, called the hydrogen bond, which involves some sort of sharing of the hydrogen atom. A very important such bond in biology is the following:

$$\diagdown\!\!\!\diagup\overset{\ominus}{N}-\overset{\oplus}{H}\cdots\overset{\ominus}{O}=\overset{\oplus}{C}\diagup\!\!\!\diagdown$$

This bond (the dotted line) can join two peptide linkages. It thereby plays an important role in the conformation of proteins, and consequently also in all the structural, catalytic, and other functional properties that depend on this conformation. As we shall see in the later part of our tour, the phenomenon of base-pairing in nucleic acids, which governs all the transfers of genetic information in the living world, depends on the same bond, and on another, similar, hydrogen bond:

$$\diagdown\!\!\!\diagup\overset{\ominus}{N}-\overset{\oplus}{H}\cdots\overset{\ominus}{N}^{\oplus}\!\!\diagup\!\!\!\diagup\!\diagdown$$

Van der Waals forces are responsible for the attraction that exists between nonpolar groups made only of carbon and hydrogen, such as constitute the hydrocarbons found in gasoline and other petroleum products. A number of amino acids have R groups that can establish connections by means of van der Waals forces.

A point of key importance with respect to these interactions is that the water molecule has an asymmetric structure, which makes it polar:

$$\overset{\oplus}{H}-\overset{\ominus}{O}-\overset{\oplus}{H}$$

Hence, it has the ability to bind electrostatically to all electrically charged or polarized groups. These are called hydrophilic for this reason. In contrast, nonpolar groups have no affinity for water; they are hydrophobic, water repellent. We are all familiar with this phenomenon: oil does not mix with water, certain plastic surfaces don't get wet; they are hydrophobic. Actually, this term is a little misleading. Hydrophobic groups do not really repel water; they are excluded by it, as a result of the strong hydrophilicity of the water molecules themselves, which keeps them huddled together by means of hydrogen bonds.

Most of the cellular milieu being aqueous, interactions with water have much to do with the kind of conformation that a polypeptide chain will tend to adopt. In very schematic language, we may say that whenever this is structurally feasible, a polypeptide chain will fold in such a way as to expose the greatest number of hydrophilic groups on its surface, or in clefts accessible to water, and to bury most of its hydrophobic groups together in internal regions from which water is excluded and in which the groups can interact with each other. Proteins that lend themselves to this kind of group segregation on the basis of water affinity are generally soluble in water. Those that are unable to bundle themselves inside a hydrophilic shell tend to home to hydrophobic regions, where their hydrophobic parts are made welcome by van der Waals types of interactions. As we shall soon find out, membranes provide the main resting place of this kind.

An important element in determining the conformation of proteins is represented by the tendency of polypeptide chains to twist into a corkscrew kind of arrangement stabilized by hydrogen bonds between nearby peptide linkages. The most common such arrangement is the α-helix, which has a pitch of 0.54 nm and contains 3.6 residues per turn. It forms a relatively rigid rod about 1 nm thick, with a knobby surface shaped by the R groups of the residues. Some amino acids cannot fit into this arrangement, and break it. Consequently, α-helical parts are usually relatively short segments. Whole protein molecules often contain several such rods, joined, usually at an angle, by less regular parts. Exceptionally, the sequence of a polypeptide may be such as to allow the α-helical disposition to be maintained over a very long distance. Such molecules are naturally filamentous. Other arrangements besides the α-helix, sometimes involving two or three polypeptide chains twisting together (coiled coils), may also provide the molecules with a threadlike struc-

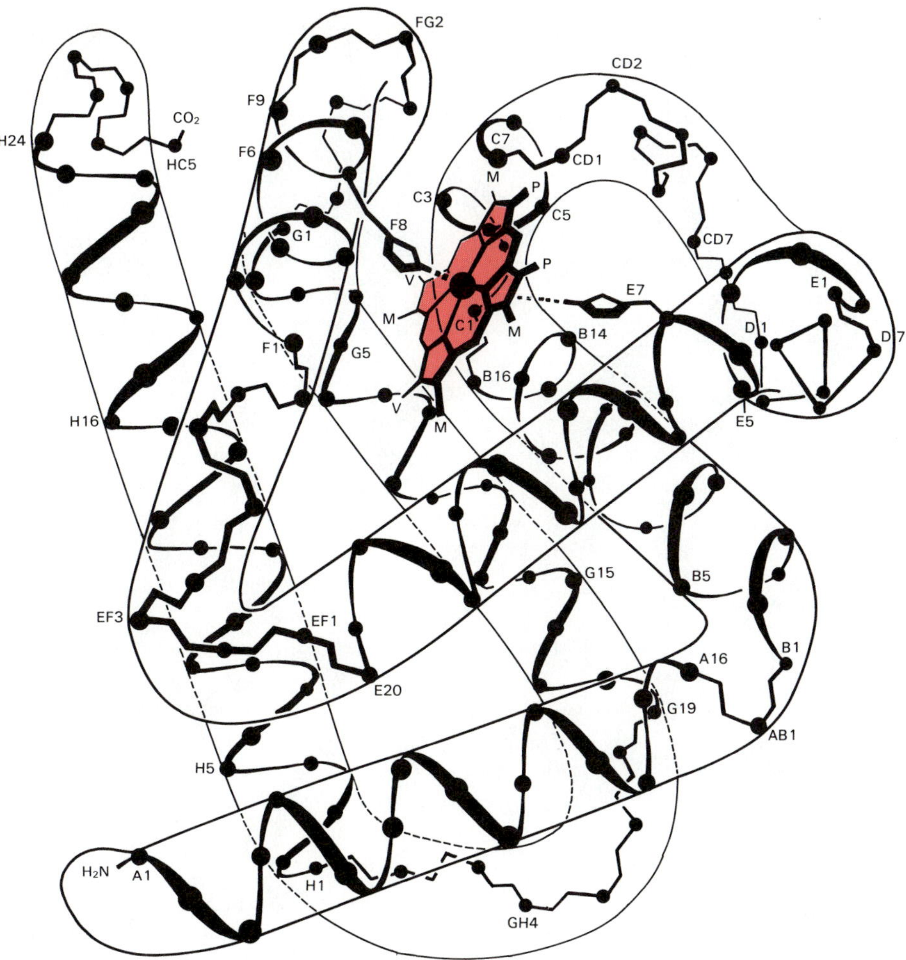

Structure of a myoglobin molecule. The protein includes eight α-helical segments folded around the inserted heme (shown in color). (Compare model on p. 29.) The letters A through H refer to α-helical segments, numbers to amino acid residues.

ture. The functions of such fibrous proteins are mainly concerned with structural support and locomotion.

To these various intrinsic factors that determine the shape that a given polypeptide chain will adopt must be added the interactions between the chain and other molecules of either protein or nonprotein nature. Many of the most important functions of proteins depend on such interactions, which often have a crucial molding effect on the conformation of the molecules involved. Sometimes a given structure is sealed by a true covalent bond between two parts of the same polypeptide chain or between two chains. The most common such bond is the disulfide linkage —S—S—, which arises by the oxidative joining of the thiol (—SH) groups of two cysteine residues:

$$2\ \text{Cys—SH} \longrightarrow \text{Cys—S—S—Cys} + 2\ \text{H}$$

Many proteins consist of a single polypeptide chain; others contain two or more distinct chains or subunits, linked together by noncovalent attractive forces, sometimes also by covalent bonds, such as disulfide bridges.

So much for this preliminary briefing. More will be said about proteins when we meet some particularly characteristic member of this immensely important, as well as varied, group of natural molecules. But now we should start our tour without further ado. A convenient approach to our goal is by way of the circulatory system, which carries most of the traffic to and from cells.

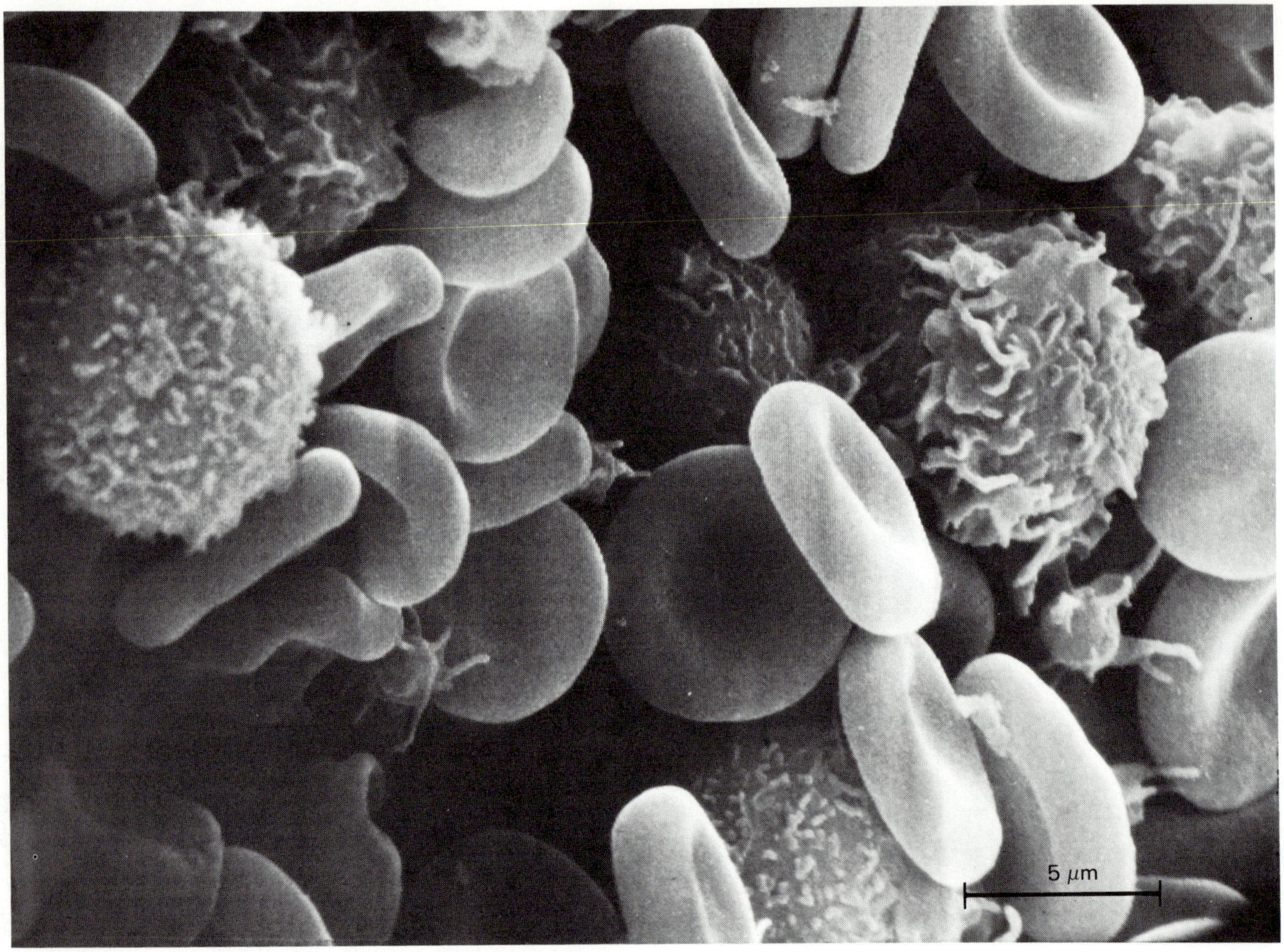

Scanning electron micrograph of blood cells from a patient suffering from Hodgkin's disease. The biconcave disks are erythrocytes; the other cells are leukocytes.

Blood Cells, Our First Companions

Blood vessels are turbulent waterways, especially the larger arteries where we find ourselves at the start of our journey, rhythmically propelled by a powerful swell at every heartbeat. All around us are cells that, like ourselves, are carried along and tossed about by the swiftly moving current of plasma, the fluid part of the blood.

Different types of junctions.

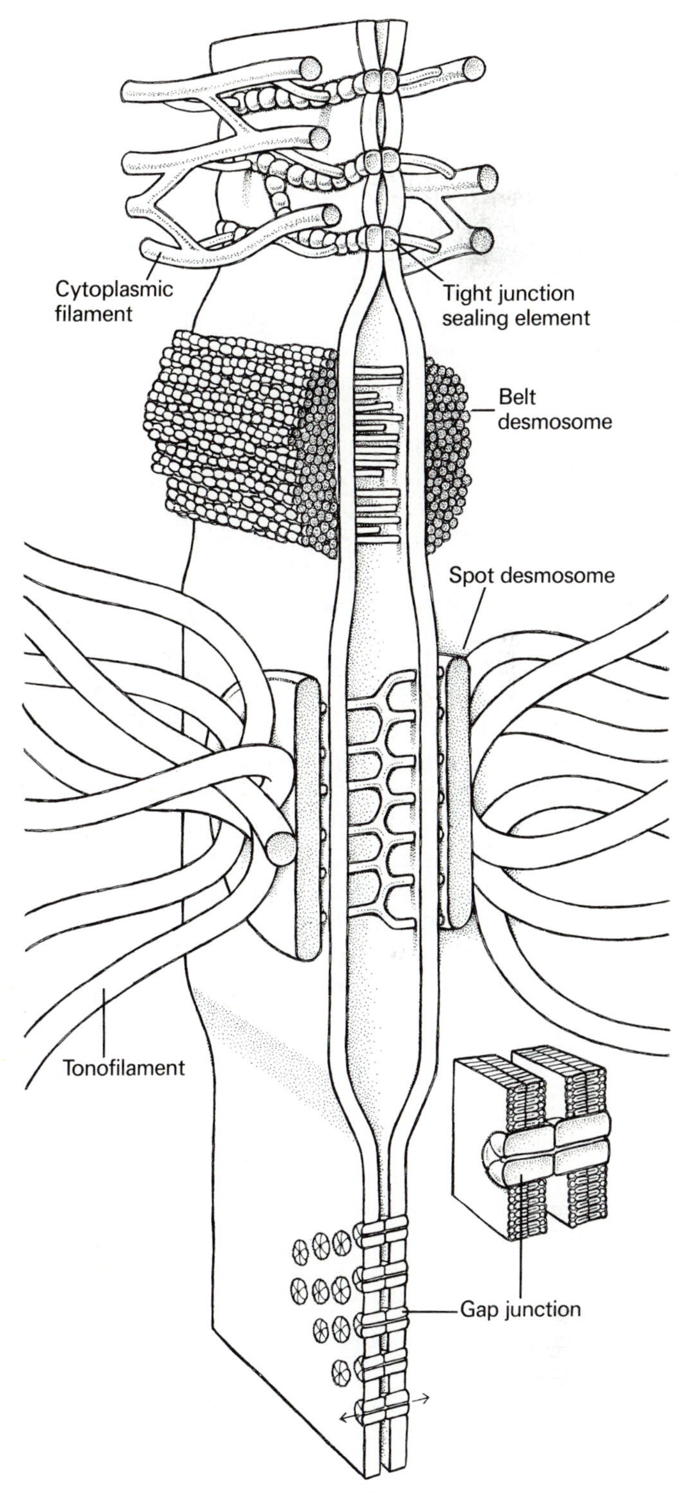

Most of these blood cells are small biconcave disks filled with a ruby-red material. They are the red blood cells, or erythrocytes (Greek *erythros*, red), whose main functions are to bring oxygen from the lungs to the tissues and to help bring the waste-product carbon dioxide back to the lungs. They are extremely valuable to the economy of the organism and contain a very important protein, the red oxygen-carrier hemoglobin. But as cells they are hardly worth a second glance, since they are completely degenerate and virtually moribund. Red blood cells have no mitochondria, no ribosomes, no intracellular membrane system, and, in mammals, no nucleus.

On rare occasions, we may bump into a white cell, or leukocyte (Greek *leukos*, white). White cells occur in different sizes and shapes. But these fleeting encounters hardly afford us a satisfactory view. We will meet some of them again later, under more restful conditions.

As the arteries become narrower and the force of the current begins to spend itself, we start catching glimpses of the cells that line the walls of the blood vessels. They are called endothelial cells (Greek *endon*, inside; *thêlê*, nipple) and form a single layer looking somewhat like an irregular tiling or pavement. They are exceedingly flat over most of their surface, so that their nuclei bulge out into the lumen of the vessel, thus giving the wall a knobby appearance. Endothelial cells are cemented together by different types of junctions to form a thin, continuous, tubular sheath that serves as an important filter and regulator of exchanges between blood and tissues. Such junctions are not peculiar to endothelial cells. Most cells are attached in this manner and organized into sheets, columns, clusters, sacs, or any other of the various multicellular arrangements of which the tissues and organs are composed. The strongest intercellular junctions are the desmosomes (Greek *desmos*, bond; *sôma*, body), in which the two cells are connected by a dense joint and riveted together by bundles of transverse fibers extending deeply into their cytoplasms; these are known as tonofilaments (Greek *tonos*, tension). Several other types of junctions exist, including one, called gap junction, in which adjacent cells are connected by closely packed rows of hollow, cylindrical pegs, about 15 nm long, and 8 nm in diameter, with a 1.5- to 2-nm bore. Plugged at each end into the membranes of adjacent cells, these structures keep the two cells separated by a narrow gap but allow the intercellular passage of electric currents and of small ions and molecules through their inner channel.

Without additional supporting structures, the multicellular connections established by these junctions would be quite unable to stand even the weakest of stresses. The junctions would resist, but not so the membranes they serve to join, which, as we shall see, are very flimsy films, hardly stronger than soap bubbles. Blood vessels are particularly in need of such reinforcement to withstand the pressure imposed upon them by the contractions of the heart. Thus the larger arteries are enveloped in a thick, resilient casing, almost impassable by even the smallest molecule. But as the vessel becomes narrower and the blood pressure decreases, the wall becomes thinner. By the time we reach the capillaries, which are very narrow, hairlike ducts (Latin *capillus*, hair), only a thin sheet, called basement membrane (see p. 40), remains to back up the endothelial lining. Here is where most exchanges between blood and tissues take place—and where we ourselves should leave if we don't want to get caught in the venous return traffic. But to do so, we must pry open a junction between two endothelial cells. As we wonder how to do this, unexpected help comes from a passing leukocyte that, before our very eyes, is undergoing an odd change in behavior.

A moment ago, this cell was just drifting along. Now it clings to the endothelial lining and forcibly worms its way between two cells in what seems to be a most frantic activity. It is, in fact, displaying chemotaxis (Greek *taxis*, arrangement), in response to exposure to some chemical. Prompted somehow by the encounter, the cell starts to follow the chemical's trail back to its source—that is, it moves up the chemical's concentration gradient, cutting its way through intervening structures by means of specially secreted enzymes. This remarkable, and still poorly understood, phenomenon is elicited by substances that are commonly released at a site of some injury or bacterial infection. It is most useful to the organism, therefore, as it automatically brings the white cells to where their defensive properties are needed. As an unexpected fringe benefit to our party, it has opened a gap for us through the capillary wall.

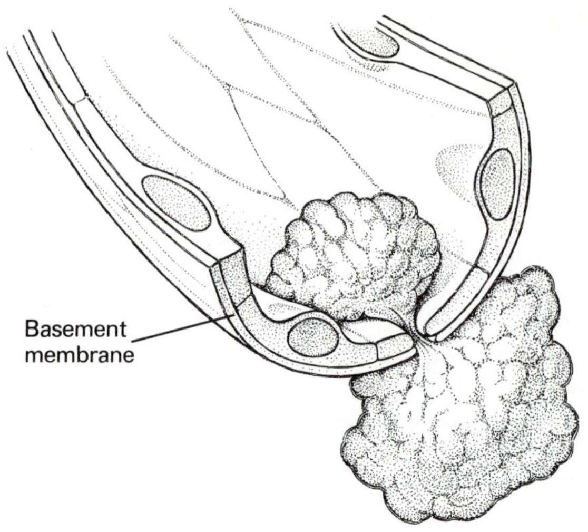

Leukocyte, attracted by chemotactic signal, squeezes through capillary endothelium.

Breaking through the Extracellular Structures

As we enter the extracellular spaces, we come upon an awesome sight, reminiscent, except for its submarine character, of those wild jungle scenes dear to the illustrators of early travel books. A thick tangle of filaments stretches like lianas between towering trunks tilted in every direction. In some areas, the mesh is almost impenetrable. In others, it is only a tenuous filigree. Such, at least, is the view that we get at the molecular level with our millionfold magnifying glasses. If we magnify our surroundings a few hundredfold less, we find that microscopic order arises out of this apparent molecular disorder in the form of fibers, sheets, plates, and other structural elements.

An awesome sight. The interior of a primeval forest on the Amazon.

Cytonaut works his way through connective-tissue jungle. Note thick collagen trunks and tangle of proteoglycan lianas.

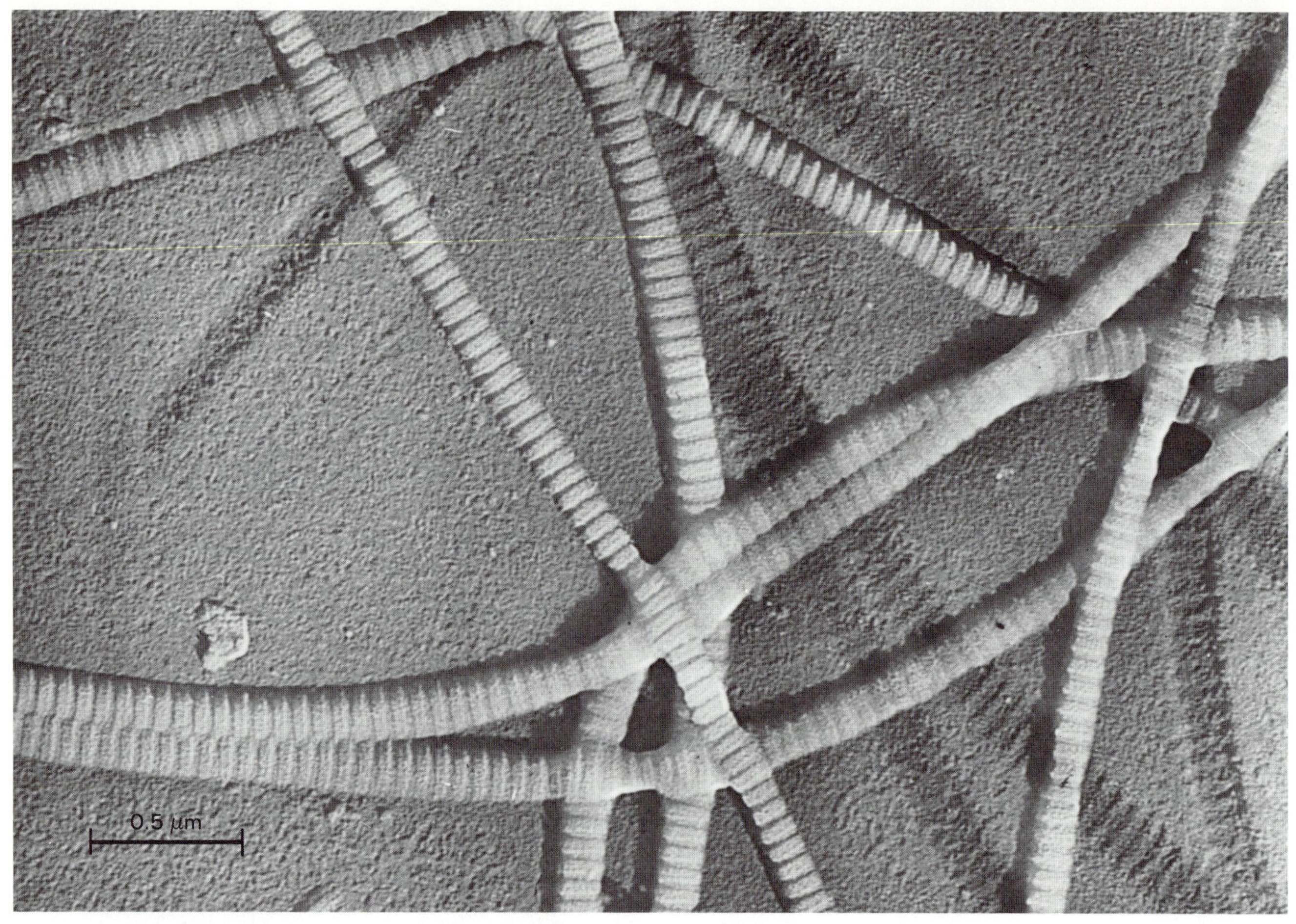

Electron micrograph of metal-shadowed replica of collagen fibers from human skin.

The main molecular component of these extracellular structures is a substance called collagen, which is extracted industrially from bones to make gelatin and animal glue; hence its name (Greek *kolla*, glue). Collagen is a protein. Its basic unit is a long polypeptide chain of 1,055 amino acids, exceptionally rich in the amino acids glycine, proline, and hydroxyproline. Note that the last one is not on the list of twenty given on page 31. It is made from proline residues after the polypeptide has been synthesized, and it is peculiar to collagen. Also peculiar to this polypeptide is its twisting into a left-handed helix of about 1-nm pitch, quite different from the α-helix mentioned earlier. Three such chains join into a right-handed helical thread about 1.5 nm thick and 300 nm long, with a 3-nm repeat (or 9-nm pitch for each individual chain).

Called tropocollagen (Greek *tropê*, turn), the triple-stranded molecule is the building block of collagen. With our molecular eyeglasses, we would see it as a short length

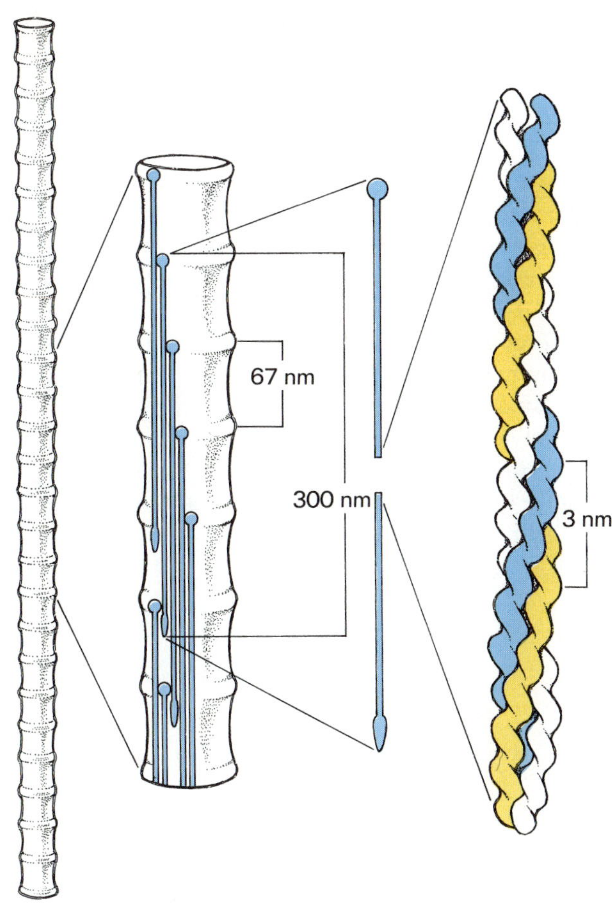

Structure of collagen fiber showing its staggered arrangement of triple-stranded tropocollagen molecules.

of string, a little over one-twentieth of an inch thick and close to a foot long. How such frail material can build the massive trunks that surround us is explained by the natural propensity of tropocollagen molecules to join side by side in staggered fashion. This property allows them to assemble spontaneously into fibers of virtually any thickness or length. The arrangement is such that the C-terminal heads of adjacent tropocollagen molecules are staggered by exactly 67 nm. Five such repeats (335 nm) are the minimum needed to accommodate the full length of a tropocollagen molecule (300 nm), leaving a gap of 35 nm between the heads and tails of consecutive molecules. Because these gaps are in register with each other across the whole thickness of the fiber, they create a characteristic cross-striation by bands 67 nm apart. Each of these bands corresponds to a region where one-fifth of the fiber thickness is free. This free space is important. In bone, it is the site of mineralization.

Tropocollagen is made as a precursor molecule, called procollagen, that possesses extra N-terminal and C-terminal pieces that prevent self-assembly from taking place prematurely. Only after the procollagen has been discharged extracellularly is it processed by enzymes that remove the interfering parts. After the fibers have formed, they undergo further changes, including strengthening by cross-links. Such modifications are believed to continue throughout life and to contribute to the progressive stiffening of connective-tissue structures with age.

Collagen fibers are the main components of the rigid framework whereby supporting structures are reinforced. They act, so to speak, like the steel rods in reinforced concrete or the fibers in fiberglass. Arranged in parallel, they serve to make all sorts of longitudinal parts, up to tendons and ligaments. Their three-dimensional intertwining forms the scaffolding of tissues and organs, including the authentic skeletal parts, such as cartilage and bone.

Several different types of collagen contribute to the building of these various structures. A special kind, called type IV, tends to make flat arrangements. Together with other proteins, such as laminin (Latin *lamina*, thin plate) and fibronectin (Latin *nectere*, to bind), and with proteoglycans (see below), it forms a resilient sheetlike material, between 50 and 100 nm thick, out of which structures known as basement membranes are made. This appellation, which goes back to the early histological literature, is unfortunate. Basement membranes are in no way constructed like the cellular membranes we are about to see; they are really walls. Sometimes they surround individual cells, as do the walls of bacterial or plant cells. More frequently, they envelop or support multicellular arrangements of various sizes and shapes. One of their functions is to provide cells with the kind of carpeting they need to creep or stick. Another is to act as molecular filters. The basement-membrane casing of capillaries plays a particularly important role in screening the substances that are allowed to gain access to the tissues from the blood. The most exacting such filter occurs in brain capillaries (blood-brain barrier).

In certain areas, as in the arterial wall, use is made of another filamentous protein called elastin. Like collagen, elastin assembles into fibrils, which themselves form more elaborate arrangements, mostly fibers and flat plates, or lamellae. Unlike similar structures made of collagen, elastin fibers and plates tend to adopt a sinuous or scalloped shape, such that they can be stretched to 1.5 times their length and spontaneously shorten again when released. As their name indicates, they provide elasticity.

As in reinforced concrete or fiberglass, the fibrous network of the supporting structures is embedded in an amorphous matrix or filler, called ground substance, which, at the molecular level, is seen as a mesh of tenuous polymeric molecules. These include a number of important polysaccharides, many of them attached by one end to a protein stalk (proteoglycans). Depending on the nature of these molecules and on their density, the resulting matrix may be no more than a viscous fluid, may behave like a jelly, or may reach the hardness and resilience of cartilage or a lobster shell. Sometimes, as in bones, teeth, coral, mollusk shells, and other biomineral structures, the matrix spaces are largely occupied by crystalline deposits of mineral salts.

As a rule, one more obstacle stands in the way of the intruder trying to approach a cell. It is a carbohydrate-rich covering called surface coat, or glycocalyx (Greek *kalyx*, husk). Its thickness and appearance vary according to cell type from a hardly detectable fuzz to a bona fide basement membrane. Bound to the cell membrane by relatively weak forces, it serves to maintain a special microenvironment around the cells, protects them against certain physical or chemical attacks, and helps them to recognize kindred neighbors and to establish connections with them and with connective-tissue structures.

3 The Cell Surface, With an Introduction to Membranes and Lipids

Most striking to the traveller who first sets eyes on the cell surface are its unevenness and changeability. With rare exceptions, cells display surface patterns of great complexity that are characteristic for each type. Some cells are ravined by deep clefts or pitted by craterlike depressions. Others are deformed by protrusions, called pseudopods (Greek *pous, podos*, foot), or studded with fingerlike projections identified as microvilli (Latin *villus*, shaggy hair) or as cilia (Latin *cilium*, eyelash). Yet others have their surface pleated by filmy veils. The variety is endless.

A Very Flexible Skin

The scanning electron microscope has revealed the rare beauty of these patterns but fails to show their constant movements and kaleidoscopic changes. Cilia beat, membranous veils wave and undulate, microvilli bend and twist, pseudopods bulge, craters erupt suddenly in the midst of a quiet surface, spewing secretory products, or, in contrast, invaginations yawn open, to be sucked deep into the cell. Sometimes, as we recently witnessed, the whole cell appears convulsed or starts crawling away, set in motion by some elusive chemical signal.

In spite of these incessant, sometimes tumultuous contortions, the cell always remains completely clothed by a tight-fitting membrane that adapts to every change in shape with what looks like effortless plasticity. This membrane is called the plasma membrane, or plasmalemma (Greek *plasma*, form; *lemma*, husk). It is a thin film, about 10 nm thick. Magnified a millionfold, this

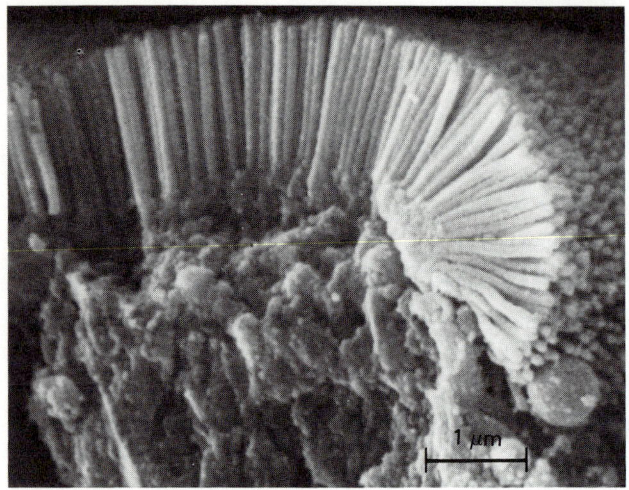

Scanning electron micrograph of intestinal mucosal cell illustrating microvilli.

amounts to no more than 1 cm, or 0.4 inch—an uncommonly tenuous casing for a structure the size of a large auditorium (which, we have seen, is the size of a cell similarly magnified). Actually, we will find out when we explore the internal aspect of this membrane that it does not alone bear the brunt of containing the cell; it is helped by a number of supporting structures that bolster it on its inside face.

Containment is only one of the many functions of the plasma membrane, which is really a major cell organ. It is a highly complex and dynamic structure that regulates virtually every interaction between the cell and its environment, including other cells. It does so thanks to a unique mode of construction that involves two main types of molecular constituents—phospholipids and proteins.

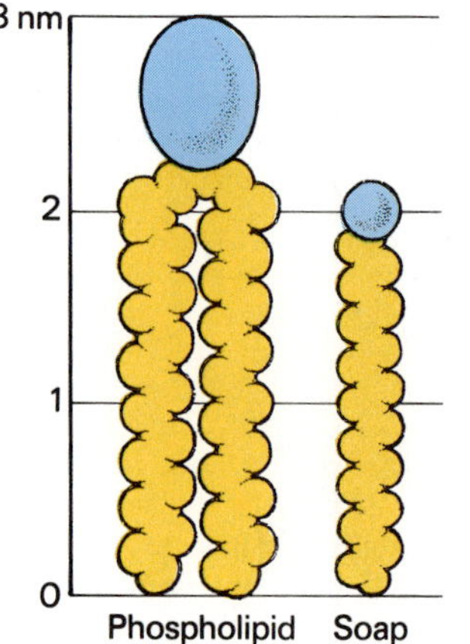

Models of phospholipid and soap molecules.

Lessons of a Soap Bubble

The secret of membrane construction—not just the plasma membrane, but all other biomembranes—is the lipid bilayer (Greek *lipos*, fat). The simplest example of a lipid bilayer is a soap bubble. Soaps, the salts of fatty acids, are linear molecules made up of only carbon and hydrogen, except for a terminal carboxyl group carrying a negative charge, COO^-. In terms of the classification that we have seen in the preceding chapter in relation to the amino acids, they have a long, hydrophobic tail and a hydrophilic head. Such molecules are called amphipathic (or amphiphilic), which is the Greek way of saying that they have two loves. The lipids of membranes are more complex than the simple soaps, but they, too, are amphipathic. They have a forked hydrophobic tail that consists of two fatty acid chains and a bulky hydrophilic head that contains a negatively charged phosphoric acid group, linked to another group that is often positively charged; they are called phospholipids.

When amphipathic substances are mixed with water, they spontaneously adopt a molecular configuration that satisfies their two conflicting requirements simultaneously. They associate in such a way as to dip their hydro-

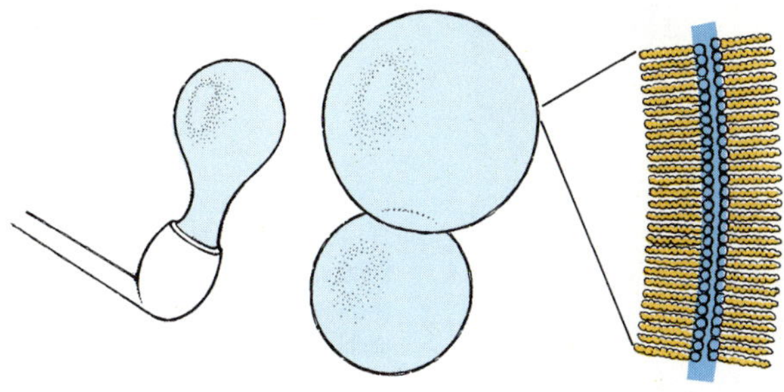

Different kinds of structures formed by amphipathic substances and air, water, water and air, or water and oil.

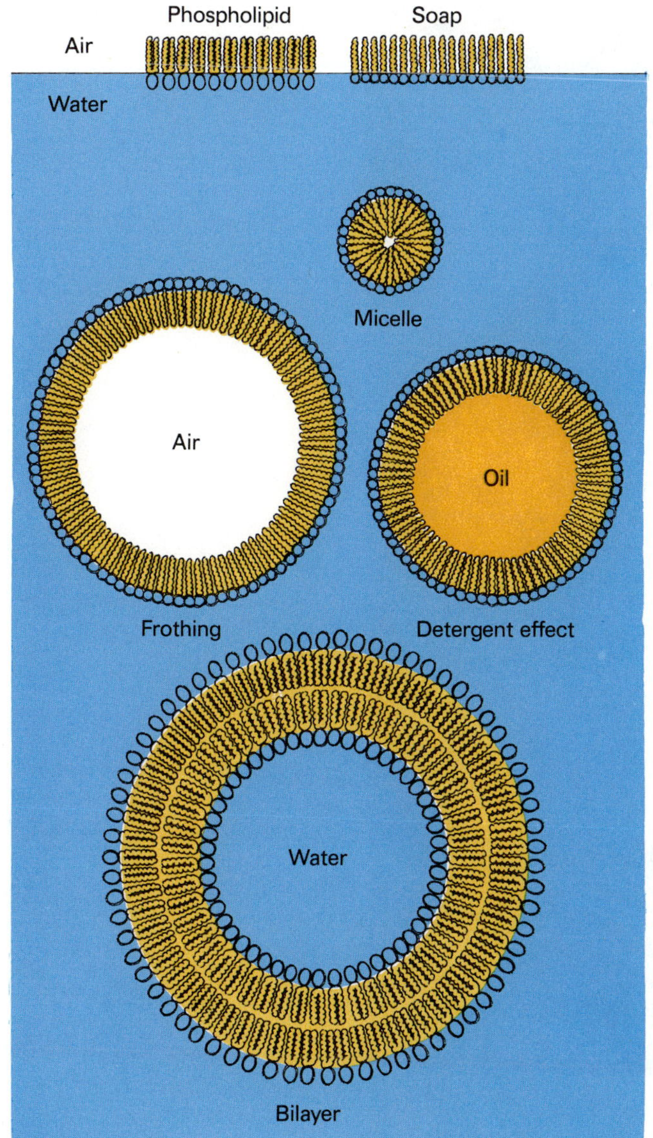

philic heads in the water, while keeping their hydrophobic tails out of it and in contact only with each other and with any other hydrophobic materials, such as oil, plastic, or air, that may be around. Essentially three structures can be generated in this way: monolayers, micelles, and bilayers. Molecular monolayers form at interfaces between water and air or some hydrophobic fluid. Micelles form in water, as spherical clusters of intertwined tails surrounded by heads. Stir a micellar solution with air, and the micelles will join and spread as monolayers around imprisoned air bubbles (frothing, surfactant effect). Stir it with oil or use it to rub a greasy surface, and the micelles will similarly change to monolayers that coat dispersed droplets of oil or grease (emulsifying or detergent effect). As to bilayers, they develop under certain conditions as partitions between two phases of similar nature. Two arrangements are possible. If the two phases are hydrophobic—air, for instance—the tails will stick outside on both faces of the bilayer and the heads will be inside, joined by a film of water. This is the structure of soap bubbles. If the two phases are aqueous, the tails will be inside the bilayer and the heads outside.

Phospholipids have a great tendency to form bilayers when they are mixed with water. Their forked hydrophobic tails are too bulky to fit comfortably inside micelles. In contrast, they are readily accommodated in bilayers. These close spontaneously into vesicles, which may consist of a single bilayer or of several concentric bilayers, depending on how they are prepared. Called liposomes, such artificial phospholipid vesicles have become an im-

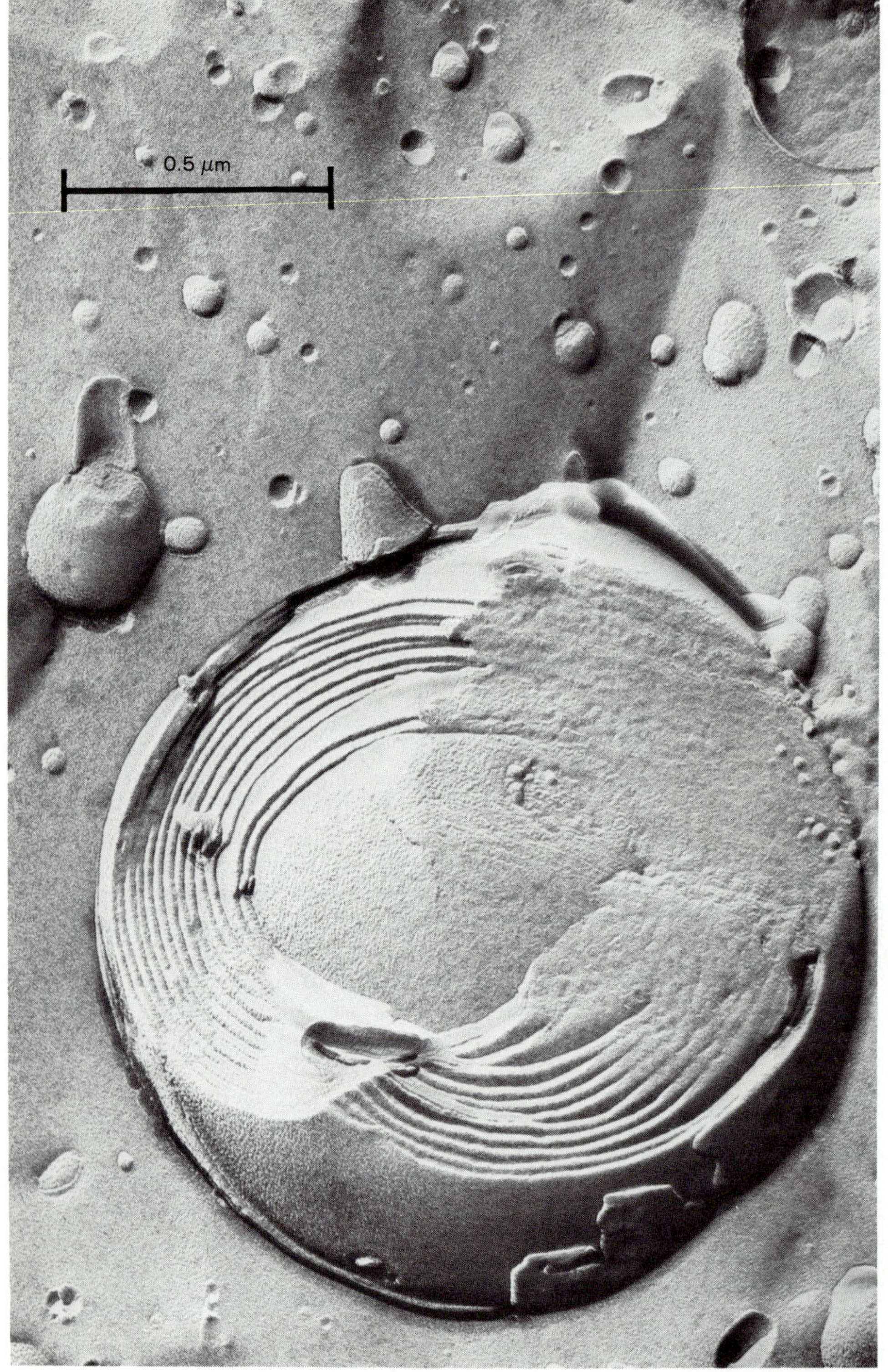

Electron micrograph of a metal-shadowed replica of a freeze-fractured multilayered liposome.

portant tool in cell research. The natural propensity of phospholipids to assemble into bilayers in an aqueous milieu lies at the heart of the structure of normal membranes, which all have as their basic fabric a phospholipid bilayer, from 5 to 6 nm thick. In addition, they contain various protein molecules, of which more will be said later.

Several important properties of biological membranes, as well as of soap bubbles, are explained by their lipid bilayer structure. One property is their flexibility. A lipid bilayer is an essentially fluid arrangement within which the molecules can freely move about in the plane of the bilayer and reorganize themselves into almost any sort of shape without losing the contacts that satisfy their mutual attraction.

This kind of flexibility is characteristic of soap bubbles, except that they are more rigid than cell membranes owing to their higher surface tension. They can suffer distortion, when blown upon, for instance, but have a strong tendency to adopt the spherical shape, which minimizes surface tension. The cell membrane, with one face supported by cytoplasmic structures and the other in contact with a watery milieu, has a lower surface tension and is therefore more flexible than a soap bubble.

The fluidity of lipid bilayers requires that the hydrophobic tails be able to slide freely past each other. This ability, in turn, is influenced by temperature. Below a certain critical temperature, called the transition temperature, which depends on the nature of the lipids concerned, the hydrophobic chains congeal into an ordered, rigid structure, no longer compatible with the functions that a membrane has to fulfill. Indeed, cells adapted to different temperatures have membrane lipids of correspondingly different transition temperatures. Bilayer fluidity is also influenced by inserted substances that interfere either with the sliding of the hydrophobic chains or with their congealing. Cholesterol, which is a key component of the plasma membrane in all eukaryotic cells, owes its importance to such interactions.

A second important property of lipid bilayers is that they are self-sealing. Again look at a soap bubble. If you cut it in half, you do not get two half-bubbles, but rather

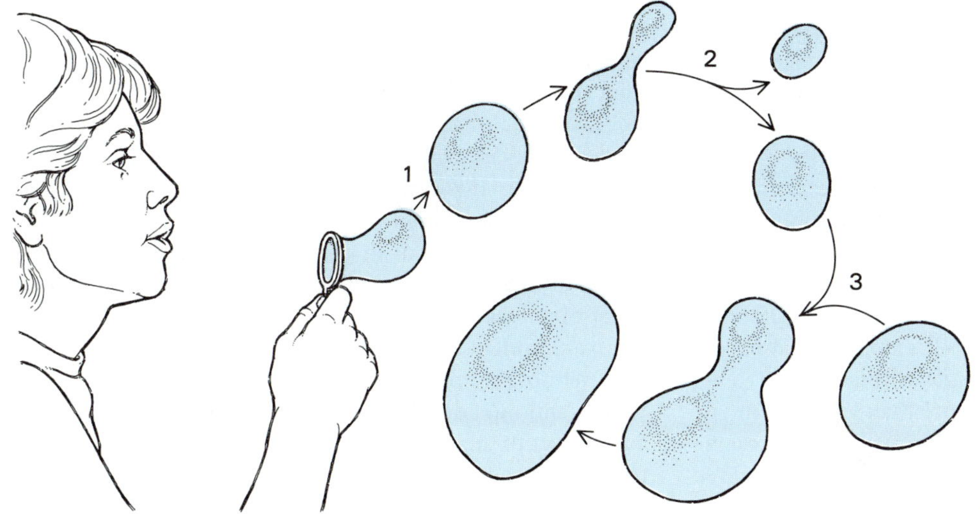

Soap bubbles (1) detach and seal; (2) divide; and (3) fuse.

two bubbles, smaller but whole. If two soap bubbles collide, they fuse, and the whole structure reorganizes to make a single and bigger soap bubble. So it is possible to have fusion and fission of lipid bilayers, always on the basis of the self-sealing properties inherent in this type of structure.

Coming back to the cell, it is possible to push a needle through a cell membrane and take it out again: the puncture site will close automatically. It is even possible to cut a cell with a microknife and obtain two pieces, each completely surrounded by a sealed plasma membrane. Conversely, cells can be made to fuse, like soap bubbles, by merger of their plasma membranes. This phenomenon happens physiologically—for instance, in muscle development—and it can be induced artificially. It is exploited for the preparation of monoclonal antibodies (see Chapter 19). Membrane fusion also plays an important role in many intracellular processes, as we will see later in the tour.

A third property of lipid bilayers (of the membrane type) is their impermeability to molecules that are soluble in water. Such molecules cannot traverse the bilayer, because in order to do so they would have to cross the oily film made by the hydrophobic tails of phospholipid molecules. To get physically through such a film, a substance must itself be hydrophobic or it must take advantage of thermal agitation and squeeze through the occasional gaps that open up in the bilayer as a result of molecular movements. This is how water and other very small molecules are exchanged between the cell and the environment. As a rule, however, the bilayer serves as a highly effective barrier, allowing the cells to retain their own constituents and ward off extracellular substances with a minimum of leakage. This is sound economy, resembling that of medieval cities, which protected themselves with a wall and a moat. But bridges and gates are needed to permit, and at the same time to regulate, the two-way traffic that necessarily must take place to support life within the city. This the lipid bilayer, which is little more than a passive moat, cannot do by itself; it requires the participation of membrane proteins, molecules endowed with a higher degree of specificity.

From Bilayers to Membranes: the Role of Proteins

As mentioned in the preceding chapter, proteins that are unable to bury their hydrophobic groups inside a hydrophilic shell find their natural residence in membranes. There they can make all their parts happy by inserting their hydrophobic domains into the lipid bilayer and allowing the hydrophilic ones to emerge into the watery outside world. Depending on the organization of these domains, the molecules will either float iceberglike on the bilayer, with their hydrophilic parts all facing the same side, or straddle the bilayer, with hydrophilic parts protruding on both faces (transmembrane proteins). Some of these arrangements are remarkably complex. Bacteriorhodopsin, the major component of the photochemical apparatus of the microbe *Halobium* (see Chapter 10), spans the bilayer seven times. The acetylcholine receptor (see Chapter 13) consists of two units, each of which is made of five transmembrane polypeptides clustered around a central channel. In these and other known examples, the protein parts that cross the bilayer consist of α-helical rods containing from 21 to 27 amino-acid residues, most of them hydrophobic.

Such proteins, which have one or more hydrophobic parts embedded in the lipid bilayer of a membrane, are called integral, or intrinsic, membrane proteins (whether they traverse the bilayer or not). Those that are attached to either side of the membrane but do not penetrate the lipid bilayer are called peripheral, or extrinsic.

Thanks to the essential fluidity of the lipid bilayer, which was commented upon earlier, membrane proteins are free to move about in the plane of the membrane by lateral diffusion. This freedom is essential to a number of functions—we will soon turn it to our own use—but there are limits to it, imposed by the interactions between the membrane proteins and other components lying within and outside the membrane. Some of these patterns are subject to modification by specific molecules that bind to membrane proteins, with consequences that may be of great functional importance.

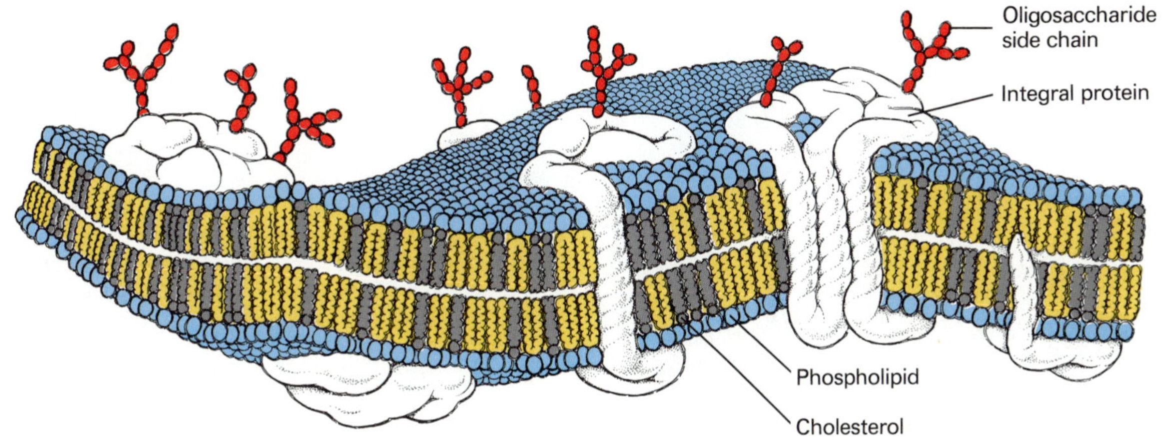

A molecular view of the plasma membrane. Integral proteins either float iceberglike on the lipid bilayer or they straddle it; they move about by lateral diffusion. Oligosaccharide side chains of glycoproteins form molecular down—or a forest of antennae—on the cell surface.

A characteristic of the plasma membrane is that many of its outside protein components are glycoproteins—that is, they bear oligosaccharidic side chains made of sugar molecules. It also contains some glycolipids. As many as a dozen or more sugar molecules of several different kinds may participate in the formation of these carbohydrate components, which often have a branched structure. They cover the cell surface with a fine molecular down. In actual fact, they are not down; they are feelers. Magnify them another hundredfold, and you will readily appreciate this. You are now reminded of one of those underwater scenes where shoals of fishes and other creatures browse peacefully among animals that look like plants, until suddenly a drooping limb turns into a deadly tentacle that snaps up an unsuspecting victim and drags it inside. In our cellular seascape, the potential prey are molecules present around the cells; the tentacles are the oligosaccharide side chains (and other parts) of the membrane glycoproteins, many of which function as receptors.

Surface Receptors

The concept of a receptor was formulated in the beginning of this century by the German scientist Paul Ehrlich, known for his contributions to immunology and chemotherapy. He used the lock-and-key analogy, introduced by his contemporary, the chemist Emil Fischer, to explain how there may be on the surface of cells certain strategically located chemical groupings—the receptors, or binding sites—that specifically bind a certain kind of molecule—an antibody, for instance (see pp. 50–51), or a drug, in general terms a ligand (Latin *ligare*, to bind). Just as there are many locks and many keys, there are many receptors and many ligands.

Modern chemistry has added to this concept that of conformational change: when occupied by its ligand, the receptor has a shape different from that when it is free—that is, the folding pattern of the polypeptide is altered. When this kind of change happens to a transmembrane protein or to a molecule capable in turn of affecting the conformation of a transmembrane protein, some sort of communication is established between the outside and the inside of the cell. Many of the most important interactions between a cell and its environment, including other cells, occur by way of receptors.

There are many kinds of receptors on the surface of any given cell, each represented by up to hundreds of thousands of molecules. Many, but possibly not all, are glycoproteins. They cover the cell with a forest of molec-

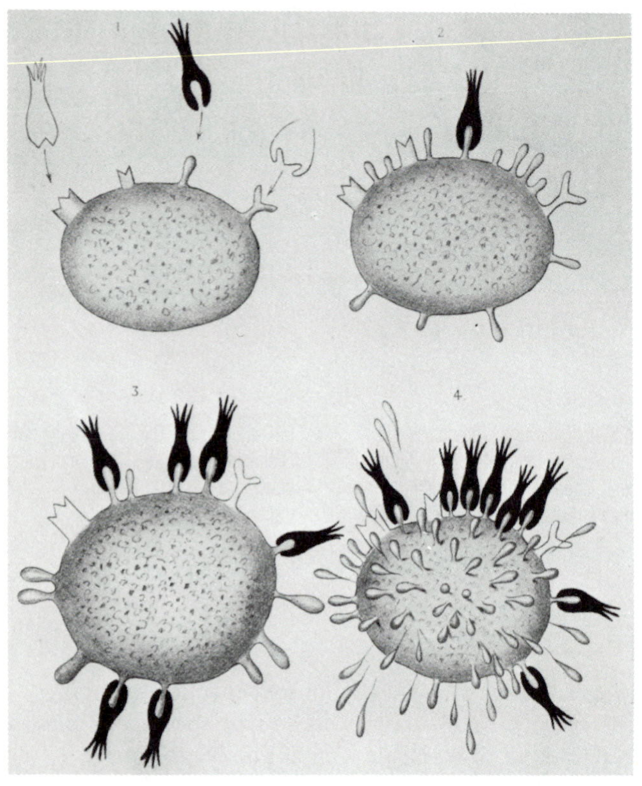

The picture at the left was drawn by Paul Ehrlich in 1900 to illustrate his receptor concept.

A receptor, when occupied by a ligand on the cell surface, changes conformation and activates an intracellular trigger, as shown below.

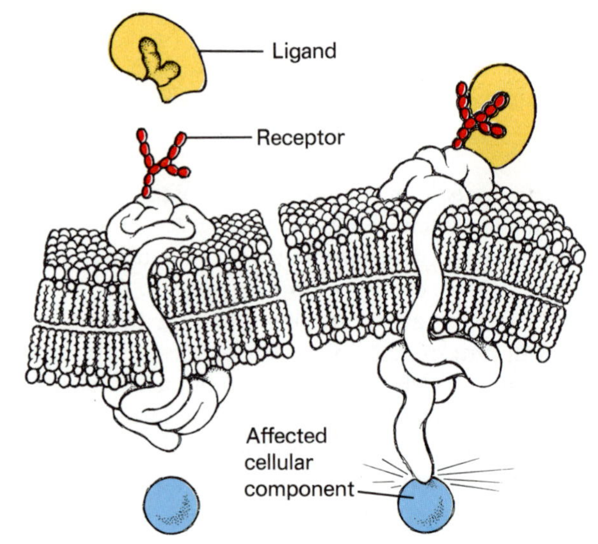

ular antennae and offer the observer a most entertaining display. As molecules of all kinds drift by, carried by the eddies and currents the cell maintains around itself, no moment passes without some spectacular catch occurring somewhere. Sometimes, more than one receptor molecule participates in the binding process. It may even happen that hundreds of receptors become involved collectively in the immobilization of a large object, such as a bacterium.

An event that often follows capture, and is readily seen from our vantage point, is intake, just as in our underwater analogy. The processes involved make up some of the most dramatic parts of the receptor show; they result in the formation of pitlike depressions into which receptors and prey disappear from our view, drawn in by an invisible force. We will take advantage of this mechanism to effect our first entry into the cell and will consider it in detail in the next chapter.

Among the substances that get caught by receptors, many are specific messengers, called hormones, that are manufactured elsewhere in the body and arrive at the cell by way of the bloodstream. Hormones that bind to a cellular receptor generally end up being taken in. Before that, however, their binding triggers some characteristic cellular response. From where we are, we can observe only the quiver of the receptor as it becomes occupied.

Not until we visit the cytosol will we be able to see the devices to which the receptors are connected on the other face of the membrane and the mechanisms whereby those devices cause the cell to respond in some specific way (see Chapter 13).

A subtle, but very important, change mediated by some receptors when they are occupied is gating—the temporary opening of a channel through which given substances or ions are allowed to enter or leave the cell. To understand the significance of such a process, we need to look more closely at the mechanisms whereby molecules cross the plasma membrane.

Molecular Traffic across the Plasma Membrane

The maintenance of cellular life depends on the continuous passage of many different substances—most of them highly hydrophilic—into and out of cells across the plasma membrane. Sugars, amino acids, and other nutrients must get in to satisfy growth and energy needs, while waste and breakdown products must get out to avoid piling up inside. In addition, ions must be moved in or out in order to maintain the ionic composition of the intracellular milieu, which is very different from that of the surrounding medium. For one thing, it is much richer in potassium ions and much poorer in sodium ions. These inequalities cause leakages, which must be compensated by transport in the reverse direction. All this adds up to considerable two-way traffic across a boundary that, it must be remembered, is made largely of a continuous phospholipid bilayer, almost impermeable to most hydrophilic molecules.

The first condition to support such traffic is a large enough surface area. In fact, the main function of microvilli, those fingerlike projections found densely crowded on the surface of some cells (p. 42), is to increase the area available for exchanges between the cells and their environment. They are characteristically found on cells, such as those that line the intestinal mucosa or the kidney tubules, that are exceptionally active in such exchanges.

A second condition is the occurrence of appropriate conduits across the lipid bilayer. These are usually provided by transmembrane proteins acting as specific carriers for the translocated substances or, in some special cases, as controllable gates or channels, operated by the binding of a specific ligand or by an electric perturbation.

The third condition is the availability of energy. The decisive factor here is whether the direction of transport is down a concentration gradient (from where the substance is more concentrated to where it is less) or up a gradient.

Different kinds of molecular transport.

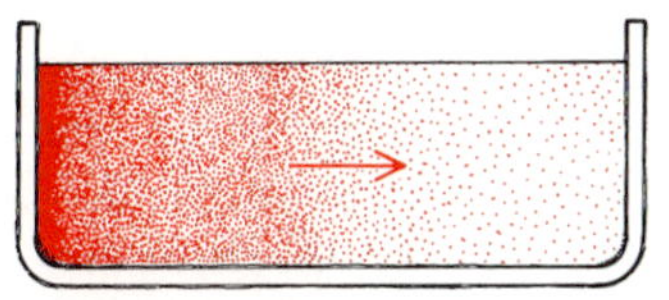

1. DIFFUSION
Source of energy: negative concentration difference in direction of movement
Rate: slow

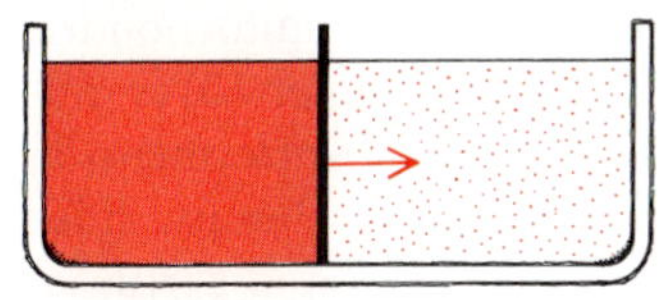

2. PERMEATION
Source of energy: same as for diffusion
Rate: very slow

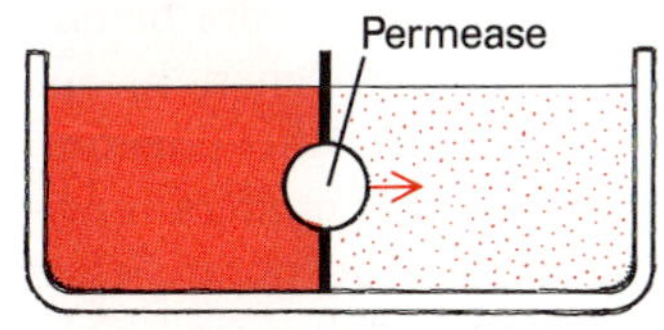

3. FACILITATED TRANSPORT
Source of energy: same as for diffusion
Rate: fast

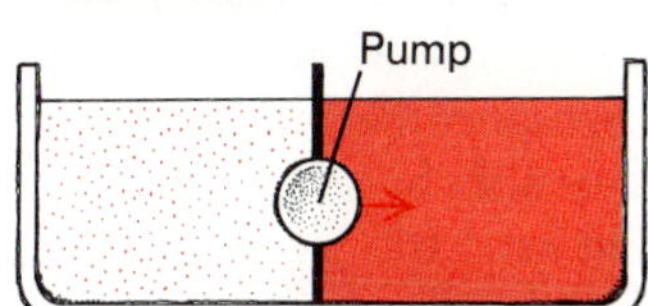

4. ACTIVE TRANSPORT
Source of energy: must be supplied (usually ATP)
Rate: fast

Downhill transport can occur spontaneously (see Appendix 2), but it tends to level the concentration gradient that supports it and thereby to exhaust its source of energy. For transport to continue, the gradient needs to be maintained through the production or supply of the substance on one side of the membrane and its consumption or removal on the other. Downhill transport is called diffusion if it is unhampered and permeation if it is restricted by a permeability barrier. The transport is said to be facilitated when it is helped by some sort of carrier or translocator system acting catalytically. Translocators that catalyze facilitated transport are called permeases. The term ionophore (Greek *pherein*, to bear) is used to designate a special group of carriers that mediate the transport of certain ions across membranes.

When transport occurs against a concentration gradient, the process must be directly supplied with energy, and the machinery involved is correspondingly more complex. The systems that carry out such active transport are generally referred to as pumps. The most important among them translocate positively charged ions, especially of hydrogen (H^+), sodium (Na^+), potassium (K^+), or calcium (Ca^{2+}). Their activity often results in the creation of an electric imbalance across the membrane (membrane potential). Energy transduction in mitochondria (see Chapter 9) and chloroplasts (see Chapter 10) is crucially dependent on this kind of charge displacement, as are all the manifestations of bioelectricity. Pumps, aided by controllable ion gates that allow very rapid perturbations of the generated membrane potentials, are behind nerve conduction, brain function, muscle excitation, cardiac rhythm, gland stimulation, and many other phenomena. They feed those hidden currents that modern medicine has learned to tap for diagnostic purposes. They support the discharges of up to several hundred volts whereby the torpedo fish and the electric eel stun their prey.

At present, the battery of transport machines that stud the cell surface reveals itself to us only by streams of ingoing and outgoing molecules and ions. But we will have an opportunity to see something of their inner works later in the tour, when we get to the other side of the plasma membrane (see Chapter 13).

Identity and Immune Recognition

One last important function of the plasma membrane is to provide cells with an identity card. This consists of a number of specific chemical groupings, known as transplantation, or histocompatibility, antigens. (The term antigen designates any molecule that can elicit an immune response.) The first such antigens to be discovered were those that determine the blood groups A and B. As is known, some of us are blood type A, others are B, and still others are AB or O. In other words, we fall into one of four classes that represent the four possible combinations (2^2) that can be achieved with two characters, depending on whether they are present or absent.

Today, many transplantation antigens have been recognized in the human species. Their number and polymorphism are such as to make it highly improbable that any two individuals should have exactly the same combination. In practice, it happens only in identical twins. Transplantation antigens are displayed more or less completely on the surface of every cell of a given individual and are characteristic of that individual. As a means of identification, they are as reliable as fingerprints.

In the organism, these chemical identification marks are continuously being inspected by a special cellular defense corps, the lymphocytes, the agents of the immune system, which have the ability to recognize by its surface identification marks any invader that might have infiltrated our defenses and to destroy or help destroy it. They circulate through blood and lymph from a number of bases, which include the spleen, the thymus, the lymph nodes, the tonsils, and various so-called lymphoid patches.

There are two types of lymphocytes, known as T and B for their main source of origin, thymus and bone marrow (originally bursa of Fabricius, a lymphoid organ of birds). Within each type, there are several subclasses. T lymphocytes—at least the main subclass designated cytotoxic—are the infantry of the immune system; they are specially equipped to kill other cells by direct contact through a "kiss of death" mechanism. The B lymphocytes are the artillery, or rather give rise to the artillery, in the form of cells (plasma cells) that discharge missiles known as antibodies, which are endowed with the ability to combine specifically with their target. These antibodies, or immunoglobulins, which are of protein nature, do not kill by themselves, but they serve as recognition devices for a number of killing mechanisms. In particular, when bound to their target, they cause it to become attached to a receptor present on the surface of white blood cells, which then engulf and destroy the enemy thus branded. This mechanism serves as a major defense against microbes and viruses (see Chapter 4). Antibodies also serve to alert a soluble killing system carried by the bloodstream and known as complement.

Obviously, it is extremely useful for us to have such a defense corps. In fact, we could not survive without it. But this advantage carries with it a danger of mistaken identity and summary execution of friends. Here is where our transplantation antigens come into play. Early in fetal life, our lymphocytes learn to recognize the specific pattern displayed on the surfaces of our cells and to treat the cells thus labeled as friendly. But their inspection is very thorough, and they will spot even a small deviation from the pattern defined as "self." For instance, it is believed that they can detect and kill some cancer cells, which have almost the same identity card as normal cells. Lymphocytes certainly have no trouble recognizing cells originating from another individual, and they thereby tend to oppose the successful surgical transplantation of a tissue or organ. When transplants are made, a detailed typing of the transplantation antigens of the recipient and of the potential donors is done to allow the best possible matching. After the operation, the patient is treated with immunodepressant drugs, which weaken the rejection mechanism but which also, unfortunately, reduce the patient's ability to withstand infectious attacks and perhaps to reject cancer cells. There is a better way of eluding immune rejection, if only we knew how it works. Fetuses use it to entice their mothers into overlooking the foreign identification marks that they carry as part of their fatherly inheritance. Some women have immune systems that refuse to be subverted in this manner; they suffer repeated spontaneous abortions caused by immune rejection.

Lymphocytes are organized like no other defense force in the world, in that each individual lymphocyte can recognize only a single type of foreign grouping, as though each member of the corps were able to grapple with only a single type of invader. Since there are millions, if not billions, of such distinct groupings, most of our lymphocytes are never called to combat, and those that are are necessarily few. Quite often, in fact, they are much less numerous than the invaders.

This may look like an extremely inefficient way of doing business, but it seems to be the only way in which immense versatility can be combined with perfect reliability of recognition. And this, of course, is of paramount importance. Just imagine the havoc that would be created by a trigger-happy corps of the kind that fired first and asked questions afterward. Apparently, efficiency and safety can be achieved simultaneously only on the basis of the one-lymphocyte-one-target principle. Such being the case, Nature has provided a neat solution to the problem of numerical weakness. When a lymphocyte encounters and recognizes its specific target, it starts multiplying. This is yet another example of a receptor-mediated response. The recognition occurs through binding of the target to specific surface receptors of the lymphocyte, and the binding results in a mitogenic response (one that stimulates mitosis). Thanks to this mechanism, a whole army, or clone (Greek *klôn,* twig), of identical lymphocytes directed against the target is generated by successive cell division. The organism becomes immunized.

This powerful mechanism has only one drawback. It takes time to be established and may arrive too late when the invading force is a strong one. Hence we use vaccination, a way of developing an appropriate defense corps by introducing a "dummy" enemy, such as attenuated viruses or killed bacteria, which can no longer cause a serious disease but still carries the identification marks that set off recognition, lymphocyte multiplication, and antibody formation. Recently, specific surface proteins or peptides extracted from the pathogen, or reproduced artificially either by genetic engineering methods (see Chapter 18) or by organic synthesis, have been used for the same purpose (synthetic vaccines).

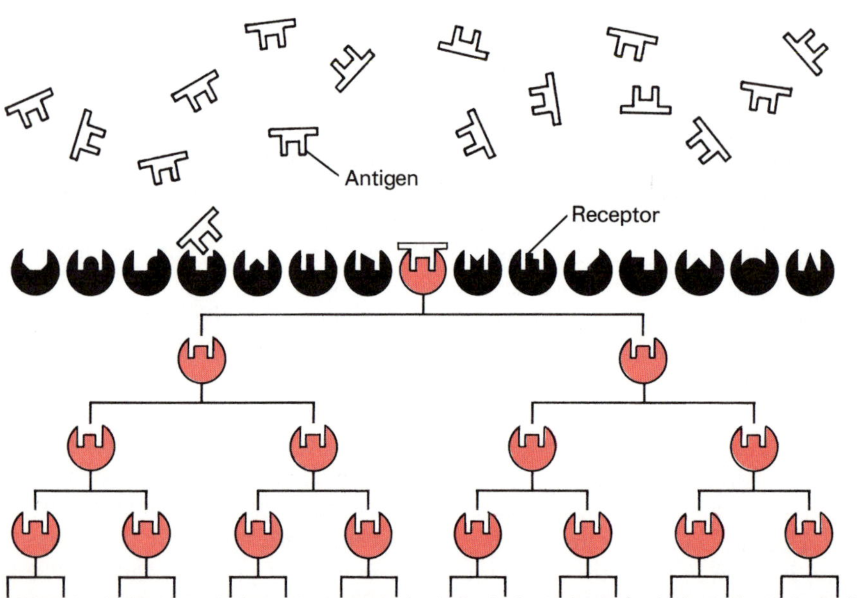

Functioning of the immune system by clonal selection.

We will return in some of the following chapters to the remarkable mechanisms whereby lymphocytes acquire their specific surface sensors in the course of differentiation (see Chapter 18) and are later stimulated to divide upon contact with the appropriate antigen (see Chapter 13). Like all living wonders, unfortunately, immune defense sometimes goes wrong. It may be congenitally defective, to the extent that the affected children must be kept constantly in a sterile environment. Or it may fail later in life, as in the recently described acquired immune deficiency syndrome (AIDS). Conversely, it may start striking friends as well as foes and launch treacherous autoimmune attacks against a patient's own liver, kidneys, joints, or other organs.

4 Entering the Cell: Endocytosis and Vesicular Transport

The various ways into a cell, whether by permeation, facilitated transport, or active transport, that were briefly described in Chapter 3 are restricted to small molecules. With rare exceptions, even single macromolecules do not cross the plasma membrane; members of our party are clearly excluded. Fortunately, there is a more accommodating means of entry, which is open not only to macromolecules, but also to much larger objects, including viruses, bacteria, and fragments of other cells. It depends on a membrane-mediated process of bulk uptake, called endocytosis.

The Endocytic Route

To understand the term endocytosis, we must go back one century. The scene is Messina in Sicily. Ilya Metchnikoff, a self-exiled Russian zoologist, is observing a transparent starfish larva through his microscope while, as he recalls in his memoirs, his "family has gone to a circus to see some extraordinary performing apes." As his eyes follow a wandering cell moving like some free amoeba through the larva's tissues, "a new thought suddenly flashes across my brain" that perhaps every organism contains a corps of mobile cells whose task it is to detect, pursue, engulf, and destroy unwelcome intruders such as microbes or viruses. This intuition made Metchnikoff one of the founders of immunology. With the help of a Hellenist friend, he coined the term "phagocyte" (Greek *phagein*, to eat) to designate the "eating cell." The act of engulfing a solid particle such as a

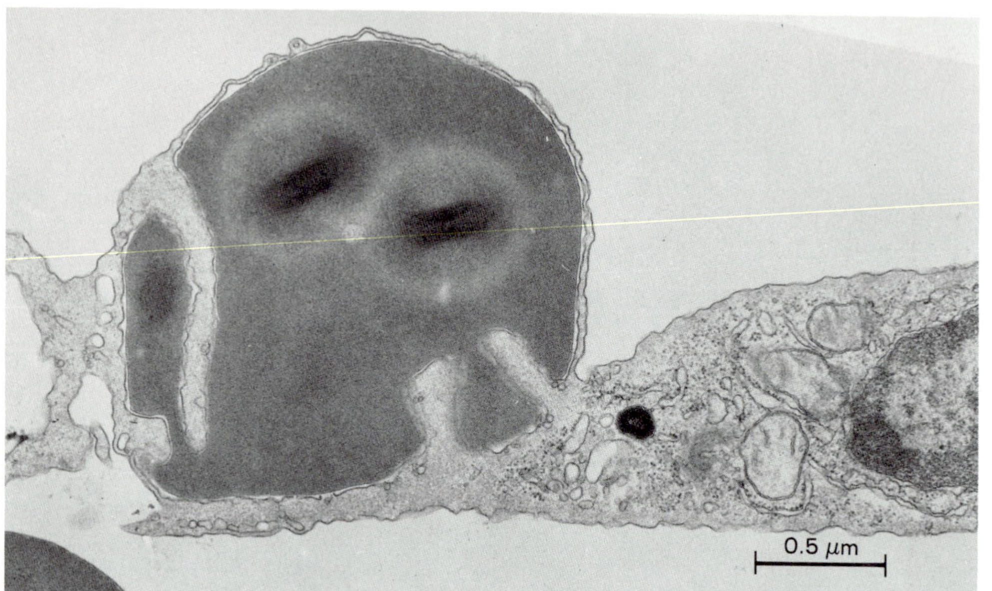

Left: Phagocytosis of a red blood cell by a mouse macrophage. Notice the thin rim of cytoplasm that completely envelops the engulfed erythrocyte.

Below: Two views of pinocytosis by a human monocyte. Invaginations of the plasma membrane are just about to close and detach as vesicles. Note the fuzzy "coat" around budding vesicle in right-hand picture. This coat is lost very rapidly after closure of the vesicle. It is not seen around the pinocytic vesicles in the cytoplasm.

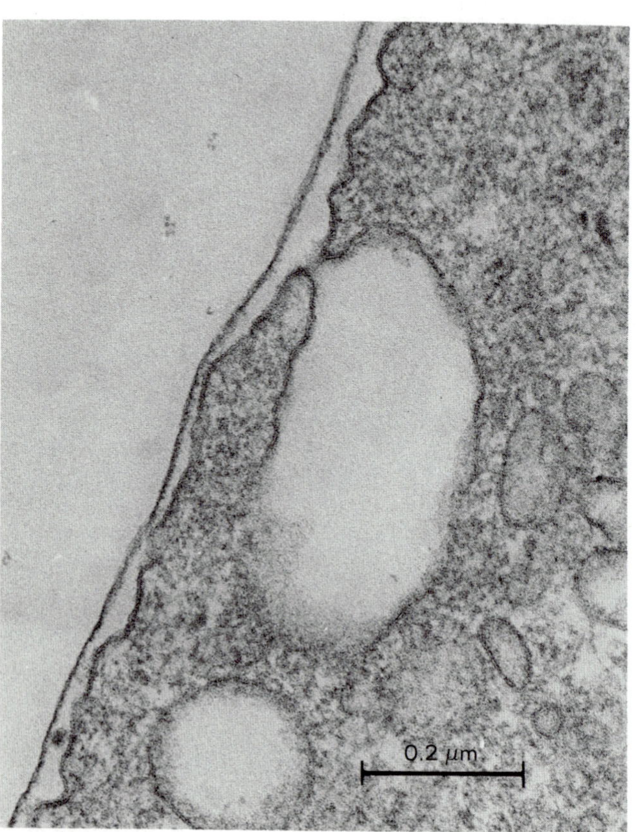

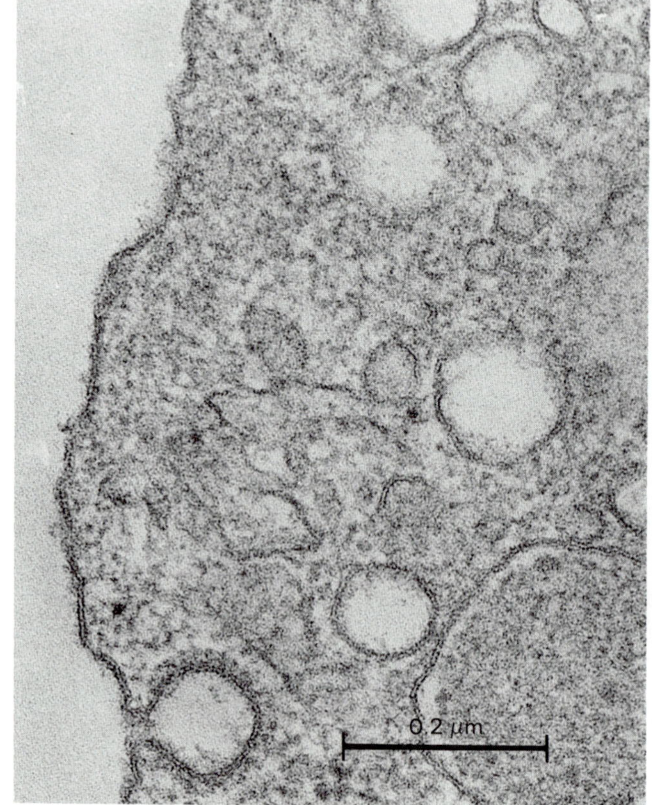

bacterium thus came to be called phagocytosis, literally the cellular act of eating. Then, in the early 1930s, the American biologist Warren Lewis discovered that cells could also engulf droplets of fluid and called this phenomenon pinocytosis (Greek *pinein*, to drink). Eventually, phagocytosis and pinocytosis were found to be manifestations of a more general mechanism of uptake, to which the name endocytosis was given.

Endocytosis can occur in many forms but depends invariably on the plasma membrane to provide the vehicle of entry. Whatever object is taken in, it always enters the cell wrapped in a sealed membranous sac that originates from an invagination of the plasma membrane. What we have learned of the fluidity and self-sealing properties of biomembranes helps us understand the more physical aspects of this phenomenon. Imagine a small patch being pulled inward from the surface of a soap bubble and eventually detaching as a miniature bubble imprisoned within the larger one. What does the pulling and how it is engineered are much more complex questions. We must wait until we can watch the process from inside the cell before we can address them.

In the meantime, some very interesting things about endocytosis can be observed from the outside. Most arresting is the participation of surface receptors. As mentioned in Chapter 3, binding of a ligand to a receptor is often followed by intake. In fact, many receptors have no function other than to select extracellular materials for intracellular intake. The process involved is called receptor-mediated endocytosis. The best way to observe this process is by introducing into the pericellular medium molecules of a ligand that is specifically bound by a given type of receptor. As ligand molecules are picked up, the occupied receptors can be seen to converge swiftly with their catch toward pitlike depressions of the membrane, and then to disappear progressively from view as the pits become deeper and their rims narrower. Eventually, only a pinhole opening remains, soon to vanish completely while the membrane regains its smooth, undisturbed appearance. Actually, a whole patch of membrane has been excised in this process and is now moving freely through the cytoplasm below in the form of a closed vesicle containing the engulfed material. Terms such as phagosome, pinosome, endosome, and phagocytic, pinocytic, or endocytic vacuole have been used to designate such a vesicle. The word endosome has acquired a special meaning as the name of the acidic sorting station in which endocytized materials undergo their first intracellular processing (see p. 58).

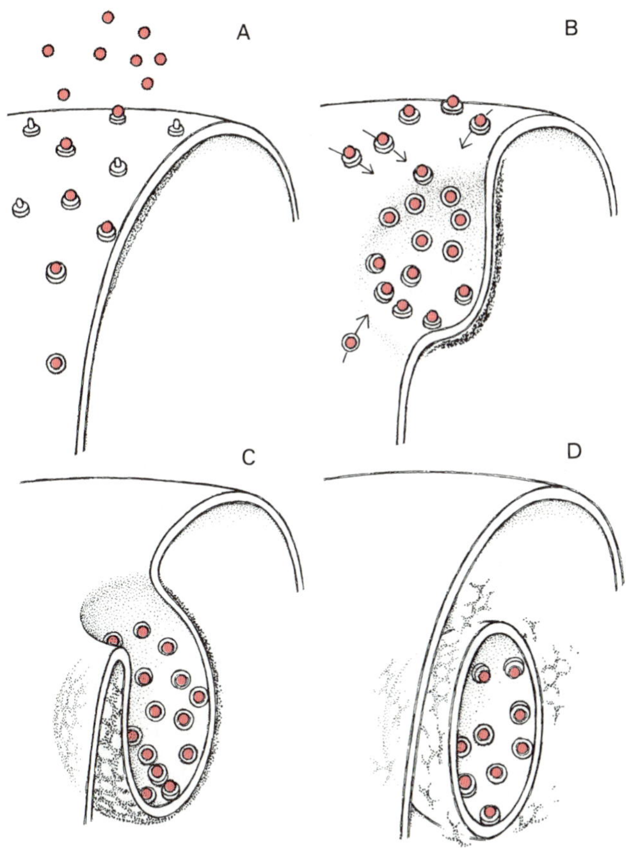

Receptor-mediated endocytosis: (A) ligands bind to receptors; (B) ligand-loaded receptors cluster in a coated pit; (C) the coated pit becomes a deep invagination; and (D) the invagination has detached as a coated vesicle, which is in the process of losing its coat.

The clustering of occupied receptors in endocytic pits strikingly illustrates the ability of membrane proteins to move freely along the plane of the lipid bilayer by lateral diffusion. What actually brings them together is not clear. Cross-linking by divalent or multivalent ligands may play a role in some cases. But, in others, the receptors seem to

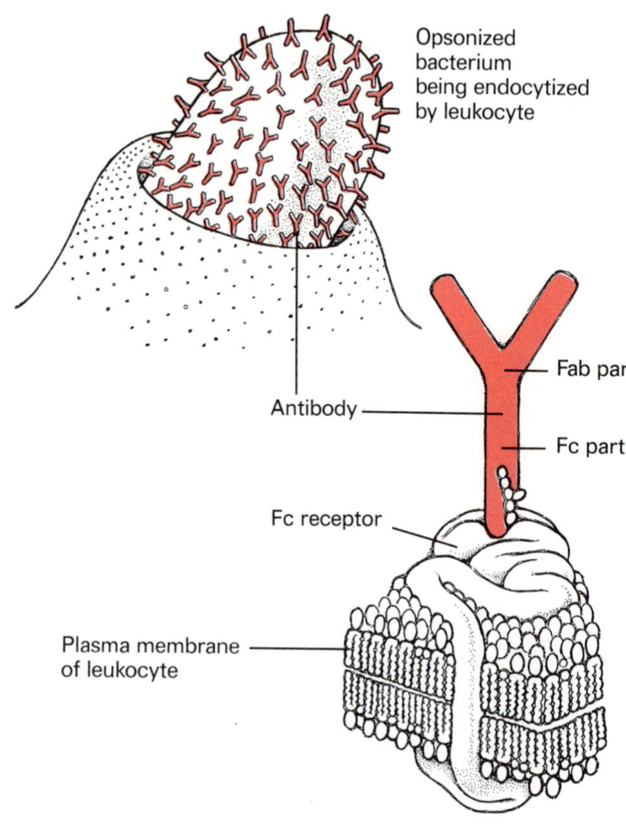

Phagocytosis of antibody-covered (opsonized) microbe mediated by Fc receptor (zippering mechanism).

be continuously on the move. Whether occupied or not, they disappear into endocytic pits and reappear on the cell surface in an uninterrupted stream, somewhat in the mode of an escalator or moving belt. As in these examples, the machinery that does the moving is hidden. We will take a look at it when we visit the cytosol. All we can distinguish from the outside is the shadow of a cagelike structure at the back of the endocytic pit; it seems to pull the membrane patch inward and to strangle it into a separate vesicle. Made of clathrin (see Chapter 12), this structure appears in cross section as a bristly coat. Hence the name "coated" given to the pits and vesicles that are thus lined.

A different kind of endocytic uptake, one that is highly dependent on cross-linking of receptors, occurs when the ligand molecules are attached to the surface of a large object, such as a bacterium. Once contact between the object and the cell membrane is established by means of a few receptor-ligand linkages, additional receptors are recruited from neighboring parts of the cell surface, and their binding causes a progressive envelopment of the ligand-coated object by the receptor-covered piece of plasma membrane. Thus, this molecular "zippering" effect actually induces the engulfment process. The name phagocytosis is generally reserved to designate this kind of uptake.

The mechanism whereby antibodies brand targets for destruction by leukocytes (Chapter 3) depends on this kind of process. Antibodies are Y-shaped protein molecules that attach to the corresponding antigens by means of their forked end, known as the Fab part (see Chapter 18). This part of the antibody carries the immunological specificity of the molecule and is therefore different for every individual type of antibody. The stem of the Y, called the Fc part, is the same for all antibodies. Now imagine a microbe entering an immunized organism and there encountering antibodies directed against some of its cell-wall components. It becomes covered with antibody molecules that all bind by their Fab forks and therefore have their Fc stems sticking out. This Fc pelt will induce zippering of the microbe by a white blood cell, thanks to the occurrence of specific Fc receptors on that cell's membrane. The early immunologists gave the suggestive name opsonin (Greek *opson*, seasoning) to such phagocytosis-inducing antibodies. They are the "dressing" that makes the objects they cover palatable to the eating cells.

We will understand the significance of endocytic receptors much better once we find out what happens to the materials that are engulfed with their help. An important point to remember, in the meantime, is that the nature and density of receptors occurring on the cell surface vary from one cell type to another, and even in the same cell, according to its functional state. This means that each cell selects its particular "menu" out of the environment by means of its surface receptors.

Endocytic uptake is not restricted to molecules and objects that are selected by receptors. Extracellular components are also taken in randomly as the dissolved contents of engulfed droplets of fluid. This is pinocytosis in the etymological sense of the term. Called fluid-phase endocytosis, it can lead to sizeable uptakes, since cells easily "drink" the equivalent of their own volume every day. Just imagine a human being guzzling from 15 to 20 gallons a day.

Even more impressive is the quantity of membrane that is being translocated by the endocytic activity. It may amount to the interiorization of as much as twice the surface area of the plasma membrane, with all its infoldings and digitations, every hour. Cells hardly make new membrane at this rate. They continuously recycle back to the surface the membrane patches used in endocytosis.

A Hazardous Start

Our visit to the cell's interior is about to start. But first we should get properly dressed. For reasons that will soon become clear, we need an acid-resistant and lysosome-proof coat if we want to avoid an abrupt—and fatal—termination to our tour. A waxy substance extracted from the cell wall of leprosy bacilli, which owe to this substance their remarkable resistance to intracellular destruction, will provide us with the appropriate material. On top of that, we still have to don an overcoat carrying a suitable ligand, so that we can be recognized and admitted by the cell we plan to visit. Should it be a white blood cell, our garment would be made of antibodies, so as to satisfy the cell's Fc receptors. But other cells may require a different overcoat. Remember that surface receptors vary from one cell to another. This means that we can, to some extent, choose the target of our visit by covering ourselves with the appropriate ligand.

With due regard to our size—still comparable to that of a large bacterium, even after shrinking a millionfold—our proper mode of entry should be phagocytosis. However, there will be more to learn if we ask the cell to stretch it a little and to accommodate us within a coated pit. There, together with a variety of receptor-bound molecules, we are about to experience endocytosis from the vantage point of its objects or victims. Even to the forewarned, it can be an unsettling experience to feel oneself being gradually enshrouded by creeping membrane folds. But worse is yet to come, after our last connection with the outside world seals shut and our flimsy craft casts off into the deepening cytoplasmic gloom, past bulky shadows that can just be discerned through the half-transparent membrane.

The first sign that our engulfment has been completed is a progressive acidification of the medium inside our membranous vessel. The hydrogen ions, H^+, or protons, responsible for this phenomenon come from the cytosol outside, delivered by an energy-dependent proton pump (Chapter 3). This pump presumably already exists in the pits on the cell surface, but the protons it ejects are diluted and washed away by the extracellular fluid. When the pits close into vesicles, the protons remain trapped inside, where they accumulate rapidly until they reach a concentration of about 20 microequivalents per liter (pH 4.7), as opposed to a cytosolic concentration near neutrality (0.1 microequivalents per liter, or pH 7.0). An important consequence of this acidification is that it decreases considerably the affinity of many ligands for their receptors, thus causing them to fall off the membrane. As it happens, our overcoat is made of ligands of this type, and we are now swimming freely in the stinging fluid that fills the endocytic vacuole, thankful for the acid-resistant coat that we borrowed from leprosy bacilli. Many other ligands have joined us, but some remain attached to their receptors.

While these events are going on, the membrane around us presents us with a spectacle strangely reminiscent of what we watched on the cell surface. Again we see receptors—some occupied, but many others free—clustering in deepening clefts and finally vanishing into the cytoplasm in the form of small, flattened vesicles that either are returned to the cell surface or are routed to some other destination that we cannot discern at present. The main function of this activity is membrane retrieval, without which, as we have seen, endocytosis could not continue

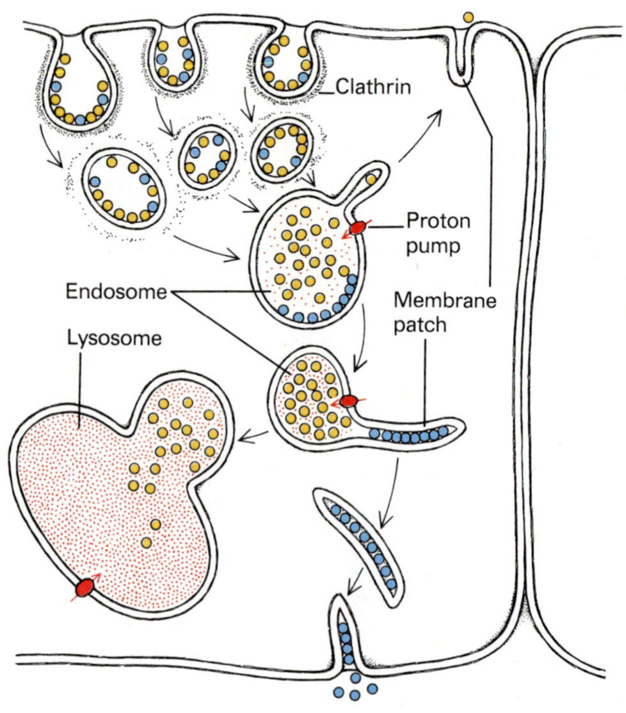

Sorting in an endosome after endocytosis. In this example, two types of ligands (represented by yellow and blue dots) are taken up together by receptor-mediated endocytosis. As a result of the acidification (shown in red) of the endosome content by a proton pump, the yellow ligands fall off their receptors and are subsequently conveyed to a lysosome by endosome-lysosome fusion. Before that, membrane patches detach from the endosome and are recycled back to the cell surface. One returns to the site of entry with some endosome contents (regurgitation). Another moves to the opposite side of the cell with a cluster of blue ligands that have remained receptor-bound and discharges them outside the cell (diacytosis, see p. 61).

very long. But the process is in no way random; it involves a lot of highly discriminate sorting and reorganization. Defined receptors and other membrane components are selectively concentrated on the excised patches and removed with them, accompanied by such engulfed materials as remain or become attached to the receptors in the acid medium created by the proton pump. At the same time, patches are also added to our surrounding membrane, brought in by newly arrived endocytic vacuoles that fuse with our own. We are in fact in an authentic organelle, the first station on the cellular import line. In it, receptors unload their catch and are recycled for a new round of duty, with the exception of a few that take their load further into the cell. Names such as intermediary compartment, receptosome, or endosome have been given to this organelle. The last is a generic term meaning endocytic vacuole (see p. 55) that now tends to be reserved for the endocytic sorting station.

All this membrane reshuffling does not go on without a certain amount of "slobbering" or regurgitation of fluid contents. By and large, however, dissolved materials and detached ligands remain inside endosomes (or are returned to this compartment by a new endocytic gulp). With them, we continue our adventurous cruise through the cytosol, somehow pulled and guided by invisible strings, parts of the cytoskeleton, as we will see in Chapter 12. Suddenly, right in front of us, looms a dark, ungainly blimp of a dirty brown color. Collision is inevitable.

The crash is mild, but its outcome is fateful, since it has caused the membrane of our vessel to merge with that of the blimp, hurling us into what turns out to be a thoroughly unpleasant environment. Everywhere we look are

Our endosome has collided and fused with a lysosome, projecting its contents into highly corrosive, digestive fluid. Scenes of destruction are everywhere.

scenes of destruction: maimed molecules of various kinds, shapeless debris, half-recognizable pieces of bacteria and viruses, fragments of mitochondria, membrane whorls, damaged ribosomes, all in the process of dissolving away before our very eyes. Obviously, we have been projected into some frightfully corrosive fluid. For any but the hardiest of beings, this would be the end of the trail. It would have been for us, but for our protective coat. In chemical terms, the kind of massive and utterly indiscriminate process of dismantlement we are witnessing is called digestion. The blimp is a digestive pocket, or lysosome (lytic body). We will spend all of Chapter 5 exploring this inhospitable, but very necessary, part of the cell. But first, let us reconstruct the events that led us into it.

The Sealed-Room Trick

In just a few minutes' time, we have been spirited through several hermetically sealed walls: first from the extracellular spaces into an intracellular endosome, and then from the endosome into a lysosome. At the same time, we have seen patches being either added to the tenuous fabric that surrounds us or removed from it. At no time, however, has this fabric suffered the slightest tear. Even Sherlock Holmes might have found this one a poser. Or, more likely, he would have regaled Dr. Watson with a discourse on the elementary properties of lipid bilayers: for the secret of our translocation lies in membrane coalescence. This phenomenon forms the physical basis of vesicular transport, the main means of mass transportation through the multiple membrane-bounded compartments that we have started exploring in this first itinerary of our tour.

Vesicular transport depends on two distinct types of events, which, in terms of containers, may be described as fusion and fission; in terms of contents, as mixing and separation; and in terms of mechanisms, as *cis* and *trans* membrane mergers. To clarify these definitions, let us look first at the second leg of our journey, when our endosome joined with a lysosome. This, clearly, was a fusion event, leading to mixing of contents. It was deter-

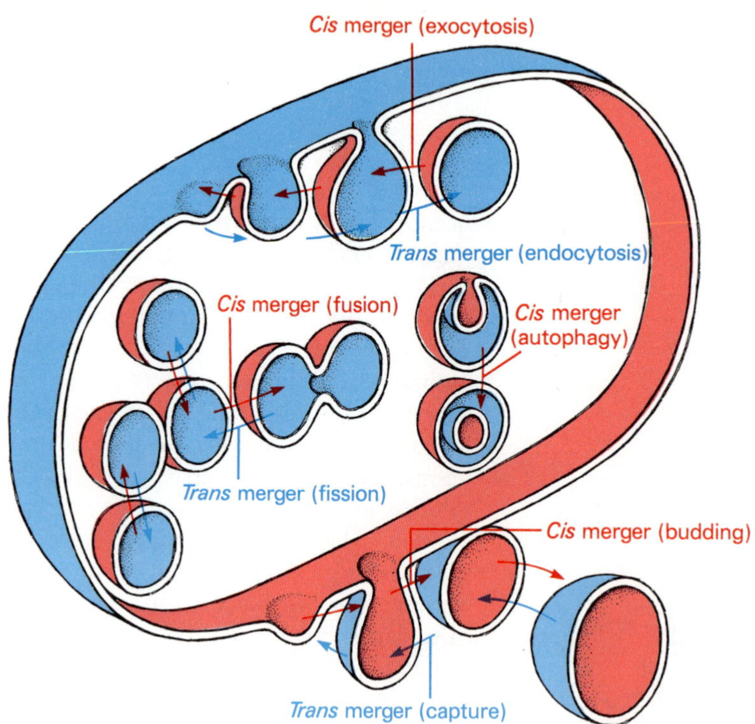

Different forms of vesicular transport mediated by *cis* or *trans* membrane mergers.

mined by the merger and reorganization of the membranes of the two bodies, approaching each other by way of their cytoplasmic faces (the faces of the membranes that are in direct contact with the cytoplasm). This, by convention, is what we will refer to as a *cis* membrane merger. Besides being alliterative with *cyt*, the definition makes sense: *cis* means on the near side; *trans* means on the far side. It is only fair that we should look at the membranes from the point of view of the cell.

It is easy to see that this type of *cis* merger is a mechanism of general significance: any two vesicles that interact in this way mix their contents. A special case of *cis* merger obtains when one of the partners is a cytoplasmic vesicle and the other is the plasma membrane. Mixing then occurs between the vesicle's contents and the outside world: the vesicle's contents are discharged from the cell. This phenomenon, called exocytosis, plays a major role in secretion. It will be examined in greater detail in Chapter 6.

Now consider the earlier event that first got us into the cell—namely, the closure of the endocytic invagination. It represents a fission event, in the sense that a small volume is cut off from the outside world and becomes isolated as the content of an endosome. If we look at its mechanism, we find that this phenomenon also depends on a merger of membranes, in the form of narrowing lips that in this case are led to approach each other by their outside, or *trans*, faces. Again, the mechanism is of general significance. Any vesicle undergoing it divides into two. In particular, the final unmooring of the vesicles we saw detaching from our endosome took place by a *trans* type of membrane merger.

There goes the mystery of the sealed room. All that we need in order to move bulk materials through any number of sealed walls, whether in import, as just experienced, or in export (see Chapter 6), are alternating fusion and fission events determined by *cis* and *trans* membrane mergers, respectively. "Elementary, my dear Watson." But is it? Only in a purely descriptive sense. When it comes to the forces that maneuver the membranes into position so that their lipid bilayers can merge and reorganize and to the factors that control the remarkable specificity of the events, we are still largely in the dark. For it must be pointed out that these mergers obey strict rules. A cytoplasmic vesicle does not simply fuse with the first membrane-bounded object it happens to bump into.

Vesicular transport also represents the main mechanism for moving membrane material through the cell. As far as is known, membrane patches always travel as closed vesicles, although these may be very flat and virtually devoid of content. This is how the piece of membrane that brought us in, or its equivalent, is returned to the cell surface. Such recycling, as we have seen, is an imperative requirement enforced by the enormous membrane consumption associated with endocytosis. It takes place largely from endosomes, but continues also to some extent after endosome-lysosome fusion. Fragile receptors are thereby rescued from lysosomal destruction. Exceptionally, materials that bind tightly to the excised membrane patch may be similarly saved. The vesicles that form in this way may travel directly to the plasma membrane and be reinserted in it by exocytosis. Or they may follow a more circuitous pathway (see Chapter 6).

The Role of Endocytosis

Finding a lysosome as the first reception room of the cell is hardly what one would expect from a friendly host. But cells do not care to be visited. They need to be fed. Endocytosis is first and foremost a feeding mechanism. For many single-celled organisms, such as protozoa, and for lower invertebrates, it is *the* feeding mechanism. Food, as we all know, must be conveyed to a stomach for digestion. And this is exactly what endosome-lysosome fusion accomplishes.

It is true that as we go up the animal scale the need for such a feeding mechanism decreases. Our cells find plenty of nourishing small molecules in the blood and extracellular fluids, which together form what the French physiologist Claude Bernard has called the "milieu intérieur." Our cells thus can afford to feed by molecular transport (Chapter 3) rather than by bulk uptake. However, the endocytic route has not been closed by evolution. It has, instead, become increasingly selective and refined and has adapted to a wide variety of different functions. Most of these involve digestion in the lysosomes and will be considered in the next chapter. But there are exceptions to this rule. For instance, most endosomes that form on the blood side of the flat endothelial cells that line the surface of blood vessels are not intercepted by a lysosome. Instead, they migrate to the tissue side of the cell, where they unload their contents by fusing with the plasma membrane. In this case, endocytosis is followed directly by exocytosis and serves to transport certain blood constituents across a cellular sheet. This process is called diacytosis (Greek *dia*, through), or sometimes transcytosis.

In some cells, endocytic feeding and diacytosis take place simultaneously. Liver cells, for example, direct some of the materials they engulf to their lysosomes, others to the bile canaliculi. Both kinds of materials enter by the same route, as the mixed contents of the same vacuole. Only after their uptake are they separated and sent to their respective destinations by means of the sorting process that we have described.

It may also happen that endosomes are altered in some manner that delays their fusion with lysosomes. They can then serve as temporary storage vacuoles. There is evidence that this occurs in the egg cells of certain insects that use endocytized exogenous proteins for the formation of yolk. Upon fertilization, fusion of the yolk granules with lysosomes is activated and digestion of the yolk proceeds, to feed the growing embryo. Exactly how endosome-lysosome fusion is delayed in such cases is not known, but there is evidence that the process can be controlled from inside the endocytic vacuole. Some endocytized substances that bind to the membrane—for instance, concanavalin A, a glycoprotein extracted from jack beans—have this ability. So have certain microorganisms; the tuberculosis bacillus is one. Thanks to this property, it is able to launch a successful intracellular invasion, remaining and proliferating within an endosome. Its cousin, the leprosy bacillus, has evolved a different mechanism. It makes a digestion-resistant coat—we have taken advantage of it—and actually resides and proliferates within lysosomes.

5 The Meals of a Cell: Lysosomes and Intracellular Digestion

A major consequence of the kind of fusion-fission events described in Chapter 4 is to allow intermittent communication and sharing of contents among all membrane-bounded pockets that can engage in this sort of activity with each other. The term space or compartment designates a set of pockets related in this manner.

The Lysosomal Space

The lysosomes, of which there may be several hundred per cell, make up a typical space. This kinship would, however, be far from obvious to a casual observer. Indeed, diversity and polymorphism are the most characteristic features of the lysosomal compartment. Lysosomes come in all shapes and sizes, and their inner structures, especially, are extraordinarily variable. This heterogeneity is reflected in the morphological nomenclature. Many different words have been coined to describe particular entities that we now know to be lysosomes. Among them are dense bodies, residual bodies, myelin bodies, multivesicular bodies, cytosomes, and cytosegresomes, to mention only a few. The pathological vocabulary is even richer, since many of the abnormal cellular inclusions seen in diseased tissues also are lysosomes.

As soon as we look at biochemical function, however, the kinship of all these different structures becomes obvious: all are sites of digestion. At the same time, their structural heterogeneity becomes understandable. The contents of lysosomes consist for the most part of materi-

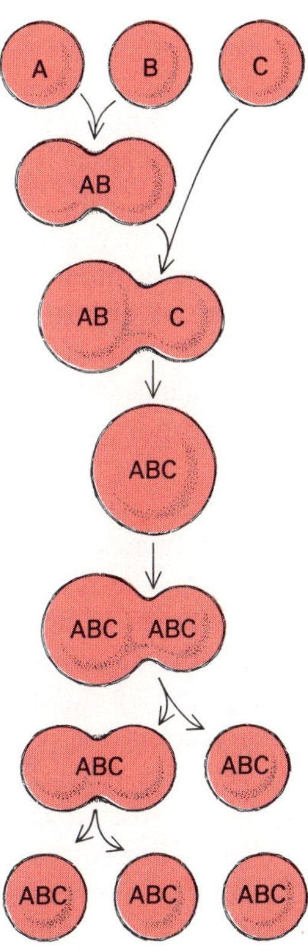

The concept of space or compartment. This diagram shows how three vesicles containing different components (A, B, and C) can be converted into three identical-looking vesicles in which the components are now mixed together, thanks to the ability of the structures to fuse with each other by *cis* membrane mergers and to divide by *trans* mergers.

als in the process of being digested and of indigestible residues. Inspect a few hundred stomachs chosen at random, and you will find that their contents, too, vary greatly, depending on when the owner had his last meal and what he ate. But, having once realized this fact, you would have no difficulty in recognizing all these organs as stomachs. The very diversity of their contents would be diagnostic. This has happened in cell biology. Nowadays, when an electron microscopist sees a particle surrounded by a single membrane and with a "messy" internal structure, he knows from previous experience that he is most likely looking at a lysosome.

In chemical terms, to digest means to hydrolyze—that is, to split with the help of water the various bonds

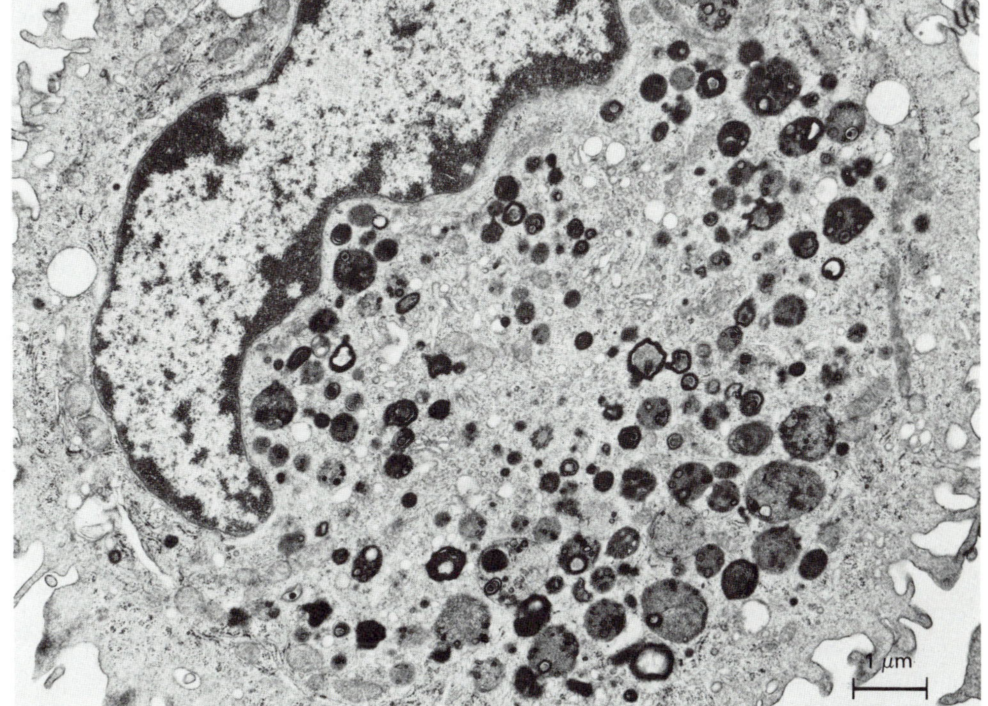

Morphological heterogeneity of lysosomes. In this electron micrograph of a rabbit alveolar macrophage, most cytoplasmic structures are lysosomes. Note the diversity of sizes, shapes, and inner structures. The crescent-shaped nucleus in the upper left-hand corner seems to envelop the lysosome-rich region, sometimes referred to by German scientists as the *Hof*, the court.

whereby the building blocks of natural macromolecules are linked. Examples are the peptide bonds, which join amino acids in proteins; the glycoside bonds, which link sugars in polysaccharides; and the ester bonds between acids and alcohols. Most of these bonds are quite stable and are broken only under harsh conditions of temperature and of acidity or alkalinity. Living organisms could neither realize nor tolerate such conditions, yet they readily digest their food. They do so with the help of special catalysts, called hydrolytic enzymes, or hydrolases, which are secreted into their digestive cavities.

Hydrolases are specific catalysts. Each splits only a well-defined type of bond. Since food is usually made of many different constituents that contain many kinds of chemical bonds, it follows that the act of digestion requires the concerted or sequential participation of many different enzymes. Indeed, the digestive juices secreted into the gastrointestinal tract contain many different hydrolases, and that is why the human organism can utilize a variety of complex foods of vegetal and animal origin. This ability is, however, limited. For instance, it does not include cellulose, as was mentioned in Chapter 2.

These basic notions are essentially valid for lysosomes also. In each lysosome, we find a large collection of different hydrolases—more than fifty have been identified—that together are capable of digesting completely or almost completely many of the major natural constituents, including proteins, polysaccharides, lipids, nucleic acids, and their combinations and derivatives. Like that of the human intestine, however, the digestive ability of the lysosomes is subject to certain limitations.

While they act on many different chemical bonds and substances, the lysosomal digestive enzymes have one property in common: they act best in a slightly acid medium. In technical terms, they have an acid pH optimum, usually between 3.5 and 5.0. In keeping with this property, lysosomes maintain an acid interior with the help of a proton pump, as do the endosomes from which they derive (Chapter 4).

The lysosomal space occupies a central position in the economy of the cell. Incoming roads bring materials to be digested not only from the outside, but also from inside the cell. The products of digestion make up the main outgoing stream. In addition, the lysosomal space is connected by a special line to a subsection of the cell's export machinery that supplies it with the necessary enzymes (see Chapter 6). As is to be expected for a membrane-limited space, this commerce uses two means of transport: permeation for all materials capable of traversing the membrane, eventually with the assistance of special pumps or permeases, and vesicular transport for all the others.

A major problem, which so far has received no satisfactory explanation, concerns the manner in which cells succeed in containing their lysosomal space. The lysosomal membrane is made, like other biomembranes, of proteins and phospholipids that are intrinsically digestible by lysosomal enzymes. Indeed, when pieces of the membrane are excised by autophagic segregation (see p. 69), they are completely broken down. Yet, they resist digestion when in the lysosomal lining. In fact, they make up an astonishingly safe shield that effectively protects the surrounding cytoplasm against the corrosive contents of the lysosomal compartment. The key to this mystery is not known. We can only surmise that the surface components, probably glycoproteins, that line the inner face of the lysosomal membrane are folded in such a way as to put their chemical bonds beyond the reach of the hydrolases to which they are susceptible. We must assume further that this special conformation is maintained with the help of metabolic systems present in the adjoining cytoplasm and is lost when the connection with the cytoplasm is severed.

Of course, this kind of explanation is really no explanation at all. It recalls the answer given in Molière's *Malade imaginaire* by a medical candidate who is asked in a mock examination why opium puts people to sleep. "Because opium has sleep-inducing virtues whose nature it is to numb the senses," the aspiring physician answers solemnly (in Latin, to sound more learned). We are no better when we state with equal profundity that the lysosomal membrane is resistant to digestion "because its surface lining has an enzyme-resistant conformation." Unfortunately, that is where we stand at the present time.

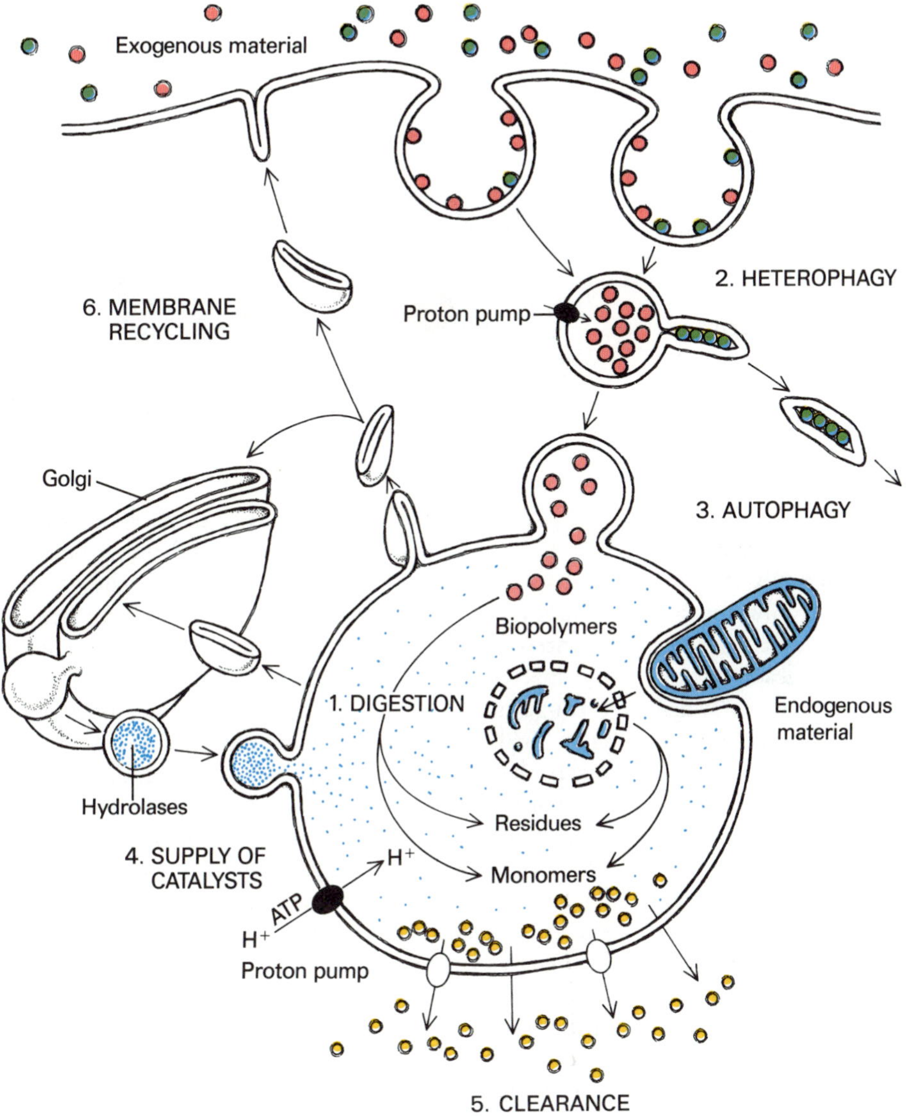

Composite view of lysosome function. The large circular body represents the whole lysosome space.

1. The main event occurring in the lysosome space is the digestion (hydrolysis) of biological polymers to monomers, catalyzed by hydrolytic enzymes (hydrolases) requiring an acid medium (protons, or H^+ ions) for optimal activity.

2. Substrates of digestion are introduced into the lysosome space from outside the cell by endocytosis (*trans* merger) followed by endosome-lysosome fusion (*cis* merger), often after preliminary sorting of contents and membrane in the endosome.

3. Other substrates enter from inside the cell by intralysosomal budding followed by severance (*cis* merger) of the bud.

4. Newly made enzymes are conveyed to the lysosomes by Golgi vesicles (*cis* merger). Protons are actively transported into the lysosomes by a proton pump, with the help of energy provided by ATP (see Chapter 7).

5. Products of digestion diffuse or are transported across the lysosomal membrane into the cytosol. Indigestible materials unable to traverse the membrane in this way remain in lysosomes as residues.

6. Much of the membrane material added to lysosomal membranes is removed by *trans* merger and recycled back to the cell surface, either directly or by way of the Golgi apparatus. A small amount of membrane is interiorized by autophagic segregation and digested. Note that the *trans* face of the lysosomal membrane is resistant to digestion when its *cis* face is in contact with the cytoplasm but not after segregation.

Physiological adaptations of heterophagy.

A. Defense: a leukocyte captures and destroys a pathogenic bacterium.

B. Cleaning of lung alveolus: a macrophage sweeps up dust particles, microbes, viruses, and other foreign objects.

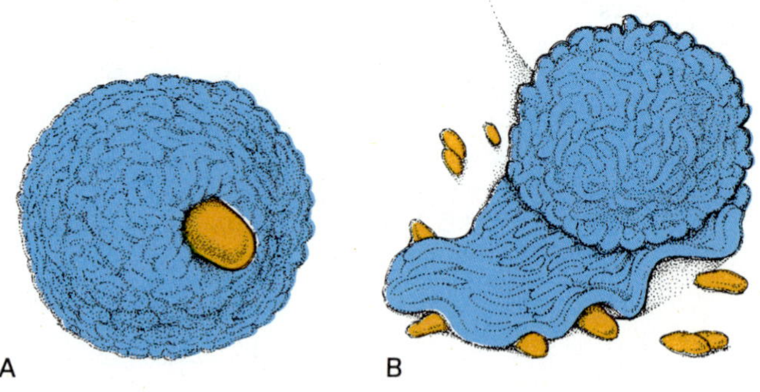

Heterophagy

Uptake and digestion of extracellular materials is called heterophagy (Greek *heteros*, other). It is a faculty that comes down to us from those early single-celled ancestors that first learned to chase after living prey and to kill and digest them intracellularly. It represents the main feeding mechanism of protozoa and lower invertebrates and persists almost unaltered in the leukocytes, whose function it is to pursue and destroy invaders.

There is, however, an important difference between a free-living protozoon and a leukocyte. For the former, endocytosis is a matter of life and death. Its survival depends on a daily catch of bacteria. The leukocyte, on the other hand, lives in an unctuously rich fluid full of sugars, amino acids, and other small molecules that can be used directly, without prior digestion. It has no need for heterophagy. On the contrary, heterophagy kills it. Leukocytes are constructed so as to consume only one big meal in their lives. They are made in the bone marrow, loaded with lysosomal hydrolases and other deadly weapons, and then sent out to search for the enemy. When they encounter them, they gobble up as many as they can. Soon after, they die, tricked by natural selection into committing mortal gluttony for the higher good of the organism. Until suffering this fate, they feed mostly on small molecules that they take in from the environment through the various transport systems with which their plasma membrane is fitted. They are essentially osmotrophic (Greek *ôsmos*, push; *trophê*, nourishment). Free-living protozoa, on the other hand, are phagotrophic.

Human cells, and in general the cells of most multicellular organisms, are essentially osmotrophic. To them, endocytosis and lysosomal digestion do not have the same survival significance as they have in the lower organisms, where these processes represent the sole means of food collection. Even in higher organisms, however, heterophagy still serves some nutritive function, because certain foodstuffs, such as iron or cholesterol, are carried by proteins that have to be taken up and processed for the nutrients they carry to become available.

In addition, in multicellular organisms, heterophagy is adapted to a wide variety of functions that serve the organism as a whole, rather than the individual cells. One such adaptation—to immune defense—has already been encountered in leukocytes. A related activity is carried out by macrophages, a group of cells that are scattered throughout the tissues and are characterized by a particularly avid and indiscriminate phagocytic activity. The macrophages also act in defense but, in addition, have a general role as "cleaners" and "garbage collectors." We find them in the lungs, for instance, endlessly scrubbing the surface of the alveoli, sweeping up dust, soot, tar, bacteria, viruses, any particulate material brought in by the inspired air. Whatever they cannot digest they deposit in their lysosomes. When these are full, the cells curl up and die, mission accomplished, and are ejected with the sputum.

Other cells are more selective and remove specific molecules from their environment by means of their endocytic receptors. As a rule, the function subserved by this activity is regulation by destruction. For instance, many hormones are endocytized and destroyed in lysosomes after binding to surface receptors and eliciting their intracellular effects. Thus, by being at the same time a sensor and an endocytic catcher, the receptor both mediates and limits hormonal activity.

C. Bone remodeling: an osteoclast breaks the bone matrix down into fragments by means of secreted acid and lysosomal hydrolases; it completes the breakdown by endocytosis and lysosomal digestion of the fragments.

D. Thyroxin production in the thyroid gland: (1) thyroglobulin is secreted into the lumen of a follicle; (2) thyroglobulin is endocytized by follicular cells and digested in lysosomes; (3) one product of digestion is thyroxin, which diffuses into pericellular spaces to be picked up by blood and lymph.

E. Renal reabsorption: proteins and other macromolecules that have filtered through renal glomerulus are endocytized by proximal tubule cells and digested in lysosomes.

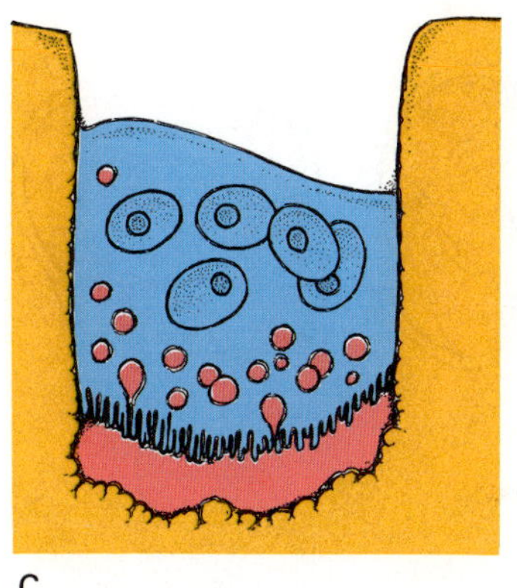

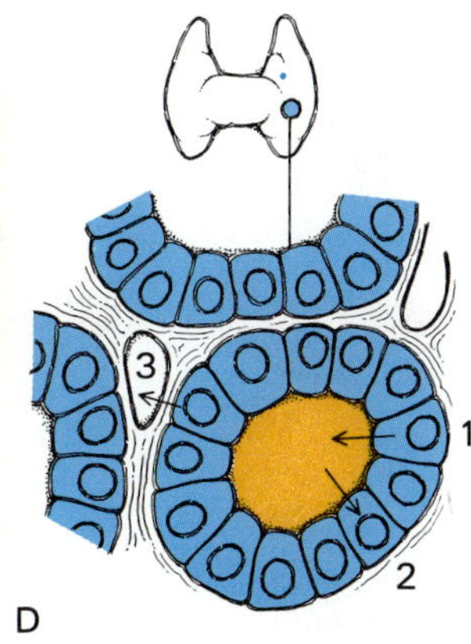

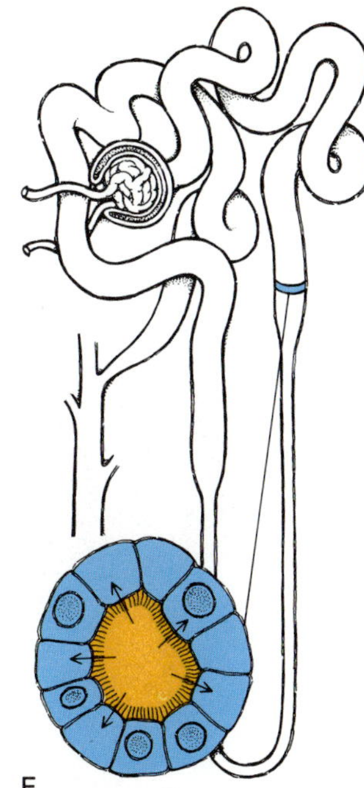

Heterophagy is also involved in the turnover (see p. 68) and remodeling of insoluble extracellular structures. Bones, for example, contain cells, called osteoclasts (Greek *klastos*, broken), that spend their time like moles, digging galleries and tunnels through the bone matrix. To do this, they secrete acid and lysosomal enzymes into the blind end of the tunnel; they thereby dissolve the crystals of hydroxyapatite (a complex of calcium phosphate and hydroxide) that make up most of the bone mineral and dismantle the scaffolding of collagen fibers that form the organic matrix of the tissue. They then complete the job by phagocytosis and intracellular digestion of the fragments. The damage is repaired by builder cells, called osteoblasts (Greek *blastos*, germ), that construct new bone elements, or trabeculae (diminutive of Latin *trabs*, beam).

Sometimes endocytosis is used for retrieval. This happens to the proteins that leak through the kidneys. While passing through the renal tubules, they are reabsorbed, digested, and recovered as amino acids. A remarkable case of retrieval is seen in certain cells—in particular the fibroblasts, which manufacture a good part of the collagen and other components of connective tissue. These cells display on their surface a special receptor for lysosomal enzymes, which are thereby both removed from the extracellular spaces, where they can do harm, and returned to the lysosomes, where they can do good. The same receptor also occurs intracellularly, as a key piece of the machinery whereby newly made lysosomal enzymes are conveyed to their destination (see Chapter 6).

It may even happen that heterophagy is geared to a synthetic activity. This occurs in the thyroid gland, where the hormone thyroxin arises as the product of the lysosomal digestion of thyroglobulin. This is an iodinated protein made by the thyroid cells and discharged by them into a common extracellular reservoir, subsequently to be taken up again and digested by the same cells upon appropriate stimulation.

These few examples illustrate the remarkable functional diversity of heterophagy. Starting as powerful nutritional aids, lysosomes have thus provided evolution with a particularly rich theme on which it has composed countless variations. And this is only one side of the coin.

Autophagy

Cells also practice autophagy; that is, they "eat" and digest little pieces of their own substance (Greek *autos*, self). They do this not only when they are deprived of food and forced to live on their own resources, but even when they are abundantly provided with nutrients. The rate at which this self-destruction

Schematic representation of autophagy. Four stages are shown from left to right: (1) interiorization by intralysosomal budding of the cytoplasm, (2) segregation by *cis* merger, (3) digestion, and (4) clearance. The size of the segregated material distinguishes macroautophagy from microautophagy. Multiple microautophagic events lead to the formation of multivesicular bodies.

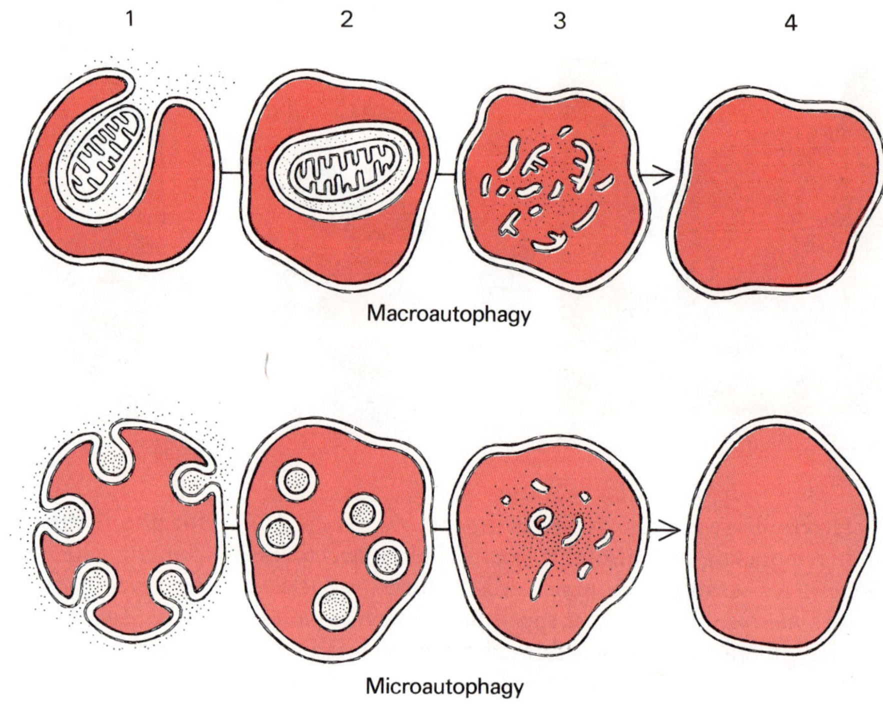

goes on is astonishing. An average liver cell, for example, destroys most of its contents in less than a week.

This you would hardly suspect from examining the cell, even with the sharpest of morphological or chemical tools. A hepatic cell may live for many years, and during all that time its structure and its chemical composition hardly change. The intense molecular turmoil that goes on behind this façade came to light only when isotopes became available, so that chemically identical molecules could be distinguished from each other by their atoms. It was then found that cells continually destroy and rebuild their constituents at a remarkably rapid rate. They are like those old houses that look exactly as they did when they were first built but, owing to multiple repairs, are left with few of their original window panes, or tiles, or even bricks or boards. But what can take centuries in a house is completed in a matter of days in a living cell.

Turnover, as this molecular renewal is called, is a finely regulated activity. Each cell constituent has its own characteristic average lifespan. Protein molecules, for instance, generally live between a couple of hours and a few days, depending on their nature. This implies that each type of protein has its own characteristic rates of synthesis and destruction and that these two rates must balance each other perfectly if the cell is to remain unaltered. The same is true for most other cell constituents, with the exception of DNA, which does not turn over as such, although it does undergo local injuries and repairs.

To destroy their own constituents, cells need digestion just as much as they do when they break down exogenous materials of similarly complex nature. Several different systems participate in this digestive activity. Among them

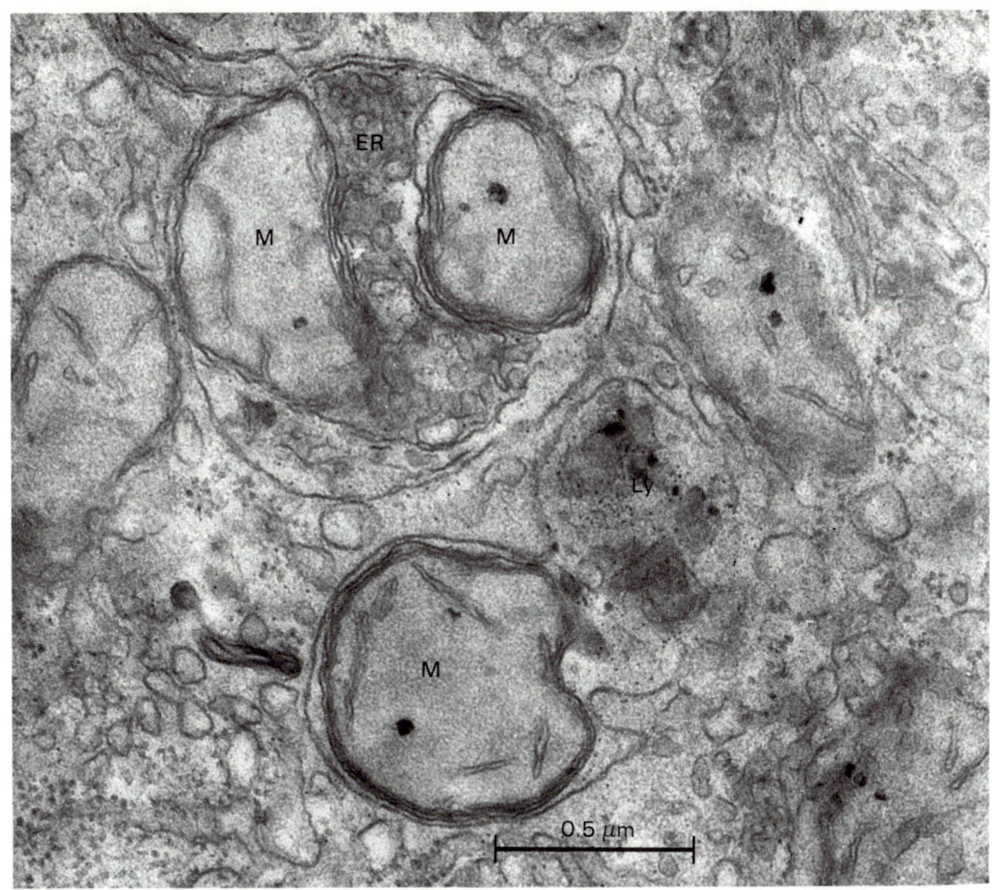

Autophagic vacuoles. Electron micrograph of a thin section through the liver of a rat that was injected with the hormone glucagon, which stimulates autophagy. The autophagic vacuole at the top contains two mitochondria (M) and endoplasmic reticulum (ER). Remnants of a second membrane are seen near the outer membrane of this autophagic vacuole. Just below it is another autophagic vacuole containing a mitochondrion. This vacuole seems to have fused with a lysosome (Ly), recognizable by its dense, polymorphous contents.

are the lysosomes, which serve as dumping ground and disposal site for a variety of intracellular materials, especially bulky objects such as whole chunks of cytoplasm, mitochondria, clusters of ribosomes, and membrane fragments abstracted from the endoplasmic reticulum. These objects enter the lysosomal compartment by a variant of the sealed-room trick (Chapter 4) that is essentially similar to the mechanism whereby cytoplasmic buds detach from cells. In budding, a piece of cytoplasm jutting out of the cell is amputated as a result of the progressive narrowing and final severance—by a *cis* type of membrane merger—of its connecting link. Let the same mechanism operate on a blob of cytoplasm that protrudes into the lumen of a lysosome and the bud ends up segregated inside the lysosomal compartment, a prey to cellular autophagy. Multiple events of this kind lead to the formation of characteristic structures described by morphologists as multivesicular bodies.

In the early stages that follow the act of autophagic segregation, the snared cytoplasmic objects can still be recognized inside their membranous trap. At first, they are surrounded by two membranes, as required by the budding mechanism just described. Later, the piece of membrane surrounding the sequestered bud degenerates and is no longer visible. Eventually, the contents of the bud become themselves unrecognizable, as their digestion progresses. The term autophagic vacuole designates all the structures that can be identified by any of the above characteristics as arising by autophagy. It has been claimed that other membranous structures besides lysosomes may serve in the initial segregation mechanism. Autophagic vacuoles arising in this manner would contain no digestive

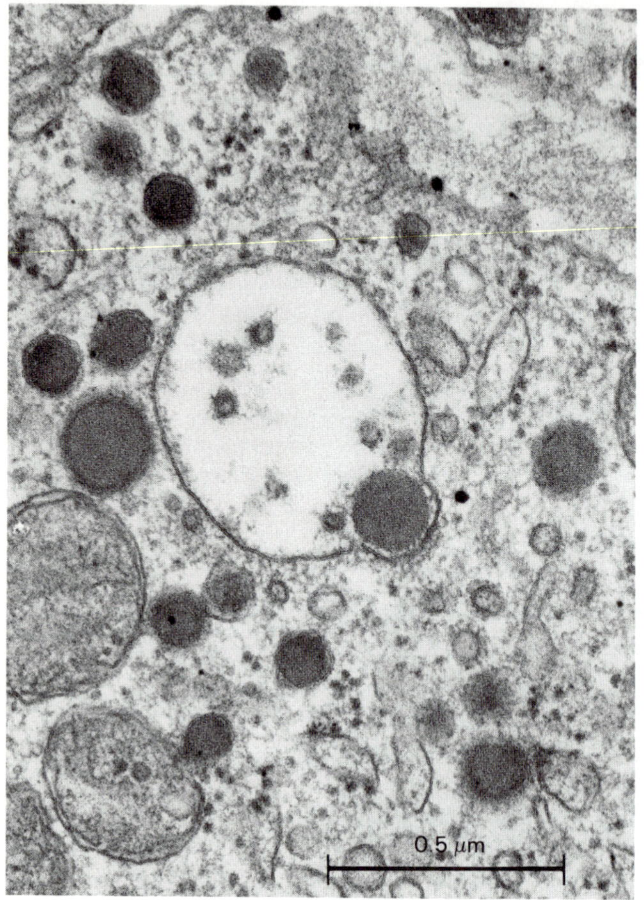

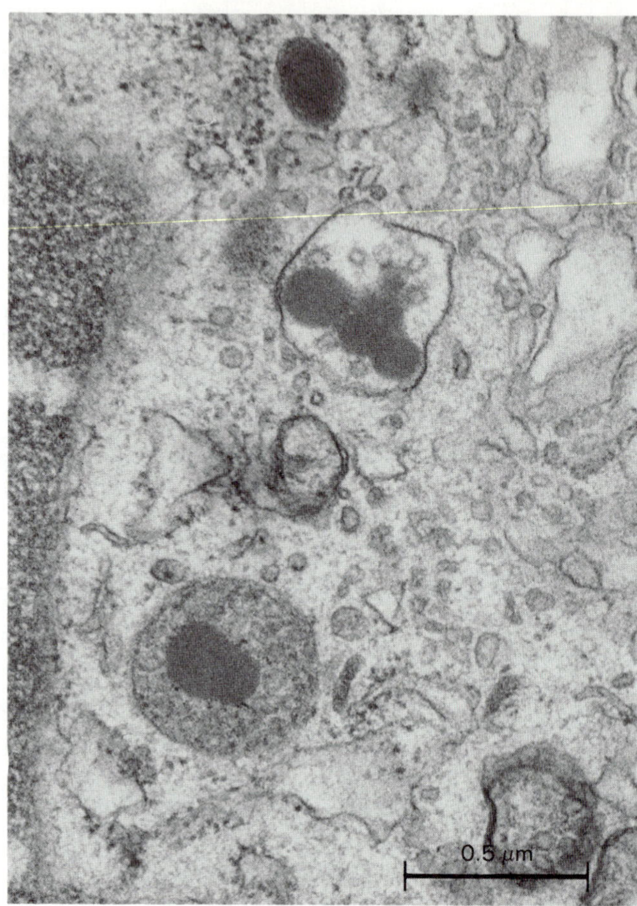

Two views of crinophagy. These electron micrographs illustrate crinophagy in hormone-secreting cells of the pituitary gland. The picture at the left shows a dense secretion granule that has just fused with a multivesicular body believed to be a lysosome. In that at the right, secretory contents are seen inside a multivesicular body and a dense body. Note that the membrane of the secretion granule is not interiorized but is added to the lysosomal membrane (*cis* merger). This fact distinguishes crinophagy from autophagy.

enzymes at first but would acquire such enzymes by subsequent fusion with lysosomes.

Faced with the extent of self-destruction that goes on in living cells, we may well wonder what advantages outweigh such a costly activity. It is likely that, in its primitive form, autophagy, like heterophagy, served an essentially nutritive function: it provided a mechanism for survival during starvation and evolved into a highly efficient process in which cellular constituents are sacrificed in their order of decreasing dispensability. Thanks to autophagy, a cell can do without food for extended periods. It uses up its own substance progressively, but in an orderly fashion, so that it remains organized and functional for a long time. This is true also of organisms. The walking skeletons that came out of Hitler's horror camps survived by consuming the proteins of their nonessential muscles to feed their more important hearts, brains, and blood. Many recovered remarkably well when supplied again with the nutrients needed to rebuild their tissues.

Autophagy remains a primary cell response to food deprivation in most present-day organisms, including the higher animals, where it may also be stimulated by hormones, such as glucagon, that stimulate mobilization of the organism's reserves. But it is obvious that self-support hardly explains the continual and intense turnover of constituents in cells that are perfectly well nourished most of the time. The main advantages that cells gain from their autophagic activity are rejuvenation and adaptability.

Thanks to turnover, cells continually replace their constituents with newly made ones and thereby achieve something very close to perpetual youth. Take the brain cells of an aged individual. They have been there for decades. Yet most of their mitochondria, ribosomes, membranes, and other organelles are less than a month old. Over the years, the cells have destroyed and remade most of their constituent molecules from hundreds to thousands of times, some even more than 100,000 times. These events do, however, leave some trace, in the form of a progressive accumulation of brown, indigestible residues in the lysosomes. This so-called age pigment, or lipofuscin, betrays the age of the cell. It could even be one of the flaws—the result of a slightly less than perfect lysosomal digestive activity—that finally defeat the cell's bid for eternal youth, if, as has been suggested, overloading of lysosomes plays a role in cellular senescence (see p. 72).

The second important advantage of autophagy is adaptability. The cell part that is destroyed can be replaced not only by an identical one, but also, if need be, by another one better suited to the current requirements of the cell. Some of these adaptive changes are evoked transiently, in response to environmental changes of one sort or another. Others are part of the fundamental processes of differentiation and development.

In gland cells, there is a special kind of autophagy called crinophagy (Greek *krinein*, to sift or separate; by extension, to secrete). It occurs through direct fusion between secretory granules and lysosomes (*cis* merger), and results in the destruction of the secretory material. It is an important regulatory mechanism (see Chapter 6).

Clearing Lysosomes

In the gut, the final products of digestion are cleared by what is called intestinal absorption. They are removed by the mucosal cells, usually with the help of active pumps, and discharged into the bloodstream. Something similar goes on in lysosomes. The various small molecules formed by the digestive process diffuse or are transported across the lysosomal membrane into the cytoplasm, where they are used by the metabolic systems of the cell. It is likely that this clearance requires the help of special carrier systems located in the membrane, but these have not yet been characterized.

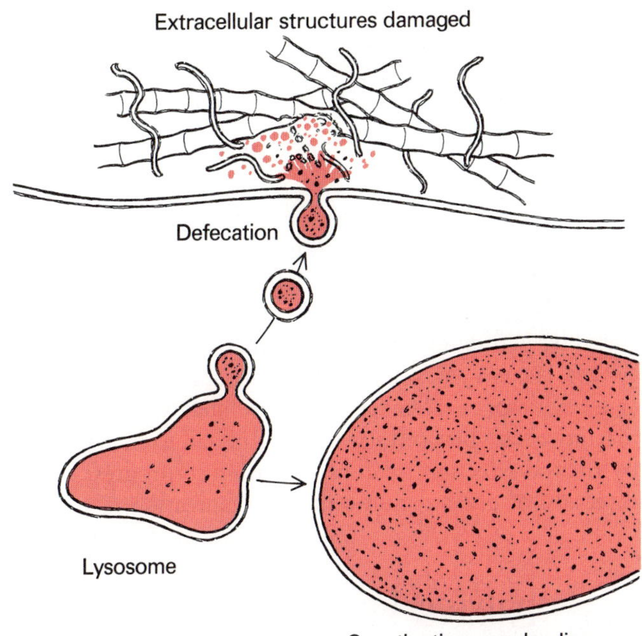

An evolutionary dilemma: what to do with lysosomal residues in multicellular organisms? If the residues are discharged by defecation, as they are in unicellular organisms, the released enzymes may damage extracellular structures. If they are kept stored within the lysosomes, they threaten to cause lysosome overloading. Between these two evils, the second seems to have been selected as less harmful, as indicated by the state of "chronic constipation" of most cells.

But what if digestion does not occur or is incomplete and does not progress to the point where its products can be cleared? What, for instance, if the cells take in some digestion-resistant material or if they suffer some digestive deficiency?

In most protozoa and lower invertebrates, accidents of this sort are of no special consequence, because these cells have the ability to unload the contents of their old lysosomes into the surrounding medium. That process, graphically named cellular defecation, depends on exocytosis. In higher animals, most cells seem to be unable to empty their lysosomes in this manner. They are, so to speak, chronically and persistently constipated. This represents a serious deficiency, which is responsible, as we shall see, for numerous pathological conditions associated with lysosome overloading. The alternative, unfortunately, is just as bad, or probably worse, as also illustrated by pathology. When lysosomes are unloaded in the tissues, the enzymes that are discharged with the residues cause widespread damage to extracellular structures. We may speculate that cellular defecation had to be repressed by natural selection to allow evolution to progress in the direction of greater multicellular organization.

Supplying Enzymes to Lysosomes

Like all enzymes, the lysosomal hydrolases are proteins. Therefore, they should themselves go the way of all the other proteins that enter the lysosomes. They should be digested by their proteolytic congeners. The fact that they do their job properly indicates that such digestion takes place slowly. This is hardly surprising, because without digestion-resistant enzymes there could be no lysosomes. As to what makes for their resistance, presumably it is their ability to keep themselves tightly bundled up in the acid environment prevailing within their natural habitat. As a rule, a protein chain needs to be at least partially unfolded, or denatured (Chapter 2), to become exposed to the clipping action of proteolytic enzymes. Many proteins are denatured by acidity. Obviously, the lysosomal enzymes are not, and this explains their survival in the lysosomal milieu.

They are not eternal, of course, and have to be replaced as they succumb to a chance hit that starts them on the road to breakdown. To find out how this replacement occurs, we have to move to another division of the cell, its export department, of which lysosomal enzyme production is a subbranch. This we will do in Chapter 6. At present, we can only watch the final delivery step. Not unexpectedly, the newly made enzymes arrive membrane-wrapped and are discharged into the lysosomal space by membrane fusion (*cis* merger). Such fresh packages of enzymes are called primary or virgin lysosomes, as opposed to the digestively active secondary lysosomes.

The Digestive Tribulations of a Cell

Dyspepsia, hyperacidity, constipation, and other digestive upsets are the common lot of mankind, the source of a great deal of personal discomfort, as well as of a very profitable alleviating industry. Yet these troubles are nothing compared with the digestive ills that afflict cells. Now that medical investigators are turning into subcellular and molecular detectives they are beginning to find out that a great many diseases are nothing but the manifestations of some digestive disturbance affecting certain cells.

As we have already seen, most of our cells suffer from constipation. It cannot be considered a disease, since it is a natural condition. But it is a serious disability, which leaves our lysosomes very vulnerable to the risk of overloading. The most dramatic examples of lysosome overloading are seen in young children suffering from a genetic deficiency of some lysosomal enzyme. More than twenty-five such deficiencies are known. They make up the group of genetic storage diseases, of which Tay-Sachs disease is a well-known example. In each of them, a lysosomal enzyme is severely deficient. As a result, the lysosomes become filled with materials that require the missing enzyme for their digestion; they swell progressively to enormous sizes and end up choking the cells to death. These lysosomal storage diseases are rare, fortunately, but they are extremely distressing. Children afflicted with them often suffer severe mental retardation and die at an early age.

Besides genetic deficiencies, many other situations may lead to lysosome overloading. It is seen in arteriosclerosis, occurs as a major complication in certain kidney diseases, and can be induced by drugs. The more we get to know about it, the more it appears to be one of the most frequent cellular illnesses, responsible for numerous diseases. In a way it is universal, and an inevitable concomitant of aging, for there is no way in which a cell can completely escape having its lysosomal compartment visited occasionally by an indigestible molecule. Once in, the

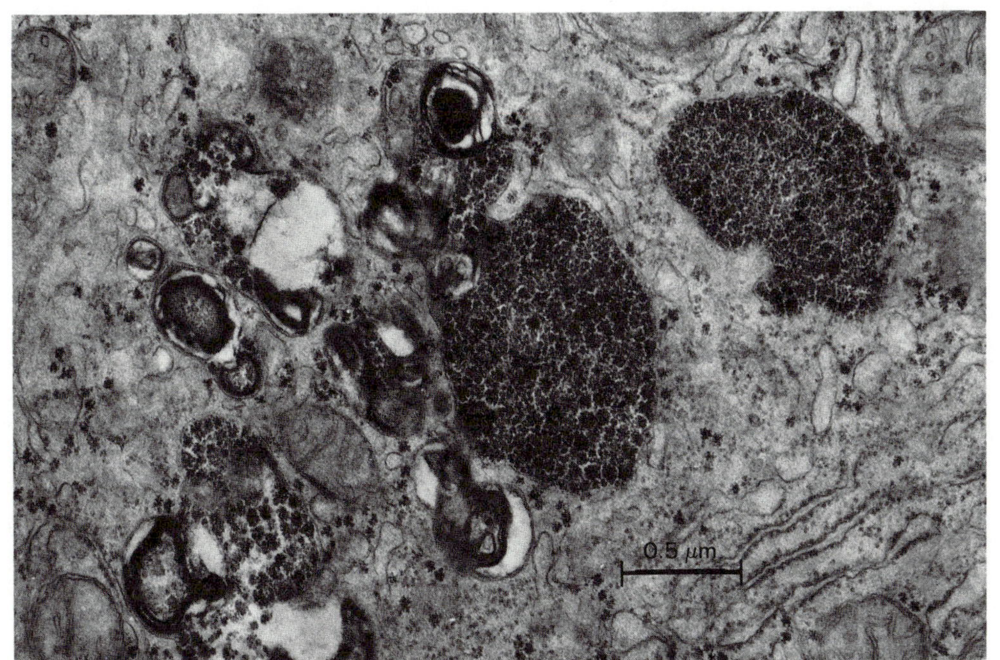

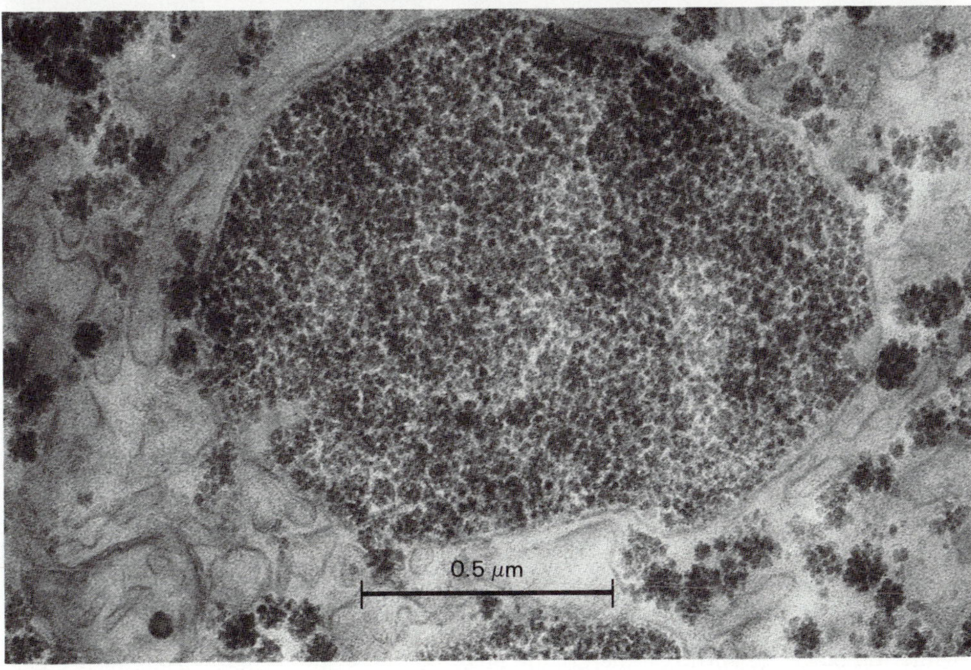

An example of a genetic lysosomal storage disease: glycogen storage disease type II, or Pompe's disease. Due to the genetic deficiency of the lysosomal hydrolase normally responsible for the digestion of the starchlike polysaccharide glycogen, which enters the lysosomal space by autophagy, this material accumulates in lysosomes. In these two electron micrographs of a fragment of liver from a child who died of the disease, glycogen particles can be recognized by their high density and star-shaped structure. The upper picture shows an early stage of incorporation of glycogen into dense lysosomes. In the lower one, a swollen lysosome is completely filled with glycogen.

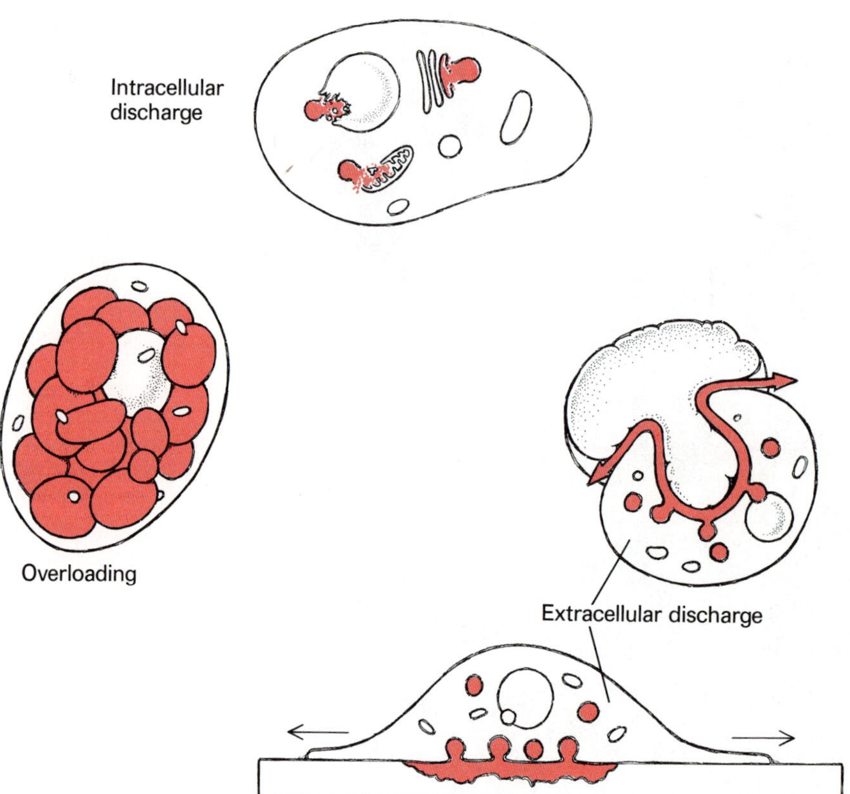

Digestive pathology of cells: the three great lysosomal syndromes.

molecule cannot get out again. So, with time, the lysosomes fill up with residues.

In certain pathological conditions, the constraints that prevent extracellular unloading of lysosomes break down. This may happen, for instance, when cells struggle to take in very bulky objects, such as antigen-antibody aggregates, or if they attempt the impossible task of engulfing a flat surface, such as a basement membrane accidentally opsonized (Chapter 4) by autoantibodies (antibodies against endogenous constituents). Endocytic invaginations then fuse with lysosomes while still open and allow the lysosomal contents to leak out into the extracellular spaces. Rheumatoid arthritis and several other autoimmune diseases illustrate the type of gruesome erosion that extracellular structures may suffer as a result of this kind of discharge, showing that there is a good reason for cellular constipation, in spite of its many drawbacks.

Another frequent lysosomal syndrome is that which results from some injury to the lysosomal membrane. Whatever magic renders this membrane resistant to digestion, it is not foolproof. Predictably, if the lysosomal membrane gives in, the neighboring cytoplasm becomes

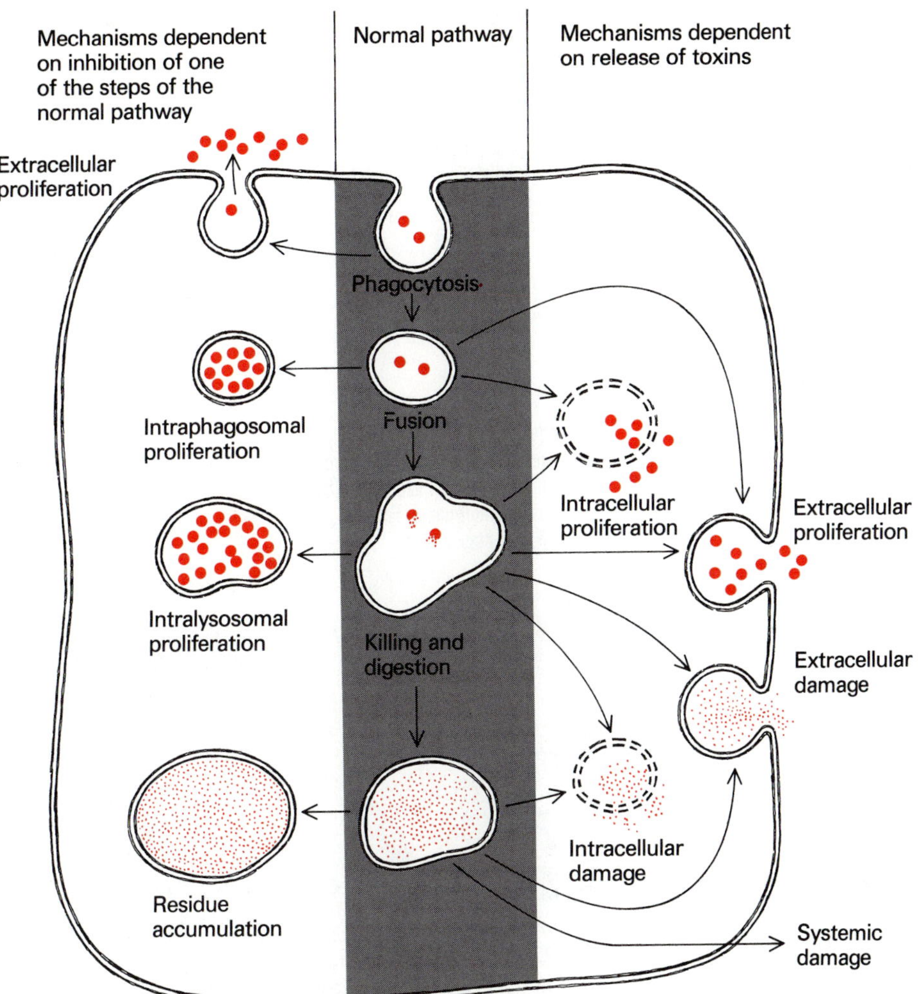

Lysosomes and infection. This diagram illustrates some of the mechanisms whereby pathogenic microorganisms evade, inhibit, or resist destruction by lysosomes, or, alternatively, turn this defense mechanism into a means of causing local or systemic injuries.

exposed to enzymic attack. Extensive cell injuries, and even cell death, may ensue. Gout, asbestosis, and silicosis (miners' black-lung disease) are examples of conditions where such injuries are believed to occur.

Finally, some mention must be made of the role of lysosomes in infection. In a way, most successful infections may be seen as some sort of failure of lysosomal defense. Or to put it differently, a pathogen is a microorganism capable of evading lysosomal destruction. Our bacterial enemies have displayed remarkable ingenuity in this respect. Some slip by our immune system without alerting it and, not being opsonized, avoid being phagocytized. Others inhibit endosome-lysosome fusion—remember the tuberculosis bacillus—or, like the leprosy bacillus, resist lysosomal destruction. Yet others escape by breaking open the membrane of endosomes or lysosomes with an exotoxin (a secreted toxic substance). The most perfidious allow themselves to be killed and then exert a posthumous revenge by poisoning the organism with an endotoxin (i.e., a toxic substance of endogenous origin) produced by lysosomal digestion of their cell wall.

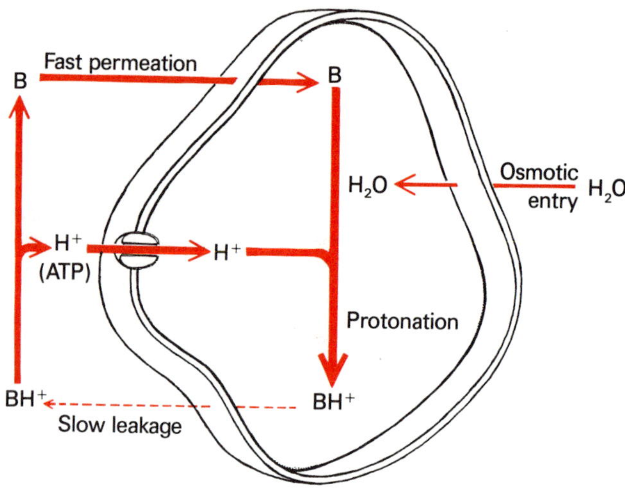

Lysosomotropism of weak bases by proton trapping. Permeant, nonprotonated form B is trapped in the lysosome as nonpermeant, protonated form BH^+. Water is attracted osmotically. The process is powered by ATP through a proton pump, which forces protons inside.

Consequence of lysosomotropism is illustrated by vacuolation. The fibroblast shown here was vacuolated by exposure to chloroquine. For vital staining, see book jacket.

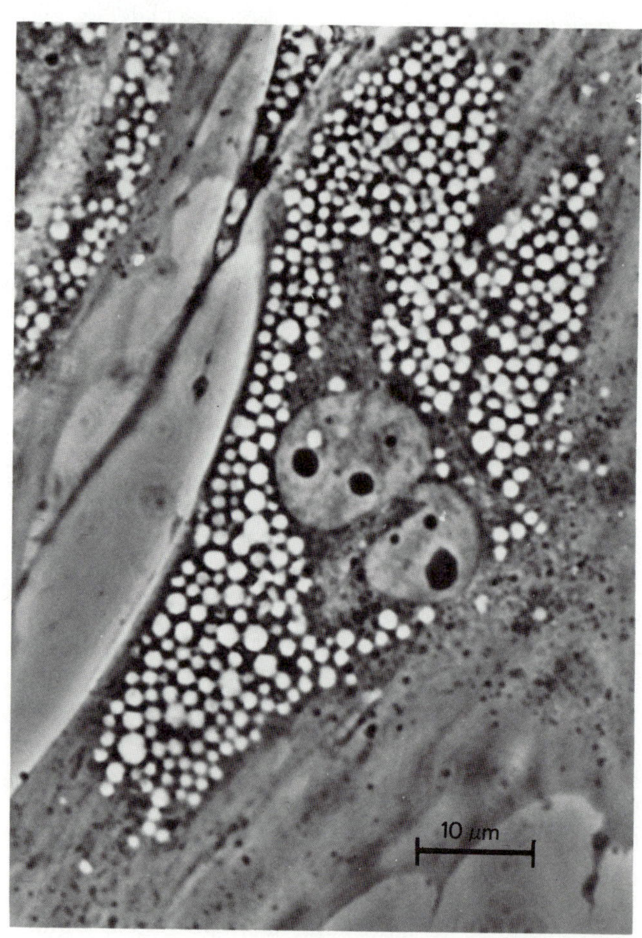

Sending Drugs into Lysosomes

There are many reasons why we may want to send drugs into lysosomes. The most obvious is to correct some local disorder, of which we have just seen there are many different kinds. But, in other instances, lysosomes are not so much a target as a tool of the therapeutic intervention. A large group of substances—among them many drugs, but also a number of dyes, such as neutral red—home spontaneously to lysosomes, where they accumulate very rapidly and reach up to several hundred times the concentration they have elsewhere in the cell. Called lysosomotropic, all these substances are weak bases. They readily cross membranes, including the lysosomal membrane, in their unprotonated form, which is uncharged and adequately lipophilic. When exposed to the lysosomal acidity, they become protonated, acquiring one and sometimes two positive charges that prevent them from crossing the membrane in the other direction.

This phenomenon of proton trapping is not limited to lysosomes. It occurs in endosomes and in any other structure (some parts of the Golgi apparatus?) that is kept acidic by the operation of a proton pump. It leads to all sorts of interesting manifestations. The most spectacular is vital staining, in which the whole endosome-lysosome space lights up like a constellation of brightly colored stars. Known for more than a century, this phenomenon has only recently been explained. Another manifestation of lysosomotropism is vacuolation of the cells, the consequence of osmotic swelling of the invaded structures. In addition, many lysosomotropic substances interfere with lysosomal digestion by neutralizing the local acidity, sometimes also by inhibiting one or more hydrolases. A number of toxic effects and therapeutic actions are explained by such mechanisms. An interesting example is that of the antimalarial drug chloroquine, which is intensely lysosomotropic and is believed to accumulate in the lysosomes of the malarial parasite and there to block digestion of hemoglobin, the parasite's only source of food in the red blood cells where it resides.

Another way of getting into the lysosomes is by endocytosis. This mode of entry holds particularly exciting promises because of its selectivity. Entrance by the endocytic route is governed by surface receptors, and these vary from one cell type to another. Now that we know something of what happens in the lysosomes, we can add a new twist to this tale. Suppose we wish to send a drug selectively to a given type of cell—a cancer cell, for instance, or some pathogenic parasite. Provided our target has receptors that do not occur, or are less abundant, on the cells we wish to protect, we are in business. All we need do is attach the drug to a carrier that is recognized by these receptors, and do so by a chemical bond that will be split inside the lysosomes. There are a few additional conditions. But, basically, this is the kind of concept that is now inspiring the design of what may well be the "magic bullets" of the future.

Lysosomotropism by endocytosis and its application to drug targeting. A drug-carrier complex is taken up selectively by cells possessing a receptor for the carrier on their surface (target cells). In lysosomes, the drug is detached from its carrier by enzyme action and moves out to its target—for instance, the nucleus.

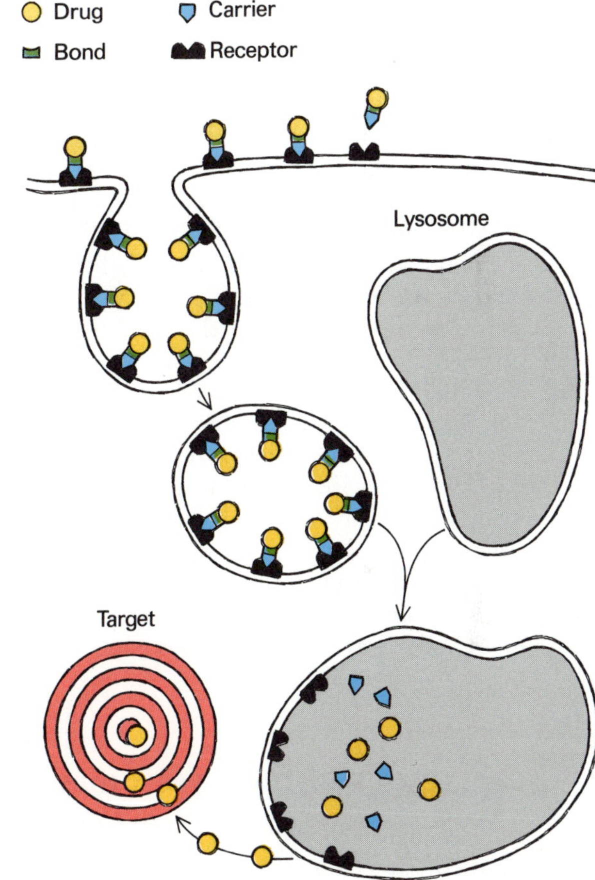

The Importance of Biodegradation

Most students of life have reserved their main admiration for biosynthesis, this wondrous ability of living organisms, shared by even the humblest of bacteria, to manufacture thousands of specific substances, many of which are so complex that even our most sophisticated chemical technology is unable to reproduce them. Degradation, on the other hand, was seen as the natural and inevitable "way of all flesh," a fate constantly menacing the "miraculously fragile" fabric of protoplasm, a threat to life rather than a vital function.

Biosynthesis is indeed a wonderful activity, and a good part of our tour will be devoted to its key energetic and informational aspects. But biodegradation, although much simpler, is equally important, as our plastic age is beginning to find out. Had not the appropriate hydrolases appeared at about the same time as the first proteins, nucleic acids, polysaccharides, and other biopolymers, there would be no biosphere, only a "plastosphere." There would be no adaptation, no differentiation, no evolution, no life. The fabric of life is really not fragile at all. It is made largely of very tough molecules that are no less stable under normal circumstances than are polystyrene or polyvinyl chloride. Digestion, as we have seen, proceeds only in the presence of suitable catalysts.

Biodegradation, therefore, is an essential, life-saving process. But, at the same time, it is by its very nature undissociably life threatening as well. Such tightrope-walking between life and death cannot be entrusted to a random downhill process. Biodegradation is really a strictly controlled activity, fenced in by an elaborate network of protective defenses. That is the lesson we take back with us from our visit to this fascinating lysosomal compartment, which somehow allows the kind of harsh and indiscriminate breakdown treatment that is demanded by biodegradation to be administered with the necessary discrimination and under remarkably safe conditions. But now we must find a way out of the lysosomes.

Escaping out of the Lysosomes

A few years ago, nobody in his senses would have dared to enter a cell by the endocytic route unless there were some strict guarantee that the lysosomal compartment would be bypassed. Once inside a lysosome—so the belief ran—abandon all hope, as in Dante's *Inferno*. Even if you escaped being burned by the acid or cut to pieces by the hydrolases, you would remain forever trapped within a membranous prison, endlessly tossed around from pen to pen by the capricious play of *cis* and *trans* mergers. Indeed, innumerable well-documented examples supported this view, including the many pathologies associated with lysosome overloading.

Yet, as we now know, the view is not entirely correct. There is a way out of lysosomes. There may even be several exits, connected to different intracellular routes that lead either to the plasma membrane or to the Golgi apparatus. All we have to do is to grab hold of some appropriately chosen membrane patch. Soon it will detach and take us out of the lysosomal compartment. The difficulty, apparently, is to hang on to the membrane in the lysosomal milieu. Most substances that bind to plasma-membrane receptors are detached by the lysosomal acidity or by enzyme action. Even some of the receptors are destroyed upon exposure to the corrosive lysosomal contents. Thus, by the time the membrane patch is severed from the lysosomal membrane to be recycled back to the cell periphery, it moves away naked and with its surface scoured. This, at least, is how we interpret the fact that so little is removed from the lysosomes in spite of the continuing recycling of their surrounding membranes. We must assume further that the membrane patches move off in a form—perhaps of very small vesicles—that offers little space for fluid transport. This may make our escape pretty uncomfortable. But we will have to chance it.

Should we accompany the membrane all the way to the cell surface, we would be back where we started. However, some recycling membranes follow a more circuitous route and stop at the Golgi apparatus, perhaps to provide an opportunity for repair of the damage the membrane has suffered while passing through the lysosomal compartment. In addition, there is a direct lysosome-Golgi shuttle associated with the supply of fresh enzymes to the lysosomes. We will take advantage of this service, which is clearly advertised by the occurrence of a specific receptor for lysosomal enzymes on the membrane patches involved. It will bring us conveniently, if not comfortably, to our next objective: the cell's export department.

6 The Cell's Export Industry: Endoplasmic Reticulum, Golgi Apparatus, and Secretion

Living cells manufacture all sorts of export materials, which they assemble, process, package, and transport in a chain of interconnected, membrane-limited enclosures and finally ship out by exocytic discharge. These materials are mostly made of protein or carbohydrate, often combined as glycoproteins or proteoglycans. They fill a considerable catalogue.

A Catalogue of Products

First, we find a variety of substances produced for short-range export and used by cells to organize or clean their immediate surroundings. The walls, the fibers, the matrices, and all the other pieces of the framework that holds the cells together and gives tissues and organs their characteristic architecture are assembled from soluble precursors, such as procollagen (Chapter 2), that are made and discharged on site. The enzymes that trim and consolidate these building blocks are similarly produced and secreted locally by the export machinery of the cells. Specialists of such construction work are the osteoblasts, which erect the scaffoldings of bone; the chondroblasts, which lay down the matrix of cartilage; the fibroblasts, which make connective-tissue fibers.

As is the rule in the living world, construction also calls for destruction. This is accomplished extracellularly by lytic enzymes that are likewise products of cellular export. From the point of view of individual cells, these enzymes serve as a means of invasion; they fragment extracellular structures and clear passages for the cells to

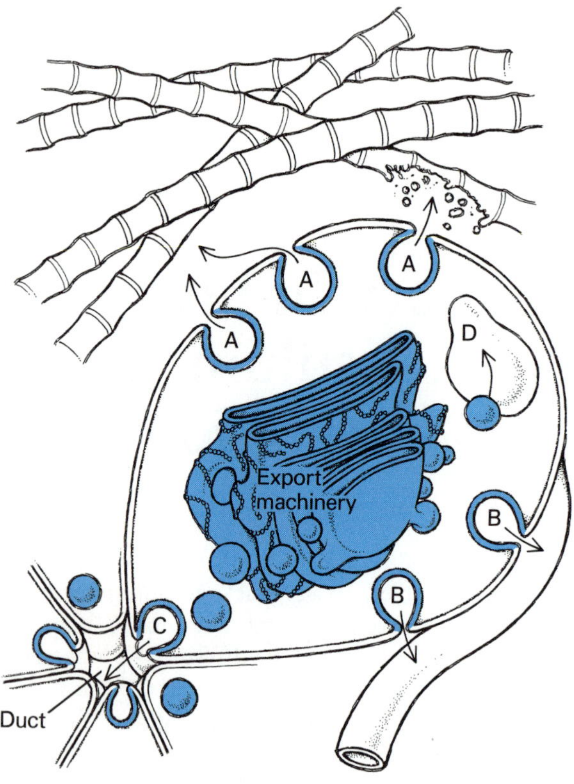

Composite picture showing how different products manufactured by the cell's export machinery are discharged exocytically: (A) into pericellular spaces (building blocks and processing enzymes of extracellular structures); (B) into the bloodstream (plasma proteins, immunoglobulins, endocrine secretions); (C) into a secretory duct (exocrine secretions); and (D) into lysosomes (acid hydrolases). Not all cells make all kinds of products, but most cells make more than one and therefore need a dispatching or sorting system.

move through. Without such a secretion, the leukocyte that led us out of the bloodstream could not have broken through the capillary's basement membrane. Cancer cells may owe their characteristic invasiveness to a particularly aggressive policy of this sort. But such demolition tools are useful also to the resident cellular societies, by providing them with the ability to clean, repair, and remodel their habitat.

While all cells tend to clean their own doorsteps, so to speak, certain specialized cells make the destruction of extracellular structures their lifetime avocation. We have already met the osteoclasts, which burrow their way through the bone matrix with the help of an acid mixture of lysosomal hydrolases (Chapter 5). Fibroblasts, strangely enough, also double as "fibroclasts." When properly stimulated, they will discharge a powerful collagenase, which specifically fragments collagen fibers, and a special protease acting on structural proteoglycans (Chapter 2). Such secretions are physiologically important, but carry grave pathological hazards, as we have seen in the unloading of lysosomes (Chapter 5).

Another major division of the cellular export industry deals with the manufacture of materials for long-distance shipment. Cells involved in this activity are usually arranged into special organs called glands, which are subdivided into two groups. In exocrine glands, the secretory products are collected by ducts and conveyed by them to a specific destination. In contrast, the products of endocrine, or ductless, glands are discharged into the pericellular spaces, from which they are spread throughout the organism by the bloodstream.

The better-known exocrine secretions, such as saliva or pancreatic juice, supply enzymes to the digestive organs. But there are many others. Glands occur in skin, in eyelids, in ear ducts, in the genital tract, and in many other areas. Their products serve for lubrication, protection, waterproofing, disinfection, scenting, and chemical communication, as well as for digestion.

Hormones make up the most important endocrine secretions. But here we have to distinguish between the polypeptide hormones, such as insulin or the pituitary hormones, which are manufactured and shipped out by the machinery that we are about to visit, and a number of other chemical messengers of small molecular size, including thyroxin, epinephrine, acetylcholine, and the sexual steroids, which either diffuse out of the cell by permeation or are released by special devices that we will not have the opportunity to see on this tour.

A special group of cellular export products is represented by the proteins of blood plasma. Many of them are made in the liver, except for the antibodies, which are manufactured by the B-lymphocyte-derived plasma cells (Chapter 3).

Finally, there are the lysosomal enzymes. Technically, these do not deserve the name of export products, because they are mostly used intracellularly. In practice, however, the space into which they are discharged is only one step away from the extracellular space, from which it is derived by the mechanism of endocytosis; its enzymes may even end up physically outside the cell, as mentioned in Chapter 5. In these respects, lysosomal enzymes have the character of secretory products. The cell obviously considers them as such, since it uses its export machinery to supply enzymes to its lysosomes.

Secretion is a specialized activity in that each product is usually made by one or a few cell types. But cells often make more than one product, sometimes for different destinations. This means that the export machinery must include some sort of dispatching system, whereby, for instance, acid hydrolases are sent to the lysosomal space, plasma proteins and hormones to the bloodstream, exocrine products to the appropriate duct, and structural elements and the agents of their remodeling into the immediate cellular surroundings.

Some cellular exports are regulated only by production. The cells dispose of their merchandise as they make it and keep no stocks. The plasma cells, which secrete immunoglobulins, behave in this way; so do the fibroblasts, the main manufacturers of structural collagen. However, other cells, such as those that secrete hormones or the components of digestive secretions, store their products in special granules and discharge them only on command. We will speak of continuous and intermittent secreters to distinguish these two types.

In many cases, the cells export incompletely finished products, leaving the finishing touch, which is usually some proteolytic clipping, to be administered at the site of destination. They do this particularly when the finished product is potentially harmful and can cause local damage. Then they make an innocuous precursor, which is converted into the final product only where and when it is needed. Many digestive enzymes, including some lysosomal hydrolases, are made in this way, in the form of inactive proenzymes, or zymogens. So are certain tissue-degrading enzymes, such as collagenase, and a number of blood-borne systems involved in the formation and dissolution of clots, in cell killing, and in other dangerous activities. Structural components destined for assembly into insoluble aggregates are similarly made as larger precursors incapable of forming such combinations spontaneously. An example is tropocollagen, which is secreted as soluble procollagen (Chapter 2). The finishing of these various precursors takes place outside the cells and is mediated by enzymes that are themselves export products. Thus the cells involved are linked by complex networks of mutual interactions.

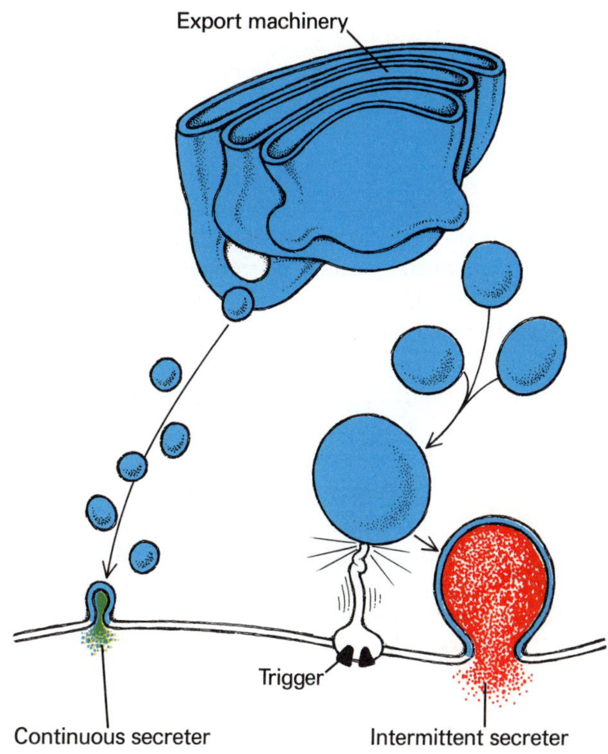

A continuous secreter discharges a product as it is manufactured. An intermittent secreter stores a product and discharges it only when triggered to do so.

Sometimes, when particularly strict safeguards are required, a whole cascade of precursor-activation reactions is inserted between the initial trigger and the final process-

After fighting their way upstream against the secretory current, cytonauts emerge into a deep gorge of impressive dimension. It is a cisterna of the endoplasmic reticulum. Tufts of growing polypeptide chains cover its walls.

ing step. Blood clotting is a typical example of such a cascade. It is the result of the proteolytic conversion of a soluble blood protein called fibrinogen into a product called fibrin, which assembles into an insoluble network. The converting enzyme, thrombin, is present as an inactive precursor designated prothrombin. When a tissue is wounded, certain substances released from the injured cells initiate prothrombin activation, but they do so by a complex chain of successive proteolytic events. Activation of fibrinolysin, the enzyme that dissolves clots, likewise depends on a cascade of such events. A particularly striking example of such a mechanism is the activation of complement, the powerful killing system that is attracted to antibody-branded cells (Chapter 3). It consists of no fewer than nine distinct circulating protein molecules, some of which are made of more than one subunit. Binding of one of these components to the Fc tails of bound immunoglobulin molecules triggers a set of consecutive interactions that end up in the construction of an elaborate, derricklike structure by which a life-draining hole is bored through the plasma membrane of the antibody-coated cell. The impression one gains from these and other similar examples is that the number of steps between a triggering event and the final outcome is a direct function of the danger associated with setting off the mechanism accidentally. Proteolytic cascading is Nature's way of constructing multiple-control, "fail-safe" devices.

A Magic Spinnery

The factories in which cells manufacture their export products are not normally accessible to visitors from the outside. The Golgi apparatus, where we landed coming from the lysosomes, lies halfway down the assembly line. If we wish to follow the industrial process from its beginning, we must first force our way upstream. Our unlawful journey is not easy and takes us through a series of narrow caves and convoluted tunnels. But this bit of cellular speleology soon comes to an end, as we emerge into a deep gorge of impressive dimension, part of the endoplasmic reticulum, or ER.

The term reticulum means network in Latin. Its choice reflects the two-dimensional view of the morphologists who observed this system in cross section as a filigree of thin lines. These were recognized as the edges of membranes cut perpendicularly to their plane. Three-dimensional reconstruction showed that the membranes form large, flattened sacs, or cisternae, completely sealed off except for the connections—permanent or intermittent—that link them with each other and with the Golgi apparatus.

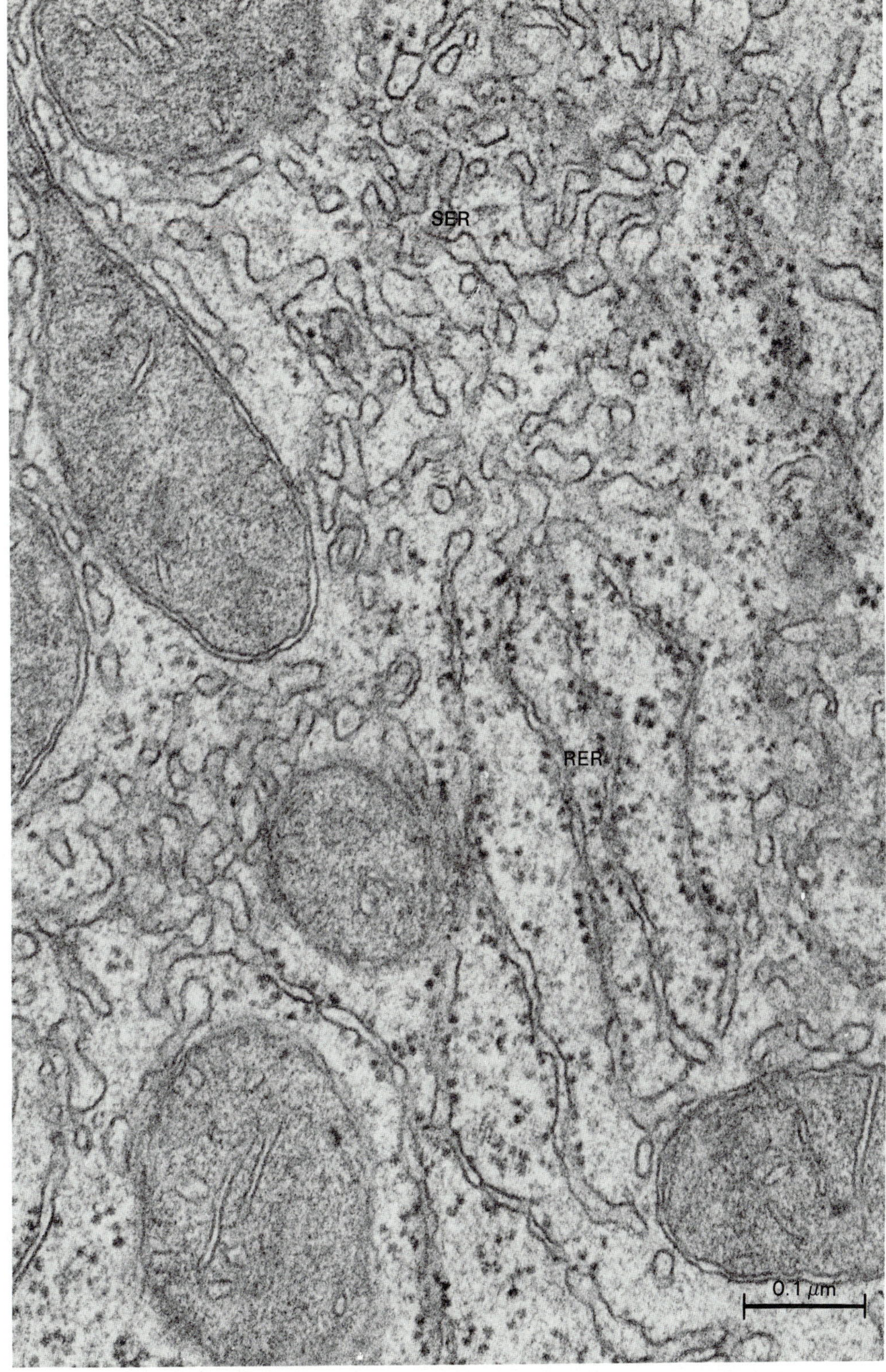

Electron micrograph of a thin section through a rat liver cell shows the endoplasmic reticulum. Narrow, elongated, membranous profiles studded with dense particles (ribosomes) on the cytoplasmic side are cuts through the cisternae of rough-surfaced endoplasmic reticulum (RER). Small, irregular, smooth profiles belong to smooth-surfaced endoplasmic reticulum (SER). Note the continuities between the RER and SER. The large bodies are mitochondria.

Seen from the inside, the flat, membranous walls of ER cisternae appear to be constructed according to the same bilayer-protein blueprint as are the plasma membrane and the boundaries of endosomes and lysosomes. But they are thinner, smoother, more flexible; the phospholipids in the bilayer have a different composition, and there is virtually no cholesterol associated with them. The proteins also are different and are largely devoid of carbohydrate side chains. These endoplasmic membranes lack the characteristic bristly appearance of the plasma membrane. They are covered instead with tufts of silky, tenuous threads, which make them look like shimmering gossamer. The shimmer is not just an optical effect; it is a reflection of movement: the silky threads grow. They grow at a visible rate: roughly 1 nm per second, which at our millionfold magnification amounts to more than 2 inches per minute. They grow steadily until they reach from 50 to 200 nm in length; then they fall off and get carried away by the current. We are enmeshed in a sea of moving silk.

What we are witnessing here is the production of export proteins. The silky threads that grow out of the surrounding membrane are polypeptide chains. They form sinuous rows comprising from ten to twenty threads, characteristically arrayed in an order of regularly increasing length. All the threads of a given set grow at the same rate but with a phase difference proportional to the distance between them. About every 15 seconds, the rearmost thread of a row reaches its full size and drops off, and the tip of a new one emerges through the membrane somewhere in front. Thus, as its individual threads grow out, the row itself moves on, crawling on the surface of the membrane like some strange, hairy centipede. Hundreds of such ghostly creatures trace ever-changing arabesques on the walls that surround us. It is an unforgettable spectacle.

If we now examine the growing threads at the point where they emerge into our cave, we will notice that they are being delivered from behind the membrane through a narrow, proteinaceous tunnel built across the lipid bilayer. At the far end of each tunnel, the outline of a bulbous root can be vaguely distinguished. About 25 nm wide (1 inch at our magnification), it seems to be the actual site where the thread is being spun. Thus, behind each array of growing threads there is a row of bulbs, separated from us by the membrane. By peering through the membrane, we can just discern one more detail: a faint, slender tape strung between the bulbs of each row, and actually slithering through them at the remarkable speed of some 250 nm (10 inches at our magnification) per minute. The movement of this tape is synchronized with the growth of the threads and with the lateral displacement of their points of emergence on the membrane. In fact, the tape, with its string of bulbs, is the real centipede behind the ghostlike creature of which we see only the growing hairs. The name of this hidden demon is polysome, short for polyribosome.

Ribosomes, which will be examined in great detail in Chapter 15, are small particles that serve as the sites of protein synthesis throughout the living world. Their business is to attach amino acids together into polypeptide chains. They receive appropriate instructions as to the order in which to do this from the messenger RNAs, or mRNAs, which are segments of ribonucleic acid (hence the abbreviation RNA) that carry pieces of genetic information transcribed from deoxyribonucleic acid (DNA). These mRNAs are thin, threadlike molecules that run through the ribosomes like a tape through a cassette player. But, instead of informing one ribosome at a time, they run through ten or more simultaneously and thus link them in a string, the polysome. Imagine a tape running through a series of cassette players, each of which is hooked to a separate speaker. The same music will be rendered by each speaker, but in the form of a canon or a fugue, with a delay proportionate to the distance between the cassette players. While the last one in the row starts on the overture, the first one is already into the coda. This is exactly what happens with polysomes, except that they render the tape's message into a specific sequence of amino acids, not of sounds. All the ribosomes of a given polysome read the same text—that is, they make the same polypeptide. They do so at the same rate, but out of phase with each other: the greater the length of tape already translated, the longer the part of the polypeptide already

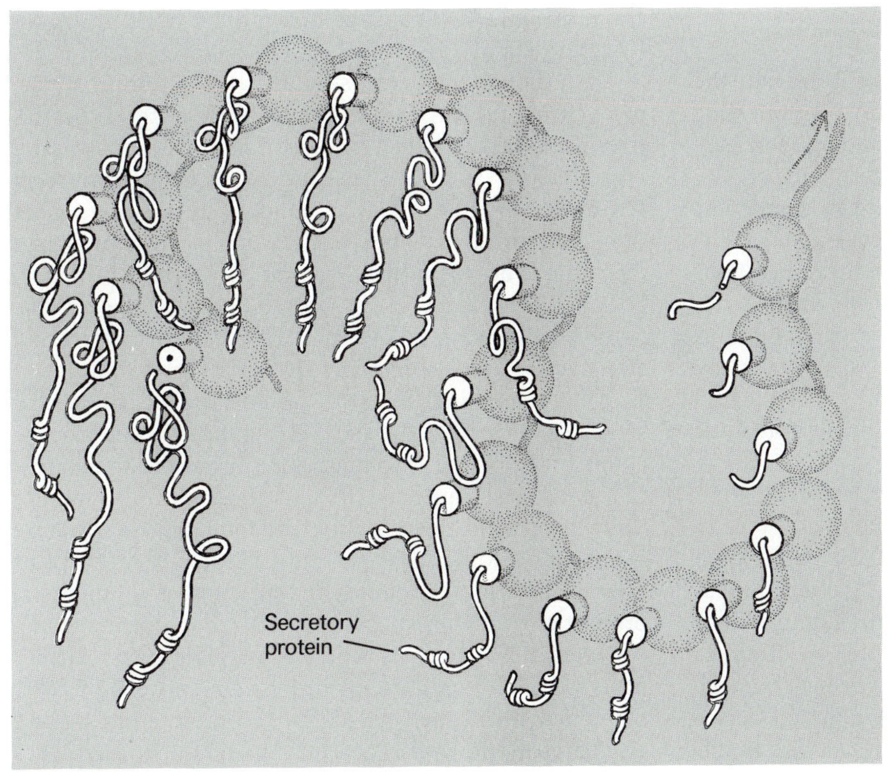

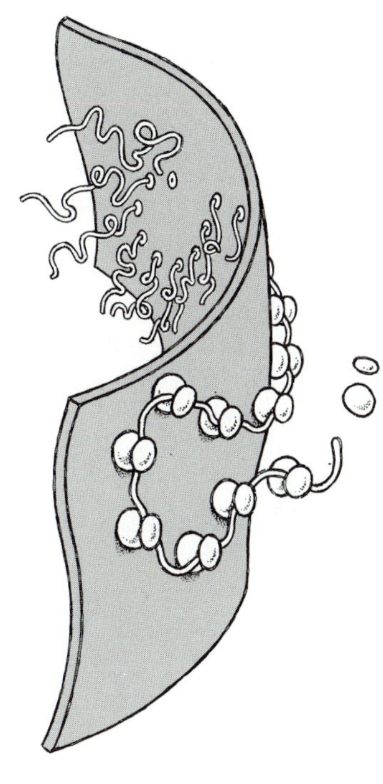

Secretory proteins grow out of the wall of the endoplasmic reticulum. The arrow shows direction of movement of the "centipede," which consists of ribosomes and mRNA. Shadows of these ribosomes and of mRNA can be vaguely distinguished through the membrane.

made. Note that polypeptides grow at their roots (i.e., on the ribosomes), not at their tips. Thus, if you compare two growing chains arising from the same polysome, you will find that the shorter one is identical with the terminal part of the longer one.

We will have an opportunity to watch all this at much closer range when we visit the cytosol (in Chapter 15). In the meantime, we can take advantage of our present location to answer a question that has evoked a considerable degree of speculation among experts. Do all the individual tufts that grow on the walls of a given cisterna consist of the same polypeptides? Or are they different, but all intended for the same final destination? Or do they make up a completely haphazard mixture? Faced with the problem of dispatching different products to different destinations, cells could have found a simple solution to this problem by segregating production, as is generally done in our chemical factories. It seems, however, that their export industry does not operate in this manner. It throws all its products into the same cauldron and sorts them out later.

Processing and Transport

In most instances, polypeptide chains are but the raw secretory products; they require a considerable amount of further processing, trimming, and garnishing before they are ready for shipment. Some are fitted with bulky carbohydrate side chains that contain as many as ten or more sugar molecules (glycosylation). Others are linked with lipids, sometimes many times their weight, as are the plasma lipoproteins made in the liver. All the chains must be folded into their proper configuration, a process that, for some molecules, is sealed by chemical bridges, usually disulfide bonds: S—S. Finally, many molecules are pared down by additional proteolytic pruning.

Some of these changes, such as glycosylation, begin to take place even before the polypeptide chains are actually completed. Others occur as the molecules drift down with the secretory current, sliding along the membranes where the processing enzymes are situated. Complex biochemical reactions are involved, but of these we often see little more than the final steps from our present vantage point inside ER cisternae. Most of the action takes place on the cytoplasmic side of the membranes, where building blocks and energy are in plentiful supply; the ER space itself serves mostly as a collecting channel and assembly line. This organization resembles that of an automobile factory, where engine, body, chassis, wheels, and other parts are made in separate shops and then simply put together on the assembly line with minimum expenditure of manpower.

This we have found to be true of the polypeptide chains, which are synthesized by ribosomes on the cytosol side of the membrane and delivered ready-made into the endoplasmic cisterns. Lipid components and carbohydrate side-chains are likewise assembled on the cytosol side and conveyed through the membrane. How the carbohydrates get through is puzzling, since sugars are notoriously hydrophilic and are not likely to pass readily through a lipid bilayer. A hydrophobic carrier molecule called dolichol is known to be involved in the process, but how it does its job is not clear. We will have more to say about this when we consider biosynthetic mechanisms (in Chapters 8 and 13).

Even though they are restricted, the reactions that take place inside the ER cavities are highly specific and selective. Each polypeptide is processed in a characteristic and reproducible fashion. That this should be so may seem self-evident—indiscriminate, random processing could only lead to chaos—but actually it carries a profound meaning. If polypeptide A, but not polypeptide B, undergoes a given change, it can only be because the enzyme involved "recognizes"—that is, binds in a catalytically effective fashion—something that exists in A and not in B. That something, in turn, can only be an amino-acid sequence, at least at the start. A subsequent step may be commanded by the preceding one, such as the attachment of a given sugar molecule. But, to begin with, the instructions must lie in the naked polypeptide, whose structure is the expression of a genetic message. It follows that this message determines not only the structural and functional properties of the polypeptide at birth, but also the whole concatenation of events whereby its molecule is further processed and modified, routed within and out of cells, combined with other molecules, incorporated in a given cell part, or otherwise handled. In other words, genes govern the whole four-dimensional history of protein molecules. We will revert several times to this all-important point (see Chapter 15).

For now, however, we must continue to accompany the secretory polypeptides on their journey. As we move on, we notice the gradual thinning of the silky growth on the membrane surface. Polypeptide tufts become sparser and eventually disappear, which means that there are no more polysomes crawling on the cytoplasmic face of the membranes. This difference can readily be seen in cross sections. Where polysomes occur, they stand out as dense dots lined up against the membrane edges and give them a rough-surfaced appearance; membranes that do not bear polysomes look smooth. Hence the terms rough endoplasmic reticulum (RER) and smooth endoplasmic reticulum (SER) given to these two parts of the ER. As pointed out, the transition from one to the other is gradual. By the time we enter the smooth part, however, the scene begins to change considerably. We are reaching the end of our roomy endoplasmic cave and approaching the tortuous passages that we traversed on our journey upstream. We are now about to reach the Golgi apparatus, this time lawfully, along with the secretory current.

The shape of the transitional elements that connect the ER with the Golgi apparatus is not easily discerned. At some stages of our journey, we have the impression of squeezing through narrow, twisting ducts; at others, of being carried in a small, sealed capsule. In other words, it is not clear whether the connection through which we are traveling is permanent and tubular in shape or transient and mediated by vesicles that use the soap-bubble trick of fusion-fission to transport secretory products from one compartment to the other. This question is still being debated, and the answer could be different for different cells. Here, again, we encounter the difficulty of having to reconstruct a three-dimensional picture from two-dimensional cross sections. Cut through a convoluted tube or through a string of closed vesicles, and you will see the same picture: a row of ringlike profiles.

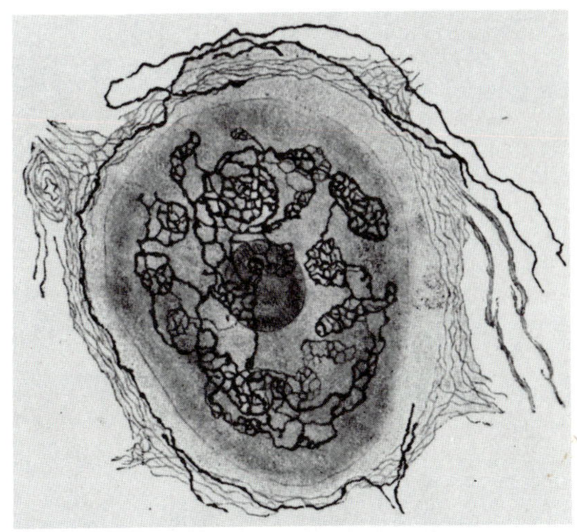

The "internal reticular apparatus," as drawn by Camillo Golgi at the beginning of the century.

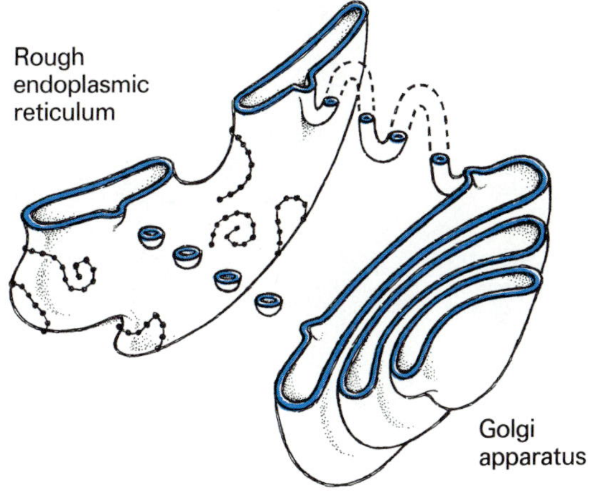

The connection between the endoplasmic reticulum and the Golgi apparatus may be either mediated dynamically by shuttling vesicles or established statically by convoluted tubules. The two possibilities are not distinguishable in thin sections.

Packaging and Delivery

Our present location is named after the Italian histologist Camillo Golgi, who, at the turn of the century, discovered this system in nerve cells that had been impregnated with certain metal salts. He saw it as a fine network and called it internal reticular apparatus.

The exact three-dimensional shape of the Golgi apparatus (often simply called the Golgi) is still being mapped. It has certain characteristic components, but structural details vary from cell to cell, and the exact boundaries of the system are imprecise. It is a sprawling complex of membrane-bounded cavities big and small, with the untidy appearance—which actually hides a considerable degree of order—often seen in warehouses and packaging centers.

The largest and most typical Golgi component is the dictyosome (Greek *diktyon*, network), so named because of the way Golgi envisioned the system. Of course, the

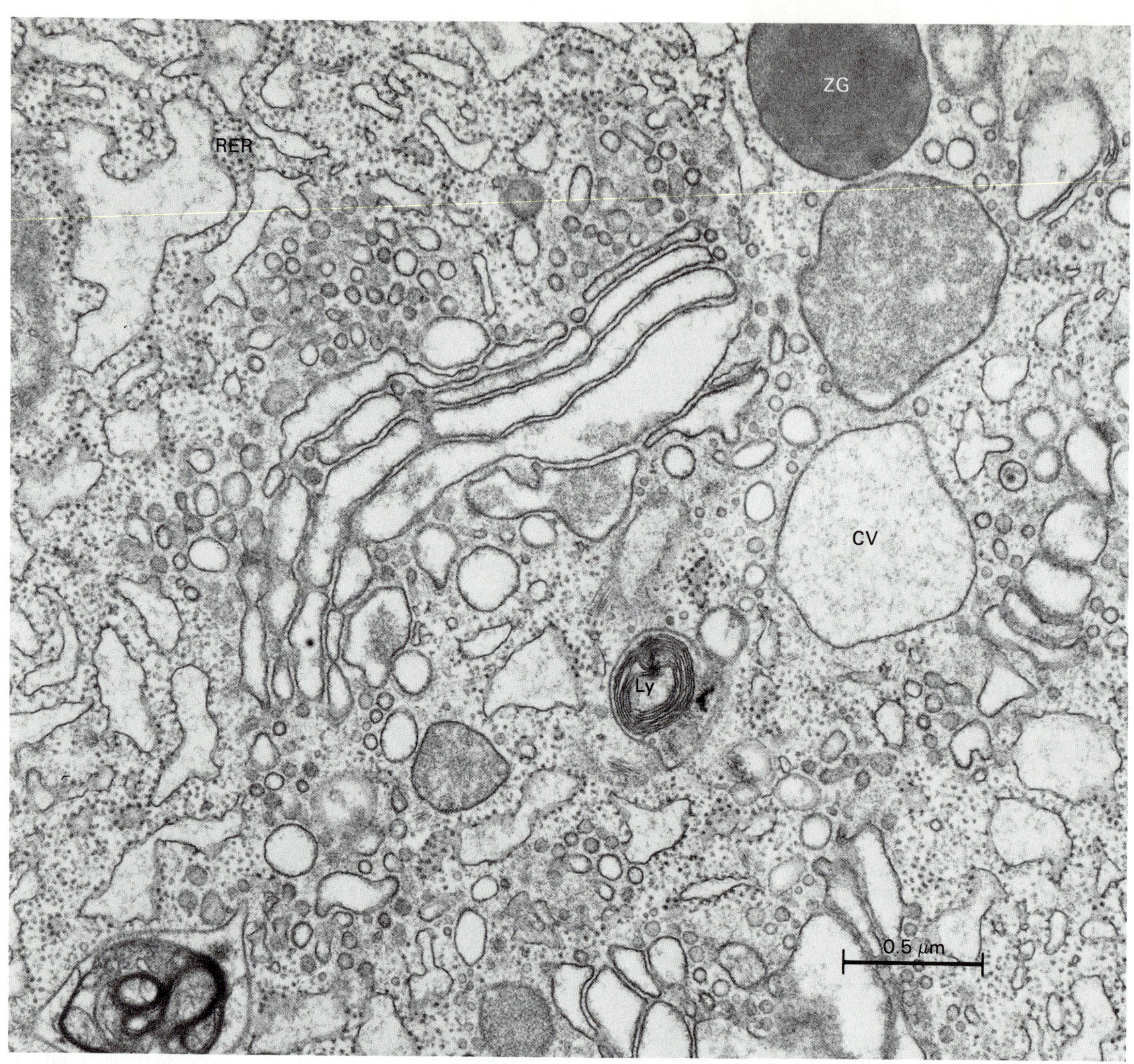

This electron micrograph illustrates the disposition of a Golgi region in a cell of rat exocrine pancreas. A typical stack occupies the center of the picture. Behind its *cis,* or endoplasmic, face (to left) are profiles of RER ending up in SER transitional elements and connecting vesicles. On the *trans,* or exoplasmic, side of the stack, the cytoplasm is filled with vesicles of varying sizes, including three large vacuoles showing successive stages in the maturation of a zymogen granule (ZG) from a condensing vacuole (CV). The small vesicle containing membranous whorls is a lysosome (Ly).

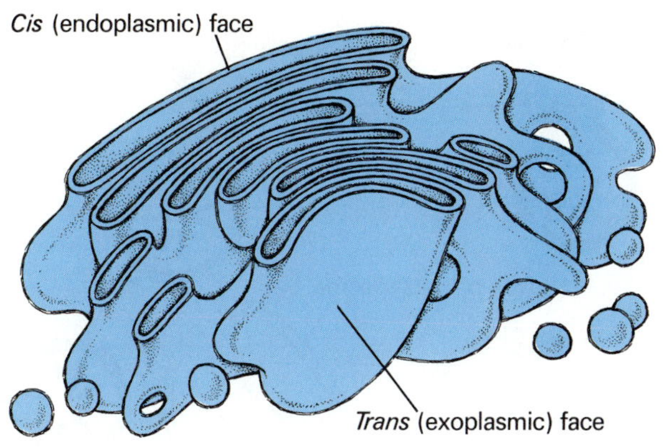

Three-dimensional representation of the Golgi apparatus.

Golgi apparatus is no more a network than is the ER; it only looks like one in cross section. The dictyosome is a stack of about half a dozen large, flat, membranous cisterns, pressed close together and looking somewhat like a pile of large, double-walled dishes. It resembles such a pile also in having curved faces—one convex, the other concave. At present, we are in the first cistern on the convex face of the stack, close to the ER. The walls of this structure do not differ greatly from those of the smooth ER. However, as we wander from cavity to cavity in the direction of the concave face of the stack, we notice that the membranes become thicker and coarser. There are increasing amounts of cholesterol in the lipid bilayer and of carbohydrate side chains on the membrane proteins. Manifestly, our surroundings become more and more plasma-membranelike. This polarity reflects the packaging function of the Golgi apparatus. As secretory products traverse the system, they move over from ER-like containers to containers that are constructed like the plasma membrane. This change is probably required before exocytic discharge by fusion with the plasma membrane can take place.

Exactly how secretory products move through the Golgi apparatus is a hotly debated question. Some believe that the Golgi cisterns move with their contents from one position to the other in the stack, undergo a progressive change of their membranes in the process, and finally fragment into vesicles that carry the products to their final destination. Holders of this dynamic view describe the Golgi stack as having a forming (from the ER) and a maturing face. Others see the Golgi stack as a more static system, through which products move either by way of permanent connections or by vesicular transport, following itineraries that may vary, depending on the product, and that may not necessarily pass through all the cisterns of the stack. In accordance with this conception, they use the purely topological *cis* and *trans* (with respect to the ER) to distinguish the two faces of a Golgi stack. We will use the terms endoplasmic and exoplasmic to avoid confusion with *cis* and *trans* membrane mergers (see p. 95).

On the whole, the static view is probably closer to the truth. The transition reflected in the polarity of the system is not the gradual change postulated by the dynamic view. It involves profound and relatively abrupt differences in which, for instance, whole groups of enzymes are replaced by others. It is clear from all the available evidence that Golgi membranes do not undergo this kind of extensive reorganization, at least at a rate in any way comparable to the rate at which products move through the system. Hence, contents and containers cannot move together, contrary to the view postulated by the dynamic theory.

Secretory proteins undergo a considerable degree of further processing as they move through the Golgi, including partial trimming of some of the oligosaccharide side chains that were assembled in the ER, followed by capping with new sugar molecules; addition of phosphate groups (phosphorylation) or of fatty acids (acylation); and further proteolytic clipping.

A second important function of the Golgi is sorting. How this takes place is known in some detail only for the acid hydrolases destined to become part of the digestive

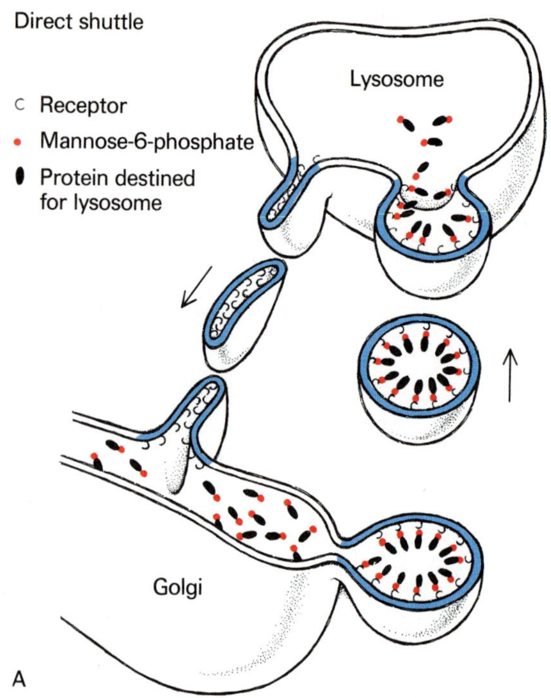

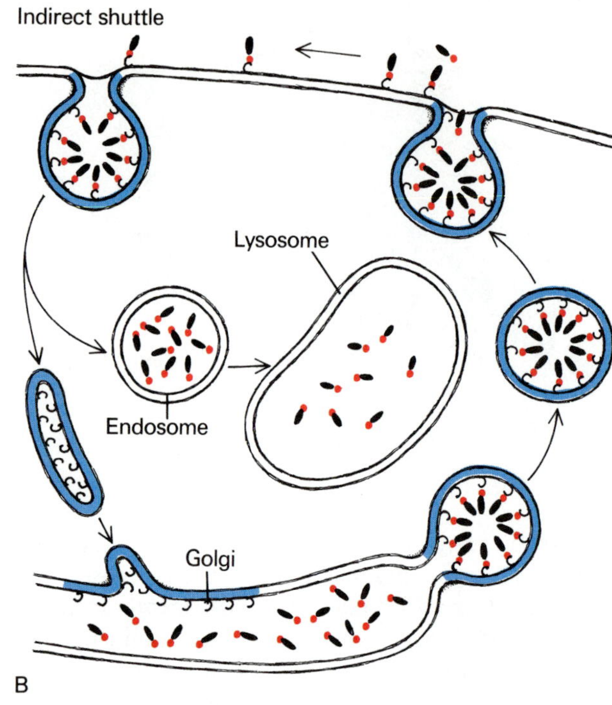

Selective transport of lysosomal enzymes from Golgi to lysosomes.

A. The structure of lysosomal enzyme proteins is so arranged as to allow them to be substrates for enzyme systems that catalyze the attachment of mannose-6-phosphate groupings. Golgi membranes have mannose-6-phosphate receptors clustered in invaginations (probably coated) that give rise to lysosome-directed vesicles. In acidic lysosomal medium, the enzymes detach from the receptors, which are shuttled back to the Golgi.

B. An alternative route takes hydrolase-loaded receptors to the cell surface, from which they return to the cell's interior by endocytosis. Enzymes are unloaded in acidic endosomes, from which the empty receptors are recycled back to the Golgi. The enzymes accompany the endosome contents into the lysosomes.

system of lysosomes. Reconstruction of the events, still partly hypothetic, is as follows. In the Golgi, lysosomal enzymes receive a characteristic addressing label, in the form of terminal mannose-6-phosphate groupings attached to some of their oligosaccharide side chains. The enzyme responsible for initiating this change does not similarly label other glycoproteins; it recognizes a specific structure—presumably an amino-acid sequence or type of sequence—typically shared by all lysosomal hydrolases. Subsequent recognition of the mannose-6-phosphate label is done by specific binding sites clustered on the inner face of certain Golgi membrane patches. These thereby fish out the lysosomal enzymes from the mixture of secretory proteins. The loaded patches then detach in the form of vesicles that selectively move toward the lysosomes (or to the endosomes) and fuse with them, unloading their catch upon contact with the local acidity. Presumably, the empty vesicles are recycled back to the Golgi afterward.

This remarkable sorting mechanism is not perfect; some lysosomal enzyme molecules accompany the main secretory stream and are discharged extracellularly. If they have been properly labeled, these molecules can still be salvaged, because, as briefly mentioned in Chapter 5, the mannose-6-phosphate receptor also occurs on the plasma membrane of many cells. In fact, the possibility that a major part of the loaded receptors may normally travel from Golgi to lysosomes via the cell surface is still being debated, although the existence of a direct intracellular route seems very probable. In any case, because of the occurrence of an extracellular pathway, true secretion of lysosomal enzymes takes place to some extent even under normal circumstances. It may be enhanced considerably in pathological situations in which transport of the enzymes to lysosomes is defective. The consequences of such leakages are, however, less dramatic than those of lysosomal unloading (defecation), because several lysosomal hydrolases, including the particularly dangerous cathepsins (proteases), are synthesized in the form of inactive proenzymes, which are activated only after they reach the lysosomes.

In addition to assigning their destination, packaging of secretory products must also make provision for their mode of delivery, continuous or intermittent. In continuous secreters, the products flow steadily out of the Golgi region, carried by small membranous vesicles that detach from the system by fission and then discharge their contents extracellularly by exocytosis. In intermittent secreters, the products are concentrated in large condensing vacuoles, which give rise to mature secretion granules, the large, densely packed, membrane-bounded structures that are the most typical feature of regulated gland cells. These granules unload their contents by exocytosis, but only upon appropriate stimulation by what is often a complex chain of nervous and hormonal relays that end in the local release of acetylcholine. This messenger unites with a receptor on the cell surface (see Chapter 13), and the resulting conformational change triggers the exocytic discharge of the secretory granules, perhaps by letting in calcium ions. If the stimulus fails to be provided and the cell goes on making more secretion granules, the granules start fusing with lysosomes instead of with the plasma membrane, and the surplus secretory product is destroyed by crinophagy (Chapter 5).

Lysosomal hydrolases are generally conveyed to the lysosomes in a continuous fashion. But there is one striking exception: the polymorphonuclear leukocyte. This white blood cell makes and stores lysosomal hydrolases over several generations in the course of its progressive development and differentiation from stem cells that re-

Exocytic discharge of secretion granules in exocrine pancreas. The zymogen granule (ZG) is approaching the plasma membrane (PM). Fusion of the two membranes by *cis* merger will produce an exocytic invagination (EX).

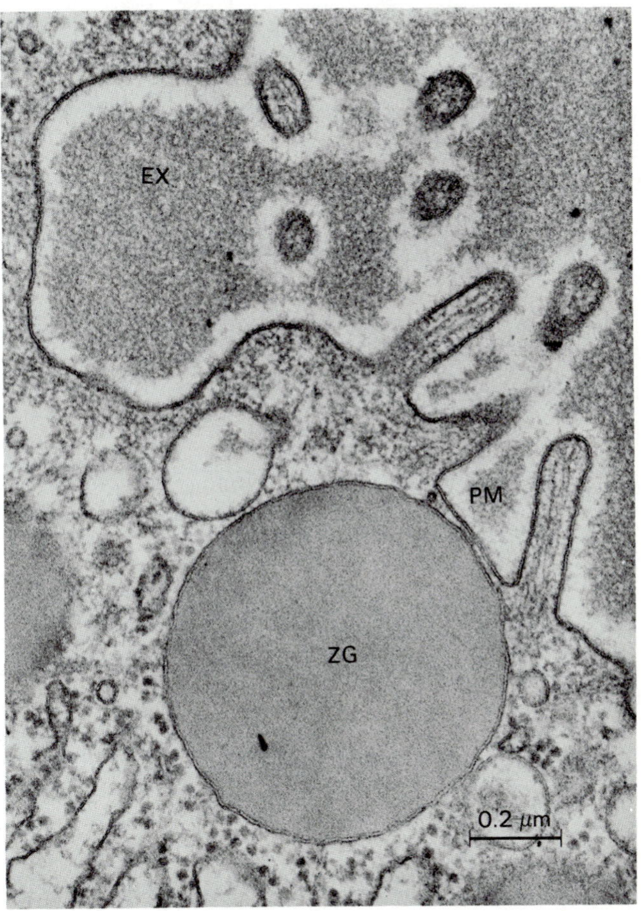

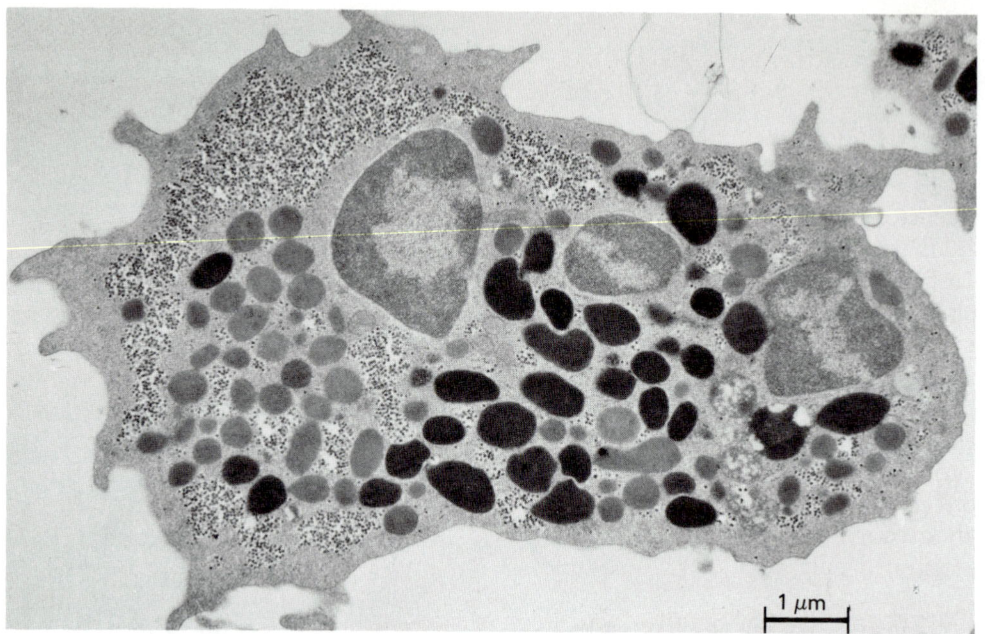

Electron micrograph of a rabbit neutrophil polymorphonuclear leukocyte. The larger, dense granules are azurophils, which are related to the lysosomes of other cells. The smaller, less dense granules, known as specific, are not lysosomes. Both types of granules are discharged massively into phagocytic vacuoles.

side in the bone marrow. It does the same for lysozyme and a number of other microbe-killing agents. It accumulates these materials in two different types of large cytoplasmic granules, known as azurophil and specific. These granules remain essentially inert in the cytoplasm until the cell is seduced by appropriate antibody seasoning (opsonization) into gorging itself with bacteria or other foreign particles (Chapter 4). This triggers a massive discharge of the contents of the granules into the phagocytic vacuoles. Killing and digestion of the prey ensue (Chapter 5). As we know, the leukocyte, unlike other phagocytic cells, does not recover after engaging in such activity. The high degree of specialization that has turned it into a controlled secreter has made it into a one-time secreter as well. At the end of its phagocytic bout, it dies.

The Birth of a Membrane

As far as is known, membranes are never born *de novo*. They always arise from pre-existing membranes by the insertion of additional constituents. This process may be almost as old as evolution. Each generation bequeaths to the subsequent one, mostly through the female egg cell, a stock of preformed membranes from which all the membranes in the organism grow out by accretion, directly or indirectly.

As we will find out when viewing it from the cytosolic side (see Chapter 13), the ER is a major site of membrane biosynthesis. Most phospholipids, as well as cholesterol, are made there. And so are many integral membrane proteins, which are synthesized by bound ribosomes, as are secretory proteins, but which, instead of falling off into the cavity, remain embroiled in the lipid bilayer by some hydrophobic sequences after they are completed. Not all ER proteins arise in this manner. Some are synthesized in the cytosol by free ribosomes and are inserted into the membranes afterward. On the other hand, many non-ER membrane proteins are made in the ER and then translocated to their final abode. By and large, such proteins travel in parallel with secretory products, but by way of the membranes themselves, rather than through the cavities. They move by lateral diffusion along physically continuous membrane domains, cluster on patches that serve in the vesicular transport of products, and thereby hitchhike from one domain to another until they reach their destination, be it in the Golgi, the plasma membrane, the lysosomes, or elsewhere. Like the secretory products, such newly made integral membrane proteins may undergo extensive processing in transit. Examples are the

The secretory pathway. The right-hand side of the diagram shows the itinerary, and concomitant processing, of secretory proteins, from their assembly by membrane-bound ribosomes to their exocytic discharge. The left-hand side shows how an integral plasma-membrane protein made in the ER uses the same pathway to reach its destination.

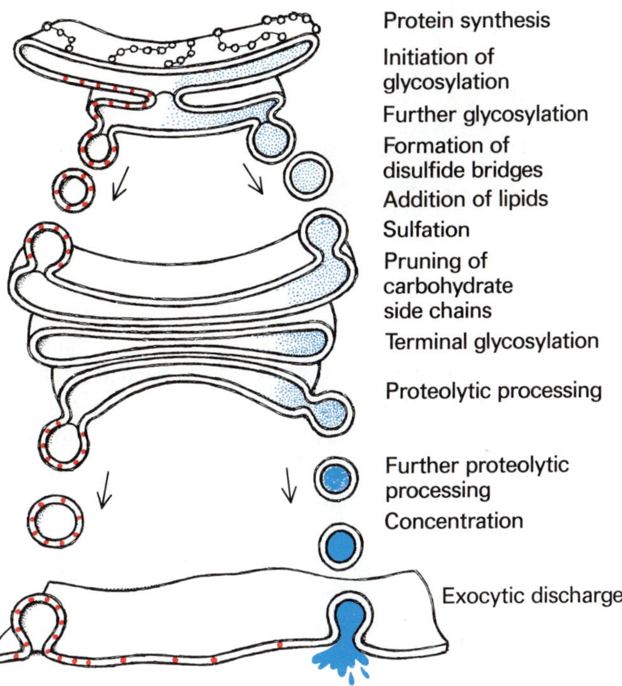

many glycoproteins that characteristically dangle their oligosaccharide side chains on the cell surface. They come from the ER via the Golgi and receive their carbohydrate components as they pass through these structures.

Closing the Circle

Several exits out of the Golgi apparatus are open to us, depending on what kind of recognition mark we wear. Not wishing to return to lysosomes, we will avoid mannose-6-phosphate. Except for that, all other outlets lead to some point outside the cell, and we might as well take the intermittent secreter line, which offers the roomiest accommodation. Even so, the experience is far from pleasant. First, we must enter a condensing vacuole and submit to an almost intolerable degree of compression, as water is being pumped out and the secretion products thicken around us to a nearly solid state. Then comes the waiting, in acute physical discomfort, and with the growing fear that crinophagy may thwart our escape and cast us back into the lysosomal compartment. When release finally comes, it does so with shocking violence. Exocytic unloading of secretion granules is an explosive manifestation. The granules jostle and push against each other, irresistibly drawn toward the periphery of the cell. Often they fuse with other granules lying in their way, creating deep chasms on the cell surface and discharging masses of secretory products. To an outside observer, the phenomenon looks like the sudden eruption of a chain of volcanoes, a veritable "cellquake." To the helpless companions of the secretory products, exocytosis brings a last moment of panic, fortunately of short duration and quickly forgotten in the sweetness of freedom recovered. Back in the balmy extracellular sea, we contemplate again the pitted, moving expanse of the cell membrane, with its twisting excrescences and waving veils. We have completed a circle, entering by endocytosis, exiting by exocytosis, and meandering in between through an endless succession of chambers and corridors.

It has been a fascinating journey, but unreal in some way, like wandering through some surrealistic maze.

The Anxious Journey, by Giorgio de Chirico.

Blank walls everywhere; endless spreads of membranes, always tightly sealed, tantalizingly opalescent, hiding from our view vast sets of machinery which, judging from their manifestations, must be of stupendous complexity. Clearly, we have hardly entered the cell as yet and will have to find a way through this ever-present screen. Before doing so, however, we should take a last, comprehensive look at the whole of the cytomembrane system.

The Vacuome

In the early part of this century, French cytologists proposed the name vacuome to designate what they saw as a complex system of vacuoles and granules occurring in both plant and animal cells. In their view, the vacuome comprised a variety of cytoplasmic structures, including the Golgi apparatus, though not the mitochondria, which they considered a separate system, the chondriome. This distinction was remarkably prescient, and the word vacuome, which never gained general acceptance, deserves to be resurrected. The term vacuolar system, which is sometimes used in the same sense today, is more ambiguous, as it has long served to designate only the import arm of the complex.

Import and export, each with its associated processing activities, represent the main functions of the vacuome. If, however, we consider the anatomical and functional organization of the system, the feature that strikes us most is its transverse division into two domains separated by the Golgi apparatus.

The ER, or endoplasmic domain, lies on one side of the Golgi—the *cis* side, which we call endoplasmic. Geared toward the manufacture of secretory proteins, it has a clear-cut polarity, stretching from the ribosome-binding sites in the RER to the transitional elements between the SER and the Golgi. Except for this asymmetry, ER membranes, which also house a number of metabolic systems unrelated to secretion (see Chapter 13), are largely homogeneous in composition. Traffic through this part of the vacuome is simple and unidirectional. The ER feeds its contents into the Golgi with little apparent reflux. According to all available evidence, secretory products are translocated through an efficient one-way lock even though their transport may depend on a two-way membrane shuttle. This also means that import stops at the Golgi barrier. Indeed, materials taken in by endocytosis do not enter the endoplasmic domain. We succeeded in doing so, but only by forcing our way through the lock.

Things are very different on the other side of the Golgi apparatus, where lies what may be called the exoplasmic domain of the vacuome (not a common appellation, but a very useful one, which we extend to the *trans* face of the Golgi). This domain is composed of a multiplicity of vesicles, vacuoles, and granules of all sizes and shapes, linked with the Golgi, with each other, and with the pericellular environment by a complex network of transport ways. The membranes surrounding these structures share some characteristic features that are also common to the plasma membrane. They are thicker than endoplasmic membranes (about 10 nm, as against 7 nm), largely because of a much greater density of oligosaccharide side chains planted on their luminal face. They also share a preponderance of certain phospholipids, such as sphingomyelin, and a high cholesterol content. Their protein composition, however, is far from homogeneous. Even the plasma membrane is subdivided into different areas.

Traffic in this part of the vacuome is remarkably dense and multidirectional. Out of the Golgi, the main secretory line transports products to their unloading sites at the cell surface, either continuously, by way of a trickle of small vesicles, or discontinuously, with an intermediate stop in secretion granules after concentration in condensing vacuoles. An important side road brings acid hydrolases from the Golgi to the lysosomes. In addition, crowding of the secretory line tends to open the crinophagic detour, which sends excess products from secretion granules to lysosomes.

Phagocytosis, receptor-mediated endocytosis, and perhaps other forms of endocytosis make up the import lines. These converge mainly in the direction of lysosomes, but on their way they may, especially when passing through endosomes, send out side branches that bypass the lysosomes and lead engulfed materials back to where they came from (regurgitation), or across the cell to a separate extracellular region (diacytosis), or to storage granules.

This network of transport lanes appears even more complex when the movement of containers is considered in addition to that of contents. Until a few years ago the magnitude of the container traffic was poorly appreciated. It was generally assumed that membrane material incorporated into the plasma membrane by exocytosis was retrieved by endocytosis, added to the lysosomal membranes by endosome-lysosome fusion, interiorized by autophagic segregation, and finally broken down by lysosomal digestion. In compensation, new membrane was believed to be made in the ER and conveyed to the Golgi to replace the material lost to the plasma membrane by exocytosis. According to this concept, which is known as the "membrane flow" theory and includes the dynamic view of Golgi function (see p. 89), newly made membranes go through a single export-import cycle and then are destroyed. In other words, the cell uses disposable containers.

This view became untenable when the extent of the container traffic was discovered. A gland cell may double its plasma-membrane area with each secretory discharge. A macrophage interiorizes as much as twice its surface-membrane area every hour. Such rates are at least one order of magnitude larger than the turnover rates of membrane constituents. Cellular containers, therefore, cannot be disposable; they must be returnable.

The pathways of membrane recycling are still being mapped out. They probably include a number of direct shuttles, as well as more circuitous routes. The hub of this traffic lies in a region sometimes designated GERL (Golgi-endoplasmic reticulum-lysosomes), of which endosomes are now recognized as a central component. Surprisingly, this rapid and complex recycling of containers does not greatly disturb the essentially unidirectional transport of contents—although a certain amount of "slobbering" does occur, as we have seen (Chapter 4). The cell evidently possesses efficient mechanisms for moving empty containers around, presumably in the form of flattened vesicles. In all likelihood, the acidity main-

Cellular container traffic. The Golgi apparatus serves as a central clearing house and channel between the endoplasmic and the exoplasmic domains: (1) ER-Golgi shuttle; (2) secretory shuttle between Golgi and plasma membrane; (2′) crinophagic diversion; (3) Golgi-lysosome shuttle; (3′) alternative route from Golgi to lysosomes via the plasma membrane and an endosome; (4) endocytic shuttle between the plasma membrane and an endosome; (4′) alternative endocytic pathway bypassing an endosome; (5) plasma membrane retrieval (a) by regurgitation, (b) by diacytosis, (c) via Golgi, (d) via a lysosome; (6) endosome-lysosome pathway; (7) autophagic segregation.

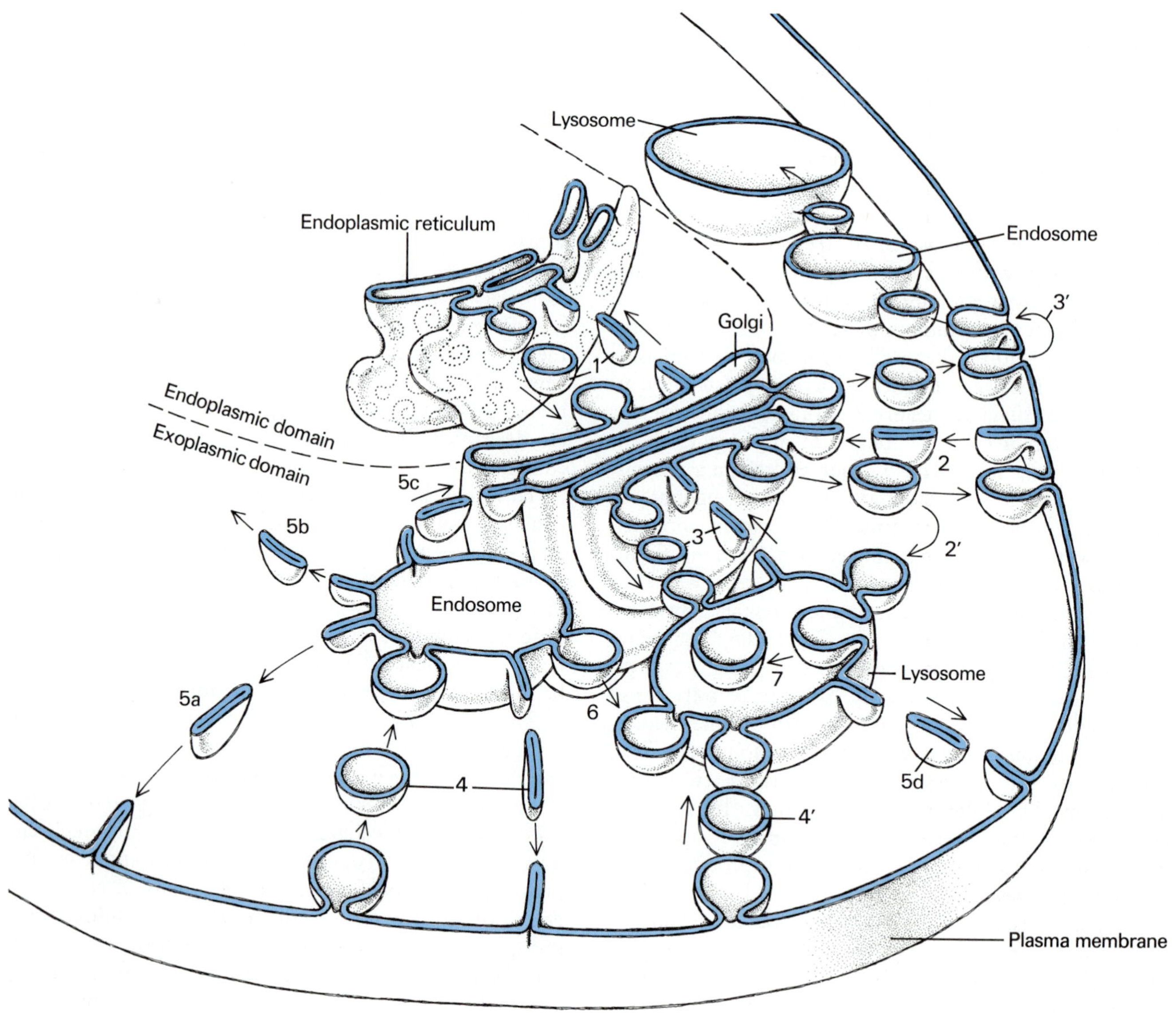

tained in the endosome-lysosome part of the exoplasmic domain plays a key role in stripping membranes of ligands before their recycling.

Even more remarkable is the ability of the cell to preserve the highly differentiated organization of its vacuome. Somehow, in spite of innumerable mergers between membranes of different composition, there is hardly any "running" of components by lateral diffusion across the junction. Sharp discontinuities are maintained, or even created, as in the clustering of occupied receptors, which plays an important role in selective transport. Wherever sorting occurs—on the cell surface, in endosomes, in the Golgi, and perhaps elsewhere—membrane patches bearing specific receptors must be involved. It is likely that the underlying constraints are imposed largely by membrane-anchored cytoskeletal elements, which also provide much of the necessary motile force (see Chapters 12 and 13).

Now that we have taken a more comprehensive look at the vacuome, we understand better our feeling of not being quite inside the cell throughout this part of our tour. Especially in the exoplasmic domain, we were actually looking at an extraordinarily shifting scene, in which a piece of membrane that happened to be on the cell surface at one moment could at another moment be part of the endosome-lysosome system, appear a little later in the Golgi, and finally be back on the surface, all in a few minutes' time. Permeases, pumps, and other transport systems present on the plasma membrane presumably continue to work when interiorized, catalyzing the same exchanges between the cytosol and the vesicle contents as they did on the cell surface between the cytosol and the extracellular medium. (However, the effects may not be the same; remember the proton pump.) We are indeed, functionally speaking, only halfway into the cell—or halfway out of it—when we are inside the exoplasmic domain of the vacuome. So are most of its contents, which either have been freshly brought in by endocytosis or will soon be discharged by exocytosis. We are somewhat deeper inside the cell when in the endoplasmic domain, to the extent that the Golgi lock protects us from invasion by extracellular objects. On the other hand, the irrevocable commitment of the ER contents to delivery into the exoplasmic domain, most often followed by discharge into the pericellular environment, clearly gives ER cavities what may be called a pre-extracellular character.

Prokaryotic Export and the Origin of Eukaryotic Cells

Possession of a vacuome is characteristic of eukaryotic cells. Most bacteria have no intracellular membranes to speak of. Yet eukaryotes arose out of prokaryotes, about one billion years ago. Consideration of the export machinery of prokaryotes may give us an inkling as to how this momentous transformation, crucial to the appearance on earth of all plants and animals, may have taken place.

Most bacteria secrete proteins into their environment. Prominent among these proteins are hydrolytic exoenzymes, which play a digestive role. It is interesting, and probably very revealing, that these exoenzymes are concomitantly made and extruded by polysomes attached to the inner face of the bacterial plasma membrane, in very much the same way as nascent export proteins are delivered through the membranes of the endoplasmic reticulum. The similarity even extends to the molecular mechanisms involved (see Chapter 15), suggesting strongly that the rough ER of eukaryotes originated from the plasma membrane of their prokaryotic ancestors, presumably as a result of some kind of progressive interiorization process. Surface flexibility would be important for such a process to occur, making it likely that our putative distant forbears were devoid of a rigid outer wall. Such naked bacterial cells can be generated by the action of lysozyme (protoplasts) and may also arise occasionally in nature (L forms). They are very fragile, but one can see how, under certain circumstances, their vulnerability might be compensated by a tremendous evolutionary advantage conferred by their surface flexibility. Should this property result in the formation of infoldings of the plasma membrane and, especially, in the intracellular vesiculation of these infoldings, such as occurs in endocytosis, the exoen-

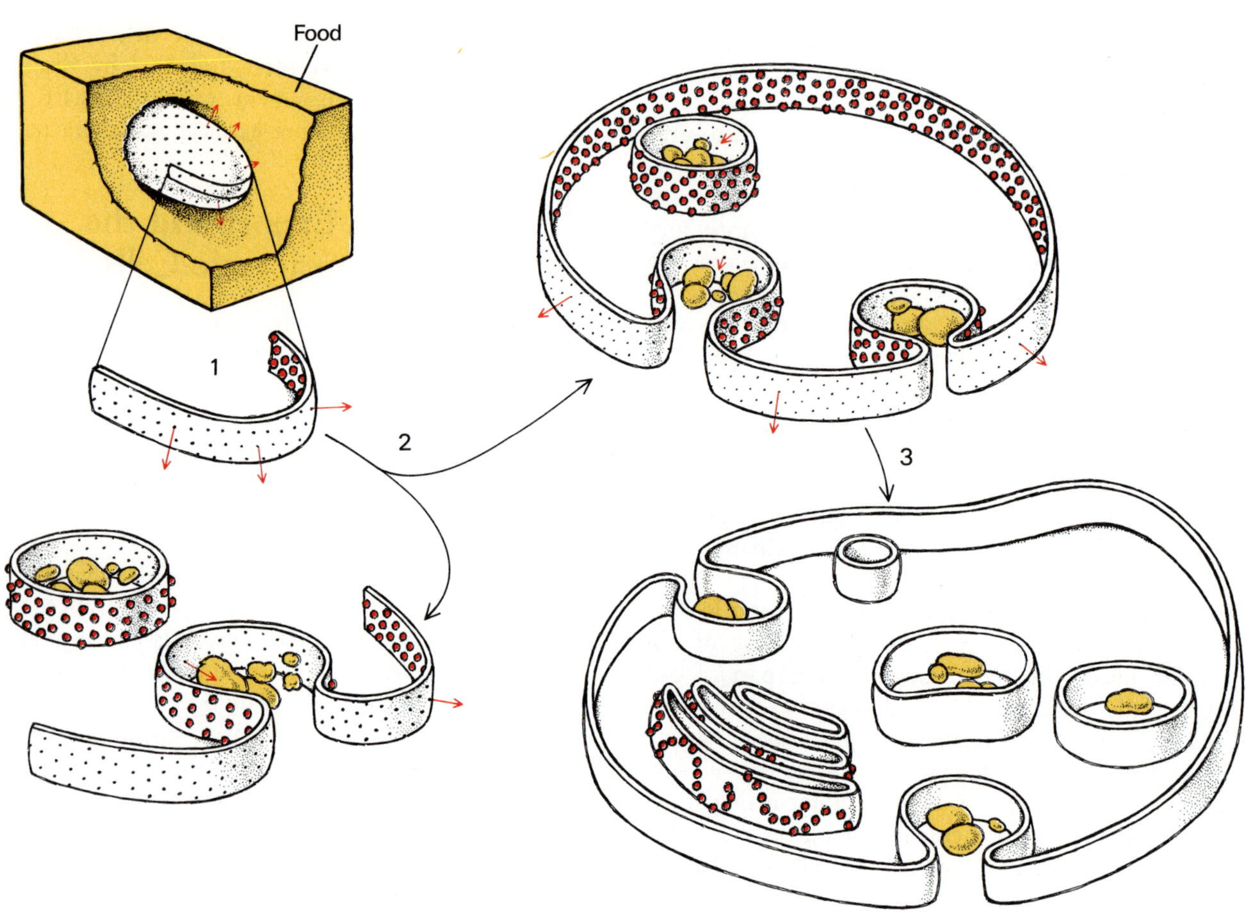

Hypothetical mechanism explaining the evolutionary origin of a primitive phagocyte from a prokaryotic ancestor: (1) like many present-day heterotrophic bacteria, the putative bacterial ancestor relies for nutrition on the extracellular digestion of food by exoenzymes that are made by plasma-membrane-bound ribosomes and discharged outside; (2) invaginations of the plasma membrane in a wall-less form of bacterium create primitive intracellular digestive pockets in which interiorized food particles (primitive endocytosis) are broken down by trapped exoenzymes; (3) differentiation of the intracellular membrane system relegates bound ribosomes to parts of the ER and leads to the formation of different membranous domains, and the cell increases considerably in size.

zymes made by the membrane-bound polysomes would remain trapped inside the interiorized vesicles, where they would be able to act on any extracellular material that got caught in the interiorization process. This would amount to the simplest possible form of intracellular digestion, within vacuoles having the combined properties of endosomes, lysosomes, and rough ER cisternae.

Even in this primitive and haphazard form, such an acquisition would provide a distinctive evolutionary advantage, likely to favor any chance mutation that would result in its further development. Until then, in order to benefit from the digestive activity of their exoenzymes, the cells had to rely on extracellular digestion. Unless they had other means of subsistence, they were practically condemned to reside inside their food supply, like maggots in a chunk of cheese. Henceforth, they would be free to roam the world and to pursue their prey actively, living on phagocytized bacteria or on other engulfed materials. This development could well have heralded the beginning of cellular emancipation.

Another important consequence of the postulated change would be the possibility of growth. Cells depend on exchanges with their environment and are limited in their capacity to grow by the surface area available for such exchanges. We have seen in Chapter 3 the importance of microvilli and other surface projections in this connection. The development of an intracellular vacuolar system, in direct continuity with the outside through endocytic interiorization and lined by a plasma-membrane type of membrane fitted with all the necessary transport systems, would enhance the exchange capacity of the cells manifold and thereby allow them to reach considerable sizes. At the same time, a progressive differentiation of the cytomembrane system might have occurred, with ribosomes becoming progressively relegated to the deeper parts of the system and the more peripheral elements evolving into plasma membrane and the smooth-surfaced parts of the vacuome of today's eukaryotic cells.

According to this hypothesis, the crucial evolutionary step from prokaryote to eukaryote was accomplished by way of a large-size primitive phagocyte. We may never know what really happened. But it is interesting that the most popular theory of the evolutionary acquisition of such important eukaryotic organelles as mitochondria and chloroplasts depends on the occurrence of just such a kind of cell (see Chapters 9 and 10). Also, if this hypothesis is correct, our picture of the vacuome as somehow extending the main boundary between cell and environment to thousands of intracellular recesses would be rooted in evolutionary history.

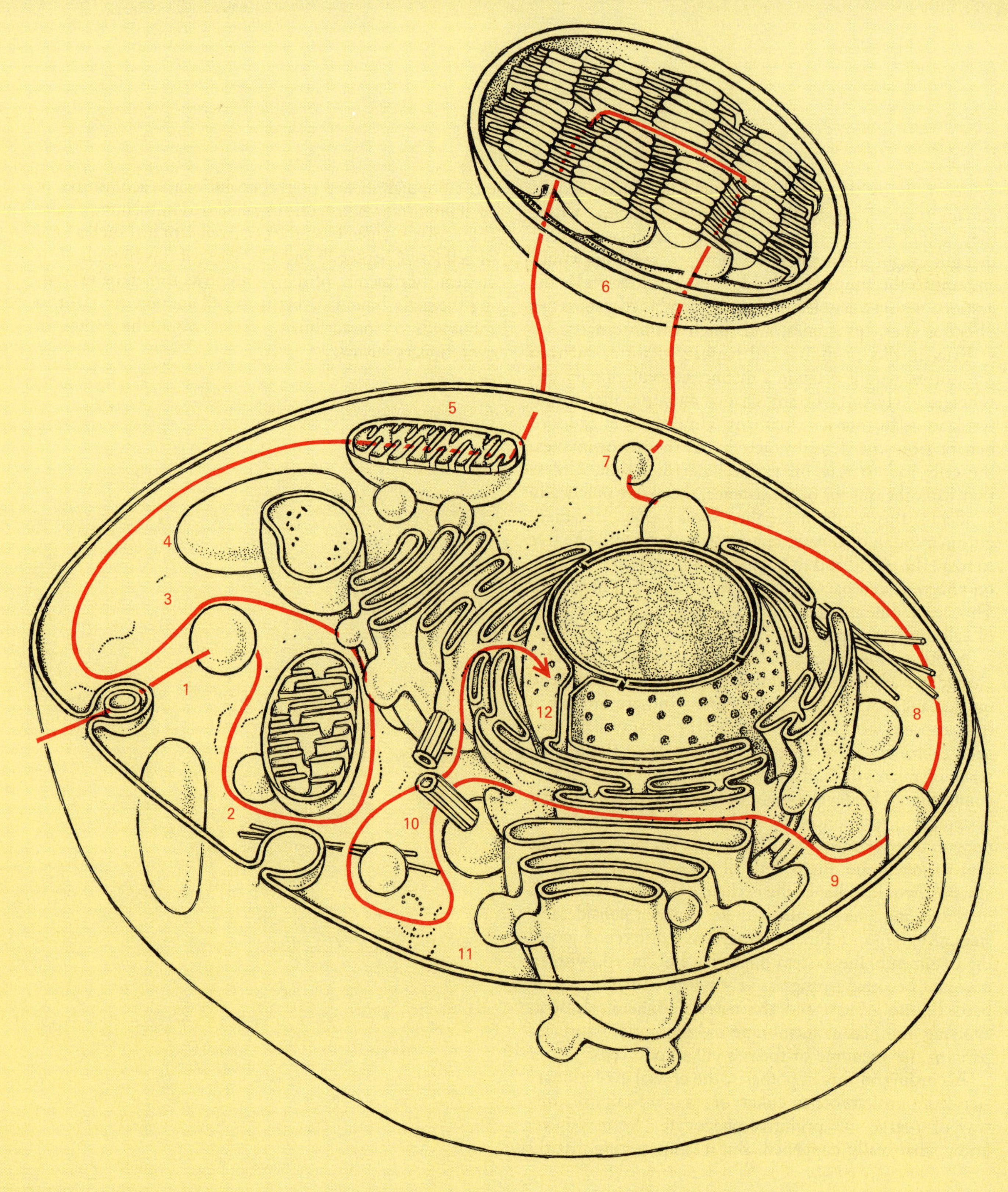
6
5
7
4
3
1
12
8
2
10
9
11

ITINERARY II

The Cytosol and Cytoplasmic Organelles

1. Use virus wrapping to enter cytosol by way of endosome.
2. Survey cytosol and its multifarious contents.
3. Watch anaerobic glycolysis and become acquainted with electron transfer and energy retrieval by phosphorylation.
4. Tour biosynthetic factories and learn about group transfer.
5. Visit mitochondria and examine their protonmotive power generators.
6. Make a side trip to the plant world and see how chloroplasts utilize sunlight.
7. Take a look at peroxisomes and sundry other microbodies.
8. Stop at various skeletal and motile elements.
9. Retrace steps through cytosol, paying special attention to the cytoplasmic faces of membranes.
10. Take last embracing view of cytosol as nexus of metabolic regulation.
11. Contemplate protein assembly on ribosome.
12. Wait near nuclear envelope for mitotic opening.

7 The Cytosol: Glycolysis, Electron Transfer, and Anaerobic Energy Retrieval

At first sight, it would seem that, even reduced a millionfold, bulky creatures of our size could not possibly get through the plasma membrane without causing the cell some irreparable injury. Yet there is ample evidence that it can be done. In a number of infectious diseases, the causal agent, be it a virus, a bacterium, or even a protozoan parasite, can be seen residing and developing freely in the cytoplasm of viable cells of the patient. It came from the outside, and therefore must have succeeded in traversing the plasma membrane without killing its host cell.

Piercing the Veil

The method whereby these intruders do their breaking and entering varies from one to another, but always relies initially on a Trojan-horse type of deception. The attacking microorganism first acts the part of a defenseless prey and allows itself to be engulfed without resistance. Then, when its too-trusting host is just getting ready to dump its catch into the lysosomal acid bath, the wily invader unsheathes its secret weapon, usually a membrane-dissolving enzyme or toxin that opens the endocytic vacuole and frees the imprisoned microorganism before it can be destroyed.

Some viruses—for instance, the causal agent of influenza—use an even craftier technique, prepared at the time the particles get ready to emigrate from their native cell to spread the infection to neighboring cells. As they leave their host cell, these viruses wrap their nucleic acid-protein bodies (nucleocapsids, see Chapter 18) within a spe-

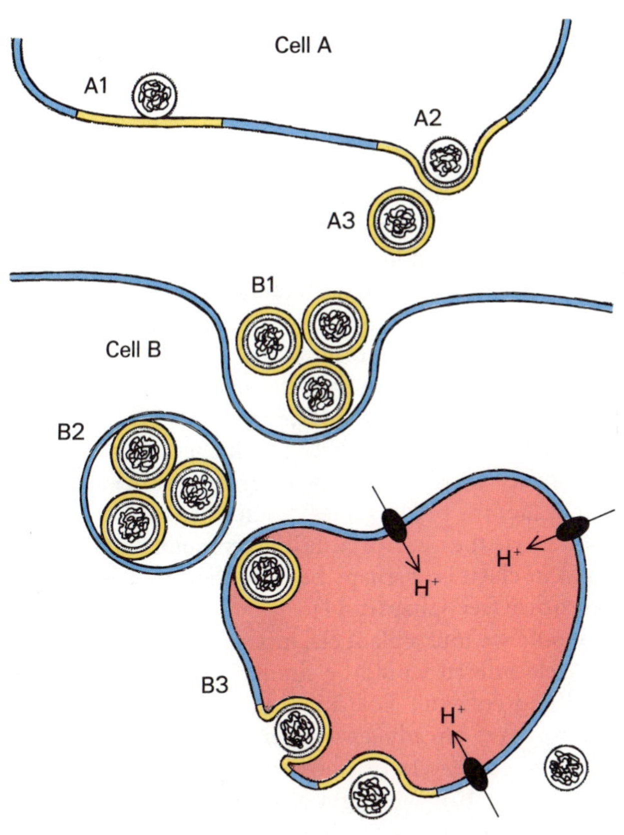

How some membrane-wrapped viruses travel from one cell to another.

A1. A piece of virus-directed membrane is assembled on the plasma membrane of infected cell A. A newly generated viral nucleocapsid approaches the area.

A2. The nucleocapsid pushes the viral membrane patch out to form a bud.

A3. The bud detaches from the cell, releasing a membrane-encapsulated virus.

B1. Viral particles are caught in an endocytic invagination by cell B.

B2. The endocytic invagination closes to form a virus-containing endosome.

B3. The endosome interior acidifies by the action of proton pumps. Exposure to acid triggers the fusion of the viral membrane with the endosomal membrane, projecting the free nucleocapsid into the cytoplasm.

cial piece of plasma membrane assembled under their instructions. Upon meeting another cell, they go through the defenseless-prey routine, with, in this case, the actual dumping into the acid bath included. Now comes the surprise: actuated in some manner by its exposure to the local acidity, the envelope of the virus fuses with the membrane of its vacuolar trap, projecting its infective inmate into the cytoplasm. This mechanism probably operates in endosomes before they fuse with lysosomes.

Once again, therefore, all we need to enter the cell safely is some magic tool borrowed from a microorganism. We have a choice between a chemical membrane-opener and an automatic acid-triggered membrane-ejection device. Let us use the latter. It requires no action on our part that could be botched by faulty timing, and it will remind us of that earlier exciting experience when we donned another type of lysosome-proof jacket and plunged for the first time into the cavernous depths of the cell's vacuolar system.

Behind the Scenes

Our viral vehicle has performed as expected, and we are now in the heart of the cytoplasm. Many strange objects meet our eyes, and it will take quite some time to examine them all in detail. But first, let us take a general look around.

Perhaps our first impression is of being backstage in some theater, seeing the reverse side of the sets with the ropes and pulleys that move them. Endosomes, lysosomes, endoplasmic cisterns, Golgi sacs and vesicles, secretion granules, all those bubble-shaped structures that so far we have seen only from the inside, we now contemplate from the outside. Together they make up thousands of balloons of various sizes and forms, ranging all the way from bulging spheres to flaccid sacs. Milk-white or coral-tinged, but for an occasional ocher-brown lysosome, they glow like opals in the cytoplasmic penumbra. Their surfaces are mostly satin smooth, except in those

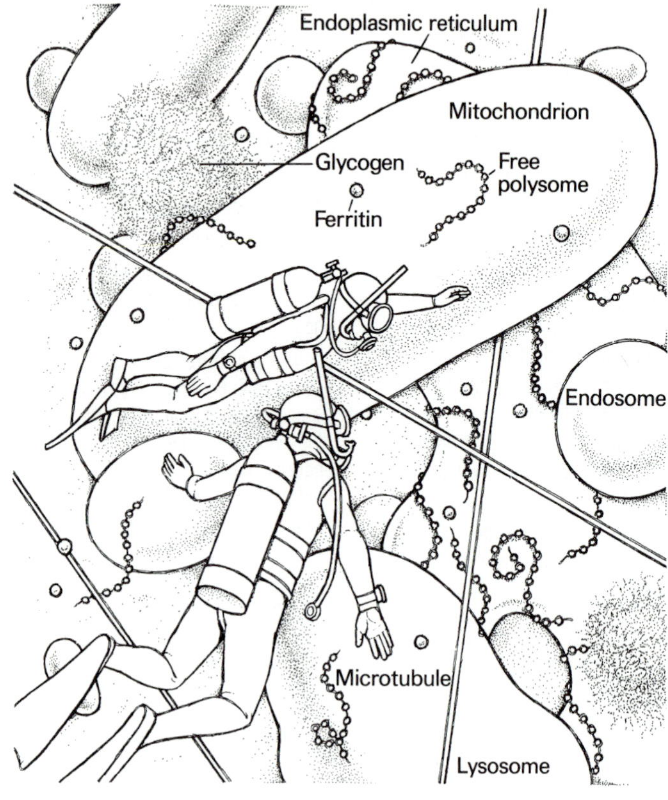

areas where they are roughened by crawling polysomes or raised into small conical or spherical mounds by a trellis of clathrin fibers. We will come back to these intriguing appendages later.

Some of the larger membranous sacs, parts of the endoplasmic reticulum and Golgi apparatus, are joined by coils of tubular connections. They make up massive systems, stolidly immobile, with only surface changes to betray their inner churnings. Others drift sluggishly by in the wake of some cytoplasmic current or jump joltingly ("saltatory" movement), as though pulled by an invisible spring. Flitting between them are shoals of small vesicles, many of them coated. Collisions are frequent in this crowded traffic. As a rule, they are of no consequence; the two partners disentangle themselves, none the worse for wear. But from time to time there results a fusion or fission event, which causes a minor local explosion, sending shock waves reverberating through the cytoplasm.

Interspersed between these now-familiar parts of the import-export machinery, a number of fat, oblong bodies of a vivid pink hue provide an arresting contrast. About the size and shape of a large bacterium, they are in a state of perpetual agitation—twisting, jerking, jostling neighboring structures with the indefatigable automatism of teenagers in a discotheque. Sometimes they split into several parts or join to form weird, hydralike structures. They are the mitochondria, the main centers of energy production in the cell (see Chapter 9).

In some cells—of the liver and kidneys, for instance—we may see yet another type of granule surrounded by a membrane: the peroxisomes. Somewhat smaller than mitochondria, they are of a dull green color, and they are found in clusters, perhaps because they are connected to each other (see Chapter 11).

Should we tour a plant cell instead of an animal cell, the spectacle offered would be similar, but even more variegated and colorful. For, in addition to the components that we have already encountered, we would meet the bright-green centers of photosynthesis, the chloroplasts (see Chapter 10). We might also run into one or more of the numerous membrane-bounded vacuoles and granules within which plants deposit their stores of starch and oil, as well as all those wonderful pigments that adorn their flowers. Such splendor, we might remember, is not an exclusive prerogative of the plant world. If man is a dull animal, relying largely for his color effects on a single black dye, melanin, packed in bodies called melanosomes, many other species—birds, butterflies, fishes—rival the most beautiful flowers.

Together, these objects occupy more than half the cell volume. The spaces between them are filled by a viscous, gelatinous matrix that forms the cytosol, or cell sap, which is the ground substance of the cell. This compartment is limited only by the plasma membrane on the outside and the membranous envelope that surrounds the nucleus on the inside. Thus we can move freely through it, having only to skirt the cytoplasmic bodies or sometimes to squeeze between them when they are pressed close together, as we might do in some aeronautical museum cluttered with multicolored exhibits.

Actually, we are not quite as free as this image might suggest. All sorts of other obstacles impede our mobility.

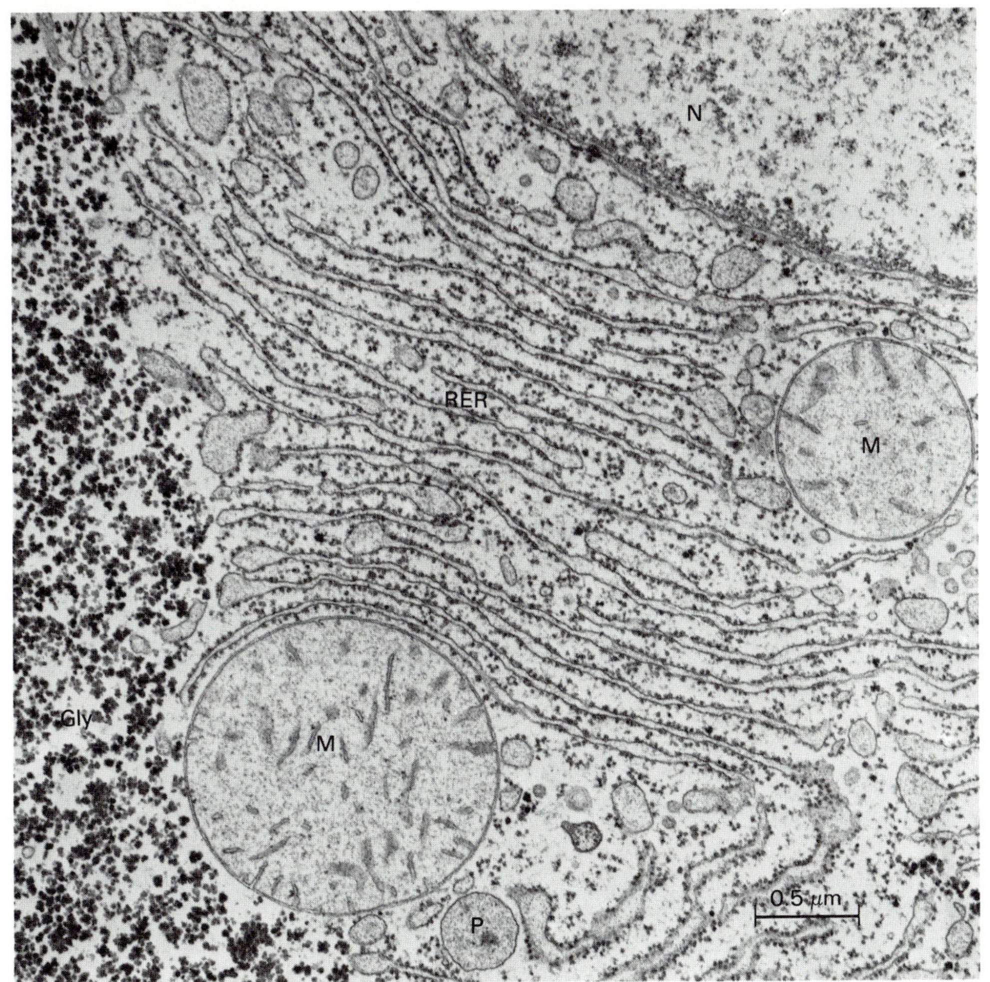

Electron micrograph of a section through a liver cell of a newborn rat. Part of the nucleus (N), surrounded by its envelope (see Chapter 16), can be seen at the upper right. Notice the pores through the nuclear envelope, and the ribosomes attached to it on the cytoplasmic side. Much of the cytoplasm near the nucleus is occupied by parallel arrays of RER cisternae. The bulging ends of the cisternae are transitional elements filled with secretory products, presumably in the process of detaching. Note also the small profiles through similar elements. Nestling between the RER cisternae are two mitochondria (M), as well as a peroxisome (P). The dense area at the left consists of glycogen particles forming a glycogen "lake."

One type we have already met and will explore again in great detail in a subsequent chapter. It is represented by the ribosomes, which generally occur as strings of ten to twenty particles held together by a strand of messenger RNA. So far, we have seen such polysomes only on the surface of ER membranes, but many also float freely in the cytosol and do their job there. With their tenuous threads of growing polypeptide chains, they pose treacherous traps, like invisible seaweeds in a swimmer's way.

In some areas—for instance, near the smooth connections linking the ER to the Golgi apparatus—we may run into an almost impenetrable obstruction: dense clumps of snow-white particles about the size of ribosomes. Looked at through our molecular magnifying glass, however, these particles are very different from ribosomes. Every one is actually a single, treelike structure, branching out into thousands of ramifications, each of which is a chain of glucose molecules linked end to end. These particles are giant macromolecules of a starchy polysaccharide called glycogen, sometimes designated "animal starch." Glycogen trees tend to congregate into small copses, which themselves assemble into forests of fairly extensive size. They are readily seen in the electron microscope as aggregates of small, dense particles, sometimes referred to as "glycogen lakes."

Another impediment in our way is the iron-storage protein, ferritin. It is a compact little particle, much smaller than glycogen, dark brown in color. Although barely discernible in the electron microscope, it is not a harmless object to stumble against: about one-quarter of its weight consists of solid iron hydroxide. Here and there, large droplets of stored lipid may oppose us with their water-repellent surfaces.

Finally, the cytosol is also crisscrossed in many places by an impressive network of girders and cables made up of various microfilamentous and microtubular elements. At any given time, these structures are assembled into a fairly rigid framework that connects the different cell parts with each other and is largely responsible for the shape of the cell and for its attachment to neighboring cells and other extracellular anchoring points. But the framework is not static. Some of its parts are continually dissassembled, to be reassembled in a different manner. Others are made to slide along each other by small molecular motor units. Movements within the cell, and of the cell itself, are largely determined by such changes. Chapter 12 will deal with these various structures, which make up the "bones and muscles" of our cells.

Discounting all these particulate and filamentous elements that clutter the cytosol, we are still left with a considerable amount of material—easily one-third of the total weight of the cell. It consists essentially of water-soluble components and forms the cytosol proper. The architecture of this compartment in the living cell is much debated. Some believe it to behave simply as a concentrated solution of randomly dispersed material. Other investigators see it as a highly organized system in which all constituents are linked by specific interactions or are immobilized by a network of "microtrabeculae." Still others accept the possibility of reversible changes between two such states, a concept that, under the name of "sol-gel transformation," goes back to the early days of colloidal chemistry. Most likely there is a grain of truth in each view, and reality is a compromise. On the whole, however, randomness probably prevails over organization. By all appearances, the cytosol permeates, in an essentially homogeneous fashion, every nook and cranny of the cell that is not occupied by something else. It is the basic filler, the true ground substance of the cell. It is no inert filler, however, but rather a feverishly busy place, where some of life's most important deals are being transacted, in particular those related to energy. It is not the only such site and, for many cells, not the main one. But it is probably the oldest in terms of geological time and, for this reason, has much to tell us about life at its most basic.

Energy, the Perennial Problem

Life is an active process, which depends on the continuous performance of various kinds of work. Consider its most typical character: growth and multiplication. To make a new cell, thousands of proteins, nucleic acids, carbohydrates, fats, and other complex substances have to be constructed, either from scratch or from relatively simple building blocks. To do this, cells need energy, which means that they must be able both to extract energy from their environment and to use it for the accomplishment of chemical work. In addition, living organisms move, manufacture electricity, remodel their surroundings, sometimes even emit light. All this requires energy. As we shall see when we visit the mitochondria and the chloroplasts, evolution has come up with some elaborate solutions to the problem of cellular energy. But those are late inventions, which took more than 1 billion years to develop. Life thrived during all that time. What, then, did it do about energy?

Our guess is that it depended on the kinds of systems that are found today in the cytosol. We have no proof of this, for we cannot go back in time to find out. Nor is there any trace in the fossil record of how early forms of life went about their daily business. What the fossil record does tell us is that as far back as 3.2 billion years ago there existed a microbe, named *Eobacterium isolatum*, that, according to the imprints it has left in some South African rocks, may have been very similar to some of the bacteria we know today. Traces of an even earlier microorganism, *Isuasphaera*, dating back 3.8 billion years, have been uncovered in Greenland. It is almost certain that in those ancient times the earth's atmosphere contained very little oxygen, which is generally believed to be mainly a prod-

Anaerobic fermentations are caused by living microorganisms. This drawing, made by Louis Pasteur, depicts what he saw through the microscope in a droplet of fermenting beet juice.

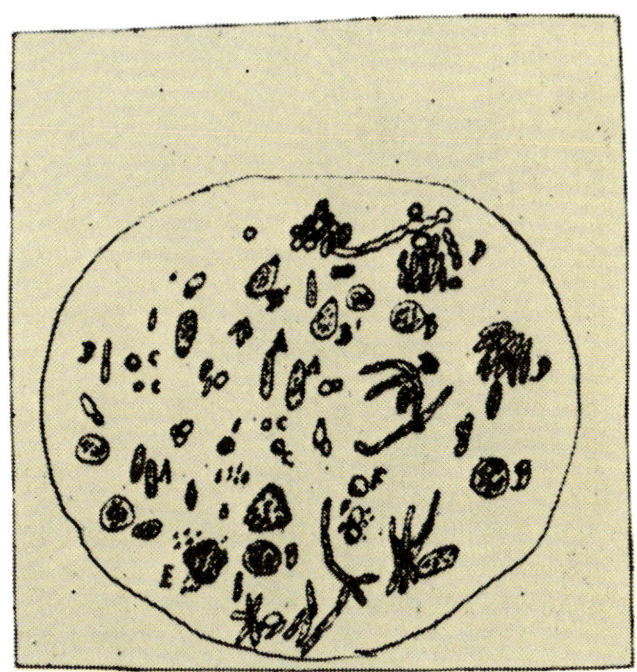

uct of photosynthesis. Thus, *Isuasphaera, Eobacterium*, and many of their descendants must have derived their supply of energy from anaerobic mechanisms—mechanisms capable of supporting life (Greek, *bios*) without (Greek negative: *a*-) air (Greek *aêr*). It so happens that what we find in the cytosol of higher cells and in the cell sap of most present-day bacteria is exactly that: an anaerobic energy-producing mechanism, which is known as glycolysis. Hence our assumption that this system comes to us, perhaps with little change, from those early forms of life that started peopling the earth some 4 billion years ago.

Indeed, glycolysis has the hallmarks of primitiveness, including the quality of relative simplicity. In this respect, the cytosol offers a good introduction to biological energy supply. Even so, for our visit to be profitable, we will need better eyesight than we have used so far. This is not so much to discern complex molecular structures—these will be kept at an absolutely strict minimum—but to apprehend certain key concepts without which we cannot possibly understand life's solutions of its energy problem.

There is another historical aspect to this part of our visit. It will also retrace in some ways the voyage of those early explorers who first discovered the principles of bioenergetics, for anaerobic glycolysis is the cradle of dynamic biochemistry. One might even say that it has been a key element in the development of human civilization. It was discovered many millennia ago in the form of fermentation and was employed by our distant ancestors for the manufacture of leaven, cheeses, and alcoholic beverages. These age-old industries remained purely empirical until 1856, when a Mr. Bigo, a distiller established in the French town of Lille, found himself suddenly threatened with ruin. For some unexplained reasons, his sugar-beet fermentation vats were all going sour, producing lactic acid instead of alcohol. He went to seek the help of a young chemist from Paris who had recently joined the local university staff and who, it was reported, had done some brilliant work. The young man obliged and eventually succeeded in rescuing Mr. Bigo's business. At the same time, he made a discovery that would change the world: anaerobic fermentations are caused by living microorganisms. His name: Louis Pasteur. Later, in 1897, a German chemist, Eduard Buchner, found that a mere "juice" expressed from yeast—none other, in fact, than the yeast cells' cytosol—could bring about the conversion of sugar into alcohol. Buchner thereby demonstrated that the function of the microorganism in alcoholic fermentation is purely chemical and that it does not rely, as was believed by Pasteur, on a special vital force peculiar to living organisms. Buchner also opened the way to a chemical dissection of the glycolytic system present in the cytosol of cells and, in so doing, launched a vast movement that has given us the detailed knowledge of metabolism that we have today.

The term metabolism comes from the Greek word for change (literally "the act of throwing about," from *ballein*, to throw). It covers the sum total of the chemical changes that take place in living organisms. It is subdivided into anabolism (from *ana*, up) and catabolism (from *kata*, down). Anabolism includes all the processes that require energy and are described thermodynamically as endergonic (Greek *endon*, inside; *ergon*, work). Its main function is biosynthesis. Catabolism is made up of the reactions that produce energy and that are called exergonic (*exô*, out) for this reason (see Appendix 2). By necessity, catabolism supports anabolism, as well as all the other forms of work carried out by living organisms (except for such reactions as are powered directly by an outside source of energy, mostly light).

When Buchner made his discovery, only the overall balance of some metabolic changes was known; mechanisms were completely unknown. Glycolysis is the first metabolic process ever to be elucidated. It took all of 40 years and the participation of many of the world's greatest scientists to accomplish this feat, which remains one of the most remarkable and far-reaching pieces of detective work of all time. Yet, all that it brought to light was a dozen chemical reactions, less than one-hundredth of the number of metabolic reactions that were to be recognized in the next 40 years. What makes the elucidation of glycolysis so important is that it was a first. Knowing only the starting point—the simple sugar glucose—and the end-products—lactic acid in one variant, ethyl alcohol plus CO_2 in another—even the most perspicacious of organic chemists could not possibly have predicted the astonishingly circuitous route taken by the natural process. Every step came as a surprise, and their identification required an enormous amount of patience and perseverance—especially with the pathetically primitive tools of the day: a few test tubes, a Bunsen burner, a balance, a light microscope. But once the pathway was clarified, it served as a beacon that illuminated the whole course of subsequent discoveries. It will do the same for us.

Glycolysis, a Power-Giving Snake

Yeast cells convert sugar into ethyl alcohol (ethanol) in twelve consecutive chemical steps that form a reaction chain, a metabolic snake:

$$\text{Glucose} \underset{1}{\longrightarrow} \text{A} \underset{2}{\longrightarrow} \text{B} \dashrightarrow \text{J} \underset{11}{\longrightarrow} \text{K} \underset{12}{\longrightarrow} \text{Ethanol} + CO_2$$

The same pathway is followed, up to the tenth step, by lactic bacilli (those that contaminated Mr. Bigo's vats), as well as by our muscles when they make a sudden effort. Only at step 11 do these diverge from yeast, converting intermediate J (pyruvic acid) into lactic acid instead of CO_2 and alcohol.

Thus, lactic and alcoholic fermentations differ only at the chain's end. Before that, they follow the same universal route, known as the glycolytic chain. This route is not laid out as a visible trail in the cytosol; the snake has no real body. If we put on our high-powered chemical glasses, all that we will see is a chaotic jumble of molecules, A, B, . . . , J, K, mixed with many others, intermediates in other pathways. What joins them, and gives the snake its substance, are the arrows, each of which indicates the occurrence of a specific enzyme (Chapter 2) that catalyzes the chemical transformation shown. The dynamic ordering of the twelve enzymes involved in the glycolytic chain follows automatically from the nature of their substrates and products. The reaction giving rise to D from C must necessarily follow immediately after the conversion of B into C and precede that of D into E. No physical channel is needed to guide the molecules toward their destination. The apparent chaos that we see hides a high degree of order, a dynamic organization generated by the properties of the enzymes present.

This lesson of glycolysis may be generalized. Behind each of the thousands of chemical reactions that take place in living cells there is an enzyme. This is commonplace knowledge today but came to be appreciated only after the glycolytic chain responsible for alcoholic fermentation in yeast began to be unraveled. Enzymology, the study of enzymes, has greatly enriched our understanding both of life and of chemistry and is now beginning to yield important practical results based on the industrial use of enzymes extracted from natural sources. The exigencies of our tour will allow only occasional references to this important branch of biochemistry. But we must at least keep in mind that every activity we observe, what-

In a chaotic jumble of molecules A, B,..., J, K, each enzyme picks out its own specific substrate and converts it into the next intermediate in the metabolic chain. Order is created out of the chaos by the specificity of the enzymes.

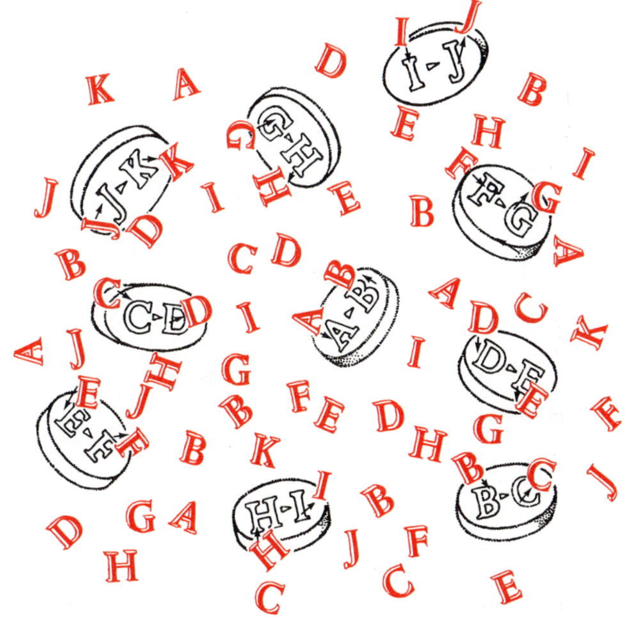

ever its nature, depends on the catalytic participation of enzymes.

Enzymes are often assisted by accessory substances called cofactors, or coenzymes. In glycolysis, two such cofactors require our attention. One is called NAD, which stands for nicotinamide adenine dinucleotide. Biochemists, you will notice, are greatly addicted to abbreviations. But they have an excuse. Most of the substances they deal with are too complex to be represented explicitly every time they are referred to. NAD is one, and we will not even bother to look at its chemical structure here (those interested will find it in Appendix 1). It is, however, worth noting that the nicotinamide part of the molecule is a vitamin known as vitamin PP, which stands for *pellagra preventiva*. Deficiency of this vitamin in the diet causes pellagra, a severe nutritional disease formerly widespread on the American continents. This is not an isolated example. Most vitamins either act as coenzymes or are part of one, which explains why an organism cannot do without them. We will see the function of NAD a little later.

The other cofactor that we must look at is designated ATP, for adenosine triphosphate. Eventually, we will have to consider the chemical structure of ATP. But right now, all we need know is that it can be hydrolyzed (split with the help of water) into adenosine diphosphate (ADP) and inorganic phosphate (P_i) and, conversely, that it can arise (provided energy is supplied) from the condensation of ADP and P_i with removal of water:

$$\text{ATP} + \text{H}_2\text{O} \underset{\text{(endergonic)}}{\overset{\text{(exergonic)}}{\rightleftharpoons}} \text{ADP} + \text{P}_i$$

The critical function of ATP in glycolysis was revealed when it was found that the breakdown of glucose is coupled to the assembly of ATP: for every molecule of glucose converted into lactic acid or ethanol, two molecules of ADP are phosphorylated to ATP. The link is an obligatory one. If ATP synthesis cannot occur, as when there is a lack of ADP, glycolysis stops.

What this remarkable phenomenon actually means became clear when the energetics of the process were considered. Glucose fermentation releases free energy: about 47 kilocalories (kcal) per gram-molecule of glucose broken down. On the other hand, the assembly of ATP from ADP + P_i requires free energy: about 14 kcal per gram-molecule of ATP formed. Thus, of the 47 kcal released by the breakdown of glucose, $2 \times 14 = 28$ kcal, or 60 per cent, are utilized to form ATP, instead of being dissipated as heat. Glycolysis powers ATP synthesis; coupling of the two processes is an energy-retrieval mechanism.

Here again, glycolysis provided a first. When other catabolic processes were subsequently discovered, they also were found to be coupled with the assembly of ATP. Not only glycolysis, but the whole of catabolism powers ATP formation: coupling is a universal energy-retrieval mechanism.

But what of ATP itself? What use its synthesis? The answer to this question, or rather an inkling of it, was first given in the early 1930s, when it was found that a muscle rendered incapable of glycolysis by a poison (mono-iodo-acetic acid) could still perform a small amount of

Obligatory coupling compels cell to use 60 per cent of the energy released by anaerobic glycolysis to manufacture ATP from ADP + P_i. Breakdown of ATP, in turn, powers various forms of biological work.

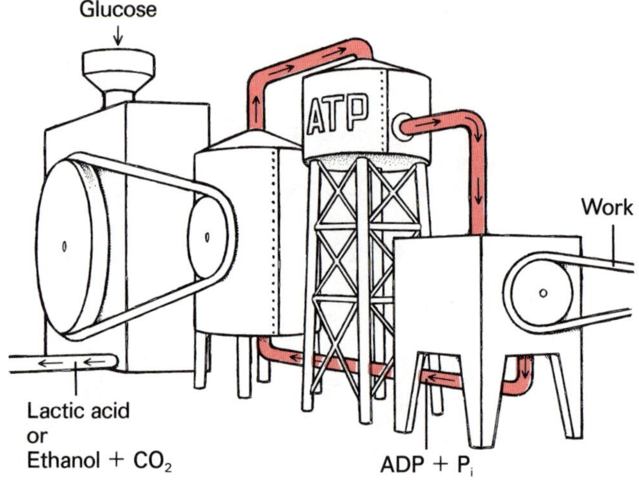

work at the expense of stored "phosphate-bound energy." Eventually the chemical reaction directly connected with the contractile machinery was identified as the hydrolysis of ATP to ADP and P_i. In this way, ATP was recognized as the missing link between glycolysis and muscular work. Glycolysis powers ATP formation; ATP breakdown powers muscular work:

$$\begin{array}{ccc} \tfrac{1}{2}\text{ Glucose} & & \text{ADP} + P_i \\ & \times & \quad\rightarrow \text{Work} \\ \text{Lactic acid} & & \text{ATP} + H_2O \end{array}$$

This was a tremendous discovery, which opened one of the main doors to the understanding of life. For not only muscular work, but virtually every kind of work performed by living organisms is powered by ATP. Scratch the surface of any kind of bioengine, be it an ion pump in a membrane, a contractile fiber in a flagellum, a light generator in a glowworm, or any of the multiple synthetic reactions whereby living organisms manufacture their own constituents: almost invariably you find ATP as a source of the required energy. It is the main fuel of life, and the function of catabolism is explained by its ability to support the restoration of ATP:

$$\begin{array}{ccc} \text{Foodstuffs} & & \text{ADP} + P_i \\ & \times & \quad\rightarrow \text{Work} \\ \text{Breakdown products} & & \text{ATP} + H_2O \end{array}$$

Epitomizing this relationship is the historical example of anaerobic yeast. This organism grows and multiplies, accomplishing tremendous feats of chemical engineering and information transfer in order to make new little yeast cells similar to their progenitors. It does all this with the conversion of sugar into alcohol as sole source of energy. Between this crude form of energy and the thousands of different processes it powers lies ATP, just as electricity lies between the burning of coal or oil and the wonders of modern technology. We will encounter ATP again and again on our tour.

The last lesson we learn from glycolysis concerns the mechanism of the coupling. Here we must look more closely at the anatomy of our glycolytic snake. This part of the visit may be a little rough for some of our fellow tourists. Unfortunately, there is no easier way. The cell is a chemical machine. Its workings cannot be understood without at least an elementary acquaintance with certain physicochemical concepts. (To help those in need of a refresher course, some basic notions have been summarized in the appendices.)

Burning without Air

From the chemical point of view, glycolysis consists simply in a halving of the glucose molecule and rearrangement of its constituent atoms, with neither gain nor loss of matter:

$$\underset{\text{(Glucose)}}{C_6H_{12}O_6} \longrightarrow \underset{\text{(Lactic acid)}}{2\ CH_3\text{—CHOH—COOH}}$$

or:

$$\underset{\text{(Glucose)}}{C_6H_{12}O_6} \longrightarrow \underset{\text{(Ethanol)}}{2\ CH_3\text{—}CH_2OH} + 2\ CO_2$$

Count the atoms on either side of the arrows, and you find the same number of carbons, hydrogens, and oxygens. From head to tail, the snake neither gains nor loses weight. In between, however, it goes through some fairly elaborate contortions that are all geared to one central function: the making of ATP. This is accomplished at steps 6 and 7 by a complex process of oxidoreductive phosphorylation.

The role of the five steps that precede this central reaction is to ready the glucose molecule for participation in it. This is quite a job, and an expensive one, since it means converting the 6-carbon glucose molecule into two molecules of a phosphorylated 3-carbon compound called phosphoglyceraldehyde. The two phosphate groups required for this purpose are supplied by ATP, so that we have an apparently paradoxical situation: a reaction destined to make ATP starts by spending it. This is not unusual. Many foodstuffs need to be activated, with expenditure of energy, before they become susceptible to degradation. It is one more of the chores the cell's maid-of-all-work, ATP, is saddled with. Of course, this initial energy investment is subsequently reimbursed with interest. Otherwise it would be useless.

With the making of phosphoglyceraldehyde, we come to the key reaction in glycolysis, which is the oxidation of this compound to phosphoglyceric acid, and the coupled condensation of ADP and P_i to ATP. We will not consider all the details of this reaction, since that would get us involved in some fairly complicated chemistry. But one aspect deserves our attention, namely the actual significance of the word oxidation. We are all familiar with the saying that we get our energy from "burning" our food. But the analogy with some sort of combustion engine that this image evokes is misleading and should be corrected. In strictly descriptive terms, the conversion of phosphoglycer*aldehyde* into phosphoglyceric *acid* consists in the addition of an oxygen atom to the aldehyde group (CHO) to make a carboxylic acid group (COOH). In this sense, it resembles the type of oxidation that accompanies combustion. Unlike what happens in combustion, however, the extra oxygen atom does not come from atmospheric oxygen; there is no such oxygen. Glycolysis is an anaerobic process, which can take place in the complete absence of oxygen. We are dealing here with a case of "burning without air."

If not atmospheric oxygen, what then is the source of the extra oxygen atom acquired by phosphoglyceraldehyde? The answer is *water,* though not regular water picked up from the medium. It is a water molecule that arises from the coupled condensation of ADP and P_i and is transferred directly to the oxidative reaction in chemically bound form. We indicate this fact by putting H_2O in parentheses; then, if we represent the rest of the phosphoglyceryl radical by R, we may formulate the reaction as follows:

$$\text{ADP} + \text{P}_i \longrightarrow \text{ATP} + (\text{H}_2\text{O})$$
$$\text{R—CHO} + (\text{H}_2\text{O}) \longrightarrow \text{R—COOH} + 2\ \text{H}$$

What this equation tells us is that the oxidative step actually consists in the removal of hydrogen. It is a dehydrogenation. But here again we must watch out. The hydrogen removed in this reaction is not hydrogen gas, H_2, which is a stable diatomic molecule. It is the reactive hydrogen atom, H, which never occurs in free form but is always transported or exchanged through the mediation of carriers, either as such or in the form of a stripped electron (e^-), with which it freely equilibrates according to the relationship

$$\text{H} \rightleftharpoons e^- + \text{H}^+$$

In aqueous media, the protons, or hydrogen ions, H^+, that participate in this equilibrium are readily available, thanks to the dissociation of water:

$$\text{H}_2\text{O} \rightleftharpoons \text{H}^+ + \text{OH}^-$$

Thus the oxidative reaction of glycolysis can be written also as a removal of electrons:

$$\text{R—CHO} + (\text{H}_2\text{O}) \longrightarrow \text{R—COOH} + 2\ e^- + 2\ \text{H}^+$$

What is true of glycolysis is true also of most other oxidative reactions that occur in living organisms. As a rule, biological oxidations take place without direct participation of oxygen through removal of hydrogen atoms or electrons, a type of reaction no doubt inherited from those early times when life first appeared and went on to evolve for hundreds of millions of years in the absence of atmospheric oxygen. When the oxidized molecule acquires oxygen, as in the example considered, the source of the extra oxygen is water or some water-generating reaction.

As to atmospheric oxygen, which is obviously essential to all aerobic organisms, including ourselves, its role is to pick up the electrons released by the oxidative reactions:

$$\frac{1}{2}\ O_2 + 2\ e^- + 2\ H^+ \longrightarrow H_2O$$

The end result is the same as in ordinary combustion. But the mechanism is different. This can be verified readily with the help of the heavy isotope of oxygen, ^{18}O, which can be distinguished from the prevalent isotope ^{16}O by means of a mass spectrograph. If glucose is burned in a furnace in the presence of $^{18}O_2$, the isotope will be found in the CO_2 formed, indicating that the glucose carbon has joined with the atmospheric oxygen. But, if the glucose is burned by a living organism breathing $^{18}O_2$, the isotope is recovered not in the exhaled CO_2, but in H_2O. This seemingly trivial difference actually makes all the difference between life and death. No organism has developed a way to retrieve energy in biologically usable form from a combustion type of oxidation. In contrast, there are countless devices for extracting energy from electron transfer. We will come back repeatedly to this important topic. Note that addition of oxygen does take place in some biological reactions. These are termed oxygenations, to distinguish them from oxidations. They are not directly involved in oxidative energy retrieval.

The reverse of an oxidation—the gaining of hydrogen atoms or of electrons—is called a reduction. Since electrons and hydrogen atoms cannot circulate in free form in an aqueous medium, neither reaction can ever happen without the other. Electrons cannot be abandoned unless they can be picked up; whenever a substance is oxidized, another is reduced. The reactions, therefore, are always oxidation-reduction reactions, electron-transfer reactions. Enzymes catalyzing such transfers are called electron transferases, or oxidoreductases.

In the central oxidative step of glycolysis, the electron acceptor is NAD^+, oxidized form of the cofactor NAD, which owes its role in this and countless other metabolic processes to its ability to act as an electron carrier:

$$NAD^+ + 2\ e^- + H^+ \longrightarrow NADH$$

The complete electron transfer reaction may therefore be written schematically as follows:

$$R\text{—}CHO + (H_2O) + NAD^+ \longrightarrow R\text{—}COOH + NADH + H^+$$

This is the reaction that is coupled with the assembly of ATP (from which you may remember it derives its hidden water molecule). Whatever the mechanism of this coupling, it means that the transfer of one pair of electrons from phosphoglyceraldehyde to NAD^+ (1) releases enough free energy to power the assembly of one molecule of ATP and (2) is subjected to a constraint such that it can proceed only if ATP is made at the same time.

Formally, the system may be seen as an electrochemical transducer that converts a flow of electrons into chemical work. We may therefore apply the general theory of electricity, which says that the maximum amount of work—the real work is always less owing to inevitable losses as heat—that can be performed by an electric machine is given (in joules) by the quantity of electricity (in coulombs) passing through the machine, multiplied by the potential difference (in volts) of the electric generator.

In the present case, we know the work. It takes 14 kcal (58,600 joules) to make one gram-molecule of ATP. We also know the quantity of electricity passing through the system: 2 electrons per molecule of ATP formed, or 2 electron-equivalents or Faradays ($2 \times 96{,}500 = 193{,}000$ coulombs) per gram-molecule of ATP. If we divide the joules by the coulombs, we find the minimum voltage of our electricity source: 0.3 volt, or 300 millivolts (mV). In other words, the transfer of electrons between phosphoglyceraldehyde and NAD^+ must occur across a potential difference of at least 300 mV. Otherwise it could not power ATP synthesis. (For additional theoretical background, see Appendix 2.)

Of course, the glycolytic snake is hardly constructed like a conventional electric generator. It has no outlets into which we can plug a voltmeter to verify our conclu-

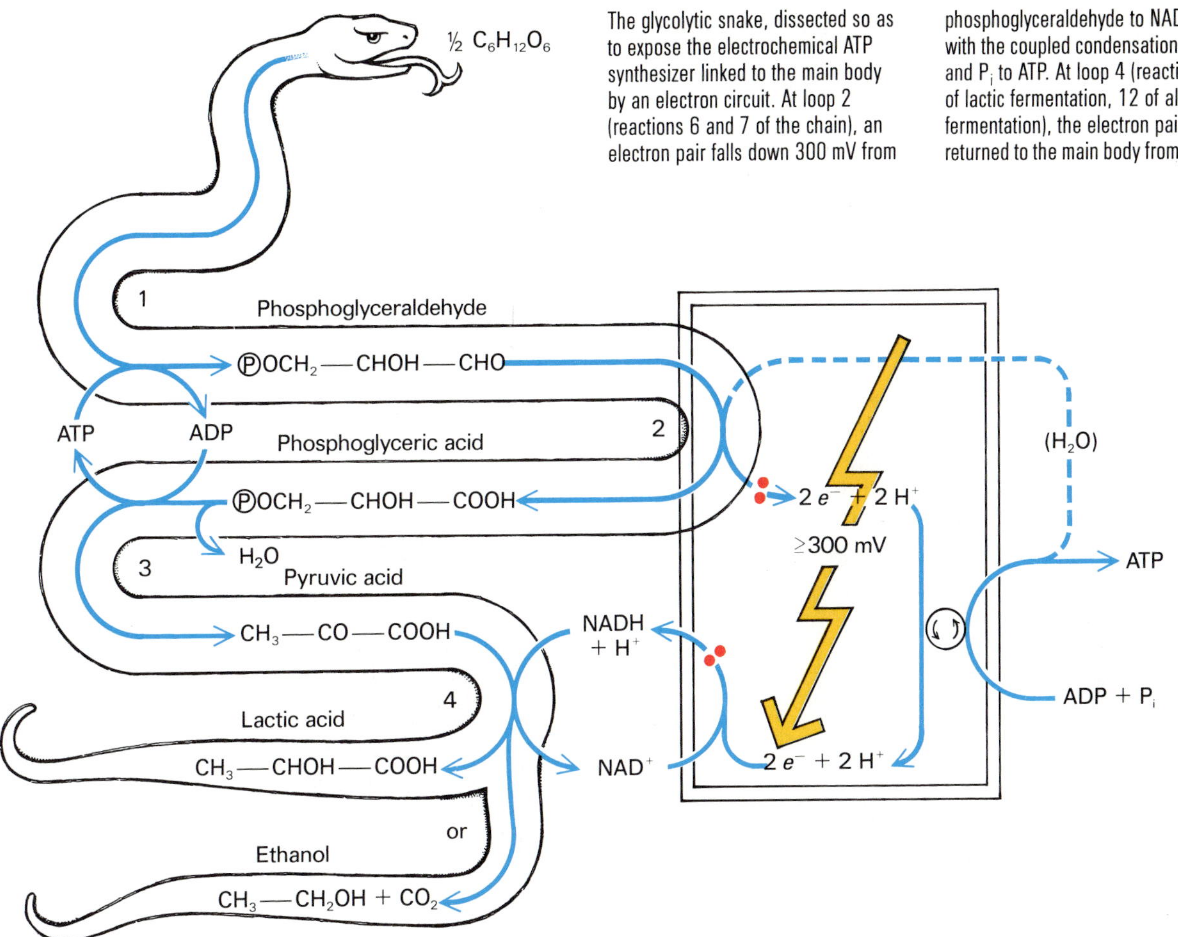

The glycolytic snake, dissected so as to expose the electrochemical ATP synthesizer linked to the main body by an electron circuit. At loop 2 (reactions 6 and 7 of the chain), an electron pair falls down 300 mV from phosphoglyceraldehyde to NAD^+, with the coupled condensation of ADP and P_i to ATP. At loop 4 (reaction 11 of lactic fermentation, 12 of alcoholic fermentation), the electron pair is returned to the main body from NADH.

sion. But it has an electron donor (phosphoglyceraldehyde) and an electron acceptor (NAD^+). For the transaction to take place, the affinity for electrons of the acceptor must be stronger than that of the donor. This affinity can be measured. It is called the oxidation-reduction potential, or redox potential, and is expressed in volts. Each redox couple (Red/Ox, e.g., phosphoglyceraldehyde/phosphoglyceric acid; NADH/NAD^+) has its characteristic redox potential. What we have established by the preceding calculation is that, if the transfer of an electron pair is to support the formation of one molecule of ATP from ADP and P_i, the difference between the redox potentials of the donor and acceptor couples must be at least 300 mV, which it indeed is for the two couples in glycolysis.

Once its main business of making ATP has been successfully completed, the glycolytic chain still has to wind up its affairs and balance its books. This is the main function of the latter part of the chain. There are three accounts to be settled: (1) phosphate, which was initially donated to the chain from ATP; (2) water, which entered the chain at the oxidative step to supply the extra oxygen of phosphoglyceric acid; and (3) electrons, which left the chain at this same step. These imbalances are now compensated. First, a water molecule is given off. Then the phosphate group is returned to ADP to regenerate the ATP that was invested at the beginning. And, finally, NADH gives back its electrons. In lactic fermentation, the electron acceptor is pyruvic acid, which is reduced to lactic acid:

$$CH_3\text{—CO—COOH} + \text{NADH} + H^+ \longrightarrow CH_3\text{—CHOH—COOH} + NAD^+$$

In alcoholic fermentation, the electron acceptor is the product of decarboxylation of pyruvic acid, acetaldehyde, which is reduced to ethanol:

$$CH_3\text{—}CO\text{—}COOH \longrightarrow CH_3\text{—}CHO + CO_2$$
$$CH_3\text{—}CHO + NADH + H^+ \longrightarrow CH_3\text{—}CH_2OH + NAD^+$$

These final electron transfers take place across very small potential differences, with no energetic benefit. They are necessary only to make the system self-contained. Cut off the snake's tail, while substituting another acceptor to collect the electrons from NADH, and the system can still perfectly do its job of generating ATP, except that it is now an oxidative system and its final product is pyruvic acid:

$$\underset{\text{(Glucose)}}{C_6H_{12}O_6} \longrightarrow 2\ \underset{\text{(Pyruvic acid)}}{CH_3\text{—}CO\text{—}COOH} + 4\ e^- + 4\ H^+$$

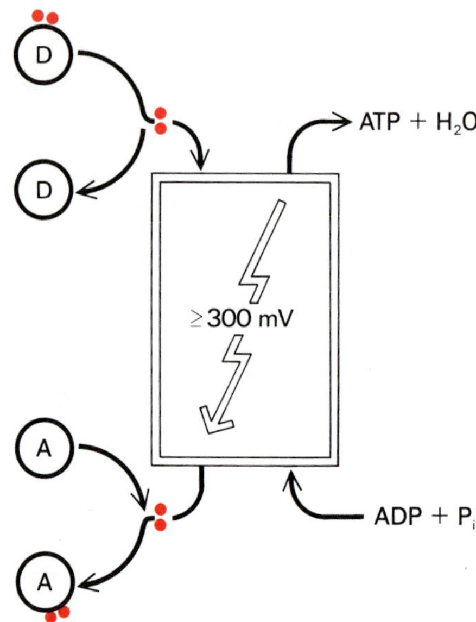

The oxphos unit is the universal energy generator. It receives a pair of electrons from a reduced donor (D) and returns it to an oxidized acceptor (A) at a potential level at least 300 mV lower. The electron pathway through the unit is so designed as to force ADP and P_i to join together into ATP with removal of water. This device is often reversible, allowing an electron pair to be lifted to a potential level at most 300 mV higher, at the expense of the hydrolysis of ATP.

Oxphos: Life's Golden Energy Gadget

Several features of the glycolytic snake are of general significance and apply to the whole of energy metabolism. The most universal of these generalizations concerns the production of ATP. Throughout nature, this central piece of energy currency arises, as in glycolysis, through a coupled electrochemical reaction that links the phosphorylation of ADP to the transfer of electrons across a difference of electric potential. There is virtually no exception to this rule. Animals, plants, fungi, bacteria, all living beings, including man, derive their ATP from the operation of such coupled reactions. As might be expected, a great many different reactions of this type occur, and their chemistry is often complex. So as not to have to go into such details, while retaining the possibility of gaining some insight into the working of the cell's main energy centers when we visit them, we will refer to systems that catalyze an oxidoreductive phosphorylation as *oxphos* units and will represent them schematically by the "boxed lightning" symbol shown above. Note that this terminology and symbolism do not belong to the standard language of biochemistry. They are introduced here for the sole purpose of helping us during our tour.

The basic design of an oxphos unit is that of an electrochemical transducer coupling the assembly of one molecule of ATP from ADP and P_i to the fall of two electrons down a 300-mV potential difference. The electrons are fed into the oxphos unit by a donor (D), which passes from the reduced to the oxidized state, and are collected at the other end by an acceptor (A), which changes from the oxidized to the reduced state. Protons may or may not accompany the electrons, depending on the nature of the molecules involved.

In glycolysis, the donor couple is represented by phosphoglyceraldehyde/phosphoglyceric acid, and the acceptor couple by NAD^+/NADH. But this is just one particular case. Other systems use other donors or other acceptors or both, allowing for a large number of different oxphos units that operate by a wide variety of mecha-

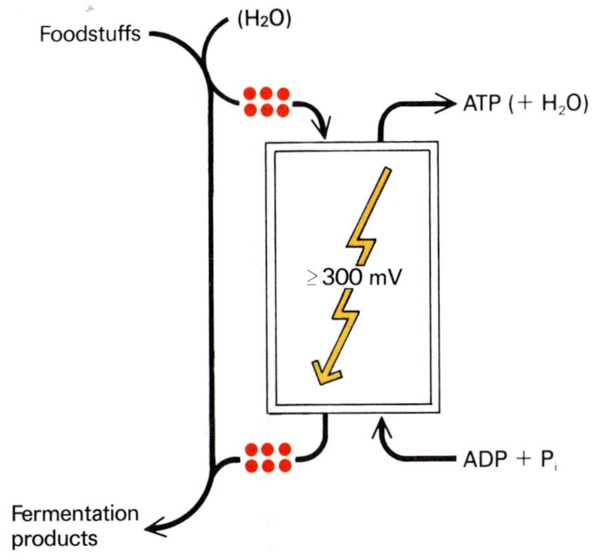

In anaerobic fermenters, a single flow of matter suffices to support electron flow and ATP assembly through oxphos units because the electron acceptor arises metabolically from the electron donor. In glycolysis, for example (see the illustration on p. 113), pyruvic acid (or acetaldehyde) arises from phosphoglyceraldehyde.

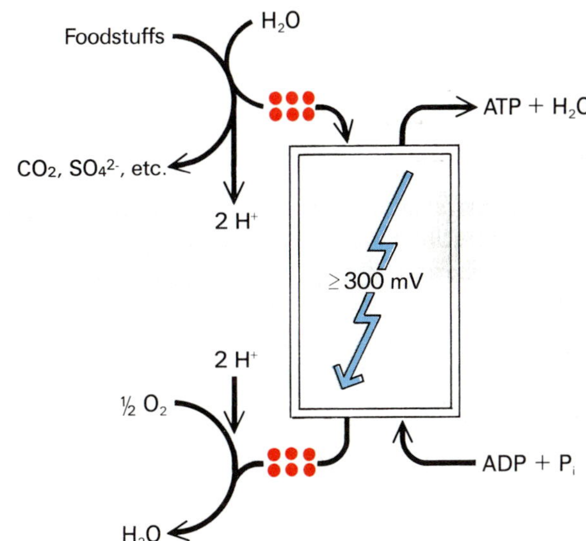

In aerobic heterotrophs, a dual flow of matter is needed to support electron flow and ATP assembly through oxphos units. One feeds electrons into the oxphos units at the expense of foodstuffs, which, with the help of water, are oxidized to their final waste products. The other, supported by oxygen, serves to pick up the electrons from terminal oxphos units.

nisms. We will encounter many in our tour. Some, like the one in glycolysis, receive their electrons from metabolic substrates and are said for this reason to catalyze substrate-level phosphorylations. Many others, including the most important ones, such as are in mitochondria and chloroplasts, are fed with electrons from NADH and other carriers. They catalyze carrier-level phosphorylations and operate by mechanisms that are entirely different from those involved in substrate-level phosphorylations. We will examine these mechanisms in Chapter 9.

Oxphos units are powered by a flow of electrons. To function, they need to be connected both to a source of electrons and to an electron collector. Glycolysis and the other anaerobic fermentations have the peculiarity that their electron flow is supported by a single flow of matter, thanks to the fact that the metabolic chain generates its own final electron acceptor. The main advantage of this kind of metabolism is that all it requires from the environment is an appropriate foodstuff—glucose, for instance. But it is tremendously wasteful, because the final products of fermentations, such as lactic acid or ethanol, are energy-rich molecules. As will be seen, they leave the cells with more than 90 per cent of the potential energy of glucose unused. This need not be so; in fact, it is quite exceptional. In the more usual situation, the electrons released by oxidative reactions are collected directly or indirectly by an exogenous electron acceptor, in which case there is no need for the cell to reject valuable materials, and catabolic degradation of the substrate can proceed further. But now a dual flow of matter is necessary to maintain the flow of electrons.

Substances capable of serving as electron acceptor abound in nature—for instance, the sulfate ion, SO_4^{2-} (which can be reduced right up to the level of sulfur, S, or of hydrogen sulfide, H_2S); the ferric ion, Fe^{3+} (which is readily reduced to the ferrous state, Fe^{2+}); the nitrate ion, NO_3^- (which will go to nitrite, NO_2^-, and further, up to ammonia, NH_3); CO_2 (which can be reduced all the way to methane, CH_4); and even the simple proton, H^+ (which will yield hydrogen gas, H_2). The most widespread and efficient electron acceptor is molecular oxygen, O_2, which is reduced to water, H_2O (or occasionally to hydrogen peroxide, H_2O_2).

Every one of these substances, and many others, have been adopted as electron acceptor by some organism through the development of appropriate enzymes. Their reduction accounts for many of life's manifestations, including the stench of sulfurous fumes, the remodeling of ferruginous silts, the recycling of atmospheric nitrogen, and the mysterious emanations that send ghostly will-o'-the-wisps flitting across the surface of marshes. Between the bacteria responsible for these phenomena and the innumerable living organisms, including man, that respire atmospheric oxygen, there is a hidden common bond: they all support their energy-yielding metabolic oxidations with the help of an exogenous electron acceptor.

Here comes another important generalization. Not just glucose, but every possible kind of foodstuff used by a living organism to support its energy needs acts by supplying electrons to ATP-generating oxphos units. There simply is no other source of metabolic energy for heterotrophic organisms—those that feed on the products of the biosynthetic industry of other organisms (Greek *heteros*, other; *trophê*, food). The autotrophs (*autos*, self), also feed electrons into oxphos units, but from other sources.

"Burning" food for energy really means breaking down the foodstuffs and enriching them with oxygen at the expense of water, in such a way as to produce electrons that are fed into ATP-generating oxphos units from which they are collected by oxygen (or by some other acceptor).

As we will see when we visit the mitochondria, life has displayed remarkable ingenuity in the exploitation of this electron flow, intercalating up to four consecutive oxphos units on the pathway of metabolic electrons, many of which cascade down potential differences of 1 volt or more. As much as 80 per cent of the free energy released by the oxidation of foodstuffs may be retrieved in this way and used for the assembly of ATP. We will come back to this topic in Chapter 9.

Electron Flow: A Generalized View

Not all electron exchanges take place through ATP-generating oxphos units. Quite often, electrons are transferred across small potential differences, with little change in free energy. Occasionally, they may hurtle down a major potential difference, but without the kind of constraint that would allow the cell to make use of the energy released. Biological electrons resemble rivers in this respect. Like waterfalls, precipitous electron falls are infrequent, and not every one of them is harnessed to a power station.

This hydrodynamic image is useful, provided we keep in mind that electrons do not actually flow inside living cells in the way they do through an electric conductor; they are exchanged in discrete steps, between a reduced donor and an oxidized acceptor. The donor is oxidized in this transaction and thereby becomes able to act as acceptor to some other donor that occupies a higher potential level. The acceptor, on the other hand, having become reduced, can now serve as donor for an acceptor of lower potential level. Electrons tumble down in this way from one carrier to another until they reach their final acceptor, usually oxygen.

Electrons, therefore, do not flow down grades of variable slope, as rivers do most of the time. They fall down a succession of abrupt steps of variable heights. The electron-flow map of living cells resembles not so much a natural network of waterways as the kind of artificial system of interconnected reservoirs that seventeenth-century engineers constructed with such relish in the gardens of the rich. Instead of exploiting the terrain for esthetic enjoyment, however, natural selection has favored energy, utilizing the contour in such a way as to generate as many electron falls as possible that have the right height to power an oxphos unit.

The hydrodynamic analogy helps us appreciate an important aspect of electron transfer, which so far has been mentioned only in passing, namely the absolute level of potential at which electrons are either donated or ac-

The electron cascade. Each reservoir contains a redox couple. Connections between reservoirs are mediated by electron-transferring enzymes. Altitude of the reservoirs determines the direction of flow and height of the falls. Many electron falls of 300 mV or more (for electron pairs) are harnessed to energy-retrieving machines (oxphos units).

cepted. In our image, it corresponds to the altitude of the reservoirs, their height above sea level. Once you have this information, you can predict accurately, from the difference between their two altitudes, the direction of water flow between any two reservoirs, as well as the maximum work that can be obtained from the fall of a given quantity of water from the higher reservoir to the lower one (or, conversely, the minimum amount of work that must be accomplished to pump a given quantity of water up from the lower reservoir to the higher one).

The equivalent of altitude for electron reservoirs is the oxidation-reduction potential (in volts) of the relevant redox couples. In possession of that information, one can readily estimate the corresponding energy changes, as explained in Appendix 2. A more convenient way of evaluating the potential energy of electrons is to give directly the result of such a calculation—that is, the free energy of the reaction—for the particular case when one pair of electron-equivalents is transferred to oxygen with formation of water. We choose oxygen as acceptor because of its universal function as final electron acceptor for all aerobic organisms. Take, for example, the $NADH/NAD^+$ couple. We consider the reaction

$$NADH + H^+ + \tfrac{1}{2}\,O_2 \longrightarrow NAD^+ + H_2O$$

The free energy of this reaction, $\Delta G_{ox\ (NADH/NAD^+)}$, expressed in kilocalories per pair of electron-equivalents, is a direct measure of the maximum amount of work that can be obtained when electrons fall from the NAD reservoir all the way down to what, for most organisms, is their zero level of energy: water. It truly expresses the electron potential (not to be confused with the oxidation-reduction potential) of the $NADH/NAD^+$ couple.

Like all free-energy changes, electron potentials vary with the state of the system (Appendix 2). In the present example, the concentrations of NADH and of NAD^+, the hydrogen ion concentration (pH), the partial pressure of oxygen, and the temperature come into play in fixing the exact value of $\Delta G_{ox\ (NADH/NAD^+)}$. Obviously, we cannot make these fine adjustments and in fact lack the information to do so in most cases. All we can do is try to approximate as best we can the conditions that prevail in living cells. The ΔG_{ox} values estimated in this manner will be called "physiological" electron potentials, the quotation marks serving to remind us that we are dealing with approximate values subject to a certain amount of fluctuation, even under perfectly normal conditions.

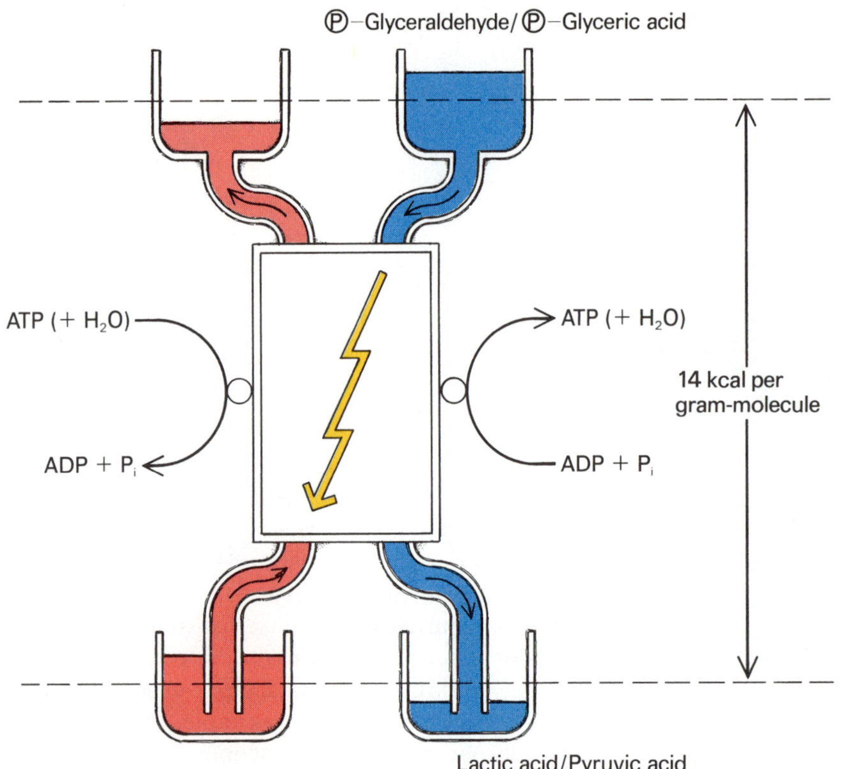

Graphic representation of electron flow through glycolytic chain. Under "physiological" conditions (dashed lines), the system is at equilibrium. The right-hand part (blue) depicts a situation in which downward electron flow, coupled to ATP formation, is elicited by an increase in electron potential of the phosphoglyceraldehyde/phosphoglyceric acid couple and a decrease in potential of the lactic acid/pyruvic acid couple. Such changes might result from corresponding changes in the ratio of the concentration of the reduced form to that of the oxidized form of each couple. The left-hand part (red) indicates a reverse situation in which upward electron flow, supported by ATP hydrolysis, is favored.

In glycolysis, the "physiological" electron potentials are of the order of −63 kcal per pair of electron-equivalents transferred to oxygen for the phosphoglyceraldehyde/phosphoglyceric acid couple and of −49 kcal per pair of electron-equivalents transferred to oxygen for the NADH/NAD^+ couple, as well as for the ethanol/acetaldehyde and lactic acid/pyruvic acid couples. These values, which are given negatively to indicate the exergonic nature of the reactions (see Appendix 2), indicate that glycolysis operates close to thermodynamic equilibrium. Between phosphoglyceraldehyde and NAD^+, the difference in potential is 14 kcal per pair of electron-equivalents, just enough to power the assembly of one gram-molecule of ATP. Between NADH and either acetaldehyde or pyruvic acid, the difference in potential is negligible. This means that the system is easily reversible and that the direction of electron flow depends on small perturbations. When glycolysis serves catabolically, as considered in this chapter, the level in the upper reservoir is somewhat higher, or that in the lower reservoir lower, than indicated, and the electrons can flow down and support ATP assembly. But, if the electron levels change in the opposite direction and ATP is supplied from another source, the flow of electrons is reversed and glycolysis has an anabolic role. This happens in liver, for instance, when carbohydrate is made from noncarbohydrate sources (gluconeogenesis), and in plants (with NADP instead of NAD), where the required ATP is provided by a light-powered mechanism (see Chapter 10).

Note that the cost of making ATP is no more constant than any other "physiological" free-energy value. It is itself subject to fluctuation, depending on the intracellular concentrations of ATP, ADP, and inorganic phosphate. If, for example, the concentration of ATP goes down and that of ADP goes up, as might occur in the course of heavy work, ATP formation will require less than 14 kcal per gram-molecule, and the equilibrium conditions of the glycolytic oxphos unit will be correspondingly altered. As we will see in Chapters 9 and 14, a fundamental regulating mechanism depends on this kind of interaction.

8 The Cytosol: Group Transfer and Biosynthesis

As ATP flows out of oxphos units, where does it go? Not an easy question to answer, for ATP rushes along hundreds of invisible trails—that is, diffuses down hundreds of concentration gradients—to wherever work is being performed and ATP consumed. Some of these trails lead to membranes, to which they bring fuel for transport mechanisms—for instance, the sodium-potassium pump. Others go to contractile fibers, to support mechanical work. Most of them, however, stop right here in the cytosol, with ATP becoming entangled with some local molecule. As a result of the scuffle, a piece of the ATP is appropriated by the encountered molecule. This liaison is usually short-lived and soon succumbs to a new collision. One or two additional affairs may follow. But eventually, this molecular group swapping comes to an end with the sealing of a stable bond between two building blocks used by the cell in the construction of its constituents.

Trailing ATP

Whatever their starting point, however circuitous their course, most cytosolic ATP trails lead to the same central biological function: biosynthesis. By charting them, we find the answer to a key question that was briefly evoked in the preceding chapter: How does a cell succeed in making thousands of different compounds with, as sole source of energy, the splitting of ATP to ADP and P_i, the central process that is repaired by the operation of oxphos units? Let us now define the problem in somewhat more precise terms.

Most biosynthetic reactions are dehydrating condensations between two molecular building blocks:

$$X\text{—}OH + Y\text{—}H \longrightarrow X\text{—}Y + H_2O$$

There are many different X's and many different Y's. They include the amino acids, which combine with each other to make proteins; the simple sugars, which associate into polysaccharides and other carbohydrate components; the mononucleotides, which polymerize into nucleic acids; the fatty acids, which join with glycerol and other alcohols to form lipids; as well as a host of other, more specialized, molecules. Putting them together correctly requires two conditions: information and energy.

Biosynthetic assemblies do not occur in haphazard fashion. They rely on the right kind of X becoming linked to the right kind of Y. As a rule, the instructions that allow the proper selection of biosynthetic partners are encoded in the specificity of the enzymes involved. When it comes to making the enzymes themselves, the instructions come from the genes, which close the circle by also providing the information for their own duplication. We will not consider this aspect of the problem further at present, because our tour will end with a detailed visit to the cytoplasmic network of information transfer and its controlling data centers in the nucleus.

The energy requirement of biosynthesis is explained by the fact that a dehydrating condensation cannot occur spontaneously in an aqueous medium. The overwhelming abundance of water drives the equilibrium of such a reaction far in the opposite direction, that of hydrolysis. To reverse the process, work must be performed, and therefore free energy must be supplied to the system from some outside source. In living cells, this source is represented in the last analysis by the hydrolysis of ATP, which, as noted, yields some 14 kcal per gram-molecule:

$$ATP + H_2O \longrightarrow ADP + P_i$$

Depending on the type of reaction, one or more molecules of ATP are consumed for every molecule of X—Y made. The overall free-energy balance is always negative, as required by energetics (see Appendix 2); most often markedly so, thereby making the biosynthetic process essentially irreversible under all conditions. But the question is: How is the energy transferred from one reaction to the other? Splitting the ATP first and then using the energy released by this process to join X with Y is not going to work. All we can get from the hydrolysis of ATP is heat—that is, random molecular motion, which cannot, under the conditions prevailing in living cells, be channeled to power a specific process. ATP splitting and X—Y formation must be *coupled,* so that one can provide the driving force for the other. The secret of this coupling is simple: never break a bond as such; always exchange one for another, by the mechanism of group transfer.

Group Transfer: Life's Second Golden Energy Gadget

Group transfer lies at the heart of biosynthesis. Its manifestations are infinitely varied and often highly involved, but its basic principle is remarkably simple. Essentially, it consists of the transfer of a chemical radical or group from a donor to an acceptor. We represent such a process schematically as follows:

$$A\text{—}B + C \rightleftharpoons [A\cdots B\cdots C] \rightleftharpoons A + B\text{—}C$$

in which A, B, and C each stand for some kind of molecular grouping, B in particular being the group transferred. As shown by this scheme, group transfer involves the participation of some sort of unstable ternary intermediate (shown between brackets), in which the group is transiently shared between its former and its new partner. It is a typical example of the eternal triangle at the molecular level: A—B forms a happy enough pair until C comes along and, after some fuzzy sharing of partners, takes off with B. Consistent with the villainous role of C, the reaction is also described as an attack by C on B or as a lysis (splitting) of A—B by C. (Example: hydrolysis, when water is the attacking agent.) The deprived victim of the attack, A, is called the leaving group. These roles are reversed when the reaction proceeds from right to left: A is

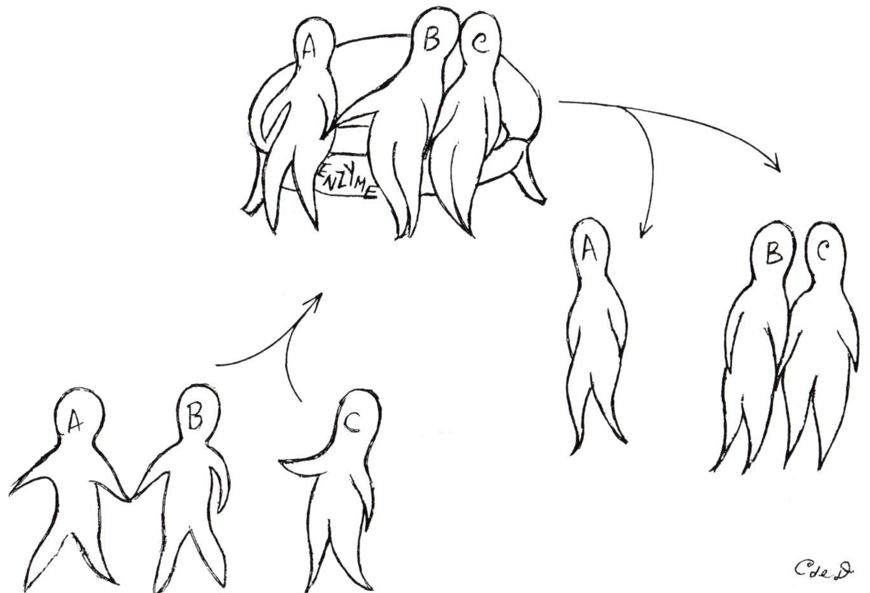

Group transfer requires two conditions: (1) the presence of an enzyme (transferase) capable of bringing molecule A–B and attacker C together in a position that allows the transfer of B; and (2) B must have more affinity for C than for A (i.e., the group potential of the B–C bond must be lower than that of the A–B bond).

the attacker or lytic agent, and C is the leaving group. The reaction remains a B transfer in both directions, making the term transfer the preferred designation.

For a drama of this sort to unfold in the human sphere, two conditions must be obeyed. First, there must be opportunity for the dramatis personae to interact in sufficiently intimate fashion. Next, B must have a greater penchant or affinity for C than for A, making the B—C bond stronger than the A—B bond. Group-transfer reactions—with due allowance for the distance between the world of humans and that of molecules—are subject to the same two conditions.

Opportunity is reflected in the kinetic condition. It generally requires the participation of a specific enzyme, or transferase, capable of bringing A—B and C (or B—C and A) close enough together to allow destabilization of the existing bond and formation of the ternary intermediate. Cells contain hundreds of such group transferases. Together with the electron transferases they make up more than 90 per cent of the total enzymic equipment of any living organism.

The second condition is thermodynamic. If the B—C bond is stronger than the A—B bond, more work must be done to break B—C than A—B, which is equivalent to saying that more free energy is lost when B binds to C than when it binds to A. On the energy scale, therefore, the B group lies lower in B—C than in A—B and, given the opportunity, will fall to the lower level. Just so in human relations; having fallen for one person does not always prevent one from falling more deeply for another. In the molecular world, however, the strength of the two bonds is not the only factor involved. The relative abundance of the four parties (A—B, C, A, and B—C) is also important. Molecular infidelities are mass events involving large numbers of individuals. A bond capable of resisting ten attackers may well yield to the assault of 10,000 because of the influence of concentration on chemical potential (see Appendix 2).

Most biological group transfers rely on what is known as a nucleophilic attack, by which is meant that the attacking agent has an affinity for positively charged radicals (the atomic nucleus is positively charged). Electrophilic attacks are rarer, except in their most naked form of electron transfer.

Nucleophilic attacks are generally perpetrated by negatively charged ions or by their protonated counterparts. In the former case, the reaction is particularly simple:

$$\text{A—B} + \overset{\ominus}{\text{C}} \rightleftharpoons [\overset{\ominus}{\text{A}}\cdots\overset{\oplus}{\text{B}}\cdots\overset{\ominus}{\text{C}}] \rightleftharpoons \overset{\ominus}{\text{A}} + \text{B—C}$$

If protonated reactants participate, protons are exchanged with the medium, as in electron transfer:

$$\text{A—B} + \text{CH} \underset{\text{H}^+}{\longleftrightarrow} [\overset{\ominus}{\text{A}}\cdots\overset{\oplus}{\text{B}}\cdots\overset{\ominus}{\text{C}}] \underset{\text{H}^+}{\longleftrightarrow} \text{AH} + \text{B—C}$$

The unstable ternary intermediate has the same structure in both formulations. It consists of two negatively charged groups vying to share an electron pair with the same positive radical B^+:

$$[\overset{\ominus}{\text{A}}\!:\cdots\overset{\oplus}{\text{B}}\cdots\!:\!\overset{\ominus}{\text{C}}]$$

The winner, as we have seen, is the one that accepts the group at the lower energy level. It is obviously very useful to know the energy level occupied by a given group in its various combinations, just as it is to know the energy level occupied by electrons, since one can then predict the spontaneous direction of the group's transfer between any two partners and, at the same time, evaluate the maximum amount of work that can be powered by this transfer (which is also the minimum needed to reverse the transfer).

The most convenient way of evaluating the energy level of a transferable group in a given combination is by the free energy of hydrolysis, ΔG_{hy}, of that combination—that is, the free energy released when the group is attacked by water or by a hydroxyl ion, OH^-. For example:

$$\text{A—B} + H_2O \longrightarrow \text{AH} + \text{B—OH}$$

or:

$$\text{B—C} + OH^- \longrightarrow \text{CH} + \text{B—O}^-$$

The free energies of these hydrolysis reactions are called group potentials. Estimated for "physiological" conditions, approximating those prevailing in living cells, the group potentials effectively measure the energy levels of transferable groups, just as the electron potentials do for transferable electron pairs. In both cases, we measure these levels with respect to the most evident natural base line: H_2O (or OH^-) as acceptor for group potentials; O_2 (to H_2O) as acceptor for electron potentials. (Note the central role of water.)

According to the convention above, the "physiological" values of $\Delta G_{hy(A—B)}$ and $\Delta G_{hy(B—C)}$ represent the group potentials of the B group in its combinations A—B and B—C, respectively. As explained in Appendix 2, the "physiological" free energy of the B transfer between A—B and C is then readily computed from the difference between the two group potentials:

$$\Delta G_{transfer} = \Delta G_{hy(A—B)} - \Delta G_{hy(B—C)}$$

If the (negative) potential of A—B is greater in absolute value than that of B—C, the ΔG of the transfer is negative: B falls from a higher energy level in A—B to a lower one in B—C; its transfer from A—B to C can occur spontaneously. In the opposite case, this transfer is endergonic, and free energy must be supplied if B is to be lifted from its lower level in A—B to its higher level in B—C.

Group potentials also serve conveniently for evaluating the free energies of coupled biosynthetic reactions. If, for example, the dehydrating assembly of one molecule of X—Y (p. 120) is supported by the hydrolysis of n molecules of ATP to ADP + P_i, the following relationship is valid, irrespective of the mechanism of coupling:

$$\begin{aligned}\Delta G_{biosynthesis} &= n\ \Delta G_{hy(ATP \to ADP)} - \Delta G_{hy(X—Y)} \\ &= n(-14) - \Delta G_{hy(X—Y)}\end{aligned}$$

As first pointed out by Fritz Lipmann, one of the founders of modern bioenergetics, the bonds found in natural substances fall roughly into two classes: high-energy bonds, represented by a "squiggle" (~); and low-energy bonds, represented by a simple line (–). The terminal phosphate bond of ATP is the archetype of high-energy bonds. Many of the bonds found in natural constituents (ester, amide, peptide, glycoside) are low-energy bonds, with "physiological" group potentials of some −6 to −8 kcal per gram-molecule. It is the difference between the two that allows the hydrolysis of ATP to power biosynthetic assemblies. As to how this is actually accomplished, the answer is: with the help of Janus.

Introducing Janus, the Double-Headed Intermediate

The ancient Romans had a god named Janus—the month of January is dedicated to him—who was believed to have two faces, one looking into the past, the other into the future. Biochemistry has rediscovered Janus as the coupling demon of biosynthesis. It arises from a nucleophilic attack by an oxygen-containing building block (X—OH or X—O⁻) on ATP or some related energy-rich molecule, which we will designate provisionally as A—B, in order to avoid going into complex chemistry. The attack is spearheaded by the oxygen atom:

$$A{-}B + H{-}O{-}X \longrightarrow A{-}H + B{-}O{-}X$$

This reaction is a transfer of the B^+ radical (B-yl group) from A—B to X—O⁻. What gives B—O—X its double-headed character is that it can also engage in a transfer of the X^+ radical (X-yl group); for instance, when attacked by building block Y—H or Y⁻:

$$B{-}O{-}X + H{-}Y \longrightarrow X{-}Y + B{-}O{-}H$$

Now watch what happens when the two reactions are allowed to proceed sequentially:

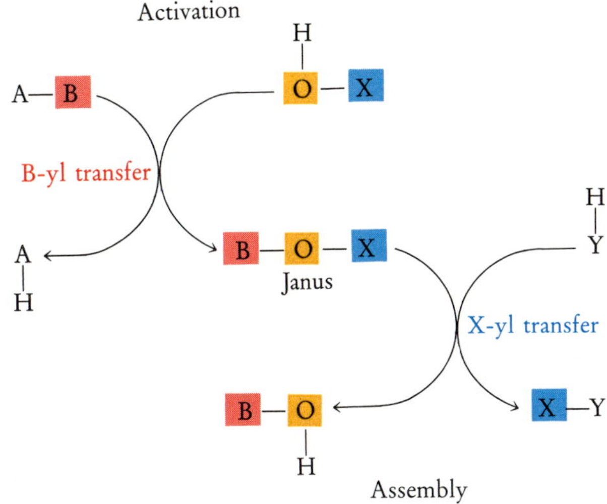

The half-reaction on the left of the above diagram reads (follow both left-pointing arrows):

$$A{-}B \longrightarrow A{-}H + B{-}OH$$

That on the right:

$$X{-}OH + Y{-}H \longrightarrow X{-}Y$$

We observe hydrolysis of A—B and dehydrating condensation of X—Y. But water is nowhere to be seen. It is transferred in hidden form from X—OH to B—OH, through the central oxygen of Janus, while simultaneously a greater or lesser portion of the group potential of A—B is retrieved in the X—Y bond. Janus thus serves both as conveyer of group-linked energy and as bearer of hidden water; it is Mercury and Aquarius all in one, if such freedom may be taken with mythology.

One can readily verify this mechanism experimentally by providing the cell with X—OH molecules labeled with the heavy isotope of oxygen ^{18}O and analyzing the products of the reaction with a mass spectrograph. The ^{18}O is found in B—OH, not in water, as it would be if the reaction were an authentic dehydrating condensation.

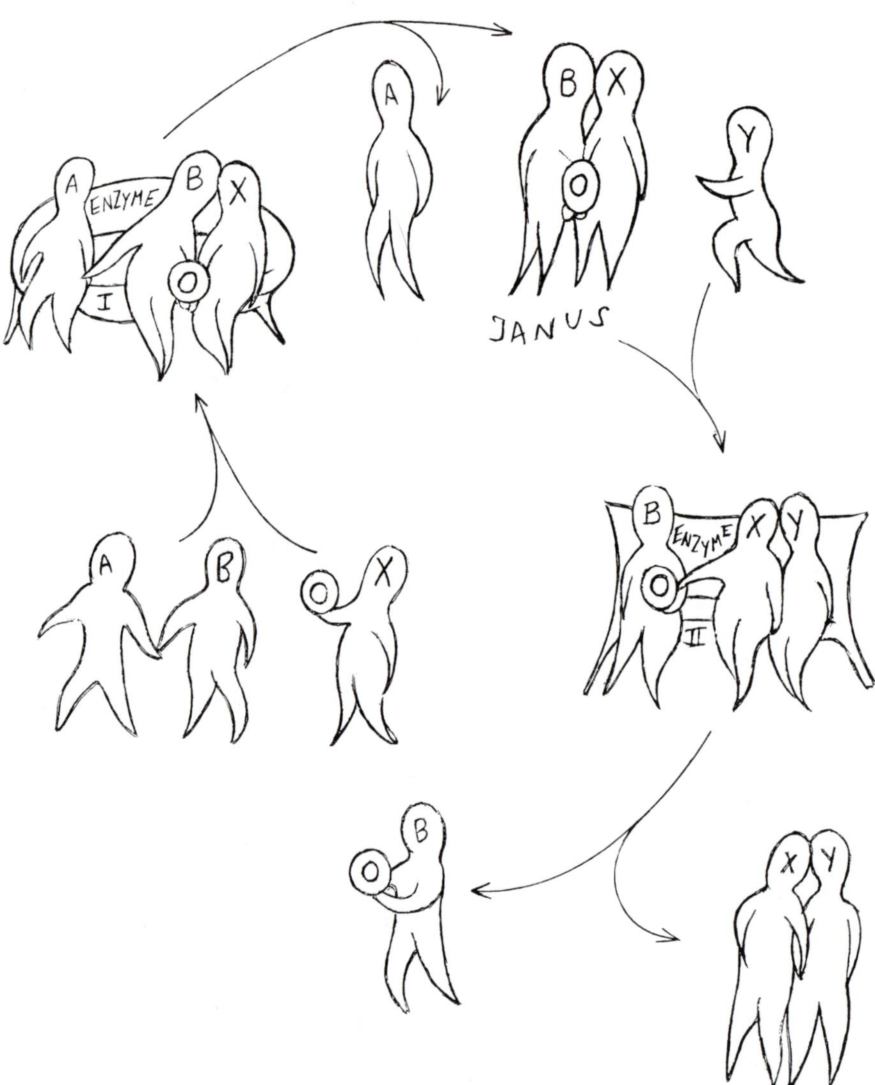

Sequential group transfer. In the first reaction, B is separated from A by an oxygen-proferring attacker X. But new attacker Y comes on the scene, removing X from B, which is left holding the oxygen by which it was tempted. The B-O-X intermediate is double-headed Janus. It consists of the two transferable groups participating in reactions I and II, joined by the oxygen atom that changes hands in the transaction.

Such, basically, is the mechanism of biosynthesis. It takes many guises. But fundamentally it always relies on sequential group transfer, linked by a double-headed intermediate. The essential anatomical features of this intermediate are two transferable groups connected by a central oxygen atom. When approached from the left, it offers an energy-rich B-yl group attached to an X—O$^-$ carrier. When seen from the right, it appears just as convincingly as an X-yl group proferred by a B—O$^-$ carrier.

According to this general scheme, biosynthesis always proceeds in at least two steps, connected by Janus. The first step, which depends on some sort of group transfer from the energy donor (ATP or some related molecule), serves to lift the X-yl group from its zero level of energy (X—OH) to the high-energy level it occupies in Janus. This step is called activation. The final step, or assembly, in which the X-yl group is transferred to its natural acceptor Y, proceeds downhill, from the high-energy activation level to the X—Y level.

In an important variant of this basic two-step mechanism, the Janus intermediate donates the activated group to a carrier, which itself transfers it to the final biosynthetic acceptor, as shown in the sequence of reactions on the facing page.

Janus, the double-headed intermediate, consists of two transferable groups linked by a central oxygen. Attacked on the left, it yields the B^+ group; on the right, the X^+ group. In each case, the remaining group is left with the oxygen.

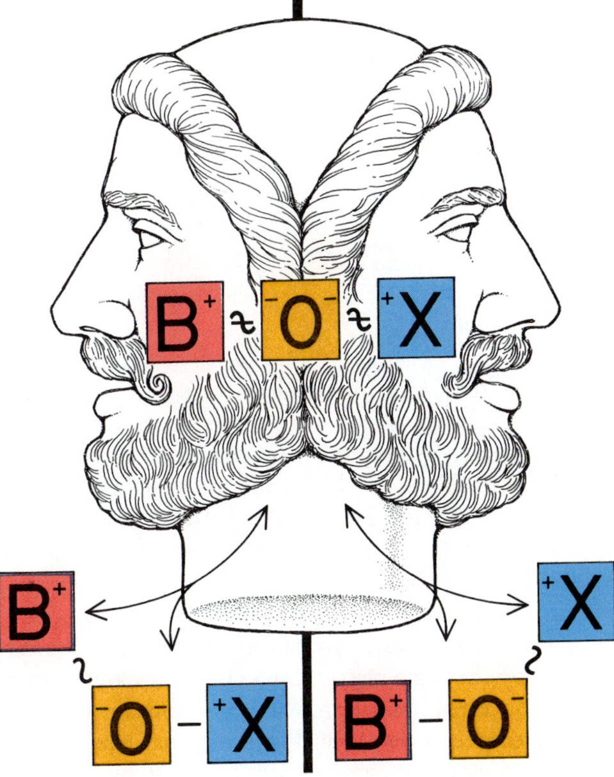

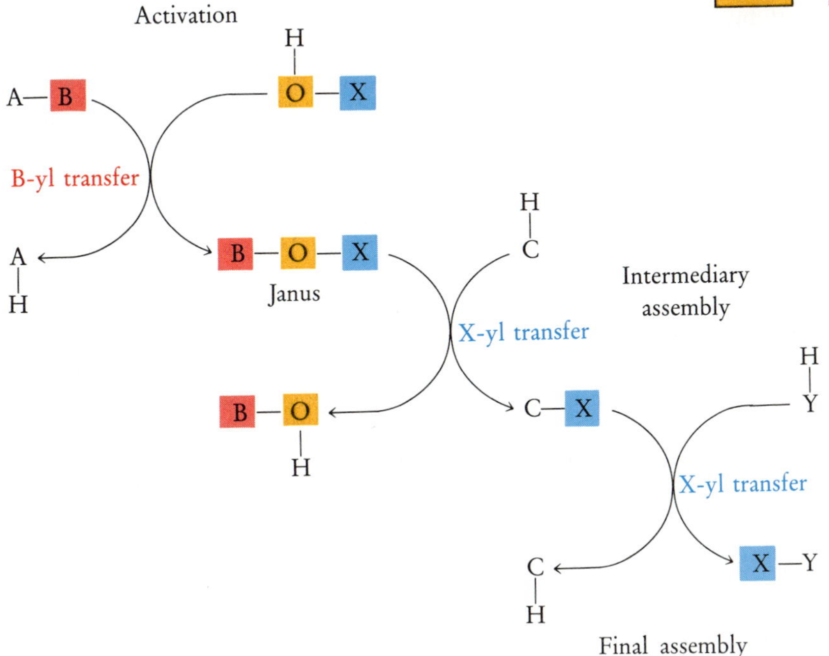

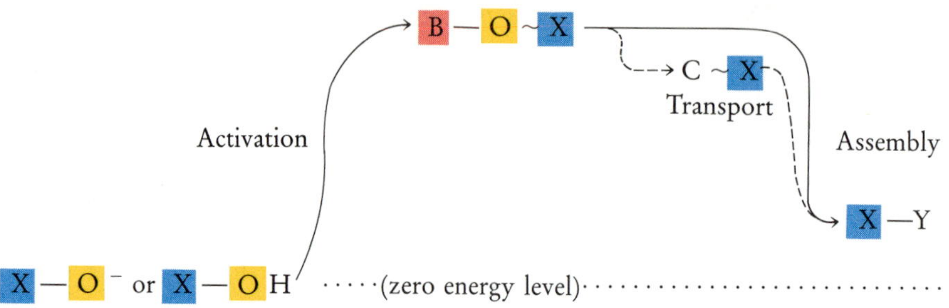

In such three-step mechanisms, the X-yl group remains at a relatively high energy level in its combination with the carrier, leaving the main energy drop to occur at the final assembly step (shown above), as befits a process in which a stable product is to be formed. The cell thus operates very much like a builder who first hoists his materials high up with a crane, moves them around in a horizontal plane, and finally lowers them into place.

In a number of cases, activation and assembly are carried out by the same enzyme, which then catalyzes some sort of concerted process in which the Janus intermediate remains enzyme-bound. Such bifunctional enzymes are called synthetases or ligases (Latin *ligare*, to bind). Without them, many biosynthetic processes would be very inefficient or even could not occur at all. Janus intermediates are often highly unstable molecules that would not survive very long if they were let loose. Frequently also, they are compounds of such high group potential that they cannot, with the coupled splitting of the group donor A—B as sole source of energy, be produced at a concentration high enough to permit efficient diffusion between two physically separated enzymes. Keeping such intermediates enzyme-bound and strategically placed so that they can immediately be trapped by the exergonic assembly process as they are formed helps overcome these difficulties.

On the other hand, it is often very useful to the cell, sometimes even indispensable, to have the two steps of biosynthesis take place at different sites. Activation requires energy and the participation of ATP; it is usually carried out in the cytosol or in some site closely connected with the cytosol and as amply supplied with ATP—for instance, the cytosolic face of a membrane. Assembly, on the other hand, frequently depends on an accurate supply of information, which is more readily secured on a structured substratum, such as is provided by the ribosomes in the synthesis of proteins (see Chapter 15) or by chromosomal scaffoldings in that of nucleic acids (see Chapters 16 and 17). Another advantage of the physical separation of the two steps is that it allows centralization. A single activation reaction suffices for each X-yl group, which can then be transported in ready-for-use form to any number of assembly sites. In actual fact, the economy is even greater: the cell often makes use of the transport phase to modify or otherwise process the X-yl group in various ways, so that a single activation reaction may serve to energize several biosynthetic building blocks that are chemical modifications of each other.

When assembly is separated from activation, a stable transport form of the activated building block is needed, and the activation step must be sufficiently exergonic in itself to produce appreciable concentrations of it. A number of Janus intermediates answer these requirements; their B—O$^-$ part then acts as carrier of the X-yl group. In other instances, the requirements are met thanks to the use of special carriers. Almost invariably in such cases, activation of the building block and its attachment to the carrier are accomplished by a single, ligase-type enzyme. Several important coenzymes function as group carriers.

The role of activation as a prerequisite to metabolic processing is not restricted to biosynthesis. In fact, a considerable part of metabolism requires prior activation of the substrate. This is even true of many catabolic reactions, as was illustrated in Chapter 7 by the example of glycolysis. The group carriers thus also serve as handles whereby the attached molecules are presented to their modifying enzymes.

But the time has come to put some chemical flesh around the bare bones of schematic abstractions. In the organization of this tour, an effort has been made to bypass chemical details as much as is feasible. But there is a limit to what can be understood of an essentially chemical

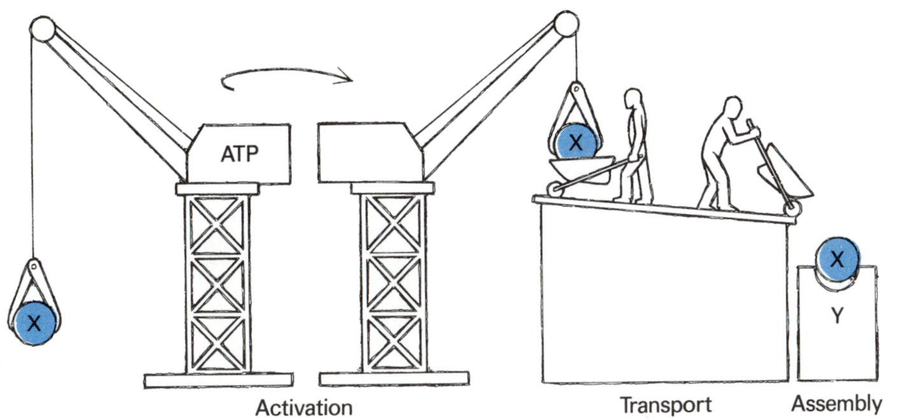

In biosynthesis, the activation step serves to lift building block X to a high-energy level (group potential) with the help of ATP. In the assembly step, the group is transferred to final acceptor Y, with a distinct drop in potential. In a number of cases, the activated group is first transferred, with little drop in potential, to a carrier C, from which it is then transferred to its final acceptor Y.

machine without the language of chemistry. Those who find the next part of the visit too arduous should, however, not lose heart. Even if they skip much of it, they should still be able to catch up with us later without too much difficulty. On the other hand, those with a better grounding in biochemistry may derive some enjoyment, and perhaps some illumination, from the kind of bird's-eye view of biosynthesis that will be provided. Let everyone tag along, therefore, and stay with the group.

The Source of Group-Transfer Energy

In the abbreviation ATP, A stands for adenosine. It is a nucleoside, which is defined as the combination of a base—adenine in the present case—with carbon atom number 1 of ribose, a 5-carbon sugar, or pentose (see Appendix 1). Three other important bases engage in similar nucleosidic combinations with ribose: guanine, which, like adenine, belongs to the group of purines, and cytosine and uracil, members of the pyrimidine family. The corresponding nucleosides are called guanosine (G), cytidine (C), and uridine (U).

We will consider the detailed structure of the bases when we look at the anatomy of nucleic acids and at the genetic language. In the meantime, let us concentrate on the other end of the nucleoside molecule, which is occupied by carbon atom number 5 of ribose (numbered 5′ to distinguish it from carbon 5 of the base). This carbon bears a hydroxyl group OH, which in most natural combinations of nucleosides carries a phosphoryl group. Such nucleoside monophosphates are called nucleotides; they are designated as adenylic, guanylic, cytidylic, or uridylic acid or by the abbreviations AMP, GMP, CMP, UMP, in which MP stands for monophosphate.

To this terminal phosphoryl group of the nucleotides, one or two additional phosphoryl groups may become attached by the kind of linkage found in pyrophosphoric acid (pyrophosphate bond) to produce the nucleoside diphosphates ADP, GDP, CDP, and UDP, and the nucleoside triphosphates ATP, GTP, CTP, and UTP.

Base	Nucleoside	Nucleoside monophosphate	Nucleoside diphosphate	Nucleoside triphosphate
Adenine	Adenosine (A)	Adenylic acid (AMP)	ADP	ATP
Guanine	Guanosine (G)	Guanylic acid (GMP)	GDP	GTP
Cytosine	Cytidine (C)	Cytidylic acid (CMP)	CDP	CTP
Uracil	Uridine (U)	Uridylic acid (UMP)	UDP	UTP

In summary, representing a nucleoside by the symbol N (which stands for A, G, C, or U) and making explicit its 5′-hydroxyl group, we have:

N—OH	Nucleoside (N)
N—O—P(=O)(O$^-$)—O$^-$	Nucleoside monophosphate (NMP)
N—O—P(=O)(O$^-$)—O—P(=O)(O$^-$)—O$^-$	Nucleoside diphosphate (NDP)
N—O—P(=O)(O$^-$)—O—P(=O)(O$^-$)—O—P(=O)(O$^-$)—O$^-$ (α, β, γ)	Nucleoside triphosphate (NTP)

These NTPs are really super-Janus types of molecules, triply double-headed. Their three phosphoryl-bond oxygens (α, β, γ) each separate a distinct pair of transferable groups. This character makes the NTPs susceptible—at least theoretically—to as many as six distinct types of nucleophilic attacks, which will be designated α_p, α_d, β_p, β_d, γ_p, and γ_d, in which α, β, γ stand for the bond under attack, and the subscripts p and d for proximal and distal (with respect to N):

N—O—P(=O)(O$^-$)—O—P(=O)(O$^-$)—O—P(=O)(O$^-$)—O$^-$

α_p, β_p, γ_p; α_d, β_d, γ_d

In practice, these possibilities (which are displayed explicitly on the facing page) are exploited very unequally. With the exception of a few rare α_p or β_d approaches, all biosynthetic attacks on NTPs are either β_p or γ_d. As far as is known, α_d or γ_p attacks are never used.

Most biosynthetic processes can be classified as regular two- or three-step mechanisms dependent on one of the above attacks. As might be expected in such a complex field, the main theme occasionally undergoes some variations. But to those who have been alerted to look for it, the theme, as in music, remains easily recognizable. One seemingly atypical variant occurs when an NTP acts in Janus capacity, as donor of a group in a final assembly process. Such single-step mechanisms disobey the two-step rule only in appearance. The group donated by the NTP still required prior activation—by group transfer from some other NTP, or by the operation of oxphos units, or by both—before it could be transferred exergonically. In such cases, the activation step coincides with the repair phase of the ordinary two- or three-step mechanisms (see pp. 130–131).

The group potentials that are brought into play in the different types of attacks on NTPs are not equivalent. As already mentioned, the "physiological" free energy of hydrolysis of the γ bond of ATP is of the order of -14 kcal per gram-molecule. All terminal phosphoryl groups in NTPs and in NDPs have the same group potential. In contrast, the "physiological" free energy of hydrolysis of the β bond of NTPs is considerably higher, partly because of a difference in standard free energy of hydrolysis (about 3 kcal per gram-molecule) and, more importantly, because most cells contain highly active pyrophosphatases that hydrolyze inorganic pyrophosphate as it arises. Therefore, hydrolysis of the β bond is followed by hydrolysis of the pyrophosphate formed:

$$\begin{array}{lll} \text{NTP} + H_2O \longrightarrow \text{NMP} + PP_i & & \Delta G_{hy(\beta\ \text{bond})} \\ PP_i + H_2O \longrightarrow 2\,P_i & & \Delta G_{hy(PP_i)} \\ \hline \text{NTP} + 2\,H_2O \longrightarrow \text{NMP} + 2\,P_i & & \Delta G_{hy(\text{total})} \end{array}$$

The same end result can be achieved by hydrolyzing the γ bond first, and then the β bond, in which case the "physiological" free energies of the reactions are known:

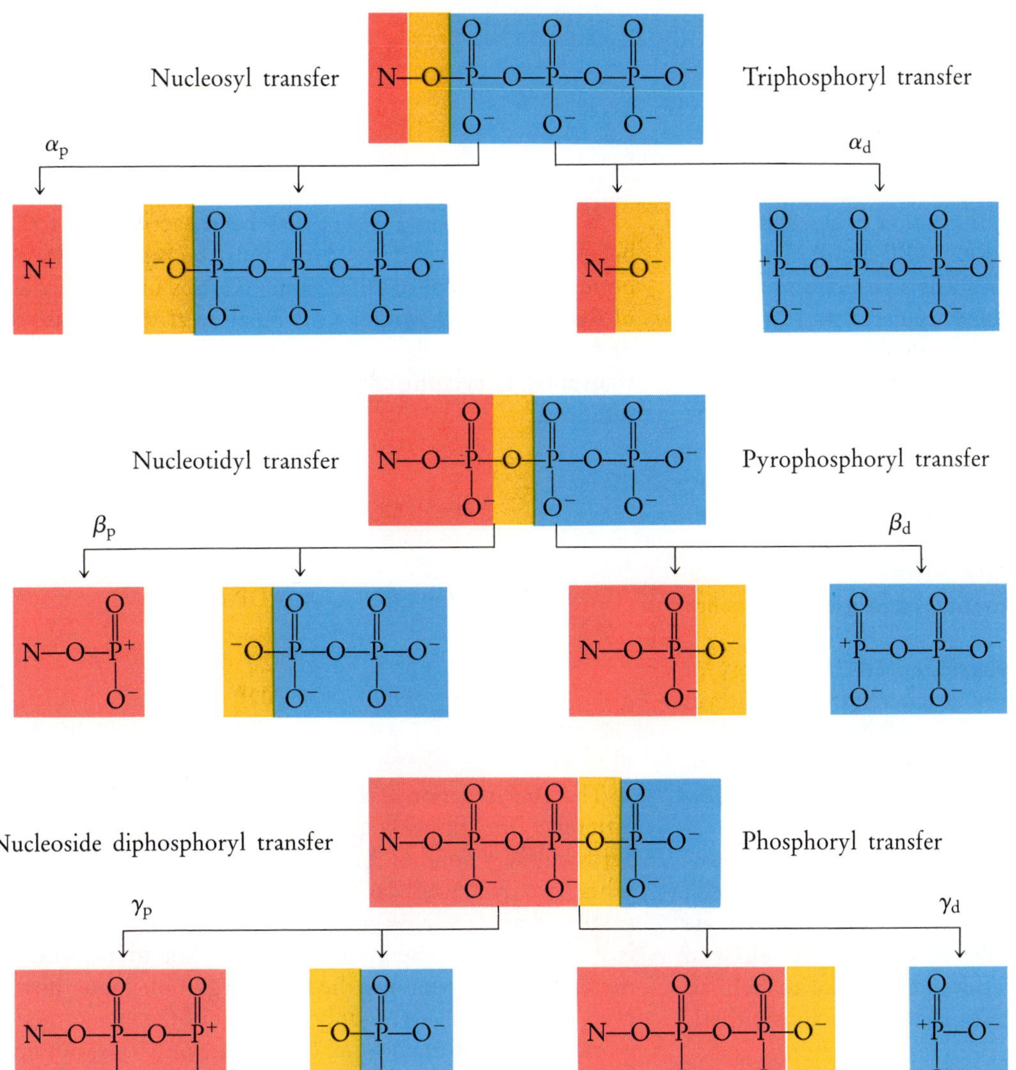

Diagrammatic representation of the six group-transfer reactions in which ATP and other NTPs may theoretically participate, depending on which of the three phosphoryl-connecting oxygens, α, β, or γ, serves as the central Janus oxygen.

$NTP + H_2O \longrightarrow NDP + P_i$	−14 kcal per gram-molecule
$NDP + H_2O \longrightarrow NMP + P_i$	−14 kcal per gram-molecule
$NTP + 2\ H_2O \longrightarrow NMP + 2\ P_i$	−28 kcal per gram-molecule

Whether we start with the β bond or with the γ bond, the total free-energy change associated with hydrolysis of the two bonds must be the same. Therefore, by elementary bookkeeping rules:

$$\Delta G_{hy(\beta\ bond)} = -28 - \Delta G_{hy(PP_i)}$$

The "physiological" free energy of hydrolysis of the β bond of NTPs depends on the value of $\Delta G_{hy(PP_i)}$—that is, on how close to equilibrium the action of pyrophosphatases maintains the concentrations of PP_i and P_i. This is not known with any accuracy. But it is a fair assumption, in view of the high activity of the enzymes, that a state very near equilibrium is maintained—in other words, that $\Delta G_{hy(PP_i)} \simeq 0$. Accordingly, we will adopt for the "physiological" free energy of hydrolysis of the β bond the (maximal) value of -28 kcal per gram-molecule. It is worth noting in this connection that some microorganisms do not maintain a very low pyrophosphate concentration but operate with a pyrophosphate-based economy instead. Apparently, the extra energy expenditure imposed on β mechanisms by pyrophosphatase action is not a vital necessity.

As to the α bond, its "physiological" free energy of hydrolysis is of the order of -7 kcal per gram-molecule in NMP. For the reason just mentioned (hydrolysis of PP_i), it approaches -21 kcal per gram-molecule in NDP, and -35 kcal per gram-molecule in NTP (extending the reasoning to PPP_i).

The table below summarizes the values of group potentials that will be used in our subsequent analyses. It will be remembered that these values are subject to fairly wide fluctuations, depending on the conditions prevailing in the cells (see Appendix 2). But they suffice to help us understand the main energetic features of biosynthetic mechanisms.

Hydrolysis reaction	"Physiological" ΔG_{hy} (kcal per gram-molecule)
$NTP \longrightarrow NDP + P_i$	-14
$NDP \longrightarrow NMP + P_i$	-14
$NMP \longrightarrow N + P_i$	-7
$NTP \longrightarrow NMP + PP_i\ (2\ P_i)$	-28
$NTP \longrightarrow N + PPP_i\ (3\ P_i)$	-35
$NDP \longrightarrow N + PP_i\ (2\ P_i)$	-21
$PP_i \longrightarrow 2\ P_i$	~ 0
$PPP_i \longrightarrow PP_i + P_i\ (3\ P_i)$	~ 0

When a bond in an NTP has been sacrificed for the benefit of biosynthetic work, it must be repaired. If the bond is the γ bond of ATP, some oxphos unit takes care of the repair. If any other, it is repaired at the expense of one or more γ bonds of ATP, thanks to the occurrence of transphosphorylating enzymes that catalyze the following reactions:

$$ATP + N \longrightarrow ADP + NMP$$

$$ATP + NMP \rightleftharpoons ADP + NDP$$

$$ATP + NDP \rightleftharpoons ADP + NTP$$

The first reaction is irreversible because of the large difference in "physiological" free energy of hydrolysis between the γ bond of ATP and the α bond of NMP. The other two reactions exchange bonds of equal energetic value and are freely reversible. The cost of these transfers is itself borne by the operation of oxphos units, which, therefore, end up paying the full energy bill. Note, however, that this bill covers only that part of the biosynthetic work—often the major one, or even the only one, but not always—that depends on group transfer. Other processes, especially electron transfer from high-potential donors, also may come into play. Biosynthetic reductions are particularly important in autotrophic organisms (see Chapter 10).

Putting the various repair reactions together, we end up with the condensed diagram at the top of the facing page, which henceforth will be referred to as the central repair machinery. This machinery also provides for the activation of such building blocks—for example, P_i or an NMP—as are donated in single-step biosynthetic processes. Note that PP_i cannot be incorporated in an NTP as such, but must first be hydrolyzed. (The same is true for PPP_i, not shown on the diagram because its appearance, if it occurs at all, is very rare and fleeting.)

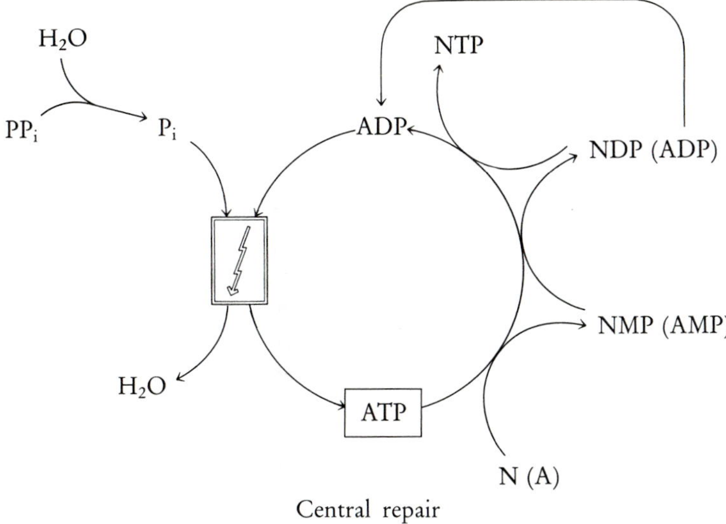

Central repair

An Optional Excursion

We have arrived at a point in our tour where many participants are probably eager for a well-deserved break. Some, however, may wish to put their newly acquired knowledge to a test. A side excursion has been arranged for these more adventurous tourists. It can be skipped by the others. It will take us browsing through a biochemistry textbook and help us recognize a few simple basic patterns behind the enormous complexity and infinite diversity of the chemical mechanisms whereby living cells manage to construct thousands of substances, most of which are still beyond the possibility of synthesis by our most advanced technology. During this excursion, the participating building blocks will be represented systematically in ionized form (X—O$^-$ and Y$^-$), unless they are known to be protonated. The movement of protons has been indicated when necessary. Detailed chemical structures will not be shown. They can be found in Appendix I.

Single-Step Processes

In these reactions, some part of an NTP is transferred to a final biosynthetic acceptor. The transferred group arises from a precursor, usually P_i or an NMP, which may be seen as the X—O$^-$ building block of our general two-step scheme, previously activated and incorporated into the NTP by the operation of what has been called the central repair machinery. The donating NTP has the character of a double-headed Janus intermediate.

Reactions dependent on γ_d transfer. The NTP involved is almost invariably ATP and acts as an NDP carrier bearing a phosphoryl group:

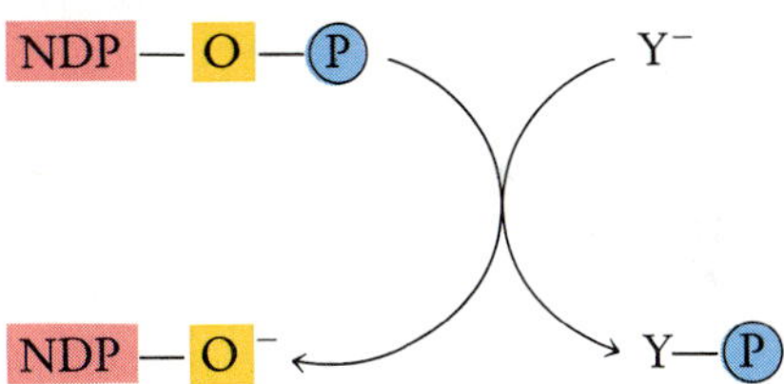

Most of the phosphorus contained in natural substances—a diversified group that includes the nucleoside phosphates, several coenzymes, the nucleic acids, the phospholipids, and numerous metabolites—first enters into its combinations by this kind of reaction. Exceptions are the terminal phosphoryl group of ATP itself, which is incorporated by oxphos action, and a number of cases in which inorganic phosphate attacks a pre-existing bond (phosphorolysis).

The cost of the biosynthetic transaction is readily computed:

$$\Delta G_{\text{biosynthesis}} = -14 - \Delta G_{\text{hy(Y—P)}}$$

Common phosphate esters, including NMPs, have "physiological" group potentials of about −6 to −8 kcal per gram-molecule. The free-energy loss associated with their assembly thus varies between 8 and 6 kcal per gram-molecule, which suffices to make the reaction essentially irreversible. A number of other phosphate compounds, however, have "physiological" group potentials of the order of −14 kcal per gram-molecule, which makes the transphosphorylation with ATP freely reversible ($\Delta G \simeq 0$). Among them are all the NDPs and NTPs, which, as we have seen, can transphosphorylate freely with ATP. Thanks to these reactions, any NDP or NMP that forms is immediately reactivated to NTP, ready for use in a new biosynthetic process (pp. 130–131). Conversely, in times of acute demand for ATP, as at the onset of muscular contraction, cells may call on their NTPs and on their NDPs (including ADP) to help restore the consumed ATP by reversal of the transphosphorylation reactions. The brunt of this responsibility, however, falls on another group of high-energy compounds, called phosphagens, characterized by an amidophosphate linkage. The phosphagen of vertebrates is creatine phosphate, which arises from creatine by a γ_d type of phosphoryl transfer from ATP:

$$\text{ATP} + \text{Creatine} \rightleftharpoons \text{ADP} + \text{Creatine}{\sim}\text{P}$$

The equilibrium of this reaction is such as to favor ATP formation. Only when the ratio of ATP to ADP concentration is sufficiently high, as it is in cells that are not subjected to an energy stress, is the left-to-right direction favored: the creatine phosphate reservoir is restored. As soon as ATP starts being consumed and the ADP level rises, the right-to-left direction becomes the favored one and creatine phosphate serves to regenerate ATP from ADP. This tides the cell over the period needed to get oxphos units in full action (see Chapter 14).

The enzymes that catalyze γ_d transphosphorylation reactions from ATP are called phosphokinases or, more simply, kinases (Greek *kinein,* to move). In addition to serving in many biosynthetic processes and in energy metabolism, phosphokinases also play an important role as primers of catabolic reactions. This is how their existence was discovered, after it was found that glucose needs to be activated by a "hexokinase" to enter the glycolytic chain (Chapter 7). In Chapters 13 and 18, we will encounter a special group of phosphokinases acting on proteins. They control a number of central regulatory processes, including those that govern cell division, and could thereby be implicated in the anarchic multiplication of cancer cells.

Reactions dependent on β_d transfer. The NTP involved in this rare type of reaction acts as an NMP carrier bearing a pyrophosphoryl group (previously assembled from P_i and activated by the central repair machinery):

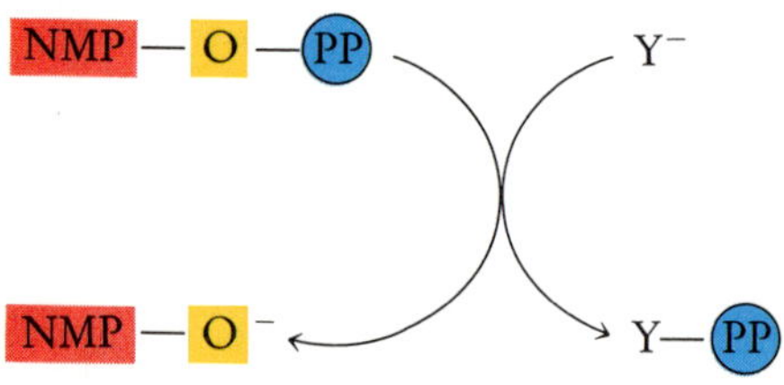

The most important acceptor of the pyrophosphoryl group is thiamine, or vitamin B_1, which happens to be the first vitamin discovered. Thiamine pyrophosphate (TPP) is an important coenzyme of decarboxylation reactions. It costs all of 28 kcal to make a pyrophosphoryl group from P_i. Part of this energy is conserved in TPP, to be dissipated only when the PP group is hydrolyzed off. There is, however, a considerable difference, of the order of −8 to −10 kcal per gram-molecule, between the free energy of hydrolysis of the pyrophosphate bond linking PP to AMP in ATP and that of its ester attachment to thiamine in TPP. This is more than enough to make the transpyrophosphorylation entirely irreversible.

Reactions dependent on β_p transfer. Here, the NTP is to be seen as an activated NMP-yl group offered by a pyrophosphate carrier, as shown at the top of the left-hand column on the facing page. The energetic contribution of pyrophosphatase is made at the initial transfer step, which therefore has available close to 28 kcal per gram-molecule for making the NMP—Y bond.

Most processes that involve the incorporation of nucleotidyl groups into stable biosynthetic products take

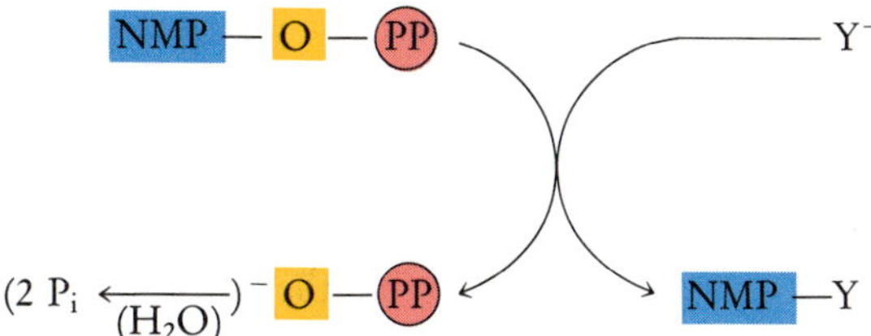

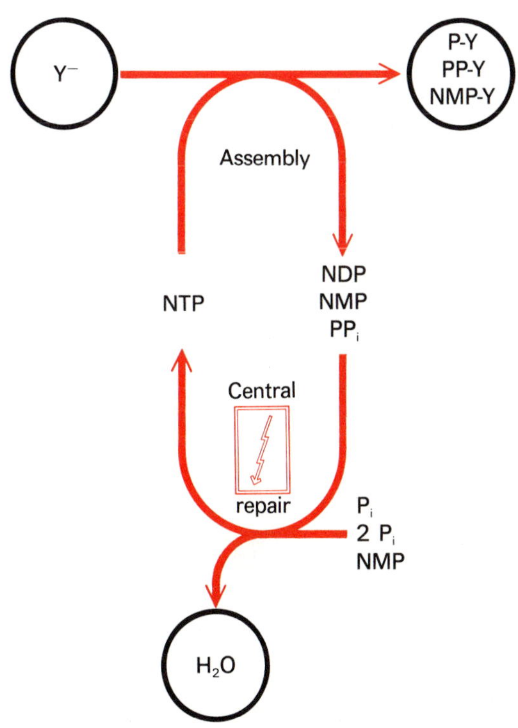

Summary of single-step biosynthetic reactions. Inorganic phosphate or mononucleotides are the building blocks. They are activated by the central repair machinery to NTP, from which they are transferred to their final acceptor in the form of a phosphoryl (γ_d), pyrophosphoryl (β_d), or nucleotidyl (β_p) group.

place by β_p transnucleotidylation. They include the fundamental processes whereby RNA (see Chapter 16) and DNA (see Chapter 17) are assembled, as well as reactions involved in the synthesis of NAD and of other coenzymes, such as NADP, FAD, and coenzyme A (see pp. 142–143), which likewise contain AMP. Similarly to phosphorylation, adenylylation plays an important role in the regulation of some enzymic proteins.

In a special, self-attacking variant of β_p transfer, the attacking agent is the internal 3′-hydroxyl group of the transferred nucleotidyl group itself. The most important such reaction, catalyzed by adenylate cyclase, leads to the formation of 3′,5′-cyclic AMP (cAMP), an important intracellular mediator of hormone action (see Chapter 13):

$$\underset{(3')\,\text{OH}}{\text{A}}\text{—O—}\overset{\text{O}}{\underset{\text{O}^-}{\text{P}}}\text{—O—}\overset{\text{O}}{\underset{\text{O}^-}{\text{P}}}\text{—O—}\overset{\text{O}}{\underset{\text{O}^-}{\text{P}}}\text{—O}^- \longrightarrow$$

$$\text{A(—O—)—O—}\overset{\text{O}}{\underset{\text{O}^-}{\text{P}}}\text{(cyclic)} + \text{HO—}\overset{\text{O}}{\underset{\text{O}^-}{\text{P}}}\text{—O—}\overset{\text{O}}{\underset{\text{O}^-}{\text{P}}}\text{—O}^-$$

$$\text{ATP} \longrightarrow \text{cAMP} + \text{PP}_i$$

Many of the bonds made by β_p transfer are high-energy bonds. Nevertheless the transfer is completely irreversible in vivo, thanks to pyrophosphatase.

Two-Step Processes

Reactions in this class conform to the basic pattern of sequential group transfer. Either they are catalyzed by a single enzyme of ligase type, and then proceed by way of an enzyme-bound Janus intermediate, or they are carried out by two distinct enzymes, often physically separated from each other. The Janus intermediate then transports the X-yl group from the activation to the assembly site, with, as carrier, the group donated by the activating NTP, in combination with the oxygen atom it has appropriated from the X—O⁻ building block.

Reactions dependent on γ_d transfer. They proceed as follows, with P_i as carrier of the X-yl group in the Janus intermediate:

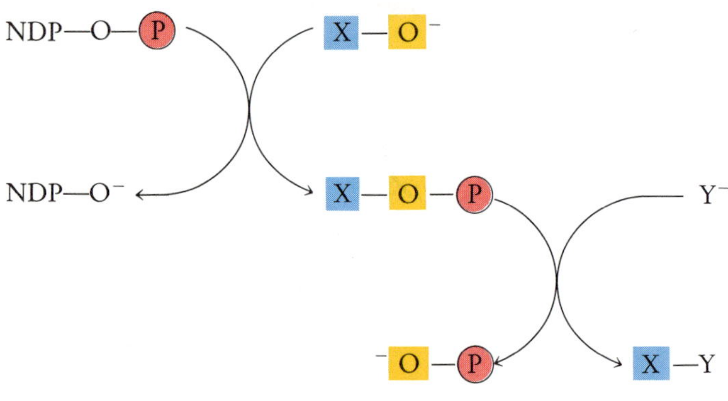

Total: $NTP + X{-}O^- + Y^- \longrightarrow X{-}Y + NDP + P_i$

In a number of such reactions, the X—O⁻ building block is a carboxylic acid, and the double-headed intermediate is the corresponding acyl phosphate:

$$R-\overset{\displaystyle O}{\overset{\|}{C}}-O-\underset{\displaystyle O^-}{\underset{|}{\overset{\displaystyle O}{\overset{\|}{P}}}}-O^-$$

Such compounds are often unstable. In addition, like many anhydrides—substances that arise by the dehydrating condensation of two acids—they are high-energy compounds, with group potentials comparable to that of the γ pyrophosphate (anhydride of phosphoric acid) bond of NTPs. These drawbacks are obviated by the participation of ligase-type enzymes. In such cases, ATP is the standard energy purveyor, and the ligases are known as ADP-forming, to distinguish them from the AMP-forming ligases, which we will meet when we look at reactions powered by the β bond.

The final acceptor of the activated acyl group is often ammonia (NH_3) or a primary amino group (R—NH_2). The resulting amide linkage (—CO—NH—) has a relatively low "physiological" free energy of hydrolysis, of the order of −6 to −8 kcal per gram-molecule. The overall process is thus sufficiently exergonic to be irreversible. The synthesis of asparagine and glutamine from aspartic and glutamic acids, respectively (Chapter 2), and that of the tripeptide glutathione from glutamic acid, cysteine, and glycine are examples of γ_d-powered, concerted two-step mechanisms. So is the synthesis of carbamoylated derivatives (R—CO—NH_2), which include intermediates in the formation of the amino acid arginine, of urea, and of pyrimidine bases. But this reaction presents us with an interesting difference: activation and assembly are catalyzed by two distinct enzymes, linked by a freely circulating Janus intermediate—carbamoyl phosphate. The thermodynamic obstacle to such a mechanism is overcome thanks to a concerted process whereby carbamate, the substrate of the activation step, arises itself as the enzyme-bound product of the condensation of bicarbonate with ammonia, also catalyzed, as it happens, by a γ_d two-step mechanism. Thus, we are dealing with a chain of two consecutive γ_d two-step mechanisms. The first three reactions in it are catalyzed by a single, trifunctional enzyme—carbamoyl phosphate synthetase—by way of two unstable, enzyme-bound intermediates (shown between brackets in the sequence of reactions at the top of the facing page): carboxyl phosphate, the Janus product of the first activation, and carbamate, the product of the first assembly step, which becomes the substrate of the second activation. The second assembly step (dashed arrows) is catalyzed by a separate enzyme.

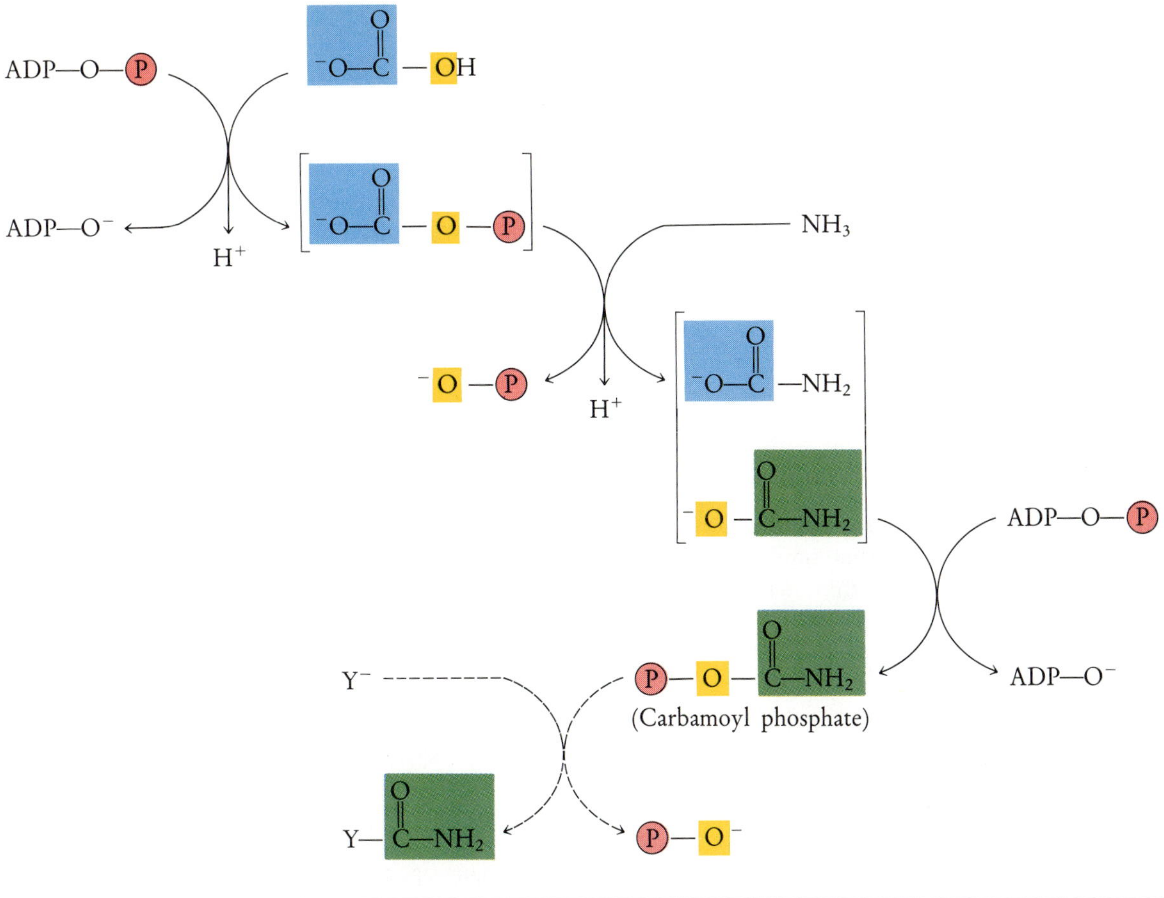

Total: 2 ATP + Bicarbonate + Ammonia + Y^- ⟶ Carbamoyl—Y + 2 ADP + 2 P_i + 2 H^+

Two γ bonds, or 28 kcal per gram-molecule, are consumed in the making of the low-energy amide bond of carbamate and the high-energy bond of its anhydride with phosphoric acid, for a total gain of about 20 to 22 kcal per gram-molecule. The overall process is strongly exergonic, thanks to the conservation of energy in the enzyme-bound carbamate. So is the subsequent transfer of the carbamoyl group to its biosynthetic acceptor, which brings it from a high to a low group potential.

It sometimes happens that the final product, X—Y, of a γ_d two-step reaction is itself a high-energy compound—for instance, a thioester (characterized by a —CO—S— linkage). Then the overall process is freely reversible and can also serve for the assembly of an NTP from the corresponding NDP and P_i, at the expense of the splitting of the X—Y bond:

$$\mathrm{R{-}\overset{\displaystyle O}{\overset{\|}{C}}{-}S{-}R' + NDP + P_i \rightleftharpoons R{-}\overset{\displaystyle O}{\overset{\|}{C}}{-}O^- + R'{-}SH + NTP}$$

Reactions of this type play an important role in the operation of some substrate-level oxphos units, in which the thioester bond is made by the oxidative condensation of the R′—SH thiol with an aldehyde (R—CH=O):

$$\mathrm{R{-}\overset{\displaystyle O}{\overset{\|}{C}}{-}H + R'{-}SH \rightleftharpoons R{-}\overset{\displaystyle O}{\overset{\|}{C}}{-}S{-}R' + 2\ \mathit{e}^- + 2\ H^+}$$

Adding the two reactions, we observe the oxidation of the aldehyde to the corresponding acid, with the coupled assembly of an NTP. This is exactly what happens in the oxphos unit of the glycolytic chain (Chapter 7):

$$R\text{—}\overset{\overset{\displaystyle O}{\|}}{C}\text{—}H + NDP + P_i \rightleftharpoons R\text{—}\overset{\overset{\displaystyle O}{\|}}{C}\text{—}O^- + NTP + 2\ e^- + 2\ H^+$$

In other substrate-level oxphos units, the substrate of the oxidation is an α-keto acid that similarly combines oxidatively with a thiol to form a thioester, with, in this case, concomitant decarboxylation.

$$R\text{—}\overset{\overset{\displaystyle O}{\|}}{C}\text{—}COO^- + R'\text{—}SH \rightleftharpoons R\text{—}\overset{\overset{\displaystyle O}{\|}}{C}\text{—}S\text{—}R' + CO_2 + 2\ e^- + H^+$$

In these reactions, the R′—SH thiol plays a catalytic role in the coupling. The thioester is an intermediate of a very rare kind, capable of acting as transducer between electron-linked and group-linked energy: it can be made at the expense of either.

Reactions dependent on pseudo-γ_p transfer. In the synthesis of glycogen, a treelike polymer made of thousands of molecules of glucose (Chapters 2 and 7), activated glucosyl units are transferred to the ends of growing branches ("tail growth," see p. 139), with UDP as carrier:

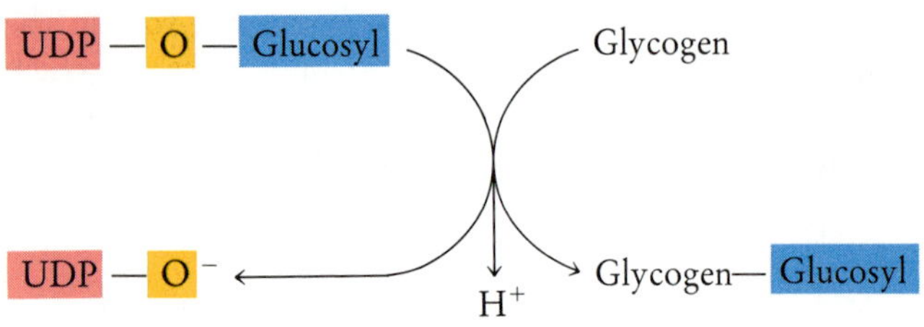

UDP — O — Glucosyl
Glycogen
UDP — O^-
H^+
Glycogen— Glucosyl

This example may serve as a paradigm of saccharide synthesis, whether they be disaccharides, oligosaccharide side chains of glycoproteins and glycolipids, or polysaccharides. Invariably, the activated sugar molecule is presented to the acceptor by an NDP, which may be UDP, ADP, GDP, or CDP, depending on the nature of the sugar. The transfer is made either directly to the final biosynthetic acceptor, as in glycogen synthesis, or by way of a fat-soluble carrier, dolichyl mono- or diphosphate, as in some of the glycosylation reactions that take place in the ER (Chapter 6). The NDP carriers also act as handles. For example, glucose molecules undergo a variety of metabolic transformations while attached to UDP.

It is interesting to note that the NDP-sugars have exactly the structure that would be expected for Janus intermediates arising by γ_p transfer (trans-NDP-ylation) on the free sugar molecule:

$$NTP + Sugar \longrightarrow NDP\text{—}Sugar + P_i$$

This, however, is not Nature's way. In reality, the NDP-sugars are made by β_p transfer with a glycosyl phosphate as acceptor. The glycosyl phosphate itself arises, directly or indirectly, by a γ_d transphosphorylation from ATP, as shown in the first sequence of reactions on the facing page.

Note the "doubly double-headed" character of the Janus intermediate. As it arises, it has an NMP-yl and a glycosyl-phosphoryl face. For assembly, it metamorphoses, so to speak, into a molecule with an NDP-yl and a glycosyl face. Because of the hydrolysis of PP_i, each glycosidic bond (from 6 to 8 kcal per gram-molecule) costs two γ bonds, or 28 kcal per gram-molecule, which is twice the price that would have been paid for a simple two-step γ_p mechanism. Why the latter was not selected for may never be known. Perhaps chance never gave it the opportunity. Or perhaps it has drawbacks that we do not perceive.

Reactions dependent on β_d transfer. In these very rare reactions, pyrophosphate is the carrier of the X-yl group in the Janus intermediate, as shown in the second sequence of reactions on the facing page.

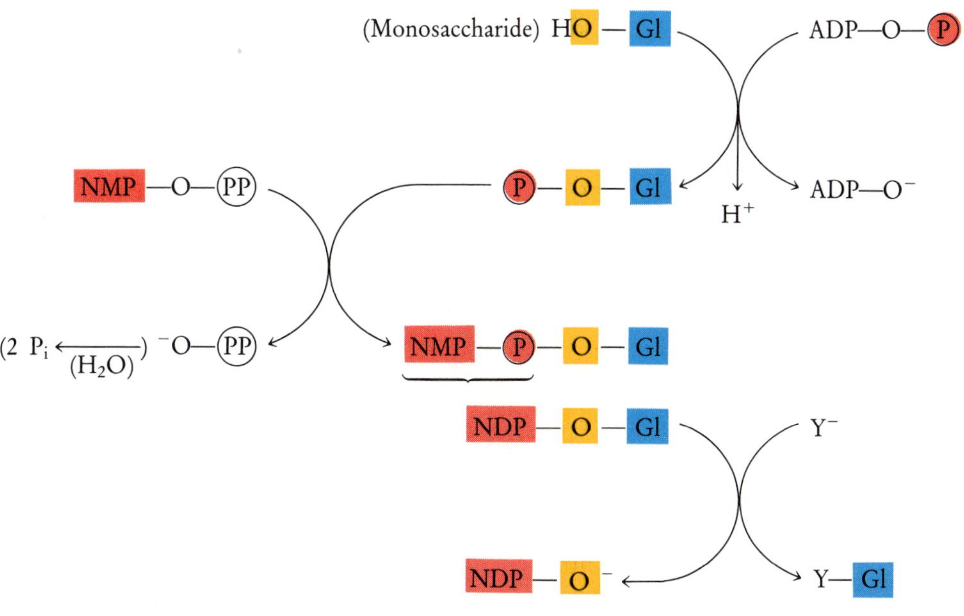

Total: ATP + NTP + Monosaccharide + Y$^-$ ⟶ Glycosyl—Y + ADP + NDP + PP_i (2 P_i)

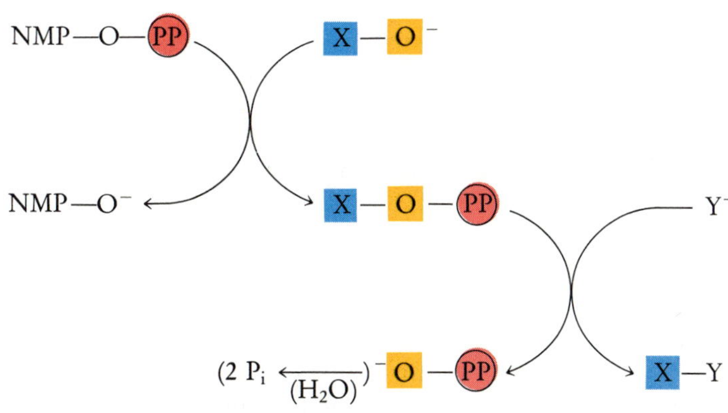

Total: NTP + X—O$^-$ + Y$^-$ ⟶ X—Y + NMP + PP_i (2 P_i)

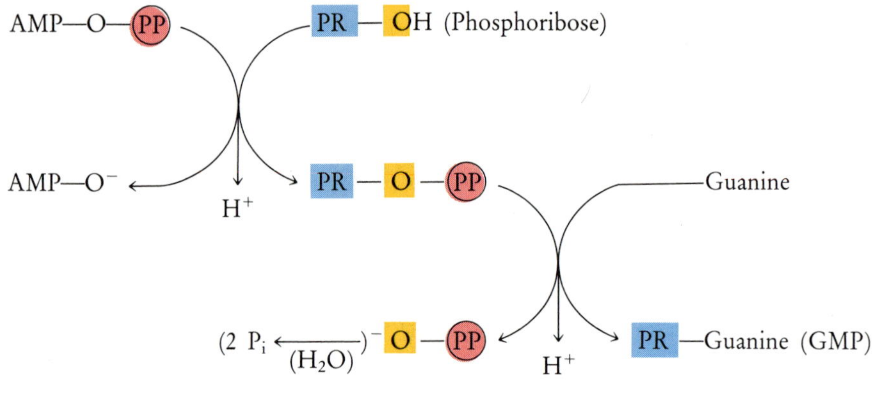

Total: ATP + PR + Guanine ⟶ GMP + AMP + PP_i (2 P_i)

The total amount of energy available for the synthesis of X—Y is 28 kcal per gram-molecule. Note, however, that much of this energy may be conserved in the double-headed intermediate, as it is the assembly step that benefits from the low PP_i concentration maintained by pyrophosphatase action. This is in contrast with β_p mechanisms (see pp. 139–140).

The most important X—O^- building block of β_d two-step reactions is 5-phosphoribose, which is the molecule that is left when the purine or pyrimidine base is hydrolyzed off from a mononucleotide. Reconstitution of some mononucleotides can then take place by a typical β_d two-step mechanism, with phosphoribosyl pyrophosphate (PRPP) as Janus intermediate. For example, guanine can thereby be rejoined with 5-phosphoribose to form GMP, as shown above.

This salvaging of bases is only one of the functions of PRPP. It is an intermediate in the synthesis of the amino acids histidine and tryptophan and in that of the purine ring.

Another key Janus intermediate with the structure X—O—PP is isopentenyl pyrophosphate, a precursor of a host of important fat-soluble molecules, including quinonic electron carriers (see Chapters 9 and 10); vitamins A, D, E, and K; sterols and steroids; carotenoids; terpenoids; latex (rubber); essential oils; and many other substances that belong to what is known as the isoprene group:

$$CH_3{-}\overset{\overset{\displaystyle CH_2}{\|}}{C}{-}CH_2{-}CH_2{-}O{-}\underset{\underset{\displaystyle O^-}{|}}{\overset{\overset{\displaystyle O}{\|}}{P}}{-}O{-}\underset{\underset{\displaystyle O^-}{|}}{\overset{\overset{\displaystyle O}{\|}}{P}}{-}O^-$$

Isopentenyl pyrophosphate

This important substance does not arise by β_d pyrophosphoryl transfer, but rather by a complex mechanism that includes two γ_d phosphoryl transfers. We are dealing here, therefore, with a pseudo-β_d mechanism. It shares with the authentic mechanism the extra boost provided to the assembly step by pyrophosphatase action and deserves to be considered here.

As a rule, two or more 5-carbon units assemble together, with release of inorganic pyrophosphate, by an iterative transfer mechanism in which the terminal carbon of isopentenyl pyrophosphate is the acceptor and the growing chain is the transferred group. This is shown schematically in the sequence of reactions at the top of the facing page. It is a typical example of "head growth" of a

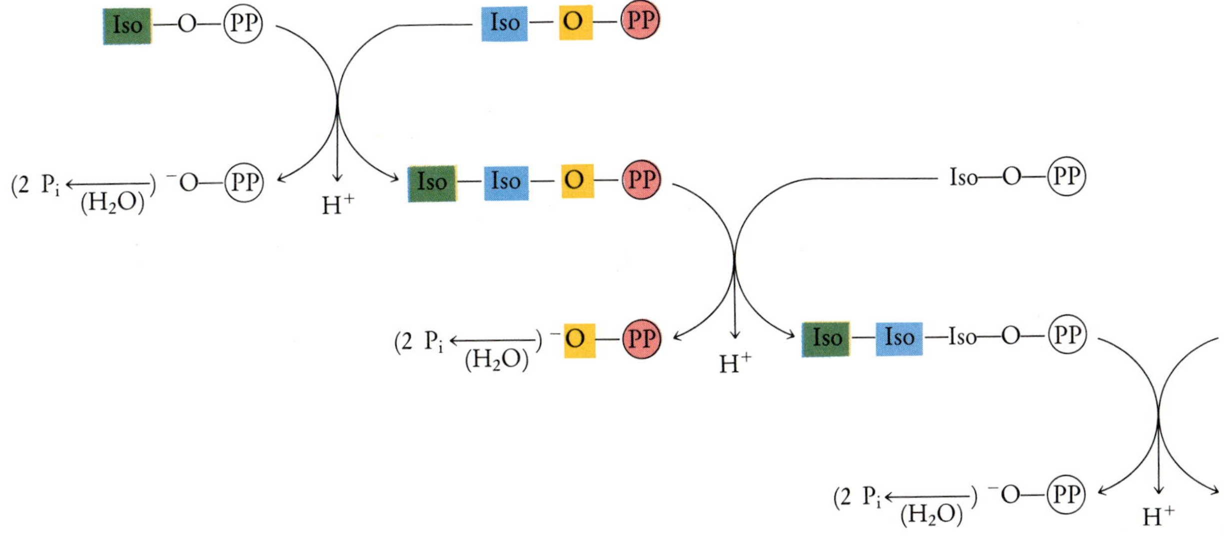

lengthening chain. Fatty acid synthesis (see pp. 145–146) and protein synthesis (see Chapter 15) are others. The characteristic of such mechanisms is that the double-headed product of the activation reaction first acts as acceptor of the part of the chain already completed, before doing its duty as group donor. In tail growth, on the other hand, the double-headed intermediate donates its group immediately to the growing chain. We saw an example of it with polysaccharide synthesis.

Reactions dependent on β_p transfer. This is probably the most widely used biosynthetic mechanism. It takes place according to the following scheme, with NMP—O—X as Janus intermediate:

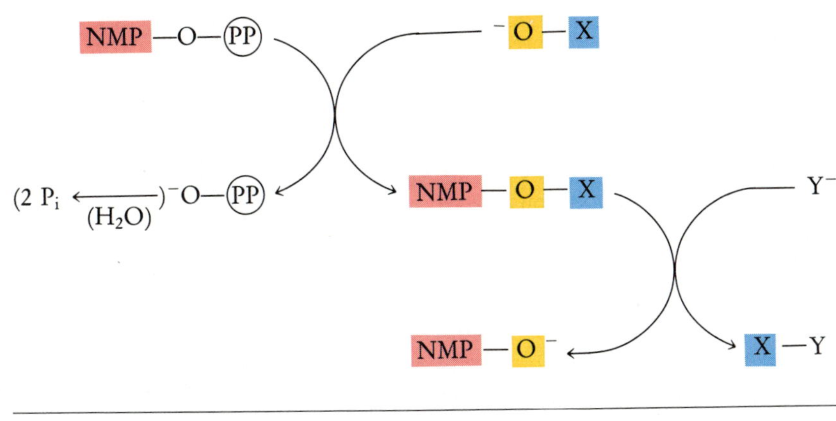

Total: NTP + X—O$^-$ + Y$^-$ ⟶ X—Y + NMP + PP_i (2 P_i)

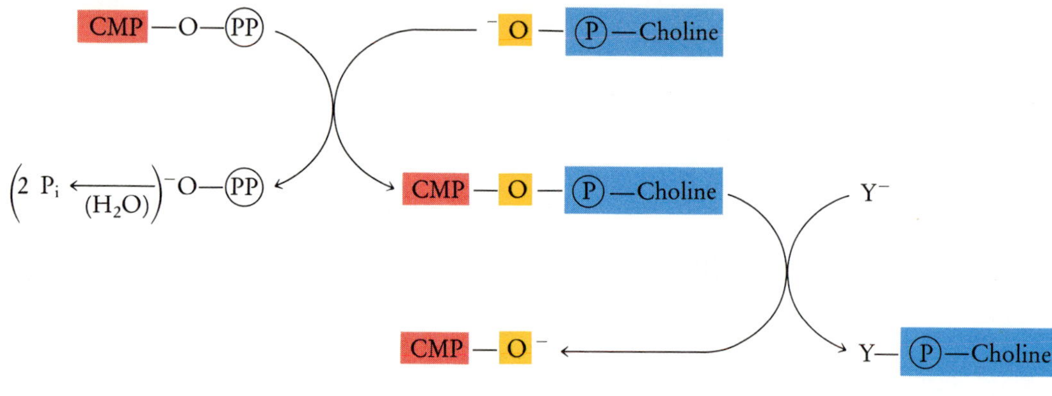

Total: CTP + Phosphorylcholine + Y^- ⟶ Y—Phosphorylcholine + CMP + PP_i (2 P_i)

The advantage of this mechanism is that it has the full strength of the β bond—up to 28 kcal per gram-molecule—available for activation. The resulting Janus intermediates are often unstable molecular arrangements that remain enzyme-bound. Activation and assembly are catalyzed in concerted fashion by a single enzyme. These ligases use ATP as energy donor, as do the ADP-forming ligases that catalyze similar concerted reactions powered by the γ bond (p. 134). An important enzyme of this group is DNA ligase, one of the main agents participating in DNA synthesis (see Chapter 17). Many other biosynthetic processes are carried out by AMP-forming ligases.

A few two-step processes dependent on β_p transfer involve the participation of two distinct enzymes, linked by a stable Janus intermediate in which the activated X-yl group is transported by an NMP carrier. NDP-sugars, it may be remembered, arise in this form, even though they behave somewhat differently upon assembly. More-orthodox Janus intermediates of the same class are generated by β_p transfer between CTP and various phosphorylated building blocks used in phospholipid synthesis, including phosphatidic acids, phosphorylcholine, and phosphorylethanolamine. These compounds have the structure CDP—R and are designated as such (CDP-choline, for example). Unlike the NDP-sugars, however, they do not act as R donors in the subsequent assembly but continue to behave as intermediates with a CMP-yl and an R-phosphoryl face, as shown at the top of this page.

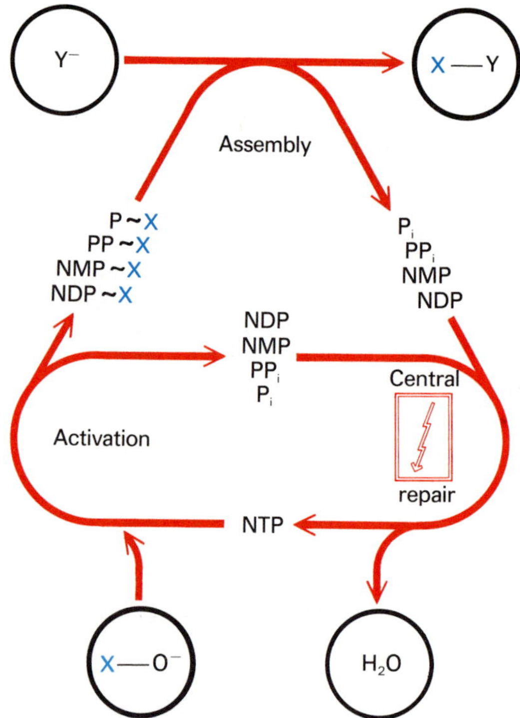

Summary of two-step biosynthetic reactions. An oxygen-containing building block (X-O⁻) is activated by an NTP to a Janus intermediate in which it is linked to a phosphoryl (γ_d), pyrophosphoryl (β_d), nucleotidyl (β_p), or nucleoside-diphosphoryl (pseudo-γ_p) group. After transfer of the activated building block to its final acceptor, the split products of the NTP used are reunited by the central repair machinery.

Three-Step Processes

These processes start systematically by a two-step concerted mechanism catalyzed by a ligase. They differ from the other processes of this kind already considered in the fact that the activated group is attached to a carrier instead of to a biosynthetic building block. Final assembly occurs in a third step by transfer of the group from the carrier to its biosynthetic acceptor.

Reactions dependent on γ_d transfer. Two important group-carrier complexes are assembled by ADP-forming ligases according to the general scheme of γ_d two-step sequential group transfer. One is carboxy-biotin, which is made very much like carbamate, with enzyme-bound carboxyl phosphate as Janus intermediate, and then serves as donor of the activated carboxyl group in a number of carboxylation reactions, as shown in the sequence of reactions below.

Biotin, or vitamin H, which plays a catalytic role in the overall three-step process, is a molecule of historical interest. It celebrates by its name an early discovery in the field of nutrition—that of a yeast growth factor—by Eugène Wildiers, a Belgian biochemist. Wildiers was so impressed by the life-giving power of his factor that he called it Bios. The name biotin was subsequently given to one of the vitamins to commemorate this event. In the performance of its carrier function, biotin is bound covalently to a flexible polypeptide arm included in a multienzyme complex that contains the ligase and the carboxylase. Its transport role is limited to the shuttling of carboxyl groups between the active centers of the two enzymes.

A similar three-step mechanism, but with enzyme-bound formyl phosphate as Janus intermediate, serves to attach formate (H—CO—O$^-$) to a carrier called tetrahydrofolate (THF), a derivative of the vitamin folic acid (Latin *folium*, leaf), one of the secret ingredients of Popeye's spinach, known for its antianemia properties.

$$\text{Total: ATP + Bicarbonate + Y}^- \longrightarrow \text{Y—}\overset{\text{O}}{\overset{\|}{\text{C}}}\text{—O}^- + \text{ADP} + \text{P}_i$$

THF not only carries the activated formyl group to various formyl transferases, but also offers this group to a number of modifying enzymes, which may convert it into a methenyl (—CH═), methylene (—CH_2—), hydroxymethyl (—CH_2OH), methyl (—CH_3), or formimino (—CH═NH) group. Each group participates in a number of transfer reactions. Thus we are dealing with a very versatile coenzyme, which acts both as carrier and as handle for activated groups. Among the many substances that depend on THF for their formation are the amino acid methionine (see pp. 146–147), the purine bases, and the pyrimidine base thymine, a constituent of DNA (see Chapter 15).

Reactions dependent on β_p transfer. Two major biological processes follow this mechanism. Both have a carboxylic acid (R—CO—O^-) as X—O^- building block, use ATP as energy donor, and depend on an AMP-forming ligase to generate a stable, soluble acyl-carrier complex by way of an enzyme-bound acyl-AMP Janus intermediate:

One such process is protein synthesis, in which the amino acids are the X—O^- building blocks, and the corresponding transfer RNAs (tRNAs) are the carriers. The whole of Chapter 15 will be devoted to this matter, and we will not consider it further here, except for pointing out that polypeptide chains grow by head growth (pp. 138–139), which means that, between the intermediary and the final assembly steps in the scheme below, there is intercalated a step in which the carrier—CO—R complex (aminoacyl-tRNA) serves as acceptor of the growing chain.

The other process that follows the three-step β_p mechanism uses a variety of organic acids, among them the fatty acids found in lipids, as X—O^- building blocks and coenzyme A as carrier. This coenzyme owes its name to the fact that it was first discovered as a cofactor of acetylation reactions. It is a derivative of vitamin F, or pantothenic acid, a ubiquitous substance (*pantothen* means everywhere in Greek), which at one time enjoyed a dubious notoriety as a hair restorer because its deficiency

$$\mathrm{AMP{-}O{-}(PP)} + {}^{-}\mathrm{O{-}\overset{O}{\overset{\|}{C}}{-}R} \longrightarrow \left[\mathrm{AMP{-}O{-}\overset{O}{\overset{\|}{C}}{-}R}\right] + {}^{-}\mathrm{O{-}(PP)} \; (\xrightarrow[(H_2O)]{} 2\,P_i)$$

$$\left[\mathrm{AMP{-}O{-}\overset{O}{\overset{\|}{C}}{-}R}\right] + \mathrm{Carrier}^{-} \longrightarrow \mathrm{Carrier{-}\overset{O}{\overset{\|}{C}}{-}R} + \mathrm{AMP{-}O}$$

$$\mathrm{Carrier{-}\overset{O}{\overset{\|}{C}}{-}R} + \mathrm{Y}^{-} \longrightarrow \mathrm{Y{-}\overset{O}{\overset{\|}{C}}{-}R} + \mathrm{Carrier}^{-}$$

$$\text{Total: } \mathrm{ATP + R{-}\overset{O}{\overset{\|}{C}}{-}O^{-} + Y^{-} \longrightarrow Y{-}\overset{O}{\overset{\|}{C}}{-}R + AMP + PP_i \; (2\ P_i)}$$

causes premature graying in rats. Coenzyme A is a complex molecule that contains several other constituents, including a molecule of AMP, in addition to pantothenic acid. The molecule has a thiol group (—SH) as reactive end and is accordingly abbreviated CoA—SH. Acyl-CoA derivatives are thioesters, which, as we have seen, are high-energy compounds, with "physiological" group potentials of the order of −14 kcal per gram-molecule. Their formation, nevertheless, is highly exergonic, as it consumes 28 kcal per gram-molecule.

The main Y—H acceptors of the acyl groups borne by coenzyme A are alcohols:

$$\mathrm{CoA{-}S{-}\overset{\overset{\displaystyle O}{\|}}{C}{-}R + R'{-}OH \longrightarrow R'{-}O{-}\overset{\overset{\displaystyle O}{\|}}{C}{-}R + CoA{-}SH}$$

The resulting esters are low-energy compounds, making the final transfer essentially irreversible. Among the molecules made in this way are the neurotransmitter acetylcholine (acetic acid plus choline, see Chapter 13), the various esters of fatty acids and glycerol found in neutral lipids and phospholipids, and many others. In addition, coenzyme A is involved in numerous other reactions of central importance—as mediator of activated acyl groups, as metabolic handle, and as a participant in some substrate-level oxphos units. (Remember the importance of thioester bonds in the operation of such units, mentioned on pp. 135–136). It is a key piece of the cell's machinery.

A Few Exceptions That Confirm the Rule

Proteins, nucleic acids, neutral lipids, phospholipids, polysaccharides, steroids, terpenoids, nucleotides, coenzymes, amino acids, purines, pyrimidines: the list of substances that we have met in our brief excursion reads like a biochemical *Who's Who*. Obviously, our simple scheme covers a lot of ground. Not surprisingly, it does not cover everything. Some of the exceptions are worth mentioning.

First, some transfer reactions do not depend on a straightforward nucleophilic attack. Among them are transamination and transthiolation, in which an exchange of groups actually takes place. Then there are the various instances where the bond to be formed has a higher free energy of hydrolysis than the NTP bond used up in the biosynthetic process. It is interesting to see where the required energy supplement comes from, or rather how it is delivered. Almost invariably, its source is ATP hydrolysis, as might be expected. As to the means, they may be classified as either "boosting the donor" or "boosting the acceptor."

An example of the former is sulfurylation, a reaction whereby the sulfate ester groups of sulfated mucopolysaccharides and sulfolipids are constructed from inorganic sulfate. The reaction is initiated by a typical β_p attack of sulfate on ATP, giving rise to the double-headed adenylyl sulfate, which is released in freely soluble form. However, the "physiological" free energy of hydrolysis of this anhydride is so high, especially at the very low concentration of inorganic sulfate prevailing in most living cells,

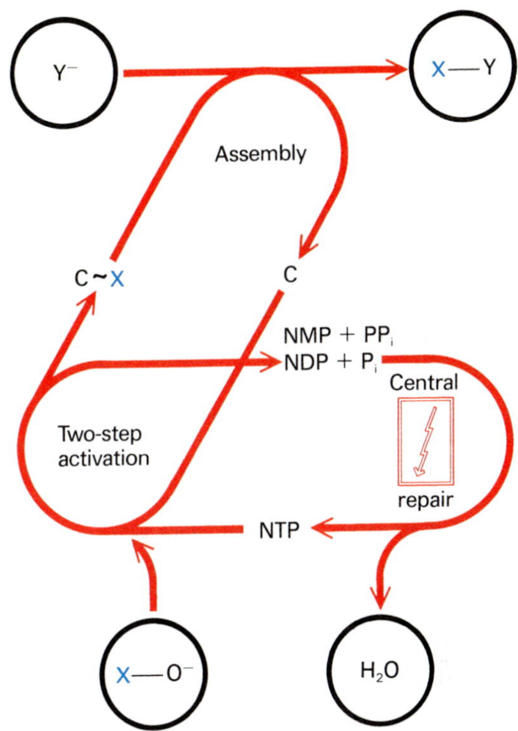

Summary of three-step biosynthetic reactions. Two-step mechanisms (see the illustration on p. 140) serve to attach the activated building block to a carrier C, from which it is delivered to its final acceptor.

that even the full complement of 28 kcal per gram-molecule made available by the splitting of the β bond of ATP does not suffice to lift its concentration to the level needed for an adequate rate of diffusion to distant assembly sites. For this reason, it cannot serve as donor of the sulfuryl group (reaction shown by dashed arrows in the diagram below). Nature's solution of this problem is phosphorylation of the 3′-hydroxyl group of adenylyl sulfate, a highly exergonic reaction capable of raising the concentration of its phosphorylated product at least four orders of magnitude above that of its substrate. The resulting 3′-phospho-adenylyl sulfate can now fulfill the role of transport form and act as donor of sulfuryl groups in a variety of assembly reactions. After it has done its duty, the 3′-phosphate group of AMP is cleaved off. The complete process adds up as follows:

$$2\ \text{ATP} + SO_4^{2-} + Y^- \longrightarrow Y\text{—}SO_3^- + \text{AMP} + \text{ADP} + \text{PP}_i\ (2\ \text{P}_i) + \text{P}_i$$

The total cost amounts to 42 kcal per gram-molecule.

Examples of acceptor activation are seen in the synthesis of fatty acids, in that of the porphyrin ring, and in a number of amination reactions. In each case, the acceptor in the final assembly reaction is fitted with a carboxyl

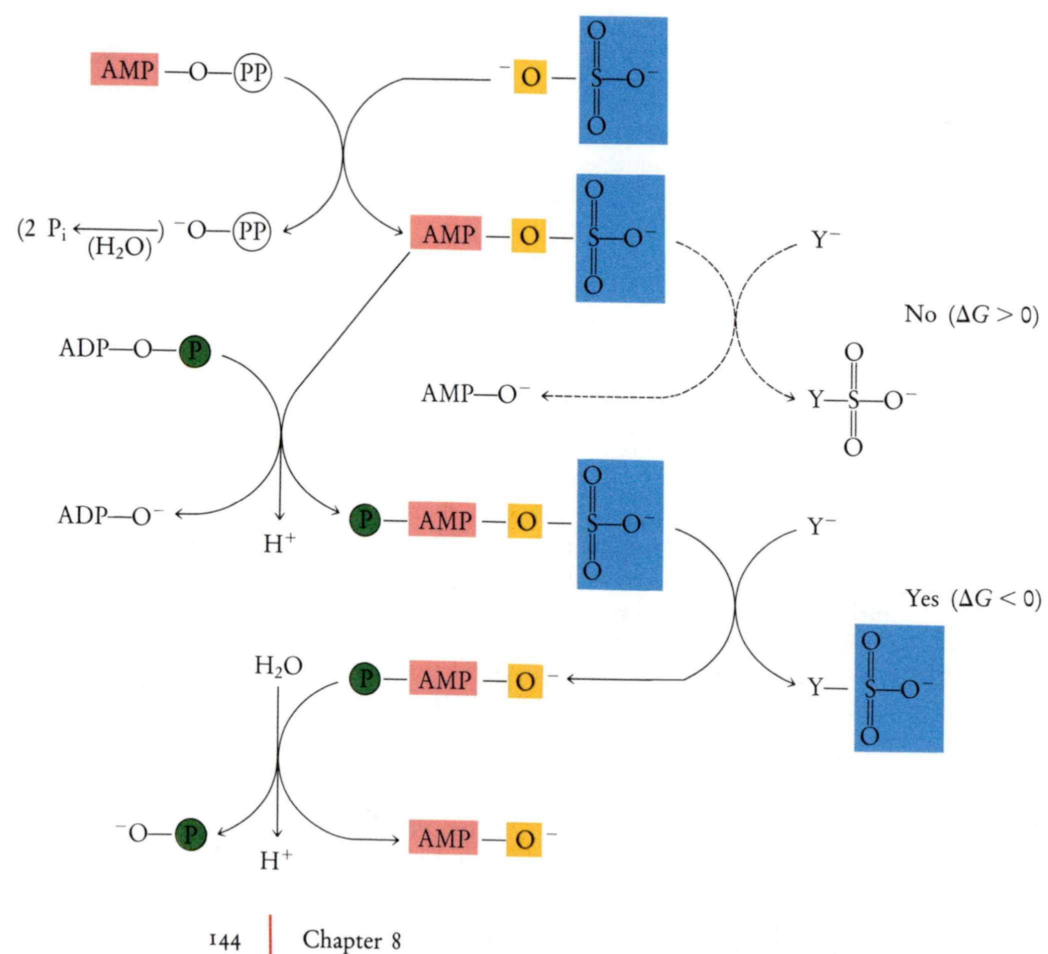

(—CO—O⁻) or acyl (—CO—R) group that falls off upon assembly, thus adding the free energy of decarboxylation or deacylation to the potential of the donated group for supporting the cost of making the new bond, as shown at the right.

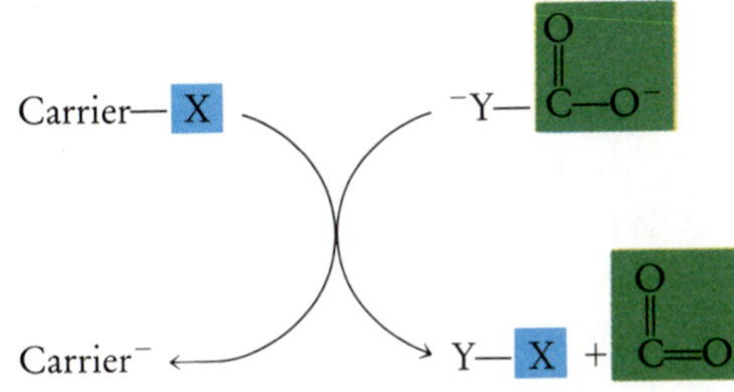

or:

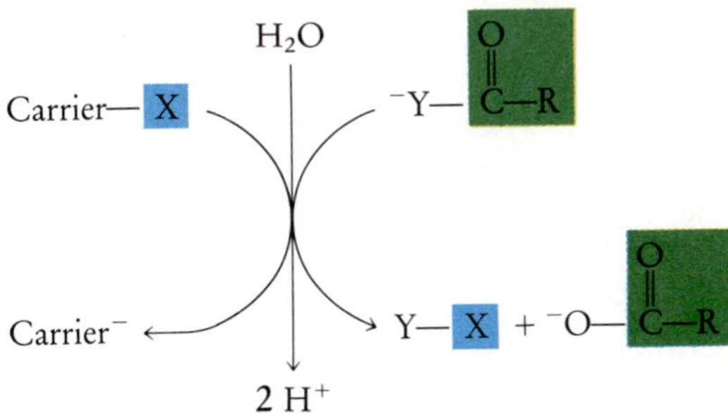

In fatty acid synthesis, the donor is a growing acyl chain borne by a protein carrier (acyl-carrier protein, ACP), which includes a piece of coenzyme A as active group. The acceptor is malonyl-ACP, derived from malonyl-CoA, which is acetyl-CoA activated by carboxylation (by a biotin-dependent γ_d mechanism of the kind described on p. 141), as shown below:

(Acetyl-CoA) CoA—S—C(=O)—CH_3 + HO—C(=O)—O⁻

ATP → ADP + P_i (biotin)

(Malonyl-CoA) CoA—S—C(=O)—CH_2—C(=O)—O⁻

ACP—SH → CoA—SH

ACP—S—C(=O)—CH_2----CH_3 + ACP—S—C(=O)—CH_2—C(=O)—O⁻ (H^+)

→ ACP—SH + ACP—S—C(=O)—CH_2—C(=O)—CH_2----CH_3 + C=O

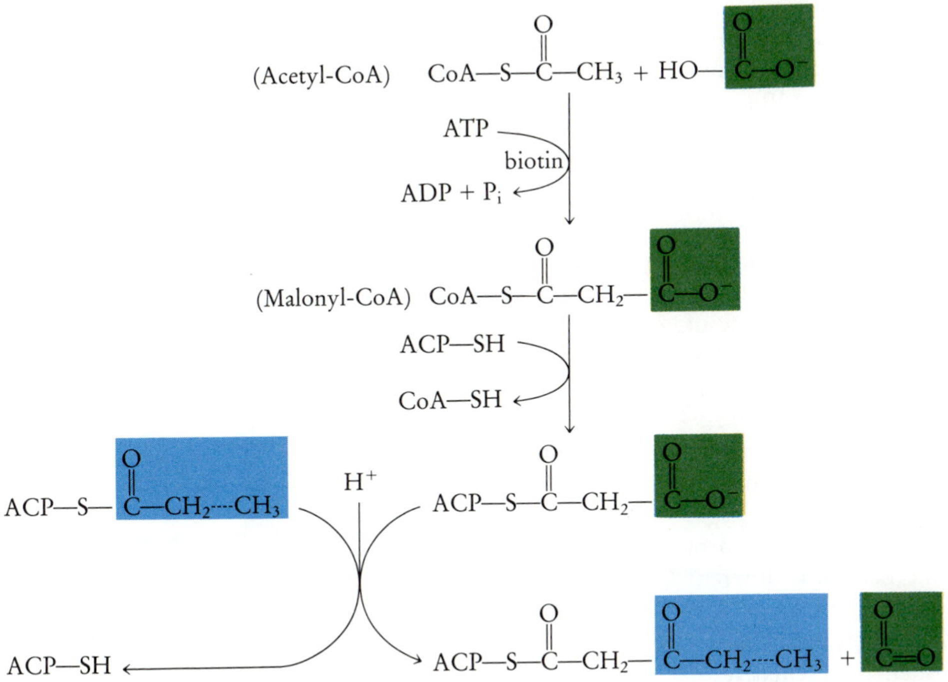

Note that the final product of the assembly reaction (a β-ketoacyl-ACP two carbons longer than the acyl-ACP donor) is exactly what would be obtained by a direct transfer of the acyl group onto acetyl-ACP. However, such a reaction cannot normally occur, because the potential of the transferred acyl group is distinctly higher in the β-ketoacyl product than in the thioester donor. This reaction does take place in living cells (with coenzyme A as carrier), but in the reverse direction (thiolysis of the β-ketoacyl-CoA derivative by coenzyme A), as part of fatty acid degradation. By carboxylating the acetyl-ACP acceptor, at the expense of an additional 14 kcal per gram-molecule, the cell provides it with about 6 to 8 kcal per gram-molecule extra energy, enough to drive the reaction in the direction of assembly. It will be noted further that the reaction consists in the addition of a growing chain to a two-carbon building block. After reduction of the β-keto group, this process will be repeated, alternating with reductive steps until the chain is completed. It is another example of head growth (pp. 138–139).

There is an analogous step in the construction of the porphyrin ring, which enters into the composition of such important molecules as hemoglobin, cytochromes (see Chapter 9), and chlorophylls (see Chapter 10). Here the donor is succinyl-CoA, and the acceptor is the amino acid glycine, which, in this reaction, behaves as methylamine ($CH_3—NH_3^+$) activated by carboxylation, as shown at the top of the facing page. The product is δ-aminolevulinic acid, a precursor of the porphyrin ring.

In amination reactions, which, with rare exceptions, are rendered thermodynamically unfavorable because of the low concentration of ammonia in cells, ammonia is replaced as acceptor by glutamine, the amide of glutamic acid, which is assembled by a typical γ_d two-step process, as was seen earlier (p. 134). In those amination reactions, glutamine has the character of a molecule of ammonia activated by acylation with glutamic acid (see facing page, middle). In some amination reactions, the activated form of ammonia is aspartic acid. The mechanism is different from that considered here.

To conclude this brief tour of biosynthetic oddities, we will take a look at the most important representative of the very rare processes that depend on an α_p attack on ATP (transadenosylation). The attacking agent is the amino acid methionine, which is characterized by a thiomethyl group ($—S—CH_3$). Exceptionally, in this case, the attack is made by a sulfur atom, not by an oxygen atom. In addition, the subsequent assembly reaction does not involve transfer of the whole activated molecule but only of its terminal methyl group, which is rendered easily transferable as a result of the positive charge acquired by the sulfur atom upon adenosylation. Inorganic triphosphate, the other product of the α_p attack, does not appear as such but is hydrolyzed to inorganic phosphate and pyrophosphate, which is itself split by pyrophosphatase action (see facing page, bottom).

This process is responsible for many important methylation reactions, including those affecting nucleic acids (see Chapters 16–18). It is, you will notice, a very expensive business: all of 35 kcal per gram-molecule. And it does not even start from scratch, with methanol (CH_3OH) as building block, but uses methionine as source of the methyl group. Methionine itself can be reconstituted from homocysteine, with, in this case, methyltetrahydrofolate as methyl donor (pp. 141–142):

$$\text{Homocysteine} + \text{THF-CH}_3 \longrightarrow \text{Methionine} + \text{THF}$$

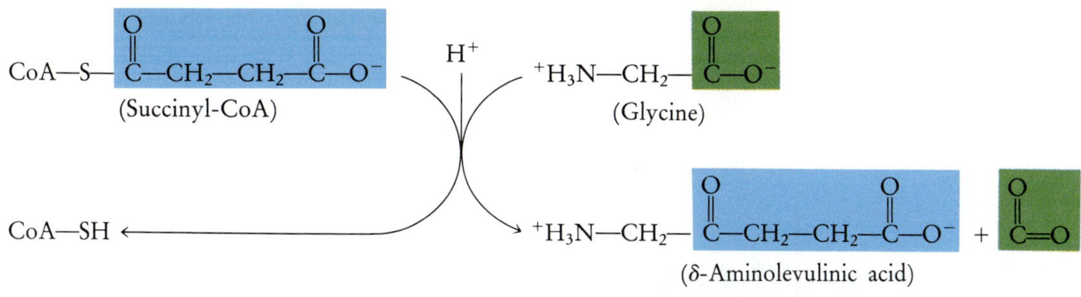

NH_3 + ^-O—C—CH_2—CH_2—CH(NH_3^+)—COO^-

(Glutamic acid)

ATP

ADP + P_i

Carrier—X

H_2O

NH_2—C—CH_2—CH_2—CH(NH_3^+)—COO^-

(Glutamine)

Carrier

2 H^+

NH_2—X + ^-O—C—CH_2—CH_2—CH(NH_3^+)—COO^-

(Glutamic acid)

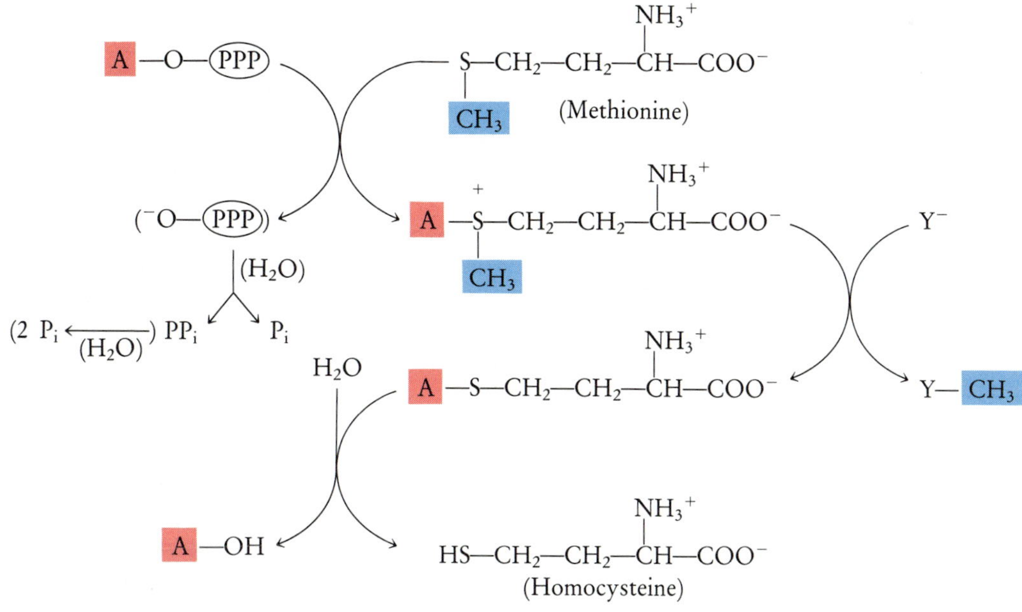

9 Mitochondria: Respiration and Aerobic Energy Retrieval

Air is so vital to us and to most living beings that we are hard put to imagine life without it. Yet, as was mentioned in Chapter 7, life almost certainly began, and thrived for a long time, in an oxygenless world. Only after the emergence of the advanced photosynthetic mechanism known as photosystem II, about three billion years ago (see Chapter 10), did oxygen start appearing in the atmosphere in significant quantity, slowly rising to its present level of 20 per cent.

This phenomenon must have posed a major threat to life and brought about the extinction of numerous species comparable to today's obligatory anaerobes. The reason for this lies in the ability of molecular oxygen to interact in a variety of ways with reducing biological molecules, giving rise to such highly toxic products as the superoxide ion, O_2^-, and hydrogen peroxide, H_2O_2. The survivors, except for those that found refuge in an oxygen-free niche, were those that developed protective enzymes, such as superoxide dismutase and catalase (see Chapter 11). Some went further and actually succeeded in taming the enemy and turning it into life's greatest ally, thanks to an adaptation of their ATP-yielding oxphos units to the use of oxygen as final electron acceptor. Their progeny now fill most of the living world.

The Taming of Oxygen

Judging from known aerobic bacteria, adaptation to oxygen must have been a gradual process, which eventually culminated in one of Nature's greatest achievements, the phosphorylating respiratory chain: a string of fifteen or more electron carriers spanning the whole 1,070-mV difference between NADH and oxygen, and arranged in such a manner as to include up to three successive oxphos units in series. A system of this sort is found in the plasma membrane of a number of contemporary bacteria, which presumably have inherited it from those remote ancestors that first acquired it well over one billion years ago. Essentially the same system is present in the inner of the two membranes that surround mitochondria, those discrete, membrane-wrapped bodies about the size of bacteria that are found scattered in large numbers throughout the cytoplasm of the vast majority of eukaryotic cells in both plants and animals, where they serve as the main centers of respiration and of oxidative energy retrieval. The link between the bacterial and the mitochondrial systems—assuming there is one, which is highly probable—makes fascinating speculation.

In the most popular version of the story, the hero is the primitive phagocyte, that hypothetical, giant, voracious, bacteria-gobbling cell that is postulated to be an intermediate between prokaryotes and eukaryotes (Chapter 6). Among its daily catch—so the story goes—were some aerobic bacteria that failed to be killed and broken down for food. They did not kill their captor either, as do many pathogenic bacteria that escape destruction. Rather, they established a permanent, mutually advantageous, symbiotic relationship with it. Their descendants have survived unto this day as the mitochondria of eukaryotic cells. Fully integrated with their host cell, as might be expected after more than a billion years of living together, these organelles have nevertheless retained the remnants of a genetic system of typical bacterial kind, together with some other vestigial properties that attest to their ancestry. As we shall see in the next chapter, the chloroplasts of plant cells may have arisen in the same way from symbiotically adopted photosynthetic bacteria.

Known as the endosymbiont hypothesis, this theory has much to commend it, including some of the phylogenetic trees deduced from molecular sequencing (see Chapter 18). Scientists, however, are a disbelieving lot, especially when it comes to reconstructing events of the

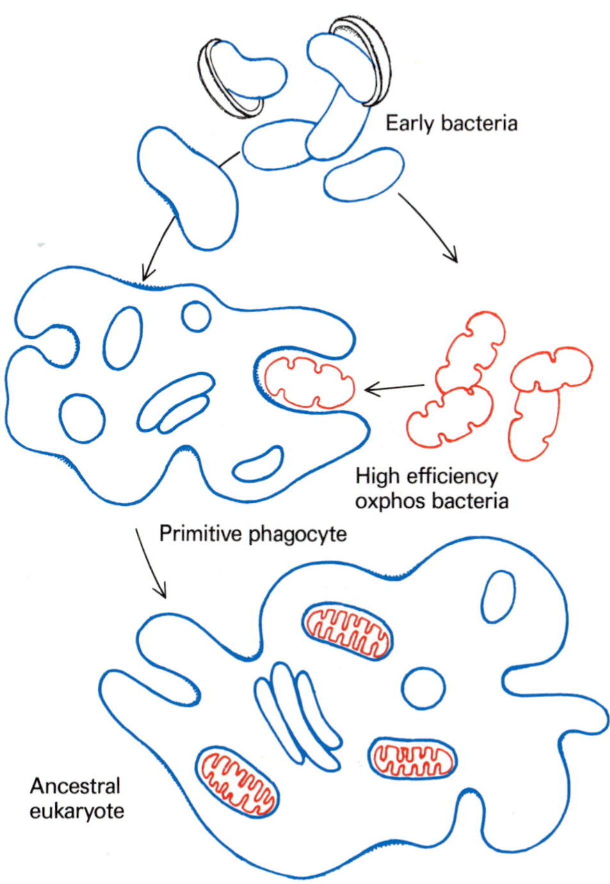

Endosymbiont hypothesis of origin of mitochondria. A primitive phagocyte (Chapter 6) engulfs and adopts symbiotically aerobic bacteria having a highly elaborate system of oxidative phosphorylation in their plasma membrane. These bacteria have developed from photosynthetic bacteria (see Chapter 10) in response to the progressive appearance of oxygen in the atmosphere. After their endosymbiotic adoption, they evolved into the mitochondria of a primitive eukaryote ancestral to all animal and plant cells.

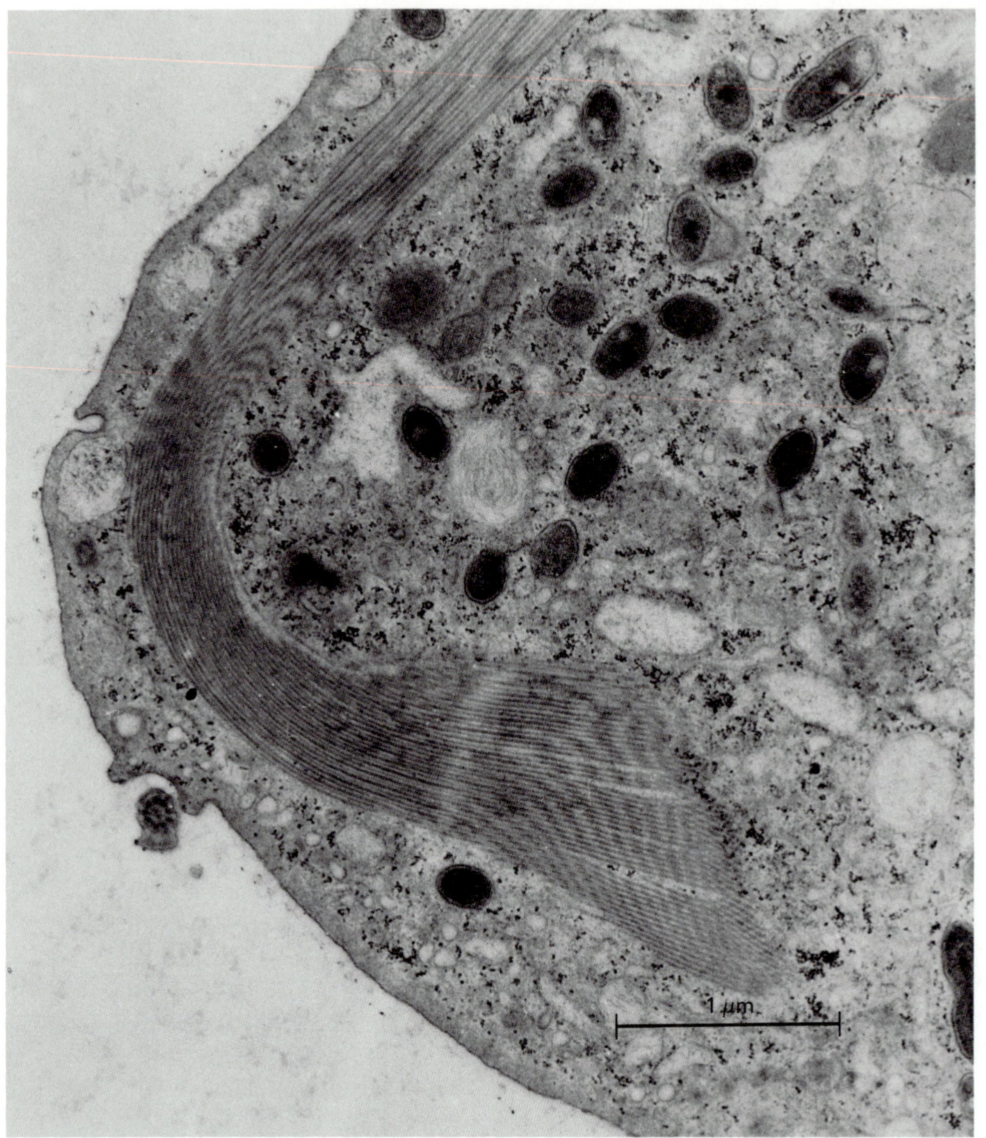

This electron micrograph shows bacterial endosymbionts living in the cytoplasm of the protozoon *Pyrsonympha*, which itself lives as a symbiont in the hindgut of the termite *Reticulitermes*.

distant past. Mitochondria, some counter, could have originated just as well from infoldings of the plasma membrane of an enlarging aerobic bacterium, much as the vacuome is supposed to have done (Chapter 6), but with a different kind of differentiation leading to segregation of the phosphorylating respiratory chain in the resulting vesicles. A number of other possibilities can be evoked, including that of an independent origin of the bacterial and mitochondrial systems through a phenomenon of convergent evolution.

It does not befit a mere guide to try to settle an argument that is still dividing experts. However, some of the features of mitochondria (and of chloroplasts) that support their origin from bacterial endosymbionts will be pointed out for the benefit of those who are seduced by the romantic appeal of the endosymbiont hypothesis.

Model reconstructed from serial sections showing a single giant mitochondrion of a yeast cell.

Reclining Figure, by Henry Moore.

Some Anatomical Details

The term mitochondrion is derived from the Greek *mitos,* thread, and *khondros,* grain. It refers to the filamentous appearance of the mitochondria in certain cells. In other cells, however, the mitochondria may be rodlike, egg shaped, or almost spherical. Sometimes several are joined together in bizarre concatenations, which, it has been claimed, may include the entire mitochondrial population of a cell, converting it into a strange, convoluted sort of open structure that could have been designed by the sculptor Henry Moore. The sizes of mitochondria are variable but tend to be large, as cellular dimensions go: 1 μm or more. Multiply this figure by one million, and you approach our own size. With a little luck, especially if we choose our target judiciously—the

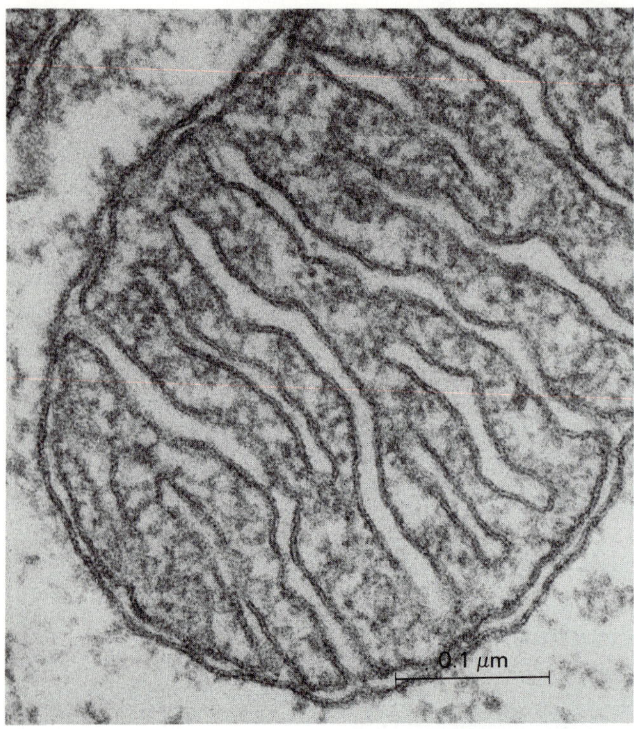

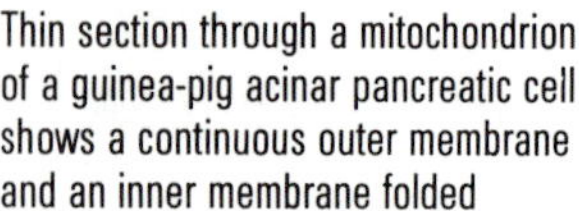
Thin section through a mitochondrion of a guinea-pig acinar pancreatic cell shows a continuous outer membrane and an inner membrane folded inward into numerous incomplete ridgelike partitions, the cristae mitochondriales.

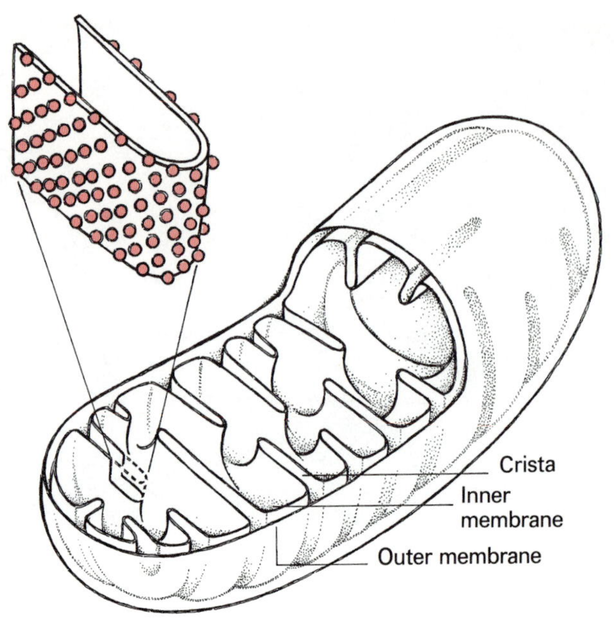

Three-dimensional reconstruction of the anatomy of a mitochondrion illustrates disposition of cristae. Enlargement shows that the inner face of the inner membrane bears small knobs (see the electron micrograph on p. 157).

bluebottle fly has particularly roomy mitochondria in its wing muscles—we may just be able to fit inside. There is no question, however, of entering without damage. We will have to resort to vivisection.

Seen from the outside, mitochondria appear encased in a thin, smooth, semitransparent envelope. Through this outer membrane can be discerned an inner one, bright pink and ravined by numerous deep clefts. Viewed from the inside, the structures corresponding to these clefts appear as ridges, or cristae (*crista* = ridge, in Latin), which form a number of incomplete partitions in the interior of the mitochondrial body. Actually, the cristae are no more than infoldings of the inner membrane, which is continuous and entirely closed. We can verify this by ripping the outer membrane and allowing the mitochondrion to take up water by applying some osmotic imbalance. It will swell to a considerable size without bursting, unfolding its inner membrane. The formation of cristae—replaced in certain cell types by tubular infoldings—represents a device for increasing the surface area of the membrane without enlarging the volume of the body. The usefulness of such an increase will become obvious, once we find out that the inner membrane is the very heart of the cellular power plant, its actual energy generator; it contains the whole respiratory chain and its associated phosphorylating systems.

From the evolutionary point of view, this inner membrane is believed to be derived from the plasma membrane of the putative ancestral bacterial endosymbiont. The outer membrane, on the other hand, presumably originates from the vacuolar system of the ancestral host-phagocyte. Indeed, it shows some kinship with the endoplasmic reticulum, with which it shares, for instance, the possession of a special pigment known as cytochrome b_5.

Energetics of Mitochondrial Oxidations

The cavity delimited by the ridged, inner mitochondrial membrane is filled with a structureless fluid, the mitochondrial matrix. It is a protein-rich sap made up largely of catabolic enzymes involved in the oxidative breakdown of all major foodstuffs, including the amino acids, which compose the proteins; the fatty acids, which are the main constituents of lipids; and pyruvic acid, which, we have seen, is formed in the cytosol as the product of aerobic glycolysis (p. 114). All these pathways join in a central metabolic vortex known as the Krebs cycle, named for Sir Hans Krebs, the British-German biochemist who discovered it in the late 1930s.

A motley collection of exotic molecules participate in these transformations, and we will make no attempt to identify them, noting only the end-products, which are very simple substances: water, carbon dioxide, ammonia or urea, inorganic sulfate—the same substances, or almost, as would be formed by combustion in a furnace. An astonishingly cool furnace, though, in which matter is burned, but little heat is produced.

The paradox, you may remember from Chapter 7, is readily explained: foodstuffs do not combine with oxygen; they combine with water and give off hydrogen atoms or electrons to appropriate acceptors. These transactions are conducted at cell temperature, and in such a way as to release little energy. On the rare occasions when they do release an appreciable amount of energy, they are directed to take place through a substrate-level oxphos unit such as that encountered in glycolysis.

Take glucose, for example. Altogether, twelve pairs of electrons are released for every molecule of this sugar oxidized in a living cell:

$$C_6H_{12}O_6 + 6\ H_2O \longrightarrow 6\ CO_2 + 24\ e^- + 24\ H^+$$

Two of these pairs arise through oxidative glycolysis (amputated snake, Chapter 7):

$$C_6H_{12}O_6 \longrightarrow 2\ CH_3\text{—}CO\text{—}COOH + 4\ e^- + 4\ H^+$$

The other ten are yielded by the further oxidation, through the Krebs cycle, of the pyruvic acid molecules produced by glycolysis:

$$2\ CH_3\text{—}CO\text{—}COOH + 6\ H_2O \longrightarrow 6\ CO_2 + 20\ e^- + 20\ H^+$$

In glycolysis, as we have seen, the electrons are transferred to NAD^+ across a substrate-level oxphos unit:

$$4\ e^- + 2\ H^+ + 2\ NAD^+ + 2\ ADP + 2\ P_i \longrightarrow 2\ NADH + 2\ ATP + 2\ H_2O$$

In the Krebs cycle, four of the five electron pairs released by the oxidation of one molecule of pyruvic acid are transferred to NAD^+, in one case through a substrate-level oxphos unit. The fifth pair is transferred to a flavin coenzyme called FAD (see below). Thus, for two molecules of pyruvic acid:

$$20\ e^- + 12\ H^+ + 8\ NAD^+ + 2\ FAD + 2\ ADP + 2\ P_i \longrightarrow 8\ NADH + 2\ FADH_2 + 2\ ATP + 2\ H_2O$$

Adding up the two processes we find:

$$24\ e^- + 14\ H^+ + 10\ NAD^+ + 2\ FAD + 4\ ADP + 4\ P_i \longrightarrow 10\ NADH + 2\ FADH_2 + 4\ ATP + 4\ H_2O$$

which gives, for the anaerobic oxidation of glucose, as it actually occurs in living cells with the help of the immediately involved electron acceptors:

$$C_6H_{12}O_6 + 2\ H_2O + 10\ NAD^+ + 2\ FAD + 4\ ADP + 4\ P_i \longrightarrow 6\ CO_2 + 10\ NADH + 10\ H^+ + 2\ FADH_2 + 4\ ATP$$

The aerobic part of the process takes place through the transfer of the electrons from the reduced coenzymes to oxygen:

$$10\ NADH + 10\ H^+ + 2\ FADH_2 + 6\ O_2 \longrightarrow 10\ NAD^+ + 2\ FAD + 12\ H_2O$$

Foodstuff	Free energy of oxidation (kcal per gram-molecule)				
		Recovered in			
	Total	NADH	$FADH_2$	ATP*	Heat
Glucose	686	490	74	56	66
Palmitic acid	2338	1519	555	84	180
Glutamic acid	478	343	74	28	33

*ATP made by substrate-level phosphorylation minus ATP used for substrate activation.

The energy balances of the anaerobic and aerobic parts are easily computed. We have already seen (Chapter 7) that the "physiological" free energy of oxidation (electron potential) of NADH is −49 kcal per pair of electron-equivalents transferred to oxygen. Considering that the "physiological" electron potential of $FADH_2$ is of the order of −37 kcal per pair of electron-equivalents transferred to oxygen, we find for the aerobic part a total of −564 ($10 \times 49 + 2 \times 37$) kcal per gram-molecule of glucose oxidized.

On the other hand, we know from calorimetric measurements that the "physiological" free energy of oxidation of glucose, with oxygen as electron acceptor, equals −686 kcal per gram-molecule. This leaves an anaerobic balance of −122 (686 − 564) kcal per gram-molecule, of which 56 (4×14) are retrieved through the operation of the substrate-level oxphos units and 66 are lost as heat.

In other words (see the table above), when living cells "burn" glucose with their own electron snatchers instead of oxygen, they conserve more than 90 per cent of the free energy that would be released with oxygen as electron acceptor. Most of this energy is stored in reduced coenzymes, and a small part in ATP. What this means, in terms of the hydrodynamic analogy offered at the end of Chapter 7, is that the electrons released by the biological oxidation of glucose are first transferred to energy-rich reservoirs situated high above the water/oxygen level. In eight out of twelve cases, these transfers take place across small differences in altitude, with little loss of energy. In the four others, the difference is substantial, but the electron fall is harnessed to a substrate-level ATP-yielding oxphos unit.

Fatty acids and amino acids suffer an essentially similar kind of cold combustion, as illustrated by the examples of palmitic acid and glutamic acid shown in the table. Their complete oxidation occurs with more than 90 per cent of the free energy of combustion stored in energy-rich cofactors.

Knowing these values, we can better appreciate the plight of our early anaerobic ancestors. Since NAD, FAD, and the other electron-carrying coenzymes are present in cells in catalytic amounts, the advantages of cold combustion can be enjoyed only if an outlet exists for the electrons stored in these coenzymes. Fermenters

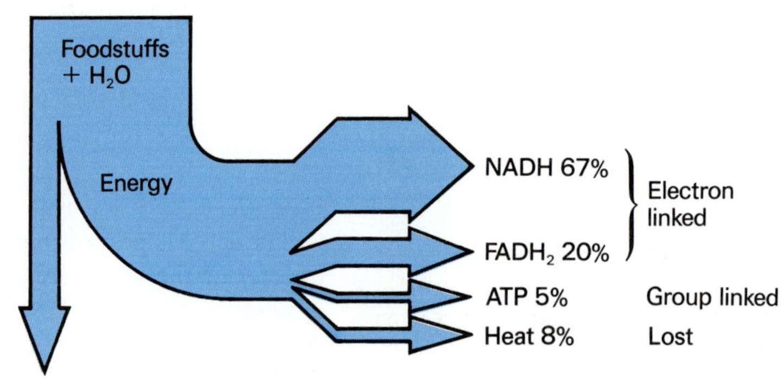

"Cold combustion," with water as oxygen supplier, brings about the complete oxidation of foodstuffs. Less than 10 per cent of the free energy released is dissipated as heat. Most is stored as high-potential electrons in NADH and, to a lesser extent, in reduced flavoproteins. A small fraction supports substrate-level oxphos units making ATP.

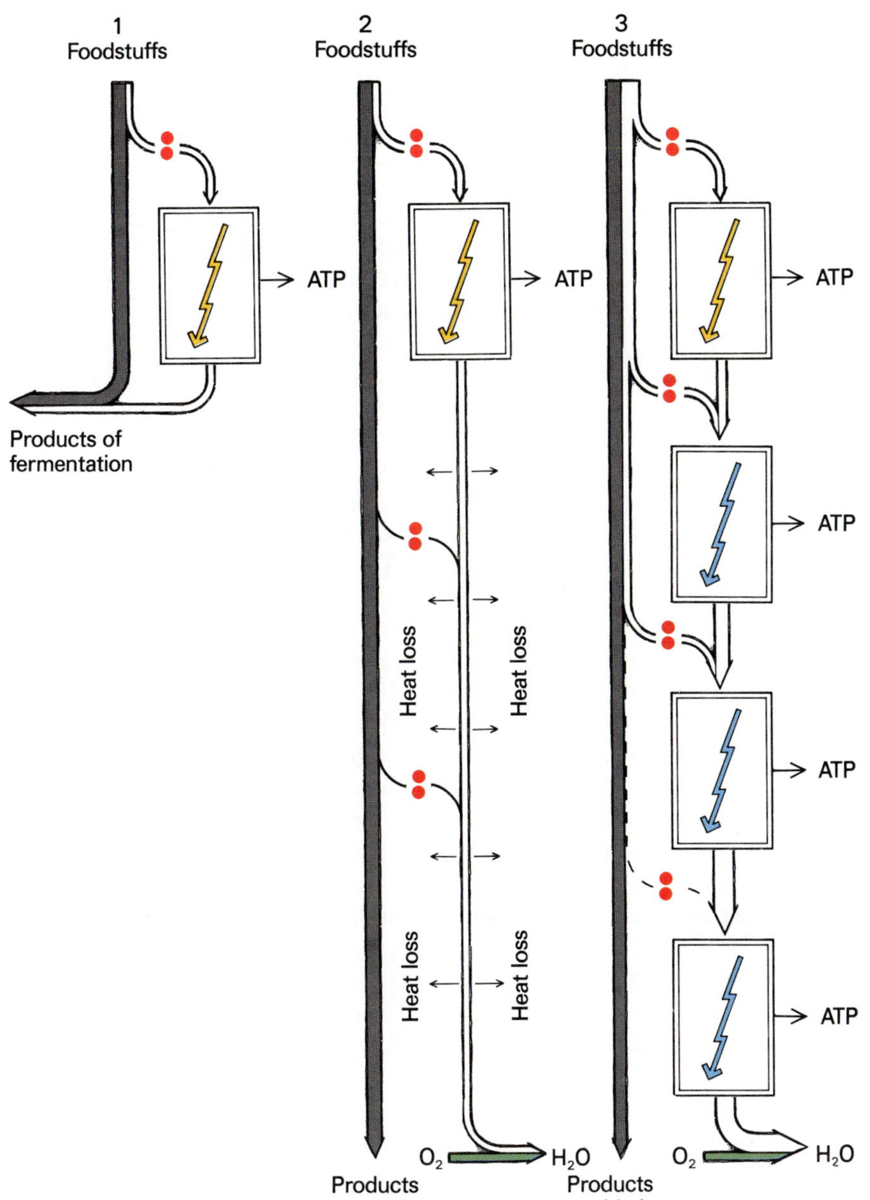

The problem of electron disposal.

Solution 1: Electrons return to metabolic flow and are excreted together with products of anaerobic fermentation. The drawbacks are poor utilization of foodstuffs and strict stoichiometric constraints.

Solution 2: Electrons are transferred freely to oxygen (or other exogenous acceptor). The advantages are greater flexibility and complete oxidation of foodstuffs, but the yield of utilizable energy (ATP) is very low.

Solution 3: Electrons are transferred to oxygen through phosphorylating respiratory chain. Foodstuffs are utilized completely, with a high energy yield.

satisfy this requirement by returning the electrons to the substrate flow. This procedure works, but it has a very low energy yield. For instance, anaerobic glycolysis, as we have seen, releases only 47 kcal per gram-molecule of glucose utilized, or less than 7 per cent of the energy that would be available by complete oxidation. It provides the cell with only two ATPs, or 28 kcal of utilizable energy.

Another drawback of fermentation is that it imposes rigid stoichiometric constraints on metabolic pathways. Only as many electrons may be released from the substrates as can be collected back into the substrate flow. These constraints are lifted if the electrons stored in coenzymes can be unloaded onto an exogenous acceptor. Then any kind of oxidizable foodstuff can be used, and used completely. The result is greater flexibility and often a better energy yield, at least from the cell's point of view.

From the point of view of world resources, however, it is shockingly wasteful if the electrons are simply unloaded without energy retrieval.

Consider again the example of glucose. As we have seen, substrate-level phosphorylation provides the anaerobic fermenter with two ATPs, or 28 kcal, for every gram-molecule of glucose utilized. The table shows that this yield is doubled in complete oxidation. But at what cost to total resources! Whereas, in fermentation, only 47 kcal are released out of the 686 that are available and the remainder is returned to the environment in the form of energy-rich molecules of lactic acid or ethanol, in unharnessed oxidation, the whole 686 kcal would be dissipated for a paltry additional gain of 28 kcal per gram-molecule of glucose used. Tolerable in a world of plenty, such profligacy becomes prohibitive when food is scarce. Yet, it is possible that some primitive aerobes operated in this way, and a number of such wasteful reactions still occur in the most advanced eukaryotes (Chapter 11).

Obviously, the only way to minimize oxidative energy waste is by making use of the energy that is released when the electrons are transferred from the reduced coenzymes to the final acceptor. This requires the successful insertion of oxphos units into the pathway of the electrons. Many different devices of this sort were developed in the course of evolution and harnessed to a variety of electron acceptors. The best results were obtained with oxygen as final acceptor, especially once the phosphorylating respiratory chain was perfected to its ultimate form, as is found today in mitochondria and some aerobic bacteria. Cells equipped with these super energy-savers can make as many as 38 ATPs with a single molecule of glucose, retrieving in usable form almost 80 per cent of the free energy released by the oxidative process.

The Phosphorylating Respiratory Chain

Physically, the mitochondrial energy transducers are situated entirely within the pink, pleated sheath that makes up the inner membrane of the body. It is a thin casing, 7 nm thick, hardly more than one-quarter of an inch at our millionfold magnification, smooth on the outside, but covered with small knobs on the inside. These knobs are about 9 nm in diameter (one-third of an inch if enlarged a millionfold) and are attached to the membrane by short, narrow stems, making the inner face of the membrane look as though it is covered with small mushrooms.

For further inspection, we will need our molecular magnifying glass. The sight it discloses is truly amazing, one of the greatest masterpieces of molecular engineering ever assembled. The whole expanse of the membrane is covered with tiny microcircuits, each made up of some fifteen to twenty different species of electron carriers beautifully put together to ensure efficient electron transfer. As many as 100,000 such microcircuits may be present on a single mitochondrial body.

Usually referred to as the respiratory chain, these microcircuits include a variety of different molecules (see Appendix 1). Prominent among them are the flavins and the hemes. The flavins (Latin *flavus,* yellow) are a remarkable class of greenish-yellow pigments derived from riboflavin, or vitamin B_2. The most important flavin derivatives are flavin mononucleotide (FMN) and flavin adenine dinucleotide (FAD). Their key property lies in their ability to act as hydrogen carriers. As such, they serve as coenzymes for numerous flavoprotein dehydrogenases.

Hemes (Greek *haima,* blood) are close relatives of the chlorophylls, the green photosynthetic pigments with which they share the characteristic porphyrin nucleus—a planar, disklike molecule made of four pyrrole rings linked by methene bridges (—CH═). In the center of this disk is a hole ringed by the electrons of four resonating nitrogen atoms. Insert a magnesium ion into this hole and you have the makings of a chlorophyll molecule. Substitute iron for magnesium and you have a heme. At the same time, thanks to the ability of iron to assume either the divalent ferrous (Fe^{2+}) or the trivalent ferric (Fe^{3+}) state, you have an electron carrier:

$$\text{Heme—Fe}^{2+} \rightleftharpoons \text{Heme—Fe}^{3+} + e^-$$

In combination with proteins, hemes make up a whole

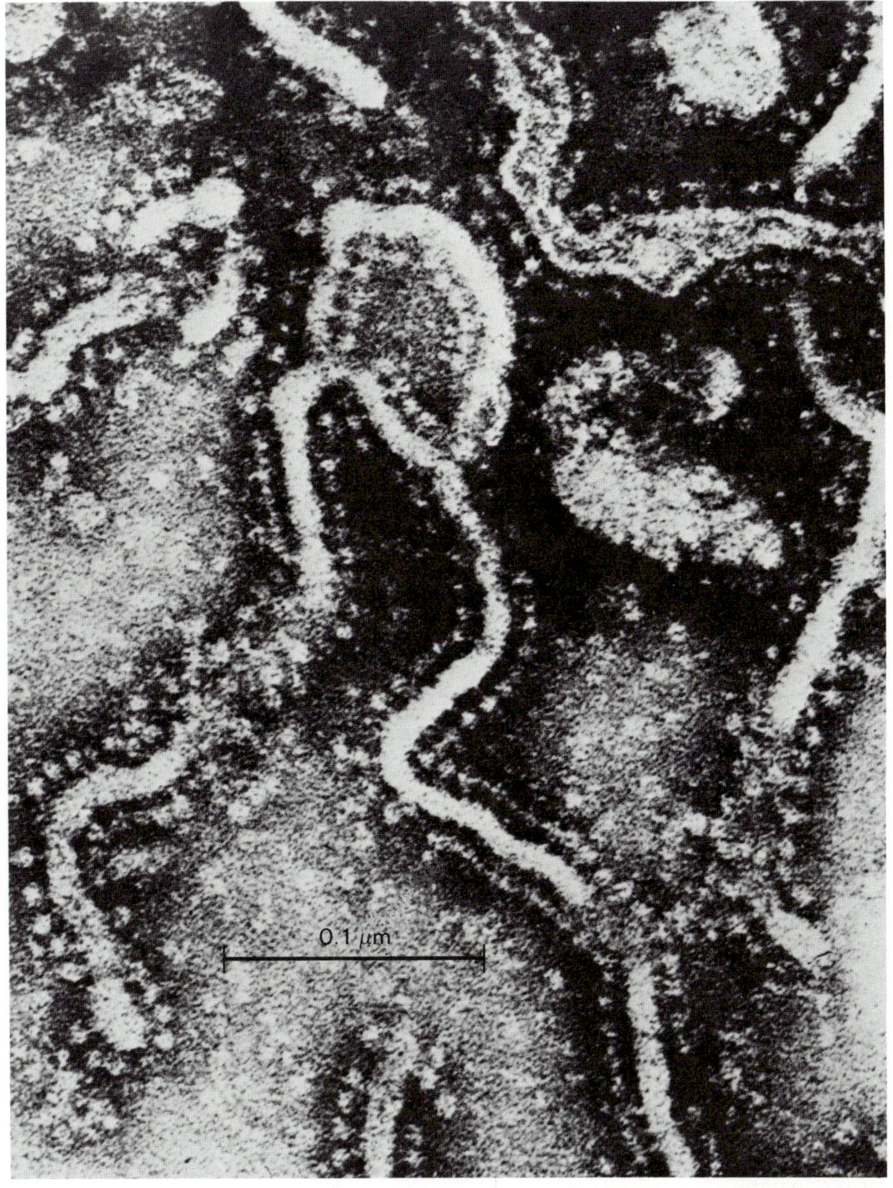

High-magnification electron micrograph of mitochondrial cristae. The preparation was negatively stained with electron-dense potassium phosphotungstate, which clearly delineates the knobs and their stalks protruding into the mitochondrial matrix. The knobs contain ATP-driven proton pumps.

panoply of colored molecules, with shades ranging from blood-red to pea-green. They serve in a variety of ways in the utilization of oxygen. Their best-known function is oxygen transport, as carried out by the red blood pigment hemoglobin. In the respiratory chain, hemoproteins are represented by a number of cytochromes, so named because they were among the earliest pigments discovered as regular cell components.

In addition to these two main classes of electron carriers, the respiratory chain also contains diphenols, which are oxidized to the corresponding quinones; iron-sulfur proteins, in which electron-carrying iron ions are encased within a shell of sulfhydryl groups; protein-bound copper ions; and possibly other components. It is noteworthy, and relevant to subsequent discussions, that some of these molecules act strictly as electron carriers (the hemoproteins and the metalloproteins), whereas the others transport hydrogen atoms (i.e., electrons in combination with protons).

Within each respiratory microcircuit, these various molecules are organized in a manner that allows them to act as a highly efficient "bucket brigade" for electrons. This means, first, that they are arranged in descending order of electron potential, so that the electrons can cascade down smoothly from one component to the next. In addition, they must be so oriented to each other that their active centers can readily exchange electrons with their partners on either side, relying only on the small displacements afforded by thermal vibrations and rotations to provide the necessary contacts. Finally, they must be fitted with appropriate inlets and outlets. As we shall see, the main inlet is provided by a flavoprotein enzyme that transfers electrons from matrix NADH to the top of the respiratory chain. A few secondary inlets also exist at lower energy levels. The main outlet is to oxygen, through an elaborate complex of cytochromes and copper known today as cytochrome oxidase, but originally designated by its discoverer, the German biochemist Otto Warburg, as *Atmungsferment,* respiratory enzyme. A most appropriate name, since most of the oxygen respired in the biosphere is utilized through the action of this enzyme.

To assemble such an electron-transport chain is certainly no mean feat. Yet these aspects of its architecture

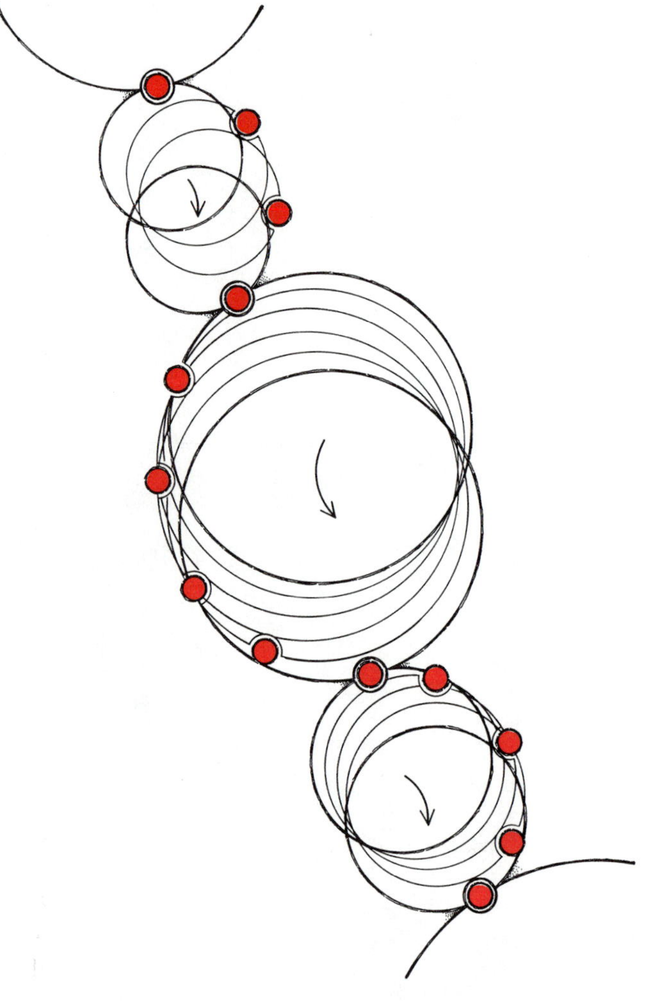

A molecular "bucket-brigade" for electrons. Members of the respiratory chain are organized so that thermal vibrations and rotations can bring their electron-carrying centers alternatively in contact with those of *two* of their neighbors.

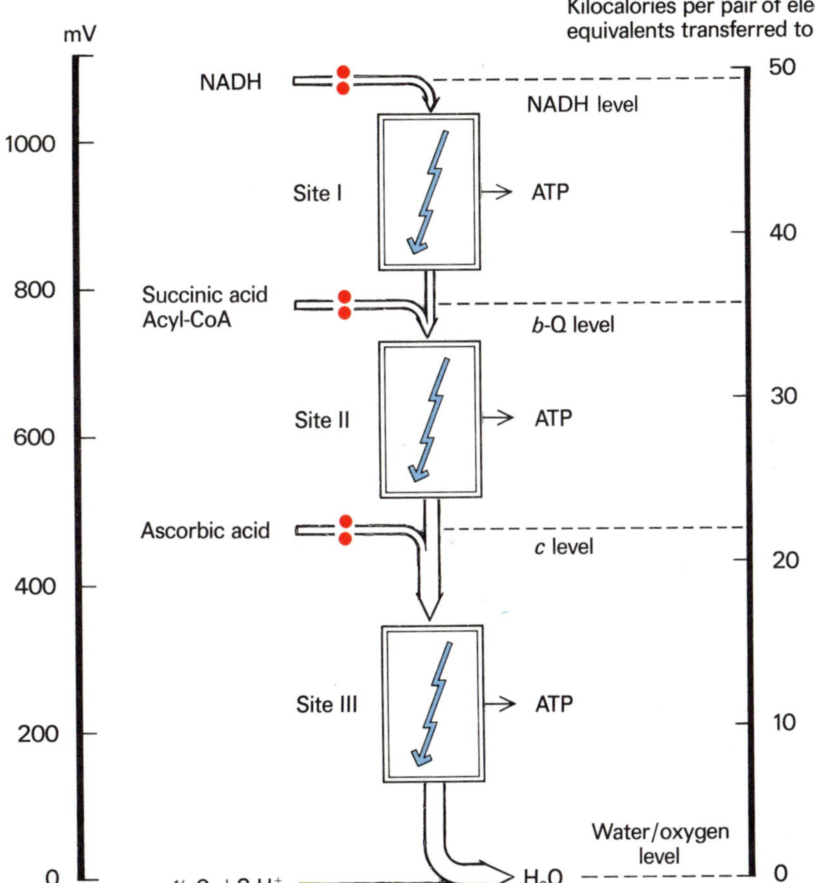

Respiratory chain phosphorylation. Electron pairs falling down to oxygen traverse between one and three oxphos units, depending on the level at which they enter.

are almost trivial in comparison with some of its other properties, for the microcircuit is not just an electron conductor. It is an energy transducer, organized so as to trap the energy dissipated by the electrons as they tumble down to oxygen and to render it usable for biological work through the mediation of ATP. In other words, the microcircuits contain incorporated oxphos units.

This introduces two additional conditions into the design of the microcircuits. First, the potentials of the carriers must be so poised as to provide appropriate electron falls. Remember, for a pair of electrons to yield enough energy to sustain the synthesis of an ATP molecule, it must fall down a potential difference of at least 300 mV. The second condition is that the electrons going down such falls must be subjected to some sort of constraint that couples their flow obligatorily to the synthesis of ATP. This is similar in principle to the kind of constraint that deviation through a hydroelectric turbine imposes on a waterfall, coupling the flow of water obligatorily to the generation of electricity.

The first condition is satisfied very efficiently. As many as three oxphos units, or phosphorylation sites, are intercalated in series in the pathway of the electrons. They are called sites I, II, and III. They are separated by platforms, which we will call the NADH level, the *b*-Q level, and the *c* level, each after some characteristic electron carrier. Each of these platforms is accessible to electrons supplied from the outside by appropriate donors. Electrons can be collected back from them with certain artificial acceptors, such as ferricyanide or methylene blue. Normally, however, the only electron outlet is to oxygen, at the zero level of energy.

Let us consider the *c* level first. It derives its name from cytochrome *c*, its principal component. Among the substances capable of donating electrons to the chain at this level is ascorbic acid, or vitamin C, the substance in fresh fruits and vegetables that prevents the nutritional disease called scurvy (scorbut), a common plight of navigators in the old days (ascorbic comes from the Greek *a*, not, plus scorbut). The antiscurvy property of vitamin C is not known to be directly linked to its ability to donate electrons, although it could be in some specialized way. In laboratory experiments, however, ascorbic acid will interact with mitochondria at the *c* level, and this has allowed the identification of the oxphos unit at site III through the finding that one molecule of ATP is synthesized for every pair of electrons transferred from ascorbic acid to oxygen.

Access to site II is through the *b*-Q level, so named because it is occupied jointly by cytochrome *b* and coenzyme Q, or ubiquinone, a quinone type of electron carrier. Several FAD-dependent flavoproteins are connected with this level, to serve as port of entry for electrons provided by certain metabolic substrates, such as succinic acid, an intermediate in the Krebs cycle, and fatty acyl-coenzyme A derivatives, the activated forms of fatty acids. When a pair of electrons enters at this level, it falls down two consecutive oxphos units: two ATP molecules are made.

The NADH level is occupied by a special flavoprotein that transfers electrons from NADH into phosphorylation site I. Altogether, three ATPs are synthesized for each electron pair that enters at this level and is collected by oxygen at the bottom. NAD^+ itself serves as electron acceptor in countless reactions. So this pathway channels most of the catabolic electrons in aerobic organisms.

The efficiency of these mitochondrial transducers is quite remarkable. With NADH as electron donor, more than 85 per cent of the free energy of oxidation is recovered as ATP: $3 \times 14 = 42$ kcal per pair of electron-equivalents out of a total of 49. With $FADH_2$, the efficiency is a little lower, but still considerable: $2 \times 14 = 28$ kcal per pair of electron-equivalents out of 37, or 76 per cent.

Combined with the high efficiency of the cold combustion reactions themselves, these values lead to overall efficiencies approaching 80 per cent, as can readily be calculated from the data of the table shown on page 154. These are staggering yields, to which no man-made combustion power plant has ever come near. They reflect the remarkable quality of the energy transducers incorporated into the microcircuits of the inner mitochondrial membrane.

How these transducers work is a question that has occupied some of the best brains in the business for more than 30 years. The answer, according to the British scientist Peter Mitchell, the father of the widely accepted "chemiosmotic theory," is: through protonmotive power. It is a somewhat offbeat topic for tourists, but let us have a go at it anyway. For, as will be even clearer after Chapter 10, we are looking at one of life's most central mechanisms. It is worth the extra effort.

Protonmotive Power, the Secret of Oxphos

The inner mitochondrial membrane is "proton tight": it is impermeable to hydrogen ions. This means that it is also impermeable to hydroxyl ions, OH^-, which otherwise could act as carriers for H^+ ions, taking them through as H_2O and returning as OH^-. In fact, the membrane is completely "ion tight": no ion, whether positive or negative, can get through it at an appreciable rate by passive diffusion. The whole commerce of ions between mitochondria and the surrounding cytosol is strictly regulated by special gates or pumps.

There is such a pump for protons—or rather, there are two types of such pumps. They are both built to pump protons forcibly out of the mitochondria with the help of a supply of energy. They are both reversible, by which is meant that they can run backward and generate power at the expense of an inward flow of protons. But their power supplies are different: one is electron driven, the other ATP driven. Such, in a nutshell, is the secret of oxphos: electron flow and ATP synthesis are interlinked by protonmotive power.

Let us look first at the ATP-driven pump. For each molecule of ATP split into ADP and inorganic phosphate,

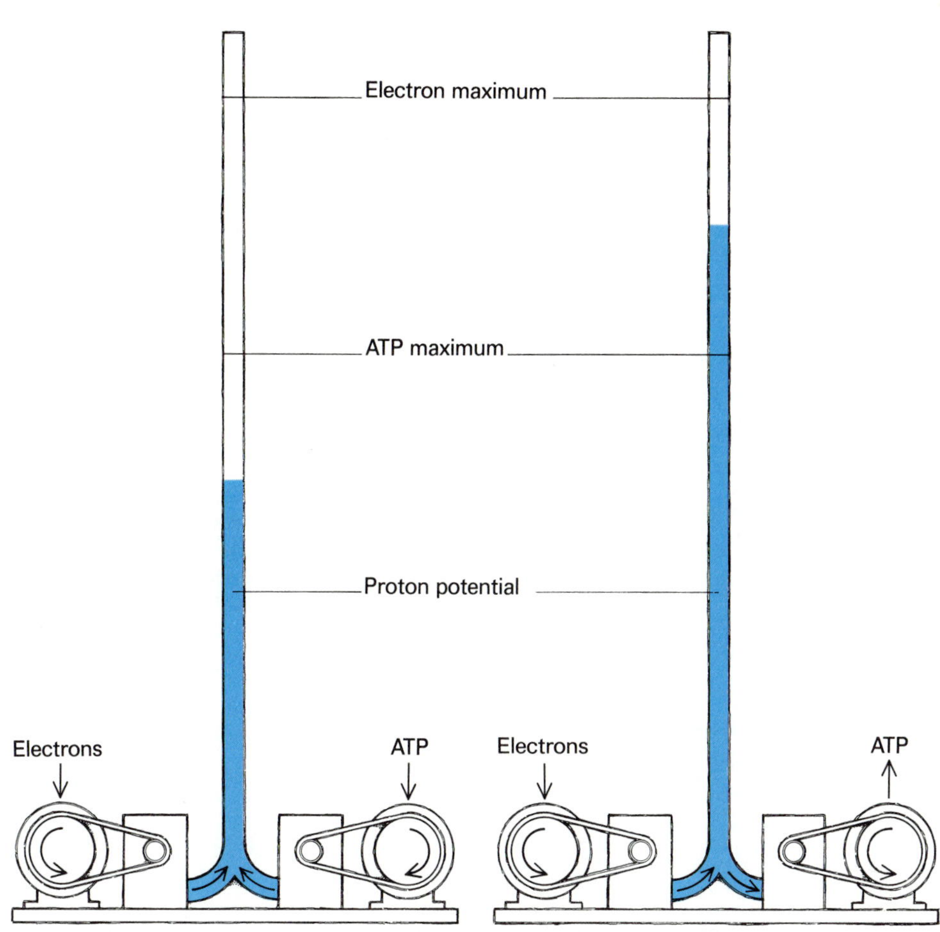

Model of chemiosmotic theory. Two pumps, one powered by electrons, the other by ATP, push protons up the same reservoir, creating a proton potential. If one pump is stronger than the other, it will raise the proton potential above the maximum power of the other pump, which will run backward. Usually, the electron-driven pump is strongest and causes the ATP-driven pump to run backward and make ATP.

it forces n protons out of the mitochondrial body. It can go on doing so as long as it is able to overcome the proton potential that builds up outside. When this potential becomes equivalent to the power of the pump, the pump will stop, just like an electric pump that can lift water only to a certain level, depending on the power of the motor. We know the power the ATP-driven pump has available; it is of the order of 14 kcal per gram-molecule of ATP split. Therefore, the maximum level to which it can lift protons is about $14/n$ kcal per proton-equivalent.

The same reasoning applies to the electron-driven pump, except that it is powered by the energy that is released by a pair of electrons falling down a potential difference. If this difference is 300 mV, the energy available will be 14 kcal per pair of electron-equivalents. Such a pump will stop, like the ATP-driven pump, when the proton potential reaches $14/n$ kcal per proton-equivalent, in which n represents the number of protons that the pump translocates per pair of electrons downgraded. As it happens—the coupled system described here would not work otherwise—n is the same for the two pumps.

If you now have the two pumps working together against the same proton potential, and one happens to be slightly stronger than the other, the more powerful pump will lift the proton potential above the limit level of the weaker pump, which, if reversible, will run backward. Just so with two electric pumps pushing water up into the same reservoir; if one pump is stronger than the other, it will bring the water level sufficiently high to reverse the flow through the weaker pump and cause it to act as a generator. In a reversible system of this sort, the electricity generated by the second pump will be equal to the electricity consumed by the first pump, barring the inevitable small losses due to the imperfections of the system.

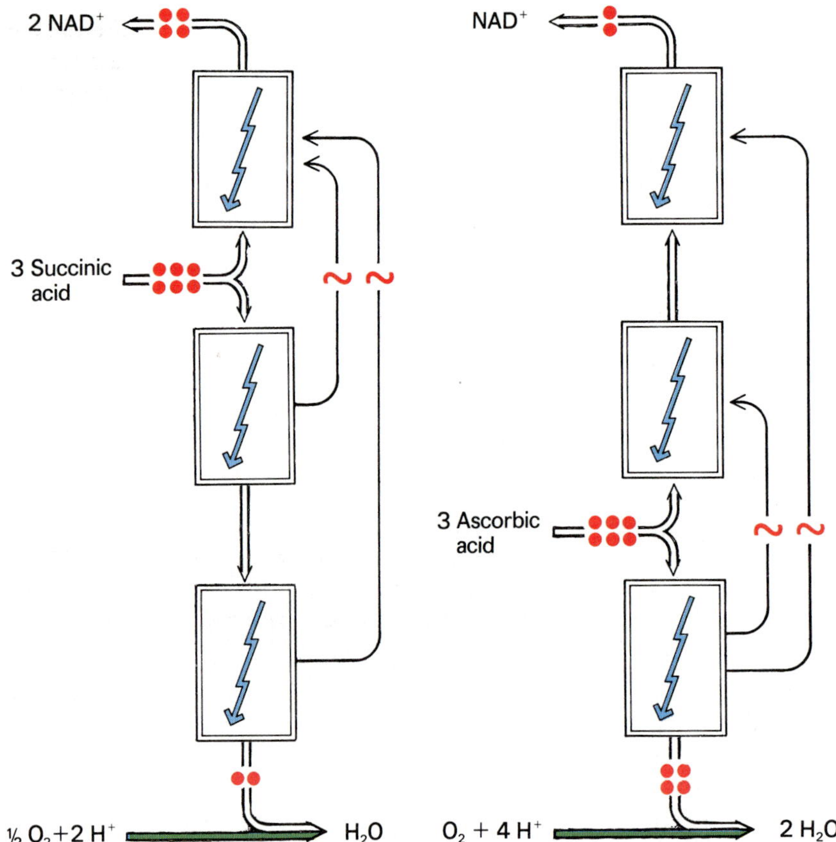

Examples of reverse electron transfer. The fall of one electron pair from succinic acid to oxygen produces two "squiggles" of protonmotive force at sites II and III. With a supply of NAD^+ as acceptor, two additional electron pairs are delivered from succinic acid and *lifted* through site I, acting in reverse with the help of the protonmotive force produced in sites II and III. Similarly, two electron pairs are transferred from ascorbic acid to oxygen through site III, producing two "squiggles" that help a third electron pair climb from ascorbic acid to NAD^+ through sites II and I acting in reverse.

Electric energy is thus being transferred by way of a hydraulic transducer.

An analogous situation obtains in the mitochondrial membrane, but with different kinds of pumps and energy. Under most circumstances, the more powerful of the two mitochondrial pumps is the electron-driven one, since it is constantly supplied with fuel by metabolism, whereas ATP is continuously consumed by the various forms of work accomplished by the cells. And so the ATP-driven pump runs backward, making ATP with the help of the protonmotive power supplied by electron flow.

Normally, this system does not run in the reverse direction, for there is no separate source of ATP to drive it. But what can happen is that one electron-driven pump has more power than the others. Remember that there are three distinct oxphos units in the respiratory chain, each connected to different electron carriers and therefore hooked onto a different proton pump. Of the three, that at site III is stronger than the other two. It has available a potential drop close to 500 mV and is essentially irreversible for this reason, whereas those at sites I and II are freely reversible, with potential differences of the order of 300 mV. Thanks to the extra drop down site III, electrons fed into the system at the *c* level or at the *b*-Q level may generate enough of a proton potential to reverse the flow of electrons through the site or sites situated above their port of entry. For such reverse electron transfer to manifest itself, a suitable acceptor—NAD^+, for instance—must be supplied to collect the uplifted electrons at the higher level. Such a phenomenon has been produced artificially with mitochondria. As we shall see, it functions physiologically in many autotrophic organisms, where it is of immense biological importance.

In trying to visualize this system, you have to realize that the mitochondrial proton pumps do not actually secrete acid in the way the proton pump in the lining of the stomach does. The reason is that the extruded protons are neither accompanied by a negative ion nor exchanged for another positive ion. As protons are forced out, an electric imbalance is created by the loss of positive charges from inside the mitochondrion; an electric potential, positive outside, is built up across the mitochondrial membrane. As the potential increases, it becomes increasingly difficult to push protons out against this potential—remember, charges of equal sign repel each other according to Coulomb's law—and eventually the pump will grind to a halt when the work required to force an additional positive charge against the electric potential becomes equal to the power of the pump. In other words, the proton potential in this particular system is expressed mainly in the form of an electric potential difference across the membrane. There is a small difference in acidity between the inside and the outside of mitochondria, but it is of minor importance.

We can calculate the mitochondrial membrane potential from the known power of the pump: $14/n$ kcal per proton-equivalent ($58{,}600/n$ joules per proton-equivalent), or:

$$\frac{58{,}600}{96{,}500 \times n} = 0.6/n \text{ V}$$

According to Mitchell, $n = 2$, which makes the membrane potential 300 mV. But not all workers agree with Mitchell. Some claim that $n = 3$ or 4, in which case the potential would be 200 or 150 mV. Others believe in an even lower value of membrane potential, not so much because they disagree with Mitchell's stoichiometry, but because they see the proton potential as being partly manifested in conformational changes in the membrane. Unfortunately, we do not have a voltmeter to settle the question. Nor is our eyesight good enough to find out whether the molecules are twisted out of shape when the microcircuits become energized.

What, now, do we know about the pumps themselves? Very little, unfortunately. Several ATP-driven proton pumps are known—remember, there is one in the membranes of endosomes and lysosomes—but the molecular mechanism linking proton transfer to ATP hydrolysis has not been elucidated for any pump so far. What is known is that the mitochondrial ATP-driven proton pump is situated in the little knobs that can be seen protruding from the inner face of the membrane and that it can be selectively jammed by a substance called oligomycin. As the name indicates, this substance is made by a mold (Greek *mykês*). We will encounter a number of other "mycins" on our tour. They have been invaluable tools to biochemists and molecular biologists in helping to unravel some of the most complex biological mechanisms. It is interesting that all these substances are in some way offshoots of penicillin. Before the discovery of that substance, few scientists showed an interest in molds. But, after it was found that a mold (*Penicillium notatum*) manufactures a substance (named penicillin for obvious reasons) capable of blocking the development of certain pathogenic microbes, a vast search was instituted for other mold products with antibiotic properties. Hundreds of thousands of strains of molds from all over the world were screened for this purpose, yielding a rich crop of active substances. A few of these were adopted for clinical use; streptomycin is an example. But most had to be rejected because they proved highly toxic to human organisms; they inhibit some essential biological process, such as DNA transcription, protein synthesis, or oxidative phosphorylation. In biochemistry, we can often learn a great deal about the working of a machine from the way it is blocked by inhibitors. And so these dropouts of drug research became instruments of key scientific advances, illustrating, as does much of the history of science, the unpredictability of discovery and the folly of those bureaucrats, unfortu-

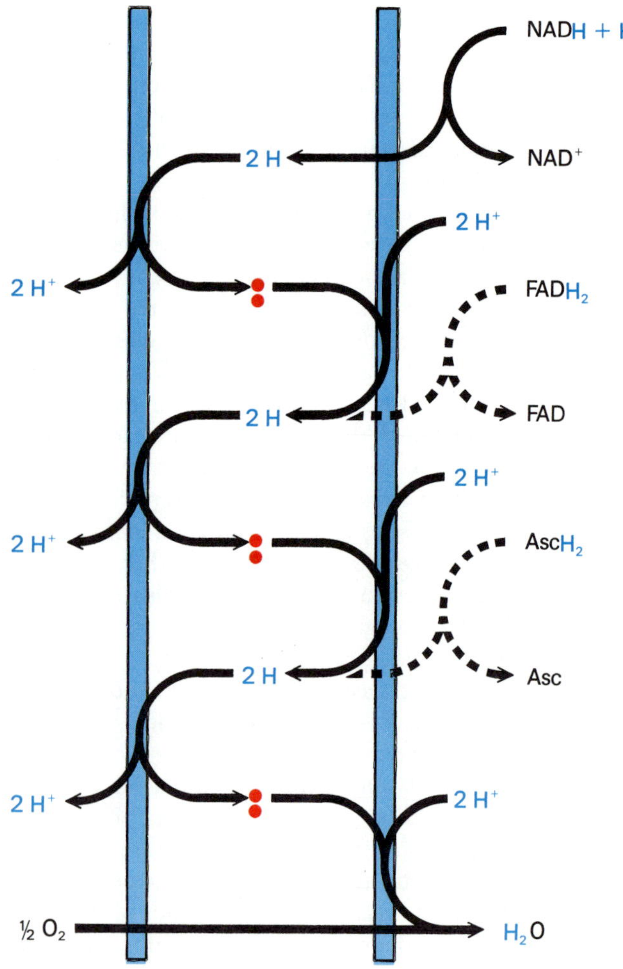

Loop hypothesis proposed by Mitchell to explain coupling between electron transfer and proton extrusion. At each prosphorylation site, electrons are assumed to carry protons from inside to outside, and to return naked.

nately more and more numerous, who would tie scientists down to rigidly predetermined programs.

Turning now to the electron-driven proton pumps, we must look for them in the molecular architecture of the microcircuits themselves, as their functioning depends on an obligatory coupling between electron transfer and proton extrusion. Mitchell has proposed an ingenious hypothesis to account for this coupling. Expressed in simple terms, it assumes that electrons enter oxphos units as hydrogen atoms supplied from inside the mitochondrion and that they leave the units naked, after having shed the protons outside. This model implies that the microcircuits are made of alternating hydrogen-carrying and electron-carrying segments and that they are organized in loops in the thickness of the membrane. This, then, would be the answer to the question, What sort of constraint does the molecular anatomy of the microcircuits impose on the flow of electrons? The constraint is that the electrons are forced along a sinuous path that takes them alternately from one face of the membrane to the other, together with protons in one direction and without protons in the other. In reality, the mechanisms are considerably more complex than might be gathered from this simplistic description, and involve subtle physicochemical and conformational changes associated with the passage of the carriers from the oxidized to the reduced state. Add these requirements to all those that were enumerated above and you have a pretty refined piece of machinery. And mind you, it is all compressed in a little "biochip" about 0.3-millionth of an inch thick and one-millionth of an inch wide!

In consequence of their built-in constraints, the mitochondrial oxphos units never spend energy unnecessarily. Electron flow and ATP synthesis are obligatorily coupled, which means that the rate of electron flow (and of substrate and oxygen consumption) is automatically adjusted to the rate of ATP consumption. At rest, little ATP is used, and consequently food consumption and respiration are low. Let there be a sudden call on ATP, as in an athlete starting a race, and electron flow immediately increases manifold, causing catabolic oxidations and oxygen uptake to increase simultaneously in the same proportion. This regulatory mechanism is called respiratory control.

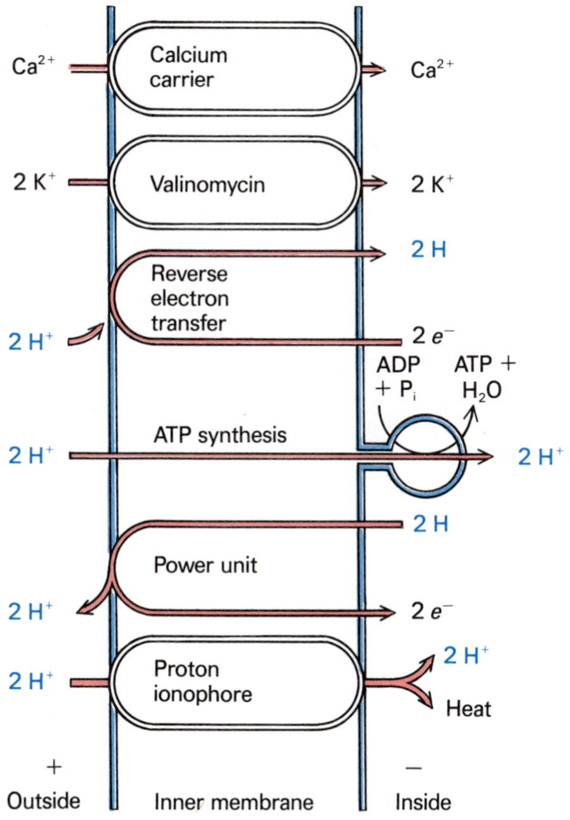

Some of the forms of work that can be powered by mitochondrial protonmotive power. Protons can *return* inside, with the coupled synthesis of ATP; or they can *return* with low-level electrons and bring them to a higher level; or they can be *exchanged* for calcium, thanks to a calcium carrier present naturally in the membrane; or they can be *exchanged* for potassium, if valinomycin (potassium ionophore of mold origin) is added; or they can *return*, with the help of an added proton ionophore (uncoupler), with collapse of the potential and dissipation of the energy as heat. Note that all these forms of work can also be powered by the ATP-synthesizing machinery, acting in reverse as a proton-extruding ATP hydrolase.

This control depends on the strength of the coupling, which is itself determined by the proton-tightness of the membrane. Any substance capable of transporting protons across the lipid bilayer of the membrane (a proton ionophore) will immediately cause the proton potential to collapse and so uncouple the system. Many such uncouplers are known. Their prototype is 2,4-dinitrophenol.

Another way of discharging the membrane potential is by providing a positive ion capable of getting across the membrane into the mitochondrion. For this to happen, there must be an appropriate carrier in the membrane. One exists naturally for calcium ions, with the consequence that mitochondria tend to keep their surroundings free of calcium ions—an important regulatory mechanism (see Chapter 12). A passage for potassium ions can be created artificially by the addition of a mold toxin called valinomycin, which has the property of a potassium ionophore. Note that, in these cases, work is still performed. Electron flow now supports the active inward transport of the added cation in exchange for the outward transport of protons.

As a last comment, it should be pointed out that substrate-level oxphos units, such as are present in glycolysis or in the Krebs cycle, do not operate by protonmotive force. They are purely chemical transducers. In fact, early understanding of the mechanism of substrate-level phosphorylations impeded for a long time the elucidation of the type of membrane-linked, mediator-level phosphorylation that is found in mitochondria, because most investigators were searching for nonexistent chemical intermediates.

Growth and Multiplication of Mitochondria

Mitochondria grow and divide, as would be expected of a symbiotic microorganism. Unlike their putative bacterial ancestors, however, they manufacture hardly any of their constituents. Most mitochondrial proteins are synthesized in the extramitochondrial cytoplasm according to instructions sent out from the nucleus. How these proteins are directed to the mitochondria and pass across two membrane layers is far from clear. The question is considered in Chapter 15.

This definitely is not how bacteria acquire their proteins. Bacteria make their proteins themselves, with their own ribosomes, reading from their own genes. One would expect their symbiotic descendants to do the same. Since they do not, should we not reject the symbiont theory? Not necessarily, for mitochondria are not entirely controlled by nuclear genes; they do, in fact, have a genetic system of their own. It is a very rudimentary system, coding for only a small number of proteins, but it is essentially complete, including DNA as well as all the pieces that are needed for the replication, transcription, and translation of the genetic information it contains. And it is operative, expressing itself in certain specific mitochondrial proteins, such as part of the cytochrome-oxidase complex. It can even suffer mutations, which are therefore transmitted by the cytoplasm and not by the usual Mendelian chromosomal mechanism. The so-called "petite" mutation of yeast is a typical example of such a hereditary change affecting the mitochondrial DNA.

The fascinating thing about this mitochondrial genetic machinery is that it has typical bacterial characteristics. The DNA is circular, resembling bacterial DNA. The ribosomes are smaller than the cytoplasmic ribosomes and about the same size as bacterial ribosomes. Their capacity to make proteins is blocked by the antibiotic chloramphenicol, an inhibitor of bacterial protein synthesis but not of eukaryotic protein synthesis, and it is not affected by the eukaryotic inhibitor cycloheximide, which also is inactive on bacteria. Most impressive of all, mitochondria do not even speak the same language as the rest of the cell; they employ a partly different genetic code.

These are some of the reasons why many scientists believe that mitochondria are descended from some distant prokaryotic ancestors that were adopted endosymbiotically and that, in the course of their prolonged process of integration, lost greater and greater control over their own fate. But a few vestiges of their erstwhile autonomy appear to have survived this attrition process to give us an inkling of historical events of immense importance that may have happened more than one billion years ago. Those events would have remained forever buried in the darkness of the past but for these telltale relics. The revelation, however, is not unambiguous and has been interpreted differently by others.

10 Chloroplasts: Autotrophy and Photosynthesis

Although our tour is concerned mostly with animal cells, it would hardly be complete without at least a passing excursion into the plant world. Plants belong to the general group of autotrophs—literally self-feeding organisms—better, though less commonly, designated lithotrophs (Greek *lithos*, stone). Indeed, the "foodstuffs" of autotrophs come entirely from the mineral world in the form of carbon dioxide (CO_2), water (H_2O), nitrate (NO_3^-), sulfate (SO_4^{2-}), and related inorganic components. The required elements are there, but in strictly "no cal" form, valueless from the energetic point of view. Autotrophs, therefore, need a separate source of energy, since they cannot, like heterotrophs (organotrophs), derive power from their food.

How to Survive in a Lifeless World

In what form this energy? There are several answers to that question, but they all boil down to a single word: electrons. That electrons are necessary for autotrophic life is obvious from the nature of its building blocks. To convert CO_2, H_2O, NO_3^-, SO_4^{2-}, and the like into carbohydrates, proteins, lipids, and other biological constituents, a large quantity of electrons must be supplied—as many, in fact, as are released in heterotrophs by the oxidation of these constituents. That electrons may also be sufficient is suggested by our knowledge of heterotrophs. Provided an organism has at least one functional oxphos unit available, it can use electrons to manufacture ATP and thereby cover all its energy needs.

The electrons must be of good quality: they must be delivered at a high enough level of potential energy. As far as ATP synthesis is concerned, this requirement is flexible and depends on the level at which the electrons leave the oxphos unit. All that is needed is that they enter at least 300 mV above exit level (if, as is usual, they travel in pairs). The requirement is, however, more stringent for reductive syntheses, which most commonly use electrons supplied at a level at least equivalent to that of NADH. The reason is obvious: that is the level from which most of the stored electrons tumble back down during catabolism.

In primitive systems, autotrophic reductions are carried out via NADH—that is, with the same coenzyme as catabolic oxidations. Animal glycolysis also functions in this way. As we saw at the end of Chapter 7, the oxphos unit of the glycolytic chain operates near thermodynamic equilibrium, so that relatively small changes in the concentrations of the participating substances may suffice to reverse the flow of electrons through the unit and to cause the cell to switch from glycolysis to gluconeogenesis.

By and large, however, Nature has assigned a special coenzyme to biosynthetic reductions, thus separating anabolism from catabolism. This coenzyme is a phosphorylated derivative of NAD and is represented as NADP, which stands for nicotinamide adenine dinucleotide phosphate. Despite its close kinship with NAD, NADP is treated differently. With very rare exceptions, dehydrogenases use either one or the other coenzyme, not both. Thanks to this distinct mode of handling, helped, if need be, by the expenditure of energy, the NADPH reservoir is maintained physiologically at a significantly higher energy level than is NADH: about 1,200 mV above water/oxygen level, or 55 kcal per pair of electron-equivalents transferred to oxygen. This is consistent with the fact that the only direction electrons can take when they move under their own power is downward. Thus in synthetic reductions, the electrons fall down from the NADPH level to that of substrate. And in catabolism they fall further down, from substrate to NAD^+, from NADH to the next acceptor, and so on all the way down to oxygen. In addition to an adequate supply of high-grade electrons, the autotrophic way of life requires an appropriate catalytic machinery. Much of this could be

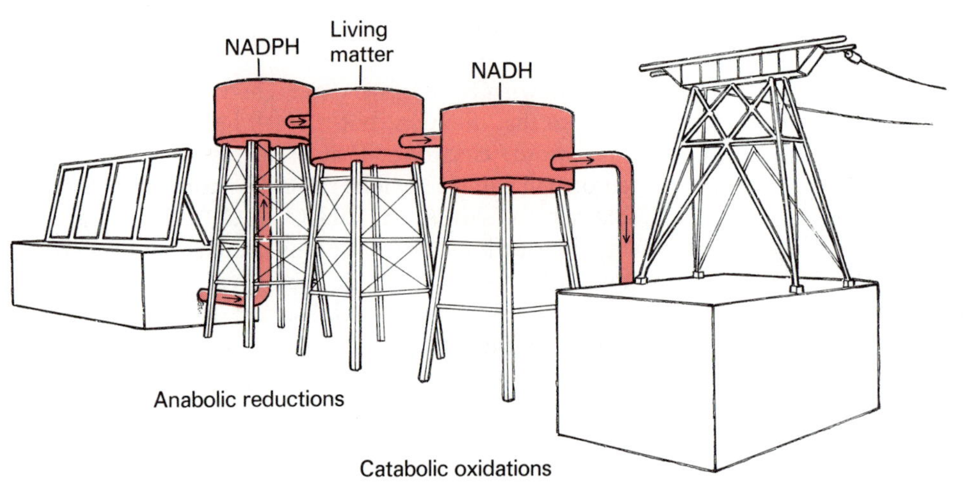

Division of work between NAD and NADP. In order to fulfill its anabolic role, the NADPH reservoir is maintained at a higher energy level than the NADH reservoir, which serves mainly in catabolism.

The dawn of life.

common to heterotrophs thanks to the ready reversibility of many of the reactions going on in these organisms. But some processes are unique to autotrophs.

To try to understand autotrophic mechanisms, we must go back in imagination to those very early, prebiotic times, some 4.5 billion years ago, when, according to most scientists, our young, cooling, and as yet lifeless planet gradually turned into a hotbed of organic syntheses. What happened in those distant days will probably never be known in any detail. But it is fair to assume that organic molecules of various kinds—among them many of the building blocks of present-day living organisms—began to be constructed spontaneously, thanks to a combination of favorable chemical and physical factors. These may have included a reducing atmosphere rich in moisture and in carbon-nitrogen gases; energy in the form of heat, ultraviolet radiation, and the discharges of electrical storms; catalytic clays and other minerals; and, finally, shallow puddles of water in which the thickening "primeval soup" incubated. It all would sound pretty fanciful, except that some of these events have already been reproduced in the laboratory through the simulation of the conditions believed by geophysicists to have prevailed in prebiotic times.

What happened next is even more difficult to imagine. Somehow, polymers of various kinds were formed, joining in all sorts of random assemblies, out of which a rare few turned out to contain the seeds of self-reproduction and self-regulation. Henceforth, natural selection had something to work on; "chance and necessity" (see Chapter 18) could consort to bring forth associations of increasing complexity and organizational stability. Eventually, structures that we would recognize today as living cells appeared. Incredible as it may seem, this whole process took no more than a billion years at most. By that time, organisms resembling bacteria existed in places as far apart as Greenland and South Africa. Life had emerged.

If you find it difficult to believe that something as complex as a bacterial cell could have arisen in this span of time, you may choose to go along with Francis Crick, of double helix fame (see Chapter 16), and prefer the alternative that the first bacteria were brought to our planet by a spaceship sent out by some distant civilization (theory of directed panspermia). Even if you accept such a far-fetched ancestry, you have, of course, not solved the problem of the origin of life. You have only shifted it to a more remote point in space and time, presumed to have offered better conditions for life to arise spontaneously than did our own planet. Considering our ignorance of what conditions were on earth those billions of years ago, the reasons for making life a cosmic import product seem anything but compelling. Note further that the modern theory that traces the origin of the universe itself to an initial "Big Bang" that took place about 15 billion years ago leaves our putative forebears hardly more time to reach their own fantastic state of development, supposedly from scratch, than we have had available to attain ours. Nor does it allow for the possibility that life is an eternal feature of the universe, present in germ form throughout outer space, as was proposed about 100 years ago by the Swedish scientist Svante Arrhenius, the father of the original panspermia theory, who is better known for working out an equation defining the influence of temperature on the rate of chemical reactions.

But back to our story. The early bacteria that appeared on earth—or elsewhere—in those distant times could conceivably have started as heterotrophs living, so to speak, on the fat of the land and consuming organic substances produced by abiotic (nonliving) mechanisms. But sooner or later, and at the latest by the time the abiotic supply of nutrients died out, at least one group of organisms must have developed some form of autotrophy. Otherwise, life would have become extinct.

The organisms most likely to have achieved autotrophic survival were those that had available high-grade electrons in their immediate surroundings. Although not very numerous, sources of "good" electrons do exist in the mineral world. Some are even found today, and they must have been more abundant at the time when a highly reducing environment prevailed. Examples are hydrogen gas, carbon monoxide, and especially several of the sulfurous fumes emitted by volcanic springs. It is probably not a mere matter of chance that so many of today's autotrophic bacteria belong to the group of thiobacteria (Greek *theion*, sulfur). They bear witness to the days when life was born in a world that, most unromantically, must have smelled strongly of rotten eggs.

Electrons supplied from the outside at a high enough level of energy—say 50 or more kcal per pair of electron-equivalents above water/oxygen level—could, with the help of suitable catalysts, support all the necessary autotrophic reductions. But ATP supply would still pose a major problem to these primitive heterotrophs if, as speculated in Chapter 7, they were anaerobic fermenters relying on substrate-level phosphorylations. To convert to "electron fuel," they needed a new kind of engine, an oxphos unit of the sort that receives electrons from an outside donor and returns them to an outside acceptor, as is found in the respiratory chain of mitochondria. How soon in the history of life such a device first appeared is beyond our knowledge. But it seems a fair guess, unless early bioenergetic mechanisms were totally different from those we know today, that such an indispensable prerequisite of autotrophic life emerged before or, at the latest, together with autotrophy. Although we have no clue as to how this primitive device operated, it is not unreasonable to assume that it depended on protonmotive force and represented, in fact, the prototype of what is now a universal energy transducer in both autotrophs and heterotrophs, whether prokaryotic or eukaryotic.

Perhaps those early autotrophs, from which all contemporary living forms presumably descend, gained this crucial acquisition originally as an adaptation to extremely acid environments. This adaptation is found today in a type of bacteria known as thermoacidophiles, which belong to the group of archaebacteria (Greek *arkhaios*, ancient), believed to have a particularly ancient history. As their name indicates, thermoacidophiles, which are found in hot sulfur springs, are adapted both to high temperature and to high acidity. They survive in this unkindly environment by actively pumping protons out and use the resulting gradient to perform various kinds of work.

Whatever truth there may be in these speculations, if today's blueprints are at all applicable to the very early forms of life, the most primitive autotroph we can imagine is one that uses high-energy electrons both for biosynthetic reductions and for fueling an oxphos unit hooked to an outside electron acceptor of suitably lower energy level (not oxygen, however, which almost certainly was not available in those days). Make this oxphos unit reversible and link it in series with one or two similar units operating at lower energy levels, as in the respiratory chain, and you have a much more flexible arrangement capable of using a new class of electron donors of moderate energy level. The lower-level oxphos unit(s) then serve(s) to cover all energy needs, including the cost of upgrading the electrons required for biosynthetic reductions. The upper-level oxphos unit does the actual upgrading by running in reverse with the help of energy supplied from below, as we have seen can happen in mitochondria (Chapter 9). Such designs, in one form or another, are characteristic of all present-day chemolithotrophs. Many of these are aerobic and must be evolutionary latecomers. But a few anaerobic chemolitho-

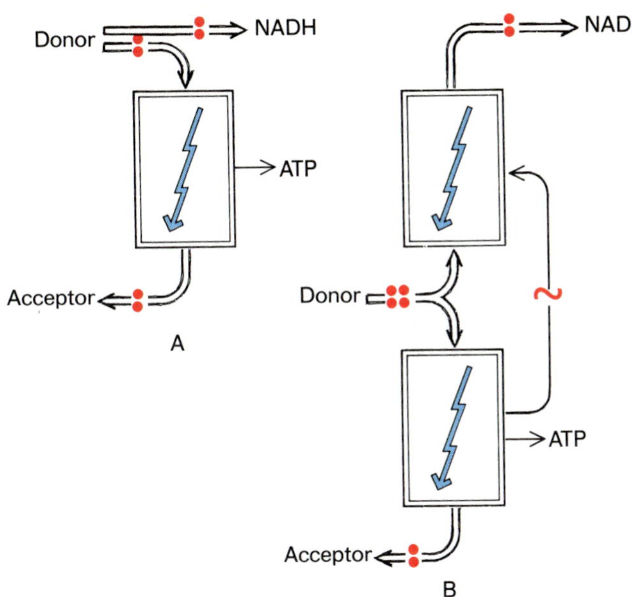

Chemolithotrophy requires: (1) an exogenous electron donor, to support both anabolic reductions and the catabolic supply of energy; (2) an exogenous electron acceptor situated at least 300 mV lower than the donor, to permit ATP formation; and (3) at least one functional oxphos unit inserted between donor and acceptor. If, as in part A, the electron donor is of high enough potential to reduce NAD^+ directly, this suffices. If not, then one or more reverse oxphos units are needed to lift the electrons up to the level of NADH, as in part B.

trophs are known—for instance, *Desulfovibrio desulfuricans* (H_2 reducing SO_4^{2-}) and *Thiobacillus denitrificans* ($S_2O_3^{2-}$ reducing NO_3^-).

Had evolutionary innovation stopped there, life, if subsisting at all, would have remained very primitive and would have clung precariously to those specialized—probably dwindling as time went by—ecological niches capable of sustaining chemolithotrophy. Another key invention was needed before evolution could truly surge forward: the harnessing of sunlight. Actually, we do not know whether this development came before or after that of chemolithotrophy. Both forms of autotrophy share the same essential requirement for an electron-flow oxphos unit. What is added in photosynthetic organisms is a light-powered device for upgrading electrons, a photoelectric unit.

The Conquest of the Sun

The magic component of this unit is chlorophyll (Greek *khlôros*, green; *phyllon*, leaf), a magnesium-containing derivative of the tetrapyrrole (Greek *tetrara*, four) porphyrin ring, which, fitted with iron instead of magnesium, we have already encountered as the active constituent of hemoglobin, the cytochromes (including the respiratory enzyme), and other hemoproteins (Chapter 9). The significance of this molecule for the success of life on earth surely needs no emphasis. But it is interesting that at least one organism has constructed a photoelectric unit with a carotene, a chemical relative of vitamin A, completely unrelated to chlorophyll. Such a device is owned today by the purple *Halobacterium halobium*, a brine-loving microbe, member of the group of archaebacteria, that grows on the surface of evaporating salterns. The carotene family of substances met with only limited success in the capture of solar energy, but it has played a key role in the development of photocommunication. Rhodopsin (Greek *rhodon*, rose; *opsis*, vision), one of the main light-sensitive pigments of the eye, is closely related to bacteriorhodopsin, the principal constituent of the photoelectric energy generator in the membrane of *Halobacterium*.

The basic function of biological photoelectric units is to accept electrons from some low-energy donor, lift them to a higher energy level with the help of light, and then donate them to an appropriate acceptor. This system may have several outlets but must in any event be linked with an oxphos unit through which the photoactivated electrons can fall back to a lower energy level in a manner coupled to the generation of a proton potential that can power the assembly of ATP or the performance of some other kind of work. Such a combination of a photounit with an oxphos unit carries out what is known as photo-

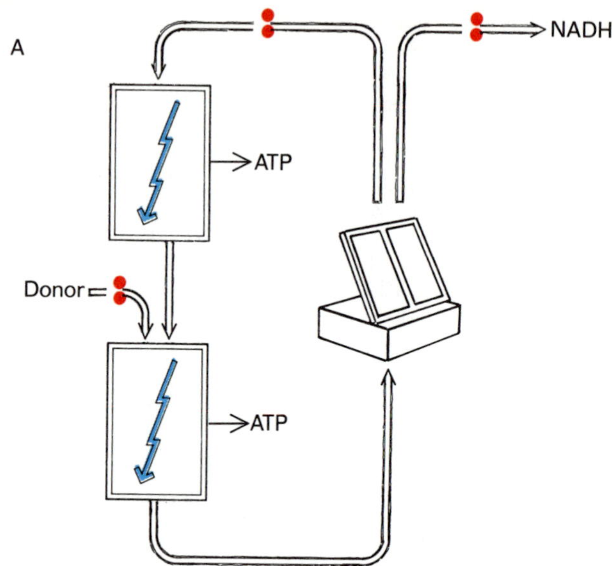

Phototrophy. In bacterial phototrophy, donor is an organic (photoorganotrophy) or inorganic (photolithotrophy) molecule of fairly high electron potential. In example A, characteristic of green photosynthetic bacteria, electrons pass through an oxphos unit and then are lifted by a photounit that either delivers them to NAD^+ or returns them to the oxphos unit (cyclic photophosphorylation), possibly by way of a second oxphos unit. In example B, characteristic of purple photosynthetic bacteria, the phenomena are similar except that the photounit is unable to lift electrons directly to the level of NADH. An additional boost is given by a reverse oxphos unit. Some donors interact directly with the photounit. In plant phototrophy (example C), photosystem II lifts electrons from water up to an endogenous acceptor that donates them to photosystem I. The latter operates like the system in green photosynthetic bacteria (example A) but with $NADP^+$ as final acceptor instead of NAD^+. The number of oxphos units inserted in the electron-transport chain separating the two photounits is unclear. Note that electrons travel singly through photounits.

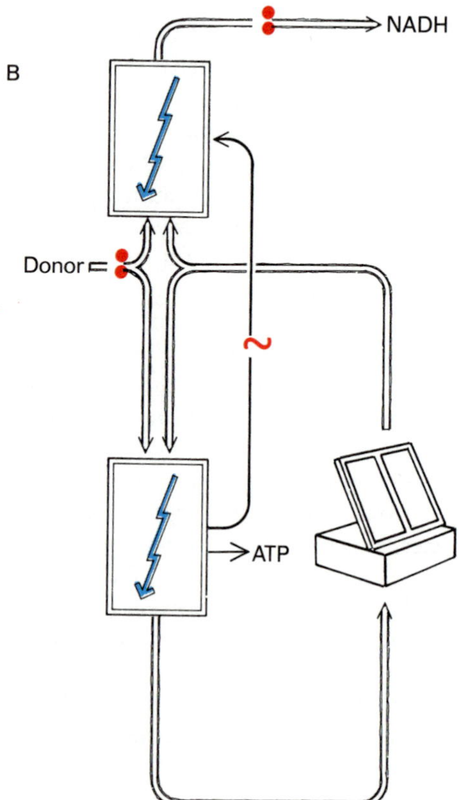

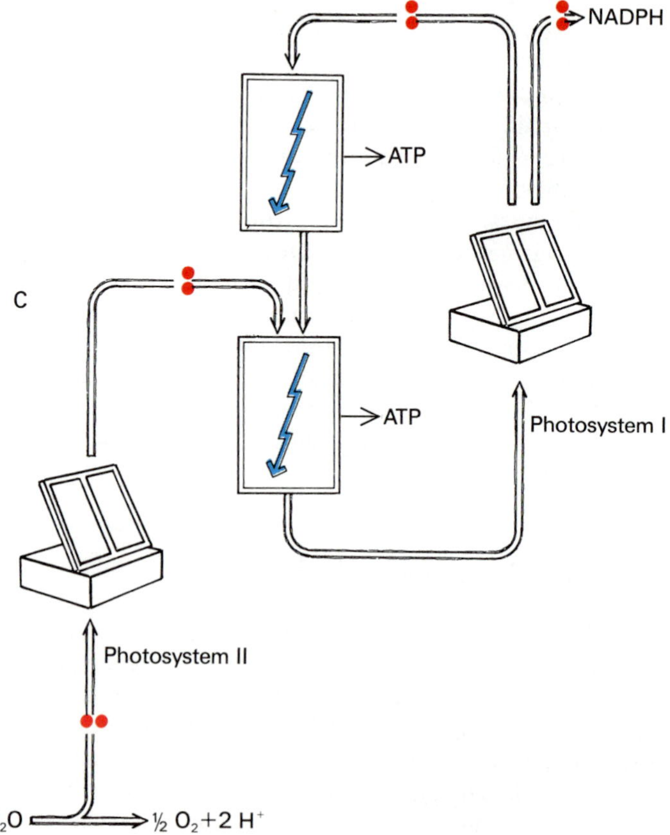

The big electron Ferris wheel of life.

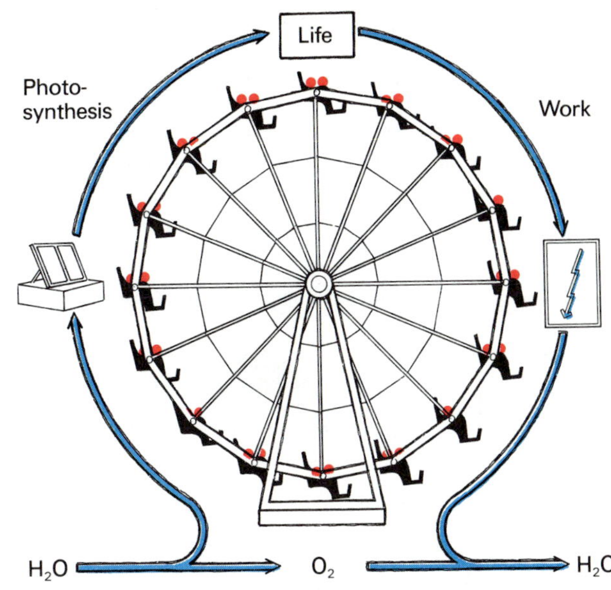

phosphorylation, which can be either cyclic or noncyclic, depending on whether the downgraded electrons are returned to the photounit or transferred to an outside acceptor.

With such a combination, the supply of electrons for biosynthetic reductions can be achieved either directly from the photounit, if its activated level is high enough, or indirectly, after an additional boost provided by another oxphos unit functioning in reverse, as in the second form of chemolithotrophy. As to the electron donors, they belong mostly to the mineral world (photolithotrophy). But cases of photoorganotrophy are also known. They may represent transitional forms between heterotrophy and autotrophy.

The most primitive chlorophyll-dependent complexes are found today in the photosynthetic purple and green sulfur bacteria. These organisms use a variety of sulfur compounds as electron donors and depend on a photochemical machinery known as photosystem I. Most of them are strictly anaerobic. A much more advanced system—actually the crowning innovation in the conquest of autotrophy—is the device designated as photosystem II. Its characteristic is that it uses water as electron donor, oxidizing it to molecular oxygen. The organisms that acquired this system achieved the highest level of energy autonomy, needing only water and sunlight to drive their engines. By itself, photosystem II cannot lift electrons all the way from water to NADPH. It does so by delivering them to photosystem I, which in turn raises them to the level of NADPH. The two photosystems are connected by a phosphorylating electron-transfer chain that allows noncyclic, as well as cyclic, photophosphorylation. The stars of this epoch-making development are the microorganisms called blue-green algae. Although unrelated to the seaweeds (*alga* means seaweed in Latin), which are eukaryotes, these organisms have been called algae because they tend to associate into various types of filaments. Their official name is cyanobacteria (Greek *kyanos*, blue); they are prokaryotes.

When did this all happen? At least 2.5 billion years ago, as indicated by clearly recognizable fossil algae in rocks of that age (Gunflint cherts) found in Canada; perhaps as long as 3.2 billion years ago or earlier, according to some imprints discovered in the South African Pre-Cambrian Fig-Tree rocks. Once they had appeared, the cyanobacteria proved, for understandable reasons, enormously successful. They invaded the entire world, covered it with an extensive mantle of solar-powered chemical factories, and, for some 2 billion years, completely dominated the evolution of life on earth. Through the abundant food they provided, they allowed the heterotrophic mode of life to thrive once again and eventually to branch out into a line of large, highly compartmentalized, voracious phagocytic cells, the putative ancestors of all eukaryotes (Chapter 6). At the same time, the cyanobacteria radically transformed ecological conditions by steadily releasing oxygen into the atmosphere. Adaptation to this change, in turn, produced aerobic life, as we saw in Chapter 9, most likely thanks to the transformation of the original photophosphorylating apparatus into a phosphorylating respiratory chain. Oxygen consumption rose progressively until it reached a steady state with oxygen production, putting in motion the big, sun-driven, electron Ferris wheel of the biosphere: from water to life, through photosynthesis; and back to water again, through respiration. Finally, if holders of the endosymbiont hypothesis are to be believed, the fateful adoption of aerobes by phagocytes, out of which mitochondria and primitive eukaryotic cells were to emerge, was accomplished.

Bacterial photosynthesis has evolved in two steps: first, through the acquisition of carrier-level phosphorylation and photosystem I; next, through the acquisition of photosystem II. According to the endosymbiont hypothesis, cyanobacteria engulfed by an ancestral eukaryote (Chapter 9) became established in their host cell and evolved into the chloroplasts of the unicellular green algae from which all plants are believed to originate.

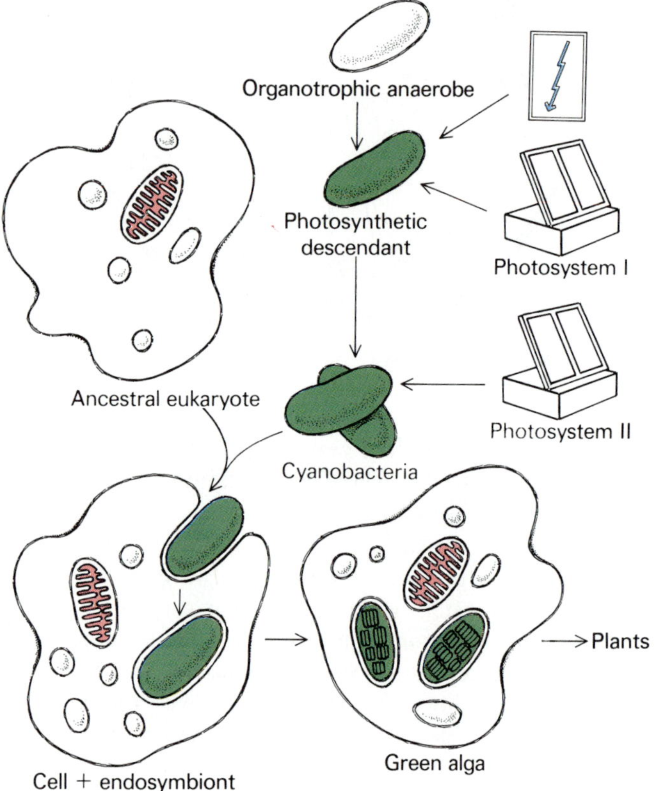

And then, between 1.5 and 1 billion years ago, the last, decisive act in this protracted evolutionary drama—or rather prologue, in view of what followed—was played. Once again, some bacterial victims of phagocytic capture managed to survive and to establish a symbiotic relationship with their captors. The guests, in this case, were cyanobacteria, and the hosts were primitive eukaryotes already equipped with mitochondria. These cells thereby acquired a second type of endosymbiont, which developed into the chloroplasts, the photosynthetic organelles of the unicellular green algae and of their multicellular descendants, the green plants. As described, the scene is, of course, hypothetical. The mechanism it invokes for the origin of chloroplasts does, however, rest on a stronger body of circumstantial evidence than does the endosymbiotic origin of mitochondria and is widely accepted.

So, 3 billion years after life first started seething on the surface of our planet, at the end of some 50 trillion bacterial generations, the stage finally was set for the fantastic development of plants and animals of ever-increasing complexity. We will consider these later parts of the evolutionary saga in Chapter 18.

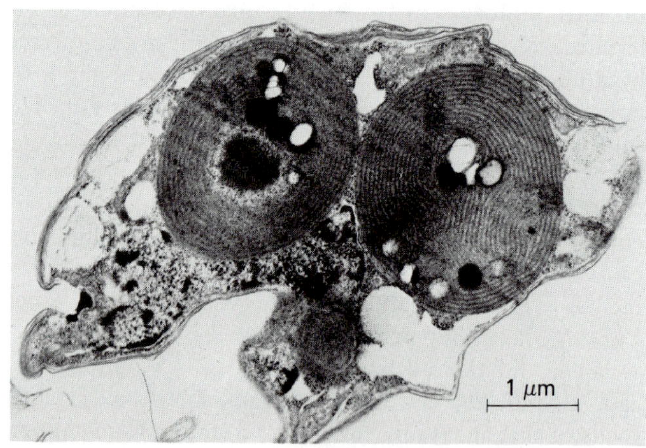

A possible missing link between cyanobacteria and chloroplasts is illustrated by this electron micrograph, which shows two photosynthetic endosymbionts closely related to cyanobacteria (cyanelles) inside a cell of the unicellular, biflagellate protist *Cyanophora paradoxa*.

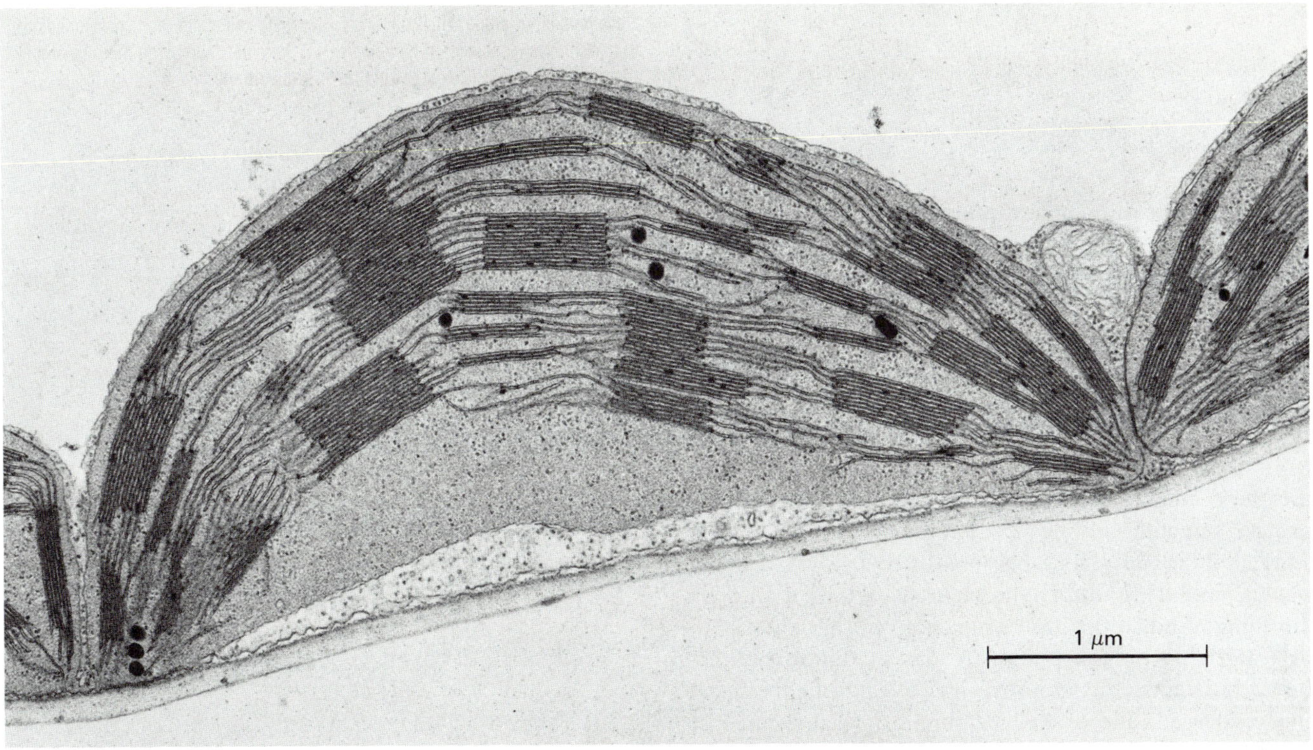

In this electron micrograph of a leaf of timothy grass, *Phleum pratense*, a chloroplast fills almost the entirety of the cytoplasm. Note how piles of thylakoids form grana. The cell is attached to its wall (see the electron micrograph on p. 25). The structure wedged between the chloroplast shown in full view and the one to the right of it is a mitochondrion.

The Green Mansions of Life

Chloroplasts resemble mitochondria in having two surrounding membranes: an outer one, which presumably originates from the vacuolar system of the ancestral phagocyte, and an inner one, supposedly inherited from the plasma membrane of the ancestral blue-green alga endosymbiont. They differ from mitochondria in both their color and their size. As a rule, they are distinctly larger than mitochondria. They measure several micrometers, so that fitting inside would not be too much of a problem for the members of our visiting team were it not for the fact that chloroplasts are generally crammed with stacks of membranes.

These membranes derive, like the mitochondrial cristae, from infoldings of the inner membrane, but with the important difference that the infoldings are severed from that membrane, forming disk-shaped sacs called thylakoids (Greek *thylakos*, pouch). Several thylakoids are stacked together into a cylindrical structure called granum. Each chloroplast contains a number of grana,

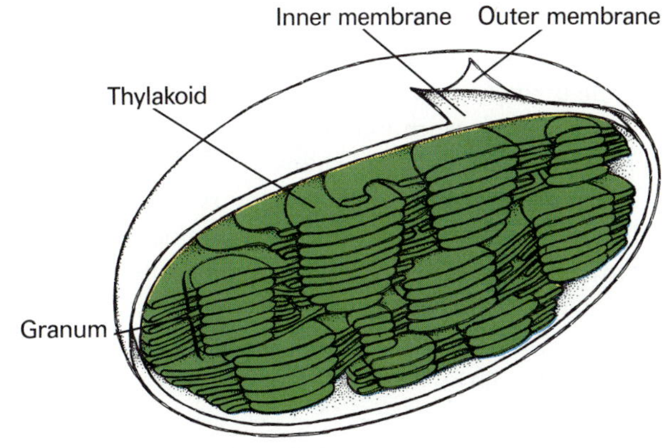

Ultrastructure of a chloroplast.

often linked by tubular connections. This system of inner membranes supports the photosynthetic machinery.

This machinery includes a phosphorylating electron-transport chain flanked by two photoelectric units. The chain resembles the mitochondrial microcircuits. It is similarly made up of a number of structurally associated electron carriers, which include metalloproteins (iron, copper), flavoproteins, quinones, and cytochromes. Electrons circulating through these microcircuits are under the same constraints as those in mitochondria, forcing protons out and building a proton potential. But "out" in a thylakoid is really "in," because the thylakoid is a sealed pouch, unlike the mitochondrial crista, which is an open infolding. And so the impression is given that mitochondria and chloroplasts pump protons in opposite directions. However, this is not really so. In each, the negatively charged side of the membrane is that facing the matrix.

Like mitochondrial inner membranes, thylakoid membranes contain an ATP-driven proton pump situated in knobs protruding into the matrix space. How many sequential oxphos units are associated with the electron-transfer chain is not entirely clear. Certainly one, but possibly two, are on the main noncyclic pathway. There is probably an additional one on the cyclic pathway.

The photosystems themselves are made up mainly of protein-bound chlorophyll, with which are associated variable amounts of accessory pigments, such as the red or blue phycobilins (open-ended tetrapyrroles, resembling bile pigments), and the yellow, orange, or red carotenes and related xanthophylls (Greek *xanthos,* yellow). When leaves start to turn, chlorophyll disappears first, and the blazing colors of the accessory pigments come through to fill our eyes with their autumnal glory.

By definition, a substance is colored because it absorbs some portion of the visible light it receives. Its color corresponds to the light that is not absorbed: it is complementary to the absorbed light. For instance, the chlorophylls are green because they absorb red light (of wavelength 680–700 nm); orange carotenoids absorb blue light (about 450 nm); and so on. In thylakoid membranes, a few hundred such molecules are assembled into

Development of grana. Thylakoids pinch off from infoldings of the inner membrane. Thus, the inner face of the thylakoid membrane corresponds to the outer face of the inner chloroplast membrane.

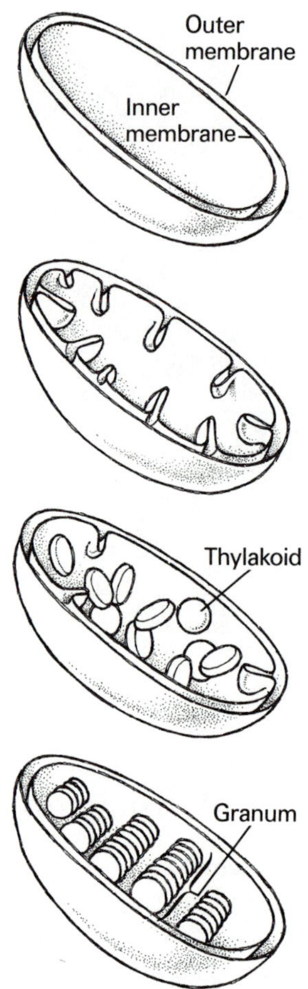

closely organized units centered around a single molecule of a special protein-chlorophyll complex, designated as P700 and P680 in photosystems I and II, respectively (from their wavelengths of maximum light absorption). To find out how such a unit operates, let us leave it in the dark for a while, and then suddenly illuminate it with a flash of light. The pigments in the unit absorb light of appropriate wavelength (depending on their absorption spectrum) and become energized, or excited—a process that involves the destabilization of an electron within the molecular structure by an amount of energy equal to that of the absorbed quantum. With most pigments, energy captured in this way is quickly dissipated as molecular agitation (heat) or sometimes is partly re-emitted in the form of light of higher wavelength (fluorescence), and the destabilized electron falls back to its resting position. Not so in the case of the P complexes, and this explains their biological efficiency. Thanks to their strategic positioning in the thylakoid membrane, they transfer the destabilized electron to an acceptor of high electron potential situated close by and fill in the resulting electron void (positive hole) with an electron taken from a low-potential donor. In this way, they serve to lift an electron from the level of the donor to that of the acceptor with the help of light energy. With P680, the donor is water and the acceptor is a carrier situated somewhere in the upper range of the electron-transport chain. With P700, the donor is a cytochrome at the exit end of the electron-transport chain and the acceptor is ferredoxin, an iron-sulfur protein of high electron potential, which transfers the electron to $NADP^+$ or eventually back into the electron-transport chain.

The P pigments, therefore, constitute the true photoelectric transducers. All the additional molecules of chlorophyll and other colored substances with which they are associated in each photosynthetic unit serve only as light collectors, or solar antennas. Their arrangement is such that they can exchange energy with each other at an extremely rapid rate (exciton transfer), thereby increasing considerably the efficiency of the system. Any quantum of light energy captured by one of the components of the unit can be conveyed to the central P pigment and utilized for photoelectric conversion, provided the connected acceptor molecule is free to receive an electron and there is a donor ready to fill in the hole.

The amount of energy made available by a quantum of light is readily calculated from Planck's formula (Appendix 2). For red light of wavelength 700 nm, the energy is 6.77×10^{-20} calories per quantum, or 41 kcal per einstein. If one quantum moves a single electron through each photosystem, which is most likely, it would take $2 \times 2 \times 41$, or 164, kcal of light energy to move one pair of electron-equivalents from water to $NADP^+$. As we have seen, the energy level in the NADPH reservoir is maintained at about 55 kcal above water/oxygen level.

The photosynthetic machinery of green plants. (Fd stands for ferredoxin.) Note that electrons travel singly through photounits.

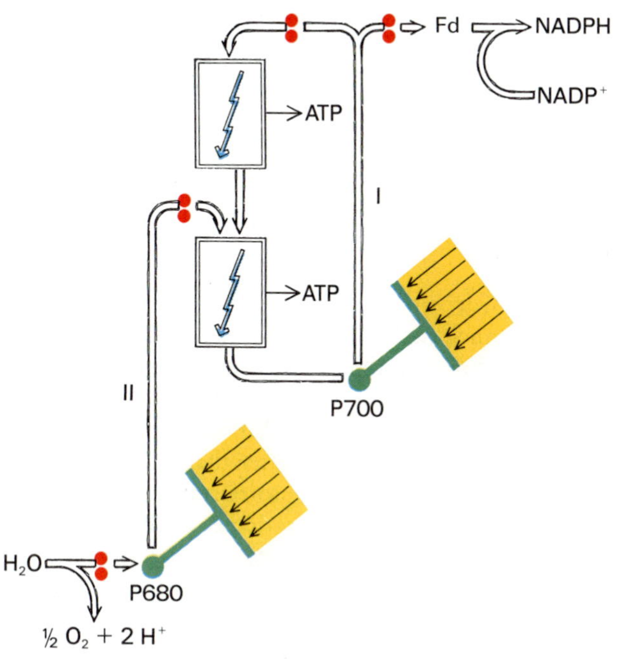

Thus the efficiency of the photoreductive process is of the order of 34 per cent. But this is not counting the gain in ATP. Electrons transferred from photosystem II to photosystem I go through at least one oxphos unit, adding at least 8 per cent to the energy yield in the form of ATP. As to cyclic phosphorylation through photosystem I and the electron-transport chain, it has a minimum efficiency of 17 per cent. But this minimum value could be twice or even three times as high, if the cycling electrons traverse more than one oxphos unit. This point is still unclear, as are many of the finer details of the pathway of photoelectrons. It has even been claimed by one school that photosystem II can by itself lift electrons all the way from water to NADP$^+$, in which case the efficiency of the process would be much higher than estimated above.

Even the lowest estimates represent excellent quantum yields, compared with other photochemical systems. They should be pondered by all those who are trying to harness solar energy. To do better than the multilayered shroud of countless microphototransducers that cyanobacteria and their descendants have woven around the earth will not be an easy task. Perhaps we should try to find better ways to make them work for us, rather than attempting to rival them.

The Dark Reaction

The only light-powered reaction in photosynthesis is the two-step photoelectric transduction that extracts electrons from water and raises their potential by about 1,200 mV. Its products are NADPH and ATP. These suffice to support all the autotrophic biosynthetic mechanisms; they can do so perfectly well in the absence of light, as they do in the nonphotosynthetic chemolithotrophic organisms.

Before we leave the chloroplasts, we should take at least a brief look at the most famous of these "dark reactions," namely CO_2 fixation. Tracing this pathway is one of the triumphs of radioisotope technology used in conjunction with chromatographic separation (Chapter 1).

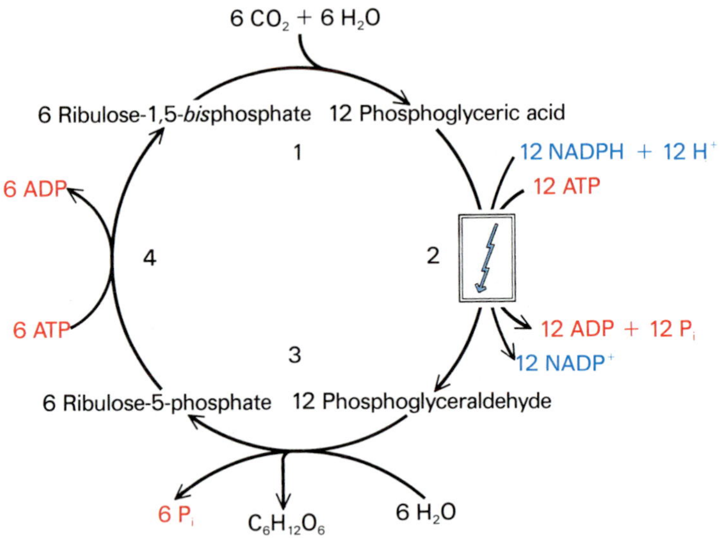

The dark reaction.

1. *Carbon dioxide fixation*, catalyzed by key enzyme ribulose-1,5-*bis*phosphate carboxylase, characteristic of all autotrophs.

2. *Reductive extraction of oxygen*, catalyzed by the substrate-level oxphos unit of the glycolytic chain (Chapter 7) acting in reverse, but with NAD replaced by NADP (in plant cells, not in bacteria). Oxygen is extracted by high-energy electrons of NADPH, with additional help from ATP hydrolysis. The resulting water is contained in the products of ATP splitting. (It will appear explicitly after regeneration of the ATP by photophosphorylation.

3. Rearrangement leading to the *formation of glucose*, the main product of photosynthesis (which is stored in the form of its polymer, starch). A complex enzyme system, known as the pentose-phosphate-pathway system, catalyzes this rearrangement, which allows twelve molecules of the three-carbon compound phosphoglyceraldehyde (36 C) to be rearranged into six molecules of the five-carbon ribulose-5-phosphate (30 C) and one molecule of glucose (6 C). This system is also present, with only minor modifications, in most heterotrophic organisms, where it supports important metabolic functions.

4. *Phosphorylation of ribulose-5-phosphate*, making it ready for participation in step 1.

Note that, for each turn of the cycle, six molecules of CO_2 are condensed reductively into one molecule of glucose, with the help of twelve electron pairs furnished by NADPH and of additional energy provided by the hydrolysis of eighteen molecules of ATP.

By illuminating leaves in the presence of radioactive $^{14}CO_2$ for shorter and shorter durations (as little as a few seconds) and then extracting and separating the labeled compounds, workers were able to identify the earliest product of CO_2 fixation as phosphoglyceric acid. With this clue, an appropriate reaction was sought and duly found. It is a remarkable process, in which a *bis*-phosphorylated five-carbon sugar, ribulose-1,5-*bis*phosphate, reacts with CO_2 and water to give two molecules of phosphoglyceric acid. The enzyme catalyzing this reaction is bound to the thylakoid membrane.

The substance made by this enzyme, phosphoglyceric acid, is not unknown to us. We met it in Chapter 7 as the product of the central phosphorylating oxidative step of glycolysis. Indeed, photosynthesis has borrowed this primeval oxphos unit for its key reductive step—first, as such and, later, with NAD replaced by NADP as coenzyme. Taking advantage of the high energy potential that it maintains in the NADH or NADPH reservoir, it causes the electrons to flow in the reverse direction through this oxphos unit, consuming ATP in the process. And, in this manner, phosphoglycer*aldehyde* is made from phosphoglyceric *acid* with the help of one pair of electrons supplied by NADH or NADPH and of the free energy of hydrolysis of ATP.

These are the key reactions of photosynthesis. They are incorporated within a complex cyclic process known as the Calvin cycle, named for the American chemist who first unraveled it. We will not look at the details, but only at the end result. A molecule of glucose is made from 6 molecules of CO_2 and 6 molecules of water. The 12 electron pairs required to extract the 12 excessive oxygen atoms are supplied by NADPH, and the additional energy needs are covered by the hydrolysis of a total of 18 molecules of ATP.

What does this make in terms of energy yield? On the debit side, we have the oxidation of 12 NADPH molecules (660 kcal) plus the breakdown of 18 ATPs (252 kcal), or a total of 912 kcal per gram-molecule of glucose made. On the credit side, we have 686 kcal per gram-molecule, the free energy of oxidation of glucose, for an overall efficiency of 75 per cent. To achieve this, a minimum of 12×164 kcal of light energy is required. Thus the maximum efficiency of photosynthesis is 35 per cent. It could be somewhat lower, depending on the number and location of the oxphos units in the electron-transport chain.

Animals lack the key enzymes of the Calvin cycle. But they do share with plants the use of NADPH as main electron donor in reductive syntheses—for instance, in the conversion of carbohydrate into fat. Like plants, they maintain a relatively high energy level in their NADPH reservoir, of the order of 55 kcal per pair of electron-equivalents, or 1,200 mV. They do this with the help of certain selected substrates acting as high-level electron donors: glucose-6-phosphate is the main one. If need be, they can send electrons from NADH to $NADP^+$ by an energy-dependent process in the mitochondria. Conversely, there is a strictly regulated overflow mechanism, whereby excess electrons in the NADPH reservoir are allowed to fall down to NAD^+.

Growth and Multiplication of Chloroplasts

What has been said about the biogenesis of mitochondria applies similarly to the chloroplasts. These particles also possess a complete genetic machinery. It is actually richer than that of the mitochondria, but nevertheless controls the synthesis of only a small portion of the total chloroplast proteins. It has the same bacterial characters as the mitochondrial system and presumably likewise represents a vestigial remnant of the ancestral endosymbiont, in this case cyanobacteria. Chloroplasts, like mitochondria, display genetic continuity and can undergo mutations that are transmitted via the cytoplasm.

The autonomy of chloroplasts is, however, limited, as is that of mitochondria. Most of their constituents are manufactured by cytoplasmic ribosomes under the control of nuclear genes. As in mitochondria, the manner in which these constituents are transferred across the particle membranes and inserted into their proper location is not well understood.

11 Peroxisomes and Sundry Other Microbodies

The first microbody was spotted in mouse kidney in the early 1950s by a Swedish anatomist, who found it to be of such nondescript character that he could think of no better name for it. Soon after, similar objects were seen in rat liver and, later, in a variety of cells in both the plant and animal kingdoms. Although widespread, microbodies remain restricted to certain cell types. In mammals, they have been detected mainly in liver and kidney.

Wherever they are found, microbodies have a similar appearance. They are roughly spherical structures, from 0.5 to 1.0 μm in diameter, which makes them slightly smaller than mitochondria. They are surrounded by a membrane and are most often filled with a fairly compact, amorphous matrix. In some cells, this matrix hides a gem—a dense, crystalloid core, or nucleoid, of beautifully delicate texture. This kind of purely morphological information leaves much to the imagination, and it did indeed provide the grounds for some rather fanciful constructions. When biochemical clues eventually put investigators on the right track, the truth turned out to be even stranger than the fictions, revealing that not one, but several, distinct types of microbodies exist, each concerned with what looks like a primitive, if not primeval, set of metabolic reactions.

Peroxisomes and Glyoxysomes

The peroxisomes are the most widespread form of microbody. They derive their name from hydrogen peroxide, H_2O_2, a key intermediate in their oxida-

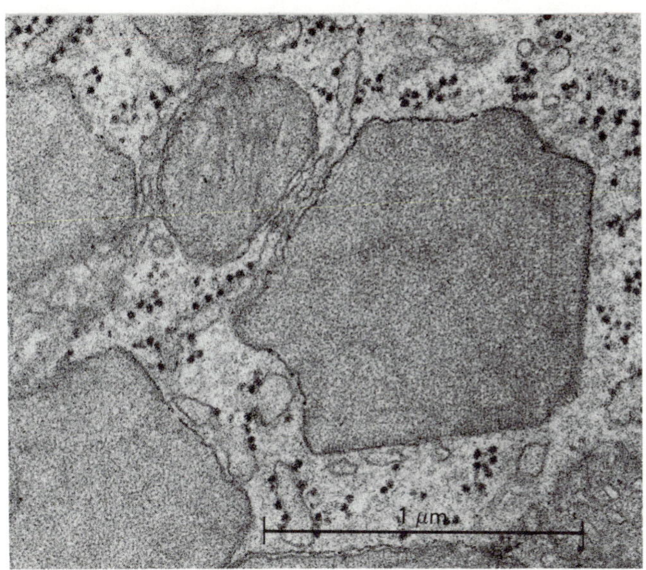

Peroxisomes in a rat kidney proximal tubule cell. Three irregularly shaped microbodies surround a small mitochondrion.

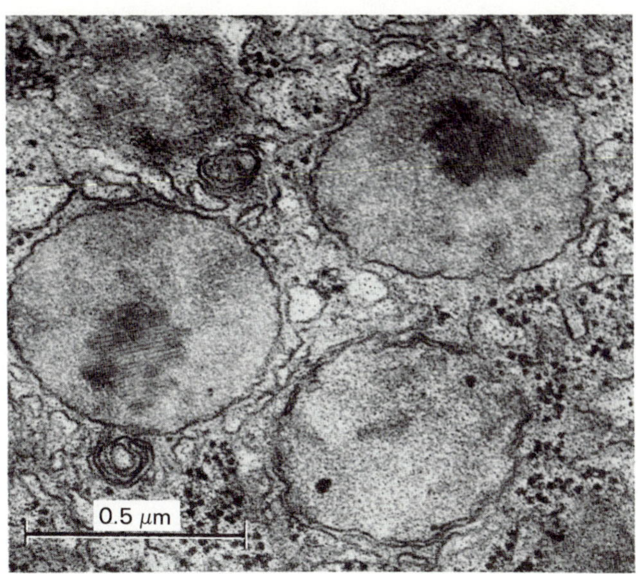

Two peroxisomes in rat liver. The regularly structured core contains an enzyme that oxidizes uric acid with the formation of H_2O_2.

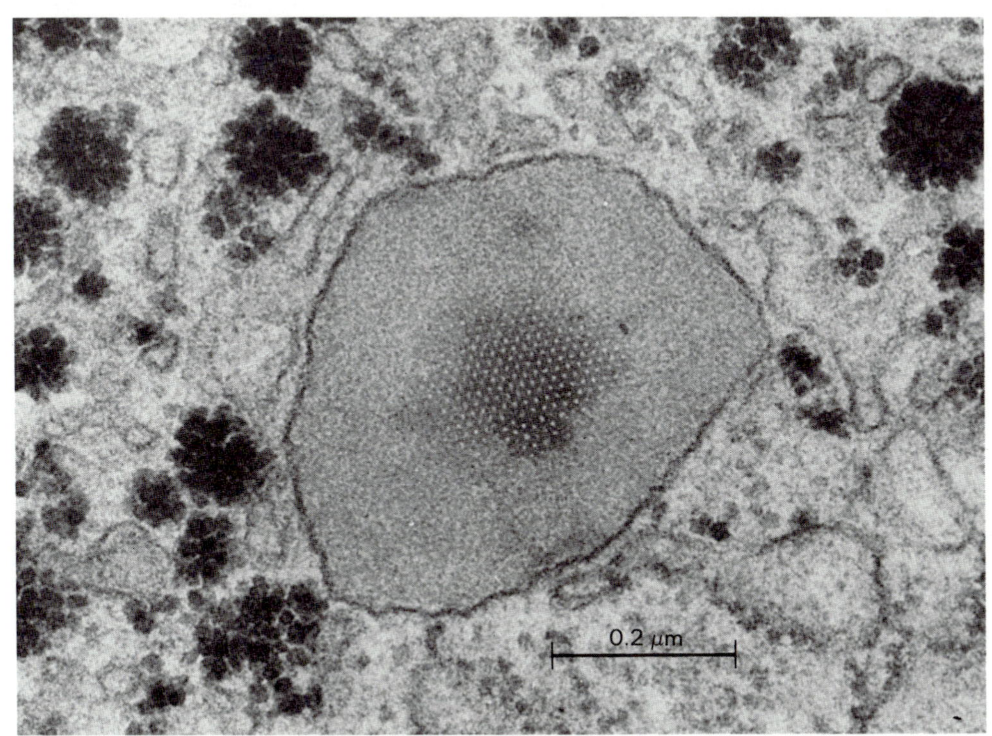

A peroxisome in guinea pig liver. The crystalloid core, which, as in rat liver peroxisomes, contains urate oxidase, shows a regular hexagonal lattice.

tive metabolism. Hydrogen peroxide is produced by a family of enzymes called type II oxidases, mostly flavoproteins, sometimes also copper proteins, that use molecular oxygen as electron acceptor and reduce it to H_2O_2:

$$RH_2 + O_2 \longrightarrow R + H_2O_2$$

The electron donors in these reactions include amino acids, fatty acyl-coenzyme A derivatives, purines, and some products of carbohydrate metabolism, such as lactic acid—in short, representatives of all major classes of foodstuffs.

The hydrogen peroxide produced in the peroxisomes is further metabolized through the action of catalase, a green hemoprotein that reduces H_2O_2 to water with, as electron donor, certain small organic molecules, such as ethanol, methanol, or formic acid, and, in the absence of a suitable donor, hydrogen peroxide itself:

$$R'H_2 + H_2O_2 \longrightarrow R' + 2\ H_2O$$

or

$$H_2O_2 + H_2O_2 \longrightarrow O_2 + 2\ H_2O$$

The latter reaction is called a dismutation. In it, one molecule of hydrogen peroxide is reduced and the other oxidized. The net result is destruction of H_2O_2 with evolution of oxygen. You can readily see catalase in action by dabbing some hydrogen peroxide on a wound: watch for the frothing oxygen. Catalase is one of the fastest-acting enzymes. It was detected as early as 1818 by the French chemist Jacques Thenard, the discoverer of H_2O_2.

Acting together, the peroxisomal oxidases and catalase allow oxidations to take place through the following mechanism:

$$O_2 \xrightarrow[RH_2 \rightarrow R]{} H_2O_2 \underset{R'H_2 \rightarrow R'}{\overset{(H_2O_2) \rightarrow O_2}{\rightrightarrows}} 2\ H_2O$$

Compare this "respiratory chain" with that of mitochondria and you have all the difference between a reckless profligate and a prudent economizer. Both achieve the same result, bringing about the oxidation of all sorts of foodstuffs, with reduction of oxygen to water. But, whereas in mitochondria much of the free energy of combustion is retrieved in the form of usable ATP, in the peroxisomes it is all dissipated as heat. This drawback is compensated by a remarkable simplicity of design. It seems likely that the peroxisomal type of respiration arose long before the delicate mitochondrial microcircuits were put together; it may represent one of the earliest adaptations of living organisms to oxygen, as was pointed out in Chapter 9.

In addition to oxidizing enzymes, peroxisomes may, depending on cell type, contain a variety of other systems. Prominent among these is the glyoxylate cycle, a Krebs cycle variant that plays an essential role in the conversion of fat into carbohydrate. This is an important biochemical process; among other functions, it makes it possible for fatty seedlings, such as castor beans, to utilize their oily stores upon germination. Hence, the peroxisomes that contain this system have been named glyoxysomes. In animals, the glyoxylate cycle was an early victim of evolution. It is present in lower forms but not in many higher vertebrates, including mammals. We can make fat out of carbohydrate—a privilege many of us would be happy to forsake—but we are unable to reverse the process because our peroxisomes are not glyoxysomes.

Microbodies identifiable as peroxisomes, while not present in every type of cell, are universally distributed in nature. They are found in a large variety of plants and animals, in molds, fungi, and protozoa. This, together with the primitive character of their respiratory machinery, has suggested that all peroxisomes may be evolutionary descendants of a single ancestral particle that was present in the first eukaryotic cell from which all plants and animals are believed to have originated. Possibly this ancestral peroxisome was already part of the primitive phagocyte in premitochondrial times and fulfilled, crudely but efficiently, the essential function of guarding against oxygen.

This hypothesis leaves the historian with two perplexing questions. First, why were peroxisomes not elimi-

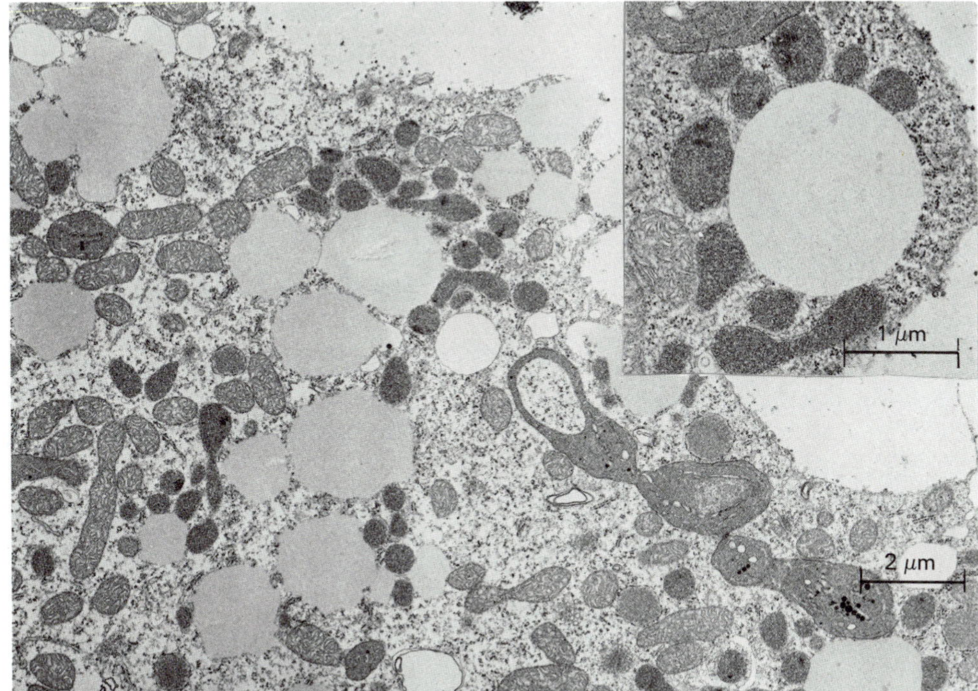

This electron micrograph of castor bean endosperm shows numerous lipid droplets (appearing as holes), mitochondria (bodies of moderate density with internal membranes), and glyoxysomes (darker microbodies with coarsely granular matrix). The inset illustrates the characteristic clustering of glyoxysomes around lipid, which they convert into carbohydrate.

nated by natural selection after they were joined by the better-equipped mitochondria? This may well have happened in many cells, although even this possibility is debatable. Many of the mammalian cells that lack regular microbodies possess tiny, membrane-bounded particles that, on the basis of their content of catalase and sometimes of other typical peroxisomal enzymes, have been identified as microperoxisomes.

In any case, in spite of a considerable attrition of their metabolic potential—the loss of the glyoxylate cycle is an example—mammalian peroxisomes are by no means "fossil organelles" of no more than vestigial interest. They undoubtedly fulfill important functions. They are involved in several ways in the metabolism of lipids, possibly also of cholesterol; in the breakdown of amino acids, including the D-amino acids, which occur only in bacteria; and in the catabolism of purines. So far as is known, they do not provide the cells with utilizable energy, but the heat they produce may sometimes be physiologically significant. It is said that the special brown-fat tissue which helps the Norwegian rat survive the rigors of winter owes part of its thermogenic capacity to peroxisomes. There are also indications that peroxisomes play an essential role in the synthesis of certain phospholipids called plasmalogens. Perhaps the best proof of the importance of these organelles in mammals is provided by pathology. There is a rare human genetic deficiency, known as Zellweger's disease, in which no morphologically detectable microbodies are seen in the liver or kidneys, where they normally are present in large numbers. Infants afflicted with this condition do not survive more than a few months.

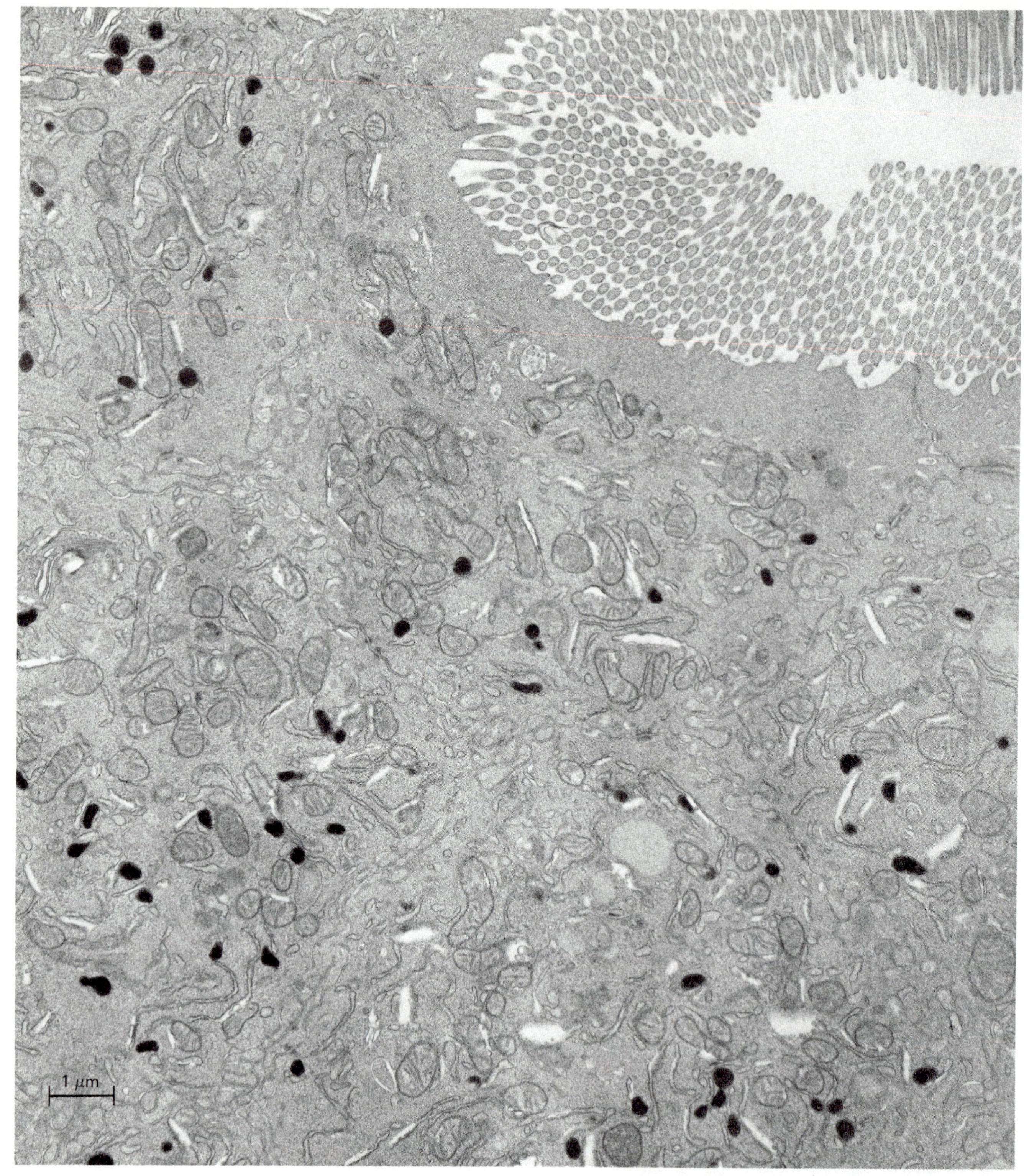
1 μm

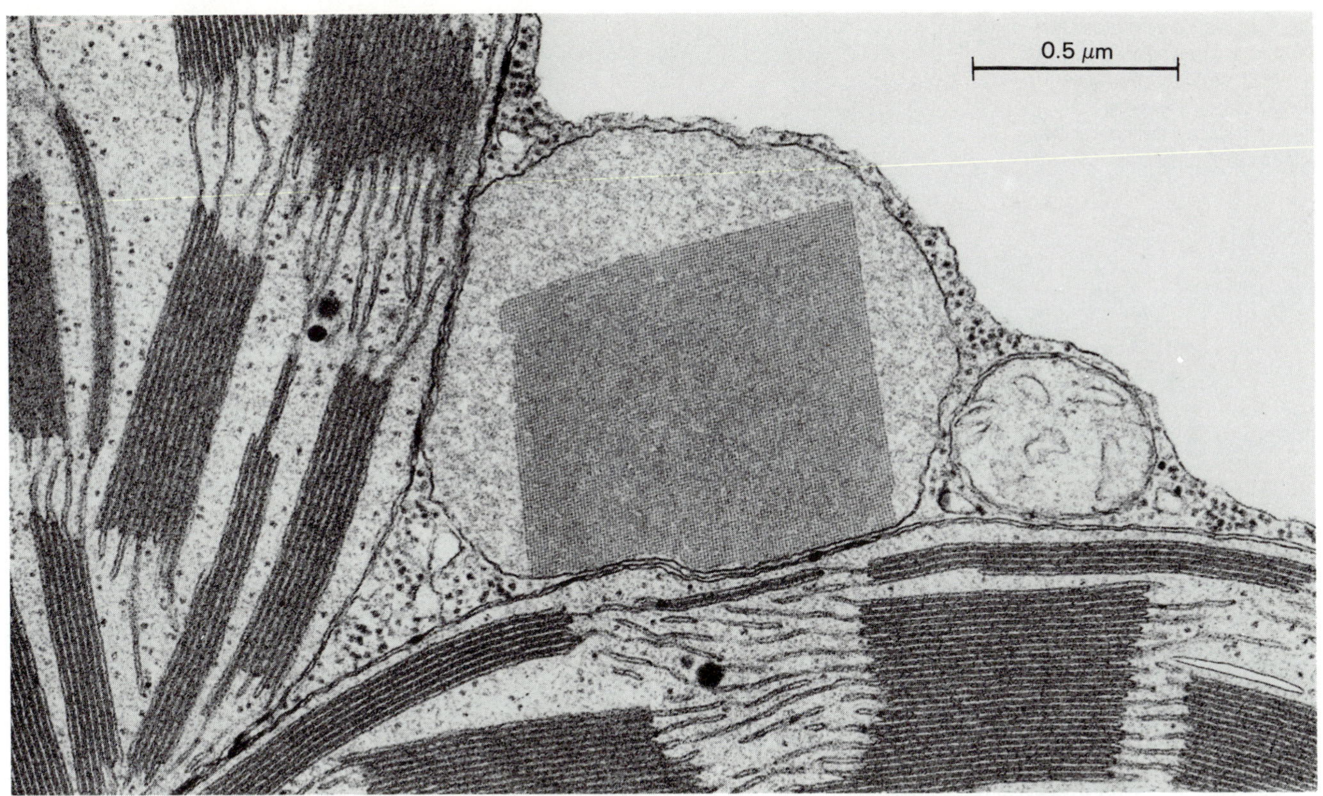

A peroxisome containing a beautiful crystalline structure believed to consist of catalase is seen wedged between two chloroplasts, and adjacent to a mitochondrion, in this electron micrograph of a tobacco-leaf cell. Localization reflects the close metabolic collaboration that exists between the three organelles.

On the facing page: Microperoxisomes in guinea pig ileum have been stained by means of a cytochemical reaction based on the oxidation of diaminobenzidine by catalase in the presence of H_2O_2. Note the numerous mitochondria in the cytoplasm and the sections through microvilli in the upper right-hand corner.

The functions of peroxisomes are even more important and varied in lower animals and protozoa, and especially in the plant world. We have already seen the key metabolic role of those peroxisomes that qualify as glyoxysomes in the conversion of fat into carbohydrate. Innumerable plants that pack their seeds with lipid reserves would be unable to reproduce without this function. Green leaves, which contain peroxisomes of uncommon beauty, depend on an intricate trilateral collaboration between these particles, chloroplasts, and mitochondria to salvage photosynthetic products that would otherwise be lost through photorespiration—a form of light-induced oxidation that results from a peculiar subversion of the central enzyme of the Calvin cycle by oxygen. These interactions are of great economic importance, as they affect the net photosynthetic yield of many crop plants. A particularly remarkable adaptation of peroxisomes is observed in certain strains of yeasts. When grown on metha-

These pictures show the remarkable structure and packing of peroxisomes in cells of yeast *Hansenula polymorpha* adapted to growing in a medium containing methanol as the sole source of carbon.

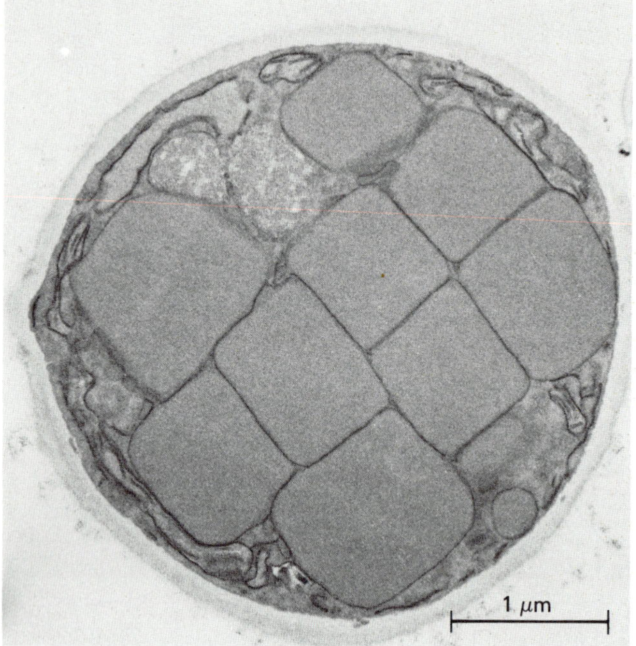

A. A thin section shows cytoplasm occupied mostly by box-shaped peroxisomes.

B. A similar view, but with relief emphasized, is offered by a carbon replica of a preparation that has been fractured in the frozen state and then shadowed with platinum.

nol, on alkanes (which are saturated hydrocarbons such as are found in petroleum), or on some other outlandish nutrient as sole source of carbon, these cells respond by an enormous increase in the level of some of the peroxisomal enzymes needed for the breakdown of those substrates. To accommodate these enzymes, the peroxisomes develop greatly in size and in number. Sometimes the cells become so packed with peroxisomes that the particles assume the shape of square boxes.

When one comes to think of it, it is not really so surprising that cells should retain peroxisomes, even after acquiring mitochondria. Right from the beginning, peroxisomes may have possessed useful attributes that were lacking in the mitochondria, and therefore their retention was favored. Or some of their characters may have become essential at a later stage—for instance, after deletion of a mitochondrial property by a mutation. Nevertheless, peroxisomes are not what they used to be—or so it seems.

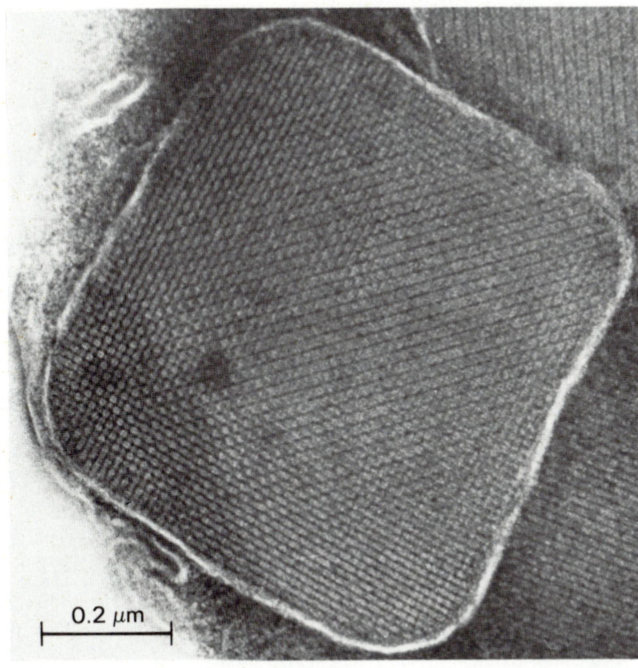

C. A higher magnification shows that a peroxisome is largely filled by a crystalline structure. This structure is made up of molecules of an enzyme that oxidizes methanol, with the formation of H_2O_2.

Putting together all the individual functions that are found today in the different types of peroxisomes, and assuming that they are all inherited from the same ancestor, one arrives at the picture of a mighty metabolizer of considerable versatility, of which today's descendants are but pale replicas.

Which brings us to the second question the historian is likely to ask: How did the primitive eukaryote—or, perhaps, its prokaryotic phagocytic precursor—manage to acquire such a powerful organelle in the first place? A few years ago, this question would hardly have been asked, since it was widely believed that peroxisomes grow out of the ER in the form of bulging buds and may even, in some cells, remain attached to that structure by membranous stalks. But this has become doubtful. Investigators looking for nascent peroxisomal proteins in ER cisternae have been disappointed. They have found instead that these proteins are made by free polysomes and released into the cytosol, from which they subsequently gain access to the peroxisomes by an unknown mechanism. As to the purported connections between peroxisomes and ER, they may have to be re-examined. We must remember that all membranes look more or less alike in the electron microscope and that investigators examining thin sections do not have our advantage of a three-dimensional view of the cellular structures. Unfortunately, the picture that we see is not as clear as we would have it. It does, however, look as though peroxisomes form clusters of interconnected particles, but perhaps separate from the ER.

If peroxisomes are not offshoots of the ER—which, incidentally, they still could be phylogenetically, if not ontogenetically—the origin of these mysterious microbodies must be sought elsewhere. Some cell detectives, their imaginations fired by the tales that circulate about mitochondria and chloroplasts, have raised the possibility that the peroxisomes may be evolutionary descendants of yet other bacterial endosymbionts, adopted at a more remote time. The primitive character of their respiratory mechanisms is consistent with this possibility. Nevertheless, it must be emphasized that, in contrast with mitochondria and chloroplasts, peroxisomes offer not a shred of evidence in support of such speculations. As far as is known, they are entirely lacking in DNA, ribosomes, or other parts of a genetic machinery. But lack of evidence does not invalidate the hypothesis. After all, if an endosymbiont can become integrated more than 90 per cent, as are mitochondria and chloroplasts, why not 100 per cent? Some day we may get an answer to these questions by interrogating the microbody proteins and extracting from their molecular structure some vestigial mementos of their evolutionary history (see Chapter 18).

Hydrogenosomes

On numerous occasions, the need for a suitable acceptor to collect electrons at their exit from oxphos units has been emphasized. But only passing mention has been made of what would appear to be the simplest solution to this problem—namely, the use of protons as acceptors:

$$2\,e^- + 2\,H^+ \longrightarrow H_2$$

Protons are available everywhere. Why, then, do we not all breathe out hydrogen, instead of having to inhale oxygen? Some organisms actually do just that. But they are very few, probably because protons are about the most unprofitable electron acceptors a cell can use. Hydrogen formation takes place at a very high potential level—of the order of 50 or more kcal above water/oxygen level. This means that, if usable energy is to be gained from the transaction through an oxphos unit intercalated between donor and acceptor, the substrate must deliver its electrons at a level of at least 64 kcal per pair of electron-equivalents. Only a handful of substances (pyruvic acid is one) can muster this kind of energy.

In the bacterial world, there is a small group of obligatory anaerobes, known as clostridia, that produce hydrogen. Among them are the pathogenic organisms that cause gaseous gangrene, an infection that develops in wounds that are not properly "aired." These bacteria seem to have an exceptionally old evolutionary history and to have developed as an independent branch in those very early days when life was still exclusively anaerobic. They never learned to adapt to the presence of oxygen. Or perhaps they lost the art after learning it. Remember, all this is conjecture.

Remarkably, the capacity to produce hydrogen is shared in the animal world by a tiny subgroup of protozoa, the trichomonads, most of which are parasites of the genital tracts of both animals and humans. They are one of the main agents of sexually transmitted diseases. These parasites can live both in the presence and in the absence of air. Anaerobically, they support their energy needs by a special reaction in which pyruvic acid is oxidized to acetic acid and CO_2 and the liberated electrons are channeled to hydrogen by way of a substrate-level oxphos unit:

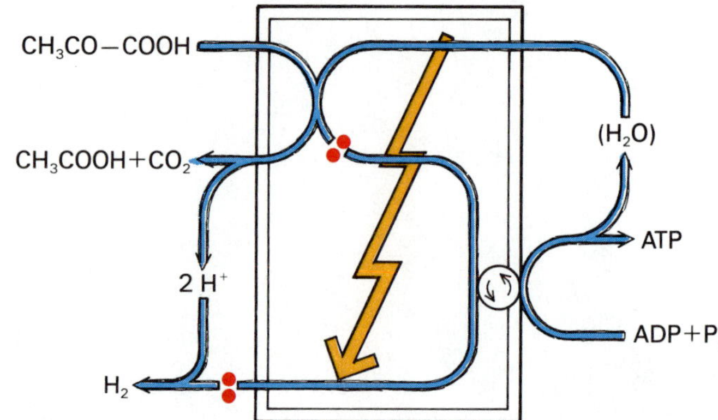

In the oxidation of pyruvic acid to acetic acid and CO_2 by hydrogenosomes, electrons are transferred to protons (with the formation of molecular hydrogen) through a substrate-level oxphos unit. As in the substrate-level phosphorylation of the glycolytic chain (see the illustration on p. 113), the oxygen added to the substrate arises from a cryptic molecule of water generated by the condensation of ADP $+P_i$.

$$CH_3\text{—}CO\text{—}COOH + (H_2O) \longrightarrow CH_3\text{—}COOH + CO_2 + 2\,e^- + 2\,H^+$$

$$ADP + P_i \longrightarrow ATP + (H_2O)$$

$$2\,e^- + 2\,H^+ \longrightarrow H_2$$

Aerobically the organisms shut off their hydrogen production and switch over to an oxygen-dependent metabolism, though one of a very simple kind since they do not contain mitochondria.

They do have prominent microbodies, however—a fact that, some years ago, raised the exciting possibility that here, perhaps, was a direct descendant of the primitive phagocyte, the offspring of a cell that had never captured a mitochondrial endosymbiont, constrained unto this day to depend on the original ancestral peroxisome for its oxidative metabolism. As often happens in science, when this possibility was put to the acid test of experi-

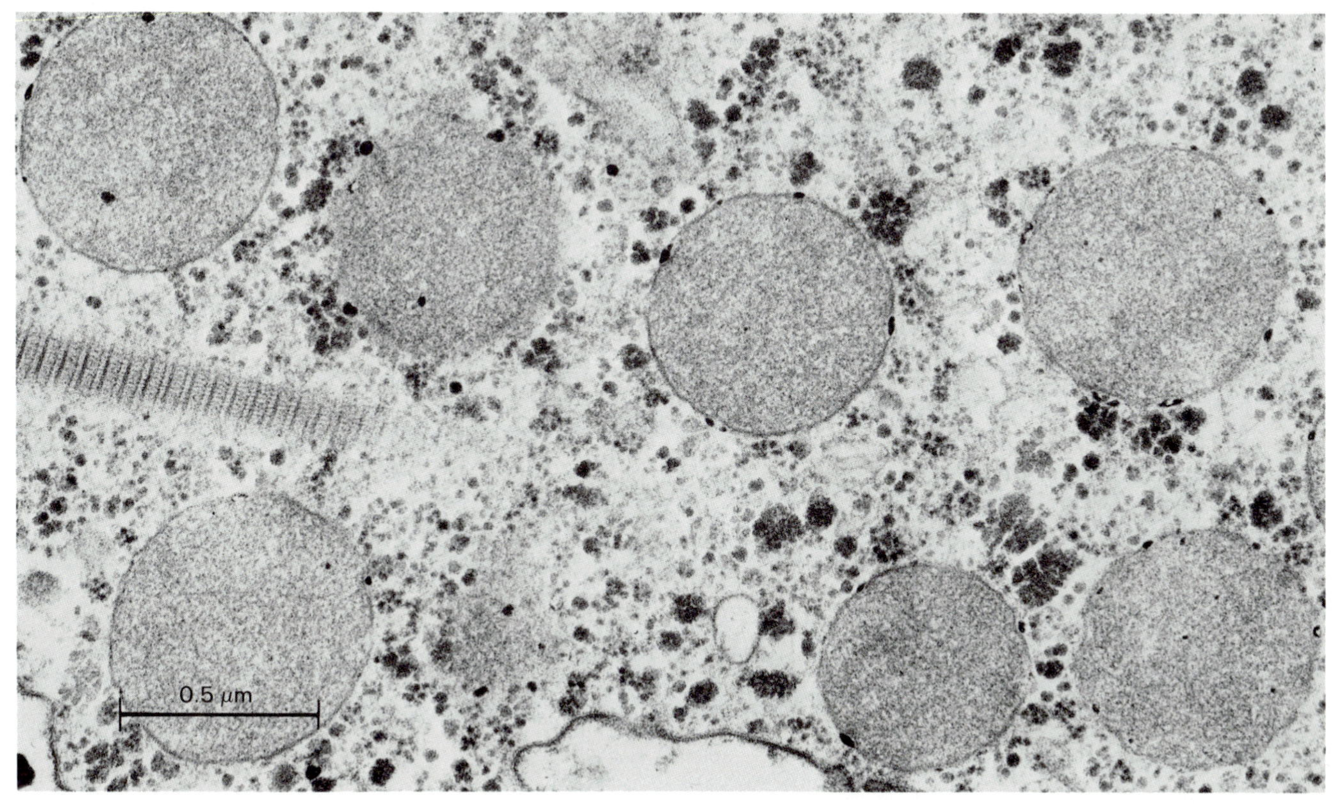

Hydrogenosomes in the human parasite *Trichomonas vaginalis.* The striated object on the left is a riblike structure, the costa.

ment, the answer was a disappointing "No." But, at the same time, it opened a door into the entirely unexpected. The microbodies of trichomonads are hydrogenosomes; they contain the whole system responsible for the oxphos-linked, hydrogen-producing breakdown of pyruvic acid.

Like all discoveries, this finding has raised new questions. The most provocative one is: Where does the hydrogenosome come from? Is it simply another kind of bud growing out of the ER? Or did it originate in some primitive clostridium that was caught and domesticated in bygone days by a member of the voracious phagocyte family? There are as yet no clues to this question. But some of the hydrogenosome proteins are being probed to reveal their ancestry.

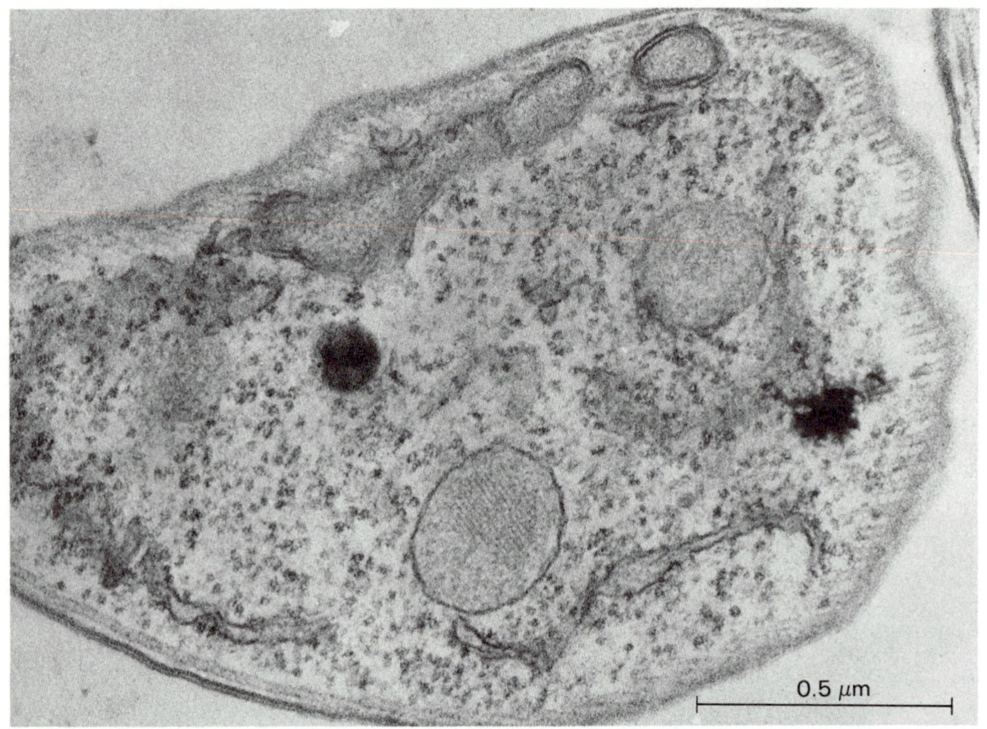

This electron micrograph shows two glycosomes in the cytoplasm of *Trypanosoma brucei*, the agent of sleeping sickness of cattle. One glycosome contains a crystalline core of unknown chemical composition.

Glycosomes

Not long ago, investigators exploring a subgroup of protozoa called trypanosomes discovered yet another species of microbody, which, amazingly, was found to contain a big segment of the organism's glycolytic chain. In every other cell type that has been analyzed in this respect, the glycolytic system has invariably been found in the cytosol. That is where we met it for the first time, earlier in our tour. How, then, did it manage to become segregated within the confines of a membrane-bounded microbody in the trypanosome cytoplasm?

Here, again, imagination is lured to fill the void created by ignorance. Could it be, we cannot help wondering, that the glycosome—as it is called for obvious reasons—is also derived from an endosymbiont, this time a primitive fermenter? Could it even be that its captor was saved by its prey from the consequences of what would otherwise have been a lethally crippling mutation of its own glycolytic apparatus? If so, the event is still making waves—and rather unpleasant ones—more than one billion years after it happened. For among the trypanosomes are some of the nastiest parasites of animals and humans, causing such severe conditions as African sleeping sickness and the dreaded South American Chagas' disease.

Whatever their origin, glycosomes are more than just glycolytic organelles. They have recently been found to contain a number of other enzymes, including some that are also found in peroxisomes. An evolutionary relation between the two particles is thus not excluded.

12 Cytobones and Cytomuscles

The shapes of living cells, including those of their infinitely varied surface folds, protrusions, and invaginations so dramatically revealed by scanning electron microscopy, are determined largely by an intricate inner scaffolding of crisscrossed solid fibers and hollow tubes, which make up what is known as the cytoskeleton. These cytoskeletal elements are the girders and cables that we noticed when we first entered the cytosol. But our view of them was obstructed by all the balloon-shaped objects that fill the cytoplasm. Should we be able to remove these objects, we would notice that the components of the cytoskeleton may reach impressive lengths, sometimes extending right across the cell, and that they are linked together in a variety of patterns of astounding beauty and often remarkable regularity. Empty a cluttered museum of its contents and only then will you appreciate the graceful and cleverly engineered architecture of the exhibition halls.

The Cytoskeleton

We cannot empty a cluttered cell in this way, but we can use antibodies directed against cytoskeletal proteins to coat in a very specific manner the structural elements of which these proteins are a part. If these antibodies are made to carry molecules of a fluorescent dye, the structures to which they attach will be as though covered with a coat of fluorescent paint. Illuminate such a decorated cell with ultraviolet light, and the whole cytoskeletal framework coated by the antibody will light up

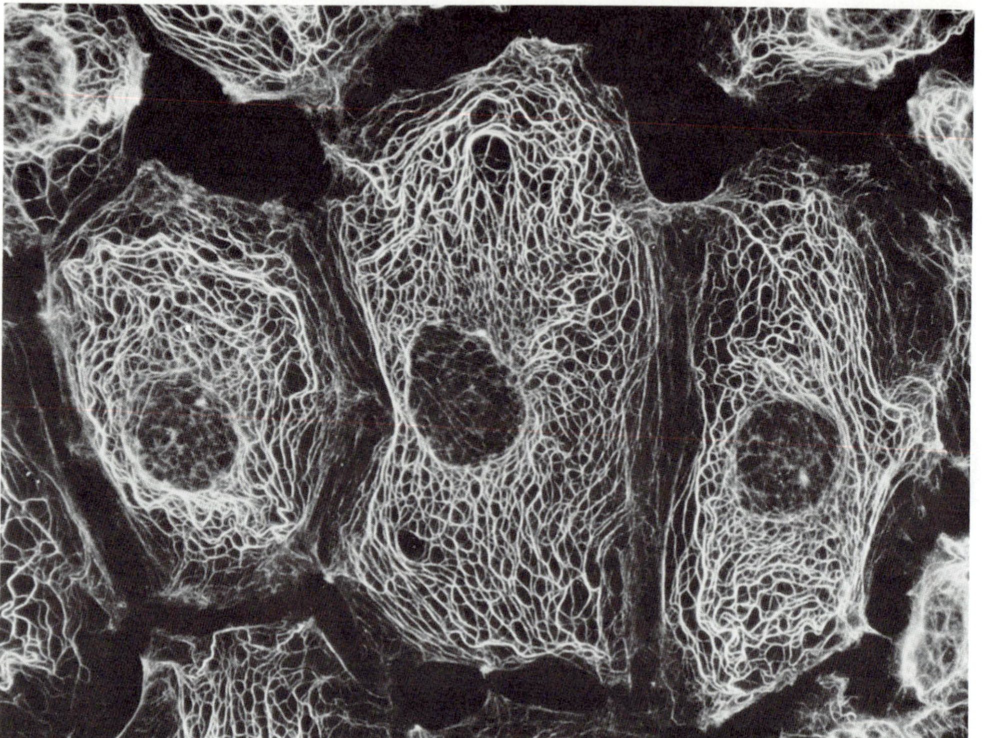

Keratin network of rat kangaroo cells (PtK_2) is revealed by immunofluorescence technique.

brightly on a dark background, as though delineated by neon tubing.

Called immunofluorescence, this elegant technique requires a great deal of prior biochemical research, since the proteins must be purified before they can be used to elicit antibodies. An additional difficulty is that several of the cytoskeletal proteins are poor immunogens—that is, they do not readily induce the production of antibodies when injected into an animal of a different species. The reason is that homologous cytoskeletal proteins, even of widely different species, have closely similar chemical structures and thereby fail to be recognized as foreign by the immune system. This kind of evolutionary conservation of structure is a strong indication that the functional properties of the proteins depend on specific amino-acid sequences that can suffer few modifications without the properties being lost. Most of the mutations that affect such proteins are therefore incompatible with maintenance of function and are eliminated by natural selection.

Beautiful as they are, the pictures revealed by immunofluorescence provide only frozen "stills" of what are constantly changing patterns. Cells alter their shapes all the time; they reshuffle their contents, generate cytoplasmic streams, propel some of their granules in saltatory motion, bend and distort membranes. They move around, creep, crawl, swim, crouch, contract, stretch out, flatten themselves on surfaces, or squeeze through narrow openings. They catch, surround, and engulf bulky objects, push out and retract pseudopods, spew out the contents of stored granules. They wave undulating veils, sweep spiraling flagella, and create currents around themselves by means of beating cilia. What of the cytoskeleton in all this frenzy of movement?

The answer to this question, as might be expected, is that the cytoskeleton is not a rigid framework. In fact, it is not even an articulated framework of the type its name might suggest. It is a much more versatile and complex arrangement of structural elements, only some of which are true fixtures. Others have the remarkable ability of rapidly disassembling into small building blocks and reassembling from them into a different shape, thus explaining the protean metamorphoses of which cells are capable.

Keratinization. This drawing illustrates the progressive differentiation of an epidermal cell as it is pushed toward the skin surface by newly generated cells arising from the germinating layer.

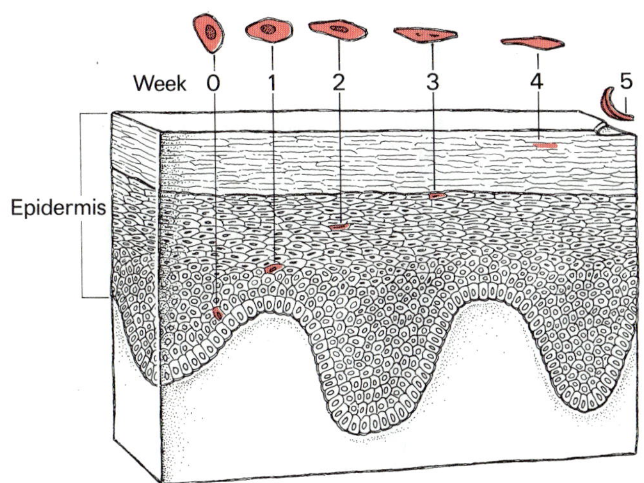

As to the more orderly forms of movement accomplished by cells or by their parts, they seem to depend mostly on the sliding of one structural element on another, elicited by ATP-powered cross-bridges between the two.

The various parts of this machinery are built entirely of protein molecules of various kinds, naturally endowed with the ability to interact either with their congeners or with other cytoskeletal proteins so as to fall together automatically into all those delicate scaffoldings and laceworks that are rendered visible in the ultraviolet light microscope by fluorescent antibodies. Only rarely does the electron microscope reveal these patterns in all their beauty and complexity. In most cases, it allows only glimpses of individual skeletal elements, which appear either as hollow microtubules or as solid filaments. The latter are divided according to their diameters into thin (6–7 nm), intermediate (8–10 nm), and thick (15–20 nm) filaments. Exceptionally, as in the fibrils of muscle cells, or in cilia and other regularly built surface projections, the molecular architecture of the cytoskeleton attains such a remarkable degree of orderliness as to appear clearly in a single, appropriately oriented, ultrathin section. Such pictures have helped considerably in the interpretation of the more random-looking patterns seen most commonly. They have also provided morphologists with some of their greatest esthetic delights.

Functionally, certain cytoskeletal elements serve only to provide the cells with an inner, essentially static, framework and to tie them together by various types of junctions. Most intermediate filaments are of this kind. In conformity with their role in structural differentiation, they tend to be different in different cell types. Examples are the keratin of epithelial cells, the vimentin of mesenchymal cells, the desmin of muscle cells, and the neurofilaments of nerve cells. Other cytoskeletal elements are present in all cells, although their arrangements may vary greatly from one cell type to another; they serve a dynamic, as well as a structural, function. Typical of these are the actin-myosin system, the microtubule-dynein system, and clathrin. We will take a quick look at each of these within the inevitable constraints of time and space that restrict the scope and depth of our visit.

Keratin, the Stuff of Toughness

From the Greek *keras*, horn, the term keratin designates a family of sulfur-rich, fibrous proteins that are the main constituents of skin, hide, hair, horns, hooves, nails, claws, scales, feathers, beaks—that is, of all the outer coverings and appendages with which vertebrates arm themselves against the assaults of the outside world. Unlike the protective coverings of plants and lower animals, which are built extracellularly from secretory products, those of the vertebrates are constructed intracellularly as a result of a very remarkable differentiation process.

The process takes place in epithelial cells. These originate from stem cells that form a layer deep beneath the skin and that divide asymmetrically into undifferentiated stem cells, which remain in the germinating layer, and differentiating daughter cells. As superficial skin layers slough off, daughter cells move slowly to replace them, pushed toward the surface by more newly generated daughter cells.

Should we visit such a young epithelial cell at the beginning of its journey toward the periphery, we would find its cytoplasm traversed here and there by sturdy fibers, about 8 nm thick (one-third of an inch at our millionfold magnification). On closer inspection, the fibers are seen to be bundles of filaments, themselves made up of

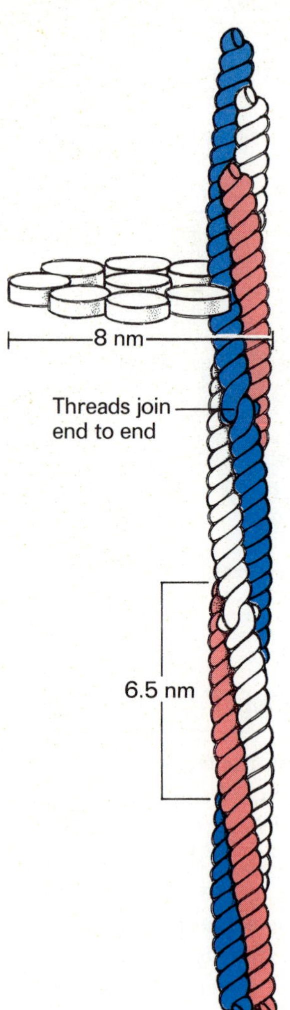

Structure of keratin filament. The basic unit is a triple-stranded, left-handed coiled coil of 6.5-nm pitch, between 60 and 90 nm long, made of three different polypeptide chains, each of which is regularly twisted in a right-handed α-helix over most of its length. Such threads join end to end and aggregate laterally to form a characteristic intermediate filament of 8-nm diameter.

thinner threads that, at first sight, remind us of the structure of tropocollagen (p. 39). Like the basic unit of connective-tissue fibers, the elementary keratin subunit is a triple-stranded coiled coil made of polypeptide chains that are themselves helically twisted. Closer inspection, however, reveals major differences between the two structures. In keratin, the triple-stranded coil is left-handed instead of right-handed and its pitch is about 6.5 nm, as against 3 nm. In addition, its three constitutive polypeptide chains are usually different, and each is twisted into a typical right-handed α-helix, of the kind seen in many proteins (Chapter 2). We will encounter remarkable samples of this structure a little later, in myosin and tropomyosin. The kind of helical structure exhibited by collagen, on the other hand, with its left-handed turn and 1-nm pitch, is unique and explained by this molecule's very unusual amino-acid composition and sequence. Keratin subunits are also shorter than tropocollagen: from 60 to 90 nm, as against 300 nm. They aggregate laterally and longitudinally to form the characteristic intermediate filaments of 8-nm diameter.

Together, the keratin bundles form a loose, three-dimensional network that envelops the nucleus within a basketlike arrangement and is strung between a number of reinforced plates distributed over the plasma membrane. As a rule, these plates are cemented by means of some sort of dense adhesive material to similar plates on the surface of neighboring cells, forming the adhering junctions, or desmosomes, that attracted our attention when we first entered a blood vessel (Chapter 2). The anchoring keratin fibers are the tonofilaments, which are seen to radiate transversely from the desmosomes into the depth of each cell and to straddle the desmosomes so as to establish direct links between the keratin networks of the two cells.

As epithelial cells move slowly from the germinating stem-cell layer to the skin surface, they devote themselves increasingly to the production of keratin, which they support by the massive autophagic destruction of their contents. At the same time, the keratin fibers become increasingly cross-linked with each other and with amorphous matrix components by means of disulfide bonds. By the time the cells reach the skin surface, they are all shriveled and dried out, lifeless and inert, but very, very tough. Firmly riveted to each other by the desmosomes, they form a single protective sheet, the horny layer, which is continuously shed by surface desquamation (flaking off) and replaced by newly differentiated epithelial cells.

If sloughing off is prevented by welding together of superimposed cell layers, the horny layer will thicken to a callus. Change the pattern of growth somewhat, modify the nature or proportion of the various polypeptides that make up the keratin, change their degree of cross-linking, and the resulting structure may be a scale, a nail, a claw, a horn, or a beak. If the differentiating cells grow into a tube instead of a sheet, they end up as all sorts of hairs and spines and, through more complex designs, as the intricate combinations of quills, barbs, and down that cover feathery creatures. There is really no end to the number of architectural variations that evolution has constructed from the central keratinization theme. For millenia, some of these variations have provided mankind with clothing and tools. Even in our contemporary plastic age, no man-made fiber can yet compete with the keratin produced by an Angora goat.

As mentioned above, keratin is a characteristic product of epithelial cells. In other cell types, the supporting intermediate filaments are made from other proteins, which assemble differently. Time does not permit their detailed inventory, and we must return to the "generalized cell" that forms the main object of our tour.

The Actin-Myosin System

Actin and myosin are two proteins that form a very remarkable locomotor combination, first discovered in muscle cells but now known to exist in all cells, in which thin actin filaments make up what may be called the "cytobones," and thick myosin filaments the "cytomuscles."

The actin filaments are usually grouped into slender bundles, which may be seen stretched like telephone wires across the cytoplasm (stress fibers) or meshed together in the form of variously structured cables, belts, felts, or webs, which most often serve as backing for the plasma membrane. They also make up the axial shafts that provide support to such surface protrusions as microvilli. Each actin filament is from 6 to 7 nm thick (one-fourth of an inch at our millionfold magnification) and consists of two intertwined threads.

Unlike the other filamentous structures that we have encountered so far, such as collagen and keratin, these threads are not built from fibrillar proteins but rather from small protein balls of the usual curled-up type, linked by complementary binding sites situated at their two poles. An actin thread thus resembles one of those indefinitely growing trains or chains that can be built out of identical interlocking pieces often put in the hands of

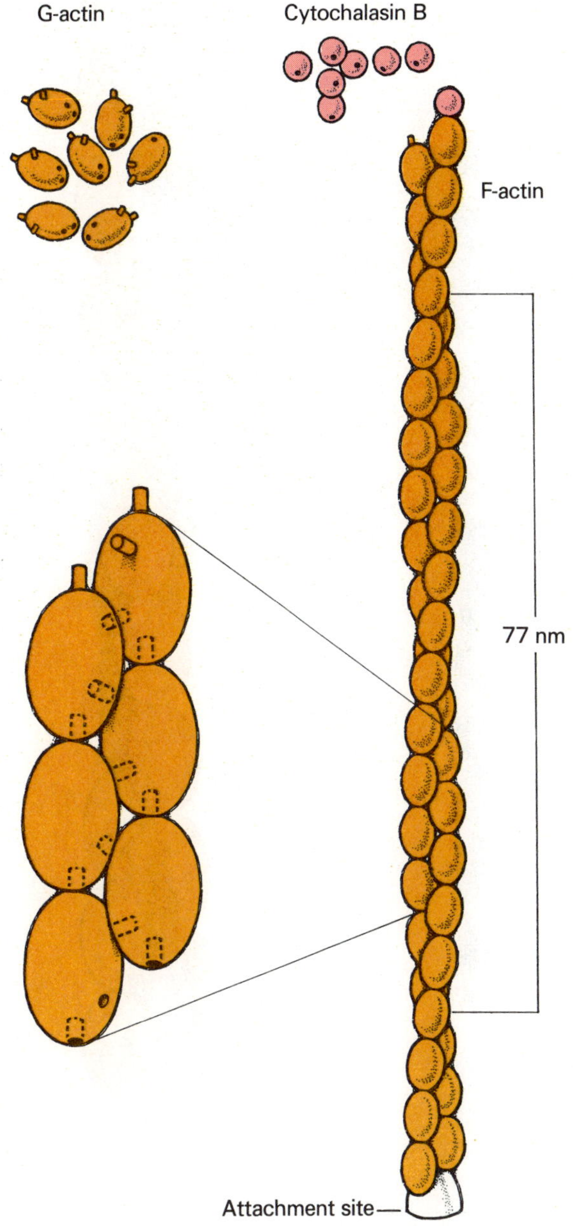

Structure of actin filament. Globular, or G, actin is an egg-shaped protein molecule fitted with two pairs of complementary (lock-key) attachment sites, one polar, the other lateral, which allow the molecules to polymerize into a twisted double-stranded string (fibrous, or F, actin) containing fourteen pairs of monomeric molecules per complete turn (77 nm). Polymerization is directional, away from the filament's attachment point, and reversible. It requires ATP and is blocked by cytochalasin B, a fungal poison.

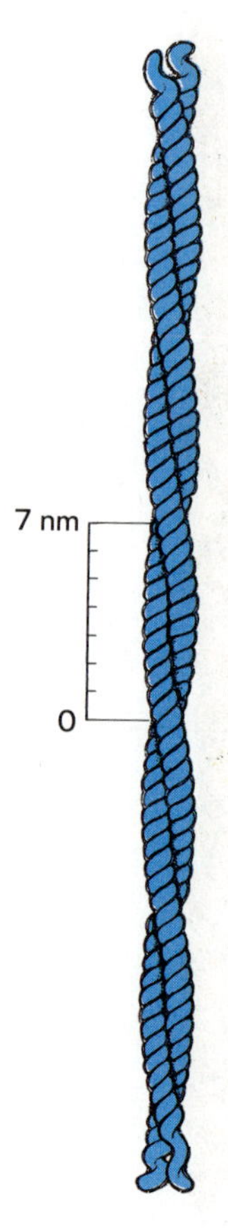

Structure of tropomyosin. This double-stranded coiled coil is 41 nm long and consists of two identical polypeptide chains, α-helical over most of their length, and twisted around each other with a 7-nm pitch.

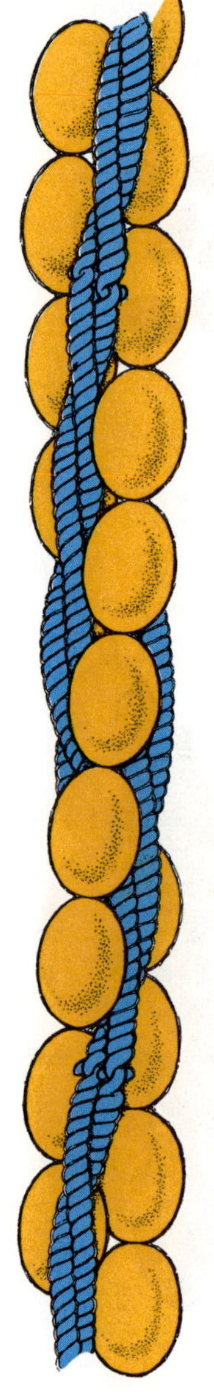

Arrangement of tropomyosin around actin filament.

young children to test their combinatorial skills. Like such a chain, it has a polarity, defined by the direction of the "lock-key" axis of the pieces. In the case of actin, the chain is not single but double, because the pieces have a second lock-key set situated laterally. The binding of two parallel threads by this set occurs with a slight right-handed twist, generating an elongated double-stranded helix with a 77-nm pitch (about 3 inches at our magnification) containing fourteen pairs of subunits per turn.

The subunits of the actin filaments are called globular, or G, actin. The product of their polymerization is fibrous, or F, actin. The interconvertibility of these two forms is what enables living cells to dismantle certain parts of their actin skeleton and to reassemble them into a different pattern. These operations are controlled by a complex set of directions, so far largely undeciphered, involving the active participation of ATP. There is a special binding site for ATP on the G-actin molecule. Upon polymerization, this ATP is hydrolyzed, and the resulting ADP remains bound to the F-actin. ADP has to be displaced by ATP before the structure can be disassembled.

As a rule, the spiral grooves separating the two strands of an actin filament are occupied by a thin thread made up of another protein molecule. Called tropomyosin (Greek *tropê*, turn; *mys*, muscle), this molecule is a left-handed coiled coil, about 41 nm long, made of two identical polypeptide chains, α-helical over most of their length and twisted around each other with a 7-nm pitch. It provides us with our second example of the α-helix, which we saw a short while ago when we looked at the keratin fibers. Note, however, that tropomyosin is double stranded, whereas keratin is triple stranded. In the actin filament, each tropomyosin molecule extends over seven G-actin subunits, or exactly one-half turn of the helix. So, for each 77-nm turn in F-actin, there are two pairs of dimeric tropomyosin molecules twisting together with fourteen pairs of G-actin subunits. In striated muscle cells but not in others, this repeating structure bears four additional molecules of another protein, called troponin, attached to the actin thread near the junction between successive tropomyosin molecules. Troponin, as we shall see, plays a key role in the regulation of muscle contraction.

Actin filaments are attached by one of their extremities—always the same one with respect to the polarity of the filament—to a sort of flat, disklike structure in which the end of the filament is firmly anchored. A protein called α-actinin has been identified as a component of this structure, but there are several others. These anchoring points remain when filaments are disassembled and provide nucleation sites from which new filaments can grow. Cytochalasin B, a poison of fungal origin, has the property of binding to the growing ends of actin filaments and of inhibiting their further elongation. It thereby interferes with all cellular activities that require remodeling of the actin cytoskeleton. It has become an invaluable tool in the identification and analysis of such activities.

The anchoring disks of the actin filaments are themselves attached to special patches on the inner face of the plasma membrane, or to each other, or to other intracellular organelles. In this way, they form the wide variety of arrangements that are seen in different cells or in the same cell in different functional states. Their attachments are mediated by a number of structural proteins that bear such suggestive names as vinculin (Latin *vincula*, bond), ankyrin (Greek *ankyra*, hook), or spectrin (so named because it was first isolated from erythrocyte "ghosts"). In muscle cells, numerous attachment disks are knit together by a protein called desmin (Greek *desmos*, bond) to form what looks like two brushes glued back to back, with their bristles pointing in opposite directions.

We will see how these bristles function in muscle contraction. But before that, it is helpful to explore the whole length of a single bundle of half a dozen actin filaments, starting from their anchoring point. At first the filaments form a rather loose skein and seem to be held together only by their common anchorage. But eventually they tighten into a cylinder in which the six filaments are grouped around a long, central shaft about 15 nm thick. This shaft extends beyond the ends of the filaments and continues over a distance of several hundred nanometers—sometimes as much as 1 μm—to end up as the core of a second bundle of filaments, symmetrically oriented. What we see at our millionfold magnification is a rigid rod, as much as several feet long and three-fifths of an inch

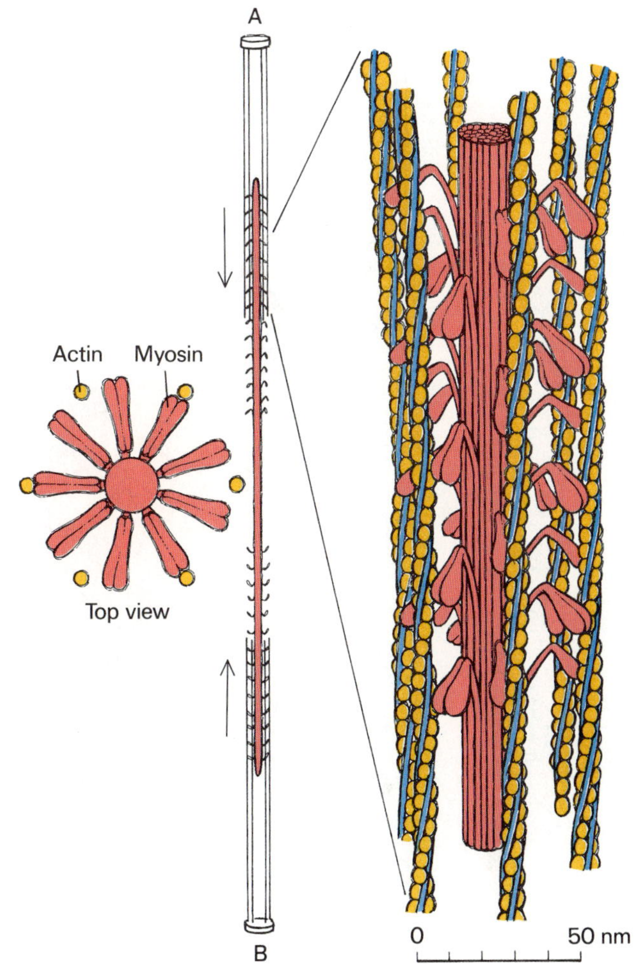

Elementary cytomuscle. A thick myosin filament connects two opposed bundles of six thin actin filaments. The forcible pulling of the actin filaments toward the center of the myosin shaft draws anchoring points A and B together.

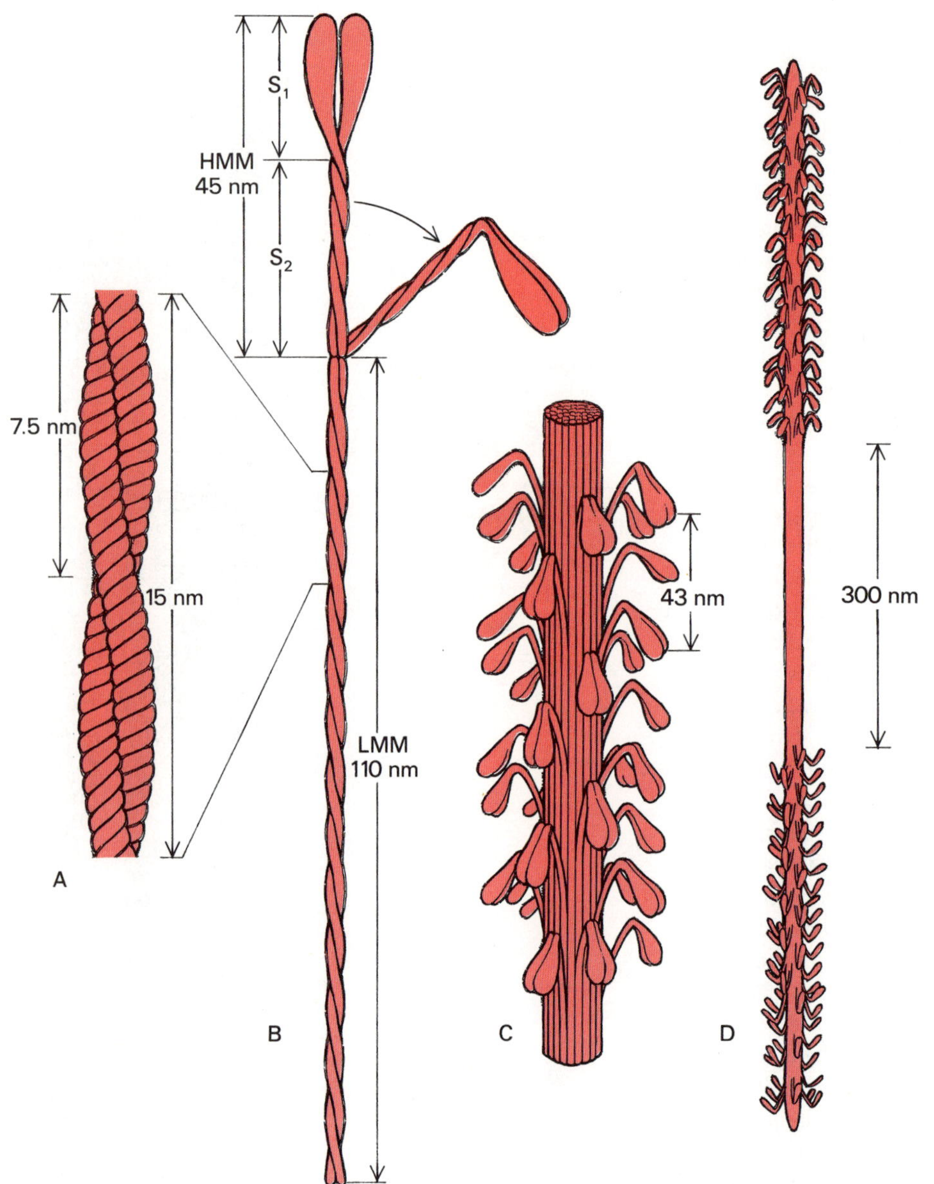

Structure of thick filament. The elementary unit is the myosin molecule (B), a long, duplex, protein molecule, consisting of a 135-nm long tail bearing twin, bulbous ends, or heads. The tail is a coiled coil of two identical, almost perfectly α-helical strands (shown in detail in part A). One irregularity in this structure provides a hinge between a hydrophobic stem, 110 nm long, and a more hydrophilic neck bearing the heads. This hinge is easily cut by proteolytic enzymes, to give heavy (HMM) and light (LMM) meromyosins. The hinge between the head and neck in HMM is also cut easily, with the formation of subfragments S_1 and S_2. Myosin molecules have the remarkable property of assembling into thick rods, from 15 to 20 nm thick and from 1.5 to more than 5 μm long. Molecules join by their hydrophobic LMM tails and have their hydrophobic HMM stalks and heads protruding on the surface of the rod in a regular helical arrangement. This arrangement is symmetrical (as shown in part D). In the middle of the rod is a bare segment, 300 nm long, made of the interdigitated stems that bear the heads of the first row on each side.

thick, connecting two opposite bundles of six identical double-stranded wires about one-quarter of an inch thick.

This connecting rod is made of myosin, a unique kind of protein that, like tropomyosin, derives its name from the Greek word *mys*. It is a long, duplex molecule, 155 nm in length (a little more than half a foot, at our magnification). Its shape has been likened to that of a golf club (with a split head) or, more romantically, to that of a long-stemmed, twin flower. The stem, about 135 nm long and 2 nm thick, is constructed very much like tropomyosin. It is a left-handed, double-stranded coiled coil, with a pitch of about 7.5 nm, made of two identical, almost perfectly α-helical, polypeptide chains. The myosin stem is the longest such structure known in nature. At its upper end, the two strands separate into distinct flexible stalks that progressively curl up into twin, globular "flowers," or "heads," each of which consists of the terminal portion of one of the stem's strands, or heavy chain, intertwined with two additional light chains. There is one irregularity in the structure of the myosin stem. It provides a hinge between a hydrophobic distal portion, about 110 nm long, and a more hydrophilic proximal segment bearing the heads. This irregularity allows cutting by the proteolytic enzyme trypsin, which sections myosin into two parts, named light meromyosin (LMM) and heavy meromyosin (HMM). A second hinge exists between the heads and the stem in HMM. It is susceptible to splitting by another proteolytic enzyme, papain, which cuts HMM into two subfragments, S_1 and S_2.

Myosin molecules have the remarkable property of associating spontaneously into bundles, which look like miniature exotic trees elegantly festooned with flowers. The "trunk" of the tree is made of the stems of the myosin molecules, joined lengthwise by their hydrophobic LMM portions and staggered helically so as to allow indefinite lengthening over constant thickness (except at the top, which is tapered). The heads, or flowers, protrude from this trunk on the more hydrophilic LMM portions and form a garland that spirals around the trunk in a regular helical fashion. The exact configuration of this arrangement is still a matter of debate among experts and may vary according to cell type and animal species. What seems to be constant is a longitudinal spacing of the heads by 14.3 nm along the trunk axis. An additional, but absolutely essential, feature of this assembly process is that it is symmetrical. Two such trees are always joined by their roots, so that each myosin filament is made of two garlanded shafts of opposite polarity, linked together by a naked central portion about 300 nm long.

Polarity of actin filament determined by "decoration" with isolated myosin heads (S_1 fragments): (below) electron micrograph; (right) model.

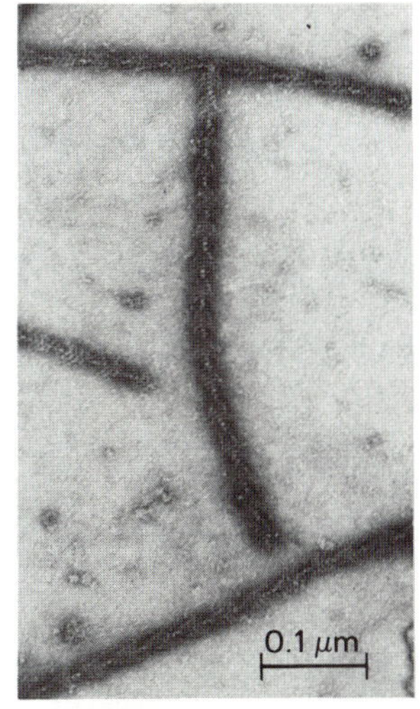

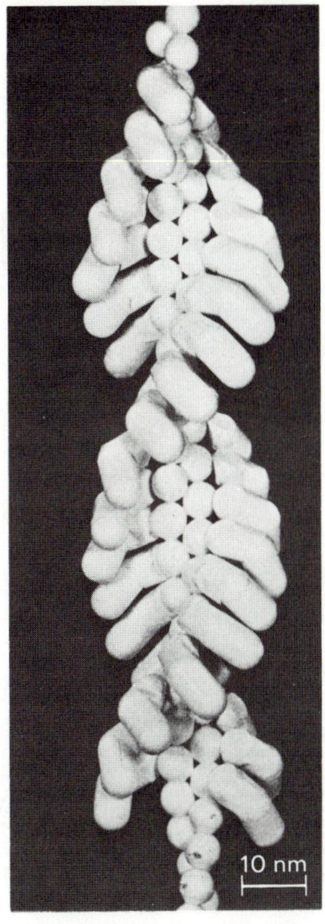

The flowers on the myosin tree—in more technical terms, the heads of the myosin molecules—are the hooks to which the actin filaments are attached. Their disposition around the trunk is such that six parallel actin filaments can bind to them, thus surrounding the myosin shaft with a sheath of hexagonal symmetry. One such sheath can form at each extremity of the myosin shaft, which thereby can connect two opposed actin bundles.

Such positioning requires the actin filaments to bear appropriately oriented binding sites for myosin heads. This property can be used to identify actin filaments in tissue sections and to determine their polarity. Myosin

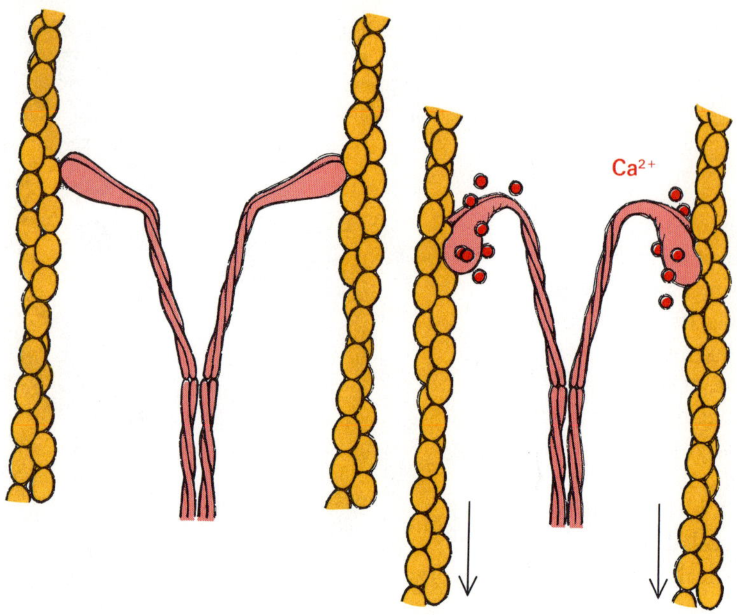

Sliding-filament mechanism of actin-myosin contractile system. In the presence of calcium ions, the ATP-bearing myosin heads bind to actin and become active as ATP-hydrolyzing enzymes. ATP hydrolysis is obligatorily coupled to the bending of the myosin heads and the longitudinal displacement of the actin filaments with respect to the central myosin filament.

heads, cut off enzymatically from their stalk and purified, are used as reagents. They bind specifically to the actin filaments and "decorate" them in a typical arrowhead pattern, with the arrows pointing in the direction of the free end of the actin filament.

So far, we have seen the myosin shaft serving simply as a connecting rod between two bundles of actin filaments. The sight is graceful but static. But let just a few calcium ions reach the system, and you will be offered one of the most dazzling displays of molecular pyrotechnics to entertain a cell tourist. With dramatic suddenness, the myosin heads come alive. Wherever they are attached to a binding site on an actin filament, they bend violently on their stalks, tugging the filament some 10 nm inward, in the direction of the middle of the shaft. This done, they relinquish their hold. By that time, however, other myosin heads have come into register with actin binding sites and exert a further 10-nm pull. This goes on as long as calcium ions (and ATP, see below) are present. With all actin filaments at each end of the myosin shaft being pulled inward in the same way, the net result is to make the two opposing actin bundles that are joined by the myosin shaft slide into each other and drag their respective anchoring points closer together. Thus the distance between these anchoring points contracts, although none of the filaments that link them actually shorten. They slide along each other by means of what may be described as a molecular ratchet.

This process accomplishes mechanical work and therefore requires energy, which, you will hardly be surprised to learn, is supplied by the hydrolysis of ATP into ADP and P_i. Myosin heads are really ATP-hydrolyzing enzymes (ATPases) of a very special kind. They can exert their catalytic activity only if (1) they are bound to actin; (2) they are activated by calcium ions (in a way that we shall consider later); and (3) they are permitted to bend at the same time. The obligatory coupling between the chemical event of ATP hydrolysis and the conformational change that forces the actin-linked myosin head to bend on its stalk and drag along the actin filament is the fundamental property that allows the actin-myosin system to convert the free energy of hydrolysis of ATP into mechanical work.

The kind of work that is accomplished depends on the cellular localization of the two anchoring points that are being pulled toward each other and on their topological relationships. Often, one point remains fixed and the pull is exerted entirely on the other. Take cell creeping, for example. A patch of plasma membrane to which an actin bundle is anchored is made to stick firmly to the substrate (adhesion plaque), after which the actin-myosin complex shortens and pulls whatever is attached to it—which

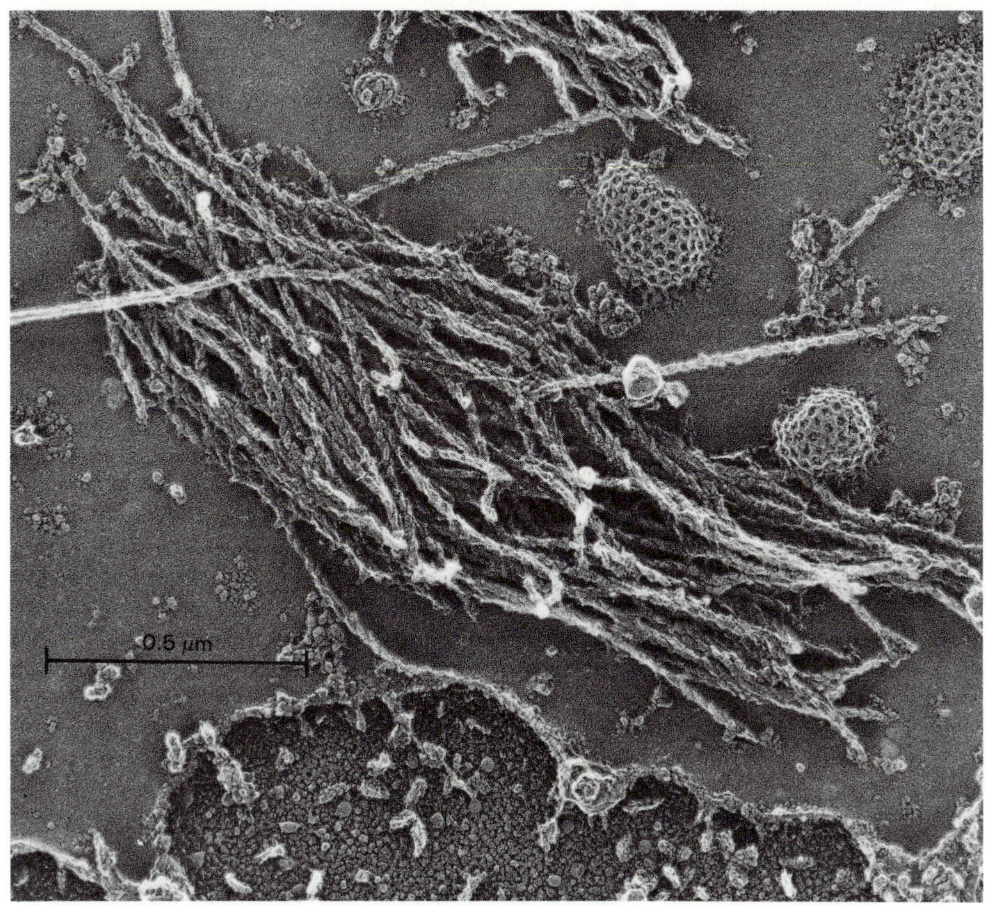

Special quick-freeze, deep-etch technique allows startling visualization of bundles of actin fibers lining the cytoplasmic face of the plasma membrane of a fibroblast. The actin fibers have been "decorated" with isolated myosin heads (S_1 fragments). Structures with a chicken-wire appearance are clathrin baskets over coated pits (see pp. 218–222).

may be the whole cell through its interconnected cytoskeleton—toward the adhesion site. When the movement is over, the adhesion plaque is "unstuck," and new ones form in front through the operation of a ruffling extension, called the lamellipodium (Latin *lamella*, small, thin plate; Greek *pous*, gen. *podos*, foot), which sends out sticky, fingerlike protrusions called filopodia (Latin *filum*, thread). This, according to some observers, could be the basic mechanism of amoeboid movement, whereby cells propel themselves toward (positive chemotaxis) or away from (negative chemotaxis) certain objects that send out chemical signals.

Actin fibers, as we have seen, often occur just below the plasma membrane in the form of belts or webs of various shapes, sometimes connected to axial strings in the inner core of microvilli and other cellular protrusions. One can see how tightening the belt may bring about cellular constriction; how retracting the web may purse, wrinkle, pucker, round up, evaginate, invaginate, or otherwise deform the surface of the cell; how pulling the strings may bend a microvillus. And, if the anchoring points are attached to intracellular organelles, one can see how these can be displaced with respect to each other and with respect to the cell surface.

Indeed, one can visualize, at least as a purely theoretical exercise, how the immense variety of movements displayed by living cells can be brought about by the coordinated operation of hundreds of tiny myosin cytomuscles busily pulling miniature actin cytobones in various directions, just as the complex movements of our own body can be explained by the manner in which our muscles pull our bones or fold our skin. But it will take an enormous amount of ultramicroscopic exploration and dissection of a kind that is still largely beyond our present technical

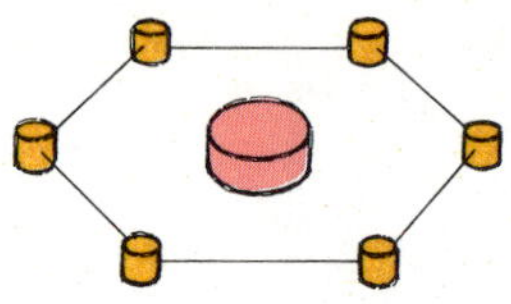

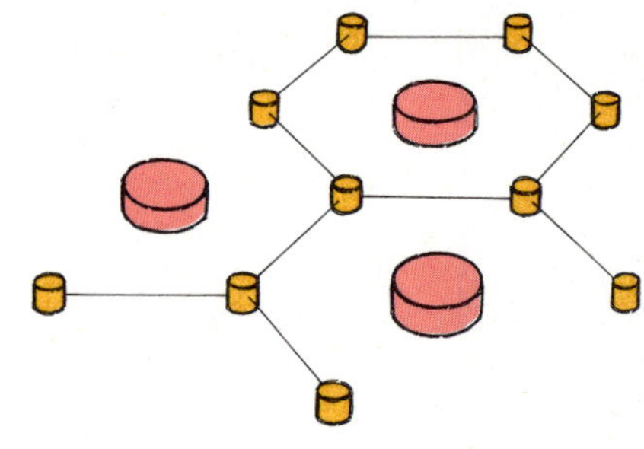

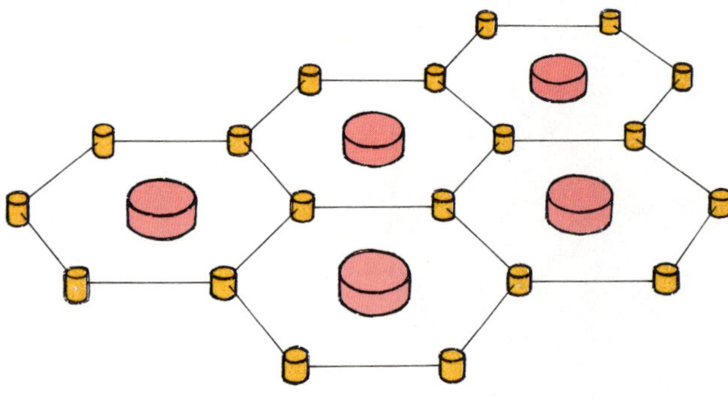

Diagram showing how the lateral joining of actin and myosin filaments leads to the assembly of a myofibril with regular hexagonal arrangement.

As actin filaments approach their points of attachment to the actinin-desmin meshwork (Z line), their arrangement changes from hexagonal to rectangular.

means before the detailed anatomy of the cell's bones and muscles can be mapped out and understood in dynamic terms. These difficulties are further compounded by the fact that the shapes and interconnections of many cytobones change, as do the sites of many cytomuscular insertions. Furthermore, as we shall see, cells contain a second cytoskeletal system made of microtubules and a second type of cytomuscle, dynein, associated with it.

The force exerted by an actin-myosin cytomuscle depends on the number of myosin heads that are pulling together at any time—that is, on the length over which filaments interact—and on the number of parallel filaments that join in this effort. If we examine the single actin-myosin connection that we have explored so far, we see that each of the six actin filaments that surround the myosin shaft has unoccupied myosin-binding sites on its exposed surface. It follows from this property that the six filaments of the first bundle can bind additional myosin shafts, which in turn can surround themselves with more filaments, and so on. Repeating such an arrangement, we end up with a structure of hexagonal symmetry, in which each myosin filament is surrounded by six actin filaments and each actin filament is flanked by three myosin shafts alternating with three actin filaments. Should you immobilize the actin filaments in such a structure by gluing together their α-actinin roots and then slide out the intercalated myosin rods, you would end up with a bristle of thin actin filaments planted from 15 to 25 nm apart at every angle of a regular hexagonal lattice, in which each hexagon surrounds a hole capable of precisely accommodating a thick myosin filament. This is what desmin, the connecting protein referred to earlier, manages to achieve in the muscle cell, except that it actually forms a rectangular network, which changes to a hexagonal arrangement upon insertion of myosin shafts into the actin bristle.

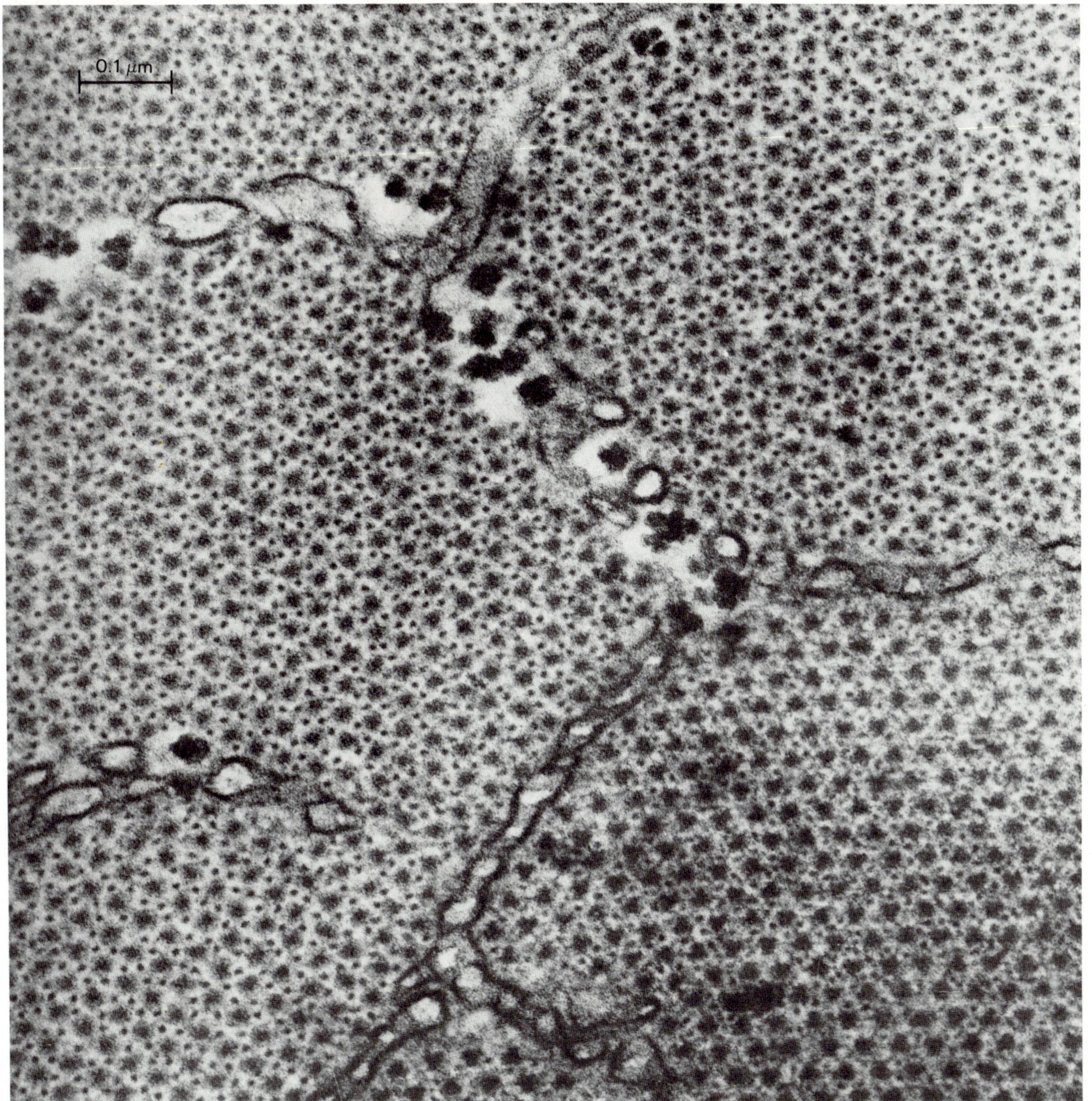

Hexagonal arrangement of thin (actin) and thick (myosin) filaments appears clearly in this electron micrograph of a transverse section through a striated muscle fiber. This picture and that of a longitudinal section shown on the next page are historical documents. They were taken more than thirty years ago and played a major role in the development of the sliding-filament theory of muscular contraction.

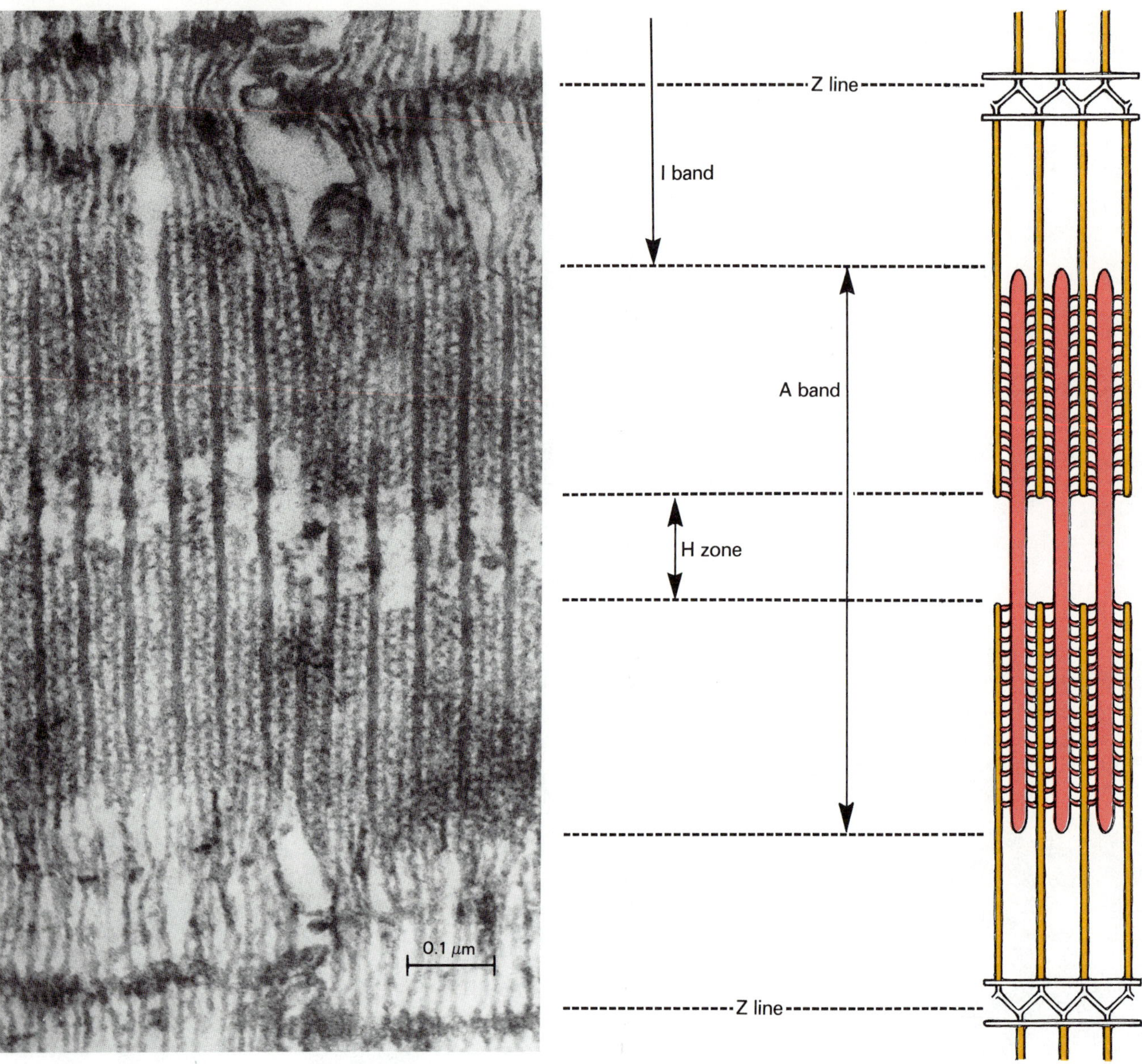

This electron micrograph of a longitudinal section through a striated muscle fiber illustrates the parallel arrangement of thin and thick filaments. The picture shows a complete sarcomere limited at both ends by a Z line (top and bottom) in which the actin filaments are joined to the actin filaments of adjacent sarcomeres by an actinin-desmin meshwork. The shape of the myosin filaments, with beveled extremities, lateral attachments (heads) to actin filaments, and bare midpiece, is clearly recognizable. This myofibril is in a partly contracted state, with thin and thick filaments having slid past each other over a considerable distance. There are two, not one, actin filaments between myosin filaments because of the orientation of the section (see the diagram on p. 202).

Diagrammatic representation at the right above illustrates the main features of cross-striation. The A (anisotropic) band, of constant length, is made of myosin filaments interdigitating at both ends with variable lengths of actin filaments, depending on the degree of contraction. The I (isotropic) band contains only actin; its length depends on the degree of contraction. The I band is bisected by a Z line, where the actinin-desmin meshwork knits the ends of two sets of actin filaments together. Midpieces of the myosin filaments make up the H zone in the middle of the A band.

In vertebrate striated muscle, all the actin filaments have the same length of 1 μm and all the myosin shafts are 1.5 μm long. The complete unit of two opposed actin bristles, connected by the central myosin shafts, is called a sarcomere (Greek *sarx*, flesh; *meros*, part). Its length varies from about 3.5 μm (actins almost entirely pulled away from the myosins) to 1.5 μm (fully interdigitated). In the latter state, the free ends of the actin filaments (250 nm) bundle up in the spaces between the bare middles of the myosin shafts, whereas the tapered ends of the myosin shafts press against the α-actinin–desmin meshwork in which the actin filaments are rooted. It would be wonderful if we could wander freely through a sarcomere, which must look like a tree planter's dream. Just imagine the veined, flower-garlanded myosin trunks alternating with the slender, helically fluted actin shafts, gracefully trimmed with tropomyosin and troponin decorations. Unfortunately, such a tour would require us to shrink at least another hundredfold. Even at our millionfold magnification, the trees of this molecular forest stand less than an inch apart, and this space is further obstructed by their flowery outgrowths. In addition, there is always the danger of a few calcium ions sneaking in and of the forest closing up on us. We must content ourselves with the two-dimensional images revealed by the electron microscope. Even so diminished, their beauty is arresting.

Structure of striated muscle. Each fiber is a giant multinucleated syncytium containing a bundle of myofibrils enveloped by sarcolemma.

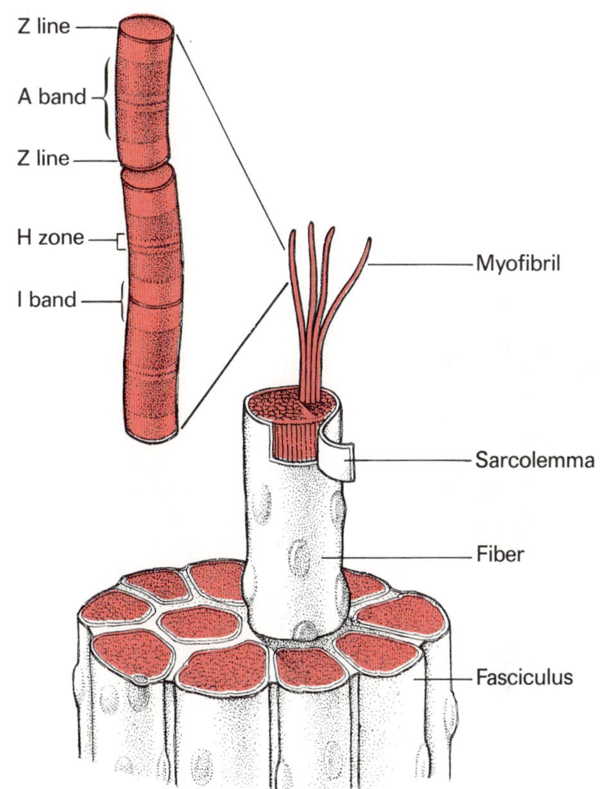

In a muscle fibril, a large number of sarcomeres are linked in series through the binding action of desmin, which not only knits actin anchoring points together to form an appropriately planted bristle, but also, as already mentioned, glues two such bristles back to back. The regular alternation of different zones gives the fibril its characteristic cross-striated appearance. Such fibrils may reach enormous dimensions at the cellular scale. They are built from hundreds of cells that fuse to form a syncytium (Greek *syn*, with). They are surrounded by an elaborate system of membranes that serve primarily in the rapid release and withdrawal of calcium ions whereby muscle contraction is regulated and by rows of mitochondria that provide the necessary ATP fuel.

Such a fibril can adopt three distinct states. In the absence of ATP and calcium, actin and myosin are rigidly

The three states of actin-myosin.

A. *Relaxation:* In the presence of ATP and absence of calcium, the system is plastic. Filaments slide freely along each other.

B. *Contraction:* The addition of calcium ions causes the myosin head to interact with the actin filament (see the next two illustrations). Actin is pulled downward, while ATP is hydrolyzed.

C. *Rigor:* The removal of calcium, with ATP absent, locks the system in a state of rigor. The addition of ATP restores the relaxed state A.

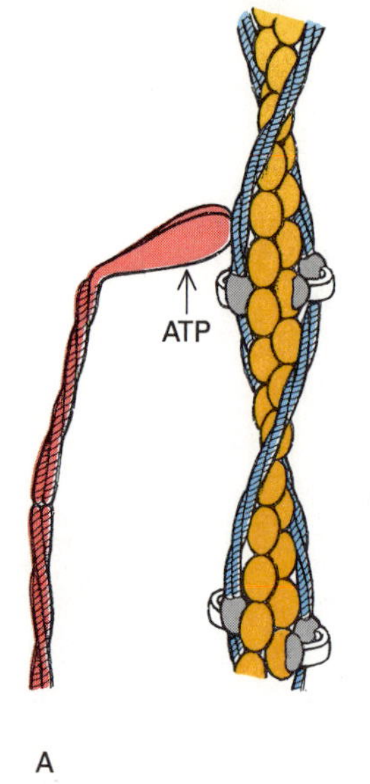

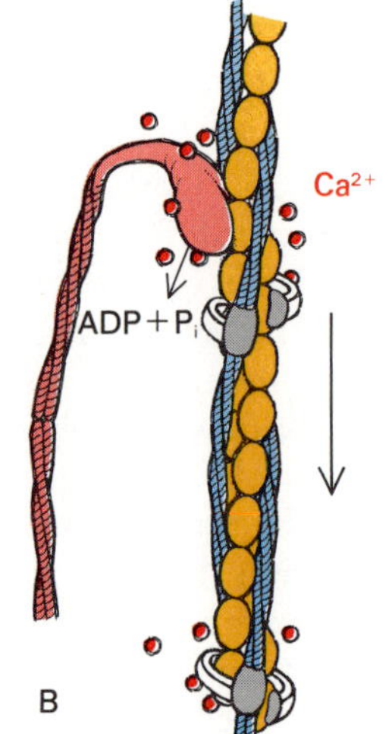

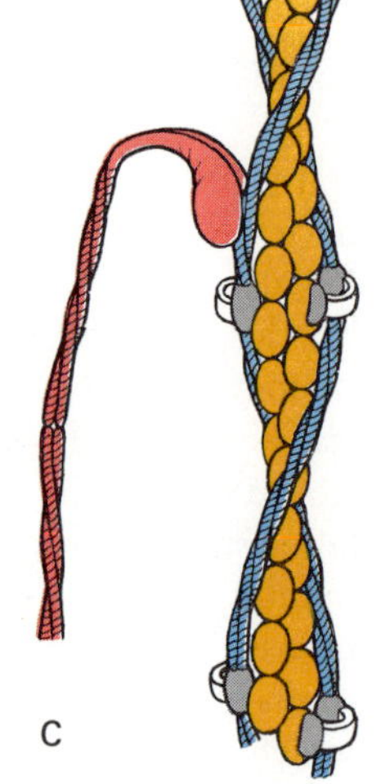

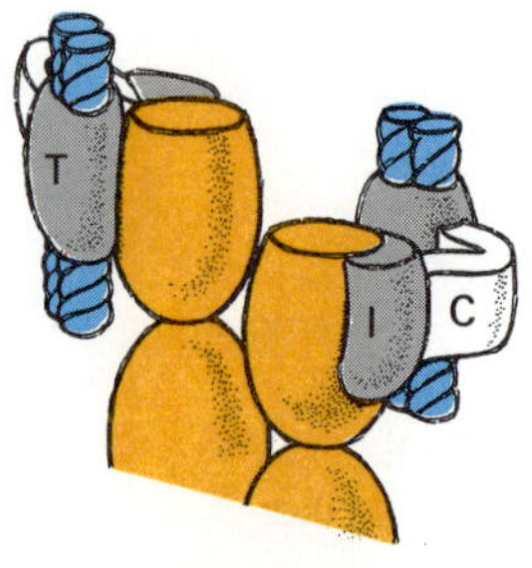

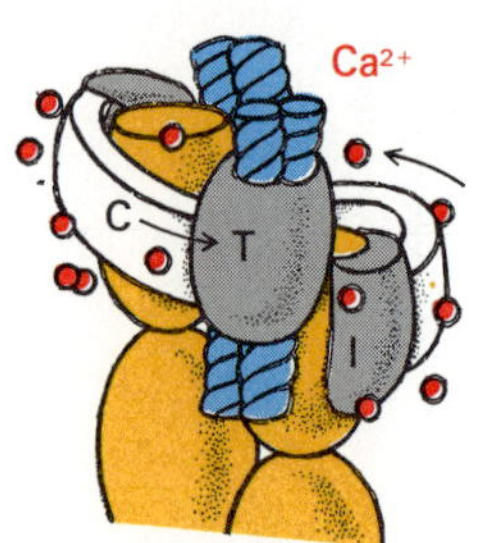

Structure of a troponin molecule. As shown at the left, it is made of a tropomyosin-binding subunit (T) and of an actin-binding subunit (I, because it inhibits the binding of myosin to actin), joined together by a calcium-binding subunit (C). The shape of the C subunit changes depending on whether calcium ions are present or absent. This conformational change alters the orientation of the T and I subunits with respect to each other. This drawing is purely imaginary. The exact shape of the molecule is not known.

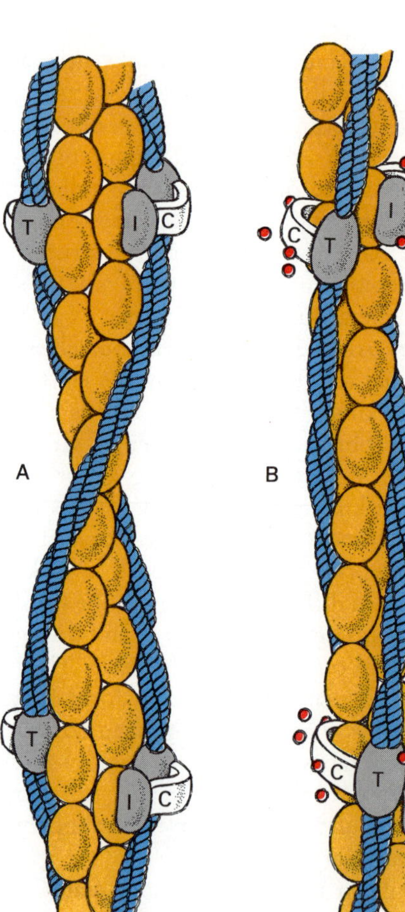

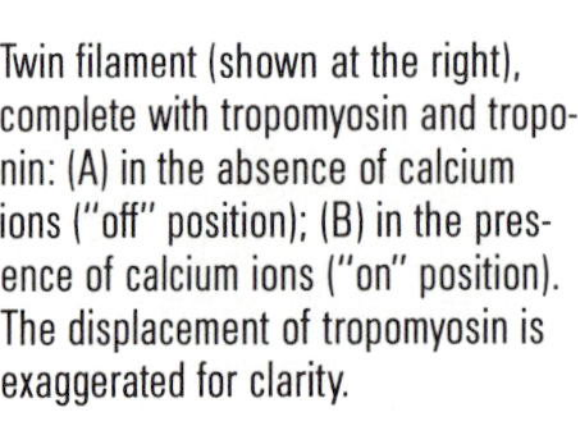

Twin filament (shown at the right), complete with tropomyosin and troponin: (A) in the absence of calcium ions ("off" position); (B) in the presence of calcium ions ("on" position). The displacement of tropomyosin is exaggerated for clarity.

interlocked. It is the state of rigor mortis that forms after death, when ATP production ceases. Add ATP and the structure becomes plastic. There is an ATP-binding site on the myosin head and, when this site is occupied, the head can no longer interact with the actin (in the absence of calcium, see below), and the filaments slide smoothly along each other. Such a fiber offers little resistance to passive stretching until other structural elements intervene to prevent its coming apart completely. This, for instance, is the state of an extensor muscle when the corresponding flexor contracts. The third state is triggered by calcium ions, which act through the troponin molecules that are attached to the actin filaments. Troponin consists of three subunits. One binds to actin and is called I, because its presence inhibits the attachment of myosin. Another, called T, binds to tropomyosin. These two subunits are linked by a third one, called C, which binds calcium ions. The C subunit is the trigger of the muscle machine. In the absence of calcium, its conformation is such that the tropomyosin threads are kept away from the grooves in the actin filament and cover the myosin-binding sites ("off" position). Occupancy of the C troponin subunit by calcium changes its shape in such a way that tropomyosin is shifted toward the grooves of the actin filament ("on" position). This allows myosin to interact with actin, which simultaneously activates ATP hydrolysis and the coupled conformational change that we have witnessed in the single actin-myosin bundle. Note that, in the fibril, each actin filament is pulled by three myosin filaments acting in a concerted fashion. Not all myosin heads are operative at the same time, however; they participate actively in the pulling only when strategically placed with respect to actin binding sites. A very smooth shortening is thereby ensured. If shortening is opposed, tension develops. This goes on for as long as the troponin C subunit is occupied by calcium and ATP is made available to cover the energy cost.

Troponin is found only in striated muscle. In other types of muscles and in nonmuscle cells, contraction is controlled by other means. But the basic principles seem universal. As far as is known, actin filaments of opposite polarity are always made to slide toward each other by a ratchet type of mechanism, powered by the ATP-splitting heads of myosinlike molecules.

Before we move on, we should pause for a last look at the actin molecule, certainly one of the most remarkable assemblages of atoms offered for our contemplation. It is not a giant molecule. It has a molecular mass of 42,000 daltons and is made of 374 amino acids. Yet, the three-dimensional ordering of this chain is such as to produce no less than eight specific, perfectly positioned, binding sites: four for the mutual association of actin molecules in the precise double spiral of F-actin; one for the ATP/ADP involved in the polymerization process; one, perhaps two, for tropomyosin; one for the troponin I subunit; and one, particularly important, for the attachment and simultaneous activation of the ATP-splitting myosin head. In addition, weaker binding sites allow actin fibers to join laterally into regularly structured bundles, or rods. How such extraordinary construction ever came to be is not known. But, once it appeared, evolution could not alter it any more. Between the amoeba and the rabbit, there is virtually no change in the structure of actin.

The Tubulin-Dynein System

More than one billion years ago, a living organism "discovered" the advantage of building scaffoldings with tubular elements, a discovery made only recently by our engineers. Presumably, the evolutionary invention was made by the usual "chance mutation–natural selection" mechanism (see Chapter 18). Perhaps some actinlike globular protein with the property of longitudinal self-assembly suffered a genetic change that altered its faculty of lateral association so that single filaments would tend to form cylindrical sheets rather than helical duplexes.

Indeed, tubulin, the constituent of the microtubules, is, like actin, a small globular protein, about 4 nm in diameter, fitted with a complementary lock-key axis that allows indefinite linear association. However, there are two tubulins, designated α and β, which are undoubtedly

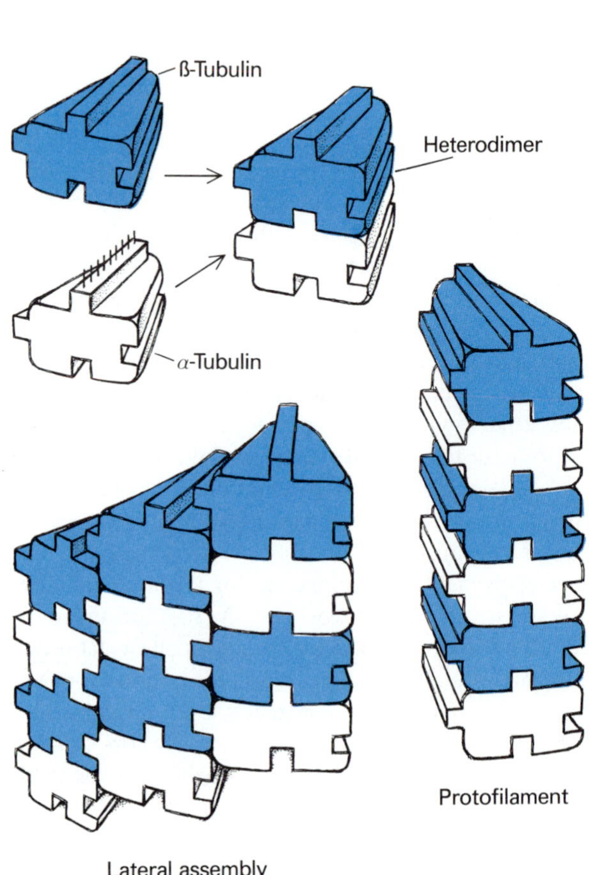

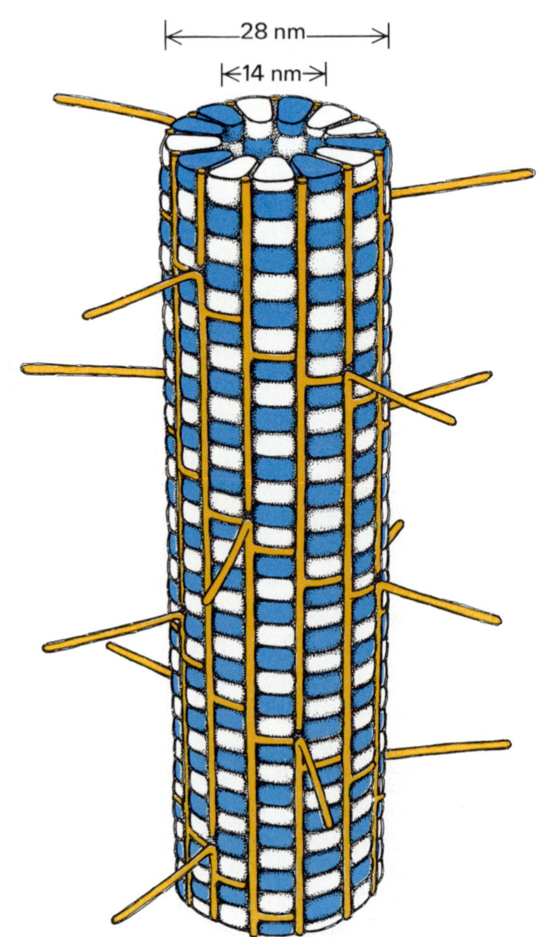

The drawings at the left above show in highly schematic fashion how very similar α- and β-tubulin molecules join preferentially to become stable heterodimers, which themselves assemble longitudinally into protofilaments, by means of a polar lock-and-key arrangement. Another such arrangement, situated laterally, allows the protofilaments to join into a cylindrical structure consisting of thirteen protofilaments, with an outer diameter of 28 nm and a bore of 14 nm. Lateral staggering is such as to create a left-handed helical periodicity of alternating α and β units, with a pitch of 12 nm per turn, and a right-handed helical periodicity of identical units, with a pitch of 40 nm per turn. Microtubule-associated proteins (MAPs) occupy the grooves.

evolutionary siblings, as indicated by extensive similarities in their amino-acid sequences. There is a preferential α–β linkage, which remains stable when microtubules are disassembled, so that the equivalent of G-actin is really an α-β heterodimer, not a monomer. These subunits assemble linearly by reversible β–α linkages to form protofilaments (Greek *prôtos*, first), which associate laterally by means of a second lock-key system, with a staggering such that each α-β subunit is flanked by a β-α couple displaced by about one-fourth of its length. The resulting sheet is not flat but curved and closes into a cylinder when exactly thirteen protofilaments are aligned side by side.

As in the transformation of G-actin into F-actin, the assembly of microtubules is associated with the hydrolysis of a nucleoside triphosphate, in this case GTP (instead of ATP for actin). There is a GTP-binding site on the α-β heterodimer. Upon assembly, GTP is split into GDP,

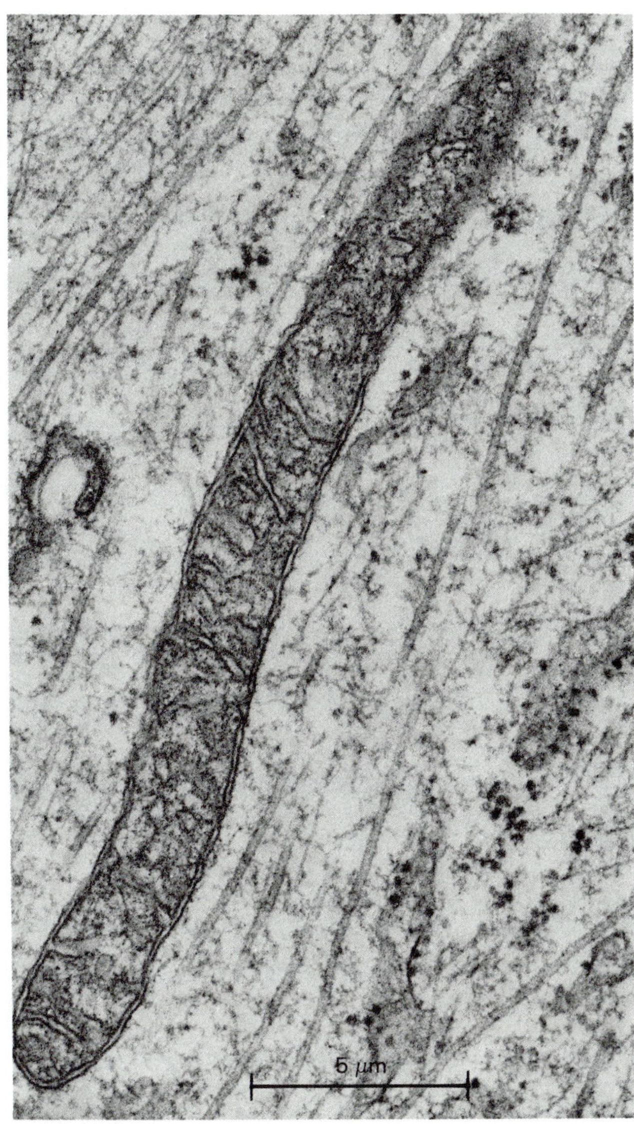

This electron micrograph of a thin section through a cultured human cell shows numerous microtubules running through the cytoplasm, on both sides of the elongated mitochondrion in the middle.

which remains bound, and P_i, which is released. In view of all these similarities, one is tempted to imagine an evolutionary relationship between actin and tubulin, with, as a possible key divergence event, the genetic change that led to duplication of a single gene into two distinct forms (ancestral to the α- and β-tubulin genes). What is known of the amino-acid sequences of actin and the tubulins does not, however, lend any support to this hypothesis.

Whatever their evolutionary origin, the development of microtubules certainly played an important role in the appearance of eukaryotes, in which they are universally present in both the plant and the animal kingdoms. According to one theory, they first appeared in some flagellate microorganism, which subsequently became adopted symbiotically by the common ancestor of all eukaryotes, to provide, among other advantages, the elements of the mitotic spindle (see Chapter 19). Unlike mitochondria and chloroplasts, however, centrioles, which are the organizing centers of the mitotic spindle and which duplicate at each cell division, are not known to contain any DNA.

To the cell tourist equipped with millionfold magnifying glasses, microtubules appear as some sort of thick-walled garden hose with an outer diameter of a little more than one inch (28 nm real size) and a half-inch (actually 14 nm) bore. They have a knobby surface; the knobs, about one-sixth of an inch (4 nm) in diameter, are arranged in thirteen longitudinal rows, in which minor differences in shape reveal the regular alternation of α and β subunits. As a result of the staggered organization of these rows, the knobs form a variety of spiral patterns around the tubing. Among these patterns are a right-handed helix of identical subunits (all α or β), with a pitch of 1.6 inches (40 nm) per turn, and a left-handed helix of alternating α and β subunits, with a pitch of one-half inch (12 nm) per turn. As a rule, this basic skeleton is further decorated with additional proteins (microtubule-associated proteins, or MAPs), partly inserted into the longitudinal grooves and partly projecting freely in the form of a hairy or fuzzy outgrowth, also arranged helically around the microtubule shaft. Altogether, microtubules offer a very pretty sight. Even more arresting are some of the structures that they serve to build.

On the facing page: Electron micrograph of a section through an axopod of the heliozoon *Echinosphaerium nucleofilum.* Note the double-helical arrangement of microtubules with twelvefold symmetry.

Many of these structures are built from labile microtubules. They are transient and changeable, the products of a dynamic equilibrium between two opposing processes that go on more-or-less continually. This equilibrium is readily displaced. For instance, disassembly of microtubules is favored by cooling, by high pressures, by calcium ions, and by drugs such as colchicine or some of the alkaloids (vinblastine, vincristine) extracted from the periwinkle, *Vinca rosea*. These drugs bind to free heterodimers and prevent them from joining with each other. Warming and exposure to heavy water, on the other hand, favor microtubule assembly.

The world of cells abounds in striking illustrations of these phenomena. A favorite of cell tourists is a group of protozoa called Heliozoa (Greek *hêlios*, sun; *zôon*, animal). These unicellular organisms owe their name to the fact that they send out long, thin, rigid spikes, known as axopods, that radiate in all directions up to half a millimeter from the cell body. Just imagine the sight, as seen through your magnifying glasses: a huge sphere, 300 feet or more in diameter, bristling with giant, rod-shaped projections some 3 to 5 feet wide and up to a third of a mile long. Should you wander through one of these projections, you would find it to be supported by a central spine made of hundreds of parallel microtubules, cross-bridged in an admirable double-helical pattern with a twelvefold symmetry. You would see a two-way traffic of particles and molecules moving briskly along this spine, linking the cell body with the very tip of the axopod, where, among other events, endocytic activity is taking place.

Should your heliozoan host be caught in an icy current, a most dramatic change would take place before your eyes. Not only would the axopodal traffic freeze to a stop, but the whole microtubular scaffolding would fall apart. In less than two hours, not a single microtubule would be left, and the axopods would be entirely retracted, leaving the sun a naked ball that has lost its rays. Not all would be lost, however, as might be realized by those who happen to recognize the heterodimeric building blocks of the dismantled microtubules among the swarms of new molecules that now fill the cell. We need only a warm wavelet, and all will be repaired. Reawakened by the gentle heat, the microtubular spine starts growing again from roots, called microtubule organizing centers, that are buried in the depth of the cell body. Soon the cell surface bulges with hundreds of new axopods, and in a matter of hours the sun again flashes its rays in all directions.

Light-micrograph of the heliozoon *Echinosphaerium nucleofilum*. Note the axopods radiating all around the cell body.

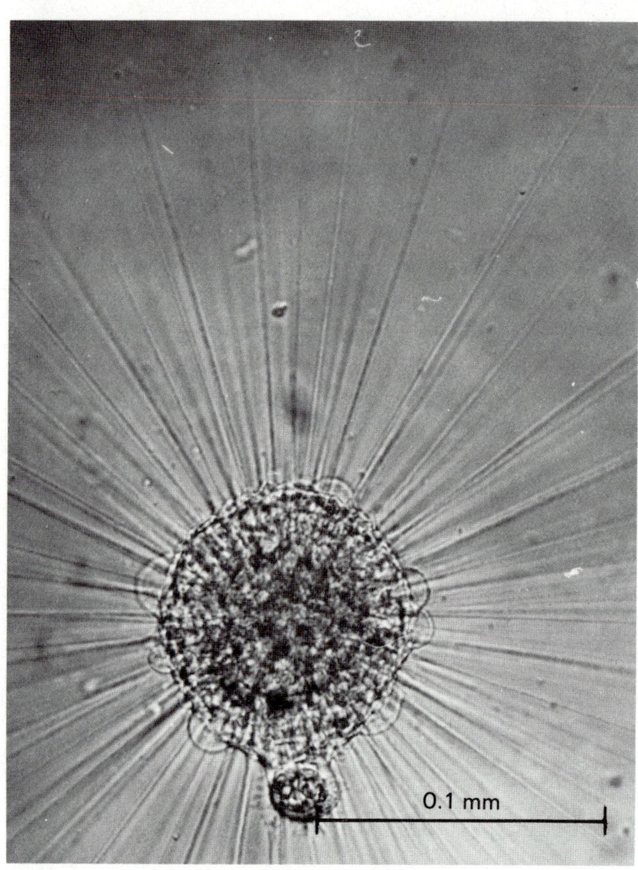

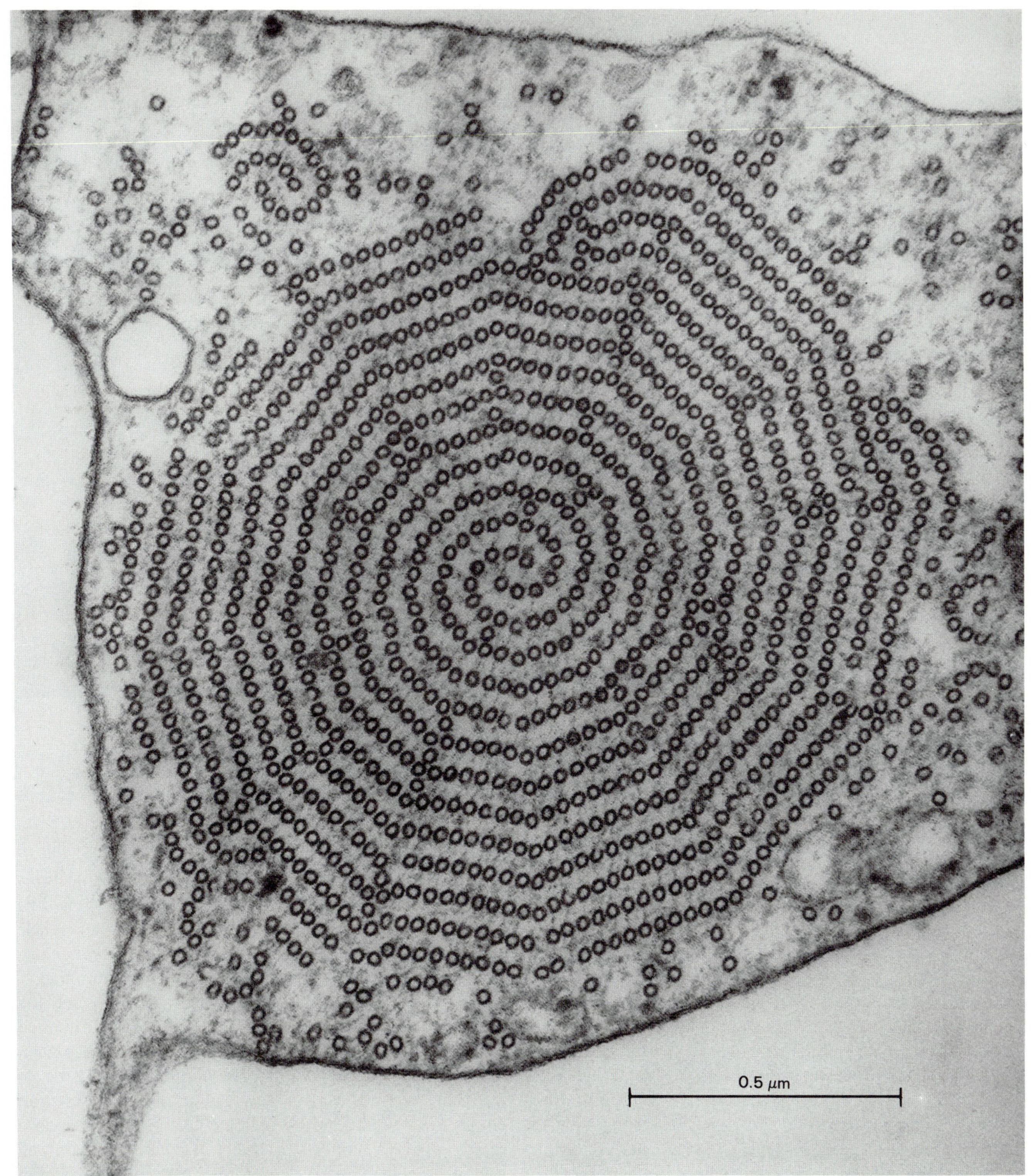
0.5 μm

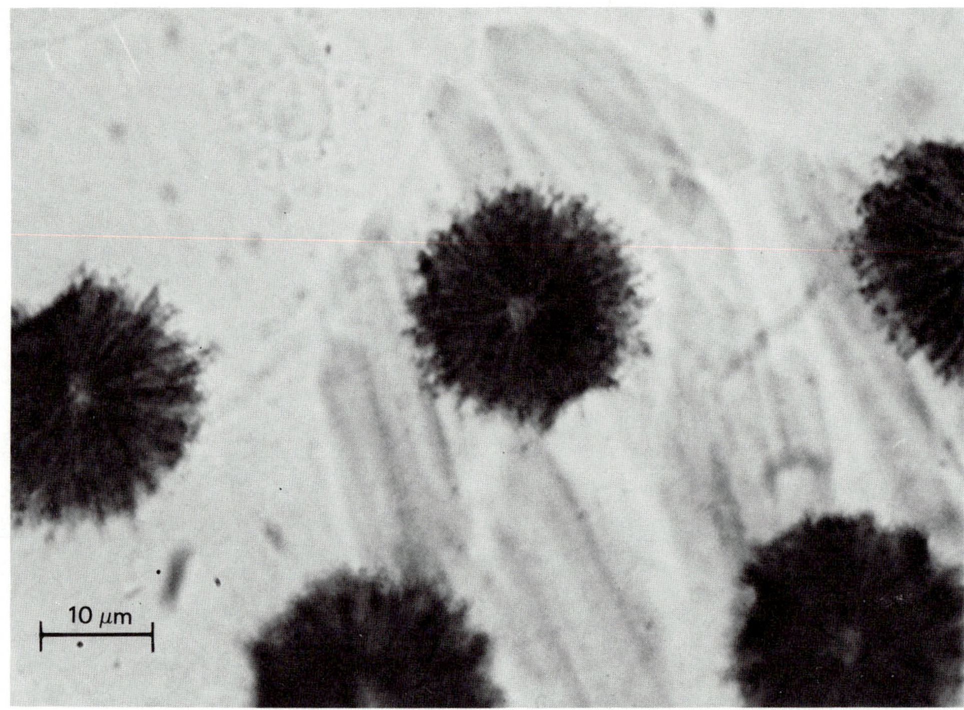

Light micrograph of a red chromatophore (erythrophore) cell of the fish *Holocentrus rufus*, with pigment granules dispersed.

Many crustaceans, fishes, amphibians, and reptiles have in their skin star-shaped pigmented cells, called chromatophores, that send out radiating processes supported by a microtubular skeleton, along which pigment granules move inward or outward with remarkable speed. Under the influence of certain hormones, whose secretion is controlled by light, the granules may, in less than a second, pack centrally and leave the cell largely transparent or invade the processes and cause the cell to darken. In some animals, two or more types of pigment granules of different color participate in these migrations independently. Herein lies the chameleon's secret. Interfere with his microtubules, and there goes his camouflaging skill.

Probably the most extraordinary cellular projections reinforced by microtubules are the axons, the threadlike extensions by which neurons (nerve cells) dispatch their signals and which may, in the nerves of the larger mammals, reach lengths of several meters. Vital communications and exchanges between the main cell body and the terminal branched endings of the axon are maintained over these incredible distances by the so-called axonal flows, of which the fastest component travels at up to 8 μm per second. Magnified a millionfold, such an axon could almost straddle the Atlantic Ocean. It could, in a matter of days, convey products made in the main cell body—say, in New York—to their terminal destination

Electron micrographs of an erythrophore of *Holocentrus rufus* (see the light micrograph on the facing page): (left) with pigment granules dispersed; (right) with pigment granules aggregated.

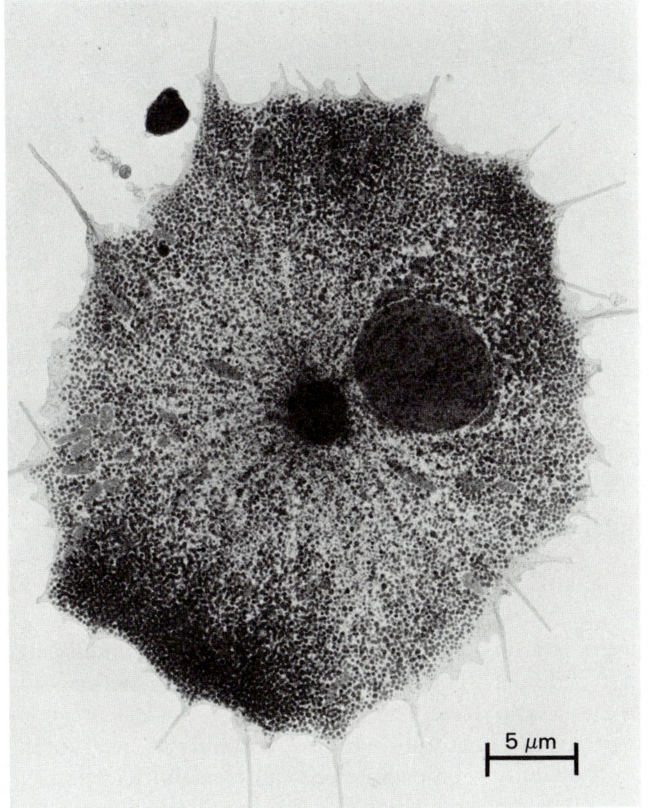

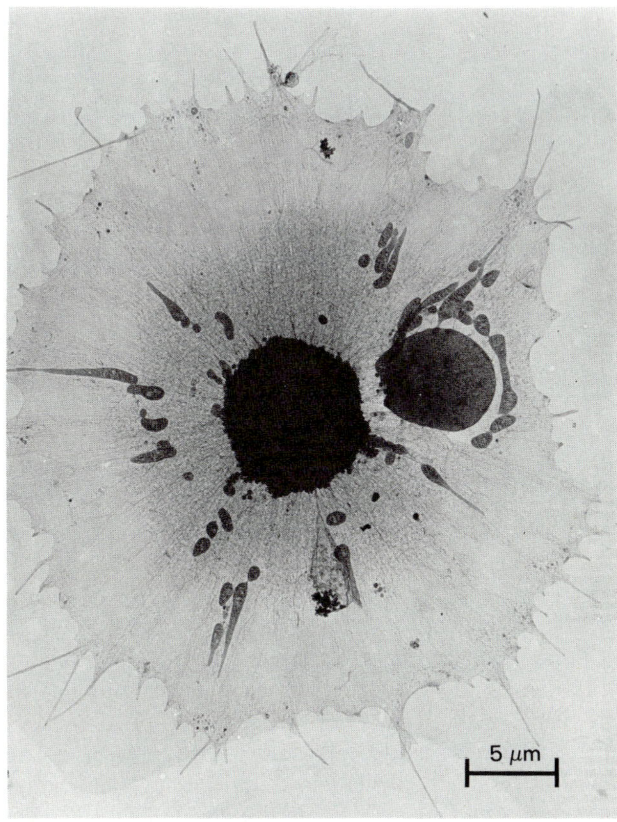

in the outer nerve endings—at some point on the Irish coast.

Microtubules also serve to induce or maintain certain structural asymmetries in cells—for instance, the elongation of growing muscle cells—and to build certain intracellular scaffoldings, generally in support of some directional transport or movement. Near the end of our tour, we will have an opportunity to witness the most grandiose of all such operations, namely the erection and subsequent dismantling of the vast, intricate machinery whereby duplicated chromosomes are disjoined and pulled away from each other to produce two separate sets in the course of mitotic cell division.

In all the examples mentioned, microtubules play an obvious structural role as cytoskeletal elements. In addition, thanks to their labile character, they may also fulfill a morphogenetic function through the alterations in shape that are elicited by their assembly and disassembly. Finally, they often provide leading tracks and bracing props for some form of guided translocation of intracellular objects or materials. The driving force of these transport phenomena could be supplied by actin-myosin cytomuscles. But there is another possibility, suggested by inspection of the more stable microtubular systems that serve to build the two main locomotor organelles of cells—cilia and flagella.

This scanning electron micrograph of the surface of the protozoon *Paramecium tetraurelia* shows rows of cilia arrested in the midst of their synchronous beat.

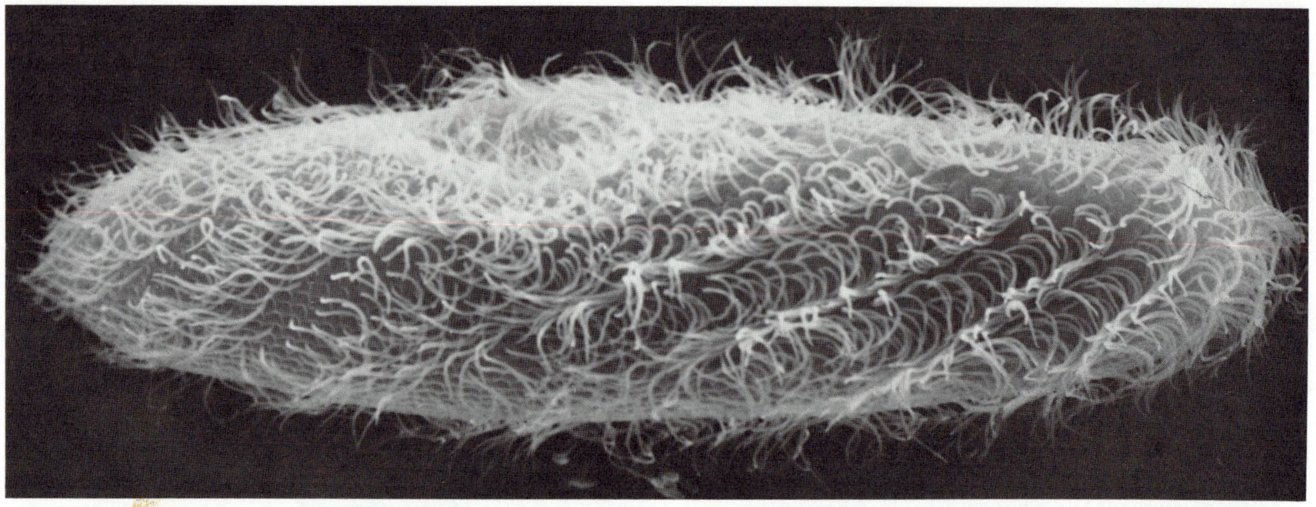

Cilia (eyelashes, in Latin) are beating protrusions. They bend and swing back, remaining essentially in a single plane, with a frequency of oscillation of some ten to forty beats per second. They are usually present in large numbers and are planted in parallel arrays, moved by what looks like waves of excitation signals. For full enjoyment of this display, you should reduce your magnification to some ten-thousandfold, instead of the customary millionfold, and equip yourself with a fast camera and a slow-motion replay system. The cell surface will then appear like a ripe wheatfield bending under a gust of wind. Ciliary beat causes a relative displacement of the cell with respect to the surrounding fluid. If the cell is free, it moves (ciliate protozoa); if fixed, it generates a liquid current. The direction of these movements can be reversed. For instance, when a *Paramecium* (a ciliate) bumps into an obstacle, it immediately swims backward. The agents of this reversal are calcium ions released by the shock.

Flagella (whips, in Latin) are longer than cilia and are usually present singly or in small numbers at the tail end of free-swimming cells, such as flagellate protozoa, gametes of algae, or animal spermatozoa. Like cilia, flagella move in a single plane, but their movement is undulating, not pendular. They serve a propelling function. (Some bacteria also have flagella, but these flagella are constructed differently; tubulin is found only in eukaryotes.)

With only minor variations, all cilia and flagella have the same molecular architecture, based on the so-called 9 + 2 pattern. Axially situated are a pair of microtubules linked laterally by cross-bridges. This central pair is enclosed by a sheath and surrounded further by a cylindrical set of nine parallel microtubular doublets joined to the central pair by radial spokes. Each doublet is made of one complete microtubule (A) of thirteen protofilaments, to which an incomplete microtubule (B) of only ten protofilaments is fused by sharing three of the A microtubule's protofilaments. The A subunit of each doublet sends out pairs of tangentially oriented side arms toward the B subunit of the adjacent doublet in a sort of ring-around-a-rosy pattern, clockwise when viewed from the tip. These side arms, which have been described as dumbbell- or lollipop-shaped, are repeated longitudinally with a spacing of 20 nm. Altogether, more than one hundred distinct proteins participate in the construction of this remarkable

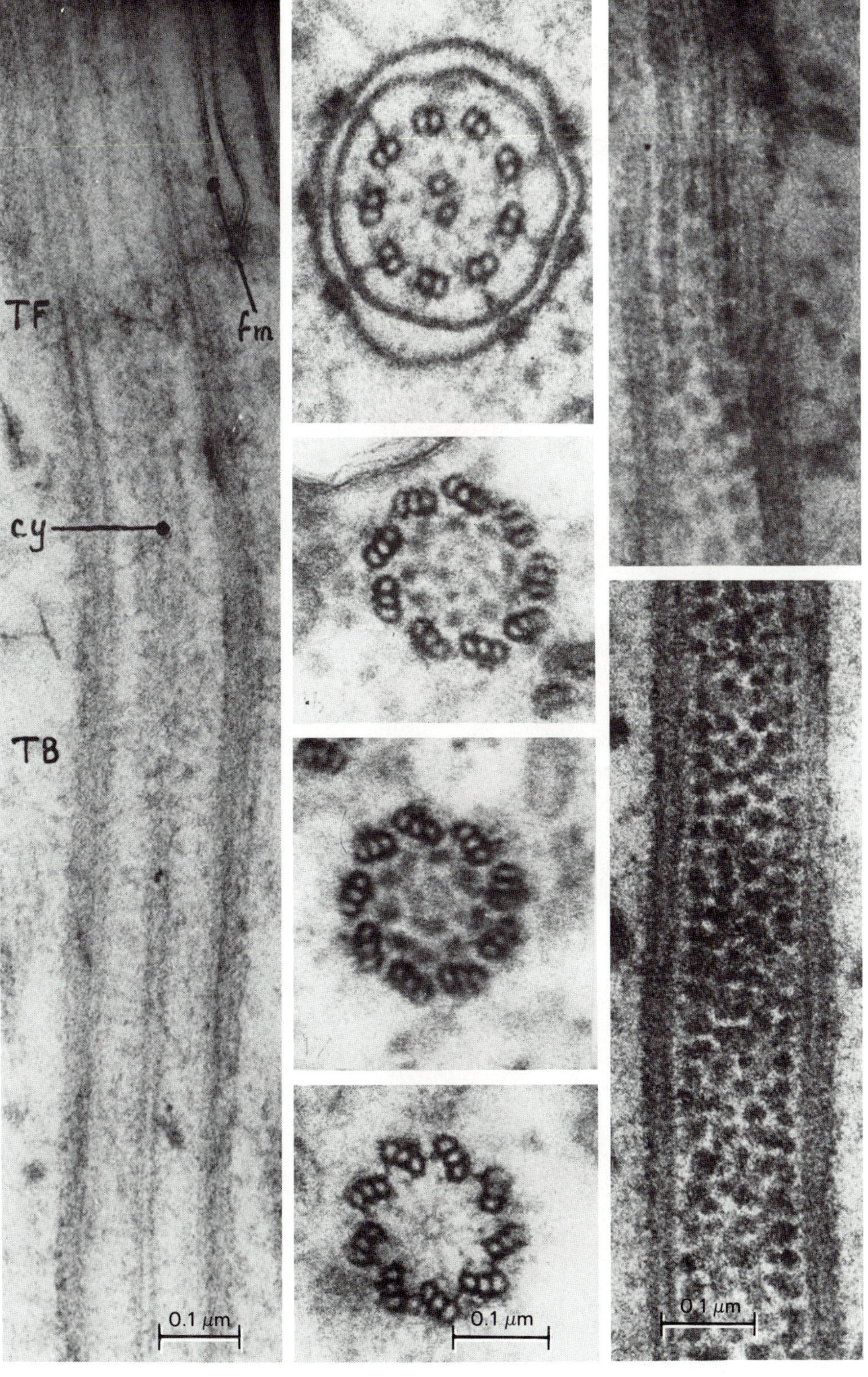

These electron micrographs, taken in the early 1960s, revealed the characteristic structure of flagella in the protozoon *Pseudotrichonympha.* The longitudinal section at the left shows the root of a flagellum on top continuing into the basal body (TF = transition between flagellum and basal body; TB = transition between proximal and distal regions of basal body; cy = central cylinder; fm = flagellar membrane). The transverse sections in the middle illustrate the structure of a flagellum and basal body at different levels (see the diagram on the next page). The longitudinal sections at the right show distal parts of the basal body.

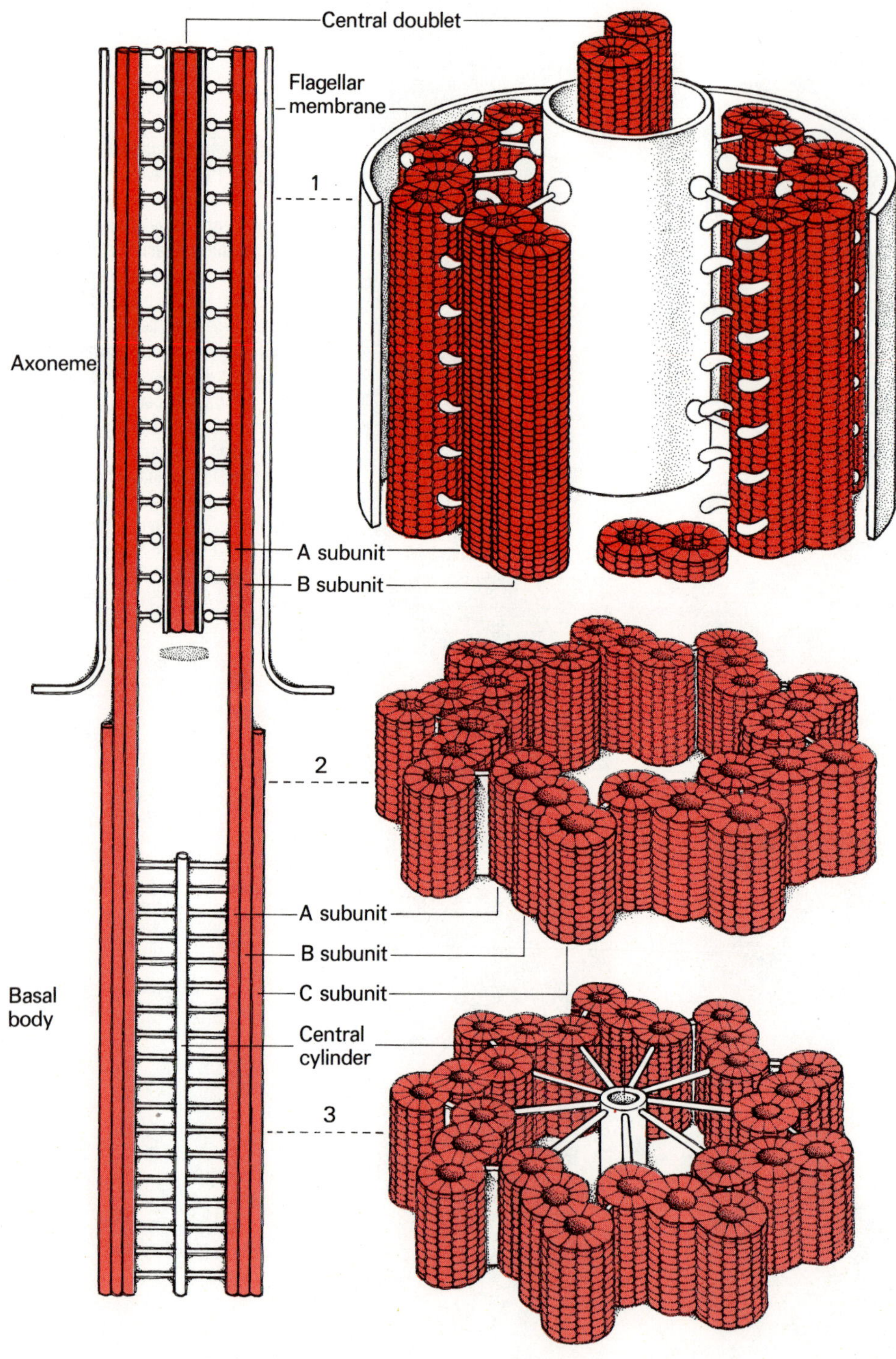

Structure of cilium. The drawing at the left shows a longitudinal section through the lower part of the axoneme and the upper part of the basal body. The drawings at the right illustrate the fine structure at three different levels. Part 1 shows the typical 9 + 2 pattern of the axoneme, with the central microtubule doublet and the ring of nine doublets linked by pairs of dynein side arms. The characteristic cartwheel structure of the basal body, with triplets attached by spokes to a central cylinder, is shown in part 3. The intermediate part connecting the axoneme to the basal body is seen in part 2. (See the electron micrographs on pp. 215 and 217.)

Longitudinal section through a cilium of the protozoon *Tetrahymena thermophilia*, prepared by freeze-etching (see the illustration on the facing page), shows four microtubule doublets connected by rows of dynein arms.

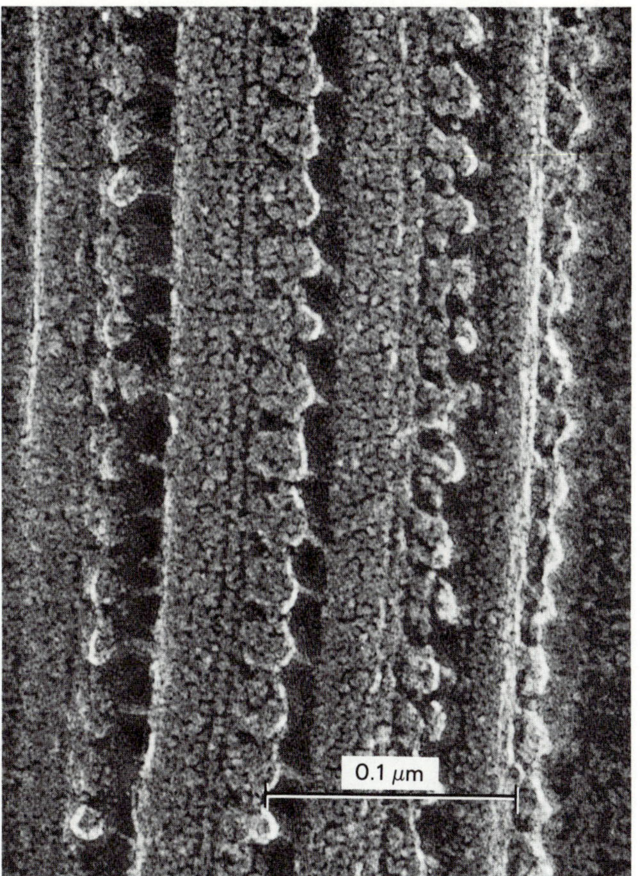

edifice. Guided by their associative properties, they combine to form a delicately sculptured shaft, known as the axoneme (Greek, *nêma,* thread). It is about 0.3 μm thick—or 1 foot at our millionfold magnification—and its beauty, like that of the sarcomere, must be enjoyed mostly in cross section.

At the point where the axoneme becomes rooted in the cytoplasm, its structure changes. The central pair of microtubules breaks off and becomes replaced lower down by a single hollow axis. The peripheral microtubular doublets, on the other hand, extend into the cytoplasm, where they are no longer connected by side arms, but instead acquire a third incomplete microtubule (C) fused to the B component of the doublet in the same way as B is fused to A. Radial lamellae connect these triplets to the central axis, replacing the axoneme's spokes. This root is called kinetosome (Greek *kinein,* to move) or basal body. It is endowed with remarkable organizing abilities. For instance, it will sprout a brand-new limb after a cilium or flagellum is amputated. As we shall see in Chapter 19, the centriole, which governs the assembly of the mitotic spindle, is a close relative of the kinetosome.

In the axonemes of cilia and flagella, the microtubules and their radial connections provide a neatly interlocked cytoskeletal framework endowed with just the degree of resilience, flexibility, and elasticity needed to permit, as well as to limit, the bending strains to which these appendages are subjected. The side arms are the power-generating cytomuscles that force the structure to bend. They are made of a protein called dynein (Greek *dynamis,* force), which shares with myosin heads the characteristic property of catalyzing ATP hydrolysis in a manner obligatorily coupled to a change in conformation. The consequence of this mechanochemical event is a relative displacement of the two microtubule doublets connected by the activated dynein side arms, in a manner that recalls the ATP-powered sliding of actin threads along myosin filaments.

This is the principle on which the molecular machinery operates. Details, of course, are considerably more complex. In particular, it is obvious that the nine rows of dynein side arms cannot be activated simultaneously. Those in or near the bending plane (which is perpendicular to the plane defined by the central pair of microtubules) must operate in seesaw fashion, so that those on one side relax their hold while the opposite ones grip, and vice versa. Side arms lying in or near the plane perpendicular to the bending plane, on the other hand, remain largely passive. How this delicate control is exerted has so far eluded the perspicacity of cell explorers.

An intriguing question, as yet unsolved, is whether dynein cytomuscles also function in other forms of intracellular movement for which microtubules provide the cytoskeletal scaffolding or whether this function is fulfilled by myosin filaments pulling appropriately anchored actin threads. In any case, a revealing parallelism is brought out by a comparison of the two main cellular motor systems: both function by a sliding displacement of cytobones under the pull of special ATP-splitting cytomuscles.

Buckminster Fuller's geodesic dome.

Clathrin

In the world of cells, wonders never cease. When the American architect Buckminster Fuller built his celebrated geodesic dome, little did he suspect that hundreds of millions of years ago Nature evolved a protein molecule that could do the same thing entirely on its own. This protein is called clathrin (Greek *klêthra*, Latin *clathrum*, lattice bar). Given the opportunity, it will, even in the test tube, assemble spontaneously into basketlike structures formed, like the geodesic dome, by a network of hexagons with a few pentagons and heptagons thrown in for good measure.

We can watch this process and, at the same time, obtain a glimpse of its function by inspecting the inner face of the plasma membrane at a patch where, on the reverse side, receptors occupied by their ligands have just congregated and are about to be engulfed by endocytosis. Alerting us to this phenomenon is the cluster of receptor-molecule tails sticking out through the membrane on our side. Apparently alerted in the same way, clathrin subunits recruited from some cytoplasmic store start polymerizing over the advertised membrane patch, covering it with what looks at first like a ragged piece of chicken wire.

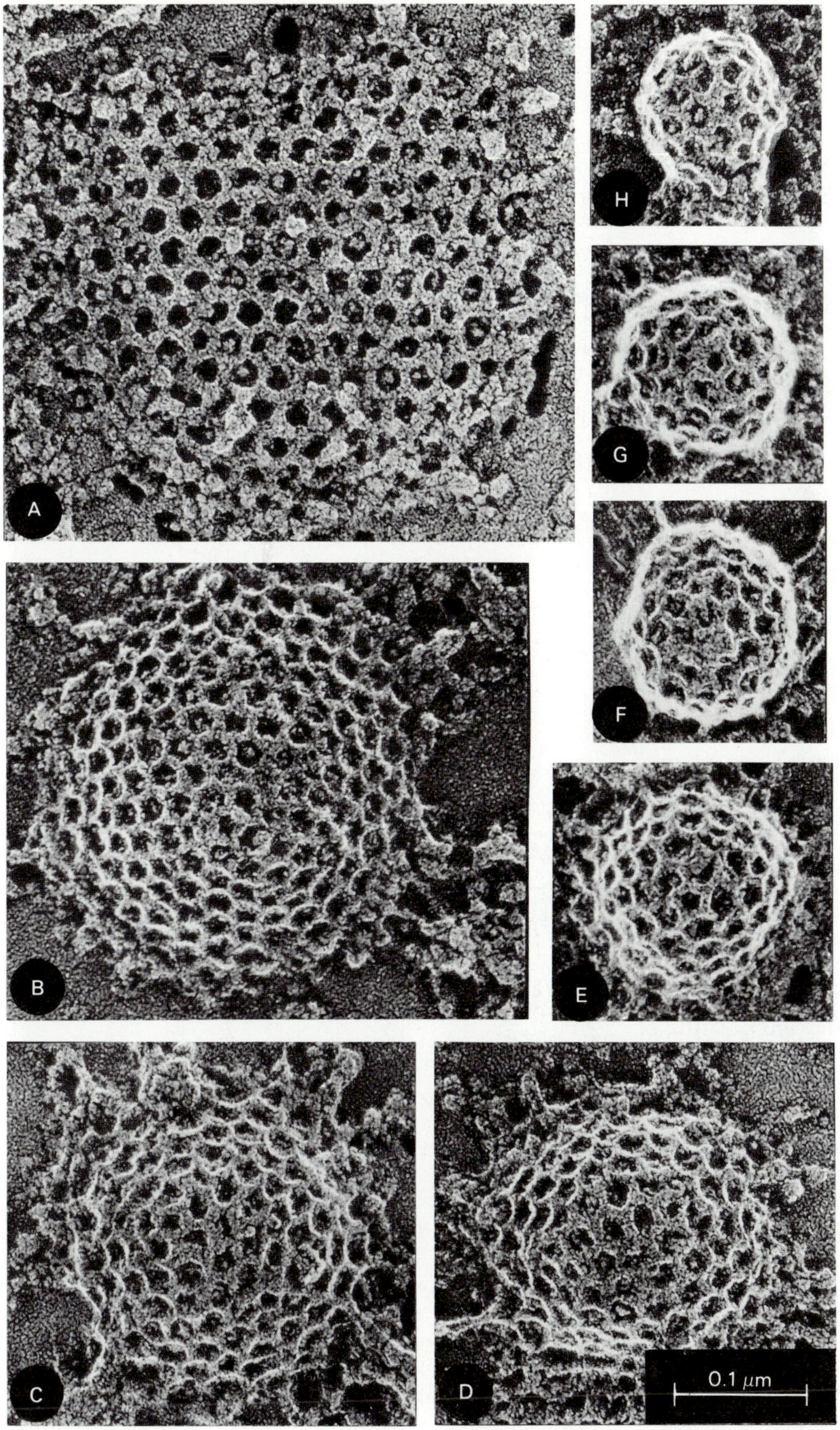

A

H

G

F

B

E

C

D

0.1 μm

These electron micrographs obtained by freeze-etch technique show clathrin baskets on the cytoplasmic face of the plasma membrane. They cover coated pits and are shown, from A to H, in order of increasing curvature, up to virtually complete closure as a coated vesicle (F to H).

Assembly of clathrin network from triskelion units.

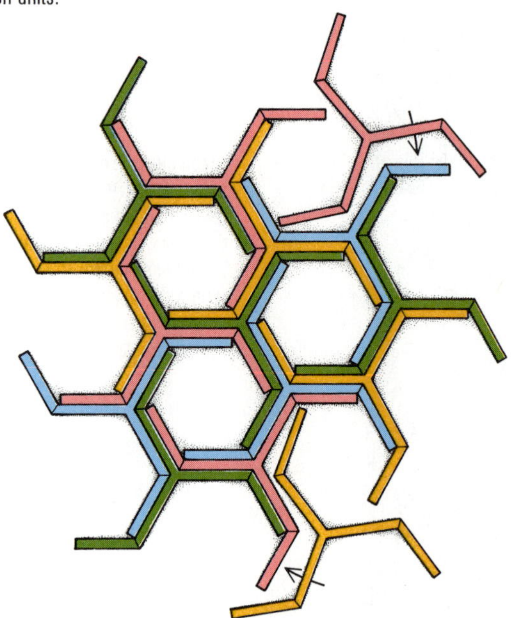

Membrane patch is pulled inward by clathrin cage in endocytosis.

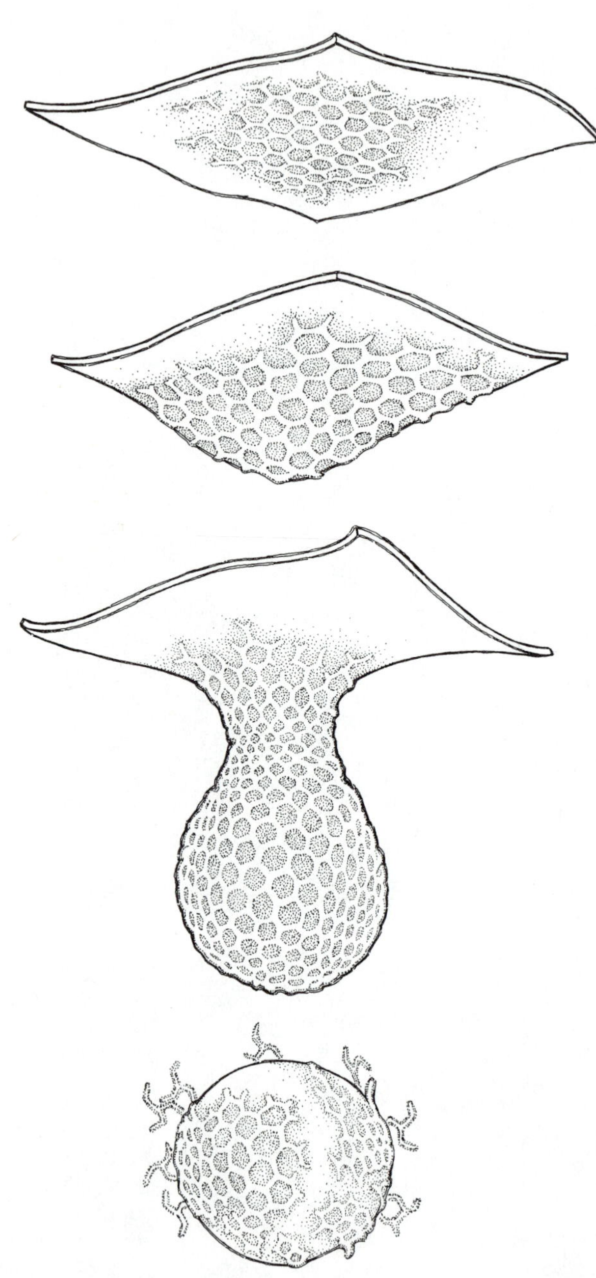

Looking closely, you will notice that each subunit is a trimeric, three-legged structure, or triskelion (Greek *skelos,* leg), with "thighs" and "legs" about 20 nm long (a little less than 1 inch at our magnification). These subunits join by lateral association of their limbs to form a hexagonal mesh in which each vertex is occupied by a triskelion center, and each strut is made of four limbs—two thighs and two legs—aligned side by side. These triskelia, however, are not flat but curved. When "running" clockwise, they fit on a convex surface. Thus, as the structure they form widens, it bulges progressively in typical geodesic-dome shape. To accommodate the increasing curvature, some triskelia fall off, and the regular hexagonal lattice of the network becomes broken in some areas by a few pentagons and heptagons. These are readily assembled by the same mechanism as the hexagons, with only minor conformational stresses of the participating triskelia. As this process continues, the curvature of the dome continues to increase while its outer rim tightens, so that its form changes gradually to that of a pear-shaped basket. Finally, the rim closes, the basket becomes a cage, and the membrane patch turns into a vesicle imprisoned inside the cage. Soon after, the cage disintegrates and disappears, releasing the vesicle.

What we have just witnessed from inside the cell is the characteristic process of receptor-mediated endocytosis.

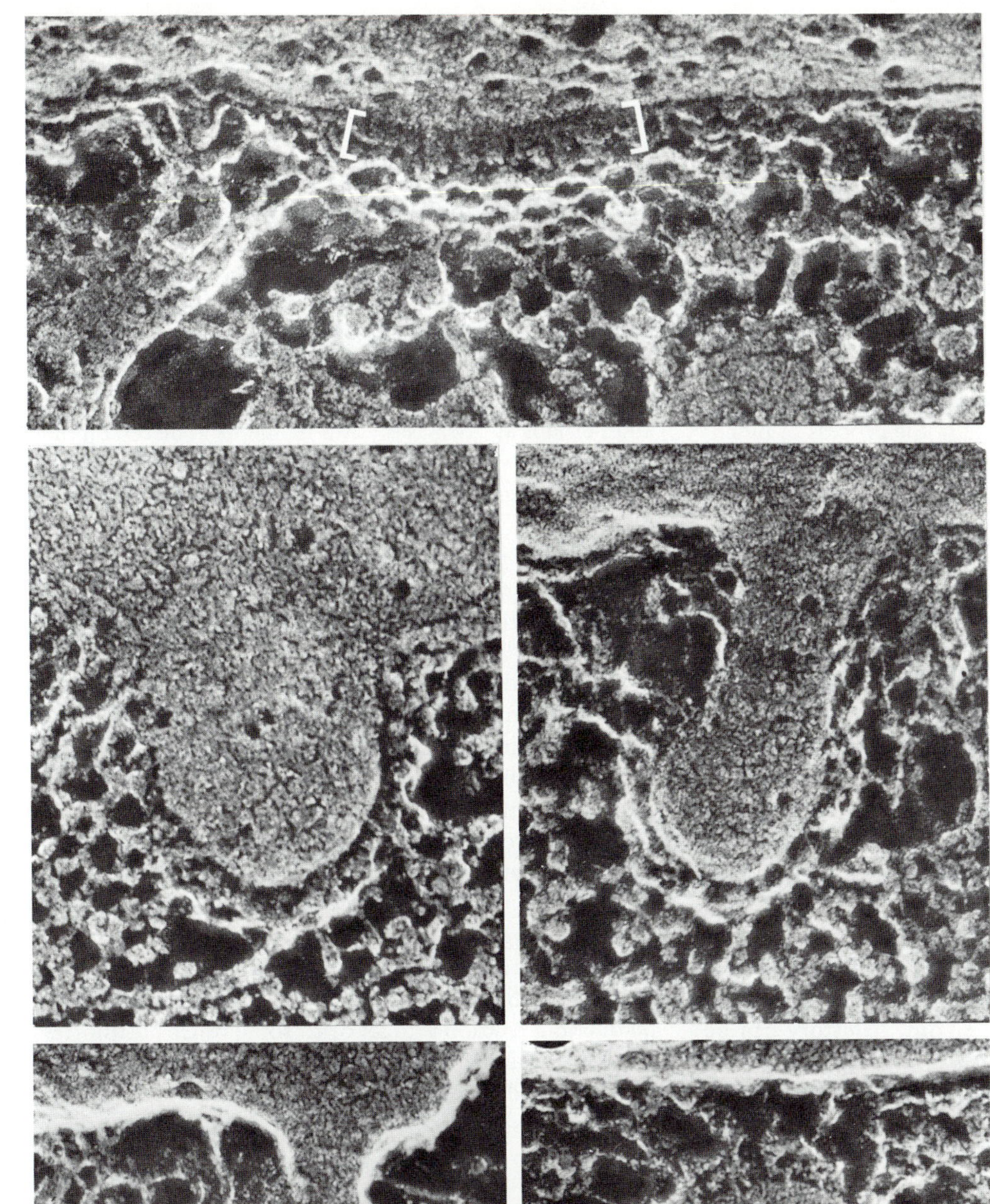

Edge views of freeze-etched coated pits show clathrin baskets at various stages of "strangling" a plasma-membrane invagination into a closed vesicle.

Early electron microscopists were given a two-dimensional hint that an organized structure might participate in this process when they noticed the presence of some sort of fuzzy contour delineating the profiles of certain membrane invaginations and closed vesicles. They invented the terms coated pits and coated vesicles to designate these structures, and put forward the hypothesis that the "coat" somehow plays a role in the process of pinching off vesicles from membranes (pp. 55–56).

Those pioneers were remarkably prescient. Indeed, it appears that clathrin baskets are involved not only in receptor-mediated endocytosis, but also in other processes of vesicular transport, such as the transfer of materials from the ER to the Golgi apparatus. How clathrin operates is, however, far from clear. As we watch the performance of this remarkable molecule, many tantalizing questions come to mind. What, on those little patches of membrane—so diverse in nature and function—where clathrin baskets form, is the nucleating signal that initiates polymerization? Where does the free energy of this process come from, and does it, as we suspect, cover the work of squeezing out a vesicle from a flat membrane? If so, how does endocytosis without the help of clathrin take place? What causes a clathrin basket to fall apart again after it has performed its function?

For the moment, we do not know the answers to these questions, as to countless others. But, if history can be trusted, we can safely predict that some day a cell tourist will find them so intriguing that he will turn into a cell explorer, not resting until he uncovers the answers. Already, there are hints that calcium ions may assist polymerization and that ATP helps in reversing this process.

Microtrabeculae

Explorers who have gone through cells with the most incisive instruments currently available, such as the million-volt electron microscope or the quick-freeze, deep-etch method of preparing samples for three-dimensional electron microscopy, have detected a fine meshwork of filaments pervading the whole of the cytosol and attached by multiple anchoring points to all cytoplasmic structural constituents present. The name microtrabeculae—literally, miniature miniature beams—has been given to the elements of this lattice. Actually, the microtrabeculae are not as miniature as their name implies: they are somewhere between 10 and 15 nm thick, which makes them definitely thicker than actin filaments and perhaps as thick as myosin filaments. The reason they were not seen before resides in their short length and random orientation in the meshwork. A minimum of three-dimensional vision was required to discern them as part of a lattice.

The pictures of the microtrabecular meshwork that have been brought back so far endow the cytosol with considerably more structural rigidity and organization than was evident to us when we first entered this part of the cell. Admittedly, our progress at that time was rather heavy and clumsy, and we could have missed this fine web. Nevertheless, there is some worry about the reality of the microtrabecular meshwork. It makes the cell look almost "overorganized," and so far has received no chemical identification, which is surprising in view of its very extensive development. The problem remains that, even with their highly sophisticated tools, the explorers who have described the meshwork have to look at cells that are dead, not at live ones. How much of the meshwork could be due to a fixation artifact or to some other postmortem change remains a debated question. On the other hand, the fact that different cell parts are often seen to move in a concerted fashion does support the view of a connecting network between them.

Calcium and Cell Motility

During our brief excursion through striated muscle, we discovered the electrifying effect of calcium ions, which, by combining with troponin C, cause the conformation of this protein subunit to change in such a way that the attached tropomyosin is displaced and no longer prevents myosin-actin interaction and ATP hydrolysis.

As it happens, this is only one of the numerous important triggering functions performed by calcium. These ions also stimulate smooth-muscle fibers, perhaps all types of actin-myosin systems; they inhibit microtubule assembly, stop or reverse ciliary beat; they influence clathrin polymerization and may be involved in the membrane-fusion events that determine endocytosis, exocytosis, and the merger and division of intracellular vesicles; finally, they activate a number of enzymes—in particular, certain protein kinases (Chapter 8). These enzymes play important regulatory roles because the biological activity of the proteins on which they act is crucially dependent on their state of phosphorylation, as will be discussed in Chapters 13 and 18. Thus, those proteins that need phosphorylation to be active are turned on by calcium. Smooth-muscle myosin seems to belong in this category. On the other hand, those that are active only in dephosphorylated form are shut off by calcium ions.

In all these functions, calcium ions act indirectly by way of a calcium-binding protein called calmodulin, which resembles troponin C in that it, too, suffers a conformational change upon binding calcium. It is this altered calmodulin that actually exerts the effects attributed to calcium. Its functional resemblance to troponin C is far from fortuitous. These two proteins are closely related, as indicated by extensive amino-acid-sequence homologies, and appear to be evolutionary descendants of a common ancestor.

Such important functions presuppose the existence of intracellular calcium reservoirs fitted with extremely efficient pumps and unloading valves. The most highly regulated of such systems are found in striated muscles—especially the flight muscles of insects, which alternate between resting and active state as many as several hundred times per second. In striated muscle, calcium is stored in a specialized form of ER, called sarcoplasmic reticulum, which is intimately wrapped around the sarcomeres, so that the distance to be traveled by the calcium ions is extremely short. Release of calcium from this reservoir can be elicited almost instantaneously, thanks to deep indentations of the plasma membrane (T system), closely connected with the sarcoplasmic reticulum in the so-called triad. When a muscle fiber is excited, a wave of depolarization is propagated along the plasma membrane and temporarily induces leakage of calcium ions out of the sarcoplasmic reservoirs as it passes through the triad system. A powerful, ATP-driven pump in the sarcoplasmic membrane sucks back the calcium ions as quickly as they are released.

Specialized ER domains comparable to the sarcoplasmic reticulum may exist in a few other cells. However, the more common calcium reservoirs are, on one hand, the extracellular fluid and, on the other hand, the mitochondria. Each is fitted with a calcium pump. That in the plasma membrane drives calcium out of the cell. That in the mitochondrial inner membrane, which, as we have seen, is operated by the protonmotive force with the help of a natural calcium ionophore, concentrates calcium within the mitochondrial matrix. Thanks to these two pumps, calcium is maintained at a very low level in the cytosol. Any local perturbation of the plasma or inner mitochondrial membranes that either slows down the calcium pump or increases calcium leakage will cause the local cytosolic concentration of calcium to rise and, by way of an enhanced occupancy of the calcium-binding sites of calmodulin, elicit some of the effects that have been mentioned. Some nervous impulses and a number of hormones and other surface ligands produce such changes on the plasma membrane. Intracellular messengers induced by metabolic changes or perhaps, as in the case of cyclic AMP, in response to surface effects act similarly on the inner mitochondrial membrane.

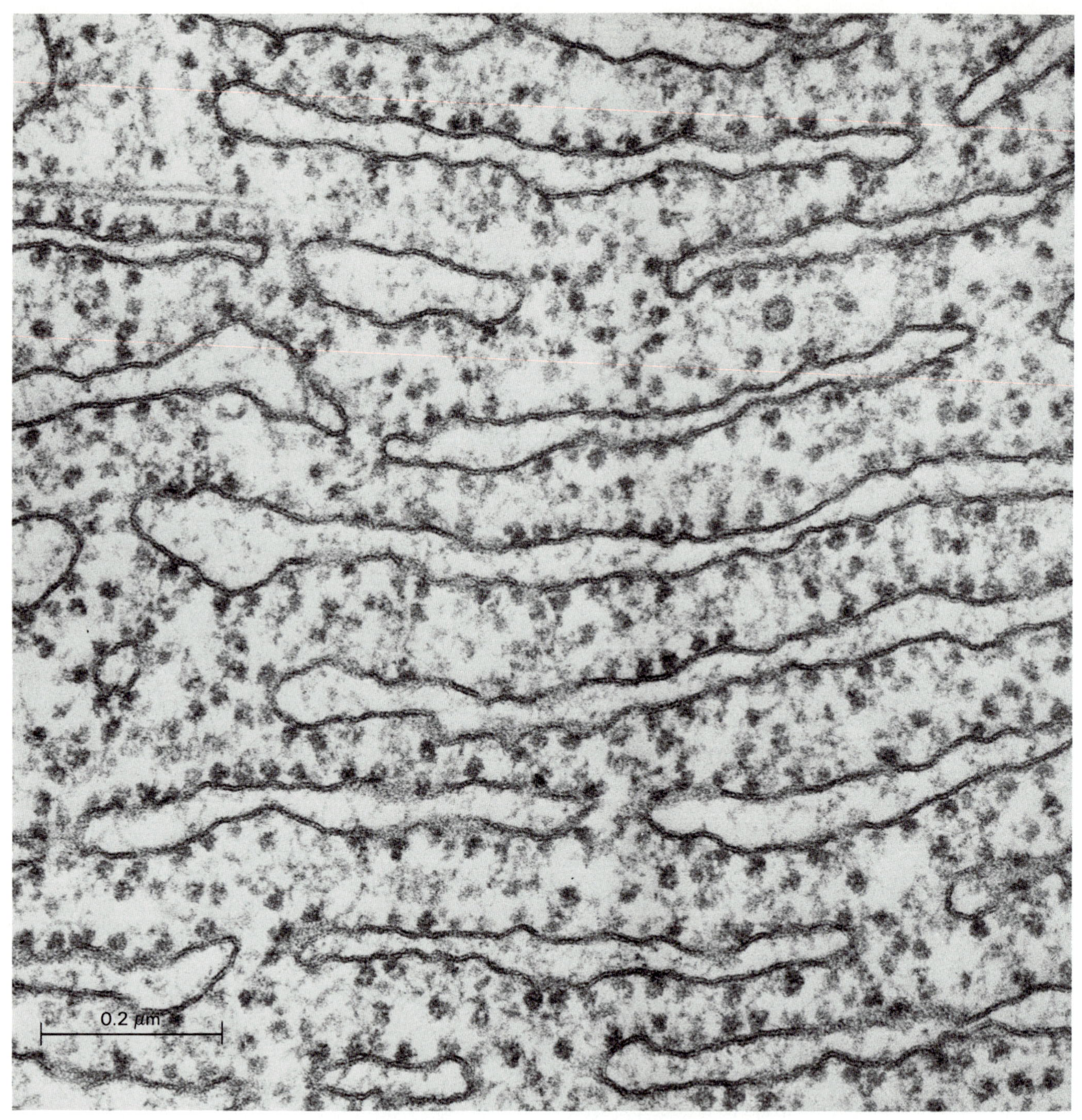

In this section through a rough ER region of a guinea-pig-pancreas acinar cell, flattened cisternae are seen packed so close together as to leave only very narrow passages between them. Note the ribosomes attached by their large subunit (see Chapter 15) to the cytoplasmic face of the ER membranes. The two ribosomal subunits are clearly discernible in a number of the ribosome profiles. The fuzzy material inside the cisternae consists of secretory proteins.

13 Membranes in Action: A View from the Cytosol

One of the most conspicuous features of many eukaryotic cells is the high degree of development of their cytomembrane system. In some cells, the total surface area of this system may exceed 0.1 mm^2, which may not seem like much until you adjust it to our millionfold magnification. Then you find that it means dividing up the space of a large auditorium with more than one million square feet of partitions. Moving through such an auditorium may be a problem, as indeed it is in the ER-rich areas of the cell, where flattened cisternae may be packed so close together as to leave only very narrow passages between them. Fortunately, the partitions are flexible and can be forced apart fairly easily, as shown by the occasional mitochondrion or other cytoplasmic granule found nestling between ER cisternae.

Membranes as Organelles

During the first part of our tour, we were given plenty of proof that these membranes are not mere partitions. But relatively little of what they do, and especially of how they do it, could be discerned on their *trans* face. And so our descriptions had to remain largely phenomenological. Now that we can inspect the *cis* face of the membranes, we get a much clearer view of the number and variety of active systems that are compressed in these tenuous films. Even more important, we can try to find out how they operate, as their works are largely exposed on the side we are facing, which is where they are supplied with substrates, receive energy, interact with

Schematic representation of a permease. This drawing shows how the binding of a substrate causes a conformational change such that a channel is opened allowing translocation of the substrate. Note that the system is perfectly reversible; transport takes place in the direction of decreasing (electro)chemical potential.

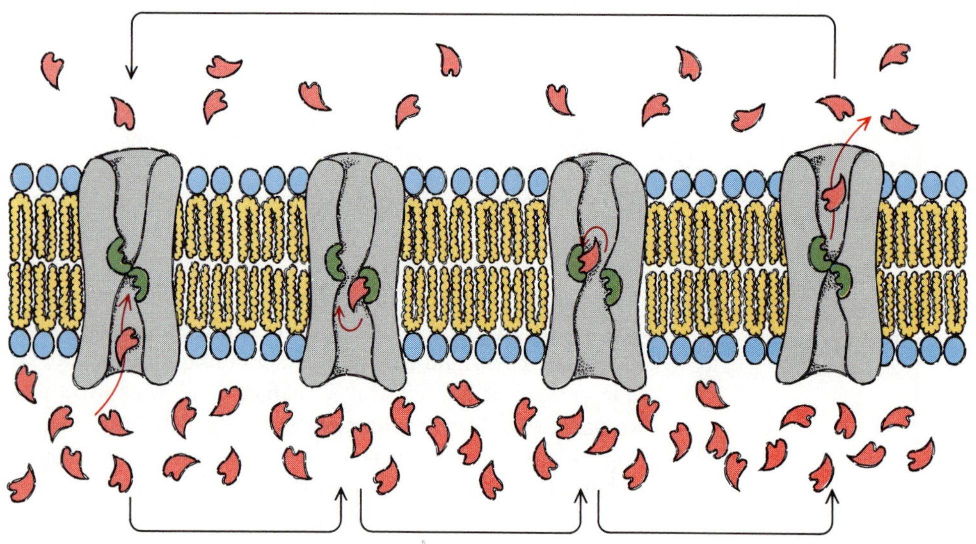

cofactors of various kinds, connect with cytoskeletal elements, and, in general, relate with other parts of the cell. Provided we can decode what is before our eyes, which is not yet possible for many of these complex machineries, we can now aim at understanding how all those displays are engineered. And understanding, not description, is the true purpose of scientific exploration.

Let us take another tour, therefore, through the meandering spaces of the cytosol and scan the surface of the membranes for evidence of the functions that they fulfill. These functions may be grouped under two main classes. One concerns all forms of exchanges—of matter, energy, information—between the two regions separated by the membrane. The other includes a variety of metabolic processes that have in common that they depend on, or take advantage of, some of the structural features provided by membranes, such as a hydrophobic milieu, a substratum for the creation of electric potential differences, or a framework on which multiprotein complexes can be assembled and disassembled with great rapidity and accuracy.

Behind the Plasma Membrane

For a start, let us find out what lies behind the formidable array of gates, checkpoints, transport systems, sensors, and antennae of various kinds with which living cells greet their visitors (Chapter 3). What these systems have in common is a dependence on transmembrane proteins equipped with binding sites that allow the recognition of some specific chemical entity. What differentiates them, in addition to the kind of ligands they bind, is the way in which they respond to ligand binding. In terms of a frequently used analogy, the binding sites are locks, and the ligands are the keys that fit the locks.

Coupled facilitated transport of two solutes.

A. Symport: Both solutes are transported in the same direction.

B. Antiport: The two solutes are transported in opposite directions.

In both examples shown, downhill transport of the red molecules drives uphill transport of the blue molecules (overall $\Delta G<0$).

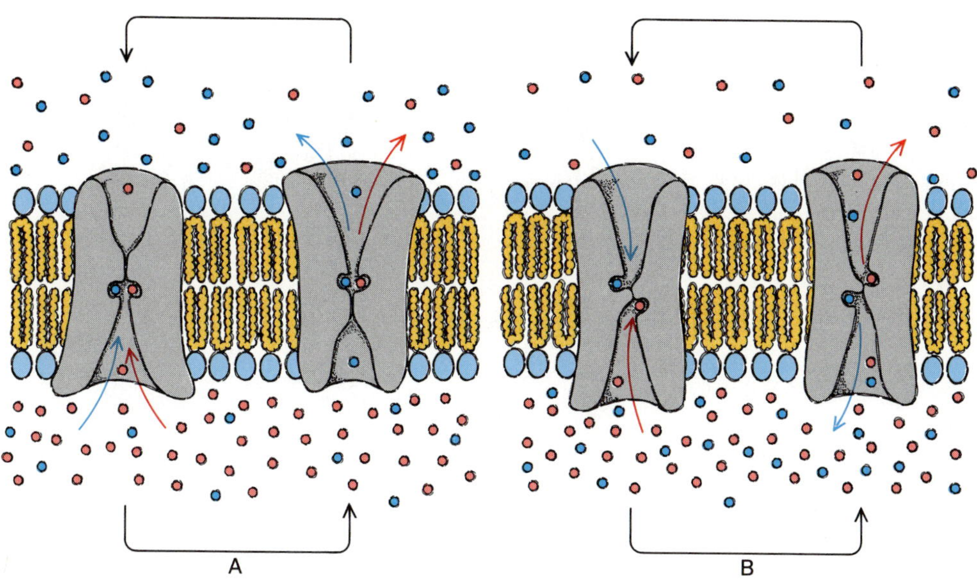

What follows the introduction of the key into the lock—the binding of the ligand to its binding site—varies from one system to another. Almost invariably, the first consequence is a change in the shape of the receptor molecule—the lock is turned—which either affects its function or activates some transducer or signaling device to which it is connected.

Molecular and Ionic Transport

One type of response is transport of the ligand across the lipid bilayer. This is what permeases do. Upon binding of their ligand on one side of the bilayer, they undergo a change in conformation such that the occupied site now faces the other side, where it can unload its molecular passenger. As was explained in Chapter 3 (see also Appendix 2), this mechanism requires no source of energy other than a difference in the chemical (or electrochemical) potentials of the transported substance across the membrane. Accordingly, the transport is always in the direction of decreasing potential. Permeases, nevertheless, are very useful because they help vitally important hydrophilic substances, such as glucose or amino acids, cross what would otherwise be an essentially impassable barrier, and do so in a very specific fashion.

A number of permeases can translocate a substance only in combination with (symport) or in exchange for (antiport) some other substance. Such systems are of interest because they can move one of their substrates up against a potential gradient, provided the other substrate simultaneously moves down a steeper gradient. They link one flow to the other, the way a rope and pulley may couple the lifting of one weight to the fall of a heavier one.

Pumps are constructed like permeases, but with the fundamental difference that the translocating shift of occupied binding sites is driven forcibly with the help of energy. In the case of the all-important sodium-potassium

Schematic representation of sodium-potassium pump.

1. The sodium-loaded pump is phosphorylated by ATP.

2. The fall of the bound phosphoryl group from a high energy level to a low one is associated with a conformational change leading to the pump's loss of affinity for sodium ions, the unloading of these ions into the extracellular medium, and the creation of two high-affinity binding sites for potassium ions.

3. Two extracellular potassium ions occupy the potassium-binding sites.

4. The potassium-loaded pump is dephosphorylated hydrolytically, with a return to the pump's original conformation; affinity for potassium is lost, potassium ions are unloaded intracellularly, and three high-affinity binding sites for sodium ions are created.

5. Three intracellular sodium ions occupy the sodium-binding sites; the cycle can start again.

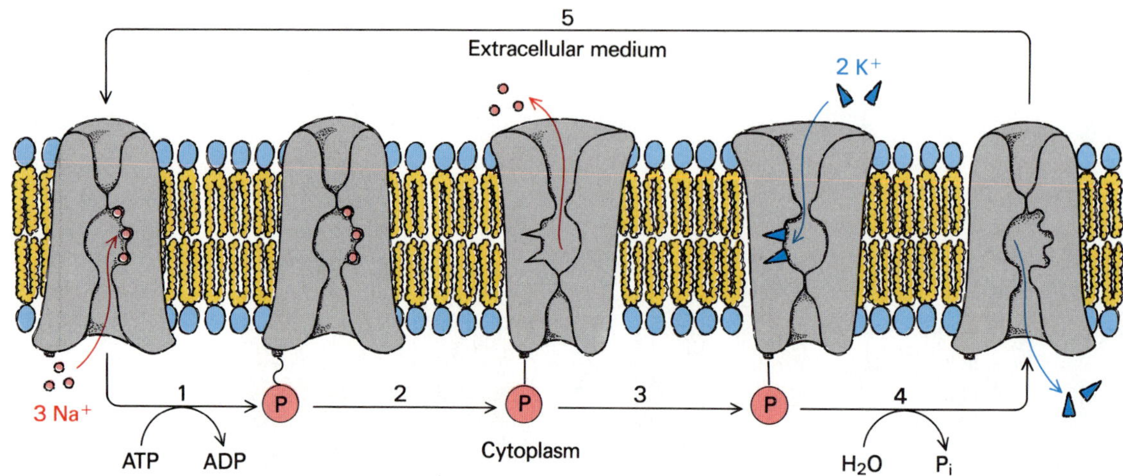

pump, the energy is supplied by ATP. When the sodium-binding sites are occupied on the *cis* side of the plasma membrane, a γ_d attack transfers the terminal phosphoryl group of ATP onto an aspartyl residue of the enzyme. This reaction conserves most of the γ-bond energy in the acyl-phosphate bond that is formed (Chapter 8). It causes the conformation of the protein to change in such a way that the sodium-binding sites now face the outside and, at the same time, lose their affinity for their substrate so that the bound ions fall off. Three sodium ions are expelled in this way against a combined electrochemical potential of the order of 3.3 kcal per equivalent (Appendix 2), at a total cost, therefore, of about 10 kcal per gram-molecule of ATP used. This energy is provided by a concomitant fall of the phosphoryl group in the enzyme from a high to a low potential level. Return to the initial state is mediated by the hydrolytic removal of this phosphoryl group. This process can take place only if two externally oriented binding sites are occupied by potassium ions and obligatorily results in the unloading of these sites inside the cell. As shown in Appendix 2, this step requires little energy, because the electric membrane potential compensates for the positive difference in chemical potential of intracellular over extracellular potassium ions. According to this description, the sodium-potassium pump may be defined in technical terms as an ATP-driven, vectorial, antiport system, which exchanges three intracellular sodium ions for two extracellular potassium ions per molecule of ATP hydrolyzed. Because the pump tends to deplete the cell of positively charged ions, it is electrogenic. Its function, however, resembles that of a bilge pump. It consists essentially in the correction of leaks and must be interpreted in the context of the various passive ion movements across the membrane. The sodium-potassium pump can even be made to move backward by high enough ionic gradients. It then powers the assembly of ATP at the expense of the exchange of intracellular potassium for extracellular sodium.

The ionic inequalities that are maintained between living cells and their surroundings by the operation of the sodium-potassium pump are of crucial functional importance. They provide the cells with an internal milieu adapted to the requirements of their enzymes. In given cell types, they may drive other active-transport mechanisms, such as an antiport expulsion of calcium ions or a symport uptake of glucose or amino acids, coupled with

Schematic representation of synaptic transmission.

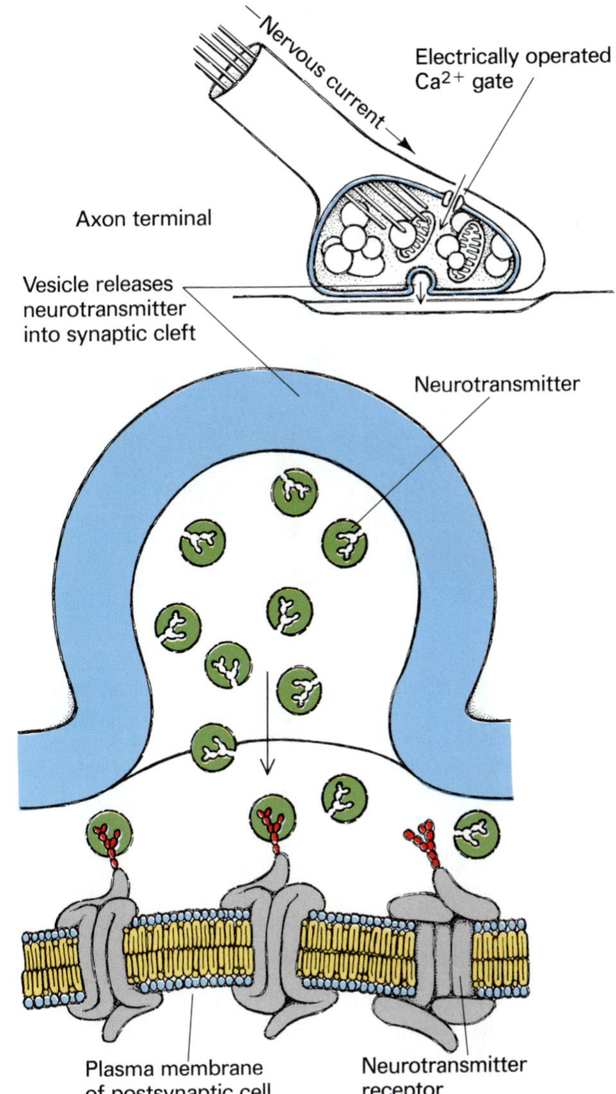

the exergonic re-entry of sodium ions into the cells. They are also responsible for the existence of a difference in electric potential at the cell surface. This resting potential usually lies between 50 and 100 mV and is created largely by the outward leakage of potassium ions, which explains why the ΔG of potassium transport is essentially zero (Appendix 2): net transport stops when the electric potential generated by outward diffusion is just sufficient to drive electrochemical transport at an equivalent rate in the opposite direction. Finally, the ionic inequalities produced directly and indirectly by the sodium-potassium pump have the advantage that they lend themselves to very rapid perturbations. Innumerable phenomena, including all the major operations carried out by the nervous system, depend on such perturbations, which are themselves elicited by means of regulated channels (gates).

Like permeases, gates allow only passive transport down concentration gradients. But instead of carrying their substrates across by means of binding sites, they let them slip through some sort of inner conduit, or channel. They do have binding sites, but these serve to control the opening or closing of the channel by regulating ligands. Gates also exist that respond to changes in certain ionic gradients or in membrane potential. The subtle interplay of chemical, ionic, and electrical factors whereby these various gates are controlled underlies much of the functioning of the nervous system, including the manner in which the ten-billion-odd neurons that make up the human brain communicate with each other and with other cells in the body to stage the most prodigious manifestations the biosphere has to offer.

Nerve Transmission

Only a detailed and extensive tour of the nervous system could do justice to even the little that is currently understood of its stupendous complexity. A fleeting visit is all we can afford; it is enough, however, to give us an inkling of what seems to be a basic functional blueprint. For this purpose, let us follow, at least in thought (we would need a fast car to do it in reality), the wave of depolarization that travels down the axon of a motor neuron that has just been stimulated. As the nervous current reaches one of the small knobby bulges whereby the axon connects with a muscle cell at the neuromuscular junction, it causes the opening of an electrically operated calcium gate. Calcium ions rush in and induce the exocytic discharge of small membrane-bounded vesicles, about 40 nm in diameter, which occupy this region of the cytoplasm. Each of these synaptic vesicles, as they are called, contains some ten thousand molecules of acetylcholine, a substance that we encountered in Chapter 8 as the product of a three-step, CoA-mediated, β_p type of assembly between acetic acid and choline.

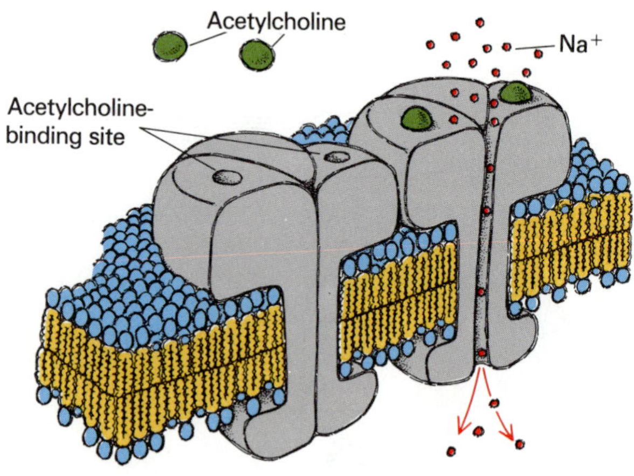

Model of acetylcholine receptor. Two identical protein subunits bearing an acetylcholine-binding site join with three other subunits (only one is shown) around a central channel, which is closed when the binding sites are unoccupied. The binding of acetylcholine (green) causes the channel to open and let sodium ions (red) through.

When released extracellularly, acetylcholine survives less than a thousandth of a second because of the presence of a powerful enzyme, acetylcholinesterase, that immediately hydrolyzes it. But this brief time is enough for significant amounts of the substance to diffuse across the narrow space (synaptic cleft) that separates the two cells and to occupy specific binding sites on the surface of the muscle cell. These binding sites belong to a protein complex, the acetylcholine receptor, which is a ligand-operated sodium gate. When the receptor is occupied by its specific ligand, the gate opens and admits sodium ions into the muscle cell. The resulting depolarization of the cell membrane is propagated by neighboring electrically operated sodium gates and, through the T-system (Chapter 12), activates another electrically operated gate, which lets calcium out of the sarcoplasmic reticulum. And so we are back to calcium, this time as trigger of actin-myosin interaction: the affected myofibril contracts.

All these events are very short and finely tuned. We have already seen the role of acetylcholinesterase in the removal of acetylcholine. Pumps, on the other hand, quickly correct the ionic perturbations. As to the synaptic vesicles, they are replaced, though more slowly, by the new formation of acetylcholine and the endocytic recovery of membrane material. The scene is soon reset for a new act.

This script, with a variety of different settings and protagonists, tells the story of virtually every communication between a nerve cell and another cell, be it another neuron, a muscle cell, a secretory cell, or some other effector cell. Except for those rare cases where electric impulses travel directly between cells, such as through gap junctions (Chapter 2), the signal is always conveyed chemically, by means of a neurotransmitter discharged by the excitatory nerve cell and acting on the target cell through the mediation of specific surface receptors. Many different neurotransmitters besides acetylcholine are known. They include norepinephrine, epinephrine, histamine, serotonin, dopamine, glycine, γ-aminobutyric acid (GABA), and a variety of peptides. The corresponding neurons are designated cholinergic, adrenergic, histaminergic, and so forth. The total number of neurotransmitters is unknown but could be well in excess of one hundred. They are particularly numerous and important in the brain, where they provide the elements of an elaborate chemical communication network.

As there are many transmitters, there are also many receptors—more than there are transmitters, in fact, as some transmitters, such as acetylcholine itself or epinephrine, act on more than one type of receptor. Some of these receptors control ionic gates, as does the acetylcholine receptor, but with a variety of effects, depending on what ions are let in or out, and where. Other receptors act differently, as we are about to see.

The extent of our dependence on neurotransmitters is truly staggering. The regular beating of our hearts, the maintenance of adequate blood flow and pressure in our arteries and veins, the continuous monitoring of blood gases, with immediate correction of any disturbance by appropriate changes in the depth and rhythm of our breathing, the whole hidden orchestration of peristaltic propulsions and digestive secretions needed for the processing of our meals are just a few among the countless silent coordinating activities that rule the functioning of our organs with the help of neurotransmitters. And this is only part of the story. Every one of our movements, voluntary or involuntary; all of the information we receive from the outside world through our sense organs; the sum total of our thoughts, impulses, emotions, and dreams, together with their substratum of unconscious and subconscious data-processing and integration—all rely critically on appropriate neurotransmitter molecules being released at the right time and in the right amounts at billions of different intercellular connections; on their interacting correctly with their receptors to produce their effects; and on their being removed or destroyed at exactly the rate needed for the effects to last long enough, but not too long.

Not surprisingly, a network of this sort is exposed to an immense variety of interferences. For each authentic neurotransmitter, hundreds of different imperfect copies (analogues) can be made that retain the ability to interact with the receptor and either mimic (false keys) or block (jamming devices) the effect of the natural transmitter. Furthermore, release of the real transmitter may be induced, enhanced, or inhibited; its synthesis or breakdown may be accelerated or slowed down. This explains the wealth of substances, both natural and synthetic, that affect the nervous system. From the deadly paralysis spread by a curare-dipped arrow to the horrible convulsions induced by extracts of the strychnos nut, from the cosmic serenity of the opium smoker to the piercing hallucinations of the LSD victim, from pep pills to tranquilizers, from knock-out drops to nerve gas, the whole gamut of neuropsychotropic drugs and poisons created by the combined ingenuity of Man and Nature is nothing but a huge collection of perturbing agents capable of interfering, in an infinite variety of ways, with chemical neurotransmission.

If so much can be done from the outside to our thoughts and moods, why not also from the inside? This question is being asked with increasing insistence by modern neuropsychiatric investigators who are searching very hard to give precise chemical identities to the demons of the past and to their more recent Freudian substitutes. Some may deplore this attempt at demythification of mental illness, but not so the patients the day they find that a simple medication can save them from protracted exorcisms on a psychoanalyst's couch or from commitment for life to the company of their fellow sufferers.

Hormonal Effects

Many receptors have more than a simple gating function and are connected to more or less complex enzyme systems that face the cytosol on the *cis* side of the plasma membrane. Best known of these systems is adenylate cyclase, an enzyme that catalyzes the formation of cyclic AMP from ATP (Chapter 8):

$$\text{ATP} \longrightarrow \text{cAMP} + \text{PP}_i \ (2\ \text{P}_i)$$

The product of this reaction is a multipurpose chemical messenger that is dispatched from its receptor-linked site of formation on the plasma membrane to various intracellular systems that are sensitive to it, in particular certain protein kinases. The important regulatory function fulfilled by this class of enzymes was mentioned in Chapter 12 in a discussion of the biological effects of calcium ions. A number of hormones, including some that double as neurotransmitters (e.g., epinephrine), act via cAMP. The selectivity of the effects of each hormone is ensured by the localization of its receptor. For example, ACTH, the pituitary adrenocorticotropic hormone, switches on cAMP-dependent steroid formation in the adrenal cortex without activating cAMP-dependent mechanisms elsewhere because its receptors occur predominantly on the surface of adrenocortical cells. Conversely, the effects of parathyroid hormone, which also stimulates cAMP production, are restricted to certain cells in the bones and kidneys that display the appropriate receptor molecules.

A remarkable characteristic of these mechanisms is their power of amplification. To appreciate this, just take a look at a liver cell that is being stimulated by epinephrine or glucagon to fragment glycogen and produce glu-

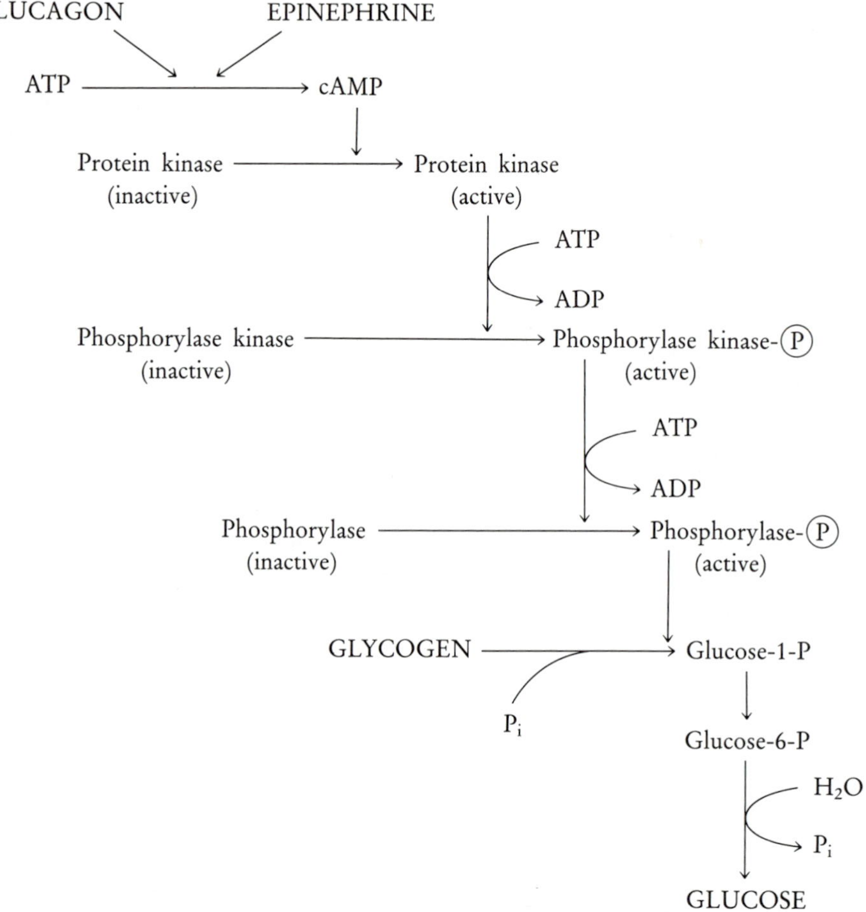

cose. Binding of either hormone to its receptor triggers the production of cAMP, which stimulates a protein kinase to activate, by phosphorylation, another protein kinase, which itself phosphorylates an enzyme called glycogen phosphorylase, which, in phosphorylated form, catalyzes the breakdown of glycogen to glucose-1-phosphate, a precursor of glucose, as shown above.

Almost every step in this cascade is catalytic. Let us assume conservatively that each catalytic step causes a hundredfold amplification: then one molecule of hormone stimulates the production of 100 molecules of cAMP; each molecule of cAMP-activated protein kinase phosphorylates 100 molecules of phosphorylase kinase; and so on. The end result is the production of 100 million (10^8) molecules of glucose for every molecule of hormone bound to its receptor. Thus, minute amounts of hormone can have dramatic effects. Note that there is a price to pay for this exquisite sensitivity: 10,101 ATP molecules (one to make cAMP, plus 100 to phosphorylate phosphorylase kinase, plus 10,000 to phosphorylate phosphorylase). This price is really negligible, however, in regard to the 38 molecules of ATP the cell can gain from the oxidation of each molecule of glucose, which means that the cell spends one ATP to get back 380,000. It is a bargain.

A more serious drawback of high amplification is the danger of things getting out of hand. It is minimized by the occurrence of numerous safeguards, including interiorization and destruction of the hormone, participation of antagonistic hormones, inactivation of the receptor, hydrolysis of cAMP to AMP by a specific phosphodiesterase, dephosphorylation of phosphorylase or of its kinase by phosphatases. But when these safeguards fail or are inoperative, tragedy may ensue. This happens in cholera. The microbe responsible for this terrible disease secretes a toxin that binds in a stable fashion to an adenylate-cyclase–linked receptor present on the surface of intestinal cells. These are thereby flooded continually with high amounts of cAMP, which, among other effects, stimulates the cells to excrete salts and water uninterruptedly. Hence the persistent diarrhea and severe dehydration that characterize the disease and used to kill most of its victims until understanding of this mechanism brought forth a simple and effective rehydration therapy.

Mitogenic Stimulation

A particularly important, as well as mysterious, class of receptors comprises those that evoke a mitogenic response. When turned on by a ligand, they start a complex sequence of events that many hours later, long after the ligand has been shed from the receptor, ends up with DNA duplication, construction of a mitotic apparatus, and cell division. The control of cell multiplication by trophic hormones or growth factors, crucially important for the generation and maintenance of each cell population in the organism, takes place through the mediation of such receptors. Somewhere along the chain of events that they control lie points through which cancer-producing agents exert their deregulating effects. There is evidence that certain special protein kinases play an important role in both normal and abnormal mitogenesis (see Chapter 18).

Mitogenic control is developed to an astounding degree of diversity and selectivity in the lymphocytes. As a result of a unique process of ontogenic diversification (see Chapter 18), the issue of this line of cells consists of literally hundreds of millions of distinct individual types, each of which has its mitogenic signaling device hooked to a different surface receptor. The number and variety of these receptors is such that virtually every possible macromolecule—except those of the organism itself, for which immunological tolerance has been built up during fetal life—is recognized and bound by at least one subclass of lymphocytes. The cells to which this happens are thereby led to multiply and to form a clone. The ligands that trigger such a mitogenic response are called antigens. In the B-lymphocyte line, the cells involved are those that manufacture an antibody directed against the stimulating antigen. (The receptor is none other than the antibody itself, anchored on the cell surface by an extra hydrophobic tail.) In T lymphocytes, the cells generated by mitogenic stimulation are killer cells that selectively destroy any cell that bears on its surface the antigenic groupings that started the multiplication process. This is how the immune system builds up specific army corps against each individual challenge by a foreign molecule or organism (Chapter 3). Certain glycoproteins, called lectins, mostly of vegetable origin, have the ability to stimulate lymphocyte proliferation nonselectively. Examples are phytohemagglutinin (extracted from kidney beans) and concanavalin A (from jack beans).

Endocytic Uptake

A general function common to many receptors is to mediate the selective endocytic uptake of their ligands, usually for subsequent delivery to the lysosomes and consequent degradation. As we saw in Chapters 4 and 5, all sorts of important regulatory mechanisms have been built around this primitive feeding process. In the framework of receptor function, it has a number of important aspects. First, there are cases where uptake conditions the cell's response to a bound ligand because the triggering change in the occupied receptor takes place only in an acidic milieu, such as is provided in endosomes and lysosomes. Epidermal growth factor (EGF), a mitogenic hormone acting on epithelial and fibroblastic cells, is said to have such a requirement. So does the invasion of the cell by a number of membrane-wrapped viruses, as we can testify from personal experience (Chapter 7), as well as the intracellular penetration of various protein toxins (see below). Another consequence of endocytic uptake is to limit the duration and intensity of a response by bringing about the destruction of the ligand. As to the receptors themselves, they may resurface, ready to function again, as early as a few minutes after interiorization, thanks to the phenomenon of membrane recycling (Chapter 6). Or, as happens with some hormone receptors, they may accompany their ligands into the lysosomes and be inactivated, with the consequence that the cells become less responsive to the hormone (down regulation).

A common characteristic of endocytic receptors is some form of connection on the *cis* side of the plasma membrane with an elaborate cytoskeletal rig that does the mechanical tasks of endocytic uptake. Clathrin baskets (Chapter 12) are the best-known such devices. But there may be others, in which actin fibers do the pulling, probably in combination with myosin. To what extent occupancy of the receptors conditions uptake is not clear and may vary according to receptor (Chapter 4). The same is true of the clustering phenomenon that causes the receptors to congregate on the membrane patches that are pulled in. Some receptors naturally home to coated pits, perhaps because they are linked on the *cis* face to clathrin or to some clathrin-associated protein. Others are brought together by divalent or multivalent ligands on the *trans* face. The most dramatic such event is capping, a phenomenon in which all the occupied molecules of a given receptor form a single aggregate, which is then interiorized in one big endocytic gulp.

Chemotaxis

Possibly related to certain forms of receptor-mediated endocytosis is the phenomenon of chemotaxis. (Remember, it was one of the very first sights to meet our eyes when we set off on our tour.) The receptors involved in this remarkable process are linked dynamically to the cell's locomotor system in such a way that its response does not depend simply on the number of occupied receptors, but also on their spatial distribution. This distribution will be uneven whenever the cell is exposed to a flux of ligand molecules diffusing down a concentration gradient: there will be a corresponding gradient of occupied receptors on the cell surface, reflected inside the cell by a similar gradient of whatever messenger molecules or other activating devices are triggered by the occupied receptors. Such a gradient seems to be all that is needed to orient the complex mechanisms whereby filopodia are pushed out, adhesion plaques attached, and actin cables retracted to move the cell toward or away from the incoming ligand molecules. Ligands capable of influencing cell movement in this way are said to be positively or negatively chemotactic. They act as cellular attractants or repellents.

Translocation

One of the strangest forms of exploitation of surface receptors is exhibited by certain toxic proteins made by bacteria (e.g., diphtheria toxin) or by plants (e.g., ricin, extracted from castor beans). These toxins consist of two parts linked by disulfide bridges: the A chain, or "effectomer," is an enzyme with the property of rendering inactive some component of the protein-synthesizing machinery; the B chain, or "haptomer" (Greek *haptein,* to bind), is a ligand with the ability to bind specifically to certain surface receptors. By itself, the effectomer is absolutely harmless because it cannot get through the plasma membrane. But, when it is brought close to the membrane by its haptomer carrier, it gets detached from the haptomer and translocated into the cytoplasm, where it does its deadly work. In several such cases, the actual piece of legerdemain takes place only after endocytic uptake and is triggered by the endosomal acidity.

This phenomenon has raised questions regarding its possible physiological significance. Several hormones resemble the toxins. In particular, the pituitary thyroid-stimulating (TSH), follicle-stimulating (FSH), and luteinizing (LH) hormones, as well as the placental choriogonadotrophin, are built of two pieces. One, common to all four, is an adenylate cyclase activator. The other carries the tissue specificity of the hormone and serves as a homing device by recognizing receptors that are uniquely present on the surface of the hormone's target cells. The analogy with the effectomer-haptomer combination of the toxins is suggestive. There is, however, no evidence that the action of these hormones requires their endo-

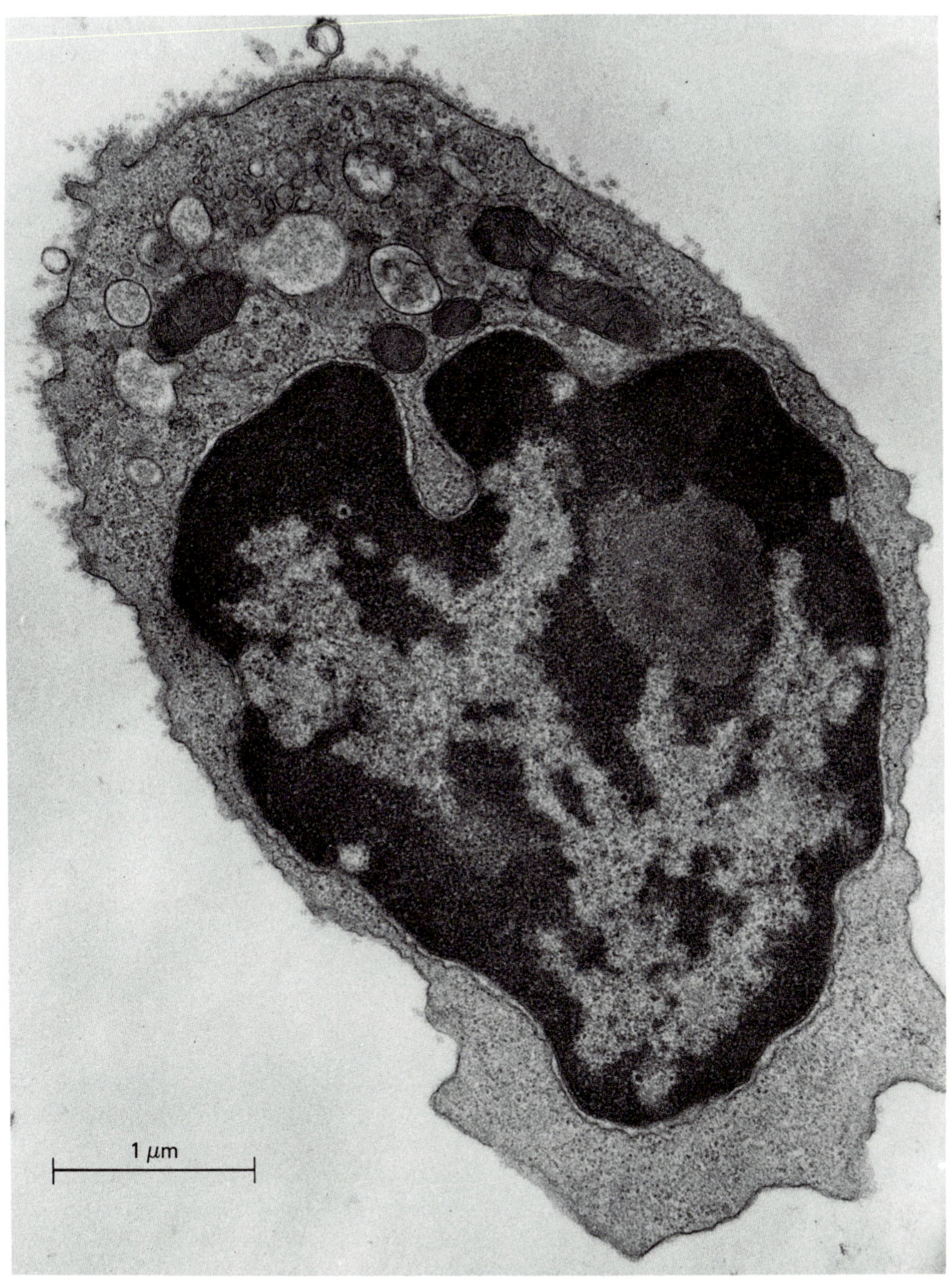

Capping. A mouse B lymphocyte, which normally has immunoglobulin receptor molecules distributed over its whole surface, has been treated with an antibody against this receptor, resulting in the clustering of all the receptor molecules at one pole (upper left) of the cell. This "cap" is made evident by the covering of small square or round markers, which are molecules of the giant protein hemocyanin (the blue copper-containing blood protein that serves as oxygen carrier in the snail) that have been attached chemically to the antibody molecules.

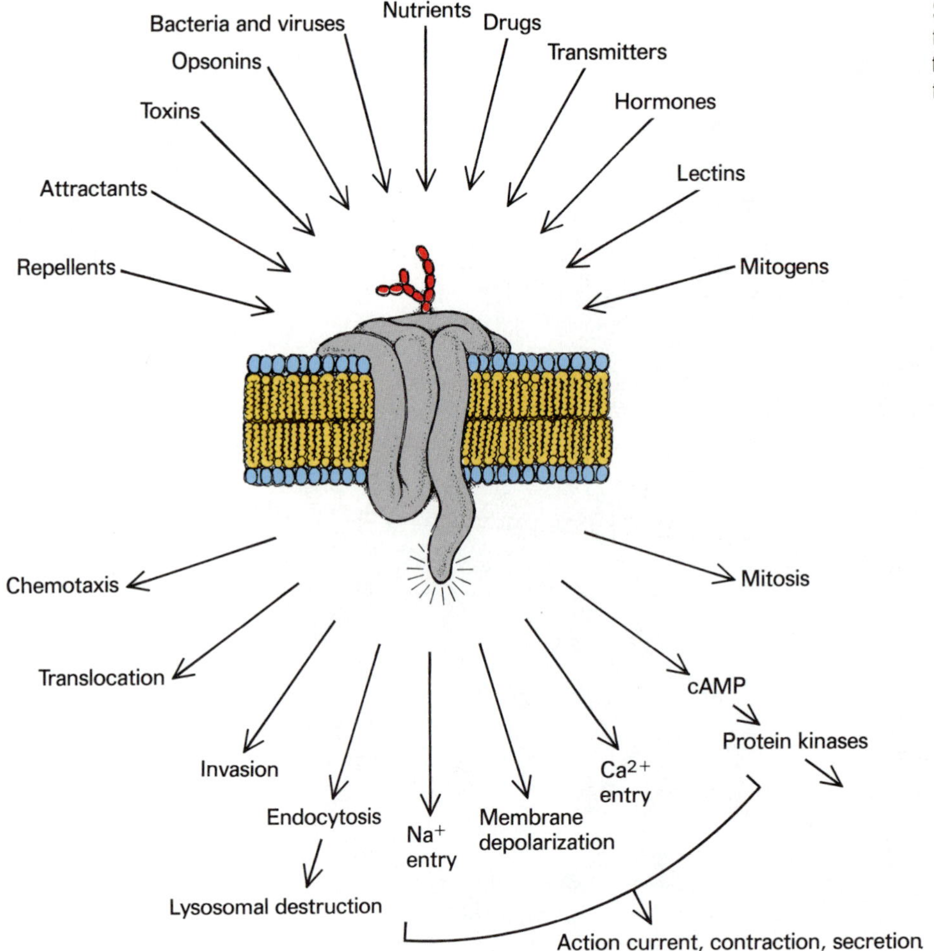

Some channels of communication established by surface receptors between the outside and the inside of the cell.

cytic uptake and introduction into an acidic milieu or involves passage of the effectomer part across the cell membrane.

So goes the dance of the receptors, a fantastic display of virtuoso acts staged on the inner face of the plasma membrane by all sorts of transducers and signaling devices and choreographed on its outer face by hosts of transmitters, hormones, mitogens, lectins, nutrients, ions, opsonins, carrier proteins, attractants, repellents, toxins, drugs, and even microorganisms and viruses, all sharing the key property of being able to bind specifically to some surface receptor. The ligands call the dance, the receptor-linked transducers command the steps, but only those cells respond that own the particular receptor capable of connecting the ligand to the transducer. Therein lies the specificity of innumerable biological responses, interactions, and effects.

A necessary corollary of this general mechanism is that each type of cell has its own combination of surface receptors. Cells, therefore, betray their identity—and thereby lay themselves open to all sorts of outside manipulations, even selective killing—to any outsider cognizant of the receptor code. No wonder, then, that therapy-motivated cell explorers who dream of "guided missiles" and other "magic bullets" as means of subduing recalcitrant cell populations expend so much time and effort scrutinizing the hundreds of different receptors exposed on the surface of the cells. It is one of the hot spots of contemporary cell research.

The Cytosolic Face of the Vacuome

In line with their role as boundaries of spaces that, directly or indirectly, make transient connections with the outside world and with their participation in the vesicular carousel that moves material in bulk into, through, and out of cells, the membranes of the vacuome share many of the properties of the plasma membrane. They bear pumps, permeases, and transport receptors and are wired to cytoskeletal elements that direct their movements. This is particularly true of the exoplasmic parts (Chapter 6), many of which are continuously recycling between the interior and the surface of the cells.

We came across some of these features in our wanderings. We saw a proton pump, possibly imported from the plasma membrane, working to acidify endosomes and lysosomes (Chapters 4 and 5) and learned of the probable participation of permeases in the clearance of lysosomes (Chapter 5). We found out in Chapter 12 that striated muscle cells subdue their myofibrils with the help of a calcium pump—a device that operates on a phosphorylation-dephosphorylation cycle, as does the sodium-potassium pump—situated in the sarcoplasmic reticulum. And, as we have just seen, they depend on an electrically controlled calcium gate inserted in the same membranes to call the myofibrils to action. In Chapter 6, the general significance of receptor clustering on movable membrane patches was noted as a mechanism of selective sorting and transport, exemplified by the mannose-6-phosphate receptor for lysosomal enzymes. Also discussed were some of the logistics of transport mechanisms that rely on an accurate mutual recognition and intimate apposition between two membranes and on the resulting fusion and reorganization of their lipid bilayers, which nevertheless occur without significant intermingling of their protein components.

Biosynthesis in the ER

Besides such general features common to all these membranes, there are important differences. Just as we contemplated many different kinds of surfaces when we toured the vacuome, so now our cruise along the cytosolic boundaries of this membranous system presents us with a similar number of different sights. Endoplasmic regions, in particular, show conspicuous signs of an intense biosynthetic activity, of which we have seen only the end result so far (Chapter 6). Most impressive is the assembly of proteins by the polysomes that cover the rough-surfaced parts of the ER. All of Chapter 15 deals with this process, and so we will skip the ribosomes for the time being. But there is much more to be seen and enjoyed, especially by those who followed our biosynthetic excursion in the second half of Chapter 8.

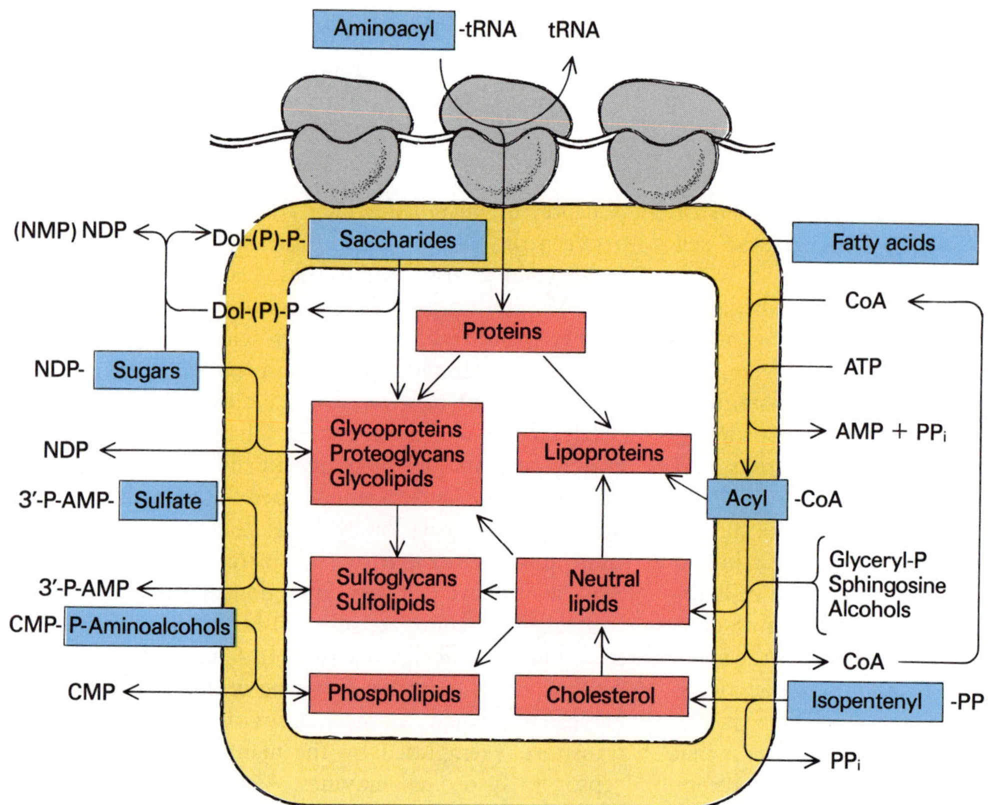

Summary of the main biosynthetic processes that take place on ER (and Golgi) membranes. Note that, with the exception of fatty acids, which are activated locally, all $X\text{-}O^-$ building blocks (shown in blue, Chapter 8) are delivered in activated form by carriers to the assembly sites. Biosynthetic products accumulate in the luminal space or, as is the case for some proteins and many lipids, remain membrane bound.

Between the attached ribosomes, a variety of nucleoside diphosphates (Chapter 8) deliver activated glycosyl groups for the assembly of the oligosaccharide side chains of glycoproteins. These chains are stubby, bushlike structures containing from eleven to fourteen sugar molecules. They are assembled on a carrier, dolichyl diphosphate, which is firmly rooted in the lipid bilayer by a long hydrophobic tail. As we saw when we were inside the ER (Chapter 6), completed shrubs are transferred en bloc onto growing polypeptide chains emerging in the vicinity. The whole process is quite baffling, as it requires the translocation, sometimes with the help of other dolichyl phosphate molecules, of the highly hydrophilic glycosyl groups across the lipid bilayer. Sulfation and phosphorylation of some of the carbohydrate side chains also take place through the membrane from carriers charged in the cytosol.

Other areas of the ER membranes are occupied with lipid biosynthesis. Some of the more hydrophilic building blocks needed for this process—for instance, glycerol, choline, or ethanolamine—are brought by special nucleotide carriers (Chapter 8) from cytosolic activating sites. Others, especially the fatty acids, are activated locally and linked to coenzyme A, from which they are transferred to their final acceptor. These are busy factories making all sorts of neutral fats and phospholipids, as well as cholesterol and its derivatives, and many other lipid-soluble molecules. Sometimes the lipid and carbohydrate factories converge to make glycolipids. Some of these lipid products are transferred into the lumen of the ER, where they are combined with proteins to form secretory lipoproteins. Others are stored in the cytosol, where they aggregate into oily droplets, immiscible with water. Yet others remain in the membranes, as do some of the newly made proteins, thus ensuring the manufacture of new membrane material.

Conjugation

Not directly related to biosynthesis as a self-building activity, but relying on the same type of mechanism, are the various phenomena known as conjugation. They can affect a variety of lipid-soluble molecules, some of them made physiologically (e.g., cholic acid), others introduced from the outside (e.g., a number of drugs); typically, they link these molecules with some hydrophilic substance such as glucuronate, sulfate, acetate, or some amino acids. The main consequence of conjugation is that it renders the affected molecules more water-soluble and favors their excretion in bile (bile acids) or urine. Most such processes occur on the membranes of the ER and use carrier-linked activated building blocks, as do the analogous authentic biosynthetic reactions.

Hydroxylation

One would think that such manifold group-transferring activities would suffice to occupy all the available sites on the membranes. But this is hardly so. Actually, a careful look will show that ER membranes appear as a quiltwork of white and pink patches and that unloading of activated groups occurs mostly on white patches. What, then, is the function of the pink patches? Color, in the biological world, is often associated with electron transport. The pink areas of the ER are no exception. They are occupied by two cytochromes, known as b_5 and P_{450}, and are involved in a very special type of electron transfer, from NADPH to molecular oxygen, that takes place in such a way that half the oxygen is reduced to water and the other half serves to hydroxylate some organic substrate:

$$\text{NADPH} + \text{H}^+ + \text{O}_2 + \text{R—H} \longrightarrow \text{NADP}^+ + \text{H}_2\text{O} + \text{R—OH}$$

This is one of the rare reactions in which oxygen is not used simply as electron acceptor but actually becomes attached to a metabolite.

A remarkable property of this system is its lack of specificity. It has its physiological substrates, the most important ones being steroid hormones and their precursors. But it will also act on innumerable artificial chemicals, including medicinal drugs and environmental pollutants. In many instances, the change inflicted on the substrate decreases its pharmacological or toxic properties and favors its elimination, either as such or conjugated. For this reason, the hydroxylating system of the pink ER membrane patches has often been referred to as a detoxicating system, and its lack of specificity regarded as a benevolent gift of natural selection, which somehow, millions of years ago, prepared our cells for the man-made chemical invasion of the world. But now we must modify this optimistic view, for it also happens that harmless molecules are rendered toxic by the action of the hydroxylating system. The kind of toxicity it conveys is a particularly treacherous one, as it often includes the ability to cause mutations and cancerous transformations. Many environmental carcinogens do not possess, as such, the property to cause cancer; they acquire it when they come into contact with the ER.

Except for the polysomes and their anchoring proteins (ribophorins), which are concentrated in the more remote parts of the ER, all these systems seem to be distributed throughout the whole membrane network, presumably because they are free to diffuse laterally along the lipid bilayer and thereby tend to occupy in uniform fashion the whole area available to them, forming only such microclusters as arise from their mutual interactions. This is indeed what we observe until we reach the junction between the smooth ER and the Golgi apparatus. There the scene changes abruptly. Enzymes characteristic of the ER disappear and are replaced by others. For instance, we find in the Golgi membranes a new family of glycosyl transferases that receive their activated sugar molecules from the same NDP carriers as do those located in the ER but transfer them to their final acceptors without the help of dolichyl phosphates. These enzymes finish off the job started in the ER. In addition to constructing new chains, they remodel those built in the ER, add new sugar molecules to them, often after their partial dismantlement by glycosidases, and cap them with special molecules (fucose, sialic acid) or groupings, including the mannose-6-phosphate tag of lysosomal enzymes. They also play an important role in fitting the membrane proteins themselves with carbohydrate side chains.

These processes are consistent with the position of the Golgi apparatus as a halfway station between the ER and the plasma membrane. But how changes in membrane composition are generated and maintained, in spite of the extensive opportunities for intermingling offered by the multiple associations between the Golgi and other types of membranes, remains a very puzzling problem.

The Mitochondrial Boundary

Like their bacterial counterparts, the energy-transducing membranes of mitochondria are authentic boundaries that mediate an intense, strictly regulated molecular traffic. The outer mitochondrial membrane has little to do with this regulation. It is an essentially open frontier of somewhat uncertain function. Virtually all the control is exerted by the inner membrane. This is understandable. To function as chemiosmotic transducer, the mitochondrial inner membrane must be tightly sealed. Hence, it alone can control such channels as have to exist to allow the cell's power plants to operate. This adds even more to the intricacy of its architecture, interspersing a network of molecular checkpoints between the protonmotive, electronic microchips that delighted us with their astonishing construction when we first set eyes on them. Most mitochondrial transport systems are of the passive antiport kind, but some are linked to protonmotive power and can drive their substrates uphill. Mitochondria, like other power stations, keep up three types of exchanges with their environment. They import fuel, export power, and undergo repair and maintenance. They do, however, differ from man-made power stations by their greatly superior versatility and efficiency. As we have seen, the mitochondrial matrix contains systems capable of oxidizing all major foodstuffs. Accordingly, ports exist on the inner membrane to admit a number of organic acids, including pyruvic acid, amino acids, and fatty acids. The last do not enter as such, but in activated form, with the help of carnitine, a special carrier that mediates the transfer of fatty acyl groups from cytosolic to mitochondrial coenzyme A. Fuel entry is complemented by an equivalent influx of oxygen, probably facilitated by the oxygen-consuming cytochrome-oxidase complex.

Systems also exist that allow the fueling of mitochondria with high-energy electrons produced in the cytosol—for instance, by glycolysis. These electrons are ferried by special redox couples that collect them from NADH in the cytosol, deliver them inside the mitochondria, and return to the cytosol in oxidized form to pick up another electron pair:

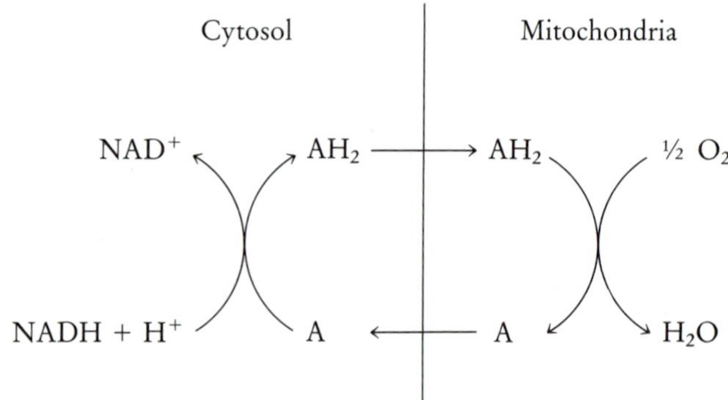

Several such systems are known. As shown by the diagram, they require two electron-transferring enzymes acting on the AH_2/A couple, one in the cytosol, the other in the mitochondria, as well as ports for the two members of the couple in the mitochondrial membrane. Their function is important because there are no passageways for NADH or NAD^+ on the mitochondrial surface. This is the only way in which cytosolic NADH can be reoxidized by oxygen.

Thanks to the existence of the Krebs cycle, mitochondria oxidize their fuel completely: they are nonpolluting power stations. The waste products flowing out of them are simple, nontoxic compounds, mostly carbon dioxide (as bicarbonate), water, inorganic sulfate, and, as sole potentially toxic substance, ammonia, which is tolerated by cells only at low concentration. In many animals and in humans, mitochondria are equipped with an antipollution mechanism that converts ammonia into urea. In birds and reptiles, ammonia is detoxified in the cytosol by conversion into uric acid.

In exchange for the fuel they receive, mitochondria export power in the form of the group potential of the γ bond of ATP assembled from incoming ADP and inorganic phosphate. A special ATP/ADP translocase allows exit of an ATP molecule in exchange for an entering ADP (or AMP, thanks to the occurrence of an AMP kinase in the space between the two membranes). Inorganic phosphate enters by an antiport in exchange for hydroxyl ions.

As already mentioned, protonmotive power can serve to move ions across the mitochondrial boundary, either by a symport or an antiport mechanism (e.g., inorganic phosphate) or by means of an appropriate ionophore or carrier with the help of the membrane potential (e.g., calcium). The key role played in many cells by the mitochondrial calcium pump was alluded to in Chapter 12. What is much more obscure is the manner in which mitochondria stock up on the materials needed to support the growth process, which, complemented by division, is the only mechanism whereby their populations are maintained in stationary cells and augmented in growing cells. Mitochondria can make many of their constituents, even proteins, from small building blocks. But, as mentioned earlier, they carry the genetic information for only a very small number of proteins. The instructions for most mitochondrial proteins are stored in the nucleus and translated in the cytosol by soluble polysomes. The synthesized proteins somehow work their way through the outer membrane and, in most cases, through the inner membrane, in spite of its hermetically sealed boundary. Most chloroplast proteins face a similar problem. So do those of peroxisomes, which also are made in the cytosol. The possible role of signal sequences, removable or not, in these posttranslational transfers will be discussed in Chapter 15.

14 The Nexus of Metabolism: Integration and Regulation

During much of our visit so far, bioenergetics has provided the main framework within which to rationalize what we saw. There were good reasons for that. Thermodynamics is an essential and trustworthy guide. It distinguishes clearly what can happen from what cannot and helps us understand why things can happen the way they do. Especially when it comes to life and its astounding combinations of apparent improbabilities, the insights given by thermodynamics have proved most illuminating. But there is an important limit. Thermodynamics tells us what *can* happen; it does not tell us what *does* happen. For example, there is nothing from the energetic point of view to prevent the whole of the biosphere going up in flames. Why it does not, and why it sometimes starts doing so, as in a forest fire, is not a thermodynamic problem but a kinetic one. This, at least, is how we distinguish them on an elementary level. At a deeper level, the two aspects fuse within the framework of statistical thermodynamics.

The Kinetic Factor

Kinetics deals with rates. It helps distinguish, among the infinite number of events that are thermodynamically possible, those few that occur in reality at measurable rates. This distinction is doubly important for living cells: first, in something of an all-or-none fashion, because most of the events that take place in living cells are of the kind that, although allowed by thermodynamics, do not take place to any significant extent without a suitable catalyst. What a cell can do, therefore, is determined by its enzymic equipment, itself a reflection of the cell's genetic endowment. In this respect, much of the diversity within the biosphere boils down to the ability, or inability, of organisms to make specific enzymes. This is true even within a given species. In humans, many congenital abnormalities result from severe deficiencies of single enzymes.

But it is not enough for a cell or organism to be able to carry out a given reaction or process. The rate at which it does the job is often crucially important as well. Regulation and adaptation are accomplished to a large extent by rate modifications. These sometimes have themselves the character of an all-or-none process. A number of the switches operated by surface receptors are of this kind, at least in first approximation. More often, the response is graded and is both mediated and modulated by more or less intricate feedback loops. We cannot leave the cytosol without taking a brief look at these mechanisms. Actually, we can see them in action everywhere in the cell. But there is no better place than the cytosol, which, in addition to being a major cell organ in its own right, containing hundreds of enzymes and up to half a cell's total protein, represents a central communication system, a unifying connecting link, the obligatory intermediate in all the exchanges and interactions that take place between the different cellular organelles.

The Simple Rules of Molecular Circulation

Much of the complex molecular traffic that fills the busy cytosolic thoroughfares is readily understood with the help of the simple law of diffusion. Soluble molecules, with no source of energy other than random thermal agitation, move spontaneously from any point where their concentration is higher to one where it is lower—down their concentration gradient. This is the direction imposed by thermodynamics (Appendix 2). Kinetically, however, diffusion is a very slow process, which becomes rapid only in situations where large concentration differences are maintained over extremely short distances—that is, where the concentration gradient is very steep. Such is the case in living cells. Indeed, once aware of the general properties of the diffusion process, we have no difficulty recognizing in the cytosol countless molecular fluxes guided by concentration gradients, often of highly complex three-dimensional geometry, and leading invariably from sites where substances are generated or admitted to areas where they are altered or let out. Some of these fluxes have been mentioned before, and the concept may now be generalized.

There are, however, cases where diffusion is too slow for the needs of the cells in spite of the short distances to be covered, because nowhere does the substance to be moved reach a high enough concentration. This may be so because of unfavorable equilibrium conditions, as is the case for some high-energy biosynthetic intermediates. Some of the devices whereby such thermodynamic hurdles are overcome have been alluded to in Chapter 8. Low solubility in water is another factor that may prevent a substance from building a steep enough concentration gradient. Most lipid-soluble substances share this drawback. Such substances are transported by carriers, generally of protein nature, which either move by simple diffusion themselves or form a chain across which the substances move by exchange, jumping from one carrier molecule to another (exchange-diffusion).

Oxygen is a special case. It enters cells by diffusion through whichever face is closest to the blood supply and then it automatically draws toward the mitochondria, each of which maintains an oxygen "sink" around itself. Fortunately, mitochondrial oxidations can function at maximal capacity even at very low oxygen concentration (owing to the high oxygen affinity of the respiratory enzyme cytochrome oxidase) and, thanks to this property, can usually generate a steep enough gradient to satisfy their oxygen requirements. An exception is the powerful, ultrafast, fuel-guzzling, red striated muscle cell, which is helped by a special pigment—myoglobin, a close relative of blood hemoglobin—that serves both to store oxygen intracellularly and to facilitate its diffusion.

In many instances, however, whether in muscle or

other cells, the limiting factor in providing oxygen to the mitochondria is not intracellular diffusion but extracellular supply from the lungs by way of the blood. If this supply is inadequate, the cell has no solution other than to switch over to anaerobic glycolysis if it is to satisfy its energy needs. During a maximal effort, such as a 100-meter run, an athlete's muscles need more ATP than can be supplied by oxidative phosphorylation with the oxygen that is made available to the mitochondria. The ATP deficit is made up by anaerobic glycolysis, which, as we shall see, is started automatically under such conditions. The runner goes into oxygen debt and accumulates lactic acid. During recovery, part of this lactic acid is oxidized, and the remainder is converted back into glycogen by reverse glycolysis with the help of the ATP generated by the oxidative process. The debt is paid back—at the cost, of course, of a small entropic levy caused by the extra run, first down and then up again, along the glycolytic chain.

Incidentally, art lovers will derive a rare form of delectation from going through a cell that is experiencing fluctuations in oxygen supply. For understandable reasons, when the oxygen tension falls, electrons jam up the respiratory chain, which means that a larger proportion of each electron carrier in the chain will be in reduced form and a smaller one in oxidized form. The opposite takes place when oxygen again becomes more abundant. Now it so happens that the various flavins, cytochromes, and other colored members of the respiratory chain have different colors or shades in oxidized and in reduced form. Thus, with fluctuating oxygen tensions, the mitochondria go through subtle changes in hue most pleasing to behold. The observer equipped with special spectroscopic lenses, especially if they reach into the ultraviolet region, will derive even greater enjoyment, as well as much detailed information on where and to what extent electron flow is hindered. Both the effects of oxygen tension and those of numerous inhibitors can be gauged and pinpointed in this way—and have been extensively by the adepts of this kind of spectrophotometric exploration.

An important property of the molecular fluxes that pervade the cytosol is that they are self-regulating. This is because the physical or chemical reactions whereby substances are caused to disappear tend to proceed faster when the concentration of their substrates increases (law of mass action). Take the simple example of a substance that is generated at site A and consumed at site B. Starting from a situation where generation, transport, and consumption occur at the same rate, imagine a sudden increase in generation rate. It is easy to see how this will cause the substance's concentration to rise in A, with, as consequence, a progressive increase in the steepness of its concentration gradient and a resulting acceleration of its rate of diffusion from A to B, leading in turn to a rise of its concentration at B and, thereby, to an increase in its rate of consumption. These readjustments will continue until new concentrations are established in A and B such that the rates of diffusion and of consumption equal the new generation rate. A situation where fluxes have been

Schematic diagram of automatic adjustment from one steady state to another.

Steady state 1: The production of the substance in A, its diffusion from A to B, and its consumption in B occur at the same rate; the concentration profile remains unchanged.

Steady state 2: After the rate of production in A doubles, the concentration profile changes until the rates of diffusion and consumption double.

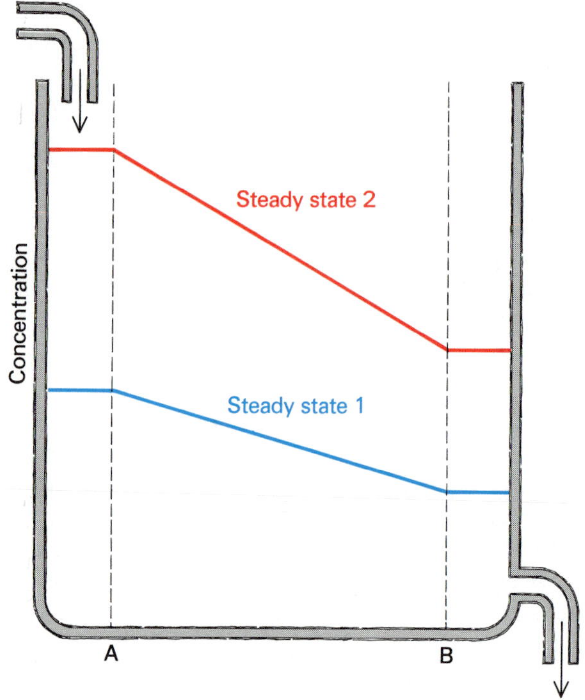

Enzyme kinetics: (1) hyperbolic relationship between the velocity of an enzyme-catalyzed reaction and the substrate concentration, according to the Michaelis-Menten equation: (2) linear relationship between the velocity of an uncatalyzed reaction and the substrate concentration, according to the mass-action law.

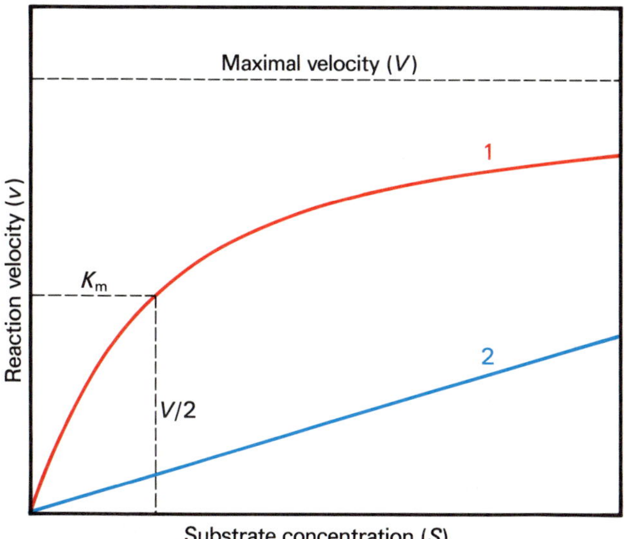

equalized in this way is called a steady state. What our example illustrates is the inherent tendency of the system to adopt a steady state and to respond automatically to a disturbance with the establishment of a new steady state. This kind of consideration may be extended to the whole complex network of metabolic processes. It goes far in explaining the homeostatic properties of living cells, which in Greek means their ability to stay alike (*homoios*). But a complication arises from the fact that most metabolic reactions are catalyzed by enzymes.

Enzyme Rules

Enzymes, like receptors, operate on the lock-and-key principle. They have specific binding sites for their substrates, just as receptors have for their ligands. The difference is that substrates bound by enzymes undergo a chemical change mediated by what is known as the active, or catalytic, site, which is closely linked to the substrate-binding site(s). It is easy to see that the number of substrate molecules altered per unit of time (reaction rate) depends on: (1) the number of enzyme molecules available; (2) their activity, as measured by the maximum number of substrate molecules each enzyme molecule is capable of handling per unit of time; and (3) their degree of occupancy. Airlines deal with the same three variables. For instance, the number of passengers an airline transports across the Atlantic per month depends on: (1) the total number of seats available; (2) the time it takes an aircraft to do a crossing and be ready for another one; and (3) the occupancy ratio. It is clear that the first two factors are characteristics of the airline's fleet and set an absolute limit to the number of passengers that can be transported. Only the third one depends on the number of potential passengers—within the limits imposed by the first two, as many a victim of overbooking can attest from personal experience.

Similarly, the rate of an enzyme-catalyzed reaction is influenced by the substrate concentration within the limits set by the number and activity of the enzyme molecules present. Unlike airplanes, however, which generally take up passengers as they come until the last seat is filled, enzymes fill their binding sites on a statistical basis. It may take tens or thousands or millions of substrate molecules dispersed in the cell volume to fill a single site, depending on the affinity between enzyme and substrate and on the number of sites remaining unoccupied. Mathematically, this relationship is expressed by the Michaelis-Menten law:

$$v = V \frac{S}{K_m + S}$$

in which v is the reaction velocity; V is the maximal velocity (the product of enzyme concentration by enzyme activity); S is the substrate concentration; and K_m is the dissociation constant of the enzyme-substrate complex (Michaelis constant)—that is, the inverse of the enzyme-substrate affinity constant (note that $v = V/2$ for $S = K_m$). Graphically, this equation is represented by a hyperbolic curve, in contrast to the straight line expressing the simple mass-action effect.

If we now refer back to the ability of living cells to readjust the rates of their metabolic reactions by simple mass-action effect, we find that the responsiveness of each individual reaction depends on the S/K_m ratio at which it operates; in other words, on how far the actual velocity v is removed from the maximal velocity V. For instance, if S equals one-hundredth of K_m (enzyme operating at about 1 per cent of its full capacity), the reaction velocity can be doubled by little more than doubling the substrate concentration. On the other hand, with $S = K_m$ (enzyme operating at half-maximal capacity), the reaction velocity cannot be doubled by any attainable increase in substrate concentration. Note that it can still be doubled by interventions at the enzyme level—for instance, by doubling the enzyme's catalytic activity or by doubling the number of enzyme molecules present (i.e., by doubling V). Most enzymes operate well below their maximal capacity, thus allowing ample scope for substrate-mediated, mass-action, regulating effects. The few that operate near maximal capacity catalyze the rate-limiting steps of metabolic pathways. They are the choice targets of enzyme-level regulatory influences.

The Three Levels of Metabolic Regulation

Within the substrate-mediated network, a particularly important role is played by the key coenzymes because of the large number of different reactions in which they participate. The most central role in this respect belongs to ATP and its hydrolysis products, ADP and P_i. They link the innumerable reactions that consume ATP to the oxphos units that regenerate it. Thanks to this link, the rate of catabolic electron flow, and thereby of foodstuff and acceptor consumption, is automatically adapted to the amount of work performed. We have already alluded to this property in reference to respiratory control in mitochondria. The diagrams on the facing page illustrate these relationships in highly schematic fashion for heterotrophs and autotrophs.

An essential feature of metabolic regulation is the existence of a loop whereby the products of anabolism can be used to fuel catabolism. The usefulness of this loop is evident. Without it, cells would have to balance their food intake and their energy expenditure from instant to instant, without any leeway—clearly a highly precarious situation, incompatible with the living conditions of most organisms. The existence of this important loop does, however, raise the question of why it does not operate continuously as a futile cycle, in which catabolism would endlessly undo the work of anabolism, with the useless squandering of energy as net result. Mythology abounds in such sad tales: Sisyphus's stone, the Danaids' pail, Penelope's tapestry. Were cells subjected only to mass-action types of regulating effects, nothing could prevent them from imitating these unfortunates.

Another problem concerns the choice of acceptor for the catabolic electrons. Oxygen has precedence, but why? This question was first raised more than a century ago when Pasteur discovered that the glucose consumption of yeast cells decreases dramatically in an anaerobic culture suddenly exposed to oxygen. To Pasteur, who was used to blowing dying embers back to life, this effect appeared paradoxical. The Pasteur effect has since been found to apply to many different types of cells; it has puzzled many investigators and inspired many hypotheses. Today we can account for most of it by simple homeostatic mechanisms. With the admission of oxygen, mitochondrial oxidations are set off and create an "electron sink" that draws (along the generated negative gradients) such available donors as pyruvate and the molecules capable of shuttling electrons from cytosolic NADH. Lactate is no longer produced. But, at the same time, the number of ATP molecules that the cell can obtain from each molecule of glucose metabolized increases nineteenfold. Therefore, to the extent that the cells do not perform more work or waste more energy in the presence of oxygen, a corresponding reduction in glucose consumption is to be expected from the ATP/ADP-mediated regulating mechanism that we have just examined. However, a problem remains. In glycolysis, the step that is controlled by the ATP/ADP ratio—the substrate-level oxphos unit—is preceded by two ATP-consuming reactions, which, as we saw in Chapter 7, are needed to convert glucose into phosphoglyceraldehyde, the actual substrate of the ox-

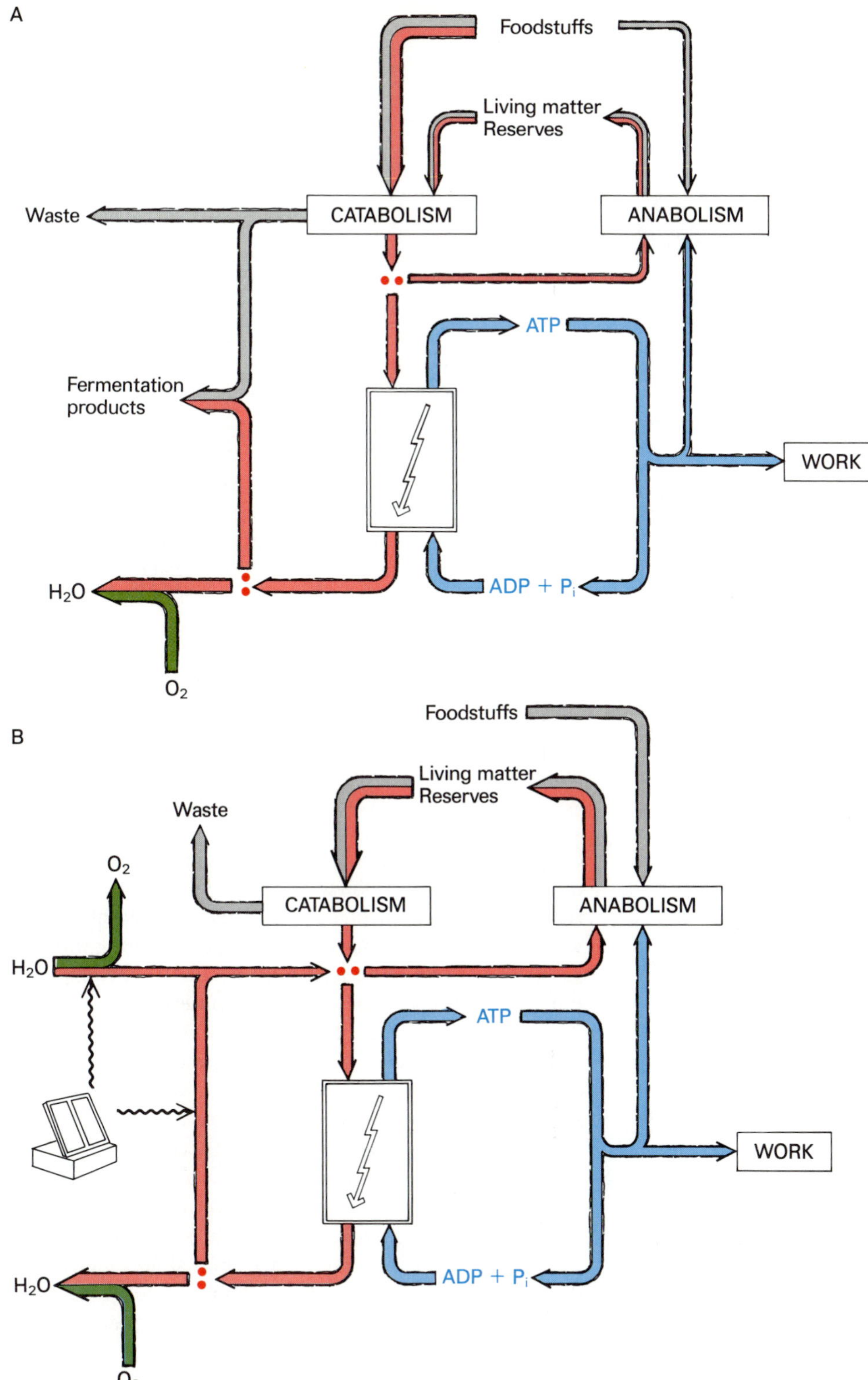

The basic blueprint of metabolic regulation.

A. Heterotrophs. Foodstuffs from the outside serve mostly to support catabolism. The main role of catabolism is to detach high-potential electrons from the foodstuffs and to feed them into oxphos units. Exiting low-potential electrons are collected by an acceptor, either supplied separately (e.g., oxygen) or generated from the foodstuff flux (fermentation). The ATP assembled by oxphos operation supports the various forms of biological work. Side fluxes of foodstuffs, electrons, and ATP support the anabolic formation of biosynthetic products, including that of reserve substances. These in turn can be mobilized to support catabolism. Self-regulation of this system is achieved principally through the [ATP]/[ADP] [P_i] ratio, which automatically adapts the rate of catabolic breakdown to the amount of work performed (including anabolic work). If the work load increases, this ratio decreases, causing the rate of electron flow through oxphos units to increase through a mass-action effect. The opposite takes place if less work is performed. Fuelling needs are satisfied thanks to the links between catabolism and anabolism. When the supply of foodstuffs exceeds needs (feeding periods), excess substrates, electrons, and ATP converge to build biosynthetic stores. In the opposite situation (fasting, starvation), stores are drawn into catabolism.

B. Autotrophs (green plants). In the dark (delete light-powered photoreduction and photophosphorylation), the organism operates essentially like a starved heterotroph, subsisting on biosynthetic reserves and adjusting catabolism to work through the mediation of the [ATP]/[ADP] [P_i] ratio. In the presence of light, a supply of high-energy electrons supports all energy needs and electron-depleted, mineral foodstuffs can be used for anabolic purposes.

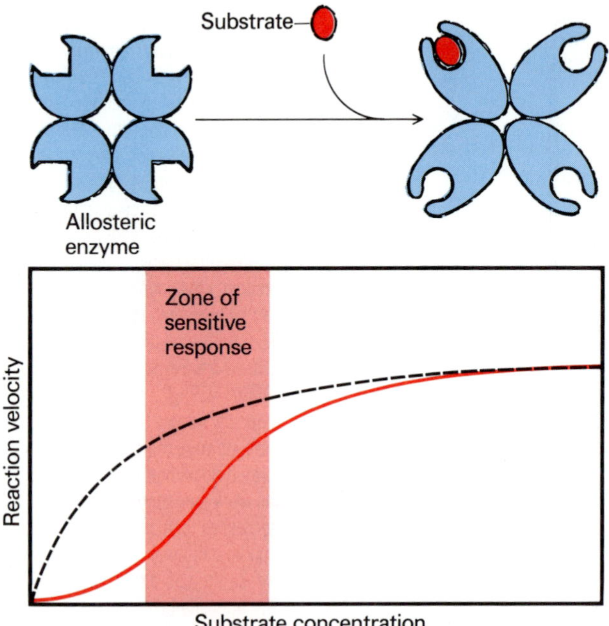

Allostery. An example of positive cooperativity. The binding of one molecule of a substrate to a tetrameric enzyme is hampered by the enzyme's low affinity for the substrate, but the binding of this first molecule causes a change in the conformation of the enzyme that greatly enhances its affinity for further substrate binding. The sigmoid shape of the kinetic curve illustrates this effect. Within a certain concentration zone, small changes in substrate concentration induce large changes in reaction velocity. Compare with normal hyperbolic curve (dashed line).

phos unit. What is it that prevents the aerobic cell from recklessly spending ATP, as long as inorganic phosphate is available, in converting glucose into various phosphorylated intermediates? Simple mass-action regulation cannot account for this.

Such examples could be multiplied. They make it abundantly clear that cells have a second set of self-regulative interactions, operating at the enzyme level rather than by simple mass-action effects. The main targets of these interventions are the enzymes that catalyze rate-limiting steps (p. 246) and thereby control metabolic bottlenecks, ideally suited for the placing of control valves. The enzymes involved are subject to a wide variety of effects, many of which depend on the simple, reversible, noncovalent binding of a low-molecular-weight substance to some site in the enzyme molecule. This can be the substrate-binding site, with as consequence an alteration in the accessibility or affinity of this site to the substrate (e.g., competitive inhibition). Or it can be the catalytic site, with consequent modification of its catalytic activity (e.g., noncompetitive inhibition). In a number of particularly important cases, it is a separate site, called allosteric (Greek *allos,* other; *stereos,* solid), whose occupancy induces a change in the conformation of the enzyme molecule that results in a modification of its catalytic activity, or of its affinity for its substrate, or, more subtly, of its affinity for other effectors. Enzymes may also suffer covalent modifications, such as phosphorylation and dephosphorylation, already alluded to (p. 232), which often have drastic effects that completely inactivate or reactivate the affected enzymes. Sometimes, the regulating agents do not act directly on the target enzyme but rather on some other molecule, itself an enzyme or not, that acts as an enzyme modifier, or as a modifier of an enzyme modifier, and so on. The possibilities are endless, and most of them are, in fact, exploited in a large variety of ways. The result is an extraordinarily intricate, multidimensional web of interactions that will challenge for a long time the sharpness of our analytical tools, as well as the power of our imagination—even of our computers.

The main agents of these interactions are normal metabolites or cofactors, internally generated. A special class comprises the actual substrates of the affected enzymes. Such substrate-regulated enzymes have more than one substrate-binding site and are constructed in such a way that occupancy of a site alters for the better (positive cooperativity) or for the worse (negative cooperativity, substrate inhibition) the substrate affinity or catalytic activity of other sites. Such enzymes have anomalous kinetics that differ from the typical Michaelis-Menten type by features that often have profound functional consequences. In a number of cases, the affected enzyme, although not directly concerned with the metabolism of the active agent,

Example of feedback inhibition. The final product of a reaction chain inhibits the first enzyme of the chain.

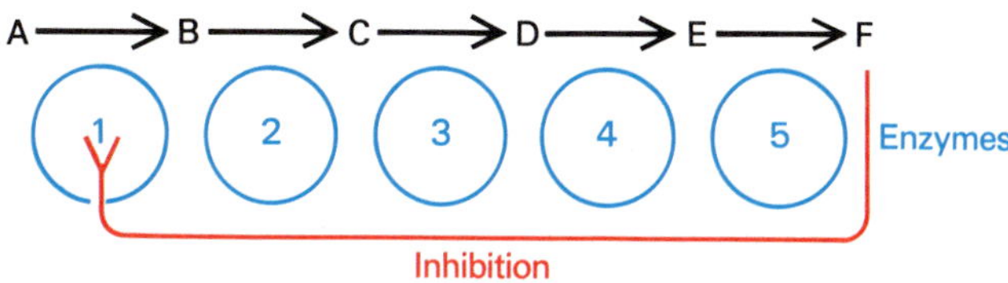

nevertheless remains in the family, so to speak, by being part of a chain involved either in the generation or in the breakdown of the agent. Such interactions are often relatively easy to interpret. A typical example is feedback inhibition, a process in which the product of a metabolic chain inhibits the first step of the chain, a straightforward regulating mechanism of the "thermostat" type. Many other interactions, however, cut across metabolic lines and are much more difficult to detect or understand.

In multicellular organisms, a number of particularly potent effects are elicited by exogenous substances that bind to surface receptors. We saw the example of the remarkable cascade of events whereby hormones such as glucagon or epinephrine can, with lightning rapidity, trigger the breakdown of glycogen (p. 232). What was not said at that time is that the cAMP-activated protein kinase that activates the phosphorylase-activating kinase acts also on glycogen synthase, the enzyme that makes glycogen (by transfer of glucosyl groups carried by UDP). But in this case phosphorylation leads to inactivation of the enzyme. One would be hard put to find a simpler and more elegant solution to the futile-cycle problem. Phosphorylation shuts off synthesis and triggers breakdown; dephosphorylation does just the opposite. Not many such examples are known, but there is little doubt that cells use regulating influences to avoid energy waste by futile cycling, except when keeping such cycles going may, like keeping a motor idling, serve some useful purpose deserving of the energy expenditure. This, of course, is not the result of calculated design; it is, like the whole network of interrelationships that give the cytoplasm its dynamic structure and organization, the product of chance trials, tested and screened by natural selection.

In addition to the two levels of regulation that have been considered, there is a third one, which operates by manipulating the number of enzyme molecules. These are generally slow effects in comparison with the others, but they are far reaching in that they may be qualitative as well as quantitative. The dominant mechanisms of this sort are found in the nucleus at the level of gene transcription. We will soon take a look at them and try to assess their significance (see Chapter 16).

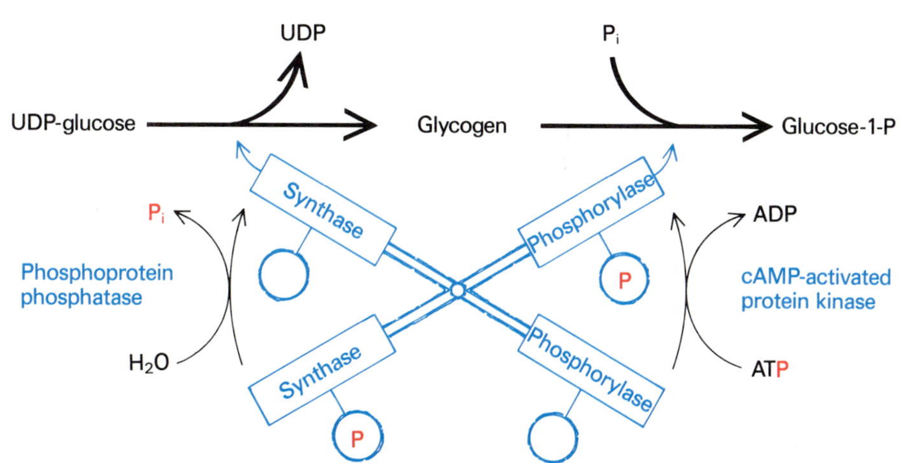

Flip-flop regulation of glycogen breakdown and synthesis by phosphorylation and dephosphorylation, respectively. The phosphorylation of enzymes by cAMP-activated protein kinase activates phosphorylase and inactivates synthase. Dephosphorylation by phosphoprotein phosphatase does the opposite.

Induced Fit, Conformation, Catalysis, and Allostery

On many occasions, in this and in preceding chapters, we have used the lock-key analogy. Born in the imagination of the great German chemist Emil Fischer, superbly adapted to immunology and therapy by his contemporary Paul Ehrlich, this analogy has been immensely useful in explaining such phenomena as antigen-antibody recognition, the interaction between receptors and their ligands, the formation of enzyme-substrate complexes, and the action of hormones and drugs. But like all analogies, especially those that refer to some aspect of the living cell, it is an oversimplification. Proteins are not constructed rigidly like locks; they are very flexible and versatile. To them, admitting a ligand is not the passive acceptance of a key by a lock. It is a real embrace, in which the protein molecule molds itself around its ligand. This does not mean that there is not a high degree of specificity in the interaction. What it means is that the fit is to some extent induced by the ligand, which is equivalent to saying that binding of the ligand causes a conformational change in the protein.

This expression has become something of a catchword, an all-purpose formula glibly invoked for want of a more-precise explanation, a learned-sounding screen behind which we hide our ignorance. But do not let the facile vagueness of the expression make you underestimate its significance. It is a real thing, of immense biological importance, which simply waits to be defined in more-precise detail for each individual protein-ligand complex as our perception of their molecular anatomy sharpens. Receptors, we have seen, depend on this effect to activate their signaling devices. It is the basis of all allosteric regulations. It also underlies all forms of cellular motility. Even in the simple binding of substrates to enzymes, it may play a role in the proper organization of the catalytic site.

Thus, not only at the cell surface, but throughout the cell, ligand-induced conformational changes of proteins represent a universal mechanism of communication, regulation, and catalytic activity, as well as a central means of performing mechanical work. They also provide infinite opportunities for chemical interference with vital processes by means of drugs, poisons, and other pharmacologically active substances that resemble physiological ligands sufficiently to interact in some way with their normal binding sites.

15 Ribosomes: The Synthesis of Proteins

Our last port of call before leaving the cytosol will be the strings of ribosomes, which, either free or stuck to the surface of ER membranes, spin out the thousands of different polypeptide threads that make up the cell proteins. What this activity means, whether in terms of importance for the cell or of intrinsic complexity, can hardly be overestimated.

The Nature of the Problem

Proteins are by far the most elaborate and versatile substances the biosphere—or, for that matter, the whole of the known universe—has to offer. Throughout our visit, we have seen them performing astounding feats as structural components, enzymes, receptors, gates, carriers, pumps, motor elements, and other functional entities. It is no exaggeration to state that one or more specific proteins lie behind every activity carried out by living cells, or, otherwise put, that life is essentially an expression of proteins. But what then of the processes that manufacture these remarkable substances? Should they not, almost by definition, be even more complex than their products?

The solution to this apparent paradox will become progressively clearer during the last part of our tour. But first let us recognize what the making of a protein molecule represents. On the basis of the chemical information summarized in Chapter 2, we know that all proteins are made of one or more polypeptide chains—that is, strings of linearly assembled amino acids belonging to twenty distinct varieties linked by peptide bonds (—CO—NH—). The innumerable structures shown by the finished products arise from the bending, twisting, coiling, and sometimes joining of such chains and from their fitting with other groups, such as oligosaccharide side chains.

The seemingly insuperable problem of constructing such an immense variety of intricate molecular edifices can be narrowed down to a single dimension, thanks to a central tenet of molecular biology, which states that all the properties of proteins—including their catalytic power, their ability to adopt given configurations and to modify them in response to outside influences, their association tendencies, their susceptibility to glycosylating and other modifying enzymes, and all the other characteristics to which proteins owe their unique biological role—are determined by the primary structure, the amino-acid sequence. In other words, the problem of making a protein is "reduced" (still a formidable job, but at least simpler to formulate) to that of assembling amino acids in the right order. All the rest will follow, provided the necessary ingredients and an appropriate environment are supplied by the cell. For some proteins, this statement needs to be qualified by adding that assembly itself must take place under special conditions. For instance, it is known that certain polypeptide chains undergo changes, such as trimming or glycosylation, while they are being put together. Their final shape is influenced by these changes.

A Mathematical Interlude

Imagine yourself an amino-acid setter. You are sitting in a ribosome and assembling amino-acid letters into polypeptide words. In front of you are 20 boxes, each containing a different kind of amino acid, or a keyboard with 20 keys, or whatever other selection device you may wish. Suppose you are assembling polypeptides containing 100 residues. Such long words do not exist in our language, but, as proteins go, this is a relatively small size, corresponding to a molecular mass of approximately 12,000 daltons. The question is, if you dip into the boxes or hit the keys at random, what are the odds that you will turn up a given sequence?

The answer is a figure not quite as small as the odds of a monkey typing the whole of *Hamlet* by chance, yet small enough to be entirely beyond the boundaries of our imagination. The computation is simple. Your chance of getting the first amino acid right is one in twenty. It is the same for the second one, so that the chances of getting the first two right are 1 in 20^2, or 1 in 400. Repeat this reasoning for every amino acid, and you find that the chances of getting all 100 right are 1 in 20^{100}, or 10^{-130}. Don't bother to visualize this figure or to translate it into familiar terms. Just forget about making polypeptides by chance. Even if the whole population of the earth should be helping you out, working day and night at the incredible speed of 1 million polypeptides per second, never making the same polypeptide twice, it would still take them close to 10^{107} years, or more than 5,000,000,000,000, 000, 000, 000, 000, 000, 000, 000, 000, 000, 000, 000, 000, 000, 000, 000, 000, 000, 000, 000, 000, 000, 000, 000, 000, 000, 000, 000, 000 times the estimated age of the universe, to rattle out all the possible combinations.

Enough; the point is made. Ribosomes cannot work at random; they must be instructed. Which indeed they are, as we shall see, by means of messages from the nucleus. But this does not settle the question entirely. What of prebiotic times, when no instructions were available? According to fossil records, it took less than 1 billion years for recognizable cells, in which hundreds of specific proteins may have been operative, to appear on the surface of our planet. Whatever the mechanisms involved in the assembly of these proteins, it is easily proved that they arose as a result of an infinitesimally small number of trials, compared with the total number of possibilities.

To demonstrate this point, let us make absurdly generous assumptions. The whole mass of the earth (6×10^{27} grams) is involved; it consists entirely of polypeptide chains of 12,000 molecular weight, each represented by a single molecular species, giving $(6 \times 10^{27} \times 6.023 \times 10^{23})/12{,}000 = 3 \times 10^{47}$ peptide molecules. Every billionth of a second the mixture is rearranged to make a new set of polypeptides, never repeating a sequence that has already turned up; this gives us $10^9 \times 60 \times 60 \times 24 \times 365 \times 10^9 = 3.1536 \times 10^{25}$ different sets made in 1 billion years, or $3.1536 \times 10^{25} \times 3 \times 10^{47}$ = about 10^{73} different molecules out of a possible 10^{130}. Even using the entire estimated mass of the universe under these incredible conditions would yield only about one-millionth of the possible sequences in 1 billion years.

Yet the facts are there: the right polypeptides needed to make a primitive cell were actually assembled some 4 billion years ago, with a total number of trials that cannot but have been vanishingly small with regard to the number of possibilities. Was it, as some biologists believe, a fantastic stroke of luck, a unique event, never to be repeated anywhere in the universe? Such a belief cannot be disproved, but it defies the laws of probability. The odds of getting one polypeptide right are small enough, 10^{-130}, to make the event close to impossible. How, then, should we qualify the probability of getting 100 different polypeptides right at the same time, or $10^{-13,000}$? To beat such odds, no less than a miracle is required.

What then? Do we have to assume that the prebiotic assembly of polypeptides was "instructed," guided by an unseen hand, so as to produce sequences possessing the catalytic properties required to put life on course? Not necessarily. There is an alternative possibility: namely, that catalytically active polypeptide sequences are the commonest thing on earth; that, in fact, almost any polypeptide has some enzymic activity. Once we make this assumption, all we need are conditions such that amino acids will arise and assemble spontaneously. If this happens, the polypeptides formed are bound to include primitive enzymes of various kinds, even though only an infinitesimally small proportion of the possible molecules has been generated. This is exactly what has been found in experiments designed to reproduce in the laboratory conditions believed to have prevailed on earth in prebiotic times. Authentic amino acids have indeed been obtained from such commonplace ingredients as methane, ammonia, hydrogen, and water. Primitive "proteinoids" have been produced from amino acids by heat or other simple physical means, and these crude artificial polymers, clearly created without the benefit of any sort of instruction, have been found to display crude catalytic properties similar to those of enzymes.

We will revert to this fascinating topic in Chapter 18. First, we must look at the protein factories of today. Whatever happened 4 billion years ago may remain forever a matter for conjecture. What we do know, however, is that it resulted in the development of a biosynthetic system capable of putting amino acids together in specified sequences that faithfully reproduce those that exist in the cell.

The Cipher of Life

Should we be asked to guess how a ribosome can possibly "know" which of the twenty available amino acids to add at each step while assembling a polypeptide chain, we would no doubt think in terms of copying. We would visualize ribosomes using the cell's proteins as templates and painstakingly reproducing them, amino acid by amino acid, with the care and accuracy of a medieval copyist. Asked to provide a chemical mechanism for such a process, we might invoke the attraction of like for like, the force behind crystallization, as did the American biochemist Felix Haurowitz not much more than 30 years ago. We would be wrong. The instructions come in code, delivered by a completely different type of molecule that belongs to the class of ribonucleic acids (RNA) and is appropriately named messenger RNA (mRNA). What happens is not copying, but translation.

Ribonucleic acids share certain features of polypeptides. They are long, garlandlike macromolecules, all with the same iterative backbone made of *n* identical molecular units and rendered uniquely informational by specific side groups, or "letters," attached to the backbone. But here the analogy ends. The backbone is made of a coarser fiber, quite unrelated to that of polypeptides. The side groups also differ from those of polypeptides. And, especially, there are only four of them. The RNA alphabet is much poorer than the protein alphabet: four letters (AGUC), against twenty.

Chemically, RNAs are defined as polynucleotides, made by the linear association of mononucleotide units. We have already encountered these building blocks, which are the 5′-monophosphates of nucleosides. In RNA, these units are linked by phosphodiester bonds that attach the terminal phosphate of each nucleotide to the 3′-hydroxyl group of its neighbor:

```
      |
      O
      |
 ⁻O—P—O—(5′)—Ribose—Base
      ‖           |
      O          (3′)
                  |
                  O
                  |
             ⁻O—P—O—(5′)—Ribose—Base
                  ‖           |
                  O          (3′)
                              |
                              O
                              |
                         ⁻O—P—O—(5′)—Ribose—Base
                              ‖           |
                              O          (3′)
```

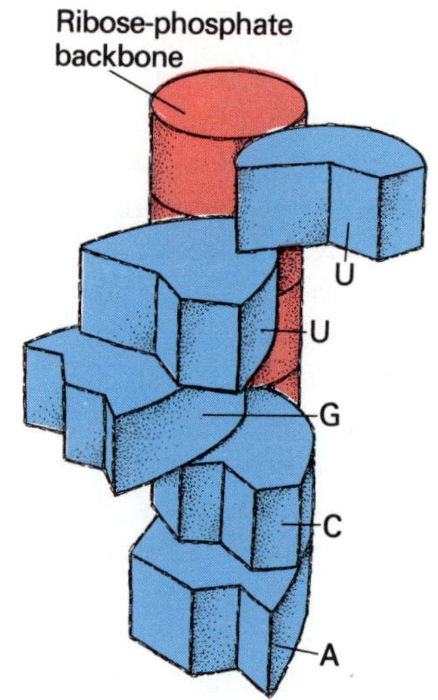

Schematic representation of a segment of an RNA molecule. The backbone is the same in all RNAs and consists of alternating ribose and phosphate molecules. Specificity is provided by the sequence of bases (from four possibilities: adenine, A; guanine, G; uracil, U; or cytosine, C) along the backbone. DNA has the same structure, with deoxyribose instead of ribose, and thymine (T) instead of uracil (see Chapter 16).

The backbone consists of 5′-phosphoryl-ribosyl units, n times repeated. At one end, called the 5′ end, the 5′-phosphoryl ester group is not engaged in a phosphodiester linkage. At the 3′ end, the 3′-hydroxyl group of the last ribose is free. The four bases are the two purines, adenine and guanine, and the two pyrimidines, cytosine and uracil, already mentioned.

It is useful to point out at this stage that DNA has essentially the same structure, except that it has a different pentose (5-carbon sugar) in its backbone—deoxyribose—and uses thymine instead of uracil as one of its pyrimidine bases; its alphabet is AGTC. But these are minor differences: deoxyribose is simply ribose without an oxygen atom in position 2; thymine is uracil with an additional methyl group in position 5 (see Appendix 1). Information transfer is not affected by this difference. The two alphabets are essentially the same: T and U are interchangeable.

Nucleic acids, both RNAs and DNAs, have one additional feature of absolutely cardinal importance that is not immediately evident unless you look at them the right way: their constitutive bases are complementary two by two, the purine adenine with the pyrimidine uracil (RNA) or thymine (DNA), and the purine guanine with the pyrimidine cytosine (both RNA and DNA). The chemically minded among you can see from the structures of the two partners (facing page) how base-pairing takes place: A binds to U or T by means of two hydrogen bonds (Chapter 2), and G to C by means of three, making the GC pair the stronger of the two.

Those who are frightened by chemical formulas may visualize the paired bases as two flat pieces, joined by perfectly fitting edges to form a rigid planar structure of roughly elliptical shape, with axes of 1.1 and 0.6 nm and a thickness of 0.34 nm. A cardinal consequence of base-

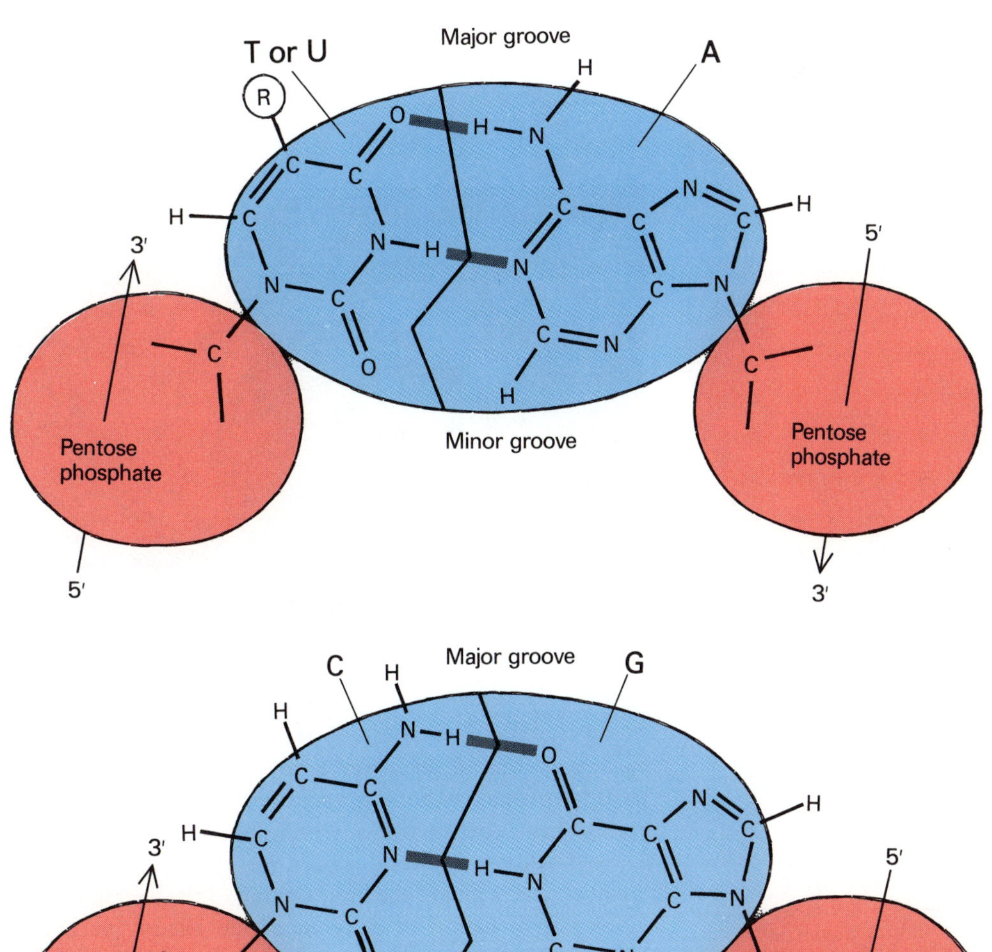

Chemical mechanism of base-pairing. The top diagram shows how the purine base adenine (A) joins with the pyrimidine base uracil (U, R=H) or thymine (T, R=CH_3) by means of two hydrogen bonds. The lower diagram shows similar joining between the purine guanine (G) and the pyrimidine cytosine (C) by means of three hydrogen bonds. Each base pair forms a flat plate, roughly elliptical in shape, with axes of 1.1 and 0.6 nm, and a thickness of 0.34 nm. The pentose-phosphate groups attached to the bases can rotate freely around the glycosidic N-C bond. But, when base-pairing occurs between two nucleic-acid chains (RNA-RNA, DNA-DNA, or RNA-DNA), these groups are forced to assume the positions indicated in the diagram, with 3′ ends pointing in opposite directions. Grooves refer to the double-helical structure of complementary chains (see the drawing on p. 256).

Regular double-helical structure generated by perfectly complementary nucleic-acid chains. Note the minor and major grooves and the double row of negative charges on the outside of the backbones. Several different structures are possible. The diagram illustrates the B form, with a pitch of 3.4 nm (or ten base pairs) per complete turn, first proposed for DNA by Watson and Crick in 1953 (see Chapter 16) and characteristic in first approximation of most of the DNA in nature.

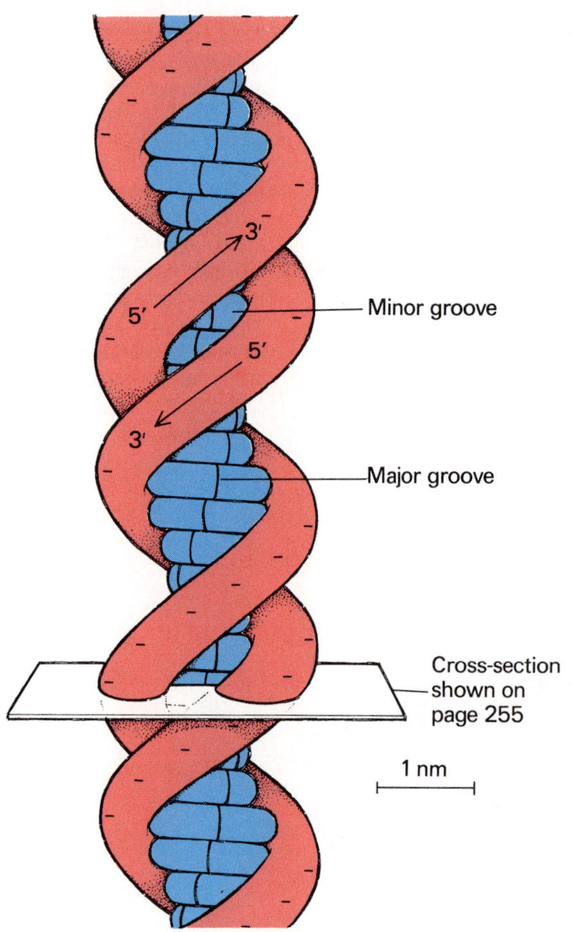

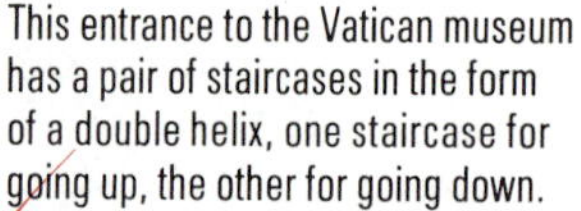

This entrance to the Vatican museum has a pair of staircases in the form of a double helix, one staircase for going up, the other for going down.

pairing is that it may serve to link two nucleic-acid strands, provided the two backbones run antiparallel to each other (one in the 5′⟶3′ and the other in the 3′⟶5′ direction). The longer the complementary sequences, the stronger the link. Owing to the characteristic molecular anatomy of polynucleotide chains, such arrangements tend to adopt a regular helicoidal shape. The paired bases pile up in parallel planes, but with their long axes forming an angle of 36°, somewhat like the steps of a spiral staircase or, rather, a double spiral staircase, one for going up, the other for going down, such as exists at one of the entrances to the Vatican museum in Rome. The two antiparallel backbones run along the outer edges of the base pairs, like the banisters of the two staircases.

When it extends over ten or more complementary base pairs, this structure appears as a twisted cord about 2 nm thick but narrowed by two helical grooves, a deep one and a shallow one. Its turn is right handed and its length is 0.34 nm per base pair, making its pitch 3.4 nm per complete turn (10 × 36°, the angle between adjoining base pairs). A double-helical row of negative charges (20 per turn), provided by the phosphoryl groups of the two backbones, lines the outside of the cord.

The most perfect double helices are found when the two strands are complementary from one end to another, as is the case with most DNA molecules and, exceptionally, with some double-stranded viral RNAs. Most RNA molecules are single stranded but include, for reasons that are obvious in view of the limited number of different combinations allowed by their alphabet, many short base sequences that are complementary in antiparallel fashion to sequences occurring further along the same strand. Pairing of these sequences causes the formation of loops closed by double-helical segments of variable length; for example:

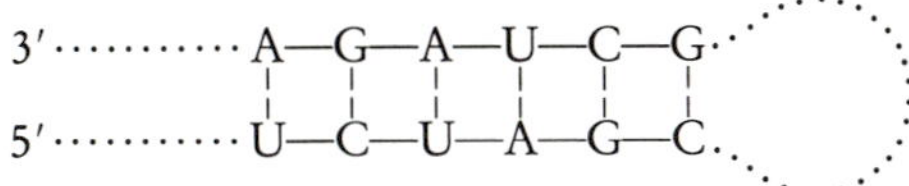

Thus, like proteins, most RNAs form characteristic secondary structures stabilized by interactions between side chains. When represented schematically, these structures resemble graceful mobiles, which have been described poetically as cloverleaves, flowers, and the like. In reality, most of them must appear as hideously twisted knots. As in proteins, such arrangements may be denatured by heat or other agents to generate a random tangle of filaments.

But the function of base-pairing is not simply a structural one. Its main role is communication. Amazingly, these two elementary relationships

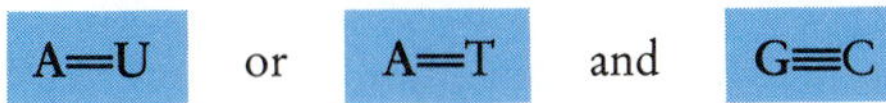

govern, through the two relatively fragile structures they embody, the whole of information transfer throughout the biosphere. They are truly the cipher of life. This will become evident as we look at ribosomal translation, and even more so when we learn about transcription and replication in the nucleus.

Diagrammatic representation of the formation of a loop in RNA. The loop is closed by the pairing of two short complementary stretches in antiparallel orientation.

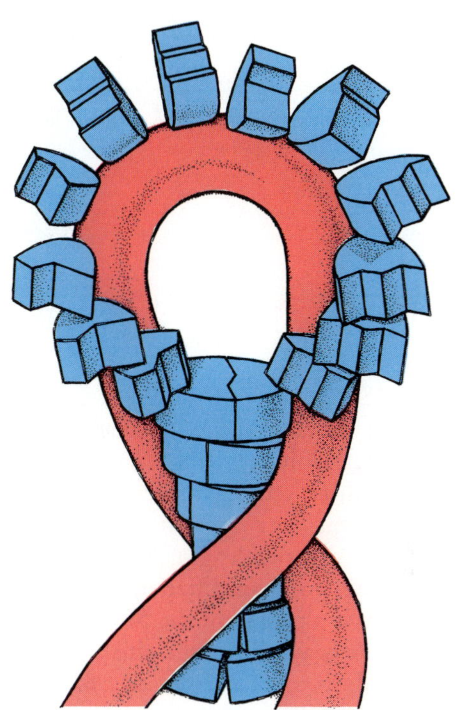

The Code

Messenger RNAs specify the sequence of amino acids in polypeptides co-linearly, with the 5′⟶3′ direction corresponding to the N-terminal ⟶ C-terminal direction in the polypeptide. In other words, the signs representing the amino acids follow each other, unbroken and unscrambled, from the 5′ to the 3′ end of the mRNA in the order in which the amino acids themselves follow each other in the polypeptide from the N-terminal to the C-terminal end. The message does not, however, start at the 5′ end and stop at the 3′ end of the mRNA; it is preceded and followed by mute sequences that are not translated. As to the message itself, since the RNA alphabet contains only four letters it is obvious that several letters must be used to code for each amino acid. How many is easily computed: with n letters in the alphabet you can construct n^m different words of m letters.

Number of letters in alphabet	Number of words in vocabulary			
	One-letter	Two-letter	Three-letter	Four-letter
2	2	4	8	16
4	4	16	64	256
20	20	400	8,000	160,000
26	26	676	17,576	456,976

By looking at the table above, you can readily see that with a four-letter alphabet (RNA language) you need words of at least three letters to represent all of the twenty amino acids (protein language), at least if you want all your words to be of equal length. You could just get by with less by using words of different lengths, as is done with the two-letter alphabet of the Morse code; but this was not Nature's way. You will also notice that with three-letter words you suffer from an *embarras de richesse*: sixty-four combinations to convey twenty meanings. You could, as in our own language, use only part of the possible combinations and reject the others as nonsense. But, again, this was not Nature's way. All sixty-four possible combinations of three bases are used in the genetic code.

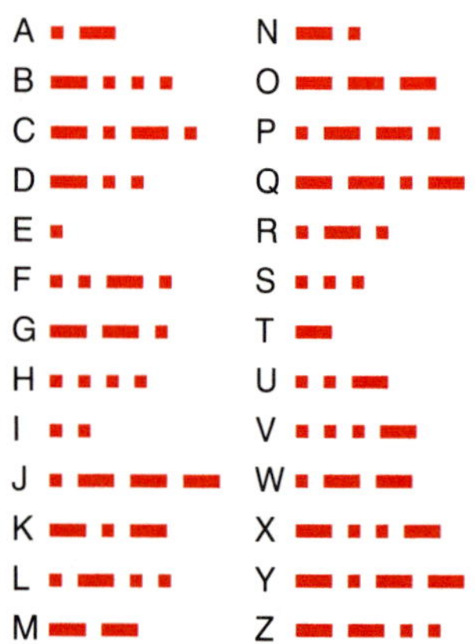

The international (continental) Morse code represents the 26 letters of the alphabet by most of the 30 combinations of two distinct signs (short-long) that can be made with codons of between one and four signs.

The breaking of this code has been one of the great triumphs of modern biology. It started in 1961, when a young American investigator discovered that an artificial polyribonucleotide containing only uracil (poly-U) induced the synthesis of a polypeptide made only of phenylalanine (poly-Phe). This yielded the first line of the genetic dictionary: UUU = Phe. Five years later the whole dictionary had been elucidated. Now we use it not only to read, but even to write, genetic messages and, with the help of genetic engineering techniques, to instruct cells to translate the message for us in as many copies as we wish (see Chapter 18). The code used by the great majority of living beings is reproduced in the table on the facing page.

As you can see, the genetic dictionary is a highly ordered system. All amino acids, with the exception of Met and Trp, are represented by more than one base triplet, or codon. Some (Leu, Ser, Arg) have as many as six synonyms. In all cases, replacing U by C at the end of the codon does not alter its significance. Often, but not always, A and G are also interchangeable in third position. The third base is entirely irrelevant for eight out of the twenty amino acids. Altogether, sixty-one triplets code for amino acids. The remaining three (UAA, UAG, and UGA) are stop signals. As we shall see, AUG, and sometimes GUG, have a special additional role as initiators.

Why this particular code, and not another? This question has already caused much ink to flow. But before addressing it, we must bear in mind that there are a few dissenters. Among them, our own mitochondria. For

The Genetic Code

First letter	Second letter: U		C		A		G		Third letter
U	UUU	Phe	UCU	Ser	UAU	Tyr	UGU	Cys	U
	UUC	Phe	UCC	Ser	UAC	Tyr	UGC	Cys	C
	UUA	Leu	UCA	Ser	UAA	OCHRE†	UGA	OPAL†	A
	UUG	Leu	UCG	Ser	UAG	AMBER†	UGG	Trp	G
C	CUU	Leu	CCU	Pro	CAU	His	CGU	Arg	U
	CUC	Leu	CCC	Pro	CAC	His	CGC	Arg	C
	CUA	Leu	CCA	Pro	CAA	Gln	CGA	Arg	A
	CUG	Leu	CCG	Pro	CAG	Gln	CGG	Arg	G
A	AUU	Ile	ACU	Thr	AAU	Asn	AGU	Ser	U
	AUC	Ile	ACC	Thr	AAC	Asn	AGC	Ser	C
	AUA	Ile	ACA	Thr	AAA	Lys	AGA	Arg	A
	AUG*	Met	ACG	Thr	AAG	Lys	AGG	Arg	G
G	GUU	Val	GCU	Ala	GAU	Asp	GGU	Gly	U
	GUC	Val	GCC	Ala	GAC	Asp	GGC	Gly	C
	GUA	Val	GCA	Ala	GAA	Glu	GGA	Gly	A
	GUG*	Val	GCG	Ala	GAG	Glu	GGG	Gly	G

*These codons have a second function as initiators of translation.
†These codons are terminators, or stop signs; UGA is read as Trp by mitochondria. For other peculiarities of the mitochondrial code, see text.

them, UGA does not mean stop, but rather Trp, like its close neighbor UGG. This peculiarity adds to all the others that support a separate origin for mitochondria, perhaps from an ancestral bacterial endosymbiont, especially since other mitochondria (e.g., of yeast and of the fungus *Neurospora*) also read UGA as Trp. Something as simple as a separate universal mitochondrial code does not exist, however. Whereas our mitochondria and those of *Neurospora* read CUA as Leu, in the orthodox way, yeast mitochondria deviate by making it Thr. On the other hand, ours have the distinction of understanding AUA as an order for Met instead of Ile.

On the whole, however, the biosphere is no Babel. It is, with the rare exceptions just discussed, linguistically homogeneous. Viruses, bacteria, fungi, plants, animals, and humans all use the same code. Does this mean that our code is the only possible code, imposed by a set of unique chemical relationships between the amino acids and their respective codons? Or, on the contrary, is the assignment purely random, one that just occurred by chance in the primitive cell from which all living organisms are believed to have originated? Neither one nor the other, apparently. Amino acids, as we shall see, have nothing to do with the reading of their codons and are not likely to have had much say in their choice, or vice versa. On the other hand, the code has an obvious structure; it is not random. It is quite clear, for instance, that chance could not possibly have grouped synonyms the way they

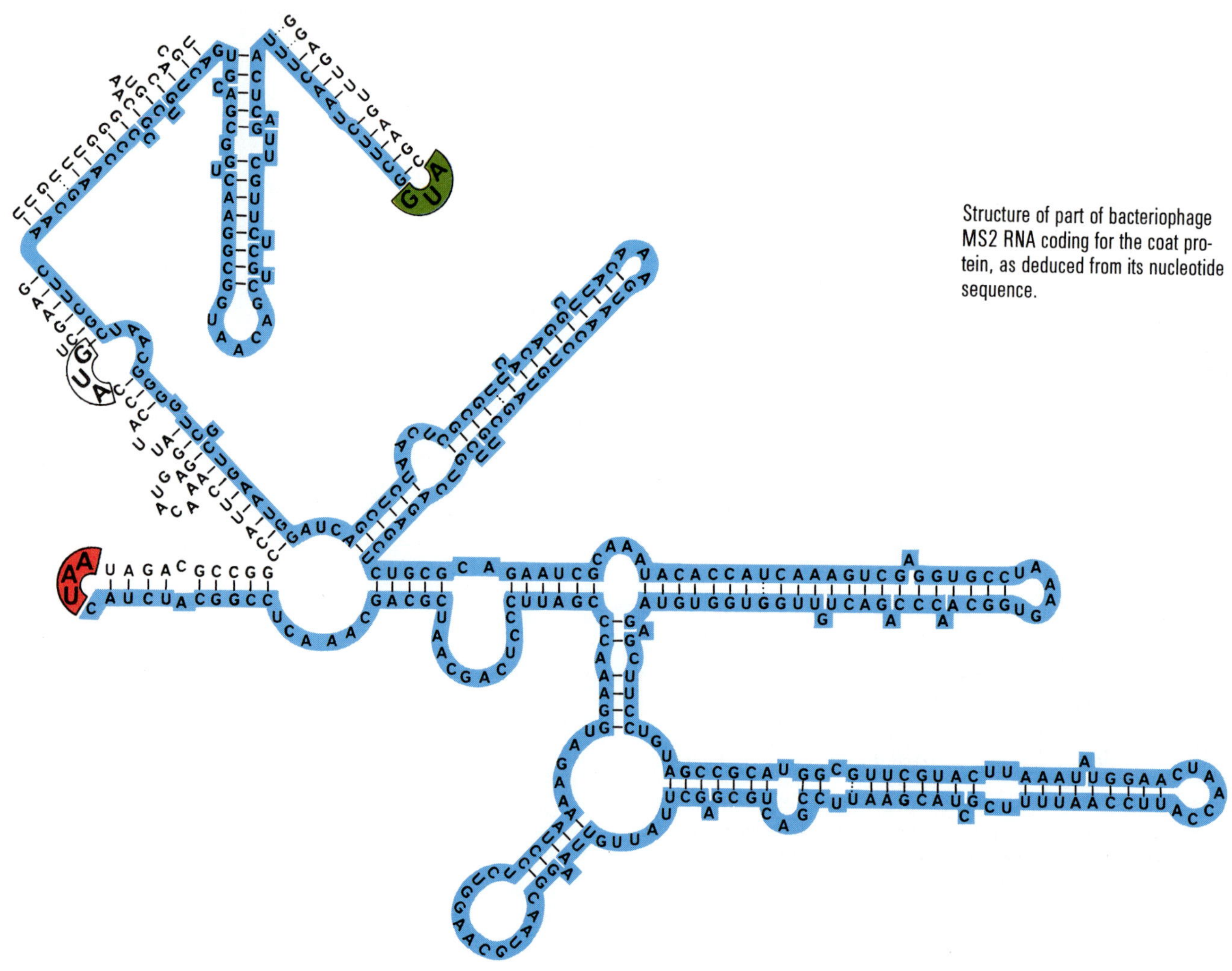

Structure of part of bacteriophage MS2 RNA coding for the coat protein, as deduced from its nucleotide sequence.

are. Their spelling similarities have suggested that in early prebiotic times there may have been no more than sixteen amino acids, each coded for by a doublet followed by an essentially meaningless third base. Some of these codons were then "borrowed" to code for the additional amino acids that appeared and to serve as stop signs. It has also been pointed out that the code is such as to minimize translational errors due to misreading, as well as deleterious effects of point mutations involving the replacement of one base by another, and may therefore be the product of natural selection. Indeed, you can easily verify that many base replacements either do not change the meaning of codons or, if they do, substitute an amino acid that resembles its predecessor in certain key properties (hydrophobicity, electric charge), so that the conformational and functional properties of the altered protein are not drastically affected. These evolutionary aspects will be considered further in Chapter 18.

When the genetic code was first deciphered, only a few proteins had been sequenced and techniques for sequencing nucleic acids were still in their infancy. The early cryptologists achieved success by indirect means through ingenious experimentation. Today, sequencing methods have been improved to such an extent that even very long messages can be "read" directly, and compared with their translation products.

One of the first natural messages to be completely decoded in this way is the RNA of a small bacterial virus, or

5′ PPP—129—GUG—1,176—UAG—23—AUG—387—UAA—33—AUG—1,632—UAG—171 3′

Protein A Coat protein Replicase

bacteriophage, called MS2. Made of 3,569 nucleotides, this RNA doubles up in more than sixty loops, stabilized by complementary base sequences, to make up a highly intricate "flower" structure, which, in three-dimensional reality, must appear as the Gordian nightmare par excellence. Genetically, this RNA codes for three proteins (sometimes for four, but we won't consider this complication): protein A, coat protein, and the enzyme replicase. The first two, but not the third, have been sequenced. The insertion of the three corresponding messages in the RNA is given schematically above, with only initiation and termination codons shown explicitly and the other sequences indicated simply by the number of nucleotides involved.

Looking at the messages first, we see that two start with AUG, one with GUG; two are terminated by UAG, one by UAA; all according to the book. In protein A and the coat protein, which have been sequenced, message and translation product were found to conform perfectly to the code. The sequence of the replicase, which was not known by direct analysis, could be deduced from that of the corresponding RNA message. The messages themselves are flanked by mute (untranslated) sequences of variable length. The 5′ end is "capped" by a pyrophosphate group attached to the terminal phosphate, reflecting that the primary acceptor in RNA assembly is an NTP (see Chapter 16). These details are not useless, but serve to lure the infected microbe into accepting the viral message and to position it correctly in its translation machinery, as though it were one of its own. This delusion proves fatal to the bacterium and highly beneficial for the spreading of the virus, since the A and coat proteins are the two protein components of the virus and the replicase is an enzyme needed for the replication of RNA. Thus, by getting its message accepted for translation, the virus actually subverts its unfortunate host into devoting its resources to making new virus particles. The multiplication of viruses always relies on some subversion of this kind, although a number of different mechanisms may be involved. More will be said about viruses in Chapter 18.

Eukaryotic mRNAs also include mute sequences. Especially important is the "leader" sequence that precedes the message at the 5′ end and ensures a proper initiation of translation (see p. 270). With rare exceptions (e.g., histone mRNAs), the 5′ end of eukaryotic mRNAs bears several added methyl groups plus a characteristic 7-methyl-guanosine triphosphate cap, and the 3′ end is terminated by a long poly-A tail of between 100 and 150 A units. Thus, a typical eukaryotic mRNA has the following structure:

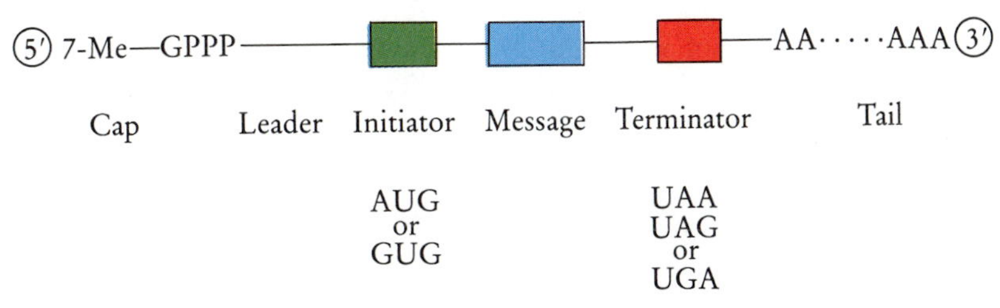

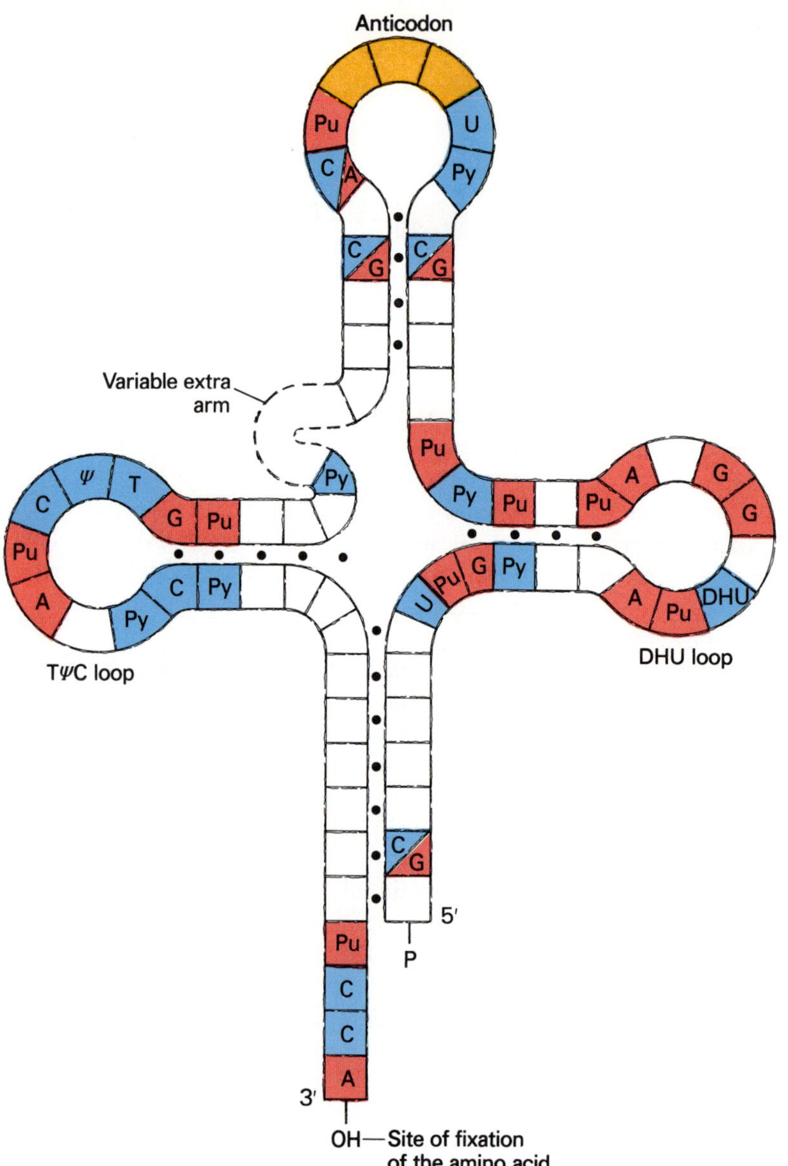

Structural characteristics common to all or most tRNAs. DHU (dihydrouridine) and Ψ (pseudouridine) are modified nucleosides. Pu stands for purine; Py for pyrimidine. C/G means C or G; C/A means C or A. The number of nucleotides in each loop may vary somewhat from one type of tRNA to another.

Reading the Message

Amino acids, we have just seen, do not read their codons. A number of attempts have been made to discover chemical affinities between the side chains of the various amino acids and the base doublets or triplets that code for them. But to no avail. The protein and RNA languages seem unrelated. How, then, is the message read? Here is where our cipher comes in. Each amino acid is attached to a special kind of RNA called transfer RNA (tRNA), which serves as reading device through base-pairing. These tRNAs are remarkable molecules, small as RNAs go (about 25,000 molecular weight) and made up of some seventy-five to ninety-three mononucleotide units, but with many of their bases modified by methylation, reduction, deamination, or transposition of the nucleosidic linkage (see Appendix 1). They all have a -CCA 3′ terminal sequence, together with certain strate-

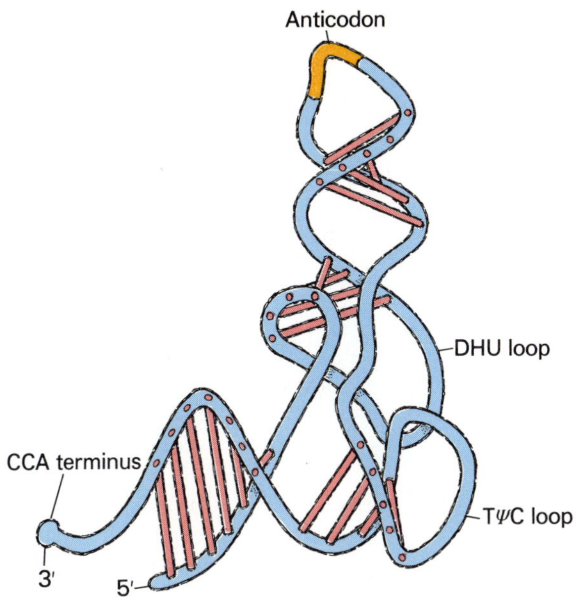

Skeletal representation of the true conformation of yeast $tRNA^{Phe}$ (transporting the amino acid phenylalanine). Compare with the schematic diagram on the facing page. Note the double-helical segments made by base-pairing.

gically placed nucleotides, and they all possess four pairs of complementary base sequences situated in such a way as to give the molecule a "cloverleaf" shape. Actually, the real structure of tRNAs (see above) is rather different from that shown in the diagram on page 262. Because of the double-helical torsion of the paired sequences that close the loops, it is screwed up into a tight knot. Even so, two features are clearly distinguishable: sticking out at the 3′ end is the typical CCA terminus with its free 3′-hydroxyl group, which serves as acceptor for the amino acid; and, diametrically opposed, occupying the exposed part of the middle loop, is the reading device, or anticodon, a base sequence complementary to the codon.

This complementarity is strict for the first two bases of the codon but may be "wobbly" for the third. For example, the anticodon of brewer's yeast $tRNA^{Asp}$ (the tRNA carrier of aspartic acid) is CUG, read from 3′ to 5′. This gives a perfect match with one of the codons of aspartic acid and an imperfect match, through the first two bases of the codon, with the other:

Codon (5′⟶3′)	—GAC—	—GAU—
	¦ ¦ ¦	¦ ¦ :
Anticodon (3′⟵5′)	—CUG—	—CUG—

The second example illustrates a case of wobble: recognition is mediated by the first two bases; the third one is loose. Thanks to the wobble, it is possible for some anticodons to recognize two or three, but usually no more than three, different codons specifying the same amino acid, allowing cells to get by with between thirty-five and forty different tRNAs to recognize the sixty-one amino-acid codons.

Recognition, you will notice, is done entirely in RNA language, a fact that has been elegantly verified by artificially attaching the "wrong" amino acid to a tRNA: the substitute is not recognized as such and is handled exactly as if it were the right amino acid. As we shall see presently, in the assembly of a polypeptide the amino acids are lined up one by one along the mRNA by means of their tRNAs; their correct insertion is determined exclusively by the accuracy and specificity of codon–anticodon interactions, and hence by the implied correctness of the amino-acid–tRNA association formed earlier.

It follows that the actual step of translation from RNA into protein language occurs when amino acids and tRNAs are matched and joined. The translators are the enzymes that do this job. They are the only bilingual elements in the cell: they can recognize both an amino acid and some part of its corresponding tRNAs, which, however, does not seem to be the anticodon itself. Interestingly, each enzyme recognizes all the tRNAs of a given amino acid. Therefore, it has imprinted in its structure one line of the genetic dictionary, with all synonyms included. There are twenty such enzymes, one for each amino acid. Together they make up the complete dictionary. But they do so in a cryptic form that relies on the tRNAs for decoding into anticodon language.

These scholarly enzymes double as energizers. They belong to the group of ligases and, in addition to their crucial role as translators, serve to provide the amino acids with the energy needed for closing the peptide bond. We encountered them before when we looked at the activation of biosynthetic building blocks. As mentioned earlier, peptide synthesis is a three-step, sequential, group-transfer process dependent on a β_p attack of ATP and on the participation of tRNA as carrier of the aminoacyl group. This is all the information we need to reconstruct the event, with the help of the general notions explained in Chapter 8.

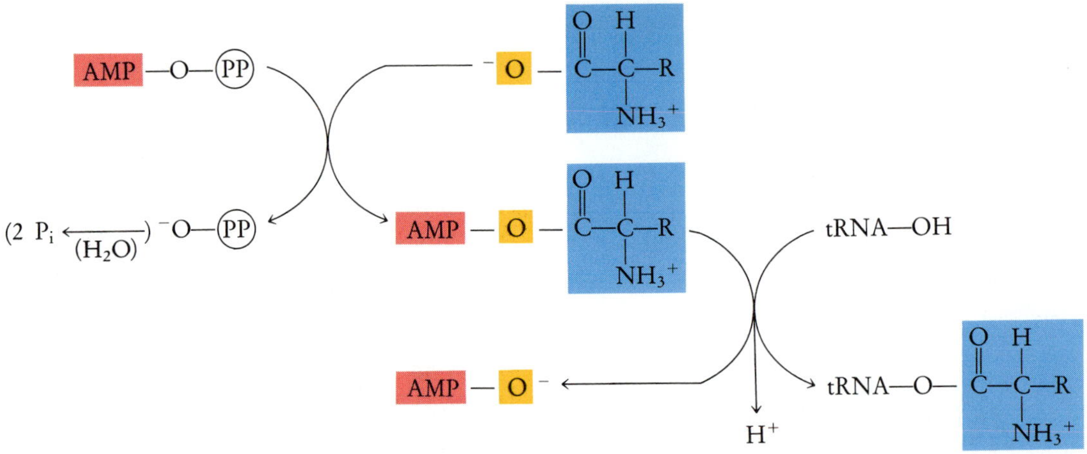

Steps 1 and 2 (activation), catalyzed by an AMP-forming ligase, occur by successive AMP-yl and aminoacyl transfer, with aminoacyl-AMP serving as double-headed Janus and oxygen transmitter from the amino acid to AMP. As is the rule in such cases, the PP_i formed is hydrolyzed to P_i, as shown above.

The third step, which takes place on ribosomes, transfers the activated aminoacyl group from its tRNA carrier to its final acceptor, an amino group belonging to an amino acid. But there is a twist to this reaction. The aminoacyl group is not, as might be expected, transferred immediately to its final acceptor. It first accepts a growing polypeptide chain on its amino group and only subsequently is transferred from its tRNA carrier with use of the aminoacyl-tRNA bond energy for the synthesis of a peptide bond. A protein grows by its C-terminal head, not by its N-terminal tail (Chapter 8), as shown schematically below.

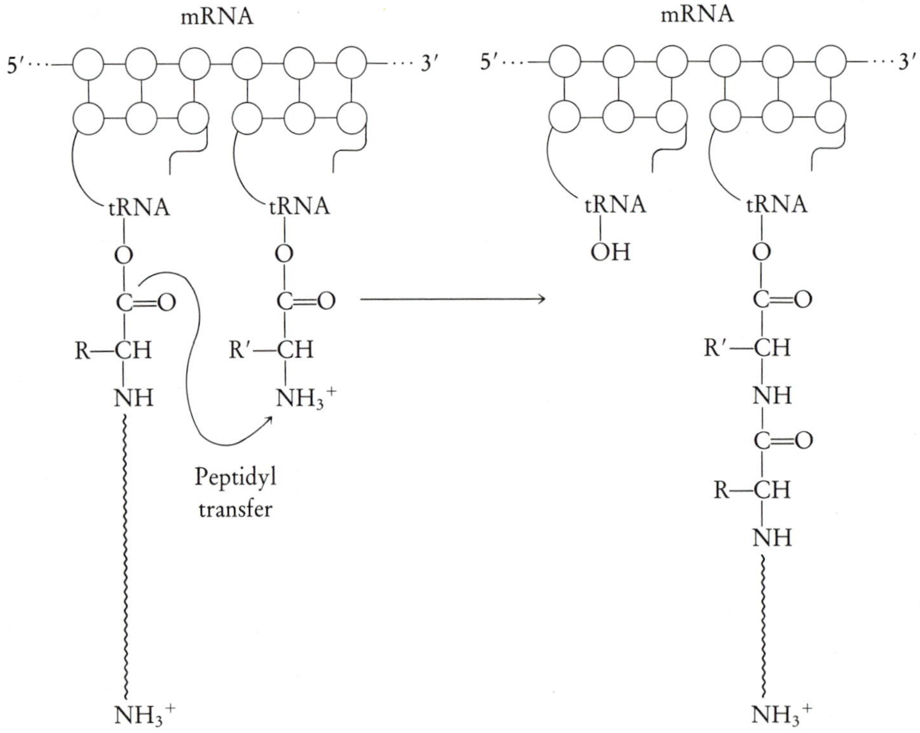

Schematic representation of a polysome.

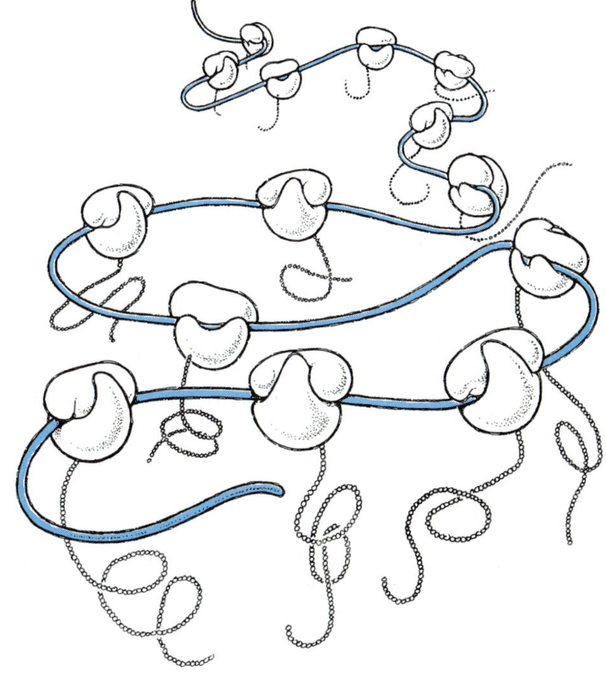

As indicated, the growing tRNA-linked chain remains attached by its tRNA to the mRNA, which is read, codon by codon, in the 5′⟶3′ direction, while the polypeptide is lengthened in the N-terminal⟶C-terminal direction.

The Perfect Robot

So much for our preliminary briefing. We can now resume our visit and start looking for some real ribosomes so that we can watch them in the act. They are not hard to find. The average mammalian cell contains some ten million ribosomes, and they are all, except for the small number of special ribosomes present in the mitochondria, directly accessible from the cytosol. A good many of them are attached to the surface of ER membranes. But others are in the cytosol, and these are the ones we will look for since they are easier to observe.

Seen from some distance, ribosomes appear as small, compact particles, slightly oblong in shape, with overall dimensions of between 15 and 25 nm—about the size of a pigeon egg at our magnification. They are not randomly distributed but are characteristically clustered by groups of ten or more. As we approach them, the reason for this clustering becomes obvious: the particles are linked, like beads on a string, by a long, slender thread. A typical 7-methyl-GTP cap at the 5′ end and a long tail of poly-A at the 3′ end identify this thread as an mRNA.

The ribosomes themselves are made of two pieces of unequal size: the large and small subunits. The large subunit looks like a stubby pear with much of its top part bitten off. The small subunit, about half the weight of the large one, is an asymmetric, bean-shaped piece lying transversely across the large subunit. The two subunits fit snugly together, except where a cleft in the small subunit leaves a channel or tunnel open between them.

The main chemical constituent of ribosomes is RNA (hence the name *ribo*somes), which makes up about 55 per cent of their weight. The remainder consists of protein. The ribosomal RNAs (rRNA) are represented largely by two unbroken strands, about 2,000 and 5,000 nucleotides long, that occupy the small and large subunits, respectively. The large subunit contains in addition two RNAs of smaller size. In biochemical jargon, the two long rRNAs are usually referred to as 28S and 18S (23S and 16S in bacterial ribosomes, which are smaller), and the short ones as 5.8S and 5S. These measurements are derived from the rate at which the molecules sediment in a centrifugal field and are given in Svedberg units, named after the Swedish scientist who developed the technique of analytical centrifugation for the determination of the molecular weight of macromolecules.

Some seventy-five to eighty different proteins (fifty-three in prokaryotes) have been identified in ribosomes, thirty in the small subunit and from forty-five to fifty in the large subunit. Many of them have been purified and sequenced, especially in the simpler prokaryotic ribosomes, and much is already known concerning the manner in which they combine with each other and with the rRNAs to form the two ribosomal subunits. Soon the complete molecular anatomy of a ribosome of the intestinal bacterium *Escherichia coli* will be known, to provide us with the most elaborate structure of a biological organelle as yet unraveled, a unique construction of more than half a million atoms. When mixed together under appropriate conditions, the sixty-odd pieces that make up a bacterial ribosome recombine spontaneously into a functional particle by self-assembly. This, no doubt, is how ribosomes arise inside the cell: their complete blueprint is contained in the primary structure of their constituents.

At the center of the ribosomal edifice lies the enzyme peptidyl transferase, whose duty it is to close the peptide bond at each successive step of polypeptide assembly. The whole intricate scaffolding around it is there to ensure the accurate positioning (and repositioning after each bond is sealed) of the mRNA and of the two tRNA-linked reactants, all the way from initiation to completion of the chain. It is assisted in this job by a number of soluble helper proteins—initiation factors, elongation factors, release factors, at least three of each—and by a liberal supply of energy provided by the hydrolysis of GTP to GDP and P_i. Let us watch this remarkable machinery as it does what it does most of the time: elongating a polypeptide chain.

Running across the ribosome, bound to the small subunit near its junction with the large subunit, lies the mRNA, the tape containing the instructions for the assembly process. Note that the specific nature of the product is entirely dictated by this tape. The rest of the machinery is strictly standard. It will turn out any kind of polypeptide, just as your cassette player will play any tune for you if properly instructed. As we face it, the 5′ end of the mRNA lies at our left, the 3′ end at our right. Therefore, the part of the message that has already been translated is to the left; what remains to be translated is to the right.

The growing polypeptide chain is readily identified by its free-floating N-terminal end. As we move up from this end along the silky filament, we recognize the amino-acid residues in the order of their successive attachments until we reach the last one, which provisionally makes up the C-terminal head of the growing chain. It is linked by its carboxyl group to the 3′ end of its tRNA, and this tRNA is itself joined by its anticodon to the corresponding codon in the mRNA. The whole bulky structure is tightly secured within a complex housing that spans the two ribosome subunits and is called the peptidyl, or P, site. This site is adjacent to the central peptidyl transferase and is organized in such a way as to put the terminal aminoacyl group of the growing chain in close contact with the active center of this enzyme. Immediately to the right of the P site lies another irregularly molded cavity of very similar shape, connected with the next codon to the right on the mRNA. It is called the aminoacyl, or A, site, and is empty at present. The stage is set for the ribosome to perform its favorite act.

Schematic functional map of a ribosome showing the site for binding of mRNA (M), the peptidyl site (P), the aminoacyl site (A), and peptidyl transferase (PT).

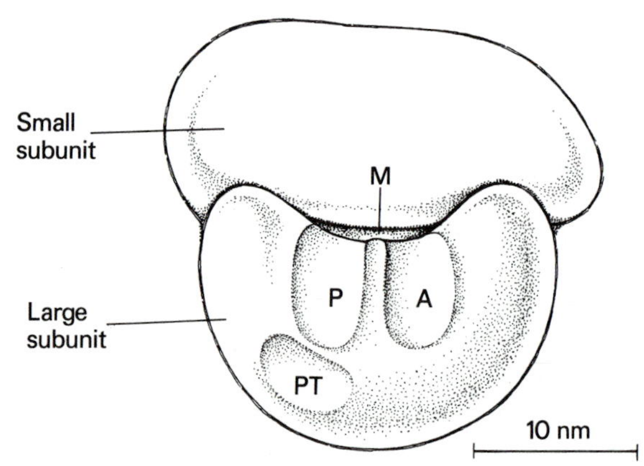

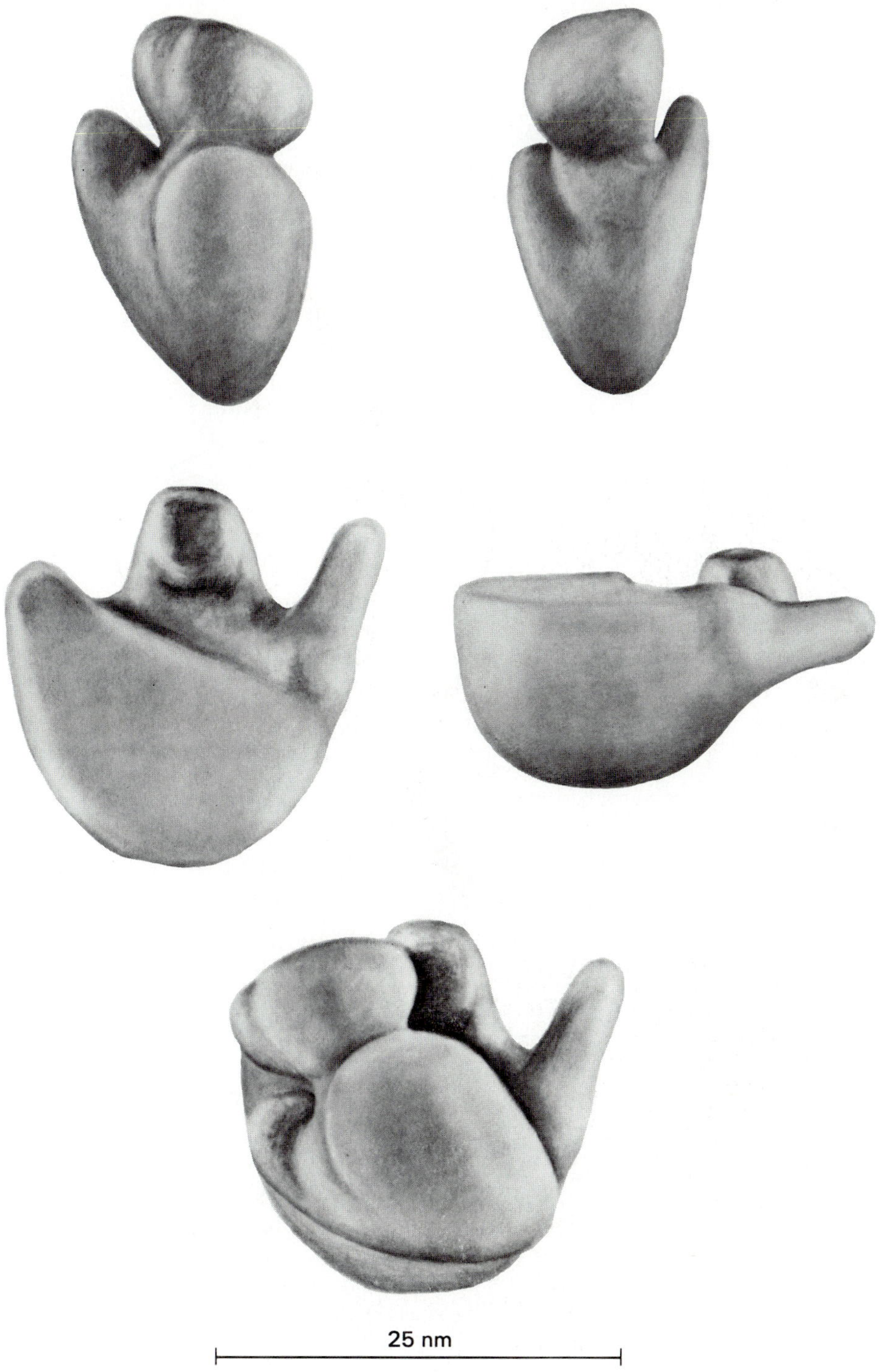

Model of *E. coli* ribosome. Top: two views of the small subunit. Middle: two views of the large subunit. Bottom: view of the assembled ribosome.

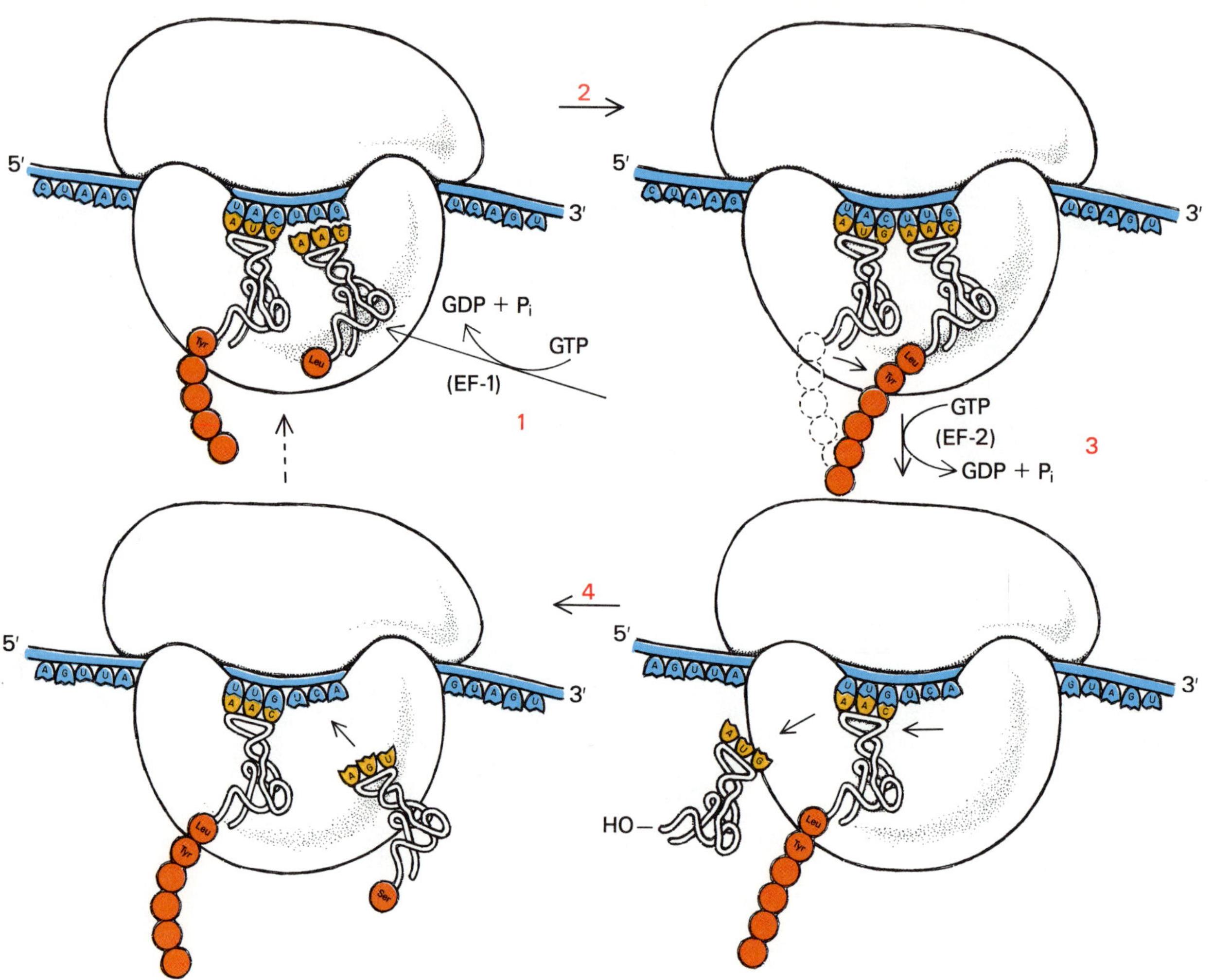

Protein synthesis: chain elongation.

Step 1: With the help of elongation factor EF-1 and the hydrolysis of one GTP molecule, the correct aminoacyl-tRNA is inserted into the A site of the ribosome (which has its P site occupied by the growing peptide chain attached by means of a C-terminal tRNA).

Step 2: The peptidyl transferase transfers the entire growing chain to the amino group of the aminoacyl-tRNA.

Step 3: With the help of elongation factor EF-2 and the hydrolysis of another GTP molecule, the lengthened peptidyl-tRNA is translocated from the A site to the P site, dislodging the free tRNA and carrying along the mRNA, which now offers a new codon to the aminoacyl site.

Step 4: The cycle starts again, with a new aminoacyl-tRNA corresponding to the codon above the A site.

For the act to proceed, the right aminoacyl-tRNA must be inserted into the A site. There may be as many as forty different kinds of aminoacyl-tRNAs swimming around, so quite a number of trials may take place before a correct codon-anticodon fit is obtained. (Remember, this fit is the only way whereby correct identification of the next amino acid to be attached can be accomplished.) Helping in this critical binding-recognition process is an important elongation factor called EF-1, which serves as carrier for the various aminoacyl-tRNAs and, by using its own binding sites on the ribosomal surface, presents the aminoacyl-tRNAs to the A site in the appropriate orientation for fitting to take place. This EF-1 factor also carries a GTP molecule. If you can adjust your eyesight to

Protein synthesis: chain termination. Upon the appearance of the termination codon UAG (or UAA or UGA) over the A site, releasing factor RF, with the help of the hydrolysis of one GTP molecule, causes the release of the completed polypeptide chain, with the concomitant liberation of the tRNA and detachment of the mRNA from the ribosome.

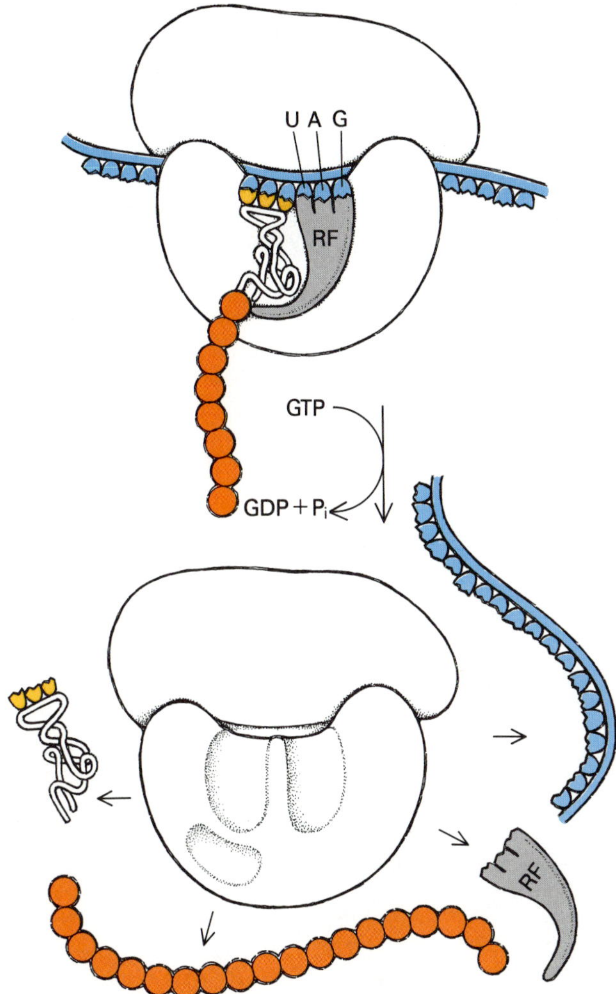

the whirl of molecular agitation, you may just be able to distinguish the multiple random trials that occur until the right anticodon aligns against the codon bordering the A site. As it snaps into place, something happens to the contact area where EF-1, GTP, and ribosome meet; the GTP is split and EF-1 floats away, leaving the aminoacyl-tRNA correctly inserted into the A site. This implies appropriate codon-anticodon pairing at the mRNA end of the A site and positioning of the free amino group of the aminoacyl radical close to the active site of the peptidyl transferase.

That is all this enzyme has been waiting for in order to do its job, which is transferring the whole growing peptide chain from its tRNA carrier to its NH_2 acceptor. As a result of this transfer, the chain becomes longer by one residue and is now attached to the A site by means of its newly acquired C-terminal amino acid and this molecule's tRNA. This is the signal for a second elongation factor, EF-2, to take over. With the help of the hydrolysis of another GTP molecule, it chases away the free tRNA that occupies the P site and causes the ribosome to translocate the whole mRNA-tRNA-peptidyl complex from the A site to the P site. We are back at stage one; the elongation act can start again.

Ribosomes go through this routine tirelessly, all ten million of them, obeying thousands of different messages and going "clickety-click" several times per second. Usually, a dozen or more ribosomes follow each other along the same mRNA strand, moving from the 5′ end to the 3′ end at the rate of ten to fifteen nucleotides per second, each trailing along a regularly lengthening polypeptide thread. We had a glimpse of this scene in Chapter 6 when we watched it through the dimly transparent membranes of the ER. There is something nightmarish about it, straight out of Charlie Chaplin's *Modern Times*, as we see it now in close detail, going on all over the cytosol and its membranous partitions endlessly, mindlessly, with metronomic regularity. But there is also great beauty in the efficiency of the machine, finely honed and perfected by natural selection, and in its excellent record of accuracy: probably no more than one mistake in ten thousand; the equivalent of one typo per five to ten pages of your average paperback. What typesetter could claim anything comparable? Mistakes in protein synthesis, incidentally, are generally of little consequence; they just cause a few faulty models of a given molecule to be produced and, since the errors vary more or less at random, no more harm than slight wastage of energy can result.

Two events break the drudgery of a ribosome's life: initiation and termination. The latter happens when one of the three stop codons—UAA, UAG, or UGA—slips into place over the A site at the last translocation, to be recognized by no tRNA (unless some mutation has modified the recognition pattern). These codons are recognized by releasing factors, which, again at the expense of one GTP, cause the successive release from the ribosome

Protein synthesis: chain initiation. Initiation factor IF, with the help of the hydrolysis of one GTP molecule, causes (1) the dissociation of a ribosome into two subunits; (2) the formation of an "initiation complex" between the small ribosomal subunit, the 5′ end of an mRNA (positioned so as to display an initiation codon AUG (or GUG) above what will become the P site), and a special tRNA bearing a methionyl group (formylated in bacteria); (3) association of this complex with a large subunit. Elongation can now start (see the illustration on p. 268).

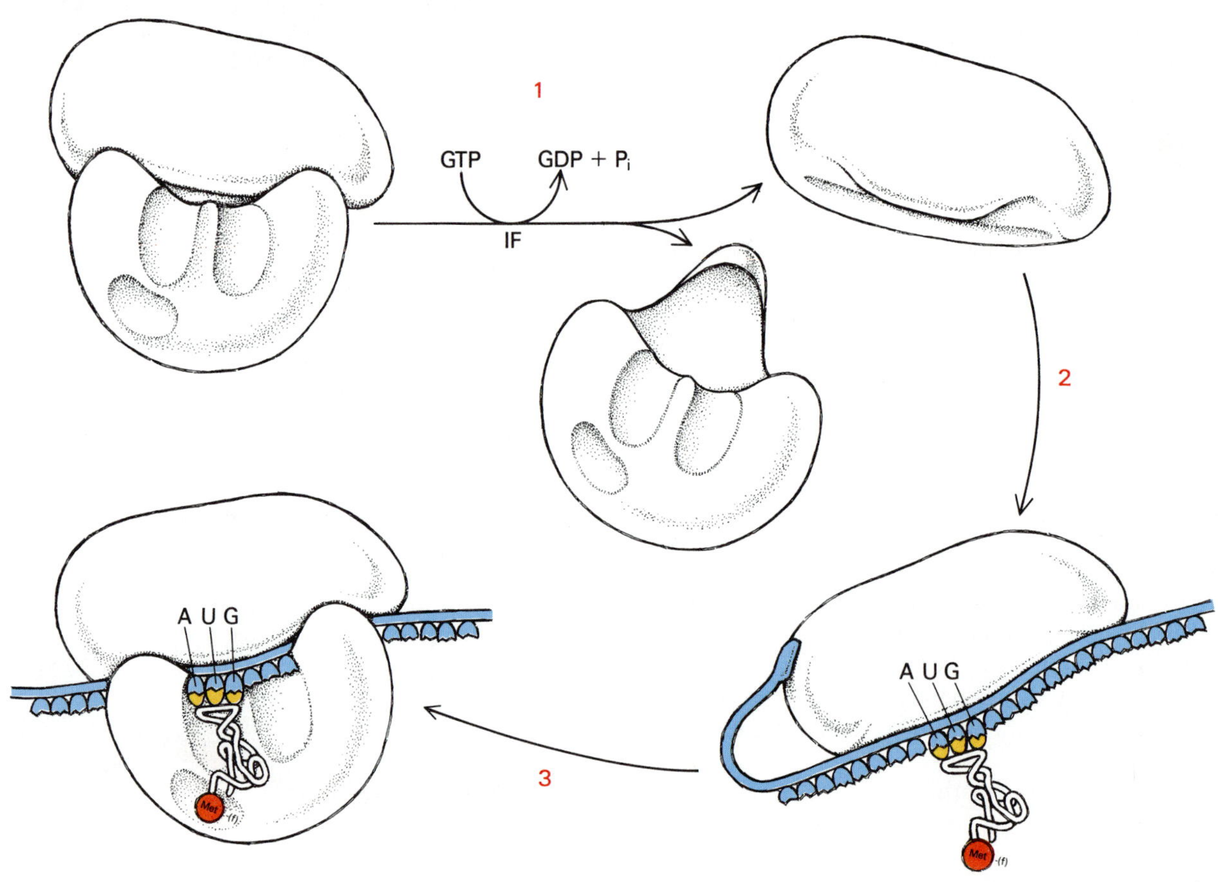

of the finished polypeptide, of its tRNA, and finally of the mRNA.

Initiation begins with the dissociation of such a free ribosome into its two subunits, which is followed by the binding of an mRNA and an aminoacyl-tRNA to the small subunit. This process, which also costs one GTP, requires very accurate matching, since it determines the reading frame. Have the fit shifted by as little as one base, and an entirely different message will be read. For a correct start, the mRNA must be bound to the small ribosomal subunit in such a way as to position an initiator AUG or GUG codon above the subunit's part of the P site. It is guided in this by its 5′ leader sequence, which recognizes special binding sites on the small subunit. The tRNA that combines with the initiator codon in this complex is different from all other tRNAs, and especially from the $tRNA^{met}$ and $tRNA^{val}$ that recognize AUG and GUG, respectively, when these codons appear inside a message: it has the unique property of joining with a small ribosomal subunit without participation of the large subunit. This initiator tRNA bears a methionyl group, which, in bacteria but not in eukaryotes, is *N*-formylated. All polypeptides, therefore, are born with an N-terminal methionyl residue, which, however, is often removed upon

Concept of reading frame. The figure shows how a message can be read in three distinct frames, to give two distinct peptide chains and an abortive runt.

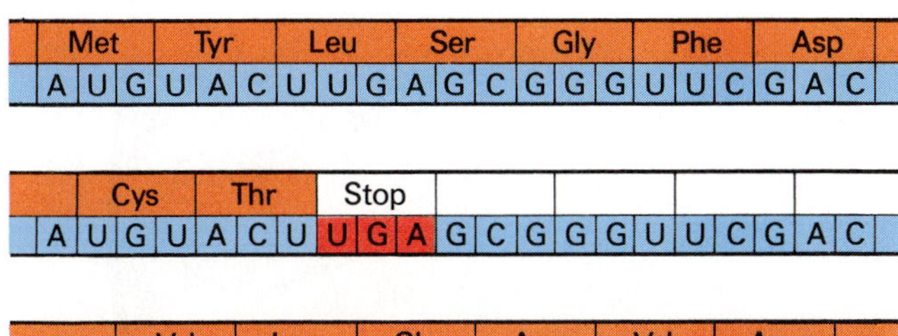

subsequent processing. In bacteria, the formyl group is removed, but the N-terminal methionine is frequently conserved. The triple complex of small subunit, mRNA, and (formyl)methionyl-tRNA is called the initiation complex. Once it is formed, it binds a large subunit; elongation can start.

Such, greatly simplified, is the sequence of events in protein synthesis. On the whole, when you come to think of all the different recognition marks, binding sites, and catalytic activities needed by a ribosome for the correct performance of its duty, the complexity of structure of this "perfect robot" becomes understandable: even half a million atoms may not seem excessive. In spite of this complexity, the system has evolved to a considerable extent. The cytosolic ribosomes of eukaryotic cells are larger than those of bacteria, and many of their recognition sites and ancillary factors are different. This is very fortunate for us, since most antibiotics (though not penicillin, which inhibits the formation of the bacterial cell wall, Chapter 2) are inhibitors of protein synthesis. Thanks to the differences between the bacterial systems and our own, the drugs thus act selectively on the bacteria without harming us. The differences between eukaryotic and prokaryotic protein synthesis do, however, complicate the job of those genetic engineers who try to get a eukaryotic message (for instance, that for human insulin) translated in bacteria (see Chapter 18). They have to tamper with the leader part of the message to make it readable by the bacterial machinery. Finally, you will no doubt remember that these differences have provided some of the most convincing arguments in support of the bacterial origin of mitochondria and chloroplasts. The ribosomes of these organelles are of bacterial type.

The Homing of Newly Made Proteins

With the exception of the rare proteins that are made in mitochondria or chloroplasts, all the proteins of a cell are initiated in the cytosol. Eventually, however, each of them finds its way to a specific intracellular compartment. The question is: How? We can obtain revealing insights into this question by taking a closer look at the "crawling polysomes," those ghostly centipedes that we discerned at the back of the tufts of growing secretory polypeptides on our tour of the rough ER (Chapter 6). We see them much better now as sinuous strings of a dozen or more ribosomes nailed to the ER membranes. We also know what they are doing, even though much of it is hidden. They are busily translating a tape of mRNA and delivering the polypeptide products of this process into the ER lumen. Let us focus our attention on the rapidly lengthening head of the centipede—the 5′ end of the mRNA, recognizable by its typical cap—as it emerges from the foremost ribosome of the string. For a while it stretches out freely, swinging and swaying with the cytosolic currents. Soon, however, a passing small ribosomal subunit gets caught by the mRNA's leader sequence; an initiation complex is rapidly built and joined by a large subunit; and translation of the message once again is set in motion. Now is when our sharpest scrutiny is needed. We have only a few seconds to watch a very remarkable event, the construction of a molecular secretory apparatus.

At first, things proceed as on free cytosolic ribosomes: the N-terminal tip of the nascent polypeptide starts growing out of the ribosome as a silky, undulating thread. But then, by the time the thread has become from twenty to

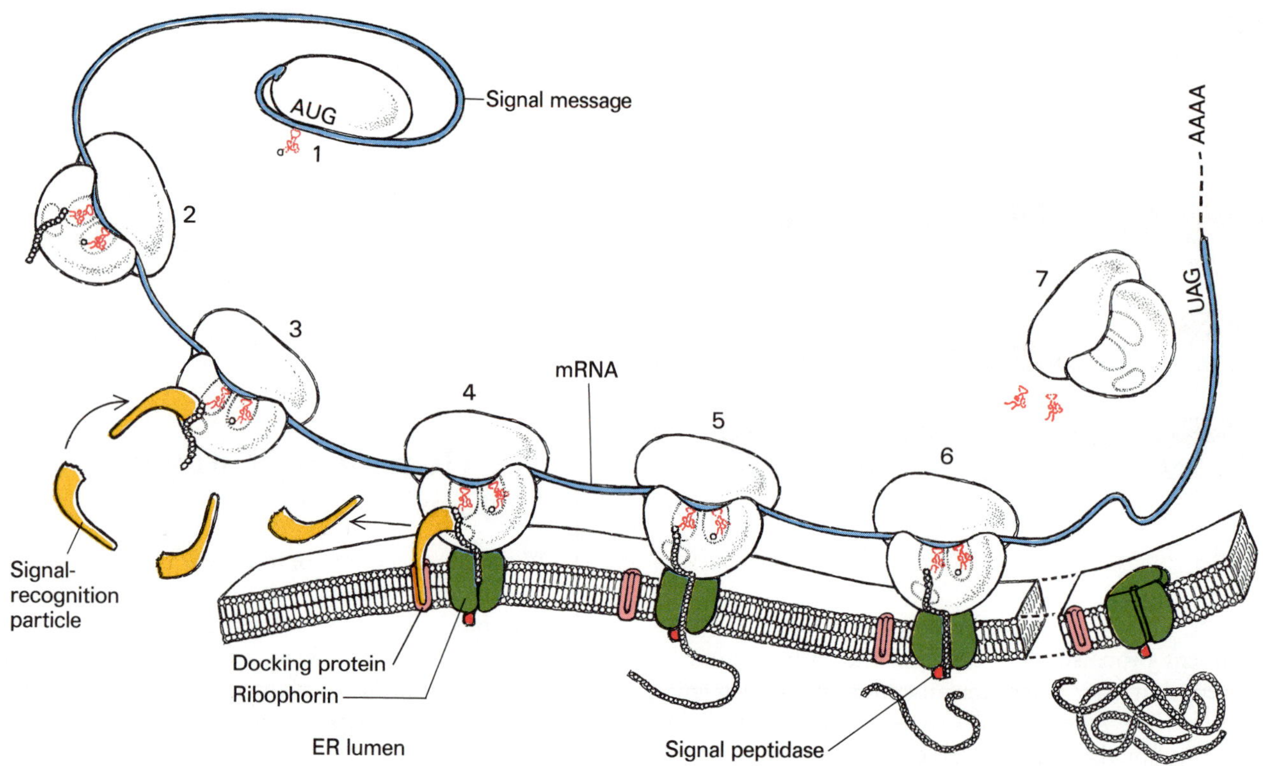

Cotranslational transfer of secretory proteins into the ER lumen, as determined by an N-terminal signal sequence.

1. Formation of an initiation complex in the cytosol.

2. Translation of the signal message has started.

3. Emerging signal peptide binds a specific signal-recognition particle (SRP) present in the cytosol. Protein synthesis is arrested.

4. Thanks to the affinity of the signal-recognition particle for an ER-linked docking protein (SRP receptor), the ribosome becomes moored to ribosome-binding proteins (ribophorins) associated with ER membranes. The SRP detaches, polypeptide assembly resumes, and the growing chain is discharged into the ER lumen.

5. The polypeptide has grown to the point that the signal peptide has been entirely extruded.

6. The signal peptide is split off by a signal peptidase.

7. Protein synthesis is completed. The ribosome detaches from the membrane.

thirty amino-acid residues long, it suddenly becomes entangled with a voluminous molecular complex, much in the manner of a surface receptor catching a ligand, and with equally dramatic consequences. For the caught complex actually stretches up to the ribosome itself, embracing it in such a way that chain elongation is arrested. N-terminal amino-acid sequences that have the ability to trap this jamming device are called signal sequences because of their directional role. The complex itself, which consists of protein and of a small 7S RNA component, is designated signal-recognition particle (SRP). A number of signaling sequences are known. They have some common features—in particular, a large abundance of hydrophobic residues. They are found at the 5′ end of all nascent secretory polypeptides. By binding the SRP complex, they activate an emergency brake that opposes the completion and delivery of such export materials in the cytosol. This state of affairs persists until the SRP-strangled ribosome bumps into a receptor, known as the docking protein, associated with the cytosolic face of ER membranes. Thanks to the specific interaction between SRP and docking protein, the ribosome becomes firmly anchored to the

ER by means of special integral membrane proteins, known as ribophorins. Its duty done, the SRP relaxes its paralyzing hold on the ribosome and falls off. Assembly of the polypeptide resumes; its product is inserted through the membrane and discharged into the ER lumen with the help of the ribophorins and, perhaps, other components, which make up the channel through which we saw the polypeptides grow into the ER lumen. Once this machinery is installed and molecular secretion is proceeding smoothly, the signal peptide is clipped off by a signal peptidase.

As a new ribosome is added to the head of the centipede by this mechanism, another one detaches from the centipede's tail, its link to the membrane severed by the detachment of the polypeptide it has just finished making. And so polysomes "crawl" on the surface of ER membranes, acquiring new ribosomes at the 5′ end and losing old ones at the 3′ end of their connecting mRNA. Occasionally, a brand-new mRNA becomes similarly involved, to start a new polysome creeping on the membrane in compensation for the inevitable wear-and-tear. These are the phenomena that take place on the cytosolic face of the membrane to produce the silky growth that made such an impression on us when we first set eyes on it.

This kind of cotranslational transfer—that is, transfer across a membrane while the message is being translated—seems to be of very general significance for export proteins, even in bacteria, where secreted exoenzymes are made by polysomes attached to the plasma membrane. It also plays an important role in the positioning of ribosomes that make integral membrane proteins for the ER and related systems, including the plasma membrane. As we saw in Chapter 6, many of these proteins, though not all, are made by ER-bound ribosomes. At first, things start very much as they do for secretory proteins, except that the signal piece is not always cut off but may remain instead as a hydrophobic anchor embedded in the lipid bilayer. If so, and if cotranslational transfer otherwise proceeds normally, a membrane protein oriented toward the ER lumen will be produced. In many cases, translocation of the growing polypeptide is blocked at some stage, as it passes through the membrane, by a specific internal amino acid sequence that acts as a stop-transfer, or halt-transfer, signal. It is often a long hydrophobic sequence that gets caught in the lipid bilayer. Chain elongation is, however, not arrested by this jamming of the delivery process; as it continues, the ribosome is progressively pushed away from the membrane with the electrostatic help of a cluster of basic residues that often follows the hydrophobic sequence. As a result of this detachment, the ribosome completes the C-terminal end of the polypeptide in the cytosol. A transmembrane protein, with its N-terminal end exposed in the ER lumen and its C-terminal end facing the cytosol, is generated in this manner. More complex mechanisms of the same general nature control the synthesis of membrane proteins with a more involved topology. Some membrane proteins that face the cytosol are made by free polysomes and subsequently become attached to the membrane by some exposed hydrophobic parts.

Much less is known about the homing of other segregated proteins. Some cell explorers claim to have seen polysomes crouching on the surface of mitochondria and other membrane-limited organelles that contain resident proteins. Most, however, have looked for such bound polysomes in vain, finding instead that the proteins of these organelles (except for the few that are assembled internally) are made by free polysomes, released into the cytosol, and transferred posttranslationally to their final abode. How this occurs is still poorly understood. One assumes, by extension of the signal hypothesis, that the proteins must include in their structure some peptide sequence or other molecular configuration that specifically recognizes, and associates with, the membrane of their host organelle. But then something must occur to force the whole bulky molecule across the membrane, and here is where imagination becomes stretched beyond the boundaries of current preconceptions. Participation of a removable signal-type sequence has been detected in some instances, though not in others. In any case, it is difficult to see how complete transfer of the molecule across the lipid bilayer of one or, more often, two or even three (chloroplast thylakoids) membranes can be accomplished by such simple means. In cotranslational transfer one can at least visualize the GTP-fueled ribosome as forcing the growing polypeptide through the membrane. No such push is available posttranslationally, and one tends to think instead in terms of an internally generated pull acting on the membrane-linked protein. Alternatively, there could be an energy-fueled protein-translocating system in the membranes. Unfortunately, explorers have not yet devised means for reaching inside a mitochondrion or chloroplast to watch what is going on on the inner face of the membranes of these organelles.

The Life and Death of Proteins

For many proteins, getting "signaled" to their location is only the beginning of a long story. We witnessed some of this on our return from the recesses of the ER through the Golgi apparatus to the cell surface. The secretory proteins that were floating downstream with us were undergoing all sorts of changes, including glycosylation, association with lipids, closure of disulfide bonds, proteolytic clipping, and paring and remodeling of carbohydrate side chains. Actually, this processing continues beyond the cell surface up to the final change that announces destruction, most often through endocytic uptake and lysosomal digestion.

Take insulin, for example, the hormone whose deficiency causes diabetes. It is manufactured by a special cell, called β-cell, that occupies small regions, called islets of Langerhans (Latin *insula*, islet), in the pancreas, a digestive gland associated with the gastrointestinal tract. In its active form, insulin consists of two short polypeptide chains, the A chain with 21 amino acids and the B chain with 30 amino acids, linked by two disulfide bridges. It is, however, made as a much longer single chain called preproinsulin. Ribosomes starting to make this molecule in the pancreatic β-cell are directed to ER membranes by an N-terminal signal sequence, which is cut off cotranslationally. This removes the "pre" part of the molecule, leaving proinsulin as the product delivered into the ER lumen. The next step is oxidative closure of the disulfide

Maturation of insulin. The total message corresponds to preproinsulin, which is converted into proinsulin upon removal of the N-terminal signal peptide. While traveling through the ER toward the Golgi, proinsulin is folded and closed in this conformation by disulfide bonds. Later in its transit through the secretory apparatus, proinsulin is hydrolyzed by a specific endopeptidase that removes the C peptide, leaving mature insulin, the secretory and active form of the hormone, to be discharged into the bloodstream by exocytosis.

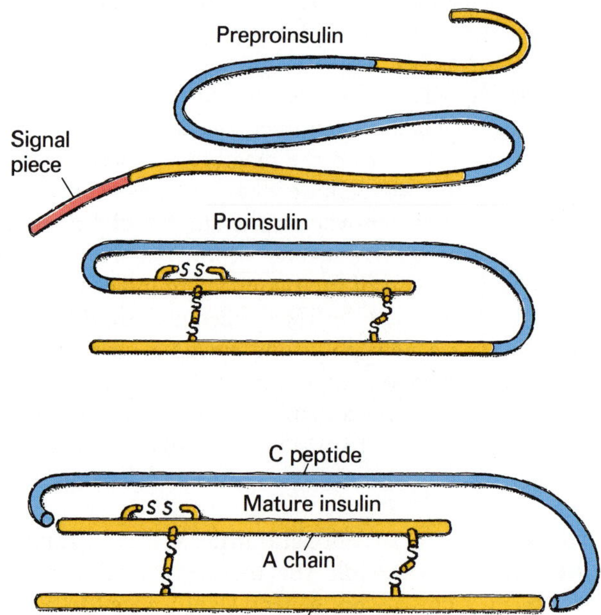

bonds. Then an internal sequence of 33 amino acids (C peptide) connecting the C-terminal A chain to the N-terminal B chain is excised by proteolysis, removing the "pro" part of the molecule.

The mature insulin is stored in tightly packed secretion granules, which, upon solicitation of the cell by an excess of sugar in the blood, discharge their contents exocytically into the bloodstream. The hormone now circulates, remaining in the blood until it happens to meet a free receptor on the surface of some cell. Binding of insulin to its receptor triggers a variety of intracellular events that result, among other effects, in an increased uptake of glucose, thus correcting the disturbance that stimulated insulin secretion in the first place. Providing a precaution against overreaction, attachment of insulin to its receptor also stimulates certain braking effects, among them endocytic uptake and digestion of the insulin in the lysosomes. All is over for this molecule. Joining with its receptor was both a fulfillment—the crowning event in its molecular life—and a death warrant, ordered by the needs of homeostatic regulation.

The fate and possible function of the C peptide are not known. But cases do exist where two or more active agents arise from a single polypeptide chain. A striking example is found in the anterior pituitary, a small, but very important, endocrine gland situated in the middle of the brain. Certain cells of this gland, called melanocorticotropic, make a signal-directed secretory polypeptide of 134 amino acids that is split into two hormones—ACTH (39 amino acids), which stimulates the adrenal cortex, and β-LPH, which acts on lipid metabolism—and a small connecting pentapeptide. Further carving of these molecules yields α-MSH (the first 13 amino acids of ACTH), β-MSH (amino acids 41 through 58 of β-LPH), and β-endorphin (amino acids 61 through 90 of β-LPH), of which the N-terminal pentapeptide forms enkephalin. The MSH factors, or melanotropic hormones, affect pigmented cells; they stimulate the changes in chromatophore color, which, as we have seen, depend on microtubule function (Chapter 12). β-Endorphin and enkephalin belong to a special group of brain peptides that mimic the effects of morphine and to which it is said we owe some of our more pleasurable sensations. These are not isolated examples. Hardly suspected a few years ago, the production of all sorts of biologically active peptides through the intracellular processing of larger precursors has now become one of the most exciting areas of research, especially in neurobiology and psychochemistry.

For a number of secretory proteins, the main carving-up occurs extracellularly: the proteins are secreted as inactive precursors, which subsequently are activated by proteolytic cutting when and where needed. A number of digestive enzymes belong to this group; they are discharged as harmless zymogens and are unleashed only after reaching the stomach or the gut. Many proteins carried by the bloodstream are similarly manacled, ready to spring open upon the right proteolytic attack to produce the powerful—but deadly, if not properly controlled—factors that cause blood coagulation, clot dissolution, enlarging or narrowing of small blood vessels, mobilization of lymphocytes and other members of the immunological police, killing of foreign invaders. As we saw in Chapter 6, some of these events are controlled by a cascade of proteolytic attacks.

Six hormones in one polypeptide. The diagram shows how different ways of carving the melanocorticotropic hormone of the anterior pituitary gland can produce as many as six different active products.

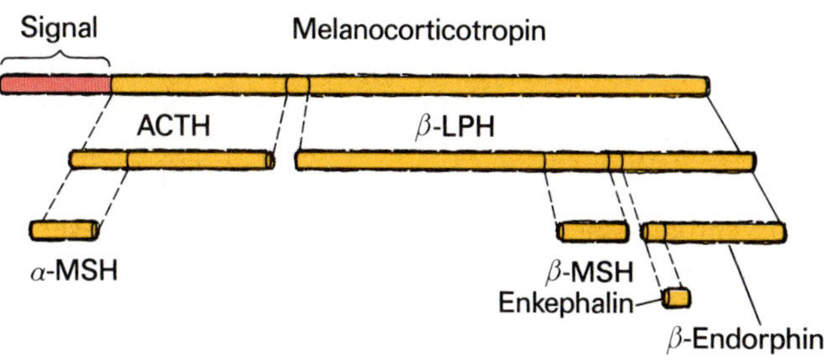

The membrane proteins made in the ER also go through many modifications, starting frequently with the cotranslational clipping of a signal piece. After completion, they start drifting by lateral diffusion in the plane of the lipid bilayer. During this voyage, they may undergo various kinds of processing on both sides of the membrane. A surface receptor protein, for instance, may acquire its oligosaccharide "hairs" in the lumen and part of its triggering mechanism in the cytosol. The itinerary and terminal destination of each protein are governed by the attachments they establish with their neighbors, with cytosolic components, and with luminal (or, eventually, external) ligands. Some proteins stay anchored in the ER, sometimes even, like the ribophorins, in the rough-surfaced parts of the ER. Others move on into the Golgi and become part of that structure. Yet others, many of them heavily glycosylated, get carried even further, either to lysosomes (with their mannose-6-phosphate-bearing ligands) or to the cell surface by way of secretory vesicles or granules. Once in the plasma membrane, they may become clustered into patches of different composition, participate in various forms of endocytic uptake, and become caught in the complex circuits described in Chapter 6. Eventually they end up in lysosomes after being segregated by autophagy.

Other intracellular proteins experience equally eventful, but different, adventures. They bind metals, acquire cofactors, assemble into multienzyme units, suffer the periodic addition or removal of groupings, such as phosphate or AMP, that affect their catalytic properties, until they, too, go the way of all mortal things.

In short, proteins have a life history. Between their ribosomal cradle and their lysosomal (or some other) grave, they go through a number of different stages, some of them associated with different locations. The whole of this molecular saga, as has been pointed out several times, is programmed in what is a protein's sole endowment at birth, its primary structure, written as a specific amino-acid sequence. Such linear sequences will then, either spontaneously or with the collaboration of some of the components offered by the complex intracellular environment, fold, unfold, and refold into defined three-dimensional configurations, which will themselves exhibit catalytic properties, display binding sites, offer targets to other enzymes, or do any of the thousands of other things proteins do in their lifetime.

Even death is included in this program, not in terms of a rigidly determined life-span, but rather in terms of a probability. The death of proteins is a stochastic process, and their life-span is a statistical notion generally defined by their half-life—that is, the time needed for half the molecules of a given species that have been synthesized at the same time to be destroyed. Half-lives are measured with the help of labeled precursors.

Just imagine that while we are visiting this cell it is exposed for a short time to a mixture of amino acids painted red. We then would be able to distinguish the proteins born during this brief period by their color and, therefore, to watch their disappearance and to measure their half-lives. The beauty of this technique is that it hardly disturbs the cell, except for the possible effect of the paint. For the red proteins that disappear before us are being replaced by newly synthesized colorless ones. The total number of molecules does not change. The cell is in a steady state; we are watching turnover without interfering with it. In reality, of course, we do not use red amino acids, but radioactive ones, at low enough dosage to keep radiation damage to a minimum. When such experiments are done, it is found that the average half-life of proteins is

The concept of turnover. At time zero, a radioactive amino acid has been provided for a brief length of time (pulse) during which radioactive proteins are synthesized. Proteins made after that to replace those destroyed are again nonradioactive. With a constant protein pool (steady state), the radioactivity in the pool falls in exponential fashion. The half-life of the protein is the time needed for the specific radioactivity of the protein to decrease by one-half.

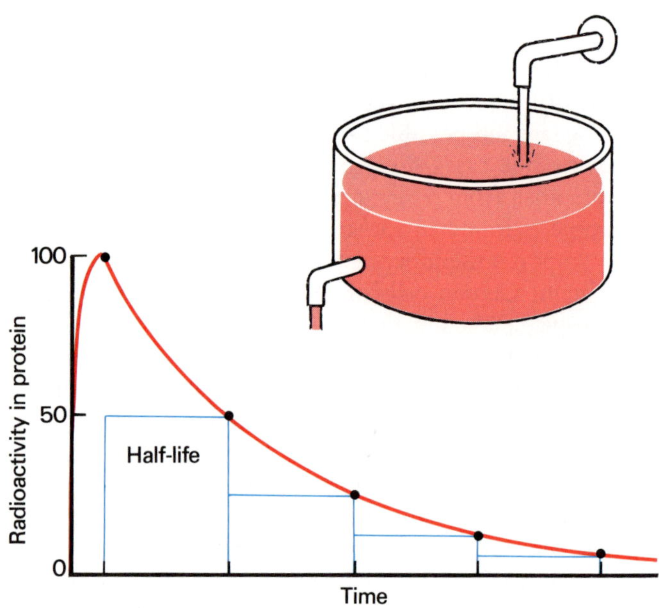

of the order of two to three days, but it varies greatly from one protein species to another. Half-lives are measured in minutes for some proteins, in hours for others, in days or sometimes even in weeks or months for yet others.

Where Is the Boss?

Many of the events that make up the life history of individual proteins are themselves mediated, directly or indirectly, by other proteins. The outcome is a four-dimensional, multifactorial network of interactions, which may, in consideration of the overwhelmingly important role of proteins, be viewed as conditioning the life of the cell itself. Different cells are different essentially because they have a different complement of proteins. When such cells associate to form an organism, a network of networks of interactions is produced. In turn, as organisms join into groups, societies, ecosystems, up to the whole biosphere, networks of higher and higher order are built up, finally to embrace, in a single interwoven unit of incredible complexity, the thin pellicle of living matter that covers our planet.

Life, in all its forms and activities, is to a very large extent a manifestation of proteins and, more precisely, of certain associations of proteins capable of generating a self-stabilizing, self-regulating network of interactions adapted to certain environmental conditions—that is, a viable system. But the proteins themselves, as we have just found out, are but the expression of the mRNAs supplied to the ribosomes of each cell. Therefore, we may reformulate the statement above and say that life is a manifestation of mRNAs. In fact, this generalization is even more correct than the first one, if we make it include all RNAs, not just mRNAs. For the mute sequences that are not translated, and especially the tRNAs and rRNAs, also play key biological functions that are strictly determined, as are the messages, by specific base sequences.

And this then raises the ultimate question: Who or what specifies the RNAs? Who or what spells out the instructions? Who or what writes the scores that are so magnificently, but quite slavishly, executed by the cytoplasmic orchestras? To find the answers to these questions, we must penetrate the inner circle, the hidden center where each cell's repertoire of information is on file. We must enter the nucleus. There the decisions are made instant by instant, or rather seem to made—there really is no boss, as we shall see in Chapter 16—about what instructions to send out and in how many copies.

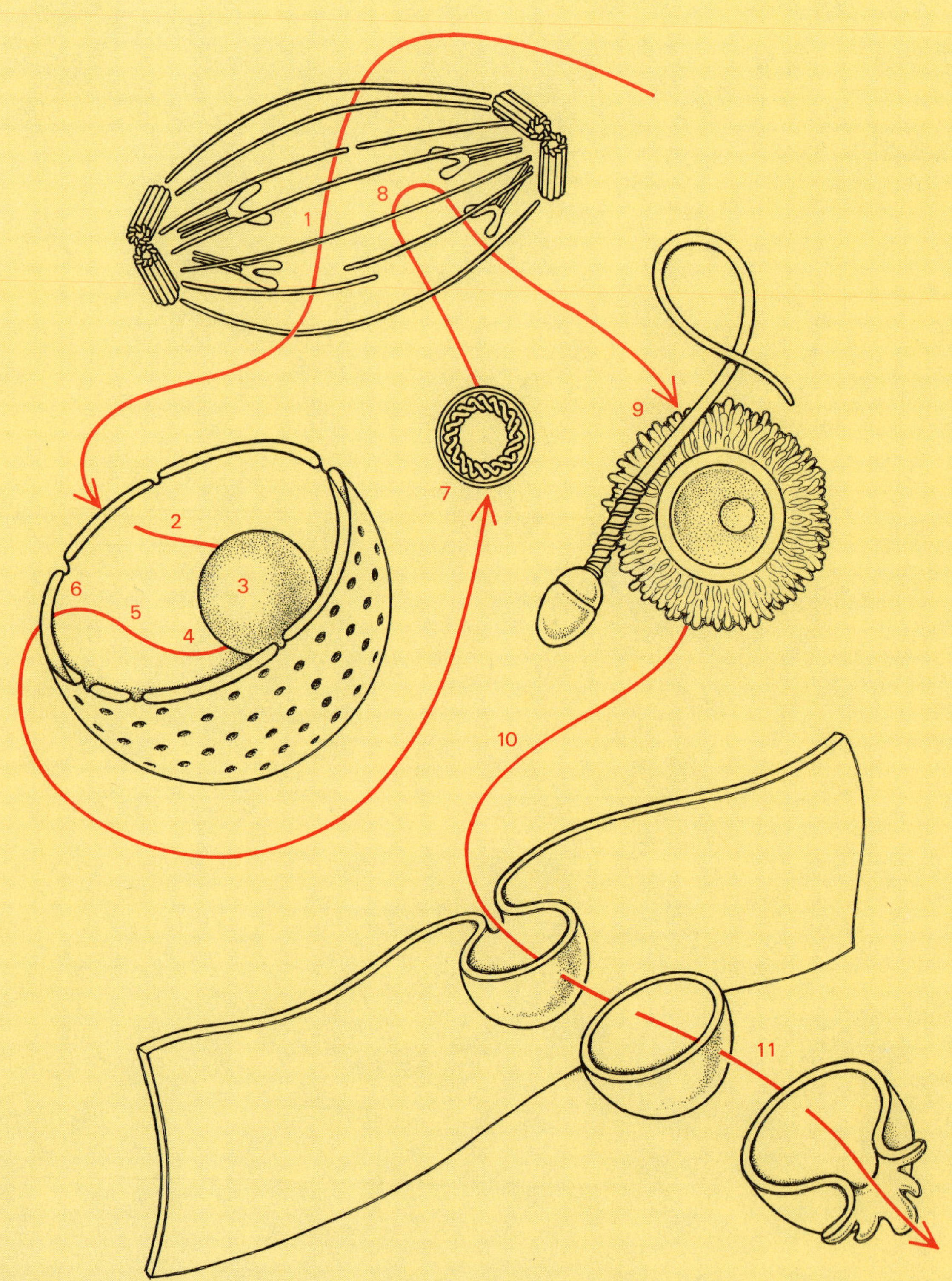

1
8
7
9
2
3
6
5
4
10
11

ITINERARY III

The Nucleus

1. Take advantage of mitotic division to enter newly forming daughter nucleus.
2. Watch chromosomes uncoil into immense lengths of DNA.
3. Visit nucleolus and observe synthesis of ribosomal RNA.
4. Move on to euchromatin area and inspect DNA transcription into messenger RNA.
5. Wait until a mitogenic stimulus initiates DNA replication and examine this process.
6. Remain in interphase nucleus to observe hidden movements of allegedly resting DNA.
7. Follow some viruses on their rounds and pause to reflect on a few problems, such as cancer, genetic engineering, evolution, and the origin of life.
8. Contemplate mitosis and separation of duplicated chromosomes.
9. Move briefly to a germ cell and take a look at meiosis.
10. Leave nuclear area and remain in the cytoplasm to watch the end of cell division.
11. Use plasma membrane to bud out of the cell and make final exit.

16 Transcription and Editing of Genetic Messages

In Latin, *nucleus* is the diminutive of *nux*, nut. The Greek word *karyon*, which has lent its root to such terms as eukaryote, prokaryote, karyolysis, karyotype, heterokaryon, likewise means nut. The nucleus derives its name from its location: it sits in the center of the cell like a stone in a cherry. Also like a cherry stone, it keeps its precious kernel enclosed within a protective shell.

A Well-Defended Citadel

The nuclear shell or envelope is not hard and rigid; it is membranous. At first sight, you might well mistake it for a rough ER cisterna; it is made of the same tenuous material and is studded with typical bound polysomes. As a matter of fact, you would not be wrong, since the nuclear envelope is indeed part of the ER. But it is a very special part, huge and completely bent and fused around the nuclear contents to form two concentric membranous spheres separated by a narrow space some 10 to 15 nm wide. This space is a true cisternal space; it receives secretory material made by the bound polysomes and transfers it by means of connecting channels to the more conventional parts of the ER. The polysomes, however, occupy only the outer wall of this remodeled cisterna, the one facing the cytosol. The inner wall is smooth; there are no ribosomes inside nuclei.

The structure enclosed by this double-walled envelope is of considerable size. It has a diameter of 8 to 10 μm and occupies one-tenth or more of the cell volume. At our millionfold magnification, this makes for a roomy cham-

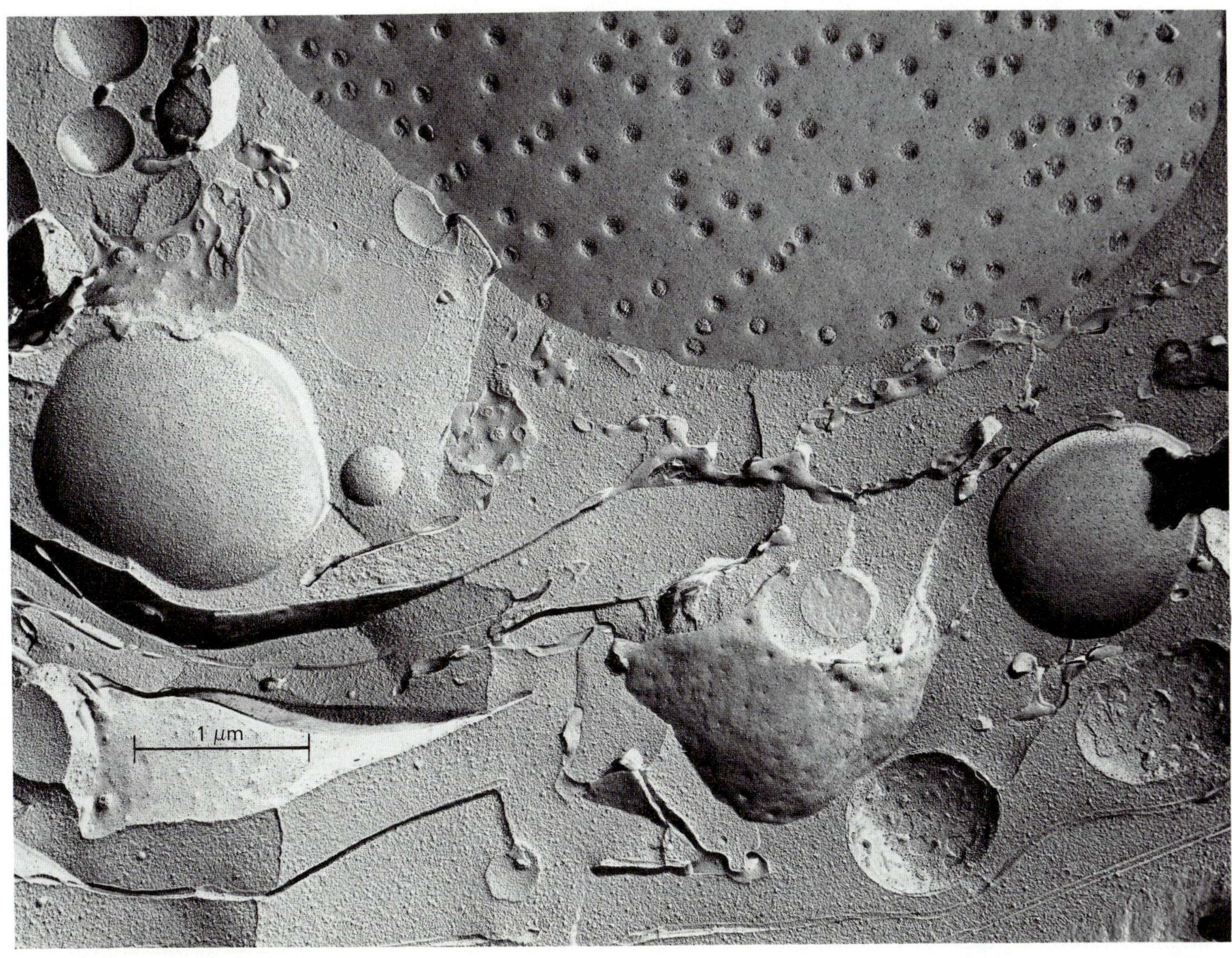

The imprint of the nuclear envelope and of its pores appears in the upper part of this freeze-etch electron micrograph of part of an onion (*A. cepa*) root-tip cell. Various fractured organelles can be seen in the cytoplasm.

ber, some 25 to 30 feet in diameter. Its casing is flimsy, hardly more than a quarter of an inch thick (7 nm, real size), which is even thinner than the plasma membrane. But it is double and bolstered by a strong protein lining called the lamina. The nuclear envelope acts as an excellent barrier between the nucleus and the cytoplasm, forcing all exchanges to take place through special channels known as pores.

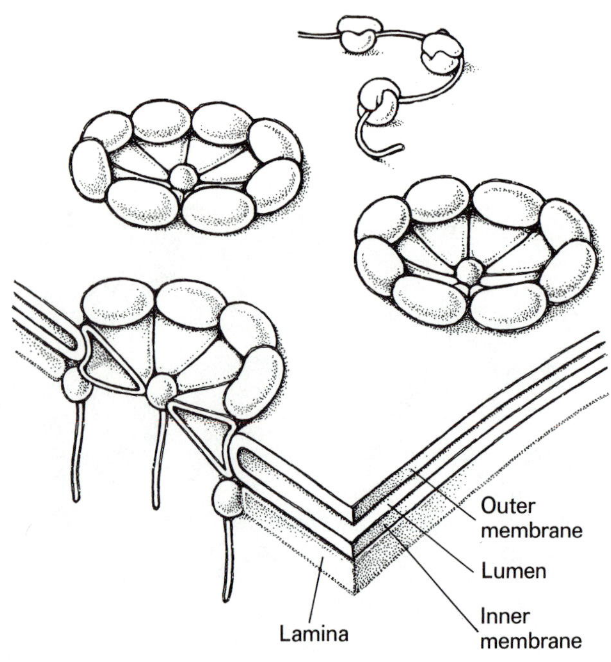

Schematic structure of the nuclear pore complex. The double-walled nuclear envelope, derived from the endoplasmic reticulum (notice the polysome on the cytoplasmic face), is pierced by thousands of round openings, or pores. Each pore is bordered by a continuous seal between the inner and outer membranes. The openings are largely closed by a proteinaceous complex made of eight conical flaps, surrounded by an inner and an outer ring (annulus) of eight globular subunits. The central hole is often stoppered by a plug. The pores are interconnected by a fibrillar web, called the lamina, apposed against the inner face of the envelope.

The nuclear pore is a proteinaceous structure with an eightfold radial symmetry inserted into a windowlike eyelet, or fenestration (Latin *fenestra*, window), bored through the two membranes of the envelope and sealed by their fusion. The outer diameter of the complex is about 120 nm, that of the hole itself, from 60 to 90 nm. But much of this is closed by a diaphragmlike structure consisting of eight conical flaps, leaving an opening of only about 10 nm, which is itself often stoppered by a plug. At our magnification, a pore would appear as a small, octagonal porthole with an overall diameter of 5 inches and a central aperture of a little less than half an inch. In our "average" mammalian cell, there may be as many as several thousand such pores distributed over the whole surface of the nuclear envelope. They give the envelope a sort of padded appearance and distinguish it clearly from ordinary ER cisternae. Judging from their structural intricacy, nuclear pores are not likely to be simple holes. They must be dynamic entities, actively mediating and regulating the transport of materials in and out of the nucleus. Unfortunately, we have as yet no inkling of how they operate. All we can do is to watch the result; a worthwhile experience in itself, which tells us a great deal about the nucleus and its functions even before we get inside.

Most striking is the steady outflow of RNA. Representing the principal product of nuclear industry, freshly made RNA threads are being extruded continually through every pore, at the rate of about 1 μm (more than 3 feet at our magnification) per pore per minute. These RNA threads come out in association with proteins, which presumably help them through the pores. Much of the RNA produced by the nucleus is rRNA, destined for the construction of the thousand-odd ribosomes that a cell must make every minute to compensate for the ineluctable attrition that affects ribosomes as it does everything else in the cell. Replacement of tRNA likewise makes a continual demand on nuclear manufacturing activity, as does the production of the small sRNAs that participate, for instance, in protein signaling (Chapter 15). Finally, a small but crucial part of the RNA that flows out through the nuclear pores is mRNA; it contains the bulk of the ordinances issued by the nucleus. In addition to RNA, a few other important molecules are made in the nucleus and delivered

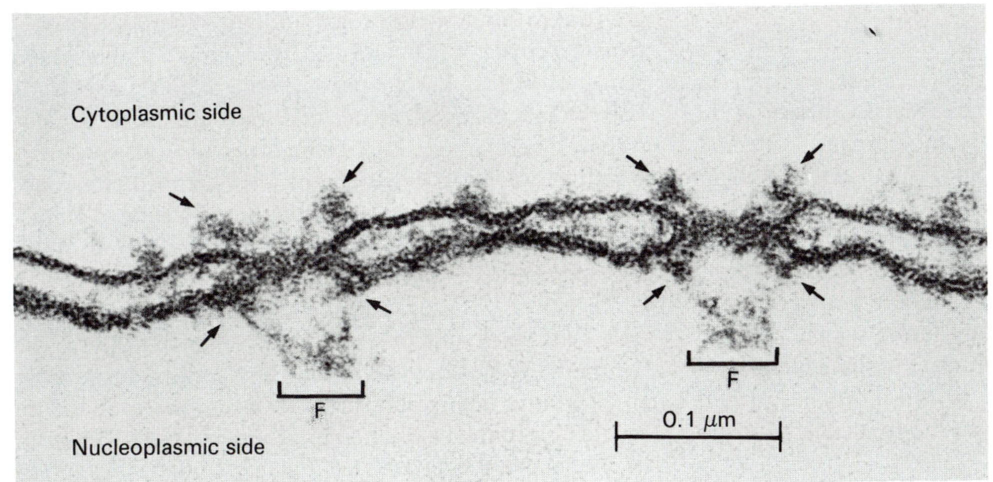

Electron micrograph of a cross-section through the nuclear envelope of a *Xenopus laevis* oocyte. Note the double membrane with its cisternal lumen and the lamina bolstering the membrane on the nucleoplasmic side. Two pores are seen: the arrows point to annular granules. Filament bundles (F) are attached to the nucleoplasmic side of the pore complex.

to the cytoplasm through the pores, among them NAD.

Flowing in the reverse direction through the nuclear pores are the activated building blocks of RNA synthesis. Except possibly for some glycolytic activity provided by cytosolic enzymes seeping through the pores, the nucleus has no autonomous supply of energy. Its biosynthetic role is restricted to assembly. It relies entirely on the cytoplasm for both raw materials and power, supplied jointly in the form of the four key nucleoside triphosphates: ATP, GTP, CTP, and UTP (or nucleoside mono- or diphosphates and a corresponding excess of ATP for their phosphorylation). When called upon to replicate its DNA, the nucleus likewise receives its energized substrates from the cytoplasm in the form of the appropriate deoxynucleoside triphosphates—dATP, dGTP, dCTP, and dTTP—or of their precursors. These supplies are complemented by a substantial additional amount of ATP that is used for purely energetic purposes and returned to the cytoplasm as ADP or AMP for recharging. The nuclear pores also allow a considerable inward flow of proteins, comprising both newly made nuclear proteins, which are all synthesized by free cytosolic polysomes, and cytoplasmic proteins that come to fetch RNA out of the nucleus. Finally, if you look carefully enough, you will distinguish in the inward stream a number of rare molecules, some of them of protein nature, others as yet unidentified, that serve to carry instructions from the cytoplasm to the nucleus.

Nuclear pores thus let many things into and out of the nucleus, though not, however, the members of our group. Try as we might, we could not possibly squeeze through one of those narrow portholes. To enter forcibly, we would have to rip apart the fragile fabric of the envelope, possibly causing irreparable harm to the delicate balance between cytoplasm and nucleus. If we wait long enough, however, our opportunity may yet come, when the cell that we are visiting enters into division. At that time, the nuclear envelope fragments into vesicles; the barrier between cytoplasm and nucleus disintegrates. All we have to do is ensconce ourselves near one of the two clusters of chromosomes that separate during mitosis (Chapter 19), and stay there. Soon a new envelope will form around us, and we will find ourselves inside the nucleus of one of the daughter cells.

Inside the Nucleus

Witnessing the reconstruction of a nucleus from the inside is a stifling experience. While you wait in the midst of the chromosomes, large expanses of endoplasmic reticulum start closing in from all sides, pressing more and more, until by the time they all come together and fuse into a single hermetic hull, you are left with hardly any space in which to move. At the same time, your main landmark disintegrates before your eyes. The chromosomes, those solid pillars under which you had taken shelter, fray progressively into interminable lengths of a ropelike material, snarling you in thousands of tangled coils. It is enough to drive the hardiest explorer into a claustrophobic panic.

There is, however, no need to yield to such an unreasonable fear; surely there will be a way out. In the meantime, let us take stock of our surroundings and see what more we can learn about the cell now that we are inside its most jealously guarded citadel. The ropes that are all around us are chromatin fibers, deriving their name, as do the chromosomes from which they originate, from the Greek word for color. Not that they are themselves colored; they are snow white, in fact. But they avidly bind a number of basic dyes. Thanks to this basophilia, they appear brightly colored in appropriately stained preparations, and were named after this property by the investigators who first saw them about a century ago. To be sure, those early explorers did not see the actual fibers. They simply recognized chromatin as the main nuclear substance and chromosomes as rod-shaped condensations arising from it during cell division.

We don't suffer the same limitations and readily see that chromatin is made of fibers some 25 nm thick. In a human cell nucleus, there are 46 such fibers, one per chromosome, varying in length between about 0.25 and 2 mm. This may not look like much until you scale it up to human size. Magnified a millionfold, it amounts to some 30 miles of hefty rope, about 1 inch thick, packed in a 30-foot ball.

Yet, this is not all. The chromatin ropes, you will notice, are actually made of a thinner, 10-nm-thick strand, coiled into some sort of solenoid, much like a telephone-receiver cord. When unrolled, this strand looks like a string of beads. Each repeating unit of this structure is known as a chromatosome; the beads are called nucleosomes. There are six nucleosomes per turn of the spiral, which means that if we completely uncoiled the ropes their length would increase sixfold, up to a total of almost 12 inches in real size for the 46 of them, or 180 miles, when magnified a millionfold.

The string of beads is itself made of even thinner twine, only 2 nm thick, and of small bundles, which make up the cores of the beads. The twine runs continuously from core to core, taking about two turns around each before moving on to the next. Should we unwind the twine from its spools, another sixfold lengthening would ensue, up to a total of about 6 feet per nucleus. Scaled up a millionfold, to the point where the nucleus forms a 30-foot sphere around us, this twine would look like ordinary string, a little less than one-tenth of an inch thick, and stretching, when all 46 pieces are strung together, over more than 1,100 miles, the distance from New York to Kansas City.

We can finish this molecular dissection by adjusting our glasses to their highest resolving power. We now see that the twine is itself double stranded and helically coiled. It is none other, in fact, than the celebrated double helix, the ultimate molecule of life, the symbol of modern biology, DNA. As to the small bundles that make up the cores of the nucleosomes, they consist of a special class of basic proteins named histones. Two molecules each of four different types of histones, designated H2A, H2B, H3, and H4, make up the core. Another histone, known as H1, is associated with the short strand of DNA that connects one bead to another. H1 molecules probably join in some sort of axial shaft when a string of nucleosomes coils into a chromatin fiber.

Most of the nuclear space is taken up by these endless coils of chromatin, which, densely packed in some places, partly unrolled in others, create a picture of utter disorder and confusion. Much that same impression might be left by the inside of a telephone switching center or a computer until we look more closely at the wiring pattern. Indeed, if we choose any part of a chromatin coil and

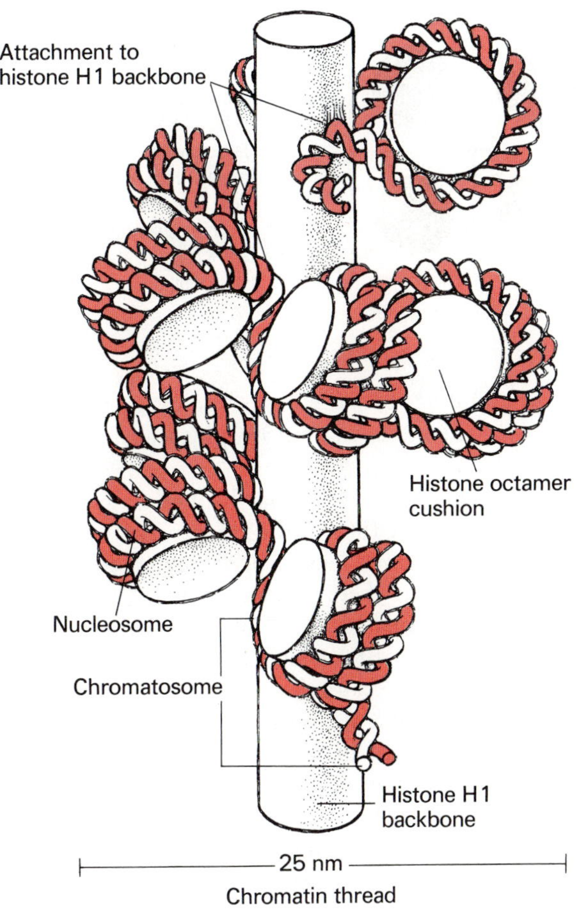

Fine structure of chromatin thread. This drawing illustrates the solenoid model, with a string of nucleosomes coiled helically, with a pitch of six nucleosomes per turn, around an axial structure believed to consist of aggregated histone H1 molecules, possibly associated with other proteins. Each nucleosome has an octameric protein core, roughly cylindrical in shape, made of two molecules each of histones H2A, H2B, H3, and H4, around which are spooled 1.75 turns of double-stranded DNA (about 50 nm). DNA runs continuously from nucleosome to nucleosome. Short linker segments of DNA between nucleosomes are attachment points to histone H1. They vary somewhat in length, depending on cell type and species. The whole repeating unit (chromatosome) contains between 160 and 240 nucleotide pairs (54-82 nm).

follow it long enough, we will find that it leads us to a protein framework called the nuclear matrix. Almost impossible to discern in an intact nucleus, the matrix is readily seen after chromatin has been selectively removed by appropriate chemical or enzymic treatment. It then appears as a three-dimensional network that pervades the nucleus. It is likely that the elements of this network originate from the protein scaffolding that supported the chromosomes during mitosis (see illustrations p. 370) and that they will participate in the rebuilding of this scaffolding at the beginning of the next mitosis. As we shall see, chromatin is anchored to the nuclear matrix in such a way as to divide each fiber into a succession of neatly separated loops, or domains. Behind the embroilment of tangled fibers, there is in fact a great deal of order.

A Bit of History

It is difficult for us, children of the Watson-Crick era, even to visualize that not so long ago most people believed DNA to be too simple a substance to have any but an accessory role in heredity. Its association with the chromosomes was, of course, known, and so was the role of the chromosomes as bearers of the genes. But only the chromosomal proteins, perhaps in the form of "nucleoproteins," were considered to have the chemical complexity and inexhaustible possibility of variation required for a genetic material.

To understand this, we must travel back more than 100 years to two events that jointly launched biology toward its greatest conquest to date, but whose fundamental interconnection remained unsuspected for a long time. One took place in the garden of an Austrian monastery, where a monk named Gregor Mendel patiently counted the types of offspring that he obtained by crossing different varieties of peas. Out of this work came the "laws of Mendel," first published in an obscure local journal in 1866, but largely ignored until the turn of the century, when they were rediscovered. From them arose the science of genetics, a term derived, like genesis, from the Greek verb *gennân*, to beget, to generate. The essence of

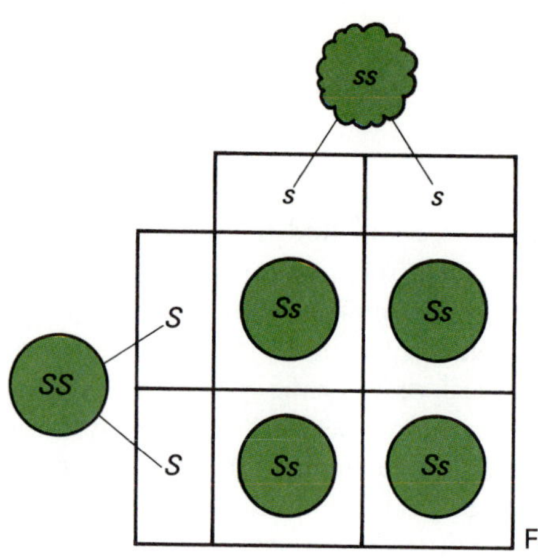

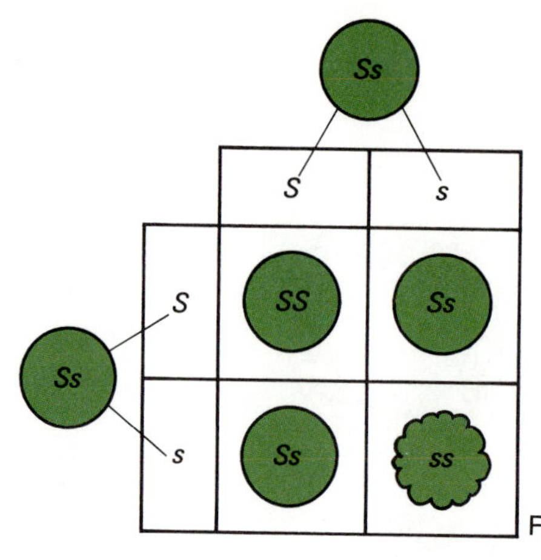

One of Mendel's crossing experiments. Cross-fertilization of two varieties of peas differing by shape of seeds gives first-generation (F_1) hybrids that produce only smooth seeds. Self-fertilization of the F_1 hybrids gives second-generation (F_2) hybrids, of which three-quarters produce smooth seeds and one-quarter wrinkled seeds. The diagram shows how these results can be interpreted by assuming that the shape of a seed is determined by a pair of "units" (alleles), each of which is either dominant (*S*) or recessive (*s*), and that allelic units are segregated in germ cells and joined at random upon fertilization.

Mendel's discovery, expressed in modern terms, is that discrete hereditary traits—as defined by the color, shape, texture, or other morphologically recognizable property of certain organs that is transmitted from one generation to the next—are determined by units, the genes, that always go in pairs, the alleles (Greek *allos*, other), of which one is provided by the male, the other by the female parent. The character of the offspring depends on what alleles, dominant or recessive, it receives from its two parents. Individuals with the same two alleles are designated homozygotes (Greek *homoios*, same; *zygos*, pair, couple). They are what we would call "pure race" (for that particular trait) and upon inbreeding continue, generation after generation, to express the character determined by their identical alleles. Heterozygotes express the dominant character, but only three-quarters of their offspring do so; one-quarter express the recessive character. Why this must be so is obvious. If *S* is the dominant and *s* the recessive allele, mating two heterozygotes, *Ss*, with random selection of the allele donated by each parent, will produce, in equal proportion, the combinations *SS*, *Ss*, and *sS*, all of which express the dominant character, and *ss*, which does not.

This is what Mendel found more than a century ago, except, of course, that he found it the other way round. He discovered the 75/25 distribution empirically and deduced the allele model from it. Inevitably, numerous exceptions and complications to this simple model were recognized subsequently. The most important one in relation to the present overview is the presence of linkage, or coupling, between genes, indicating that genes are not independent units, but are physically associated as sets.

The other major event that happened in those early days took place in 1869 in the laboratory of the German physiological chemist Felix Hoppe-Seyler, at the University of Tübingen, where a young Swiss physician, Friedrich Miescher, isolated a previously unknown substance from nuclei. His starting material consisted of pus cells isolated from septic bandages. In addition to carbon, oxygen, hydrogen, and nitrogen—the four elements commonly found in proteins and other biological materials—the new substance was found to contain large amounts of phosphorus. Miescher named his discovery nuclein, a word that was later changed to nucleic acid after the strongly acidic character of the substance was recognized. It took more than sixty years and the efforts of many celebrated chemists—in particular, the German Albrecht Kossel and the American Phoebus Levene—to unravel the structure of the new substance and of its components. In the meantime, the presence of a similar material in the cytoplasm was recognized. The two nucleic acids were first known as thymonucleic (from thymus, one of the richest sources of animal DNA) and zymonucleic (from the Greek word for yeast, an early source of RNA). Later, when the nature of their respective constituent pentoses was uncovered, they were renamed deoxyribonucleic and ribonucleic acids.

What brought Mendel's units and Miescher's nuclein together was the work of the cytologists who discovered

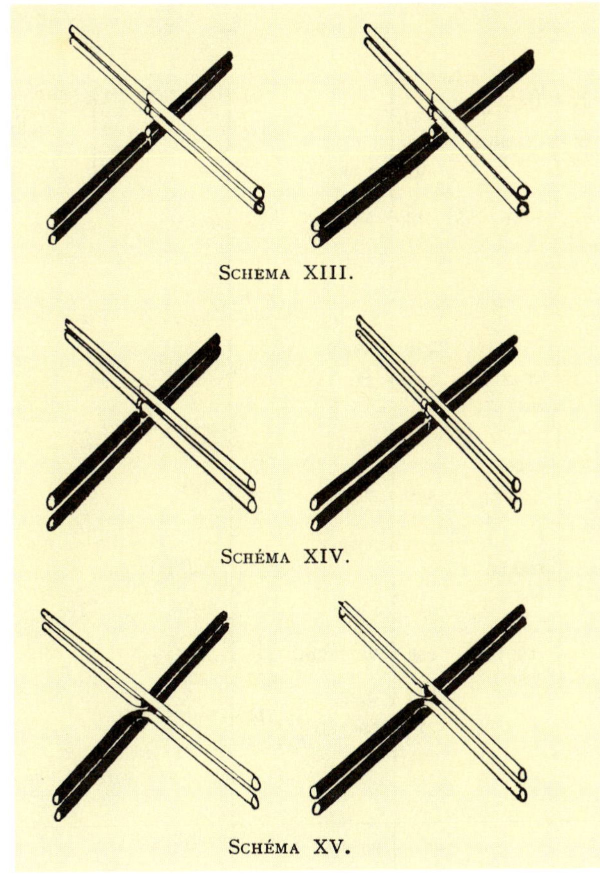

Crossing-over, or chiasmatypie. The drawing by Frans Janssens reproduced here was originally published in *La Cellule* in 1909. It illustrates the author's interpretation of the chi figure formed by two homologous pairs (dyads) of duplicated chromosomes (chromatids). After one chromatid of each pair has been broken at the point of overlap, the segments recombine crossways (for a description of meiosis, see Chapter 19).

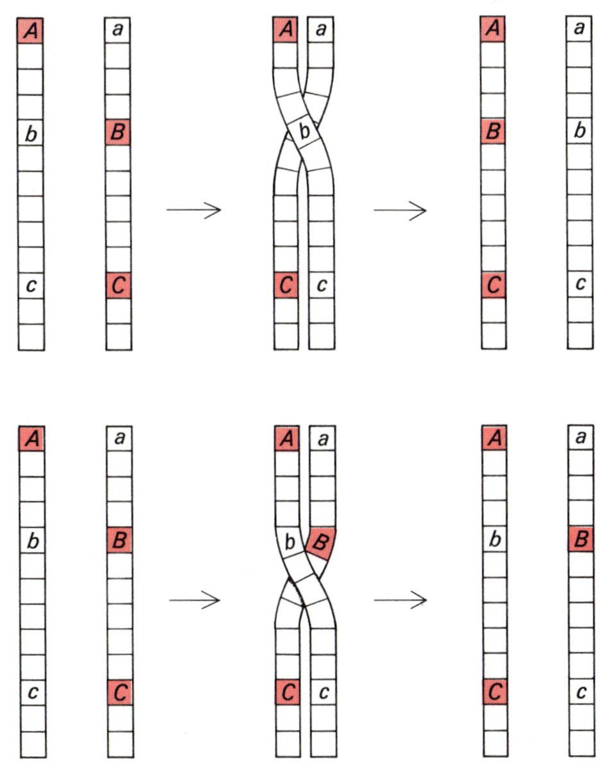

The genetic implication of Janssens's theory of chiasmatypie, according to Thomas H. Morgan. Genes situated on distinct homologous chromosomes become recombined by crossing-over. In both cases, dominant allele *C* recombines with *A*, whereas *B*, which is situated closer to the *A* locus, recombines with *A* only in the first case. Thus, the probability of two genes situated on distinct homologous chromosomes recombining on the same chromosome by crossing-over is proportional to the distance separating the loci that they occupy on the chromosome. Hence, the measurement of recombination frequencies can be used for chromosome mapping.

the chromosomes and described the behavior of these structures in mitotic division and, especially, their numerical halving in meiosis. As we will see in Chapter 19, chromosomes go in pairs, as do the postulated alleles. They duplicate at every cell division to provide identical sets for the daughter cells, as should bearers of heredity. And, during maturation of the germ cells by meiotic division, only one of each pair is retained in either spermatozoa or ovules, so that, upon fertilization, fusion of the male and female nuclei recreates a full set of chromosomes in which every pair has one member contributed by each parent, which is exactly what alleles are expected to do.

Furthermore, the size and shape of chromosomes are consistent with the concept of a linear association of many genes, indicated by the linkage phenomenon. This became even more evident after a Belgian biologist, Frans Janssens, observed that during meiosis paired chromosomes often "cross over" each other and exchange large segments before separating again. This process, called chiasmatypie by Janssens (Greek *chiasma*, X-shaped cross), and later renamed crossing-over, provided an explanation for the puzzling fact that linkage is not an all-or-none phenomenon, but a graded one, depending on the genes involved. One can readily imagine that the probability of two linked genes becoming unlinked by crossing-over must increase with the distance separating them along the length of the chromosome. Further details on crossing-over will be given in Chapters 18 and 19.

The major architect of the grandiose synthesis between genetics and cytology was Thomas Hunt Morgan, an American scientist who turned the quantitative measurement of the tightness of linkage between two genes into a method for estimating the distance that separates them on their bearer chromosome. Using this technique of genetic mapping, Morgan and his co-workers succeeded, between the years 1910 and 1922, in locating hundreds of genes on the four chromosomes of the small fruit fly *Drosophila melanogaster*. Later, these maps were given a true physical basis through correlation with the fine structure of the giant chromosomes that are in the salivary-gland cells of the fruit fly. Genes could thereby actually be localized to transverse bands on these chromosomes. Real flesh was put on Mendel's "units," but the nature of that flesh remained a puzzle.

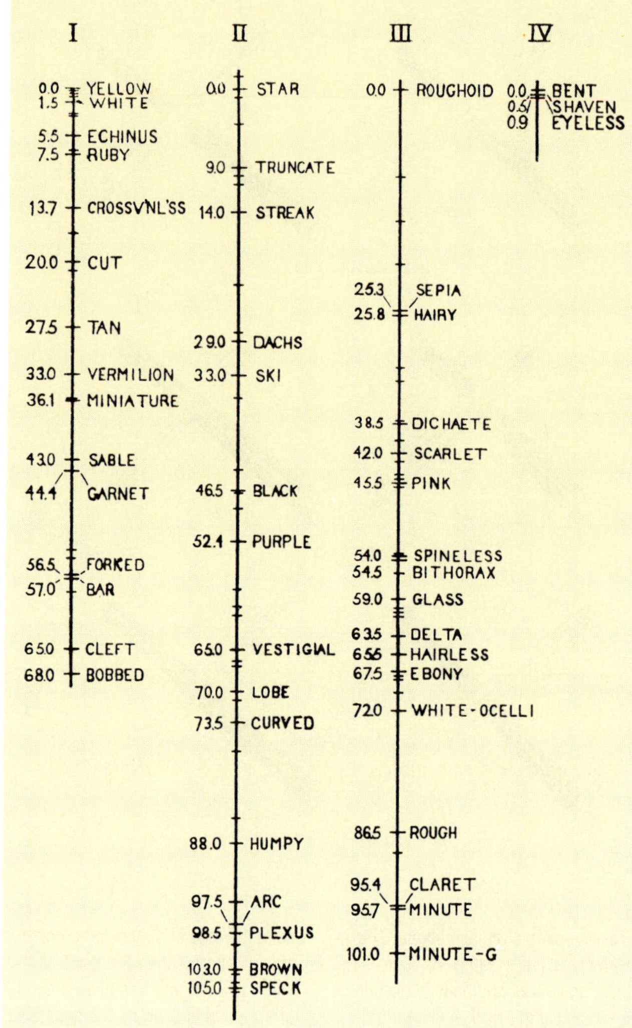

Partial genetic map of the four chromosomes of the fruitfly *Drosophila melanogaster*, as charted by the Morgan school. Genes are named according to known mutant character. The numbers at the left are distances inferred from recombination frequencies.

Here, again, cytology provided the necessary bridge, but this time with chemistry. By using various stains, enzymic attacks, and other cytochemical means, investigators attempted to characterize chemically the structures that they saw through the microscope. This search demonstrated that the chromosomes consist mainly of DNA and protein. But it took several decades before it was recognized that the DNA component, and not the protein, is the actual bearer of genetic information. The final turning point was provided by what has become one of the greatest classics of scientific literature—a paper published in 1944, with, as authors, three scientists of The Rockefeller Institute for Medical Research in New York, Oswald T. Avery and two young colleagues, Colin MacLeod and Maclyn McCarty. The origin of Avery's work was an observation made in 1928 by Fred Griffith, a medical officer of the British Ministry of Health. Experimenting with pneumococci, the agents of bacterial pneumonia, Griffith had found that mice inoculated with living bacteria of a nonvirulent strain (called R, for rough, because it forms rough-looking colonies) and with dead bacteria of the virulent S (smooth) strain would die of pneumonia and yield at autopsy a culture of live S-type pneumococci. This remarkable phenomenon, called bacterial transformation, was subsequently reproduced in vitro by treating R-type bacteria with extracts of S-type bacteria. It was known that S-type pneumococci owe their virulence to the possession of a specific capsular polysaccharide, which is absent in the R type. Transformation, however, did not represent restitution of the pathogenic polysaccharide itself, as virulence persisted through countless bacterial genera-

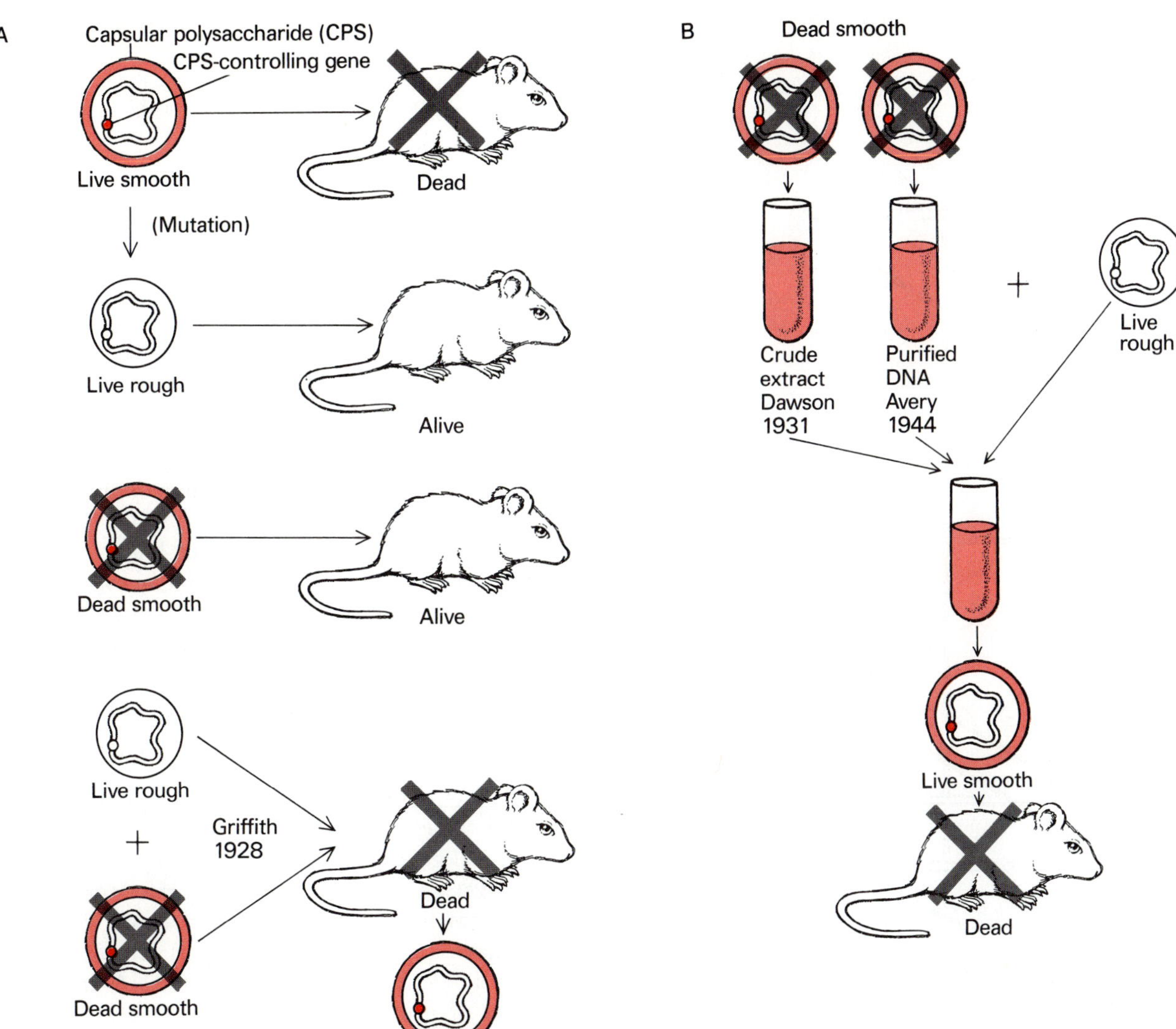

Bacterial transformation.

A. S pneumococci form smooth colonies, are surrounded by a polysaccharide capsule that protects them against destruction by the immune system, and are therefore highly virulent. R mutants form rough colonies, lack the gene for making the polysaccharide capsule, and are nonpathogenic. Heat-killed S bacteria are likewise nonpathogenic. However, as shown by Griffith in 1928, the inoculation of live R bacteria together with heat-killed S bacteria results in a lethal infection. Live S bacteria are recovered from the bodies of the infected animals.

B. Transformation of R into S pneumococci can also be effected in vitro with crude extracts of S bacteria, as first observed by Dawson in 1931. Purification of the transforming principle and its chemical identification as DNA were first reported in 1944 by Avery, Macleod, and McCarty.

tions. What was restituted or transmitted, therefore, was the ability to make this polysaccharide and to pass on the property to progeny. It had to be viewed as the insertion into the R bacteria of the corresponding gene, borrowed from the S type. Hence the immense interest in the chemical nature of the transforming factor. The famous 1944 paper was the climax of a long, patient attempt at purifying this factor. It established beyond reasonable doubt—except that of the critic willing to go to any length brandishing Avogadro's number—that the transforming factor was a DNA, not a protein.

The first schematic representation of the double helix, as it appeared in a paper published by Watson and Crick in the April 25, 1953, issue of *Nature.*

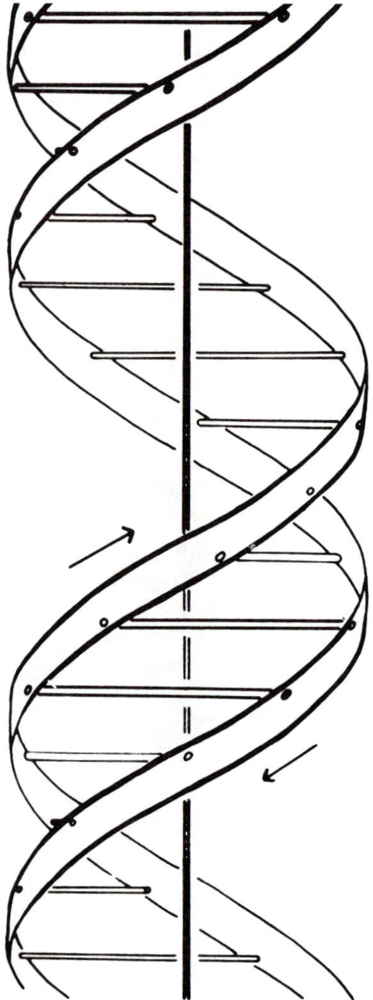

By 1950, virtually everybody was convinced, and the race for DNA was in full swing. The winners: the American James Watson, 25, and the Englishman Francis Crick, 37, with a small, unobtrusive paper occupying less than two pages in the April 25, 1953, issue of the British magazine *Nature* and titled "A Structure for Deoxyribose Nucleic Acid." It introduced the double helix to the world.

When Watson and Crick made their epoch-making proposal, the structure of DNA as a very long polynucleotide chain was common knowledge. In addition, they had available three types of facts:

> Fact one, physical, was provided by X-ray diffraction data indicating a helical structure for the fibrous molecule, with parametric dimensions suggestive of double-strandedness rather than single-strandedness.
>
> Fact two, chemical, came from the determinations of the Austrian-American biochemist Erwin Chargaff, who had found that, in samples of DNA of widely different origins, the adenine content was always equal, within limits of experimental error, to the thymine content, and the content of guanine to that of cytosine. Hence "Chargaff's rules": A = T; G = C.
>
> Fact three, biological, was the necessity that the molecule be capable of inducing its own duplication.

These facts were known to the other runners in the race, who were all busily trying to build a molecular model of the genetic molecule. The flash of inspiration that put Watson and Crick on the correct track comes down to two words: *base-pairing.* You will remember that we have already encountered this phenomenon as playing a key role in the structure of RNA and in the translation of genetic messages. Historically, however, base-pairing was first discovered through a consideration of DNA, where it explained Chargaff's rules and at the same time suggested a mechanism for self-duplication. Each molecule would consist of two complementary strands, representing, so to speak, the same information in positive and negative. As in photography, the positive could serve as template for a new negative, the negative for a new positive, leading to the formation of two identical double-stranded molecules in which, incidentally, one strand would originate from the parent molecule and the other would be newly made (semiconservative replication). Once this intuitive concept had taken shape, it remained for the two investigators to find, by empirical model-building, a molecular arrangement that fitted base-pairing into a helical structure of appropriate dimensions without imposing any

Space-filling model of the double-helical DNA molecule.

undue strain on the distances and angles known to characterize each interatomic bond. The thrill and drama of this momentous discovery, and the climate of intense competitiveness and some of the less pleasant aspects of human behavior that surrounded it, are best appreciated in Jim Watson's own candid account, *The Double Helix*.

The Double Helix

The main structural characteristics of DNA have already been described. Like RNA, DNA is a polynucleotide, made by the linear association of 5′-mononucleotides linked by 3′-phosphodiester bridges. It differs chemically from RNA in two respects: (1) the pentose is 2-deoxyribose instead of ribose—that is, ribose without an oxygen in position 2; (2) one of the two pyrimidine bases is thymine instead of uracil—that is, uracil with an additional methyl group in position 5. These differences have little effect on a number of structural properties. In particular, they do not affect base-pairing, which is essentially similar in DNA and RNA because the A—U and A—T pairs have the same shape and strength. But certain chemical properties are greatly influenced by the differences between the two nucleic acids: RNA is much more sensitive to alkaline hydrolysis than is DNA; the latter, in turn, can be selectively deprived of its purine bases by mild acid treatment, a property that is made use of in the cytochemical Feulgen reaction, whereby DNA can be selectively stained in tissue sections.

The most important difference between DNA and RNA lies in their strandedness. Apart from some viral forms, all the RNA in the world is single stranded; it base-pairs mostly with itself, to form loops closed by short double-helical complementary segments. It is just the opposite for DNA. With the exception of a few small viruses, all the DNA in the world is made of two complementary antiparallel strands. It is double helical over its whole length and, in its most common form, known as B, appears as a regular, right-handed, spirally twisted thread with a diameter of 2 nm and a pitch of 10 nucleotide pairs per turn, each pair occupying an axial distance of 0.34

nm (see the illustration on p. 256). Thus, 1 μm of DNA has about 300 turns, 3,000 nucleotide pairs, and a molecular mass of circa 2 million daltons (the average molecular mass of a nucleotide pair, sodium salt, is 660 daltons). These relationships have been verified directly, because techniques are available for visualizing DNA threads with the electron microscope. DNA can also adopt other shapes, including a newly discovered left-handed Z form, rare but possibly of biological significance.

Another difference between DNA and RNA lies in their length. In contrast with most RNAs, natural DNA threads are polycistronic. (The cistron is the molecular equivalent of a unit-gene, a sequence coding for a single polypeptide or RNA molecule.) Quite often, in the world of viruses and of prokaryotes, all of the genome is carried by a single, circular DNA thread. The length of this circle varies between 1.5 and 80 μm in viruses, and reaches 1 mm in bacteria. In eukaryotic cells, as we have seen, the DNA threads are considerably longer, up to nearly 10 cm. There is one such thread per chromosome, where it is coiled around small histone cushions to form a string of chromatosomes. The length of DNA per chromatosome varies somewhat from one species to another, with an average of about 200 base pairs (68 nm), of which a constant stretch is coiled toroidally around the histone cushion (nucleosome) and the remainder serves as internucleosomal link. As we have seen, the chromatosome string is itself twisted into a helical chromatin fiber. Depending on the state of the nucleus, this fiber is variously unrolled and dispersed. At the time of cell division, it coils and supercoils into the compact, rod-shaped structure of the chromosome. Unlike prokaryotic DNA, eukaryotic DNA is not circular. Those who like figures may be interested in learning that each of us is the possessor of more than 400 billion billion (4×10^{20}) nucleosomes and of about 20 billion miles of DNA—enough, if completely stretched, to run back and forth 100 times between the earth and the sun!

For obvious reasons, double-stranded DNA is more rigid than single-stranded RNA. Nevertheless, a DNA thread can assume all sorts of contorted shapes. The way it coils around nucleosomes and twists into chromosomes makes this abundantly clear. But these contortions generate strains and create topological problems that make even the most intricate combination of Möbius strips look like child's play. These problems are real, not just conceptual. They are encountered by the enzyme systems that transcribe and duplicate DNA and by the proteins that package it. They call for the participation of special enzymes for loosening structures, unwinding coils, and disentangling knots. We will run into several of these topoenzymes as we watch DNA actually imparting genetic information.

The stability of a double-stranded DNA thread depends on two types of forces: (1) the electrostatic hydrogen bonds that determine base-pairing; and (2) hydrophobic interactions between the planes of the stacked base pairs. Opposing these stabilizing forces are the electrostatic repulsions between the negative charges of the peripheral phosphoryl groups and the jostlings caused by thermal agitation. At high enough temperature, the disruptive forces prevail and the two strands separate. By analogy with proteins, this phenomenon is called denaturation. An alternative term is melting. It is readily detected, for instance, by a sudden decrease in viscosity of the solution, explainable by the decrease in the rigidity of the threads. The temperature at which DNA denatures is called the melting point. It increases with the GC/AT ratio, because the GC pair is stronger than the AT pair. It is of the order of 85°C for human DNA.

Denaturation of DNA is a reversible phenomenon. If a solution of heated, denatured DNA is cooled somewhat below its melting point and kept at this temperature long enough to give complementary strands an opportunity to find each other, the molecules will rejoin. Understandably, the time required for this depends on the length and variety of the different molecules present. But even highly complex mixtures that contain thousands of different genes can be reconstituted under appropriate conditions.

We cannot watch this phenomenon in the cell, because no cell can stand a temperature at which its DNA starts denaturing. But just imagine yourself inside one of those cooling DNA solutions, observing the rebirth of beautifully undulating, semirigid, double-helical threads from the jumble of billions of intertwisted single strands. It is a mind-reeling spectacle. Innumerable temporary connections form by partial base-pairing, only to be torn apart again immediately by the violence of thermal agitation until chance brings together two complementary sequences that belong to authentic molecular partners. With lightning rapidity, the two strands spiral around each other, cementing their association before another turbulent collision comes to break it again. And so, one after the other, the double strands reform.

Were it not an experimentally established fact, we would never believe that this kind of molecular reunion by blind, random groping could ever take place with threads of such length and diversity. But it does, in a matter of hours, showing that, however hard we may try to penetrate it, the molecular world must necessarily remain entirely beyond the powers of our imagination owing to the incredible speed with which things happen in it.

In practice, reversal of denaturation has produced a valuable application known as hybridization. In principle, this method is designed to allow two different DNAs or, alternatively, a DNA and an RNA, to renature together. To the extent that long enough complementary sequences exist in the two species, they will join to form hybrid duplexes containing one strand of each. Assuming that such hybrid molecules can be recognized and evaluated—and several techniques allow this—hybridization can be used for estimating the degree of kinship between two DNAs or for fishing out a gene with the help of its transcribed mRNA, and so on. For example, an early experiment of this sort established that human DNA is very close to monkey DNA, less so to mouse DNA, even less so to fish DNA, differing more and more as evolutionary distance increases. It was one of the first applications of molecular biology to the study of phylogeny. In association with electron microscopy, hybridization has been turned into a tool of almost uncanny incisiveness and simplicity. Imagine two chains, A and B, complementary over their whole length except for a one-kilobase segment that exists in A—say, between bases 3,000 and 4,000—and has been excised from B. In the electron microscope, the A-B hybrid will appear as a double-stranded thread bearing a single-stranded loop 0.3 μm long appended to it 1 μm away from one of its ends. Thus we can, with a simple ruler, find out how much has been removed from a chain and where (see pp. 308–309: mRNA splicing).

Transcription in the Nucleolus

From where we are, we can discern little more than the fine structure of the tangle in which we are caught. As luck would have it, when the nuclear envelope reformed around us, we found ourselves in one of the densest parts of the nucleus, a zone of what is called heterochromatin. The DNA in it is all bundled up and is functionally inactive. If we want to see some action, we must try to crawl out of this thicket and find one of the clearings marked on the nuclear maps left to us by those explorers who have looked at thin sections. They have described two kinds of openings: euchromatin (Greek *eu*, good), irregular areas where chromatin is dispersed instead of condensed as it is in the "other chromatin"; and nucleoli, large, intranuclear organelles rich in protein and RNA. It will be helpful to move first to a nucleolus.

The term nucleolus is the Latin diminutive of nucleus, which, you remember, is itself the diminutive of *nux*. Thus, nucleolus literally means "small small nut"; it is a nucleus within a nucleus. It may have looked small to the early observers but, at our newly discovered molecular scale, the nucleolus appears quite large. It is a roughly spherical structure, irregularly outlined, about 2 μm in diameter—almost 7 feet at our millionfold magnification; we can just fit in. It has been known for more than a hundred years, and all sorts of structural details in it have been described. Observations were also made that suggested a major role of the nucleolus in growth and biosynthesis. The most impressive example is that of the egg cell of certain amphibians, in which the nucleus measures as much as 0.5 mm (0.02 inch), and as many as one thousand nucleoli may be seen in a single nucleus.

But nothing that classical observers were able to distin-

Fine structure of the nucleolus as first seen in 1952. This thin section through a rat-liver cell shows a portion of the nucleus (N), with a dense nucleolus near the nuclear envelope (M).

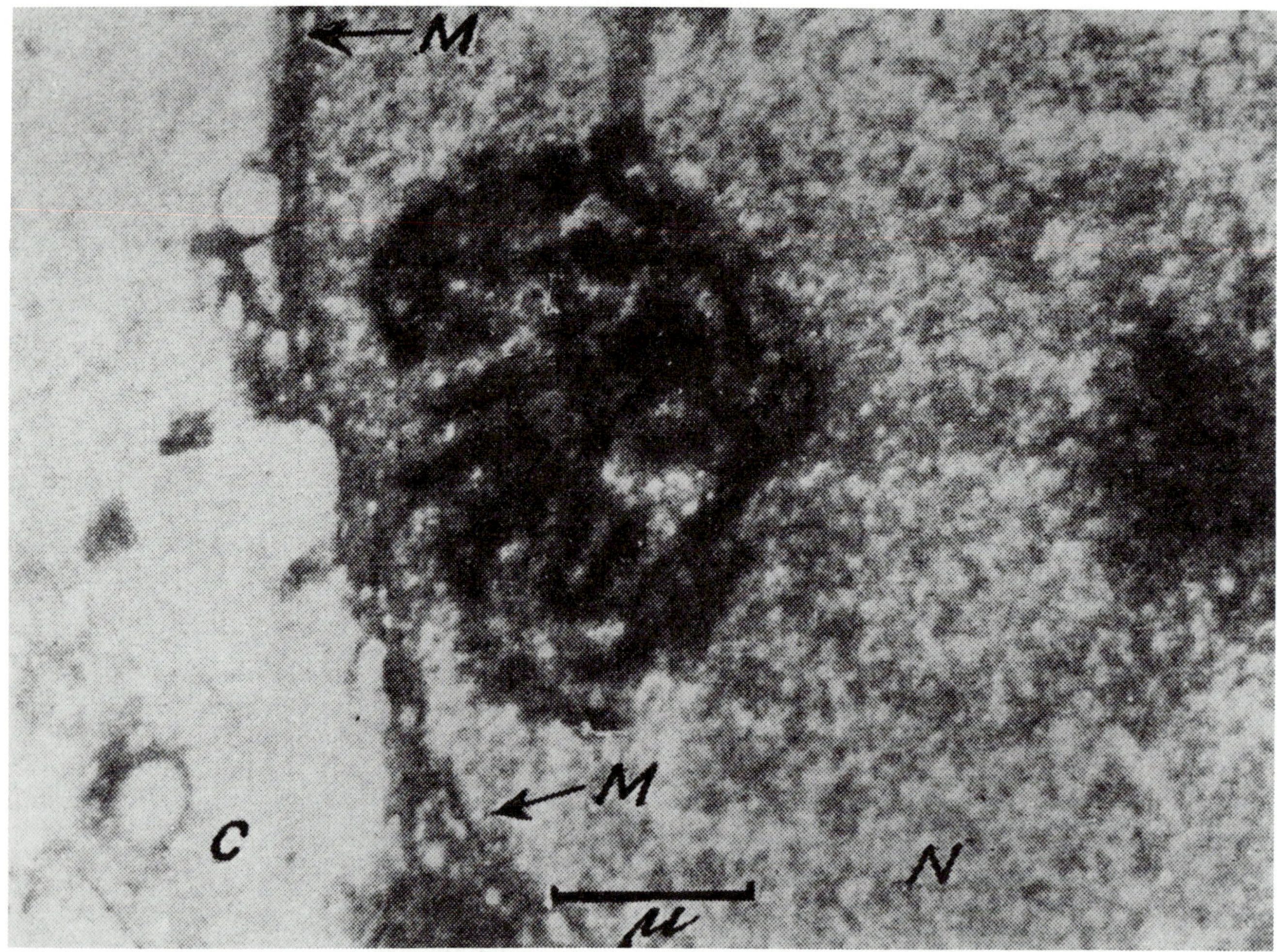

guish could have prepared us for the spectacle revealed by our molecular magnifying glasses. At first sight, it seems as if we have landed in the middle of some strange rain forest, overgrown with huge ferns garlanded by lianas. Or it could be the feather-lined nest of a giant bird. The whole space is crammed with long, delicately sculptured structures made of a central stalk, some 6 to 8 μm long (20–30 feet at our magnification) covered over a substantial part of its length with slender lateral digitations, or whiskers. Planted from 0.04 to 0.05 μm apart, these whiskers are of regularly increasing length, thereby giving the structure its fern-frond appearance. Each whisker is connected to the stalk by a knoblike bulge.

As many as a hundred or more such "fern fronds" may be found compressed within a single nucleolus. They are all strung together head-to-tail and joined by their stalks, which make up a single, uninterrupted thread, whose ends disappear in the dense heterochromatin brush that

Electron micrograph of dispersed fibrillar material from nucleoli of *Xenopus laevis* oocytes.

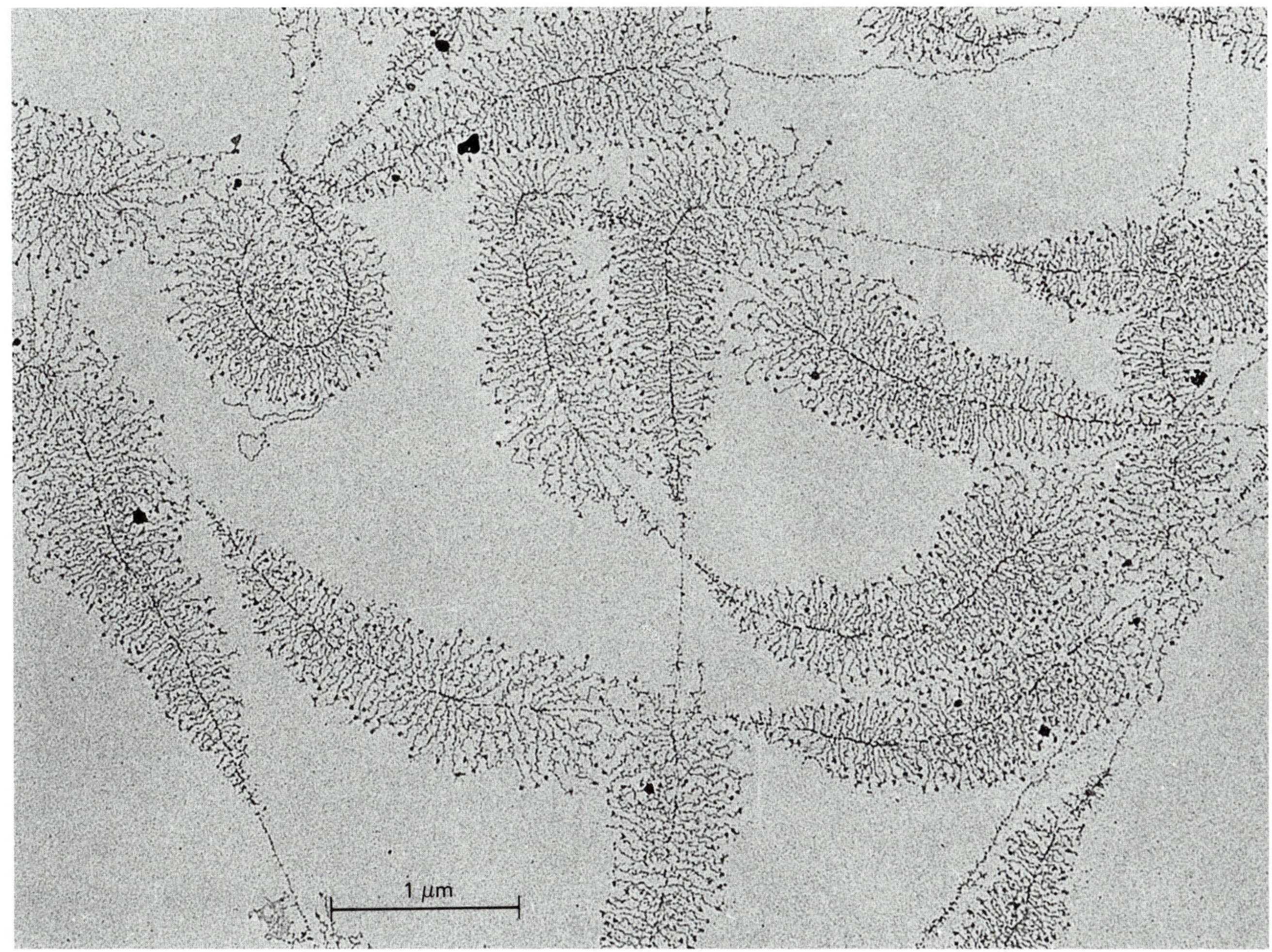

borders the nucleolus. Bewhiskered and naked segments alternate regularly along this garland. To fully enjoy the beauty of this remarkable structure, we should really open the nucleolus, as some explorers have done, and allow the garland to uncoil freely. As it presents itself to us in the intact organelle, it forms an inextricable tangle. Count about half a mile for the main stalk, plus more than 10 miles total whisker length. It needs a lot of coiling to pack all this within a 7-foot box. The nucleolus may look like a clearing to somebody crawling out of the surrounding heterochromatin. But it is a far cry from what one might call an opening. Even so, it provides us with our best opportunity for witnessing a phenomenon of absolutely central importance, and we must take full advantage of it. Look closely, and you will see that each fern frond is throbbing with activity. The knobs move, the whiskers lengthen, the stalks rotate, all in perfect synchrony.

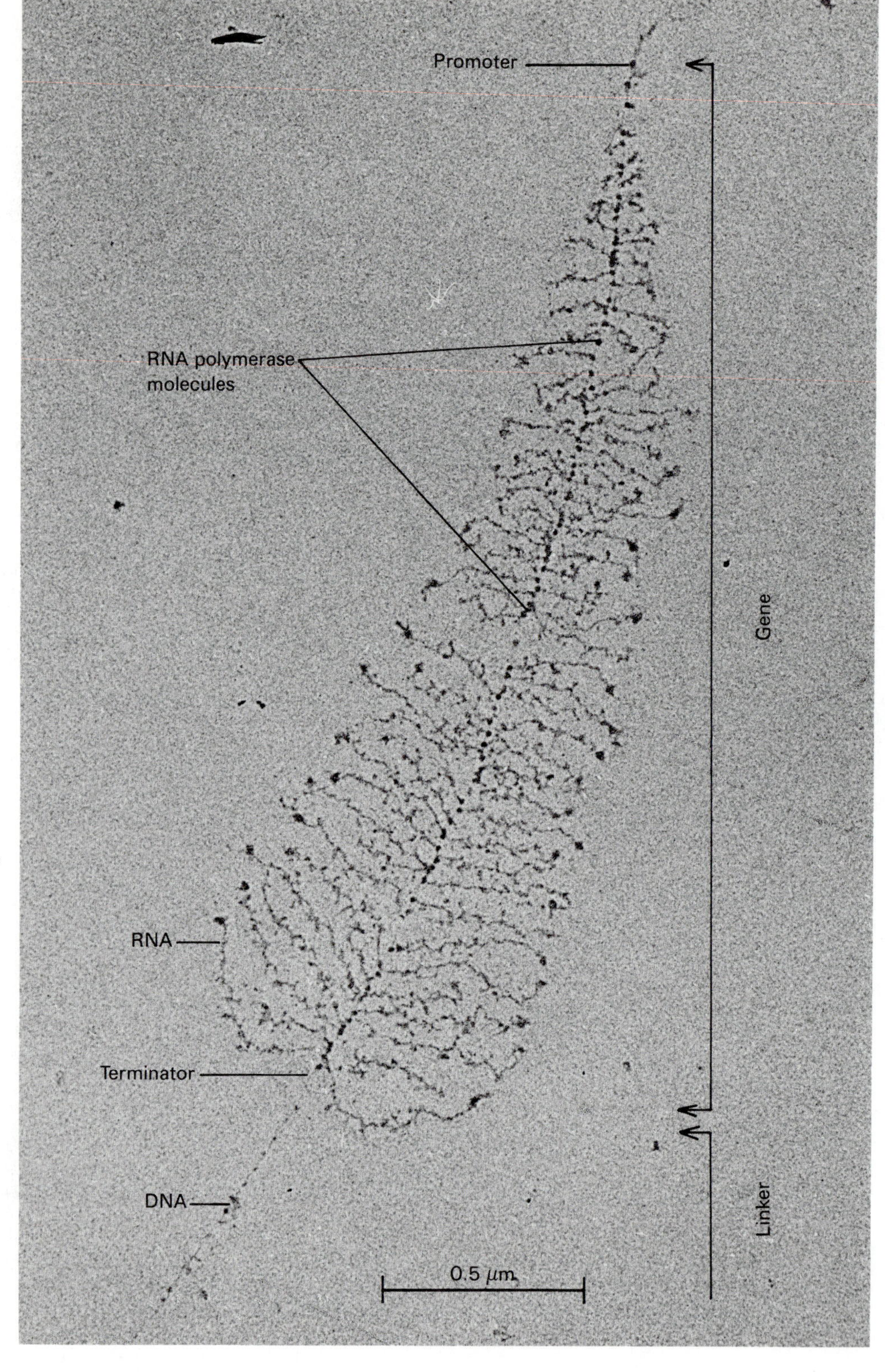

Functional significance of the "fern frond." The knobs on the central stalk are molecules of RNA polymerase in the process of transcribing a gene. The stalk is DNA; the progressively lengthening whiskers are growing RNA transcripts. The bare part of the stalk is an untranscribed linker between genes.

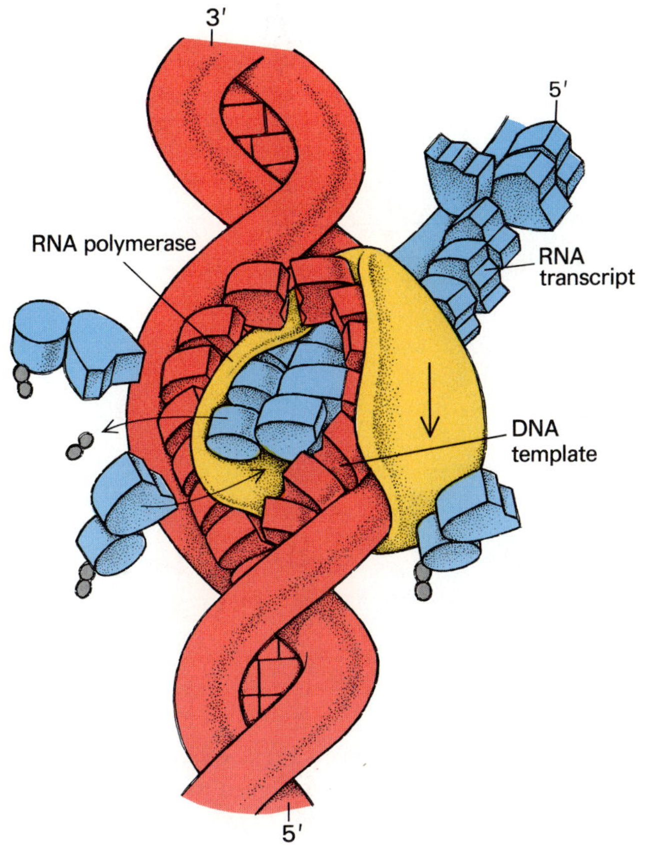

A view of the transcription process. The regular double-helical structure of DNA is disrupted locally by the insertion of RNA polymerase. This enzyme wraps around one of the DNA strands and uses it as template for the assembly of a complementary RNA chain from nucleoside-triphosphate precursors. The enzyme moves along the template DNA in the 3′ to 5′ direction. RNA lengthens in the 5′ to 3′ direction (tail growth). DNA resumes its double-helical configuration as the enzyme moves on.

We can gain some clues to the meaning of these concerted motions by watching the knobs. They course along the central stalk in the direction of increasing whisker length at the appreciable speed of about 1 μm per minute, which, magnified a millionfold, amounts to two-thirds of an inch per second. As the knobs move, the whiskers attached to them increase in length in a manner that is exactly commensurate with the distance traveled along the stalk. When a knob reaches the base of the fern frond, it falls off and releases its attached whisker, which by then is from 4 to 5 μm long. At the same time, a fresh knob binds to the stalk at the tip of the frond and starts sprouting a new whisker. What we are watching, therefore, is some sort of *perpetuum mobile* of knobs along the stalk of each fern structure and, associated with it, the growth and release of whiskers. Each fern frond turns out some 20 to 25 whiskers per minute, to give a total for the whole nucleolus of 2,000 or more per minute. In length, this adds up to about 1 cm (5–6 miles at our magnification). This material fills the spaces between the bundled-up fern fronds, where it undergoes a complex form of processing before being conveyed to the cytoplasm.

With better adjustment of our focus, we have no difficulty recognizing the main components of this remarkable factory. The central stalk, semirigid and 2 nm thick, shows the characteristic double-helical structure of DNA; it is a fully extended DNA thread. The lateral whiskers are thinner and more flexible; they are made of single-stranded RNA. The knob connecting the RNA whisker to the DNA stalk is a voluminous protein, wrapped collarwise around the DNA. This protein projects an inner spur between the two DNA strands and keeps the 3′ end of the RNA whisker closely aligned against one of the strands. The process we are looking at is none other than the biosynthesis of RNA. The knob at the growing end of the chain is the enzyme that catalyzes this process. Its association with DNA reflects the fact that it is guided in the performance of its biosynthetic job by instructions supplied by the DNA. We have, in fact, just been afforded our first glimpse of transcription.

Chemically, the growth of an RNA chain is a simple process. It takes place by successive nucleotidyl transfer

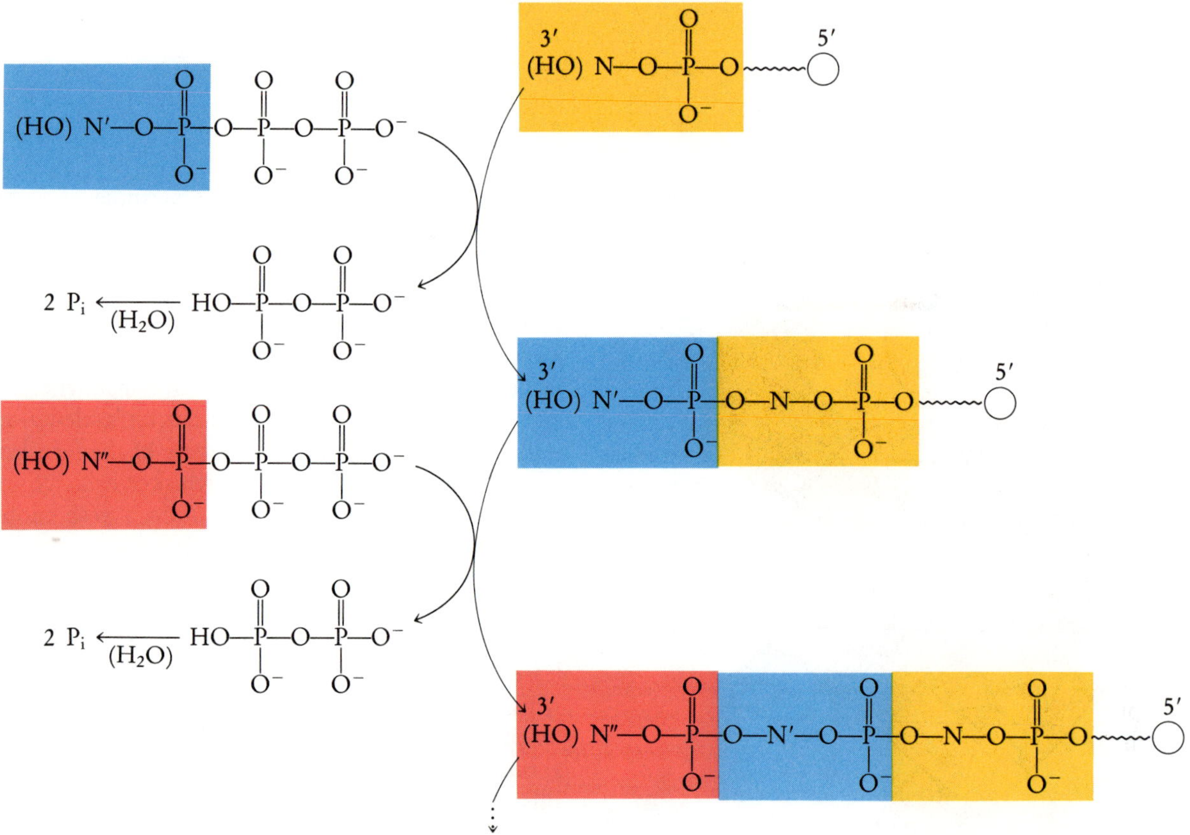

from a nucleoside triphosphate to the 3′-terminal hydroxyl group of the growing chain, as indicated in the scheme shown above. It is an instance of a single-step biosynthetic process dependent on a β_p attack (Chapter 8). Note that we are dealing here with a case of tail growth: activated building blocks are added one by one to the growing chain. This is in contrast to protein synthesis, which takes place by head growth—that is, by repeated transfer of the activated growing chain to (activated) building blocks.

The enzyme that synthesizes RNA is called RNA polymerase. Its substrates are ATP, GTP, CTP, and UTP, which may be considered the corresponding nucleoside monophosphate building blocks AMP, GMP, CMP, and UMP, activated by two successive γ_d phosphoryl transfers from ATP:

$$\text{NMP} + \text{ATP} \longrightarrow \text{NDP} + \text{ADP}$$
$$\text{NDP} + \text{ATP} \longrightarrow \text{NTP} + \text{ADP}$$

The biosynthetic process releases pyrophosphate, which, as is customary, is hydrolyzed to two P_i. In other words, to add an NMP to the 3′ end of a growing RNA chain, the cell uses two terminal pyrophosphate bonds of ATP, or 28 kcal per gram-molecule.

So far so good. But there remains the problem of which of the four available nucleotidyl groups to add to the 3′ end of the growing chain at each step. Instructions to this effect come from the DNA by way of the magic cipher. The RNA polymerase is constructed in such a way as to align its two substrates, the growing chain and the donor NTP, in antiparallel fashion alongside a DNA strand. It will catalyze the transfer only if perfect complementarity obtains between the paired bases. As soon as the right NTP comes along and its nucleotidyl group is added to the growing chain, the enzyme moves one notch, to bring its active center back in the right strategic position with respect to the new 3′-hydroxyl end of the growing RNA chain.

In this process, the DNA instructions are read, base by base, in the 3′⟶5′ direction, while the RNA strand grows in the 5′⟶3′ direction. Transcription proceeds until the enzyme encounters a terminator on the DNA—a specific base sequence that somehow causes the enzyme to stop. A special factor, known as rho, helps the enzyme off the DNA and releases the RNA, with the concomitant hydrolysis of ATP, which presumably provides the necessary energy. To start a new chain, the enzyme needs the assistance of another factor, called σ, which guides it to a special initiating binding site on the DNA, signaled by a particular base sequence called promoter. Unlike DNA polymerase, which will be considered in the next chapter, RNA polymerase has no need for a preformed chain to serve as first acceptor. It can use a single nucleoside triphosphate: in most cases, ATP or GTP.

If we put all this information together, we may now identify the DNA segment that is being transcribed by RNA polymerase as a gene—a chemically coded element of genetic information. Transcription is the mechanism whereby this information is copied in RNA language. It is the obligatory first step in the expression of the gene into what, in classical genetics, is called a phenotypic character (Greek *phainein*, to appear). Our nucleolar gene has a length of from 4 to 5 μm, or from 12,000 to 15,000 base pairs. It is flanked on its 3′ end by the promoter sequence and on its 5′ end by the terminator sequence. Beyond the terminator sequence is a stretch of untranscribed DNA—the naked part of the stalk—up to several thousand base pairs long, which ends with the promotor sequence of the next gene. This segment is called a spacer or linker. It may play a variety of roles in gene regualtion in addition to bearing the promoter and terminator sites.

Nucleolar transcription is a unidirectional, one-sided process: only one of the two DNA strands is transcribed. This, we will see, is a general, though not absolute, property of transcription. Its rate of progress is of the order of 1 μm per minute, which means that the RNA chain grows at the rate of fifty nucleotides per second, or that it takes RNA polymerase no more than two-hundredths of a second to try various nucleoside triphosphates for fit, find the right one, substitute one bond for another, move on

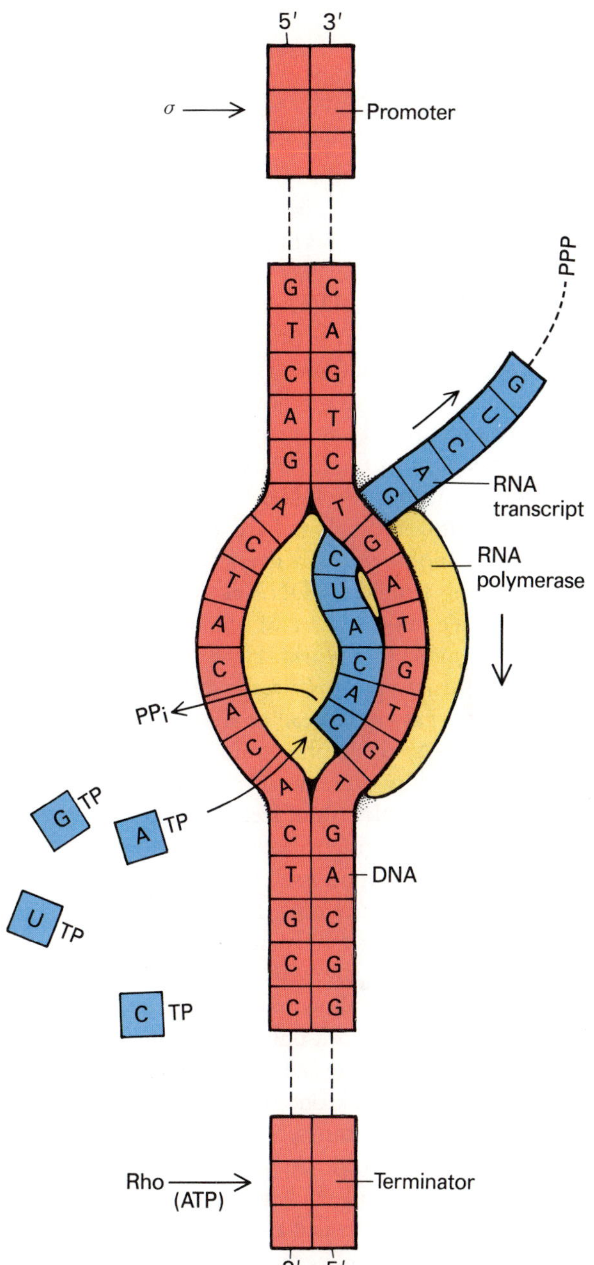

Schematic representation of transcription. RNA polymerase separates the two strands of DNA and assembles an RNA chain complementary to one of the strands (and therefore identical in sequence with the other strand except for the replacement of T by U). The initial insertion of the enzyme is determined by a special promoter sequence and is helped by the sigma factor. A terminator sequence signals the detachment of the enzyme, which is assisted by the rho factor and requires ATP. Note that the 5′ end of the RNA strand bears a nucleoside triphosphate that has served as the initial acceptor.

Topology of transcription. The progress of RNA polymerase causes the positive supertwisting of DNA in front and its negative supertwisting behind. A nicking-closing enzyme relieves the strain on DNA.

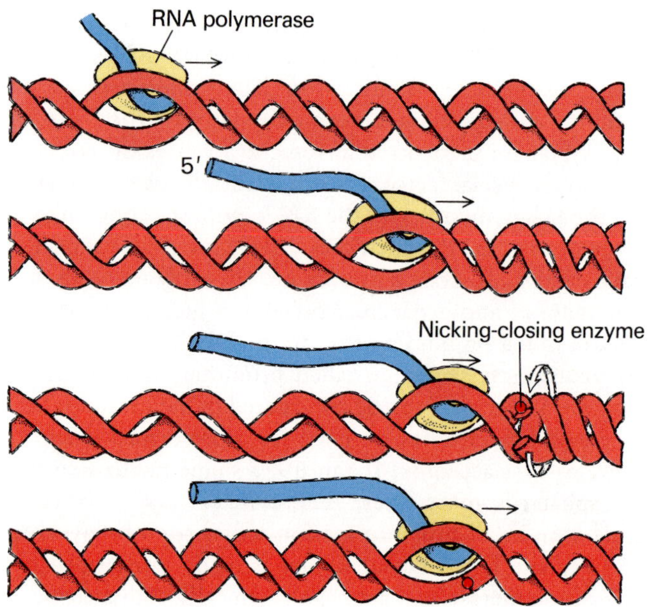

through the next base pair in the DNA double helix, and get ready for another round of the same kind. Once again, the speed of molecular events leaves us baffled. But wait until the next chapter. DNA replication is much faster than its transcription, at least in prokaryotes.

Such speed is all the more astonishing in that the polymerase has to follow a helical path around the DNA thread. You can readily appreciate that, with ten nucleotide pairs per turn of the DNA double helix, the enzyme must revolve around the DNA three hundred times per minute as it moves on. It is hard to imagine the whole bulky set of knobs, with their trailing hairs of growing RNA, whirling around the DNA at that speed. Except, of course, that we must beware of our imagination when it comes to the molecular world. How many times has that lesson been thrust upon us already during our visit. But there is an alternative. Instead of the polymerase circling around the DNA, the DNA could rotate inside the polymerase. Here, however, another topological problem is encountered: the DNA is not free to rotate. It is part of a very long loop belonging to a chromosome and, in addition, it is all twisted upon itself to fit within the exiguous space offered by the nucleolus. Rotation of such a coiled-up thread can take place only in separate short segments provided with appropriate swivels. Several of the bonds in a polynucleotide chain allow free rotation and can serve as swivels. But this will work only with a single strand, not with a double strand. The cell's way of getting out of this quandary is simple: nick one of the strands, and the remaining, intact strand can serve as swivel for the whole double-stranded structure. As long as the nick is on only one strand, the molecule runs no risk of breaking apart; the loose strand is kept coiled around the intact one by the various forces that stabilize the double helix.

In practice, if we were able to watch transcription in slow-motion replay, we would see the RNA polymerase molecules cutting through between the two strands of the DNA, causing "positive supertwisting" of the double helix in front, and "negative supertwisting" (undertwisting) behind. These two twists cancel each other between any two polymerase molecules and are, therefore, manifest only at the two ends of the gene. A nick on one of the strands near the promoter and terminator regions suffices to relax the resulting constraints, except for what is needed to provide the DNA with its rotation energy. RNA polymerase is remarkably economical in this respect. It supports the whole of its mechanical, as well as its chemical, work with the 28 kcal it has available at each step. It seems also to be able to carry out initiation, which requires slipping through about six base pairs, without any extra supply of energy. Extricating it again does, however, cost energy. Rho, as we have seen, consumes ATP in the performance of its job. Another remarkable property of RNA polymerase is its low record of mistakes, which has been estimated at about one wrongly inserted nucleotide in ten thousand. Apparently, cells can afford this level of "noise" in transcription, which, unlike duplication, does not seem to be subject to proofreading.

Now that we have some idea of how nucleolar transcription works, let us take a look at its product. It is a long RNA chain of about fourteen thousand nucleotides, 4.5 million daltons in molecular mass, 45S in Svedberg units. The first striking thing about it is that all molecules are identical. This means that the hundred or more genes that are strung along the nucleolar DNA specify the same message. What we are encountering in the nucleolus is a typical example of genetic amplification, or multiple-copy genes. It is a device used by the cell when the demand exceeds the transcription capacity of a single gene. Assume that the cell needs a thousand molecules of this particular kind of RNA per minute, which is the order of magnitude for our "average" mammalian cell. If it takes 5 minutes to make an RNA molecule, five thousand RNA polymerases must be put to work. They cannot possibly all crowd onto a single gene. The only solution, therefore, is to multiply the number of genes. In certain animal species, this process is itself genetically controlled. At certain stages of extremely heavy demand, the cells rush their nucleolar RNA production by selectively turning out numerous extrachromosomal copies of the genes involved.

If we now examine the structure of the RNA made in the nucleolus, we will have no difficulty recognizing some familiar sequences. We have encountered them before in ribosomes. In fact, the nucleolar RNA contains the whole of the ribosomal RNA except for the 5S component, but in the form of a single large chain. This chain must be processed to yield the 18S piece of the small ribosomal subunit (2,000 nucleotides) and the 28S (5,000 nucleotides) and 5.8S (160 nucleotides) pieces of the large ribosomal subunit. Like proteins, RNA molecules have a history, which is programmed in their sequence.

We can get some idea of how this programming operates by watching a growing RNA chain while it is still being elongated. It is, you will notice, assailed by a swarm of enzyme molecules that zero in on certain specific sites, no doubt attracted there by some signal. These are transmethylases. They transfer methyl groups from the activated methyl donor, *S*-adenosylmethionine (Chapter 8), onto certain ribose molecules in the growing RNA chain. As it happens, methylation affects only those segments of the RNA chain that are destined to participate in the construction of ribosomes. After the 45S RNA is completed and released, the nonmethylated parts are bro-

Processing of ribosomal RNA (rRNA) in the nucleolus. While transcription of the corresponding gene (rDNA) proceeds, portions of the newly made RNA transcript destined to form rRNA are progressively methylated. After detachment of the completed transcript as a 45S precursor, ribonuclease action cuts off the unmethylated segments, leaving mature rRNAs. These mature molecules then combine with proteins and are delivered into the cytoplasm, where they are joined by additional proteins and by a small 5S rRNA not made in the nucleolus, to form the two ribosomal subunits.

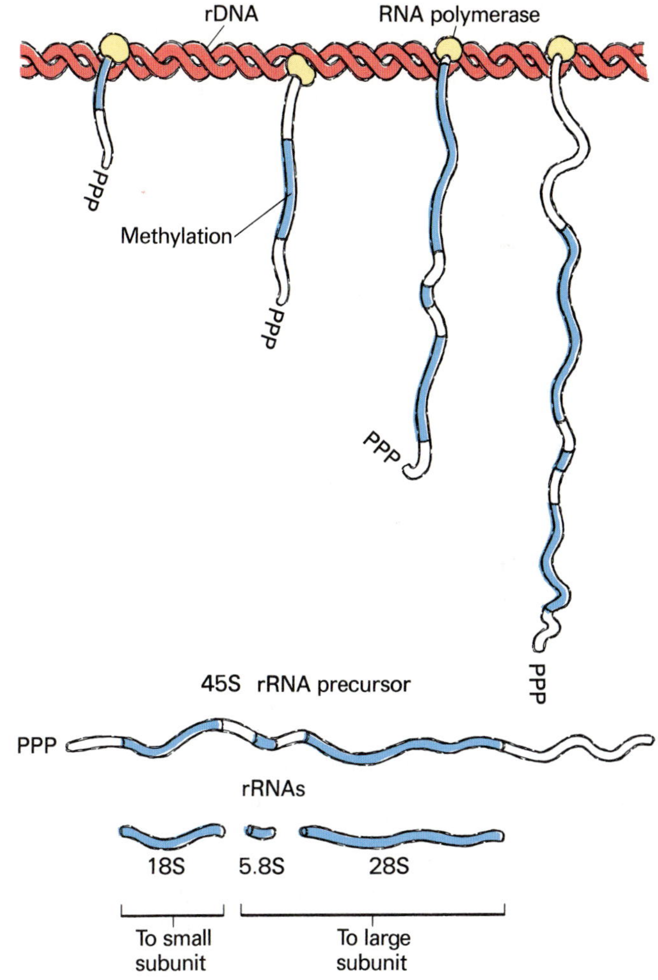

ken down by ribonucleases. But the methylated segments remain intact, protected by their methyl groups. Thus, the signal that attracts the transmethylases provides a mechanism for guiding the process whereby the various rRNAs are cut to size and completed. It is likely that this signal lies in certain base sequences that serve as binding sites for the transmethylases. These base sequences are themselves transcribed from the corresponding DNA, in which, therefore, the future of the molecule is genetically encoded.

But time once again is pressing us. We must move on. Before we do so, however, let us cast a last look around. It is remarkable how our perception has changed. When we first arrived, we were impressed with shapes and forms—fern fronds, feathers, knobs, stalks, whiskers—to the point of having a strain put on our morphological vocabulary. Early explorers had the same problem. They used other words, such as nucleoloneme, nucleolar organizer, fibrillar zone, granular zone, but their purpose was the same: descriptive. And so was the effect: a mixture of wonder and puzzlement.

How different our present molecular view. The nucleolus is simply a large ribosomal RNA factory, built by tens of thousands of molecules of RNA polymerase, transmethylase, ribonuclease, and other enzymes, gathered around a long, unfolded loop of DNA that bears multiple copies of the same gene and supports thousands of distinct transcription units operating simultaneously. Every second, this factory assembles, and methylates in certain specific sites, some fifteen to twenty RNA strands about fourteen thousand nucleotides long. These are then cut and trimmed into three pieces totaling about half this length: the 18S, 28S, and 5.8S rRNAs. Newly made ribosomal proteins, synthesized in the cytoplasm and admitted into the nucleus by the pores, come and bind these RNA pieces and return with them to the cytoplasm to participate in the assembly of an equivalent number of new ribosomes. Wear and tear of ribosomes is thereby compensated, and the protein-synthesizing potential of the cell is maintained unimpaired. In times of rapid growth (e.g., in the amphibian egg cell) this manufacturing potential may be stepped up many times by amplification of genes and multiplication of nucleoli to the point of turning out as many as one million or more new ribosomes per minute. When a cell prepares for division, the nucleolar factory packs up its gear and folds back into its chromosomal shelter. In the completed daughter nuclei, it emerges again from the same chromosomal region, which corresponds to the nucleolar organizer of the early observers.

Transcription in Euchromatin

Visiting the nucleolus has greatly simplified the next part of our tour. As we worm our way through the loosened coils of euchromatin, we have no difficulty recognizing what is going on. Everywhere, our eyes fall on the same familiar sight of RNA polymerase molecules trailing nascent RNA threads along rapidly rotating DNA fibers. Euchromatin is simply a conglomerate of sites where nuclear DNA is being transcribed into RNA. In most nuclei, the organization of these transcription sites is almost impossible to unravel. But, as is often the case, Nature has favored us with an exemplary model in the so-called lampbrush chromosomes that are seen in many oocytes at the diplotene stage of the first meiotic prophase (see Chapter 19), where paired, duplicated chromatids, still joined by their chiasmata, are actively transcribed on exactly the same sites. These transcribed sites make fluffy lateral loops—perfectly symmetrical, owing to the close fit between the duplicated chromatids—that extend from the axial stem of the lampbrush. This stem is made of untranscribed DNA segments and of the protein skeletons of the chromatids. It is clear from this structure that each of the two chromatids consists of a single DNA thread, of which certain stretches are transcribed and others are not. This is, in fact, what we would see, but without mirror image, should we have the possibility of following a chromosomal DNA thread over some distance in the usual interphase nucleus. Lampbrush chromosomes also give us an interesting glimpse of the manner in which the DNA is anchored as loops to the protein framework. However, we distinguish only the loops that are unfolded

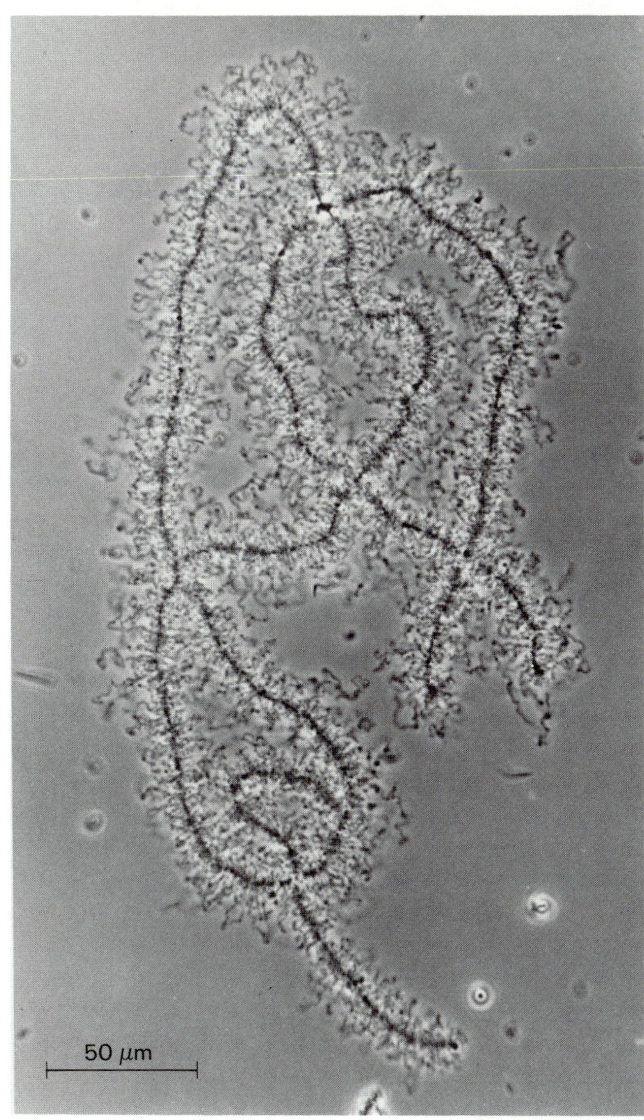

Phase-contrast micrograph of a lampbrush chromosome bivalent from an oocyte of the newt *Notophthalmus viridescens.* Hundreds of transcriptionally active loops project from the main axis of each chromosome. The two homologous chromosomes are joined by several chiasmata.

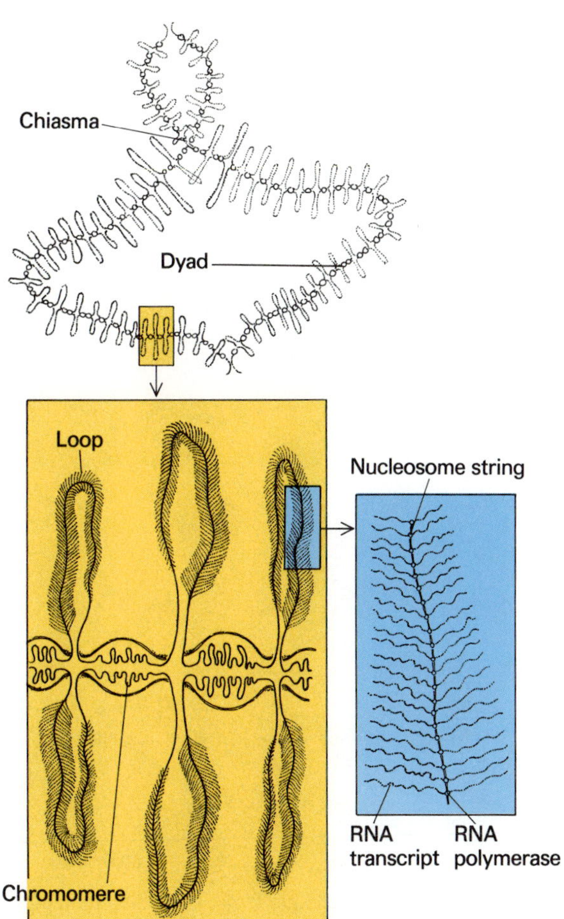

Diagrammatic representation of the structure and function of lampbrush chromosomes characteristically found in oocytes during the extended prophase of the first meiotic division (see Chapter 19). The drawing at the top shows the typical appearance that can be discerned in the light microscope, as shown at the left: fluffy, symmetrical loops alternate with thicker condensations called chromomeres, recalling the shape of a brush used to clean the chimneys of kerosene lamps. Two identical-looking chromosomes are joined together by one or more chiasmata, which are the sites of recombination during crossing-over (see the illustrations on p. 287). As shown in the yellow enlargement, each chromosome is really a dyad, made of two identical sister chromatids resulting from the duplication process that has preceded the onset of meiosis. Each chromatid consists of an uninterrupted thread supported by a protein framework. Portions of this thread are fully extended to form the loops; the others are bundled up into the chromomeres. Special treatment allows the loops to be visualized in the electron microscope (see the micrograph on p. 304, which shows part of a loop) as possessing the typical fern-frond appearance of an uncoiled string of nucleosomes in the process of being transcribed (blue enlargement). Note the change in direction of transcription indicated in the yellow enlargement. It illustrates the fact that either one or the other strand of the DNA duplex may be transcribed, but rarely both.

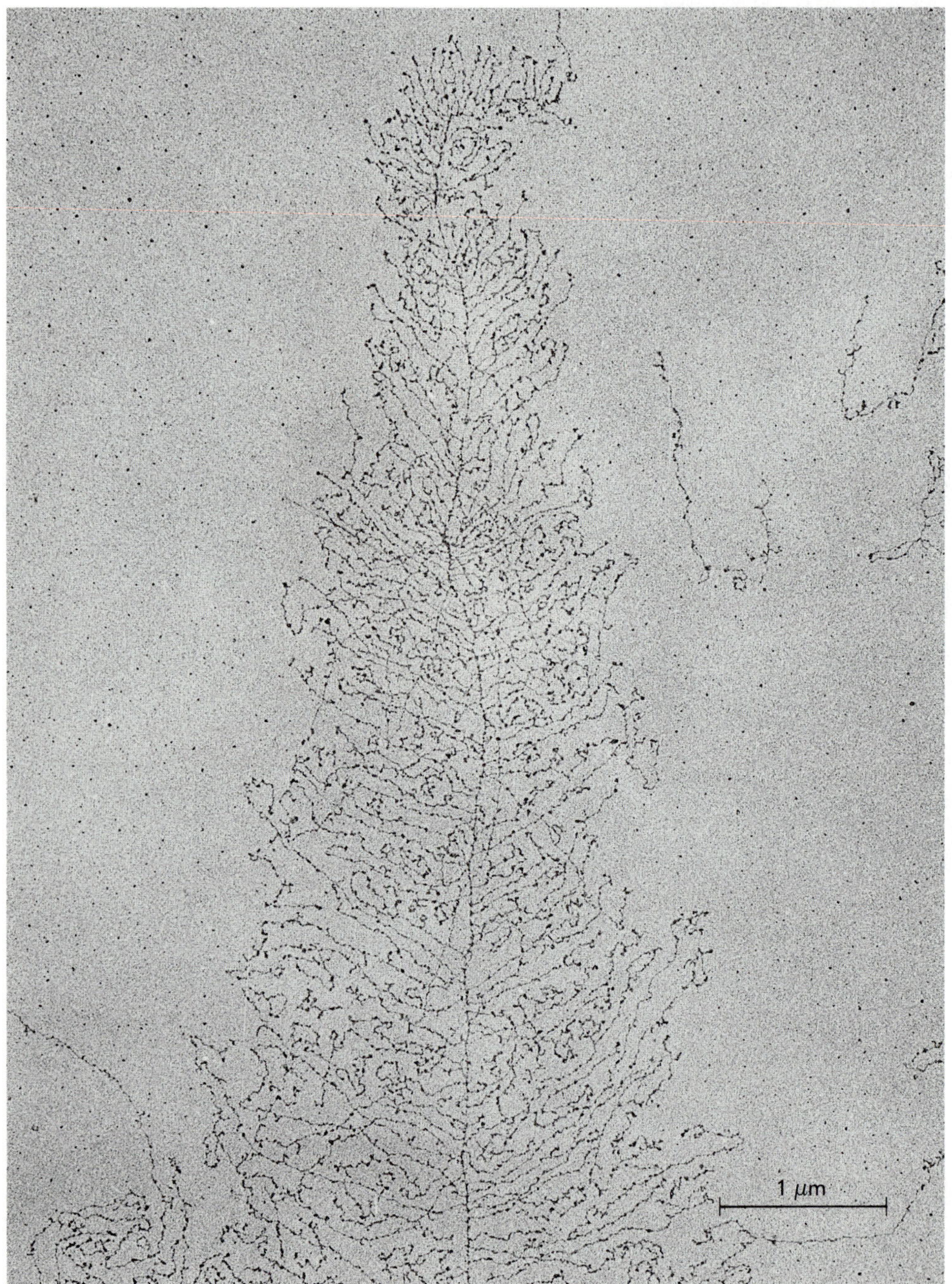

Electron micrograph of dispersed fibrillar material from lampbrush chromosome of *Triturus viridescens* oocyte. Compare with the micrographs on pages 295 and 296.

and transcribed, not those that are compacted into the stem. We will see more details of chromosome organization in the subsequent chapters.

Part of euchromatin transcription complements nucleolar transcription in the upkeep of the protein-synthesizing machinery. Here, in a corner, a string of small, identical genes turns out 5S rRNA for the large ribosomal subunit. In another area, tRNA molecules are being made, readily recognized by their shortness, by their peculiar cloverleaf folding, and especially by the number of different enzymes clinging to them and producing the numerous chemical modifications so characteristic of this type of RNA. Also made in multiple copies are certain small RNA molecules (sRNA) that are beginning to emerge as cofactors in such important processes as protein signaling (Chapter 15) and mRNA splicing (see p. 310). Much of the RNA made in euchromatin, however, is mRNA. It is made by a special RNA polymerase programmed to obey a particular set of signals, which includes a TA doublet, known as the TATA box, some thirty base pairs upstream from the starting point, as part of its promoter sequence.

By definition, messenger RNAs are tapes in which the amino-acid sequences of given polypeptide chains are written in code and are complemented by various signals that ensure proper positioning and reading by the protein-synthesizing apparatus. We have already noticed that mRNA molecules flow out continuously through the nuclear pores and therefore must be made in the nucleus. We are now finding out that they are made by transcription of nuclear genes. Ultimately, therefore, the structures of the cell's proteins are inscribed in DNA sequences. And this is how the genes (apart from the small number that specify noncoding, functional RNAs such as rRNAs, tRNAs, and sRNAs) control hereditary characters. They do so by way of their translation products, the proteins, which themselves, as we have seen, mediate most of the manifestations of life.

As in nucleolar transcription, only one of the two DNA strands is usually transcribed into mRNA. This is understandable. If both strands were transcribed, the product would consist of two complementary RNAs, which, at the first opportunity, would lock each other into a double helix prevented from participating in further information transfer. Even if this mutual block could be overcome, there is little chance that such long, complementary RNA strands could both be "meaningful"; that is, would each translate into a functional protein molecule. Two-sided transcription, therefore, would not be just highly cumbersome; most of the time it would also be useless. This does not mean that the same strand of DNA is always being transcribed. Transcription templates occur sometimes on one, sometimes on the other of the two strands of a DNA thread, and the direction of transcription therefore changes from one part of a chromosome to another. Note also that the ban against two-sided transcription is not absolute. Especially in bacteria and in viruses, several examples are known of genes facing each other on the same piece of DNA. Our own mitochondria, in a remarkable feat of economy possibly inherited from their putative prokaryotic ancestors, have succeeded in cramming transcribable information on both strands of their DNA and even in eliminating most of the spacers.

An Unexpected Editorial Mechanism

In bacteria, which do not possess a fenced-off nucleus, but have only a single, circular chromosome in direct contact with the cytoplasm, startling evidence of the relationship between DNA, RNA, and protein can be seen in the form of what may be called "composite fern fronds," structures in which the RNA whiskers produced from the DNA stalk by RNA polymerase are themselves read off by protein-synthesizing ribosomes. This can happen because both the growth and the reading of mRNA are in the 5′⟶3′ direction. Thus, it is possible for reading to start before the chain is completed. Such structures illustrate in striking fashion the principle of co-linearity between the DNA gene, or cistron, its transcription product (mRNA), and its primary translation product (unprocessed polypeptide).

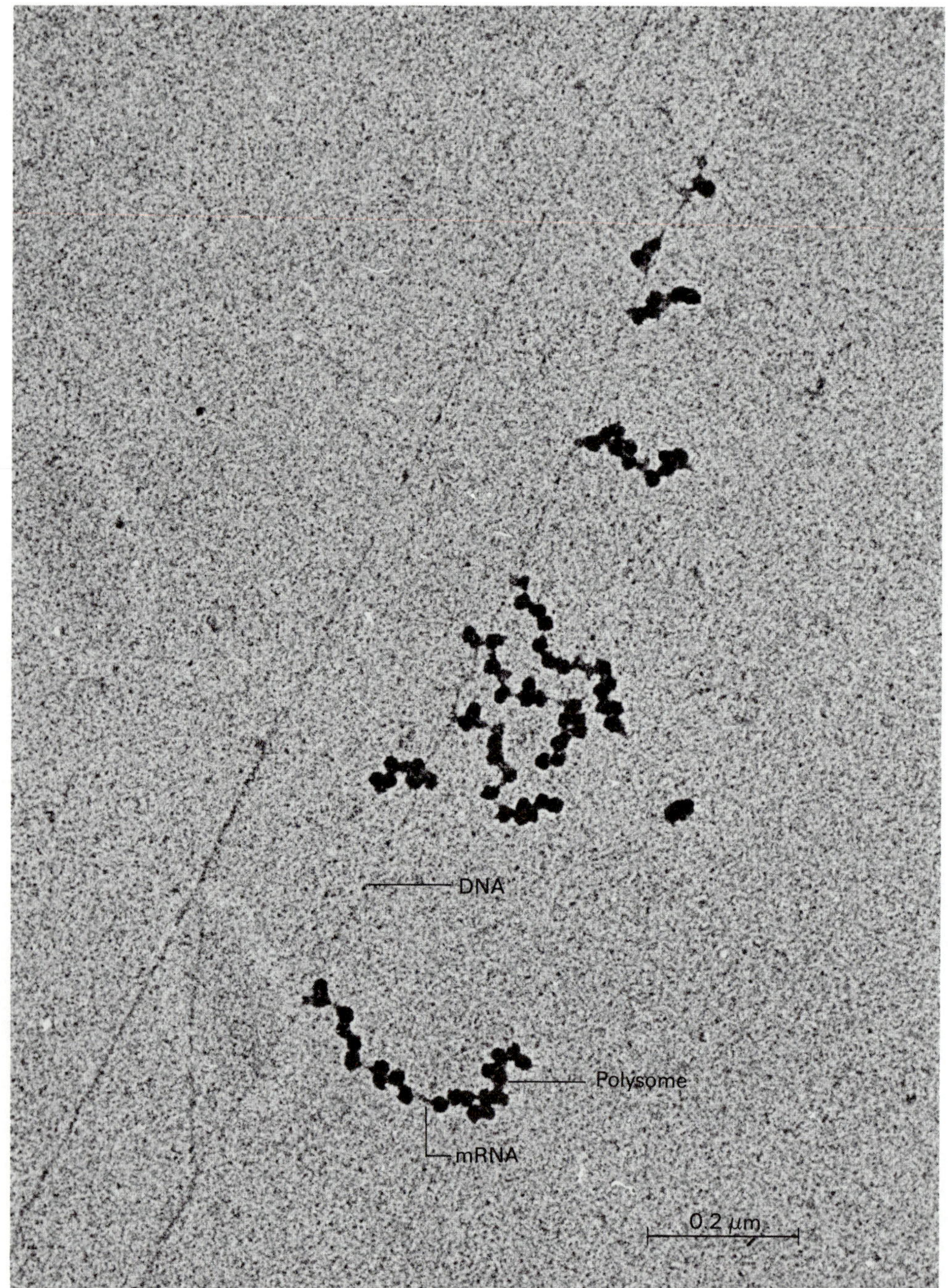

Coupled transcription-translation in bacteria. This electron micrograph of dispersed fibrillar material from *E. coli* shows ribosomes strung on RNA strands that grow from DNA in the process of being transcribed.

Schematic interpretation of the electron micrograph on the facing page. The growing polypeptide chains are beyond the resolving power of the technique used and are not visualized in the electron micrograph.

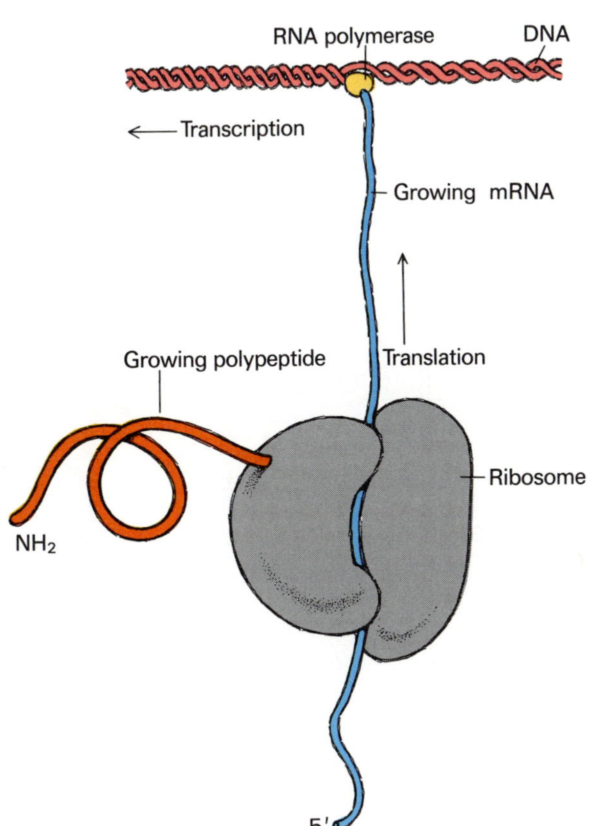

Co-linearity hypothesis.

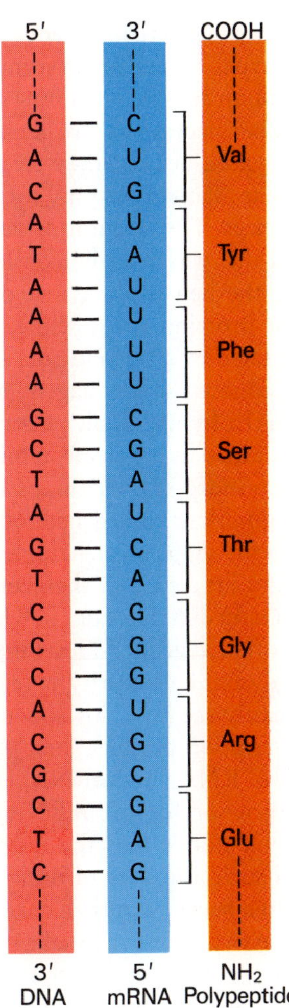

Similar figures cannot be seen in eukaryotic cells, where transcription and translation are separated by the nuclear envelope. But it never occurred to anybody to doubt for a single moment that co-linearity applied equally well to eukaryotes, so perfectly logical did it appear to be. To most people acquainted with the mechanisms of transcription and of translation, co-linearity was a self-evident necessity. A scrambled message was unthinkable, for where would the unscrambler be?

But once again—molecular biology is fertile in such surprises—the unthinkable has proved true. If you watch what happens to most primary transcription products made in euchromatin, you will see that they undergo a considerable degree of surgery before being delivered into the cytoplasm. In a process reminiscent of cutting a movie film or editing a tape, a number of segments are excised from the RNA ribbon and the remaining ones are stitched back together. The pieces that are kept in the mature mRNA are called exons because they are expressed; those that are removed are named introns, for intermediate or intervening sequences. The extent to which genes are split into fragments in this way may be considerable. For instance, a gene in the hen oviduct that codes for conalbumin, one of the proteins of egg white, consists of seventeen exons separated by sixteen introns; as many as fifty exons have been counted in a collagen gene. Introns are readily seen when the hybrid duplex between the mRNA and its gene is examined in the electron microscope. They appear as single-stranded (DNA) loops of various sizes appended to the double-stranded heteroduplex.

As indicated by the shape of such duplexes, cutting and splicing mRNA is a little simpler than cutting a movie in that the exons are united in the order in which they follow each other in the original gene. At least the message is not entirely scrambled, and co-linearity is respected to some extent. You can easily see how this is done by looking at the primary transcription product (first described as heterogeneous nuclear RNA, or hnRNA, and now designated mRNA precursor, or pre-mRNA) as it is being processed. You will notice that the introns form loops that are closed in such a way as to bring together the 5′ and 3′ ends of two adjacent exons. Excision of the intron and suturing of the two exons then take place in immediate succession. There is no opportunity for the severed exons to separate and become scrambled.

There is, however, something strange about the manner in which these loops are closed. They are not joined by complementary sequences, as are the usual RNA loops, for the RNA parts involved are not complementary, as they should be if they served to close the loop (p. 257). Therefore, these parts must be recognized by "something else" acting as a pincer or vise while the splitting and splicing enzymes do their jobs. Indeed, the terminal sequences of many introns are, if not identical, at least closely similar, sufficiently to be recognized by the same structure. Unfortunately, we cannot distinguish

Schematic representation of mRNA processing. Exons (E) are segments that are expressed. Introns (I) are intermediary segments that are excised. Note that exons are spliced together co-linearly, as shown by the mRNA-DNA hybrid. DNA introns appear as loops on the duplex between mRNA and DNA exons.

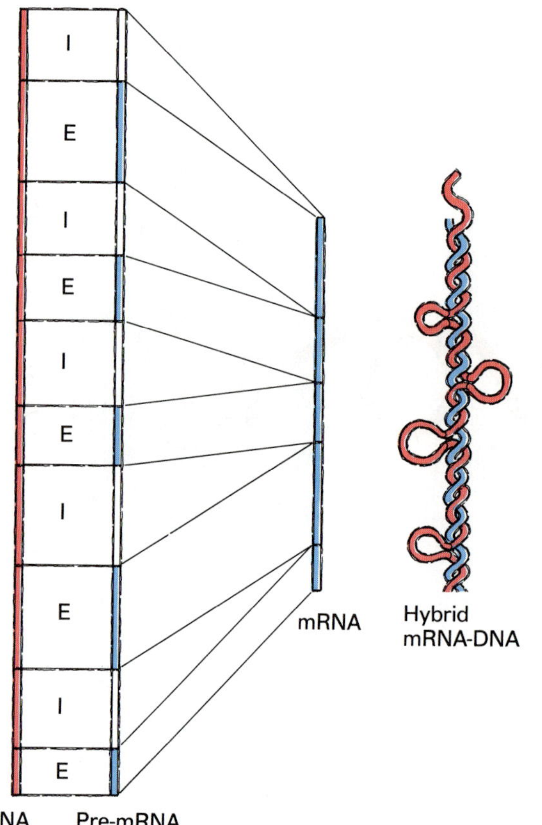

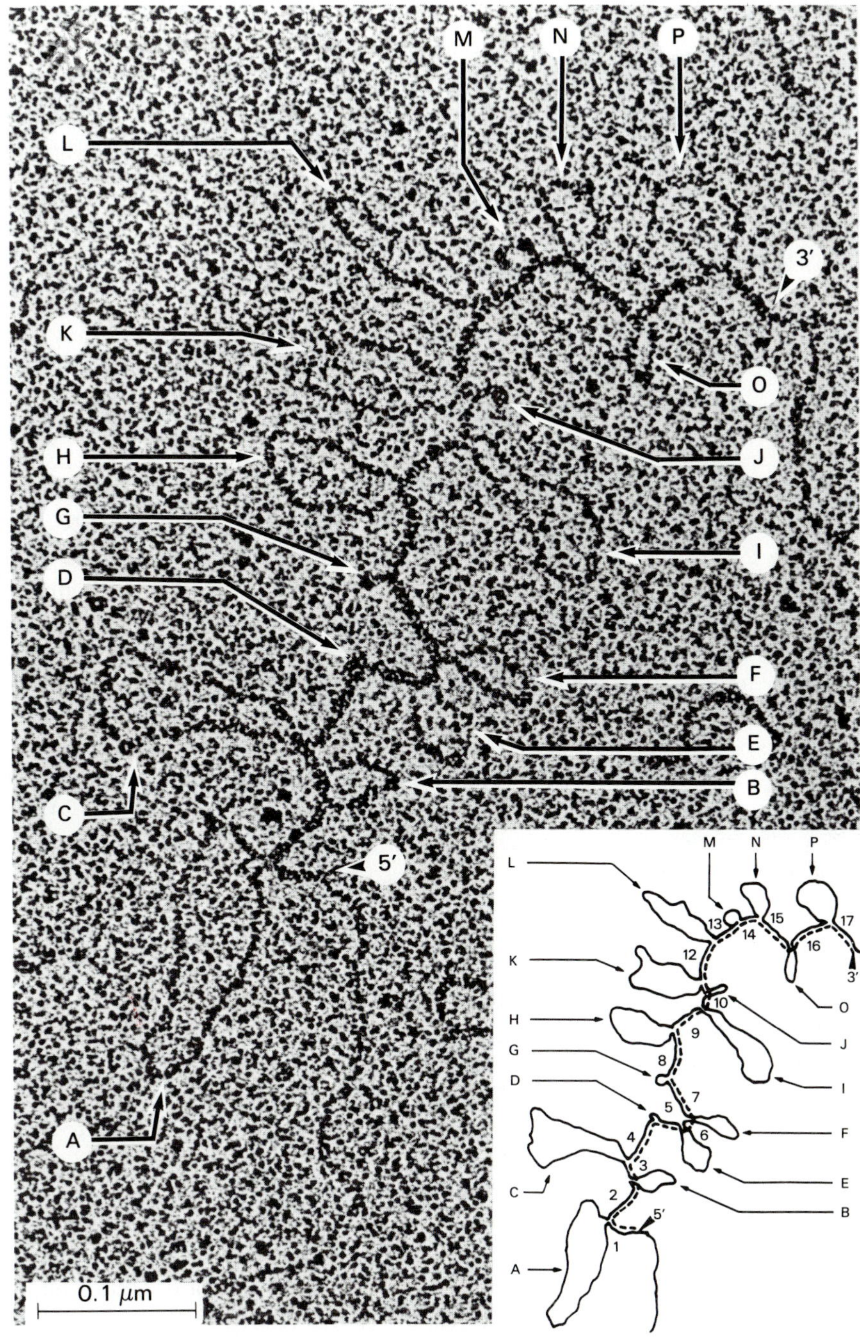

Electron micrograph of mRNA-DNA hybrid between conalbumin (from egg white) gene and mature mRNA. The thick lines, identified by numbers, are hybrid regions (exons). The thin loops, identified by letters, are DNA segments corresponding to introns that have been excised from pre-mRNA upon splicing. (Compare with the diagram on the facing page.)

Postulated mechanism of mRNA processing. The diagram shows how U1a small nuclear RNA (snRNA) may lock together the end sequences of an intron in pre-mRNA, to allow excision of the intron and splicing of the two exons.

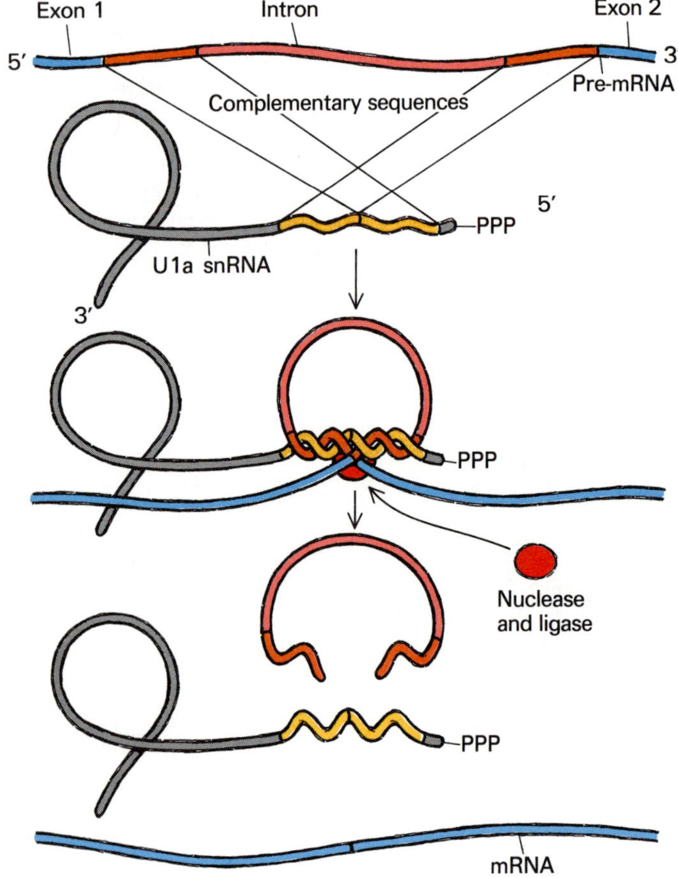

clearly what does the joining. It could be a special kind of RNA, known as small nuclear RNA (snRNA), a group of short-chain molecules from 90 to 220 nucleotides long, endowed with several characteristic structural features. One of these snRNAs (U1a) possesses a continuous sequence of bases, of which one part is near-complementary to the 3′ end, the other to the 5′ end, of introns. It could serve to align these two ends against each other in such a way as to position the exon-intron connections accurately for splitting by a ribonuclease, followed by splicing of the exons by a ligase (an enzyme similar to the DNA ligase that we will encounter in Chapter 17). Some thirty different proteins seem to be involved in these complex operations, in addition to the putative snRNA vise. They form little bundles that can be seen strung all along the pre-mRNA threads. They can also be isolated as 30S ribonucleoprotein particles (RNPs) after selective cutting of the threads.

Some other RNAs in addition to nuclear pre-mRNAs undergo splicing, apparently by different mechanisms. Most amazing is a recently discovered pre-rRNA that seems to do its own editing, including the shedding of an intron and the subsequent splicing, without any outside help. Some mitochondrial mRNAs, as we shall see, rely on their introns to generate the necessary help.

Besides splicing, mRNA molecules must also be fitted with their characteristic 7-methyl-GTP cap, additional methyl groups, and poly-A tail. The latter addition is usually preceded by substantial exonucleolytic trimming of the 3′ end. These reactions likewise take place in the nucleus, probably before splicing. Capping may even start before transcription is terminated. The finished mRNAs are then delivered into the cytoplasm through the nuclear pores in combination with special carrier proteins.

So far, split transcripts have not been encountered in prokaryotes. On the other hand, not all eukaryotic transcripts are split. For instance, the histone mRNAs represent primary transcription products, as do bacterial mRNAs. They also resemble the latter in lacking a 7-methyl-GTP cap and poly-A tail. Surprisingly, at least one mitochondrial gene (in yeast) has been found to be split.

The discovery of gene splicing has given molecular biology one of its rudest jolts so far. When light first fell on the genetic organization of primitive forms of life in the late 1950s and early 1960s, the picture glimpsed was one of satisfying orderliness. Explorers were led to think of genes as forming a neatly catalogued library equipped with retrieval mechanisms that could, in spite of their complex network of interacting feedback loops, be accommodated within an intuitively simple conceptual framework. Today, with more books read and, especially, with exploration extended to the eukaryotes, this

Relationship between products of spliced and unspliced messages when pre-mRNA includes a facultative intron.

A. The intron is fully translatable. The two proteins have the same N-terminal sequence. They also have the same C-terminal sequence if n, the number of nucleotides in the intron, is a multiple of three (1). If not, the C-terminal sequences are different (2).

B. The intron includes a termination codon. After splicing, reading continues until the next termination codon. The two proteins have identical N-terminal segments. The T and t proteins of SV40 illustrate this case (see the illustration on p. 312).

C. The intron includes an initiation codon. Its removal promotes an AUG (or GUG) codon situated downstream to function as initiation codon. The resulting protein and the C-terminal part of the product of translation of the unspliced mRNA have identical sequences if m, the number of nucleotides separating the two initiation codons, is a multiple of three—that is, if the reading frame is unchanged (1). If not, the two sequences are entirely different (2). The VP proteins of SV40 illustrate these two possibilities.

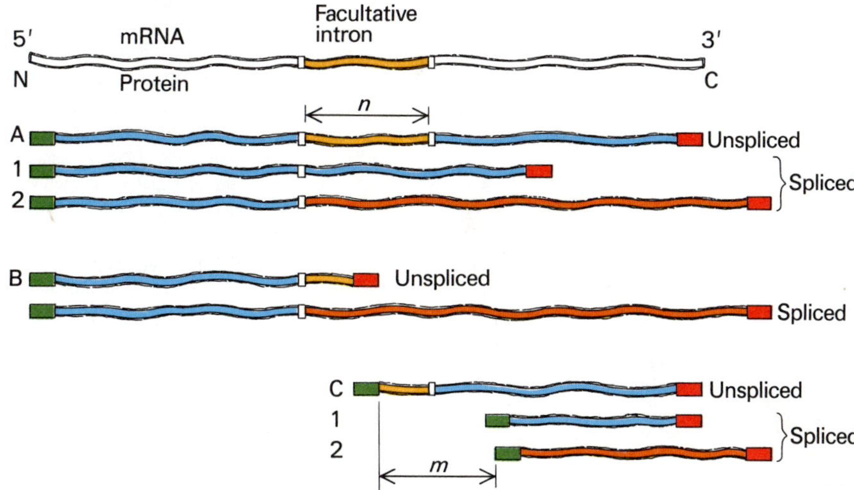

vision has been shattered. No longer a librarian's dream, it could even become a cryptographer's nightmare. It now appears that most eukaryotic genetic messages are written in pieces separated by stretches of untranslated DNA. If we have no splicing key, the messages are completely ununderstandable, as many an explorer has found after searching the nucleus directly for information. But cells can hardly derive much evolutionary advantage simply from keeping snoopers out. Surely there must be more substantial benefits to compensate for such a cumbersome, hazardous, and costly way of storing information. Otherwise, this mechanism could hardly have resisted the selective pressures that must have worked against it.

One major advantage of splitting genes, it has been pointed out, is that it enormously enhances the scope of evolutionary experimentation by recombination (see Chapter 18). Exons can be reshuffled at a high rate by means of their appended introns, and it matters little where or how exactly the joining is done. As long as the critical sequences that govern splicing are not altered, the final edited product will consist of neatly connected exons. What makes such a mechanism particularly valuable is that exons do not seem to be just randomly cut segments of the genetic message but may well correspond to a defined structural or functional domain of a protein molecule, to a "miniprotein," as it has been called. The odds of coming up with something interesting by reshuffling such pieces, even in a haphazard fashion, are appreciable, just as is the probability of generating a meaningful sentence by restringing complete words or phrases.

Another, more immediate, advantage of gene splicing derives from the possible modulation of the splicing mechanism itself by means of what might be called facultative introns (intervening sequences that may be either spliced out of a pre-mRNA or left in it). If such an intron contains the initial part of a message, its removal will promote another AUG or GUG codon, situated downstream, to the rank of initiating codon. Depending on whether this new start signal belongs to the same reading

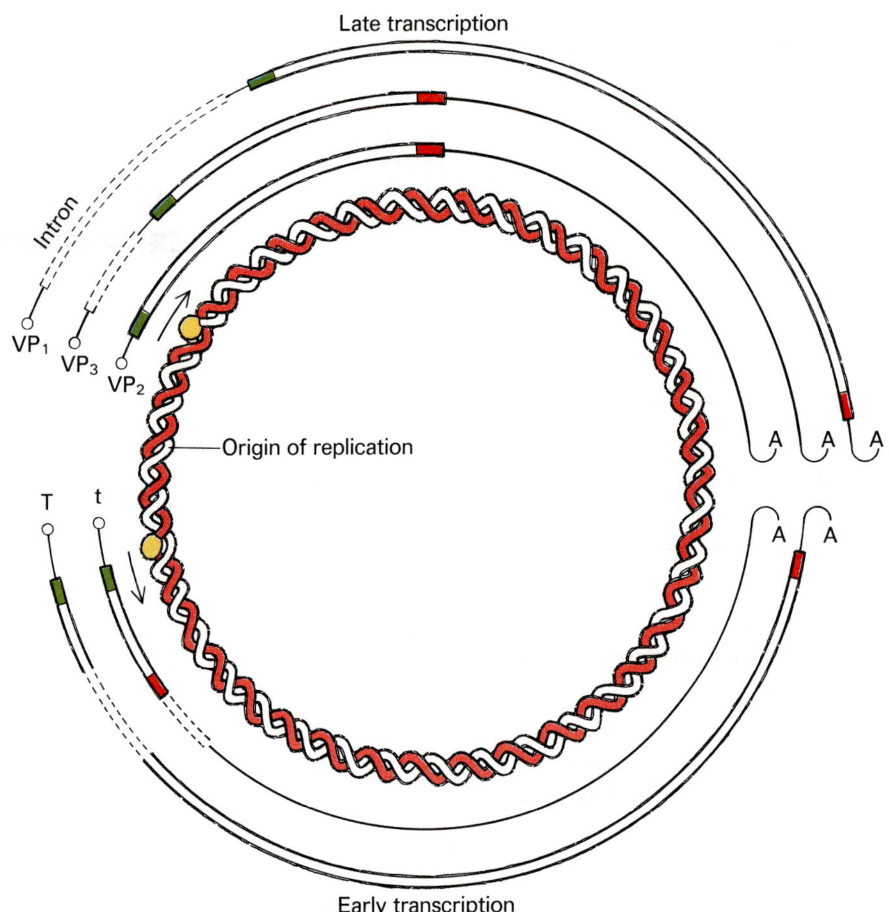

The small virus SV40 (see Chapter 18) consists of a circular double-stranded DNA, 5,224 nucleotide pairs long. Its transcription takes place in two stages. Thanks to distinct modes of splicing, the two resulting pre-mRNAs give a total of five translation products. First about one-half of the genome is transcribed from one strand. The other half is transcribed later from the other strand. The early pre-mRNA is spliced in two different ways to give two translation products, t and T, which have the same N-terminal sequence of 100 residues. T is much longer than t because the stop codon that terminates translation of the t mRNA has been spliced out of the T mRNA, allowing translation of the message up to the next stop codon, situated almost at the extremity of the molecule. The late pre-mRNA is also spliced in two different ways; in addition, it is translated in unspliced form. In VP_3 mRNA, the initiation codon of VP_2 is removed with an intron: translation starts further down, actuated by an initiation codon that happens to be in phase with that of VP_2. Therefore, the VP_3 protein is identical with the C-terminal part of the VP_2 protein. In VP_1, the two initiation codons are spliced out; translation is initiated by a codon that is out of phase with the others, so that the VP_1 protein has nothing in common with the other two VP proteins.

frame as the excised one, the resulting protein will be identical with the C-terminal part of the product of translation of the unamputated message or it will be completely different. If, on the other hand, the facultative intron includes a stop signal, its removal will allow read-through until the next stop sign, with synthesis of a protein that now shares its N-terminal sequence with the product of the intact message. Finally, if the facultative intron is fully translatable, in register with the initial part of the message, the two proteins will also have the same N-terminal sequence and they may possess a common C-terminal sequence. This will happen when the reading frame of the subsequent exon is not altered by removal of the intron—that is, when the length of the intron, measured in number of nucleotides, is an exact multiple of three.

It is remarkable that, with only a few messages decoded so far, examples of all the possibilities mentioned have already been found. A small DNA virus known as SV40 (see Chapter 18) generates as many as five different proteins from only two primary transcripts. Immunoglobulins of the IgM class are made in two forms, one soluble, the other membrane-bound thanks to an additional hydrophobic C-terminal sequence that results from the removal of an intron containing a stop codon. A particularly intriguing example of alternative splicing of the same pre-mRNA has been observed in yeast mitochondria. These organelles apparently manufacture self-adulterating

Schematic representation of fertilization illustrates the formation of a diploid nucleus from two haploid nuclei.

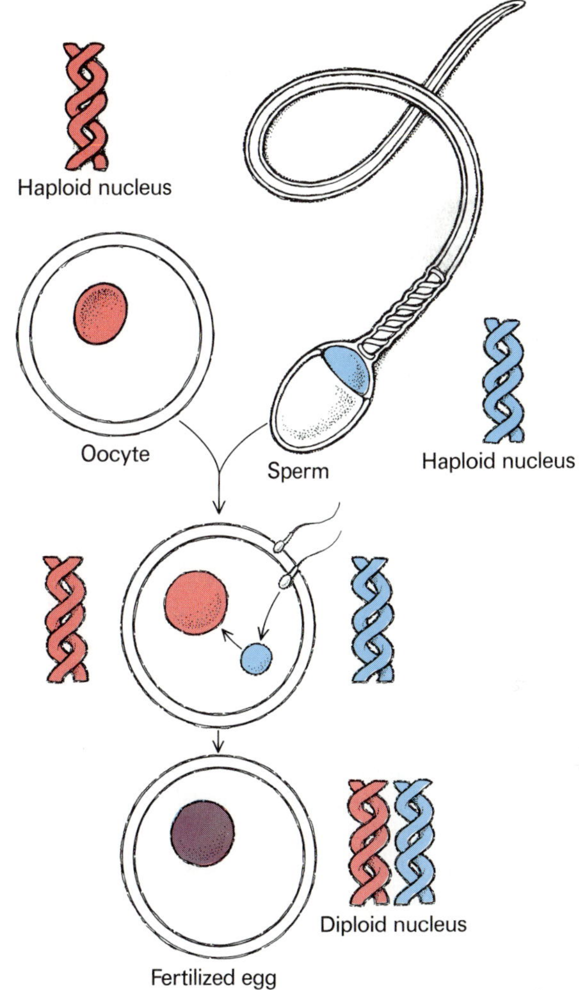

messages whose primary translation products (maturases) induce further splicing of the mRNA to a form coding for a piece of the respiratory chain.

Even in the prokaryotic world, things are not as simple as they were thought to be. Overlapping genes seem to be quite common, and at least one case is known (in a small bacterial virus) of three genes overlapping in such a way that the region of overlap in the DNA is actually translated in all of its three distinct reading frames.

So far, transcript splicing has not been observed in bacteria, whereas it seems to take place in all eukaryotes, even the most primitive. Apparently, therefore, the development of this process coincided with the great prokaryote-eukaryote transition and may even have played a key role in it. But we must beware of the obvious. It could be argued—and has been quite cogently—that genes are more likely to have started in the form of small pieces that were continuously rearranged in a more-or-less haphazard manner than as the highly compact, often polycistronic, sequences characteristic of present-day prokaryotes. If such is the case, then eukaryotes could be viewed as having tamed this ancient process and turned it into a mechanism of great precision while conserving its versatility, and the prokaryotes as having squeezed it out in the course of evolution, thereby gaining generation rapidity but losing the genetic flexibility that has allowed eukaryote development.

How to Pull Genetic Strings

This is one of the truly fundamental questions facing the explorer of the living world, the key to how cells control their fate. For cells change, adapt, differentiate, transform by turning on certain genes and shutting off others. Even the simplest bacteria do this. Yeasts, protozoa, fungi, plants, and animals—all regulate and modulate the activity of their genes. The most notable examples of this are found in the development of higher animals, ourselves included.

Take the human egg cell just before fertilization. It is a large cell and its cytoplasm is stocked with reserves and with organelles ready to get into the act. Its nucleus is haploid (Greek *haplous*, single), which is to say that it contains a single set of 23 chromosomes, totaling about 3 feet of DNA selected from the mother's genetic endowment through the lottery of meiosis and crossing-over (see Chapters 18 and 19). Upon fertilization, it receives a second set of similarly selected chromosomes from another haploid cell, the male spermatozoon, which is little more than a DNA torpedo equipped with a flagellar motor at the back and an invasion apparatus in front. This is the acrosome (Greek *akron*, tip), a modified lysosome. Through fusion of the two nuclei, the fertilized egg cell becomes diploid (Greek *diplous*, double). It now has 46 chromosomes, containing a total length of about 6 feet of DNA, or 5.8×10^9 nucleotide pairs, one out of $10^{3,480,000,000}$ possible combinations, guaranteed to be unique against all conceivable odds. (It would take more than five thousand average-size books just to print that figure: 1 followed by more than 3 billion zeros.) In this string of some 6 billion nucleotide pairs is written the whole predetermined history of one particular individual from conception to death, as well as much of what is environmentally actuated but depends on the kind of response the individual offers to an outside challenge.

Soon after fertilization, the egg cell starts to divide, producing two daughter cells, which themselves divide into four cells, then eight, sixteen, and so on. All these cells have the same 6 feet of DNA, the same identical sequence of 5.8×10^9 nucleotide pairs. Nevertheless, they soon cease to look alike. A polarity sets in within the little bunch of cells that has formed (*morula*, small mulberry, in Latin), and the cells start becoming different from each other, depending on where they are situated in the structure. This process of differentiation continues, leading in a few weeks' time to the development of a characteristic embryo in which a number of different cell types are readily recognized. And so event follows upon event, enacting with remarkable fidelity a script of enormous complexity, to produce at the end of nine months the miracle that is a newborn baby: nerve cells, muscle cells, skin cells, retinal cells, liver cells, and many others, uniquely organized to form a brain that is already sending orders in all directions, eyes that blink in the light and are moved by tiny muscles, a beating heart, lungs that fill with air and expel it in the baby's first cry, a mouth that is already seeking the life-giving nipple, a stomach and intestine that are getting ready to run their first digestive trials on the mother's milk, small arms and legs that flail, miniature fingers and toes that wriggle . . . altogether about 1 trillion (10^{12}) cells, each equipped with the same 6 feet of DNA, but nevertheless doing all sorts of different things and forming all sorts of different associations.

And this is only the beginning of a long saga, which normally will take the individual over a span of some 80 to 100 years, first up from babyhood to childhood, through puberty to adolescence and young adulthood, and then slowly down to middle age, old age, senescence, and death. It is all preordained, written into the 5.8×10^9 nucleotide pairs of the nuclear DNA. The egg cells of a mouse, mare, or chimpanzee do not look very different from those of a woman. Their cytoplasmic machineries are virtually indistinguishable. Nevertheless, each unerringly unfolds its own story, produces its own specific kind of miracle: a newborn mouse, a foal, a baby chimp. They follow different programs; their DNAs are different.

Not only life cycles and basic developmental processes are DNA-dependent in this way. So are many short-term changes and adaptations. Think, for instance, of the female sexual cycles and of all the modifications that accompany pregnancy, delivery, lactation; of the training of athletes; the healing of wounds; the adaptation to climate; the many cycles and rhythms that attune our body functions to the hour of the day. Mind you, not all adaptive changes operate by way of DNA. In general, only slow responses that take hours, if not days, to manifest themselves, do so. Rapid responses, such as are evoked by many hormones or by nerve impulses, rely on metabolic circuits, feedback loops, and other regulatory devices that are built into the existing cytoplasmic machinery (Chapter 14). DNA-dependent changes are those that alter the machinery itself, generally by adding certain proteins to it or by ceasing to supply others.

How all this complexity operates is formally extremely simple. In any given cell, only a small fraction of the genes are turned on—that is, are actively transcribed. This is how cells endowed with the same DNA manage to be different. All they need do is transcribe different parts of their DNA and thereby make different sets of proteins. Cells are, to an overwhelming degree, the structural and functional manifestations of their proteins. The ultimate control of what a cell is to be or do, therefore, belongs to whatever pulls the levers that turn genes on and off inside the nucleus.

This control is itself genetically controlled and forced to follow a precise four-dimensional program. Whatever happens at step *n* of embryological development determines what levers will be pulled in what cells at step $n + 1$. Nevertheless, the program is not written into the nuclear DNA as some sort of wound-up tape that slowly unrolls in irreversible fashion. This has been dramatically demonstrated by experiments in which the nucleus of an unfertilized egg cell has been replaced microsurgically by a nucleus removed from a fully differentiated cell, originating, for instance, from the intestinal wall. Such eggs can develop into embryos that rarely develop further but whose nuclei can be similarly transplanted to produce fully mature animals. Such development occurs in spite of the fact that the originally implanted nucleus has already played out most of its developmental program and closed, in the process, all but one of the multiple avenues that are open in the pluripotential egg-cell nucleus. It has thereby become possible, by serial transplantation of nuclei, to clone an individual—that is, to produce a number of genetically identical copies of the donor of the original nucleus. So far, such experiments have been successfully performed only with amphibians and lower forms of life. In spite of an imaginative novelist's claim, the cloning of mammals, let alone man, has not yet been done, which does not necessarily mean that it cannot be, and will not be one day.

In any case, these experimental achievements do not derive their meaning from the more freakish and sensational applications that they may allow. What they are telling us is a vitally important piece of information: the nucleus of a differentiated cell is not necessarily committed irrevocably to the expression of a single gene set. Its genome can be reawakened, either in the nucleus itself or in nuclei derived from it by mitotic division, by means of messages originating in the cytoplasm. It is obviously the egg-cell cytoplasm that orders the intestinal-cell nucleus or its offspring to re-enact the whole developmental program of the species.

Not all nuclei conserve their pluripotential property in the course of differentiation. As we will see in Chapter 18, maturing lymphocytes offer a striking example of gene processing. Other such cases may well exist. But the very fact that cloning has been achieved should suffice to correct any exaggerated idea we might have formed about

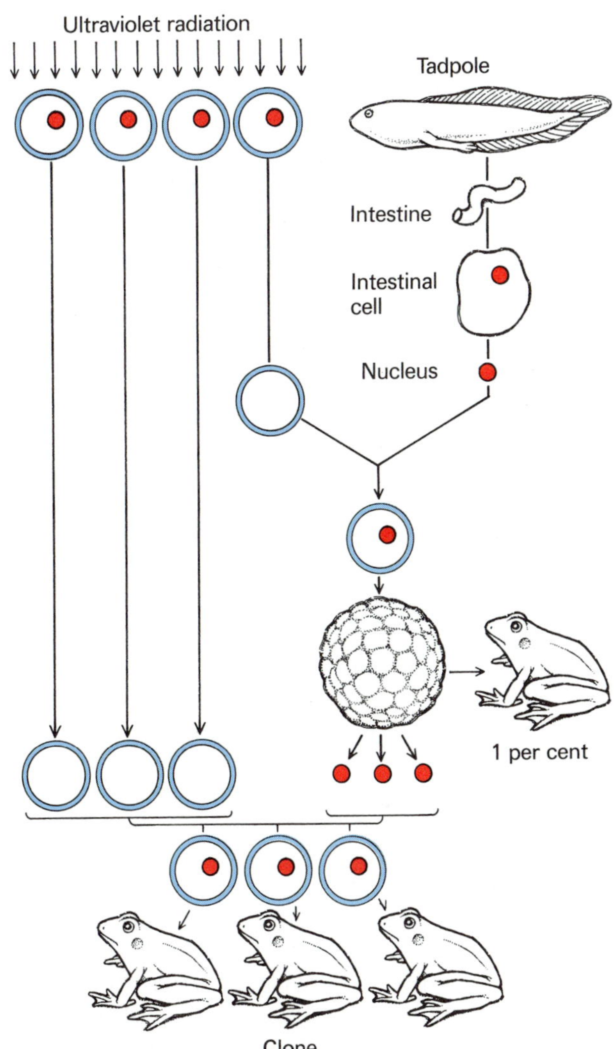

Cloning experiment demonstrating the pluripotential character of the nucleus of a fully differentiated cell. The nucleus of an intestinal cell of a tadpole is implanted into an unfertilized egg cell whose nucleus has been inactivated by ultraviolet irradiation. Some of the renucleated egg cells develop into early embryos (blastulas), of which a small fraction (about 1 per cent) continue their development to the adult frog stage. A second-generation clone of frogs with identical genetic endowment derived from the single original intestinal nucleus can be obtained by similar implantation of a number of nuclei separated from the first-generation embryo.

Jacob-Monod model of the lac operon.

A. A repressor prevents the transcription of the operon.

B. Galactose or another inducer binds to the repressor and renders it unable to bind at the promoter site of the operon. Transcription proceeds and translation of the resulting mRNA leads to the synthesis of three enzymes coded by the genes in the operon.

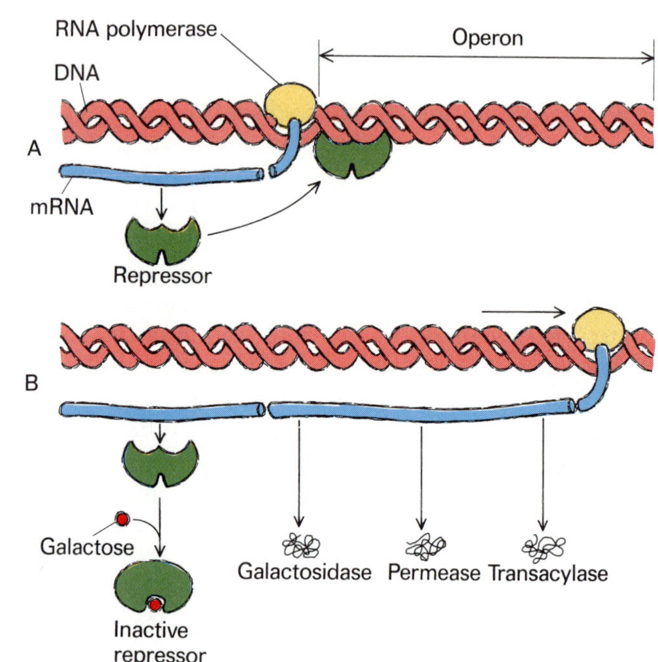

the power of the nucleus. In spite of its central location in the cell and ultimate guardianship of the organism's genetic endowment, the nucleus is not the autocratic despot that we might have suspected. It is not much more than an articulated puppet, admirably constructed and programmed, to be sure, but nevertheless manipulated continuously by the very objects of its control. When a nucleus turns certain genes on or off, it does so in response to orders received from its surrounding cytoplasm, or sometimes from even further afield through messengers produced by other cells, or through drugs, pollutants, or other substances coming from the outside. The cytoplasm, however, is no more boss than is the nucleus; its messages are conveyed or produced by proteins synthesized according to nuclear instructions. In other words, nucleus and cytoplasm do no more than interact with each other in a reciprocally coordinated fashion. The cell is a cybernetic system (Greek *kybernêtês*, steersman). And so, through its superimposed network of cell–cell interactions, is the organism.

As mentioned, the nucleus sends out instructions to the cytoplasm by means of RNA molecules transcribed from certain genes. How, now, does the cytoplasm tell the nucleus what genes to transcribe? Unfortunately, we have a fairly clear answer to this question only for certain elementary bacterial systems, to which the model proposed by the French scientists François Jacob and Jacques Monod is applicable. The main feature of the Jacob-Monod model is the presence of repressor molecules, specific proteins that bind to DNA just downstream of a promoter binding site for RNA polymerase and thereby impede transcription of the particular DNA sequence controlled by this promoter site. In bacteria, such a sequence often includes several genes that are transcribed into a single polycistronic mRNA. Such a set of genes is called an operon. Activation—or, more precisely, derepression—of this set of genes is effected by small molecules, called inducers, that combine with the repressor and thereby modify its conformation in such a way as to cause it to detach from the DNA. The first operon to be recognized was the lactose operon of the bacterium *E. coli*. Bacteria grown in the absence of lactose (milk sugar) lack the ability to utilize this sugar, for want of some key enzymes. But if they are exposed to lactose, they soon acquire the necessary enzymes and thrive on the new substrate. It took the French workers, together with many other investigators, some fifteen years of arduous and ingenious experimentation to discover that this remarkable instance of metabolic adaptation is explained by the ability of lactose to detach from the DNA a bound factor of protein nature (the repressor) that blocks transcription of a set of three genes (the lactose operon) coding for three enzymes of lactose metabolism.

Since these celebrated experiments were performed, a number of other operons have been discovered and much detail has been added to our understanding of their control. Unfortunately, these interesting bacterial systems may not tell us too much about the manner in which eukaryotic genes are regulated. So far, no operon has been identified in a eukaryotic nucleus; neither has a typical

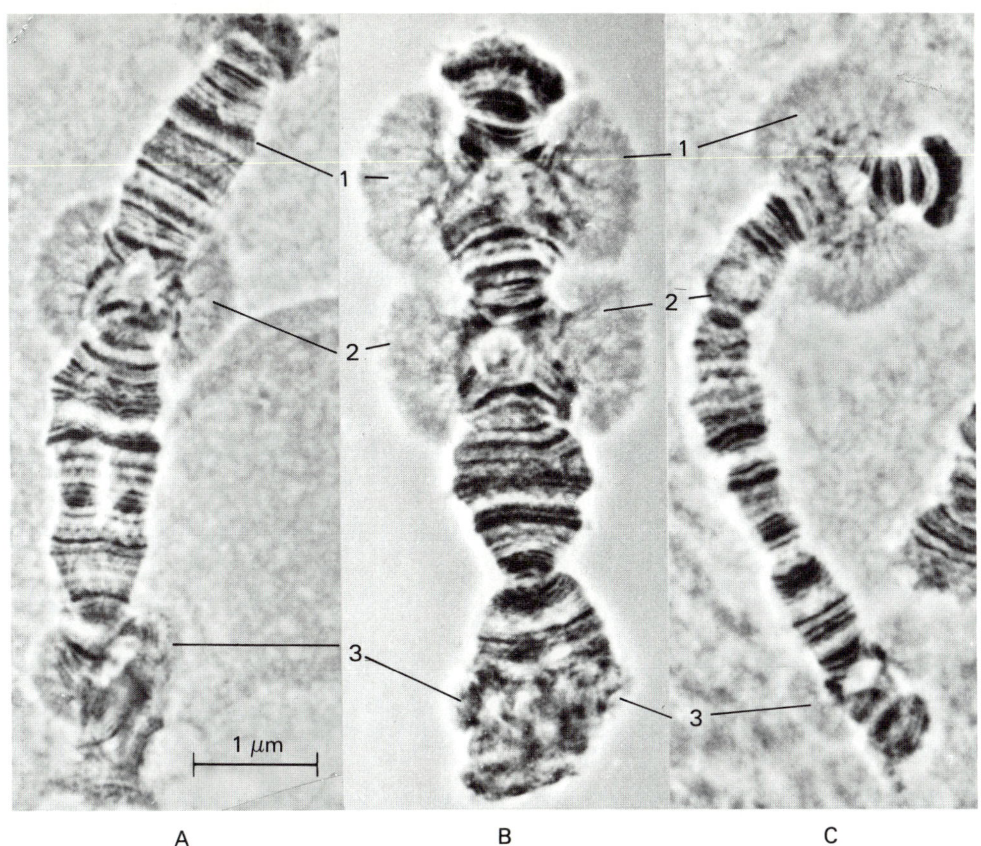

Giant RNA puffs, known as Balbiani rings, on polytene chromosome IV from salivary gland of *Chironomus tentans.*

A. Animal in natural habitat. Region 1 is unexpressed; regions 2 and 3 are fully expressed.

B. Larva exposed to high titer of ecdysone. Regions 1 and 2 are fully expressed; region 3 is poorly or not expressed.

C. Larva fed on galactose. Regions 1 and 3 are fully expressed; region 2 is completely repressed.

repressor protein. On the other hand, some information specifically relevant to eukaryotes has been uncovered, largely through the study of thyroxin and especially of steroid hormones.

The steroids are a family of molecules derived from cholesterol. They include several hormones that act on developmental processes. For instance, androgens control the production of sperm and the expression of various male characters such as the growth of facial hair; estrogens regulate the menstrual cycle and induce such female characters as the development of mammary glands. It has been well established that these effects are due largely to the production of specific proteins by the target cells, a phenomenon which is itself the consequence of the issue of specific mRNAs out of their nuclei. Steroids, in other words, act by turning on certain genes. How they do this is known in roughly schematic fashion. In contrast with most other hormones, which bind to surface receptors, the steroid hormones, being small and strongly hydrophobic, cross the plasma membrane readily and are picked up in the cytosol by soluble receptor proteins. The hormone-receptor complex then moves into the nucleus through one of the nuclear pores and binds to certain parts of heterochromatin, causing them to unroll into actively transcribed euchromatin. Unless you have a means of taking very precise bearings inside the nucleus, you may have difficulties in recognizing this change. But certain insects offer a made-to-order situation in which the arrangement of chromatin is easily perceived, even by rudimentary means. It occurs in the giant polytene (Greek for multistranded) chromosomes found in the salivary-gland cells of diptera larvae. In these cells, chromosomal fibers are duplicated manifold (polyploidy), up to thousands of times, outside of any mitotic activity. Identical fibers are accurately aligned against each other, forming thick, extended chromosomes in which individual genes, or small sets of genes, appear as alternating bands of variable density. As mentioned earlier in this chapter, the Morgan school actually located a large number of genes on these chromosomes. When such larvae are treated with ecdysone (Greek *ekdyein*, to unclothe), a steroid hormone that induces molting, certain condensed bands of the polytene chromosomes are seen to blow up into typical "puffs," which can be shown by special techniques to be centers of active mRNA synthesis.

So much is clear. Not so the mechanism whereby hormones or hormone-receptor complexes disperse DNA fibers and stimulate their transcription in certain selected areas of the genome. One of the prerequisites of transcription is the uncoiling of the chromatin threads and loosening of the nucleosomes. Understandably, many cell explorers are at present scrutinizing the histones around which nucleosomes are built, looking for changes, such as acetylation or phosphorylation, that might underlie the structural changes that accompany transcription. But all nucleosomes contain the same set of histones. There must, therefore, be sites on the DNA itself that are recognized by the hormones or their receptors in such a manner as to attract and put to work the loosening enzymes and RNA polymerase.

The selection of certain genes for transcription is only the first step in a process that also involves extensive subsequent editing of the primary transcripts, sometimes in more than one way. No doubt a number of additional controls are exerted at this level. Some explorers even claim these to be the main controls. They believe gene expression in eukaryotes to be regulated, not so much transcriptionally, as it is in bacteria, but posttranscriptionally, through selection of the messages that are to be edited and issued to the cytoplasm. Furthermore, as we shall see in Chapter 18, extensive pretranscriptional editing of genes may also occur, though probably in only a few exceptional instances.

A remarkable example of genetic regulation of a particularly drastic kind takes place in response to fever, and perhaps other stressful situations. Known as the "heat-shock" phenomenon, it is characterized by the sudden rapid synthesis of a set of five or six proteins, with the concomitant inhibition of the formation of most others. According to some reports, temperature may act by dismantling the RNP complexes that are involved in RNA splicing; pre-mRNA would continue to be made at close-to-normal rate but would no longer be processed and transferred to the cytoplasm. Transcripts that contain no introns would continue to be delivered and translated, and would be so even at a considerably faster rate, perhaps because they appropriate the protein-synthesizing machinery left idle by the slowing down of mRNA production. The resulting "heat-shock proteins" are believed to play a role in the defense against stress.

17 DNA Replication and Repair

While we were struggling through the dense nuclear underbrush and seeing its tangle of fern fronds, lianas, and other strange molecular vegetation turn into thousands of transcription centers, all busily copying and processing instructions for delivery to the cytoplasm, a momentous decision was being weighed in this very cytoplasm: to divide or not to divide?

A Fateful Commitment

The question confronts every daughter cell after it has emerged from the so-called M phase (for mitosis) and entered the G_1 (gap 1) phase of its cycle. If the answer is yes, the order is relayed to the nucleus, where preparations are started for duplication of its information content. Within a few hours, the cell enters the S phase (for DNA synthesis), which can be conveniently detected by timing the moment when the cell incorporates added radioactive thymidine into DNA. From 6 to 10 hours later, the entire genome has been replicated and the cell goes into the G_2 (gap 2) phase, where it stays for the time it needs to get ready for mitotic division (2–6 hours). Mitosis itself takes about 1 hour. The length of the complete cycle, for the cells of higher animals, is of the order of 20 to 24 hours. Bacteria, which have no nucleus and a much simpler life cycle, may start duplicating their DNA even before separating from their twins. Their doubling time can be as short as 30 minutes.

The rapid growth of bacteria is of great practical importance in both laboratory research and industry. Given

The cell cycle. After its birth by mitotic division (M), a cell enters the Gap-1 (G_1) phase, in which it may remain indefinitely (G_0). Commitment to division occurs at some stage of the $G_1(G_0)$ phase, leading after a few hours to duplication of the DNA load of the nucleus by new synthesis (S). A short Gap-2 (G_2) phase separates the end of DNA duplication from the onset of cell division. The approximate duration of each phase in a 24-hour cycle is 1 hour for M, 11 hours for G_1, 8 hours for S, and 4 hours for G_2.

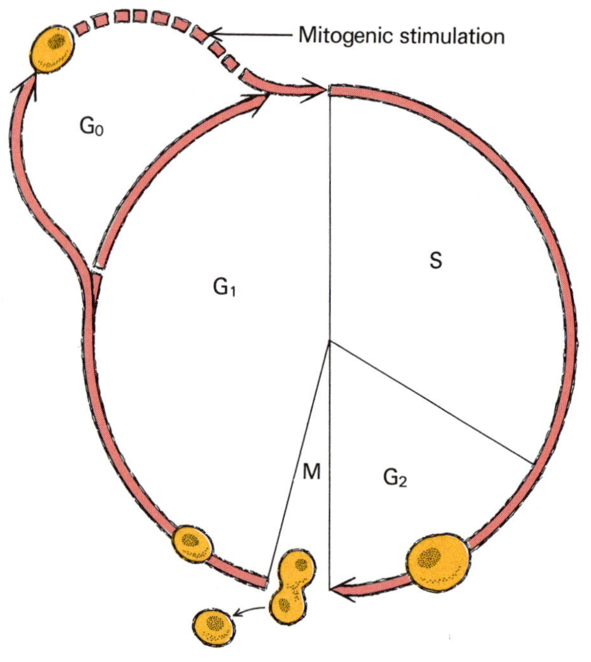

an adequate supply of nutrients, a single bacterial cell can generate 280,000 billion ($2.8 \times 10^{14} = 2^{48}$) individuals in a single day. During that time, the average human cell has barely doubled. But don't let the word "barely" delude you. If allowed unrestricted, exponential multiplication, such a cell could produce the equivalent of an adult human body in 6 weeks and of the volume of our planet in just about 4 months. Even an embryo does not grow that fast; in fact, the net growth rate of the developing organism decreases steadily until it becomes essentially zero in the adult. Cell death accounts only minimally for this limit to growth. With the exception of a few cell types—primarily those of blood, skin, and mucosae—most cell populations in the adult organism turn over very slowly or not at all (in contrast to their constituents, which we have seen turn over at appreciable rates). This is because cells have the ability to interrupt their growth cycle and to remain indefinitely in the G_1 phase, in which case they are said to be in G_0 phase.

What orders this all-important shunting is still very poorly understood. Crowding is one factor. Fibroblasts cultured on a flat dish stop growing when they reach confluence. Somehow, when the cells touch each other (contact inhibition) or reach a certain density (density-dependent inhibition), surface receptors are activated that switch the cells into G_0 phase, perhaps with the help of cyclic AMP. Something similar occurs in vivo. If one removes part of the liver from a mouse, the remaining liver cells immediately resume their growth cycle. They go on dividing until the original weight of the organ has been restored, and then they return to G_0. Similarly, after nephrectomy, the remaining kidney undergoes "compensatory hypertrophy." These phenomena, however, can hardly be explained simply on the basis of contact inhibition or lack of it, since the cells appear to sense at a distance the total number of their congeners that are present and to switch from G_1 to G_0 or back according to the information they receive. Presumably, some humoral factor is involved. The discovery of cell-specific growth factors is beginning to shed light on this central problem of growth regulation.

These examples, together with many other known instances of mitogenic stimulation, make it clear that the switch to G_0 is not irrevocable. Cells can be reawakened from their quiescent state by a variety of chemical calls. Many important physiological mechanisms depend on this property, which, however, will serve its purpose only if the return to G_0 can again be enforced stringently at some appropriate time. Let a single cell discover a means of escaping this constraint and of passing on the secret to its progeny, and an ineluctably fatal exponential growth process will be initiated, even if the loophole is of such poor quality that one might be tempted to brush it off as of no consequence. Imagine, for instance, that a cell has found an inheritable way of getting back illegally into G_1 only once in every hundred days. No cause for alarm, you might say. Indeed, it will take the lawbreaker's offspring five years to reach the tiny weight of one milligram and eight years to grow to the size of a small pea that might possibly be detected if favorably situated. After that, however, things suddenly start moving precipitously, leaving very little time to act. Two years later, the pea has turned into a one-pound tumor, which would itself require only another two years to reach the size of the body were not its growth interrupted by the death of its exhausted host. It is the whole drama of cancer. No wonder so many cell explorers now focus their scrutiny on the fateful G_0-G_1 switch.

What long-range future lies in store for the particular cell that we are visiting is hidden from us. But its immediate fate is clear. Several unmistakable signs tell us that the cytoplasm has reached, and relayed to the nucleus, the decision to move full steam ahead from G_1 into S. Cohorts of unfamiliar proteins are infiltrating the nuclear sap. And everywhere, even in the darkest corners of dormant heterochromatin, faint tremors are beginning to stir the curled-up DNA threads. It is obvious that some portentous event is about to take place.

The Secret of Faultless Copying

The event we are about to witness, as no doubt you have guessed, is DNA replication. To make sure of a front seat, let us move to one of those points where a chromatin fiber is anchored to the nuclear matrix. It is an irregularly shaped knob, made of a number of different proteins. As we watch it, this structure is clearly undergoing some sort of reorganization, apparently triggered by the advent of new proteins. Suddenly, it comes alive; it starts reeling in its attached chromatin arms from both sides and extruding them through its center in the form of a widening loop. You might well wonder at the utility of this molecular cranking until you look more closely at the chromatin loop that is coming out. It consists not of one loop, but of twin loops, identical in every respect. The machine we are watching is not a simple translocator; it is a replicator, sometimes called a replisome.

While this is going on, neighboring anchoring points have also turned into replisomes. In less than 1 hour, the long chromatin stretches strung between them have been completely dragged in and replaced by the duplicated loops that come out of their centers. In the end, adjacent replisomes, drawn together by their opposed pulls on the tightened chromatin between them, mingle in a final convulsion to join their products into a continuous duplex. Up to several thousand replisomes operate in this way all along each of the 46 chromatin fibers. They work as clusters of from 25 to 100 replisomes, but the different clusters are not all synchronized. Some start early in the S phase, others get into action late, so that it takes them a few hours altogether to finish their job. By that time, each chromatin fiber has been completely duplicated from end to end. This means not only replicating the DNA, but also doubling the number of nucleosomes with the help of new histone molecules provided by the cytoplasm, constructing a second matrix framework, and redistributing the two daughter chromatin fibers around the scaffolds in a way that will allow their subsequent separation as distinct chromosomes. Many of these remarkable events are known to us only through their outcome. They are hidden in the darkness of impenetrable tangles, like the secret life of a jungle, and no explorer has yet succeeded in setting eyes on them. But there is one important exception. Replisomes have been opened and dissected into their main component parts. We now have a fairly good idea of what goes on inside them.

Basically, DNA replication takes place according to the same tail-growth mechanism as does its transcription, described in Chapter 16. Deoxynucleotidyl groups are transferred from the corresponding dNTPs to the 3′-terminal hydroxyl group of the growing chain, the choice among dAMP, dGMP, dTTP, and dCTP being dictated by the nature of the base—T, C, A, or G—facing assembly on an antiparallel DNA strand that serves as guide. Besides these similarities, a number of notable differences distinguish replication from transcription. The most important ones are:

1. Replication affects both strands of the DNA duplex.

2. The product of replication remains associated with the instructing strand as a duplex consisting of one parental and one daughter strand (semiconservative replication).

3. Replication affects the totality of the genome.

4. Replication is several orders of magnitude more accurate than transcription.

5. Unlike RNA polymerase, DNA polymerase, the main agent of replication, cannot start anew with a dNTP as first acceptor; it needs a primer, a short, preformed DNA or RNA chain.

The unit of replication is called a replicon. It represents the entire stretch of DNA that is replicated by a single replisome, starting bidirectionally from the midpoint of the replicon. This central initiation point is called the origin of replication. Bacterial and viral DNAs consist of single replicons, which, in bacteria, may exceed 1 mm in length. In contrast, eukaryotic chromosomes contain hundreds of replicons, sometimes as many as several thousand. Their sizes vary from 50,000 to 300,000 base pairs (from about 15 to 100 μm of DNA, or 0.3 to 2 μm of chromatin).

If the view of replication that we have just been granted is correct—remember, facts are still scanty and imagination much in demand—then each chromatin region attached to the nuclear matrix is a replication origin, and each intervening chromatin segment is made of the halves of two contiguous replicons. At least, this is the disposition when the replisomes have been completed and replication is set in motion. So far, we know the structure of the replication origin only for relatively simple viral systems. The impression gained from such systems, in both the prokaryotic and the eukaryotic worlds, is that certain "palindromic" structures play an important role in initiation. (A palindrome is a word or phrase that reads identically in both directions). For instance, in the small animal virus SV40 (p. 312), the origin of replication contains the following sequence of 27 base pairs, which, except for the central G—C pair, reads identically as written or turned around 180°:

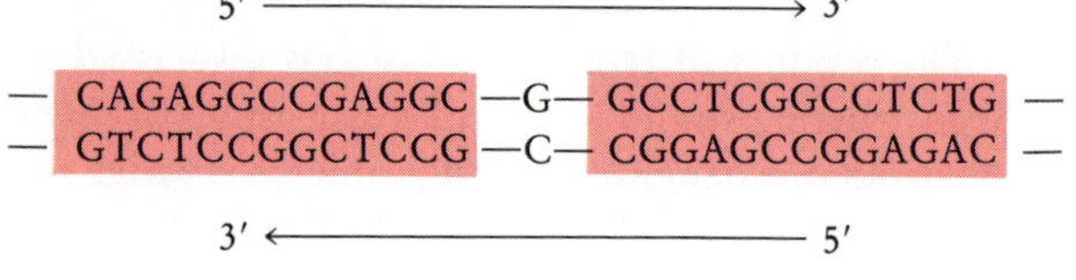

This kind of structure makes sense for a process that is expected to start symmetrically on the two strands and to share their antiparallel polarity. Furthermore, it is also capable of rearranging into opposed "hairpin loops," or cruciform structure, which may help in initiation:

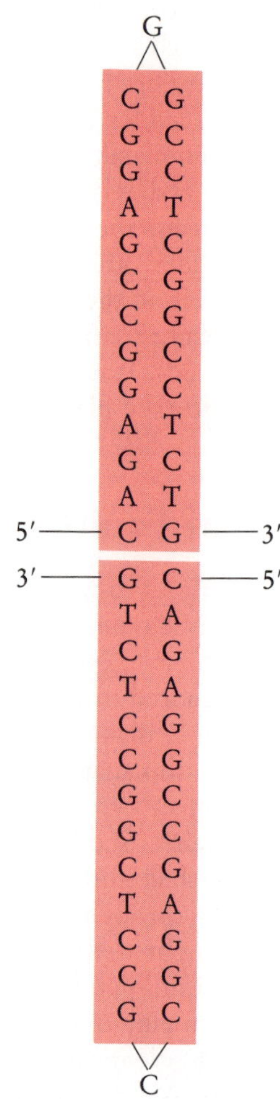

Initiation involves the participation of special proteins known as initiators. It brings into play two distinct events: strand separation and priming. Strand separation requires uncoiling a stretch of DNA at the expense of compensatory supercoiling at both ends. There are enzymes, called helicases and gyrases, that can wring DNA

DNA replication.

1. A replisome has assembled at an anchoring point of chromatin loops on the nuclear matrix (the spooling of DNA in nucleosomes is not shown for simplicity's sake). After strand separation by topoenzymes, replication is initiated in both directions by two primases, each of which assembles the RNA primer of a leading strand and opens a replication fork. The DNA of the replicon controlled by the replisome begins to be reeled in from both sides. A bubble forms between the two replication forks.

2. DNA polymerases have replaced the primases and started the synthesis of DNA on the leading strands. The bubble widens as more DNA is reeled in.

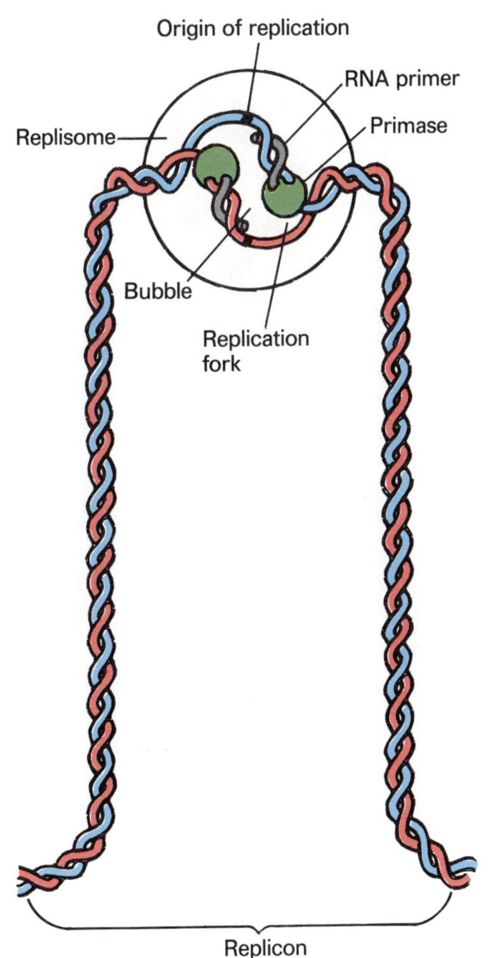

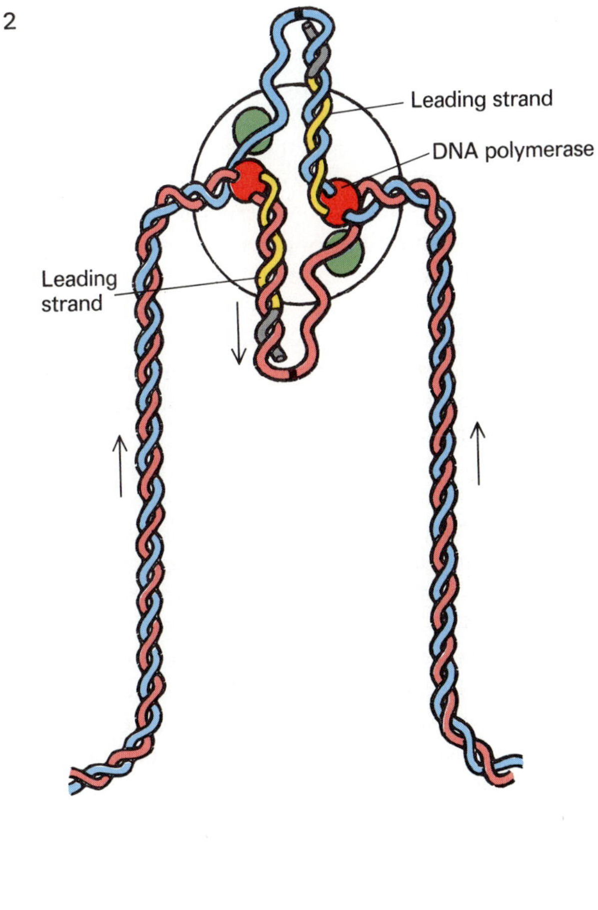

forcibly in one or the other direction with the help of ATP. Certain DNA-binding proteins, called unwinding, melting, or helix-destabilizing proteins, can do the same cooperatively by linking up all along each strand. Finally, stresses in DNA can be relieved by topoisomerases (nicking-closing enzymes), which have the ability of nicking one of the DNA strands, while conserving the bond energy, and of closing the bond again after the supercoiled duplex has relaxed by rotating around the swivel provided by the unnicked strand. We saw these enzymes at work to help transcription. Some topoisomerases even manage the extraordinary trick of nicking both strands, letting a loop pass through the opening, and closing the gap correctly afterward.

Priming is carried out by a special kind of RNA polymerase, called primase, that builds a short stretch of complementary RNA on each DNA strand, with ATP as obligatory starting acceptor at the 5′ end. Only nine or ten nucleotides long, on average, this priming RNA stretch plays an essential role because, unlike RNA polymerase, DNA polymerase cannot start a DNA chain de novo. The priming RNA is removed later.

3. Primases start the RNA primers of the first Okazaki fragments of the lagging strands while leading strands continue to lengthen.

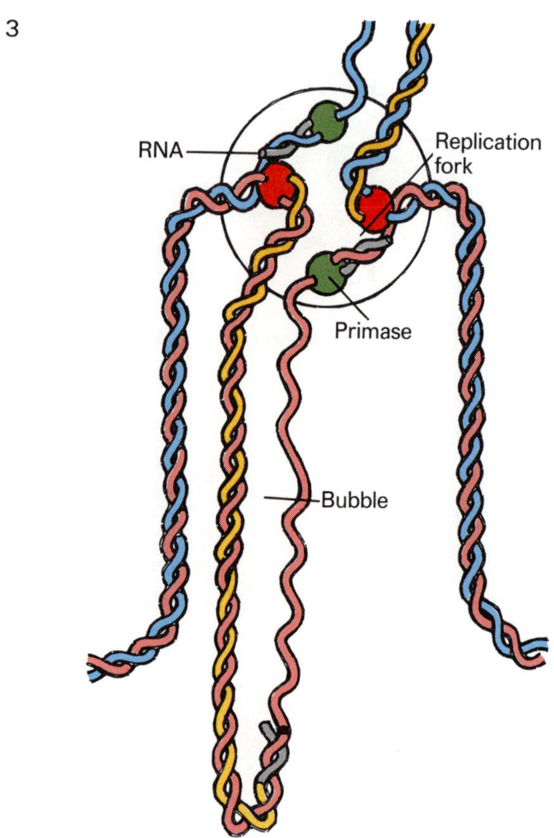

As soon as the priming RNA has been assembled, DNA polymerase takes over and actual replication starts. In this process, the polymerase cuts through the DNA duplex, much as does the RNA polymerase, with the same topological restrictions that require it to be preceded by a nicking enzyme to allow rotation of the template strand. The term replication fork designates the Y-shaped structure at the leading edge of this movement, where the two strands separate from the duplex. Because of the characteristic symmetry of the replication origin and of the replisome, these events take place bidirectionally and generate two opposed replication forks, with the help of two primases and subsequently of two polymerases, assisted by the indispensable topoisomerases.

In transcription, as we saw in Chapter 16, the polymerase—more often a train of polymerases—courses along the DNA following the instructing strand in the 3′⟶5′ direction; the disjoined DNA closes up directly behind the advancing enzyme. Things are different in replication. Each of the two polymerases remains a single and, as far as we know, stationary part of the matrix-bound replisome complex. It is the DNA that does the actual moving, under the pull of the replication process. Relative displacement and dynamics are the same in replication and in transcription. It is simply a question of what moves and what stands still with respect to the nuclear framework.

A more fundamental difference is that the separated DNA strands do not rejoin behind the polymerase, and the gap between them therefore widens to a point at which it can be easily seen in the electron microscope. Called a bubble, or eye, this gap is symmetrical in shape and limited at each end by an advancing replication fork. But watch out: the symmetry is not perfect. Each fork has a thick and a thin branch. The thick branch is the semiconservative duplex between daughter DNA and parental template; the thin branch is the other parental strand, as yet unreplicated. This is because polymerases are one-way enzymes. They can read only in the 3′⟶5′ direction and assemble in the 5′⟶3′ direction. For the same reason, the asymmetry of the forks is inverted: the thick branch of one is continuous with the thin branch of the other, and vice versa, because each polymerase is replicating a different strand, consistent with the direction of its movement.

This mechanism could conceivably suffice in the case of circular DNA if the two polymerases were somehow allowed to complete a full circle along their respective template strands until they bump into the 5′ tails of their own primers. In practice, however, this does not happen, even with short circular DNAs. The thin strands left behind by the two leading polymerases are replicated separately by other primase-polymerase systems. Owing to the obligatory polarity of assembly, this process has to move backward from the direction in which a single-

4. DNA polymerases have replaced the primases and have started the synthesis of DNA on the lagging strands. The arrows show the direction of movement of the template DNA (pink) as it is being reeled in through the replisome.

5. DNA polymerases have almost completed the first Okazaki fragments of the lagging strands and are beginning to displace the RNA-primer ends of the leading strands. The displaced RNA is broken down by 5′-3′-exoribonucleases.

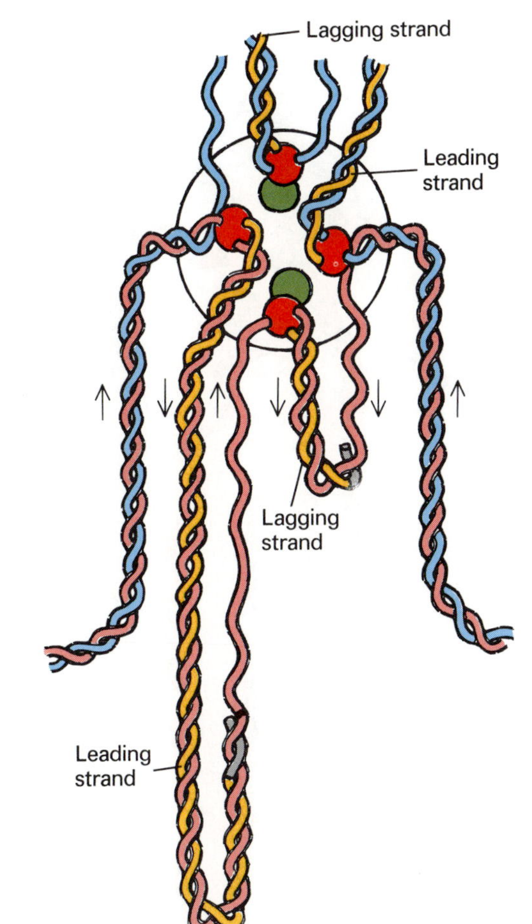

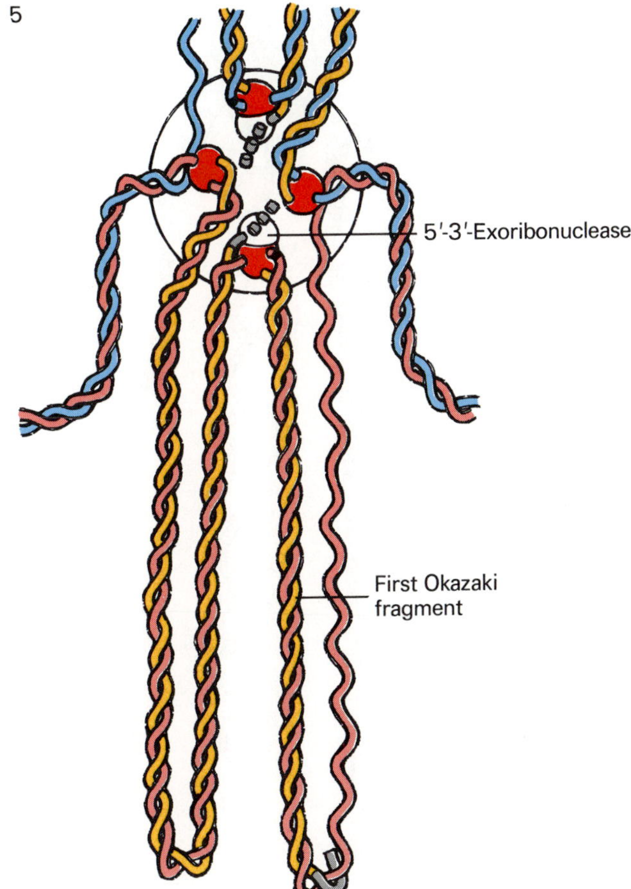

stranded template is made available by the advancing replicating fork. To meet this requirement, replication of the thin strand takes place by sudden forward leaps of the primase system, followed by backward, gap-filling elongation, to produce a series of short DNA stretches that will have to be stitched together later. Thus, whereas each strand is replicated continuously in the 3′⟶5′ direction from the point of origin, it is replicated discontinuously in the other direction. The whole process is called semidiscontinuous. Because discontinuous replication starts only after a certain length of single-stranded DNA has been liberated by the continuous process, the strand it generates is called the lagging strand. That which is synthesized continuously is called the leading strand. The terms forward arm and retrograde arm are also used.

Discontinuous replication functions by the same mechanism as does continuous replication. Primase starts the process and DNA polymerase then takes over until it runs into the 5′ end of another primer. Without further action at this stage, the product consists of a series of discrete DNA fragments (with a short 5′-RNA tail) coiled around the template strand. These fragments are called Okazaki fragments, from the name of the Japanese investigator who discovered them (and later died of leukemia: he had

6. The first Okazaki fragments of the lagging strands are completed and the RNA primers of the leading strands have been removed. Ligases now take over and attach the 3′ end of each Okazaki fragment to the 5′ end of each leading strand.

7. Primases have started the second Okazaki fragments of the lagging strands, while the leading strands (with first Okazaki fragments attached to their 5′ ends) continue to grow. Steps 4 through 7 are then repeated as many times as is required to reel in the DNA of the replicon completely.

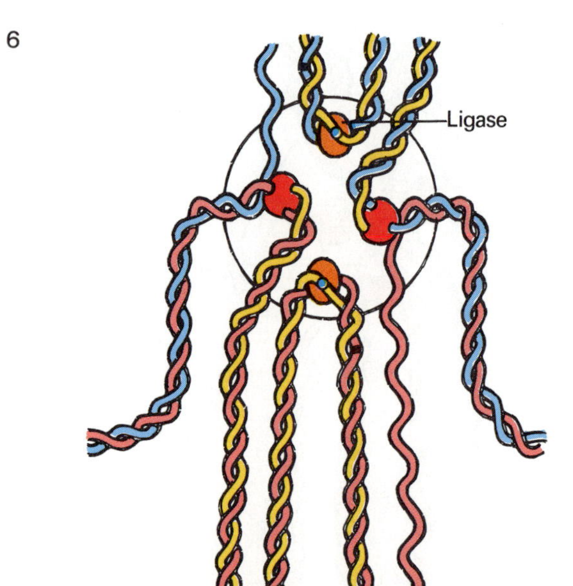

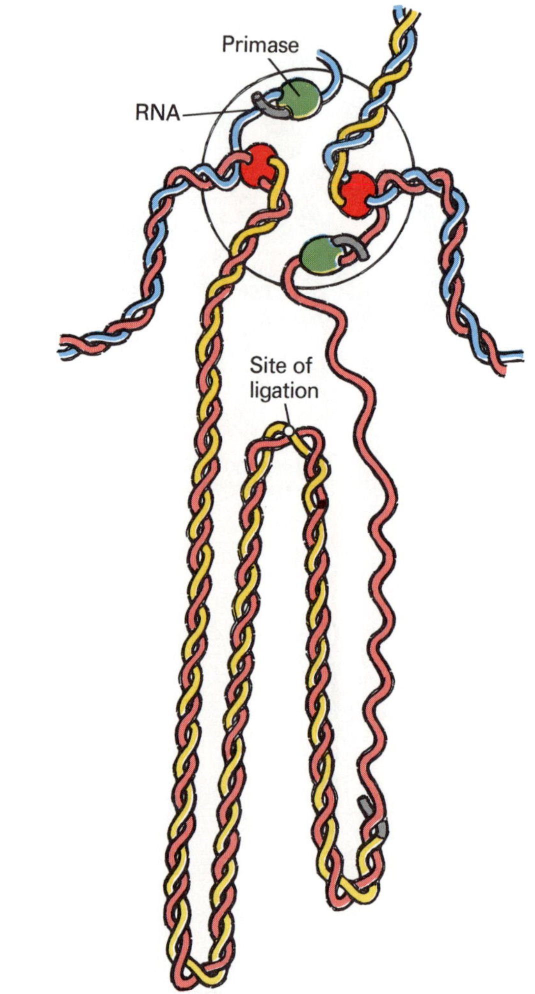

the misfortune as a child to grow up in Hiroshima). These fragments are some 1,000 to 2,000 bases long in bacteria and only some 180 to 280 bases long in eukaryotic cells. The latter periodicity is believed to be related to the organization of eukaryotic DNA into nucleosomes. It certainly indicates that primase is not finicky as concerns its attachment site when presented with a single DNA strand.

Completion of the discontinuous process requires the excision of the RNA primer that blocks the further progress of the DNA polymerase, the pursuit of replication until the 5′ end of authentic DNA is reached, and, finally, the splicing of this 5′ end with the leading 3′ end of the elongating fragment. The first step is catalyzed hydrolytically by a special 5′-3′-exoribonuclease (so named because it attacks RNA at its 5′ end, removing 5′-mononucleotides in succession). In bacteria, this activity is an intrinsic property of the main DNA polymerase and is situated at the front of that enzyme so as to clear the way automatically for replication to proceed. The final splicing step is catalyzed by an AMP-forming ligase (Chapter

8. When replication of two adjacent replicons is almost completed, the DNA polymerases that elongate converging leading strands are drawn together by the shortening DNA loops between them and start to reel in the single-stranded DNA supporting the neighboring lagging strands.

9. The RNA primers of the last Okazaki fragments of the lagging strands are displaced and broken down as DNA synthesis proceeds to completion.

10. Ligases join the fully replicated neighboring replicons.

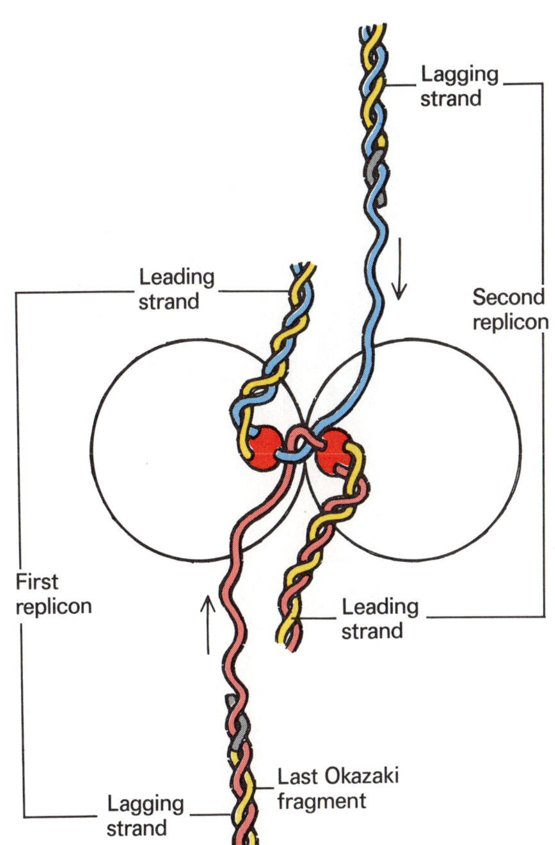

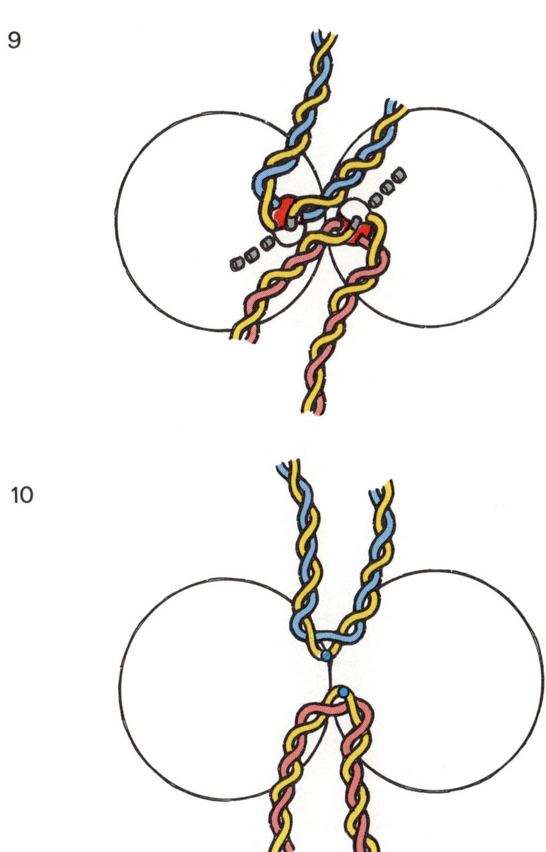

8). This enzyme activates the 5′-phosphoryl end-group by AMP-yl transfer to make an adenylyl-DNA Janus intermediate, which is then attacked on the other side of its central oxygen atom by the free 3′-hydroxyl of the joining partner, with liberation of AMP.

In circular bacterial chromosomes, the two forks meet about halfway round. Their orientation is such that the 3′ head of each leading strand encounters the 5′ tail of the corresponding lagging strand, and vice versa. These ends need only to be joined by the mechanism just described to produce two identical, semiconservative, circular duplexes. Such rings are sometimes linked and must be disentangled by special topoisomerases. In eukaryotic chromosomes, exactly the same kind of head-to-tail joining takes place at the point of meeting between two forks that belong to adjacent replicons. Replication of the whole chromosome is completed when all its replicons are joined. Such are the basic mechanisms of DNA replication. There are many variants, especially in the viral world, but we will not go into those.

Schematic view of how the multiple bubbles shown in the electron micrograph on the facing page are believed to be looped on matrix-affixed replisomes. Unduplicated DNA is shown in blue, duplicated DNA (bubbles) in red.

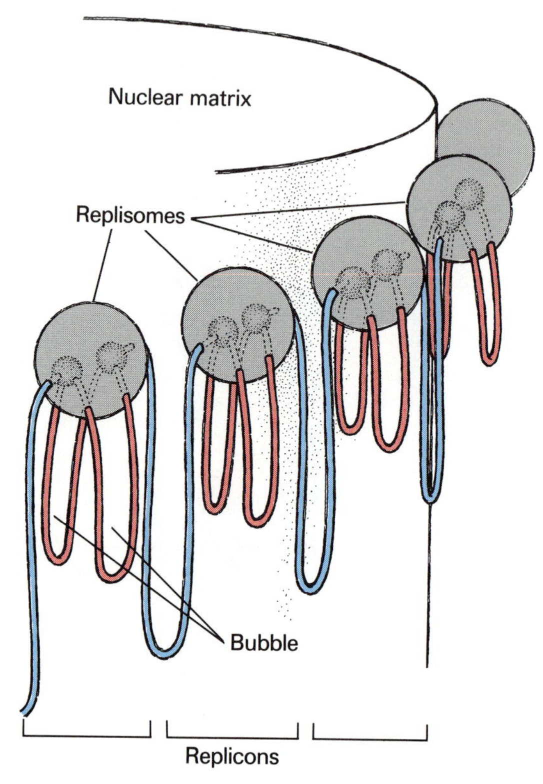

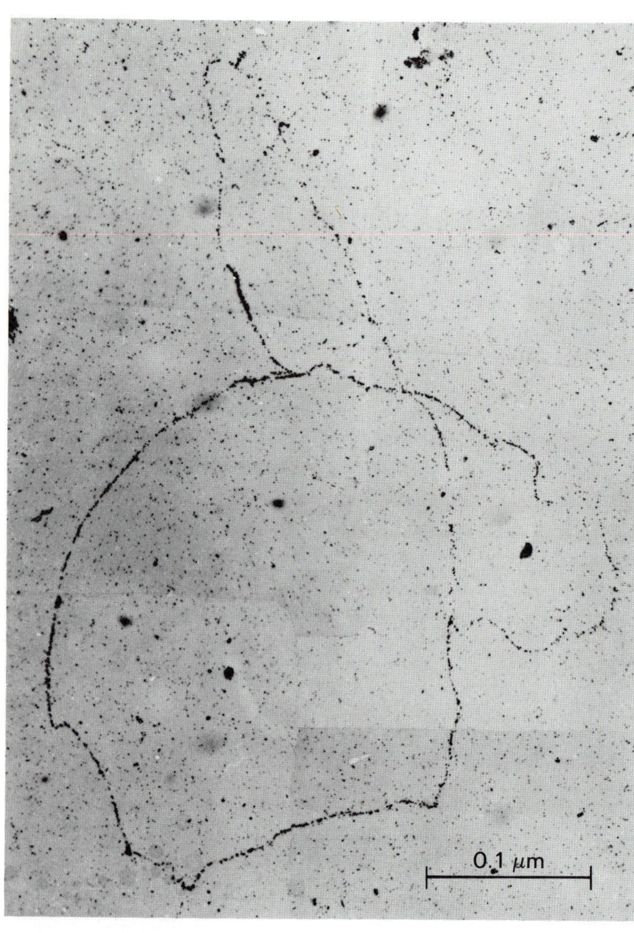

Image of a bacterial chromosome that has been detached intact from the replisome while in the process of duplicating. The two replicating forks are clearly visible. The bubble is twisted into a figure of eight. This picture was obtained by autoradiography. The bacteria were grown on a medium containing radioactive ^{3}H-thymidine to label the DNA. The cells were then killed, and their DNA was extracted by a gentle procedure designed to minimize breakage of the fragile molecules and was spread in close contact with a sensitive unexposed photographic film. Radiation emitted by the DNA impressed the film, tracing the shape of the molecule.

The problem of accuracy does, however, deserve our attention. As mentioned in the preceding chapter, RNA polymerase has an error frequency of the order of 10^{-4}. This is tolerable, considering that the mistakes are likely to be randomly distributed and that many of the resulting base substitutions will not be expressed in functionally defective proteins (owing to the properties of the genetic code or because the substitutions occur at strategically unimportant points in RNA stretches that are not translated). Therefore, transcription mistakes may cause some waste but hardly ever any serious harm. The situation is entirely different for replication. There, the mistakes are made in a master copy, whose defects will be imparted to all of its RNA transcripts and to all of its DNA progeny. They are real mutations. Their consequences may not always be dramatic. But the probability of their being harmful is such that no cell line could survive very long with a 10^{-4} frequency of replication mistakes. Yet, DNA

This electron micrograph shows multiple bubbles on a length of eukaryotic DNA (from the fruitfly *Drosophila melanogaster*) in the process of duplicating. The DNA has been extracted, spread on a surface, and revealed by shadowing with metal vapor.

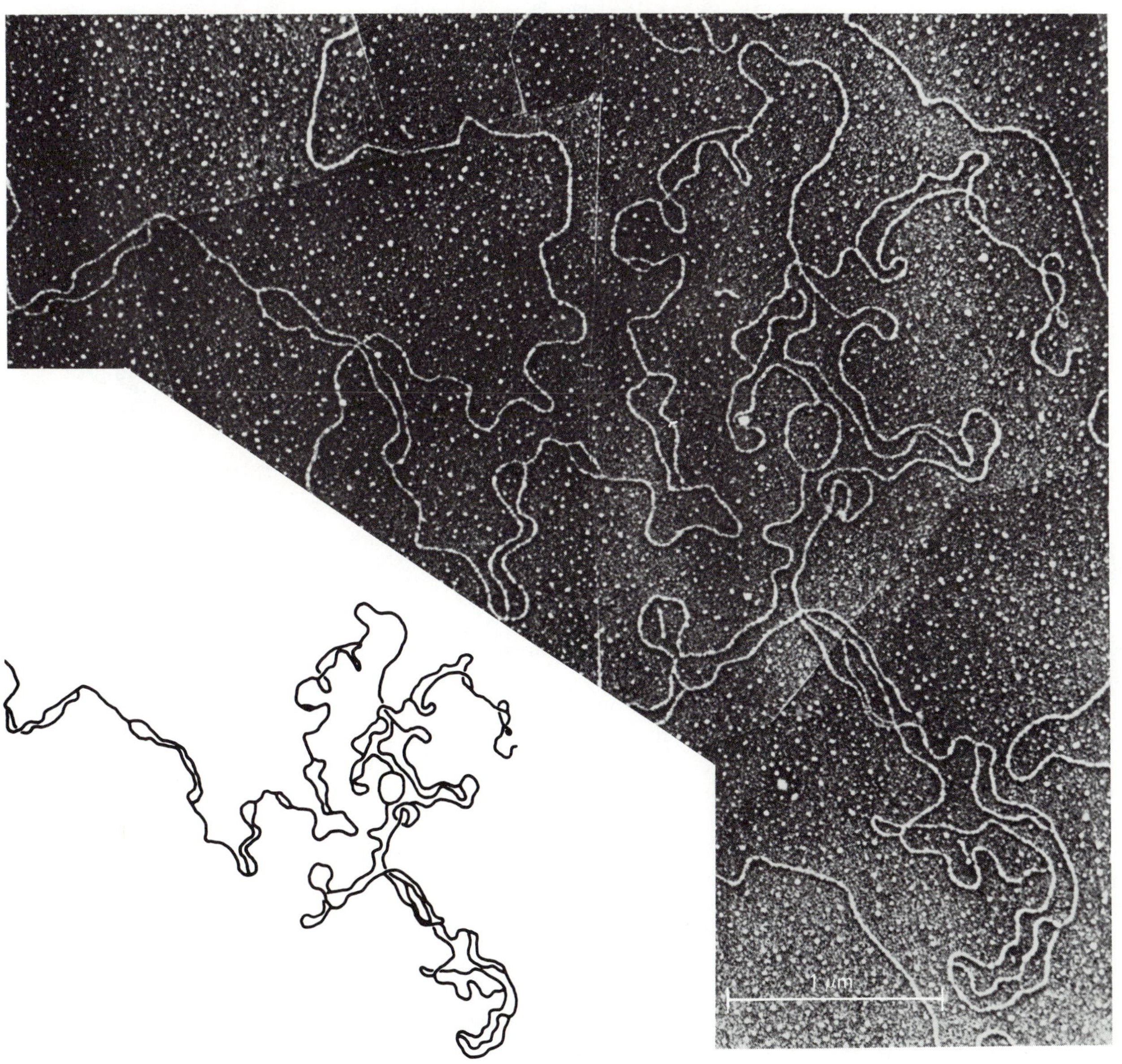

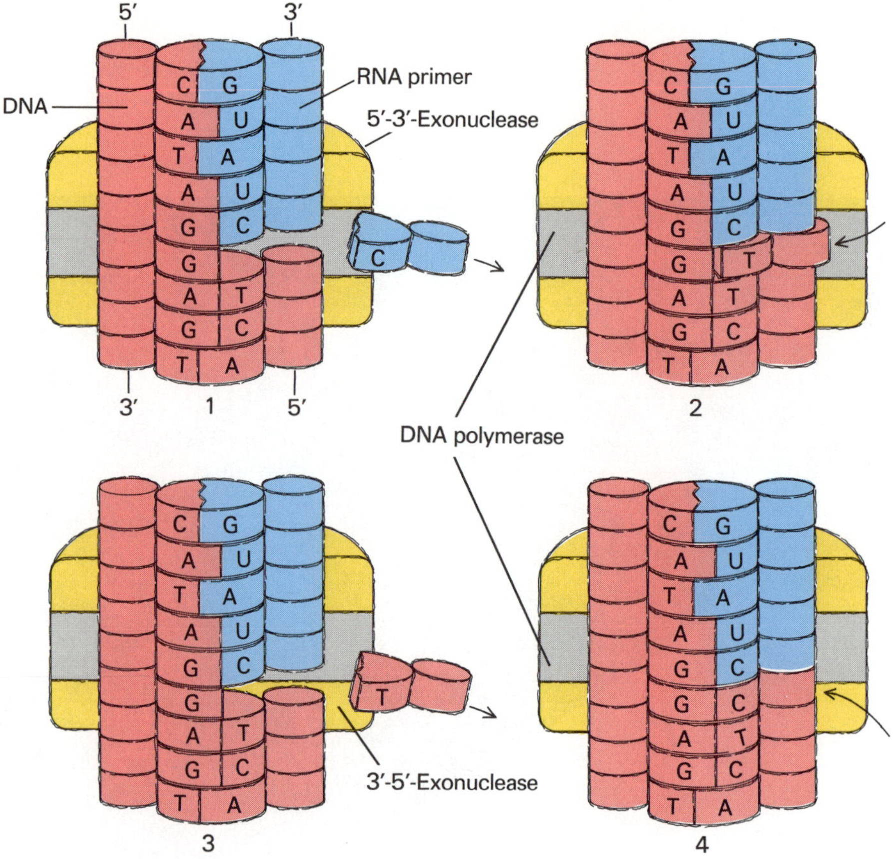

E. coli DNA polymerase I, an enzymic combine shown in action: (1) 5′-3′-exoribonuclease situated in front removes CMP from the 5′ end of the RNA primer; (2) polymerase situated in the middle erroneously attaches a dTMP group to the 3′ end of the growing DNA chain; (3) 3′-5′-exonuclease situated at the back splits off mismatched dTMP; (4) polymerase correctly inserts a dCMP group.

polymerase in itself does not seem to be more accurate than RNA polymerase.

The answer is proofreading, an operation carried out mainly by a special 3′-5′-exodeoxyribonuclease that has the ability to recognize a mismatched base at the 3′ end of a growing DNA chain, and to excise the wrong mononucleotide before it is covered by the next one. Then polymerase can try again, with 9,999 chances in 10,000 of not repeating the mistake. Thanks to this and a few other accessory safeguards, replication mistakes are reduced to a level of some 10^{-8} to 10^{-10}. This is obviously tolerable, but it is far from negligible in regard, for instance, to the number of nucleotides (about 1.2×10^{10}) that are assembled every time a genome the size of the human genome is duplicated.

In bacteria, proofreading is actually carried out by the DNA polymerase itself, which possesses an incorporated 3′-5′-exonuclease activity placed so as to "inspect"—and remove if necessary—each nucleotidyl group immediately after its transfer. This same enzyme, you may remember, also bears on its front end the 5′-3′-exonuclease activity needed to clear away any obstructing RNA primer. So far, no such remarkably coordinated molecular "combine" has been spotted in eukaryotic cells. But this may be simply because its pieces fall apart more easily upon isolation. Certainly there is no reason to believe that the efficiency and fidelity of replication are of lower quality in eukaryotes than in prokaryotes.

Considering the many steps involved, the speed of DNA replication is remarkable. In eukaryotes, it proceeds at the rate of about 1 μm per minute, comparable to that of transcription. In prokaryotes, it can be as much as 30 times as fast, which means the addition of 1,500 deoxynucleotides per second and rotation of the template DNA at the fantastic speed of 9,000 revolutions per minute. Presumably, eukaryotes owe their slower rate of DNA replication largely to the organization of their DNA into nucleosomes. However, they more than compensate for this drawback by the smaller sizes of their replicons and Okazaki fragments, which allow a much

Except for the necessary loosening of their structure, nucleosomes are not dismantled during replication. Pre-existing histone spools remain with the leading strand. Newly made histones form spools for the lagging strand. Whether they bind the strand before replication (as shown in the diagram) or after it is not known.

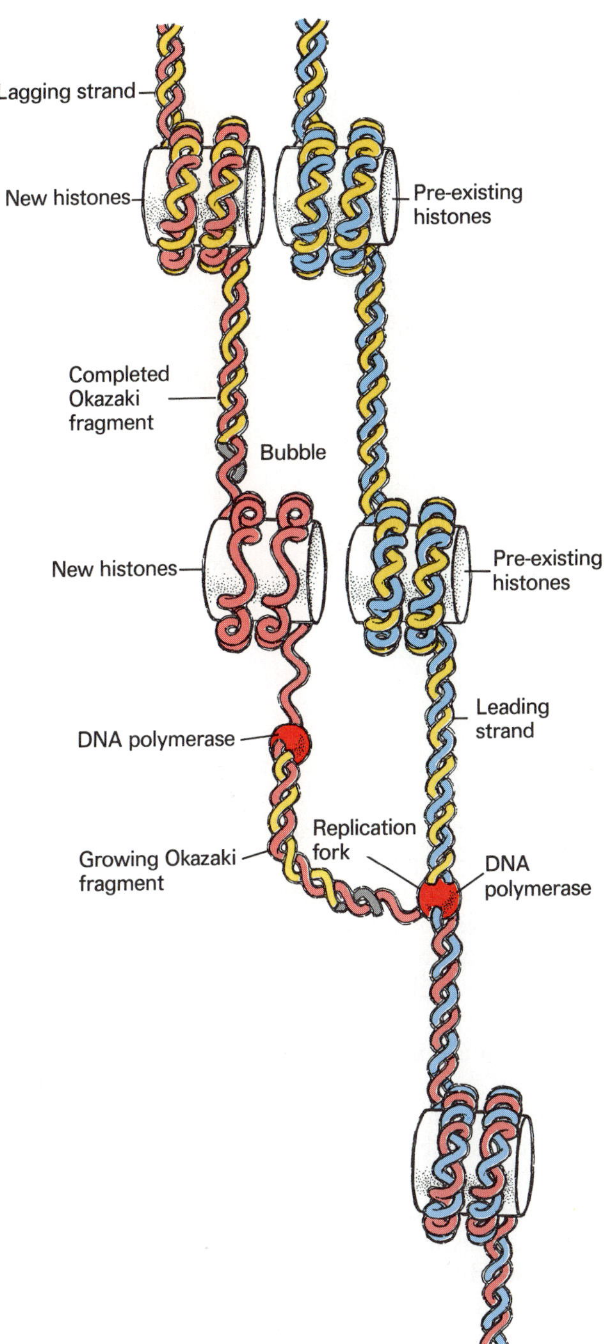

larger number of replicating systems to be put to work on a given length of DNA. In consequence, a human cell will take only ten times as long as does a bacterium to duplicate a genome that is 2,000 times the size of the bacterial genome.

We are unable to appreciate what additional topological problems the nucleosomal organization creates in eukaryotes or how they are solved. But one thing seems clear. There is no sharing of histones between new and old nucleosomes. Apparently, the leading strand hangs on to the pre-existing histone cushions, and newly made histones sent in from the cytoplasm join the lagging strand. The similarity between the length of DNA per nucleosome and that of the Okazaki fragments suggests that priming on the lagging strand may be related in some way to nucleosome spacing.

DNA Maintenance and Repair

They are the silent patrol, and they never sleep. Day and night, as much in the broad expanses of nucleoli and euchromatin as in the densest recesses of heterochromatin, they remain tirelessly on the beat, looking for flaws, possible mistakes, accidental blemishes in the cell's irreplaceable store of genetic information, ready to pounce on the slightest hint of ambiguity to set the record straight. They are the cell's DNA repair crew.

As often happens, we first learned of the existence and importance of this invaluable service from the consequences of its breakdown. There is a rare group of patients to whom sunlight is fatal. They suffer from a genetic disease called xeroderma pigmentosum, in which the skin (Greek *derma*), when exposed to ultraviolet light, develops pigmented hardenings (Greek *xêros*) that transform into cancerous lesions. Overexposure to sunlight (watch out, fanatics of the bronzing cult) can produce similar lesions in normal individuals. What happens in these sun victims is a progressive deterioration of their skin DNA due to ultraviolet irradiation injuries. The cause of the disease is not, as was first thought, hypersensitivity to UV rays, but rather faulty repair of lesions that

Excision-repair mechanism.

1. A special excision endonuclease recognizes a lesion and cuts the affected strand on the 5′-end side of the lesion, creating a free 3′ hydroxyl group.

2. DNA polymerase lengthens the nicked strand and displaces the segment bearing the lesion.

3. The displaced piece bearing the lesion is excised by 5′-3′-exonuclease action.

4. Ligase replaces DNA polymerase and stitches the 3′ end of the new segment to the 5′ end of the old strand.

5. DNA is repaired, with a short stretch of newly made DNA replacing the piece bearing the lesion. This stretch could be recognized if a radioactive precursor (for instance, ^{3}H-thymidine) had been offered to the cell between stages 1 and 3: it would be "hot," as opposed to the rest of the DNA which would be "cold." The extent of DNA repair going on in a resting nucleus (in which no duplication occurs) can thus be gauged by the amount of radioactivity incorporated.

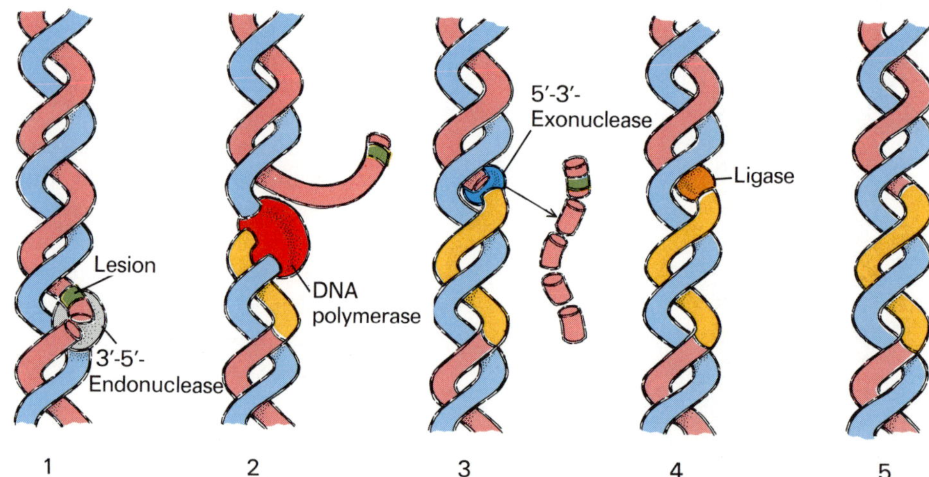

are not in themselves abnormally severe. Similar deficiencies have been observed and investigated in detail in certain bacterial mutants.

UV light is but one of the many agents that can alter DNA. X-rays, radioactive emissions, free radicals, many different kinds of chemicals, certain enzymes, and, as we have seen, replication errors that escape proofreading—all participate in what is really a constant assault on the genetic orthodoxy of living cells. Modern technology has added a sizeable contingent to this attacking army, but Nature is not as benevolently innocuous as some of our Dr. Panglosses would like us to believe. If it were, cells would not have developed the multiple defenses that they have, and, of course, the evolutionary lottery could never have taken place. The lesions that DNA may suffer are of many different kinds. Strands may be broken or bases removed by hydrolysis. Alternatively, bases may be chemically damaged or mismatched or joined covalently, either directly—the main effect of UV light is to cause dimerization of thymine—or by means of cross-linking agents. So-called intercalating substances may slip in between adjacent base pairs and dislocate the DNA helix.

As is to be expected, the nature of the repair mechanisms depends on the nature of the lesion. Sometimes relatively simple correction of the damage is possible. For instance, a hydrolytic nick is easily repaired by ligase action. Or sunlight may be coaxed into undoing its own damage with the help of a special photoreactivating enzyme that disjoins the UV-induced thymine dimers. In most cases, however, the lesion is irreversible and has to be removed. This is done by glycosylases that split off altered bases or by endonucleases (attacking bonds inside the DNA chain) that cut out a small oligonucleotide segment containing the adulterated part. A new, correct sequence is assembled by a polymerase, and its attachment is completed by a ligase, as in DNA replication.

For such an excision-repair mechanism to function correctly, the enzymes that do the surgery must be capa-

Correlation between life-span and intensity of DNA repair, as assessed by the speed of incorporation of radioactive thymidine into the DNA of resting nuclei in cells that have been exposed to UV irradiation: (left) in different mammalian species; (right) in different primate species.

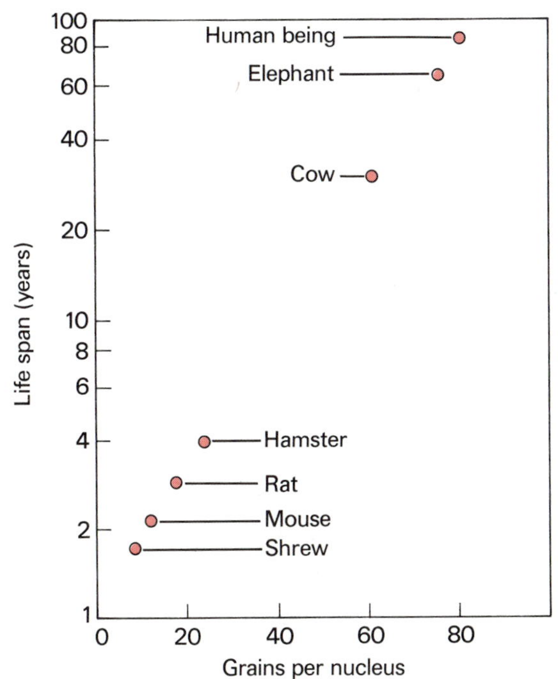

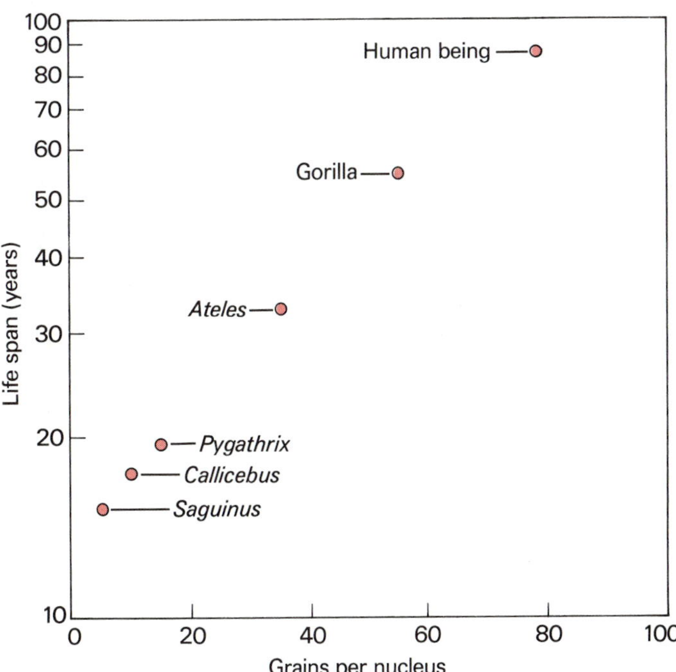

ble of recognizing the lesion. If one strand is obviously abnormal and the other normal, one can visualize, at least theoretically, how such recognition may take place. But what if the two strands are normal but simply mismatched—say, a thymine paired with a guanine? How is the enzyme to decide which of the two is wrong? Indeed, there are cases of genuine ambiguity, where repair has only a 50 per cent chance of being right. But it also happens that the excision enzymes are provided with secondary clues. Methylation is one. If one strand is methylated in the neighborhood of the mismatch and the other is not, the unmethylated strand is identified as the younger of the two and therefore most likely the faulty one, a product of incorrect assembly overlooked by the proofreading system. Excision enzymes are programmed to favor unmethylated strands.

A second prerequisite of correct repair is that the genetic information not be irretrievably lost. One strand must retain the authentic base sequence, correctly associated with the faulty strand, so as to give the right instructions to the replicating polymerase. For this reason, a double nick at the same level is generally fatal. Also, single strands in replicating forks may be difficult to repair. Even lesions of this sort may occasionally be repaired correctly if adequately splinted—for instance, by DNA-binding proteins. You may remember that certain topoisomerases cut a double strand in such a way that a DNA loop can actually pass through the cut, and yet are able to repair the damage correctly afterwards.

Although not infallible, our DNA repair crew is certainly remarkably efficient, as well as constantly in demand. The sorry plight of xeroderma patients makes this abundantly clear. It may even be that the secret of old age lies in the quality of this crew, as suggested by the evidence of a striking correlation between life span and speed of DNA repair. As illustrated by the charts at the top of this page, such a correlation has been observed among different mammalian species, as well as within the order

These two relatives (top: house mouse, *Mus musculus;* bottom: the white-footed mouse, *Peromyscus leucopus*) differ strikingly in life span: *Mus* lives at most 3.5 years, *Peromyscus* as much as 8.3 years. It so happens that *Peromyscus* cells repair their DNA two and one-half times as fast as do those of *Mus.*

of primates (with humans the winners). It has even been found to exist between two kinds of mice.

To enjoy such excellent service is clearly an invaluable boon, but there are circumstances where it is tiresome. Cancer therapy is one. Many of the agents used in our fight against this disease—irradiation, intercalating drugs such as daunorubicin and Adriamycin, cross-linking drugs such as mustard gas and its derivatives—act by harming DNA and thereby blocking or derailing its transcription and replication. One of the reasons these efforts are so often thwarted lies in the ability of the target cells to correct the injuries inflicted on their DNA. And, of course, natural selection sees to it, in its utterly blind and insensitive way, that the cells best equipped to resist the therapy remain to spread and proliferate.

Another kind of service that cells maintain to protect the purity of their genome involves the recognition and destruction of foreign DNA. This is obviously very difficult if there is nothing special about the foreign DNA, and there are, in fact, innumerable occasions on which cells are misled into accepting foreign DNA brought in by viruses or other agents, sometimes to the point of yielding their whole genetic machinery to the invader. But some defense is possible, thanks to the restriction enzymes, so named because they restrict the compatibility between host and invader. Restriction enzymes are endodeoxyribonucleases that recognize and split certain specific sequences, usually of self-complementary palindromic structure. We have already encountered a similar kind of structure at the origin of replication. One of the first known restriction enzymes is EcoRI, extracted from the bacterium *E. coli.* It has the following specificity (arrows indicate site of splitting):

```
      ↓
5′--------GAATTC--------3′
3′--------CTTAAG--------5′
               ↑
```

Restriction enzymes have become of paramount practical importance because they allow DNA to be cut into well-defined segments, limited by the so-called restriction sites. These segments are used in DNA analysis and for genetic-engineering purposes. When, as in the case of EcoRI, the splitting is asymmetrical, the resulting segments have "sticky," or cohesive, ends. They tend to join by base-pairing with any other segment that has been produced by the same enzyme and, therefore, possesses the same ends. As we shall see, this property is widely exploited in genetic engineering. For these reasons, the discovery of the first restriction enzymes has spurred a massive hunt for similar enzymes with different specificities. Today, more than two hundred different restriction enzymes are known. All, so far, have been isolated from bacteria.

But what about their physiological role? Restriction sites are so simple that they are hardly likely to distinguish foreign from indigenous DNA. Rather, the opposite would be expected, since the invading DNA is usually

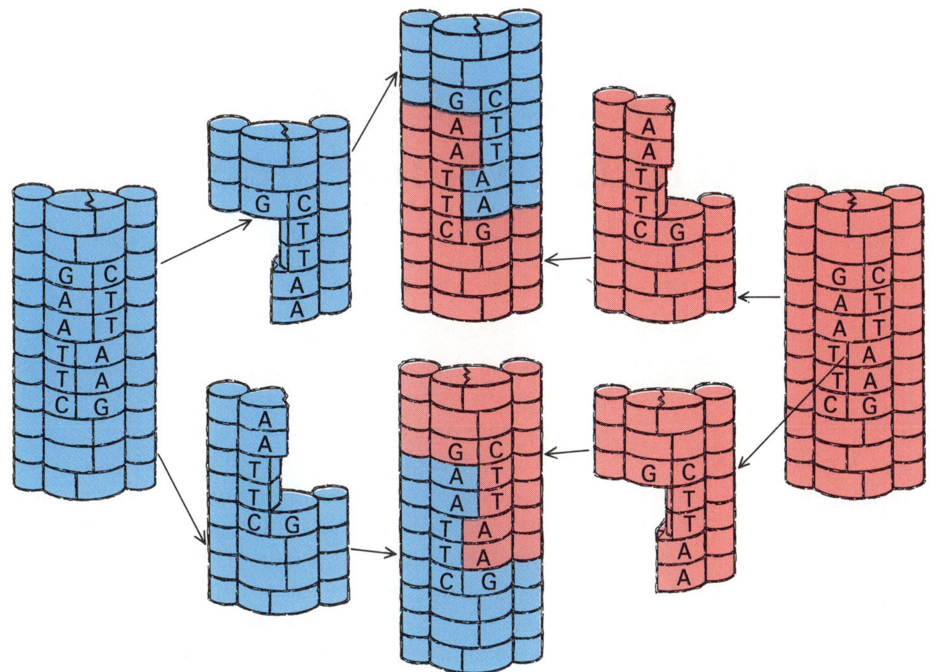

Hydrolysis of two different DNAs possessing the appropriate restriction site (GAATTC palindromic sequence) by the restriction enzyme EcoRI produces fragments with "sticky ends." Treatment with ligase reassociates the pieces at random, generating hybrids of the two DNAs, among other products. This technique is widely used in genetic engineering (see Chapter 18). Note that, if the restriction site is methylated (on central adenines), endonuclease splitting by EcoRI is prevented. *E. coli* uses this device to protect its own DNA against the action of its EcoRI.

much smaller than the host-cell DNA and may therefore be expected to lack—and indeed often does lack—restriction sites that are present in the host DNA. The recognition trick, once again, depends on methylation. Bacteria that use a given restriction enzyme protect the corresponding restriction sites in their own genome by methylation. Restriction enzymes act only on unmethylated sequences. Of course, if the invader is clever enough—that is, if it has been retained by natural selection owing to that property—to methylate its own restriction sites, it will not be recognized as foreign. And so the battle goes on. Note that this is the third time we have encountered methylation as a protective device against nuclease action. Only moments ago, we met it as a means of avoiding excision of the correct sequence in a mismatch. In Chapter 16, we saw how it serves to guide the processing of rRNA. In Chapter 18, we will meet it in a different capacity: as a means of silencing DNA and of inhibiting its transcription.

18 Recombination and Other Genetic Rearrangements

All that we have seen so far is geared to favor the strict preservation of genetic orthodoxy. In no state or church has information been policed with such rigor and deviation been prosecuted with as much severity as they are in the nucleus. Looked at from the point of view of the cell, such literalism makes sense. A living cell has a good thing going, and the dangers of change are so much greater than its possible benefits as to make heresy almost inevitably self-purging. Yet, diversity is the spice of life, variability the source of innovation in the biosphere. They are introduced into the genetic game by two kinds of accidents: some are truly accidental and fortuitous; others are made to happen by a genetically programmed enlistment of chance.

Shuffling Genetic Cards

It has been known since the beginning of this century that chromosomes may exchange segments. This happens during the meiotic maturation of germ cells, when, after duplication of their DNA content, paired chromatid dyads line up closely against each other and exchange homologous pieces by crossing-over before they separate and divide to form distinct haploid complements in the four daughter cells (Chapter 19). This phenomenon enormously enhances genetic diversification through sexual reproduction. Thanks to it, the haploid set carried by each spermatozoon or ovum does not simply consist of whole parental chromosomes selected at random from pairs of homologues, as would be the case

A mechanism for legitimate recombination.

1. A nick in a strand allows DNA polymerase to start elongating the 3′ end freed by the nick.

2. As the polymerase progresses, it displaces the nicked strand, as it does in a normal replication fork (Chapter 17).

3. The displaced strand becomes entangled with the complementary strand of the closely apposed allelic DNA, in turn displacing its counterpart as a loop.

4. At some stage the polymerase drops off, the displaced loop is cut off by endonuclease action, and the loose ends are stitched together by ligase action (arrows). The two alleles are linked by a typical chiasma.

5. Rotation of the upper part of the chi-form is followed by the cutting of their connecting strands (arrows).

6. The cut strands are attached together in reciprocal fashion, completing the exchange of segments between the alleles.

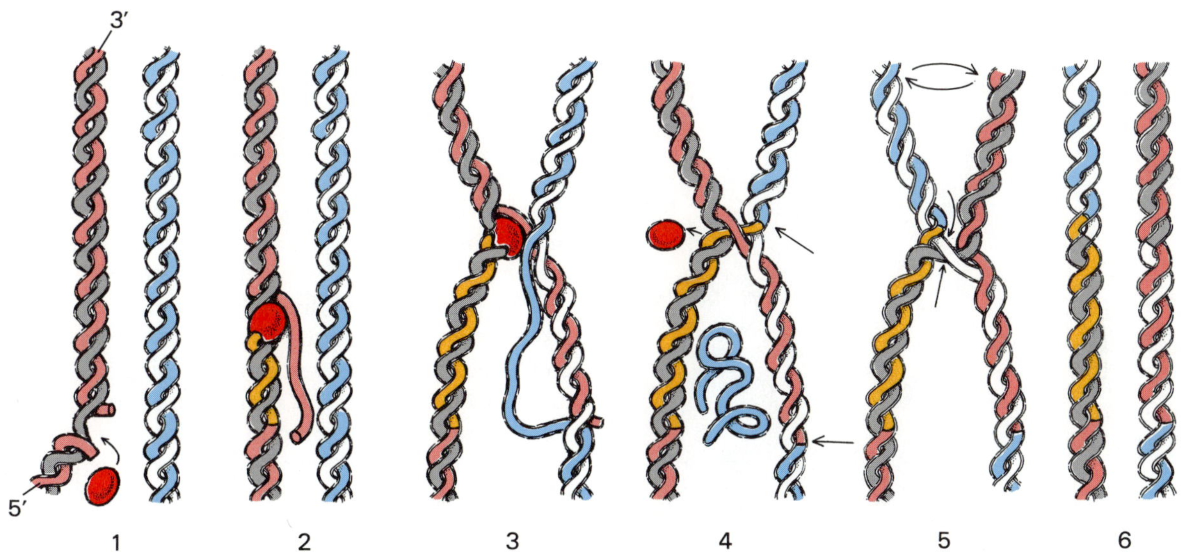

without crossing-over; but each individual chromosome is itself made of a number of fragments copied either from the father's or from the mother's contribution to the individual's genome.

A key characteristic of this recombination process, as it is called, is its faithfulness and precision. Barring accidents—the inevitable hazard of any natural process—exactly homologous pieces of the two chromosomes are exchanged. By looking at the way a DNA fiber is curled up and compressed into a stubby chromosomal rod, you can readily imagine that the exchange of segments between two adjacent chromosomes must be a complex process. But at least you can visualize its main molecular basis and, at the same time, understand why it is so extraordinarily precise. At some stage, preceded, as we shall see in Chapter 19, by parallel positioning of the two homologous chromosomes in close register with each other, some sort of hybridization must occur between a single-stranded piece of one chromosome and its complement in the other chromosome. Most likely, what causes the generation of a loose single strand is nicking followed by polymerase action, as happens in excision-repair (p. 332). Instead of being broken down, however, the displaced strand becomes entangled with the neighboring homologous chromosome. A possible mechanism leading to the formation of a chiasma and to the exchange of segments between the chromosomes is illustrated by the diagram shown above. This mechanism, it will be noted, depends entirely on familiar enzymes. Except for the initial embrace, which requires the participation of special recombination factors, subsequent events rely essentially on the same nuclease, topoisomerase, polymerase, and ligase activities as are involved in DNA replication and repair. More important, whichever the exact mechanism involved, it is clearly dependent on, and its precision guaranteed by, the close structural similarity

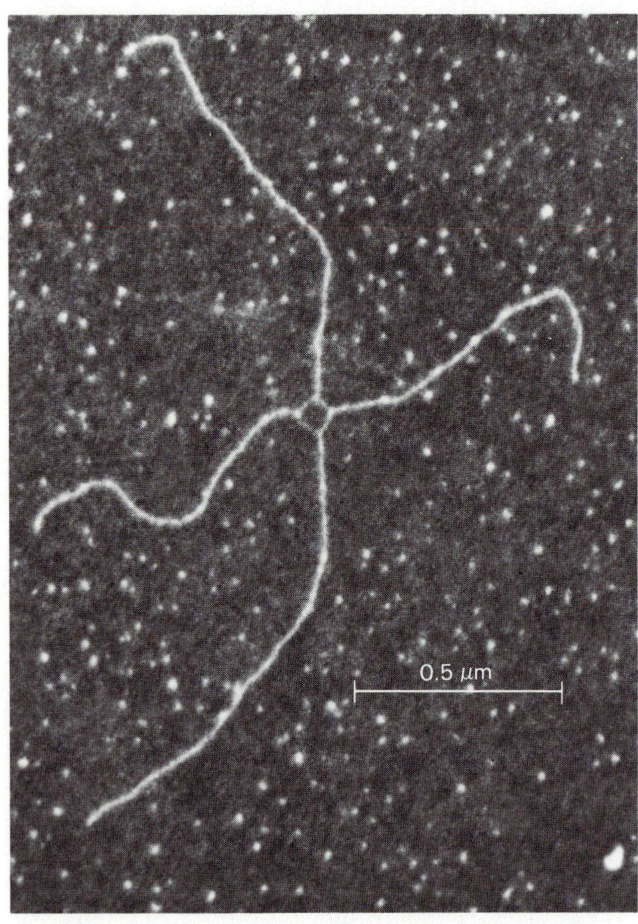

Chi-form created by two *E. coli* plasmids in the process of recombining. The central square consists of single-strand connections between homologous segments of the two plasmids (see the drawing on the preceding page). The connections are separated because the double-stranded DNA has become partially unwound in the cross-over region by the method of preparation.

between the hybridizing DNA segments. All homologous germ-line chromosomes of a given species carry the same genes in the same order, with only occasional differences (mostly base-pair replacements) that distinguish two alleles from each other.

Bacteria, although they have only a single chromosome, can engage in similar recombination events whenever they receive a piece of homologous DNA by transformation, conjugation with another bacterium, or transduction (DNA transfer mediated by a virus). In most of these cases, recombination occurs by exchange of closely homologous segments, as in crossing-over. Events of this sort are now categorized under the general designation of "legitimate" recombination.

This term implies that "illegitimate" recombinations also take place. Suspected for some time, this fact has become one of the latest and most thrilling causes of excitement in a field that seems to have an almost inexhaustible store of surprises awaiting the explorers who dig deeper than did their predecessors. Apparently, certain genes or sets of genes can wander through the genome, accounting for an immense variety of evolutionary and developmental phenomena, probably including the production of cancer. In actual fact, the word illegitimate is not entirely deserved, for homology remains the basis of the recombination. But it is an internal homology, which concerns only the two ends of the wandering DNA segment. In all known cases, these two ends are identical, or nearly so, either as such (direct repeat) or in palindromic form (inverted repeat).

An example of a direct repeat would be:

5′----ATCGCTC--------ATCGCTC----3′
3′----TAGCGAG--------TAGCGAG----5′

And an inverted repeat:

5′----ATCGCTC--------GAGCGAT----3′
3′----TAGCGAG--------CTCGCTA----5′

This kind of homology may concern only a short sequence, as shown in the example above. But quite often it

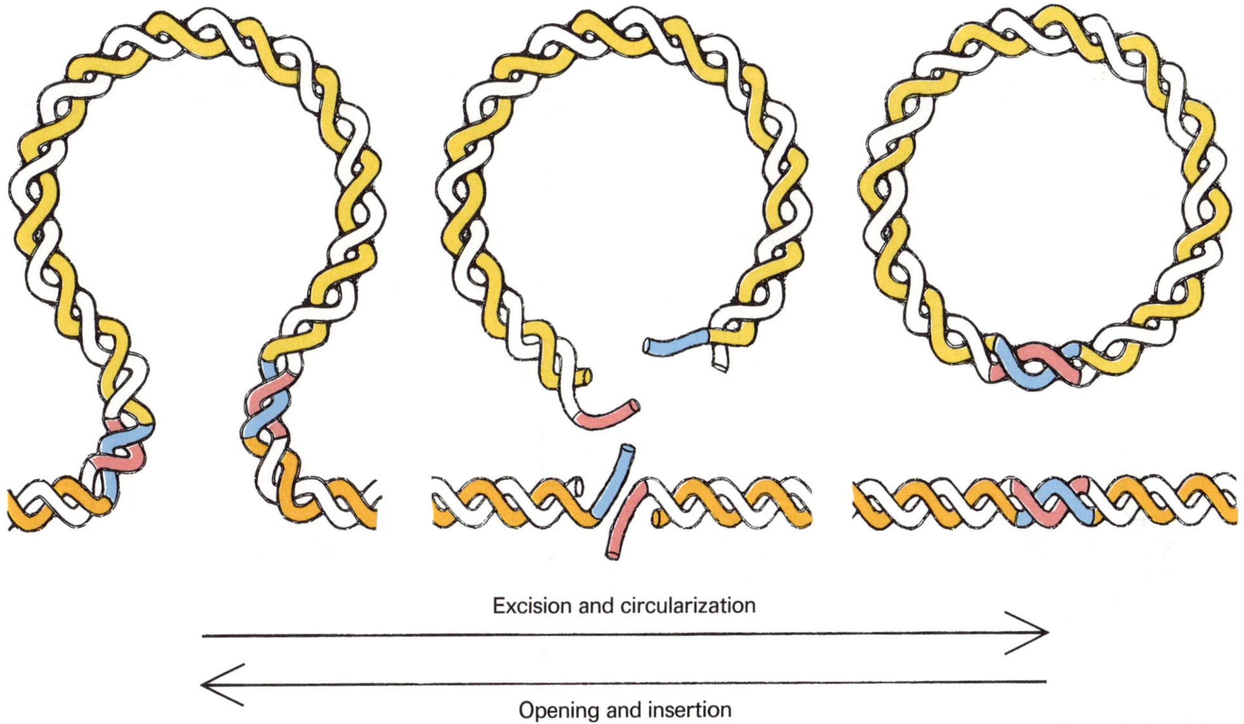

Conservative illegitimate recombination, with direct repeats, allows excision (with circularization) or insertion (with linearization) of a segment of DNA.

extends over much greater stretches, sometimes more than one thousand base pairs long.

Illegitimate recombinations are either conservative or duplicative. In the conservative type, the moving piece of DNA is translocated without duplication by what can be understood, at least in principle, in terms of an asymmetric nicking of the terminal repeats and their reclosing in a different way. If the repeat is direct, this mechanism will lead to excision and circularization of the piece of DNA flanked by the repeats or, conversely, to the insertion of such a circle into an homologous acceptor site. In essence, such a mechanism can account for the kind of "double life" led by certain viruses, either as independent, self-reproducing entities (circular viruses or phages) or as integrated parts of the host genome (proviruses or prophages). In practice, the excision and insertion mechanisms used by viruses are usually more complex.

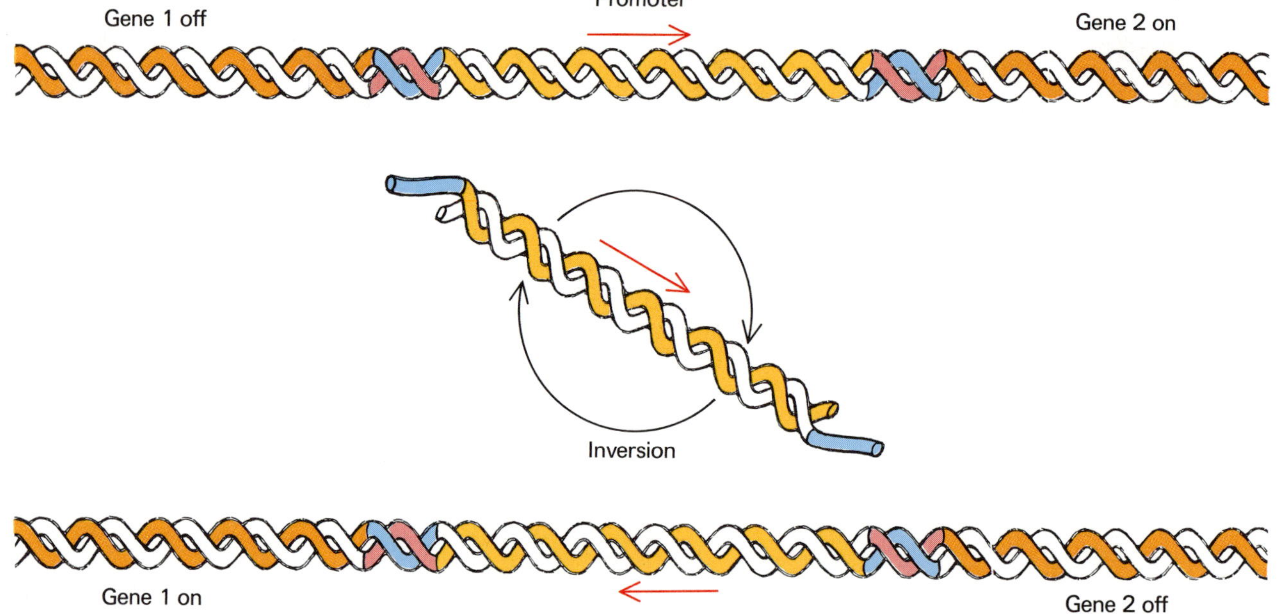

Conservative illegitimate recombination, with inverted repeats, allows inversion. A promoter on the inverton switches gene 1 or 2 on, depending on the position of the promoter.

With inverted repeats, the only possible conservative translocation is inversion. If the invertible segment, or inverton, bears an appropriately placed promoter on one of its strands, the position of this genetic "flip-flop" switch determines which of two genes that can be controlled by the promoter is to be transcribed. Such switches exist. Certain salmonellae, which are microbes that cause gastrointestinal infections, use this system to change from one to another flagellar protein. If the host becomes immune to the "flip" variant, switching over to the "flop" position will give new life to the infectious microorganism. We see here a simple example of antigenic variation. Trypanosomes, the protozoan agents of sleeping sickness, have developed this evasive defense trick to a fine art. They can change their surface coat protein many times in succession by switching on different genes. How they do this is not entirely clear but seems to involve du-

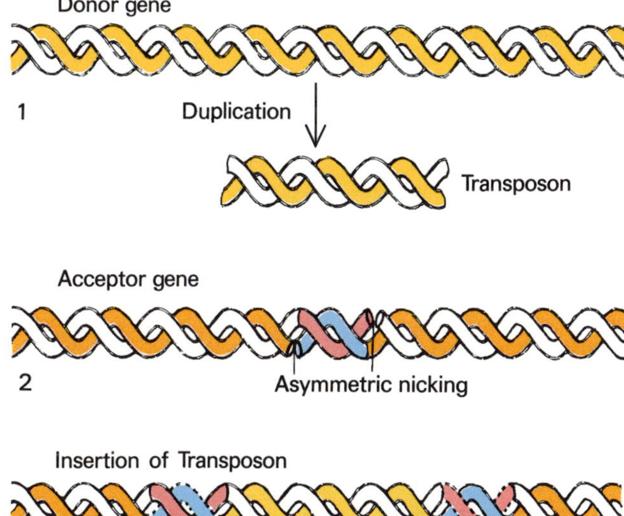

Duplicative illegitimate recombination, or transposition.

1. A transposable segment of DNA is copied by duplication.

2. An acceptor site is split asymmetrically to receive the transposon.

3. Stitching requires gap-filling replication of the single-stranded ends of the acceptor site. The inserted transposon thus becomes flanked by direct repeats.

plicative transposition of genes from mute sites where they are stored, but not transcribed, to expression sites.

As its name indicates, duplicative transposition involves making a copy of the movable DNA sequence and inserting it into a new site while the original remains in place. Sequences that are translocated in this way are called transposons if they contain genes. The name insertion sequence designates shorter transposable sequences not known to contain genes. However, the distinction between the two is fuzzy. The term transposon will be used for all transposable sequences that move duplicatively.

The mechanisms of duplicative transposition are complex and imperfectly understood. From what is known, it appears that much of the information needed for transposition lies within the transposon itself, in particular in the terminal repeats that exist in all known transposons. In addition, a number of important transposons in the bacterial world include a gene that actually codes for a special enzyme needed for transposition (transposase). So far, nothing special has been detected in the insertion site or target, except that transposition invariably leads to duplication of a stretch from four to twelve base pairs long in such a way that the inserted transposon is flanked by direct repeats of this stretch. Asymmetric nicking of the target, followed by replicative gap-filling, is the obvious explanation. The target sequences themselves are highly variable, suggesting that transposons may enjoy considerable freedom of movement.

Transposition accounts for a large number of genetic exchanges mediated within and between bacteria by such extrachromosomal, self-replicative entities as phages or plasmids. In particular, as we shall see (p. 347), it plays a major role in the spread of antibiotic resistance, since many of the genes that code for enzymes that render their owners resistant to some antibiotics are included in transposable elements carried by plasmids.

In both prokaryotes and eukaryotes, transposition is an important cause of evolutionary change. Transposons can create all sorts of disturbances merely by their insertion in a certain site. They can cut in the middle of a gene and so kill it, block or promote the transcription of neigh-

Genetic map of human globin genes. Hemoglobin, the oxygen-carrying protein of red blood cells, is a tetramer of general structure $\alpha_2\beta_2$. Each subunit resembles myoglobin (Chapter 2) and bears a heme group (Chapter 9). Early embryos, fetuses, and infants have different hemoglobins adapted to the manner in which they are supplied with oxygen. Different α and, especially, β subunits, controlled by different genes, make up these different hemoglobins. The location of these genes on their bearer chromosomes corresponds to the order in which they are expressed in the course of development. Note that all these genes (including the pseudogenes, or dead genes, $\psi\beta_2$, $\psi\beta_1$, and $\psi\alpha$) originate from a single ancestral gene that, it is estimated, underwent duplication some 500 million years ago. Since then, a large number of events—including duplications, transpositions, deletions, and other changes—took place to produce the complex situation shown on the map. Sequencing of the globins and their genes in different species made it possible to time many of these events in terms of the number of point mutations suffered by the genes since the event took place (see pp. 353-355). Such point mutations are responsible for additional diversification within a given species. Analysis of human hemoglobins has disclosed the existence of more than 270 variants, many of which are functionally abnormal. The most widespread of these abnormal hemoglobins is hemoglobin S, a β-subunit variant responsible for sickle-cell anemia (see the electron micrograph on p. 352).

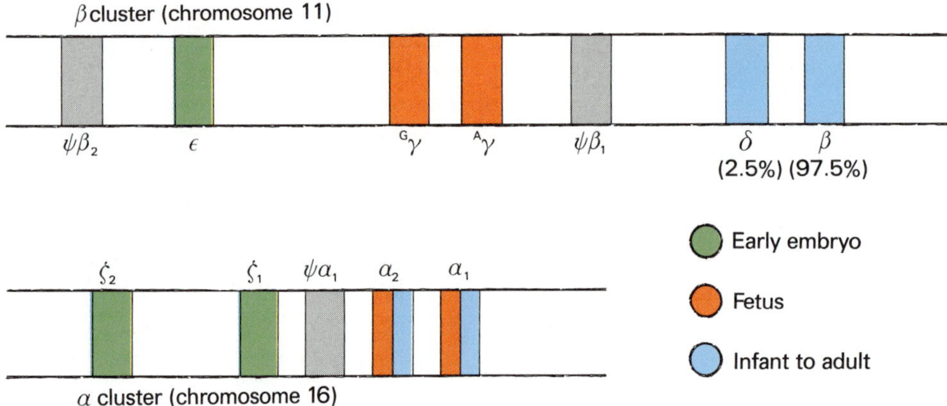

boring genes, or, alternatively, have the transcription of their own genes blocked or promoted by neighboring sequences. In addition, they have the unexplained ability of creating havoc around themselves in the form of deletions, duplications, or inversions of nearby DNA stretches.

In itself, the phenomenon of gene duplication that is an intrinsic part of all duplicative transpositions is of major evolutionary significance, because, once genes have been duplicated, their twin copies may subsequently evolve separately into two progressively different genes. The kinship between these genes can, however, be detected by a comparison of their sequences or of those of their protein products. Examples of such evolutionary siblings abound. We encountered one with the two tubulins and another with calmodulin and troponin C.

An interesting example of gene duplication and evolution is provided by globin, the protein part of hemoglobin, the oxygen-carrying pigment of red blood cells. Globin is a tetrameric protein, made of two α and two β chains. The α and β genes are evolutionary offshoots of the same ancestral gene. But this is not all. Our genome contains several different α and β genes, which code for the different hemoglobins that are found at different stages of development (embryo, fetus, adult). These genes are situated on the same chromosome in the order in which they are expressed during development. Interspersed between them are a couple of "dead genes," or pseudogenes, siblings mutated to the point that they are no longer expressed. Finally, the evolutionary history of the globin genes has varied among different human subgroups, resulting in the appearance of numerous mutant hemoglobin molecules. These all conserve the ability to carry oxygen—otherwise, the mutation would, of course, be lethal—but some of them are functionally impaired. The best-known genetic disease caused by a disabling alteration of a globin gene is sickle-cell anemia, the result of a mutation widely distributed in people of African descent. The molecular basis of this mutation and the mechanism of its spreading will be examined later in this chapter (see p. 352).

The most remarkable known instance of gene shuffling occurs in the course of B-lymphocyte differentiation. As we saw in Chapter 3, this group of cells includes a very large number of distinct individuals, each of which recognizes a different antigen. Upon contact with a given antigen, the few cells that have a receptor for it on their surface are induced to multiply (the phenomenon of mitogenic stimulation was considered in Chapter 13) and give rise to a clone of identical plasma cells that manufacture an antibody directed against the inducing antigen.

As illustrated on the next page, antibodies are Y-shaped proteins made of two pairs of identical polypeptide chains, known as L (for light) and H (for heavy), joined by disulfide bonds. Outer parts of the branches of the Y, which include the N-terminal ends of both chains, make up the variable regions that determine the specificity of the antibody. The C-terminal parts of the two chains make up the constant region. Those of the heavy chains extend beyond those of the light chains to form the stem of the Y. The antibody repertoire of an organism encompasses an enormous number of different variable regions, each made, we must remember, by a distinct type of B lymphocyte. This number is believed to run into hundreds of millions, if not billions, to account for the fact that we can make antibodies against virtually any kind of antigen, natural or artificial, to which we are exposed. In contrast, the constant regions are the same for all antibodies in a given class, irrespective of their immunological specificity. But they vary in the H chain according to the class of antibodies (IgG, IgA, etc.). Certain H-chain constant regions are fitted with an extra C-terminal hydrophobic tail that allows anchoring of the immunoglobulin molecule in the plasma membrane to serve as mitogenic receptor.

Each antibody molecule needs at least 1 μm of DNA for its genetic encoding. Clearly, with its 2,000,000-odd μm of DNA, the genome of the stem cells (or of their embryonic progenitors) from which lymphocytes originate cannot possibly accommodate the huge number of genes that code for antibodies. These genes must arise during differentiation of the lymphocytes, so that, starting from a relatively small common stock present in the stem cells, each mature cell ends up with a different gene. The new cell explorers who are now cutting their way through the dense thickets of eukaryotic nuclei with the help of restriction enzymes, cloning vehicles, and other genetic-engineering tools have identified this stock as consisting of seven distinct sets (three for the L chain and four for the H chain) of between half a dozen and a few hundred interchangeable parts, from which mature L and H genes are assembled by a special recombination process (see p. 344). A conservative estimate puts at 18 million the number of variants that can be generated by this mechanism. This number is further increased, perhaps as much as a thousandfold, by slight irregularities in the joining process and by mutations in certain highly vulnerable regions. Special splicing mechanisms at both the gene and the mRNA levels account for switching of classes in the H-chain constant region.

This example shows clearly that development is not simply a matter of turning certain genes on or off at the appropriate time and that the genome itself may undergo programmed rearrangements. A lymphocyte nucleus could definitely not be used for the kind of cloning experiment discussed in Chapter 16. We do not know to what extent gene maturation occurs in other cell types.

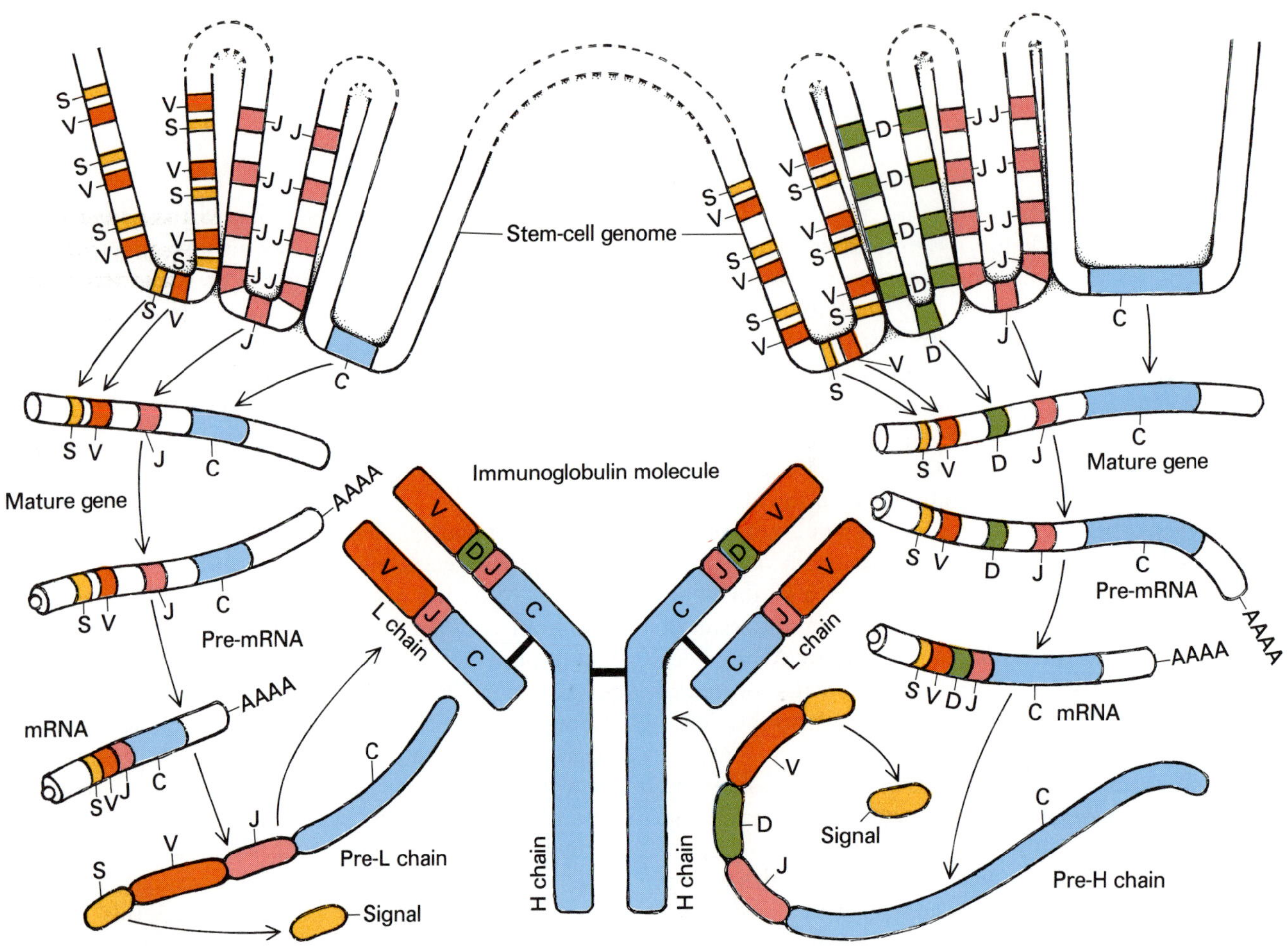

Genetic differentiation during the maturation of B lymphocytes. An immunoglobulin molecule consists of two heavy (H) and two light (L) chains, linked together by disulfide bonds. These chains consist of an N-terminal variable part (V) and a C-terminal constant part (C), connected by a joining piece (J), with, in the H-chain only, a small additional piece D (for diversity). The stem-cell genome contains multiple variants of the L-chain V and J genes and of the H-chain V, J, and D genes. The V genes are preceded by a small S segment coding for an N-terminal signal piece (the two chains are secretory proteins made by ER-bound ribosomes). As lymphocytes mature, each differentiating cell constructs a particular L gene and a particular H gene of virtually unique structure by a special recombination process, dependent on strategically placed complementary sequences, that randomly selects one out of each set of gene segments and assembles them together with a C gene. Transcription of the gene produces a pre-mRNA that is spliced to give a mature mRNA, which is translated to the corresponding immunoglobulin chain with cotranslational removal of the signal piece (Chapter 15). The immunoglobulin molecule is then further put together by oxidative formation of disulfide bridges from cysteine residues, a process that takes place inside the ER lumen (Chapters 6 and 13). This illustration does not show how different C genes are used for the H chain to make different classes of antibodies. It is not known whether recombination occurs within a single chromosome, as illustrated, or by asymmetric exchange between two alleles during mitosis.

Ballast DNA

Should we stroll through a nucleus—assuming the terrain lent itself to such a desultory activity—and try, with the eagerness of neophyte scholars, to decode all its inscriptions, we would be astonished to find out how few of them make sense. Large chunks of DNA, made of millions of simple, repetitive sequences, completely fill certain regions, which correspond mostly to the centromeric parts of the chromosomes (see Chapter 19). Known as constitutive heterochromatin, they are permanently condensed and almost certainly instruct no process other than their own replication. Then there are the long, untranscribed linkers between the genes; the dead genes; the introns, which often greatly exceed the exons in length; the untranslated leader and tail sequences of the mRNAs; the large pieces that are excised in the maturation of ribosomal and other functional RNAs. Repeats also occur in those unexpressed pieces of DNA that are interspersed between coding parts. Best known is the *Alu* family, a characteristic of human DNA, so named because its members can be neatly carved out of the genome by means of a restriction enzyme called Alu I. It includes some 300,000 to 500,000 identical, or nearly identical, sequences of about 300 base pairs. Other repeats, found in fewer numbers, are of the type that flank transposons.

Taken together, all the DNA in the nucleus that is not overtly expressed as RNA or protein may amount to more than 90 per cent of the total. The very size of the genome makes this clear. With a haploid content of almost 3 billion base pairs, we have enough DNA in our nuclei to code for 1 million different proteins of about 1,000 amino-acid residues. In fact—not counting the immunoglobulins, which represent an exceptional case, as we have just seen—we manufacture less than one-tenth this number throughout our existence, from conception to death. From which it follows that more than 90 per cent of the DNA that we diligently replicate at each cell division never becomes translated into an actual mediator of phenotypic manifestation. The puzzle is compounded by the fact that closely related species may have very different DNA contents. This is called the C-value paradox. (The term C-value goes back to the days when it was first found that different cells in a given species have the same—constant—amount of DNA in their nuclei.)

The interpretation we may put on these intriguing facts depends on what particular Darwinian sect we belong to. We may, with the "strict selectionists," put our faith in natural selection and refuse to believe that the nucleus would keep perpetuating a lot of rubbish with no selective advantage to the species. Or we could join the "genetic drifters" and argue that natural selection has no power over what is simply useless; it eliminates only the harmful. And so functionless DNA could, whatever its origin, be carried along passively and just drift, for a very long time, before turning into more than an insignificant inconvenience.

Truth probably lies somewhere in between these two extreme views. There is no doubt that DNA contains a great deal of information that, although not expressed in a transcription or translation product, nevertheless plays an essential role in the operation and regulation of nuclear activities—in the form of replication origins, promoters, and terminators; of binding sites for enzymes, repressors, or other regulatory agents; and of many other, as yet undeciphered, code words that are used in the intricate, four-dimensional communication networks of living cells and organisms. Many of these signals may be expected, like road signs, to fall into a limited number of characteristic classes and to be interspersed throughout the genome, thus accounting for at least part of the repeats. The *Alu* family, for example, is believed by some to contain replication origins, which must, indeed, number in the hundreds of thousands in a human nucleus. The highly clustered repeats of constitutive heterochromatin obviously do not function as road signs, but they could still play an important role in chromosome organization. As we shall see in Chapter 19, the centromeric region with which constitutive heterochromatin is preferentially associated may well sleep during the whole of interphase, but it wakes up to become a star performer when the mitotic drama is enacted.

Similarly, there must be a variety of regulatory signals among the DNA sequences that are transcribed but do not end up as part of a functional RNA or protein. Specific signs are needed to attract and direct the numerous factors that are involved in RNA processing and translocation. Others serve to guide the finished RNAs to their terminal destination in or on ribosomes, or elsewhere, and to instruct the systems with which they interact. Here, again, we may expect to find repeats—for instance, at the 5′ ends of mRNAs. Down at the polypeptide level, finally, we have seen that those sequences that are discarded in the fashioning of a protein are often removed only after they have carried out an important function, such as signaling (Chapter 15).

Altogether, these regulatory sequences may well add up to several times the size of the overtly expressed part of the genome. But it is not likely that they account for all the noncoding DNA. A frequently invoked criterion of functional usefulness is the degree of evolutionary conservation. Sequences that have not changed much in the course of evolution, as indicated by close similarities between distant species, are seen as important. On the other hand, those that have changed rapidly are considered not to be meaningful. Many introns fall in the second category.

As often happens when facts are scarce and beliefs strong, hardliners of either one or the other creed refuse to take evolutionary conservation as an absolute criterion, each for different reasons. The strict selectionists point out that meaningless is not necessarily synonymous with functionless. Noncoding DNA could play a role by its mere bulk, as ballast. As pointed out in Chapter 16, the flexibility conferred on the eukaryotic genome by the existence of long intervening sequences may well have been essential in the evolution of plants and animals. The drifters, on the other hand, will tell you that conservation does not automatically imply usefulness; that it is in the nature of genes to ensure their own replication, and even multiplication thanks to transposition; and that natural selection has no control over these phenomena as long as they do not affect a phenotype's ability to reproduce the genotype. This view has become known as the "selfish gene" or "parasitic gene" concept. Dead genes, of which several have already been recognized (in globin, for instance, as mentioned on p. 342), are typical examples of selfish genes.

As tourists, you are not expected to take a stand on these matters. Nor does your guide command enough expert authority to advise you. It would seem, however, that, until more facts are known, we would be wise to take the ecumenical position that there is probably some truth in each point of view.

Viruses, Phages, Plasmids, and Other Flying Genes

A virus may be defined as a piece of genetic material, either DNA or RNA, surrounded by a protein coat, or capsid (Greek *kaps*, box), and, in some cases, by a membranous envelope. The main function of these wrappings is to mediate the penetration of the genetic material into some recipient cell in which multiplication of the virus will occur. If the recipient cell is a prokaryote, the virus is often called phage. This term is derived from bacteriophage or "bacterium eater," the word originally proposed by the French bacteriologist Félix d'Hérelle, who discovered these entities through their ability to kill bacteria. A plasmid is a small piece of circular DNA that is not integrated in a chromosome and can be exchanged between cells. Plasmids are found mostly in bacteria, but they are known to be present also in eukaryotes, as, for instance, in yeast.

Long known as agents of major diseases—among them smallpox, measles, poliomyelitis, hepatitis, influenza, and rabies—viruses and their counterparts in the prokaryotic world, the bacteriophages, have become the darlings of molecular biologists. Thanks to their relative simplicity (the term "relative" is appropriate, because even the simplest of viruses is highly complex), they have served in an invaluable manner to initiate cell explorers into the mysteries of the gene. They make up an extraordinarily diversified group, which our other commitments allow us to consider only in the most general of terms.

Viruses travel light, according to a strictly economical policy. They rely almost entirely on the enzymic machinery and metabolite supply of their host cells and carry in their genetic baggage only the blueprints of their capsid and envelope proteins, together with those of such enzymes and other protein factors they may need to subvert their hosts and to ensure their own replication. The smaller phages and viruses require only half a dozen genes (or fewer) for this purpose, whereas the larger ones may code for more than a hundred enzymes, including a complete DNA replication kit and, sometimes, a rather fiendish system that radically shuts off DNA replication by the host and leaves the virus in full control. All these proteins are synthesized by the host's protein-manufacturing machinery from transcripts of the viral genome.

Viruses have their genetic information encoded in either DNA or RNA. Most viral DNAs are double stranded and are replicated and transcribed by different variants of the general mechanisms described in Chapters 16 and 17. A few small DNA phages are single stranded. The first thing that happens when such a phage enters a cell is the formation of a complementary strand. The resulting duplex, or replicative form, serves in transcription and in the production of new viral strands.

DNA viruses may mount two types of attacks. In the virulent kind, the host cell is completely taken over by the virus and used for viral multiplication until it can no longer survive. The infected cell eventually disintegrates (lysis), and the viral progeny are released to infect other cells. In the nonvirulent, or temperate, kind, the viral DNA becomes integrated in the host cell's genome by some recombination or transposition process and replicates with it. This occurs in such a way that the genes that control lysis are shut off. Such integration may have a number of fateful consequences. It may serve to spread a latent infection (lysogeny) in a whole population of cells without a detectable symptom until some stress, such as exposure to ultraviolet light, causes the viral DNA to be excised from the genome and to initiate a catastrophic virulent infection. Another possible effect, if the viral DNA has the properties of a transposon and can wander through the genome, is multiple mutations. In animal cells, DNA viruses that become integrated in the genome tend to cause cancerous transformations. A particularly simple example of such an oncogenic virus is SV40, already encountered in Chapter 16. This virus is normally found in certain Asian monkeys (SV stands for simian virus), in which it grows without causing any apparent illness. In other species, or in cultured cells, SV40 may, according to circumstances, cause a lethal infection or induce a cancerous tumor. It develops inside the nucleus of infected cells, where it behaves very much like a small, local chromosome. In particular, it combines with histones to form nucleosomes, and its transcription products are spliced, capped, and polyadenylated, as are most eukaryotic mRNAs. We saw in Chapter 16 how different modes of splicing allow five distinct proteins to be made from only two transcripts of the SV40 DNA. Of them, the VP proteins are viral proteins, and the t and T proteins mediate the cancerous transformation.

Plasmids differ from viruses by lacking a capsid and being largely uninfective for this reason. But they can be exchanged between bacterial cells during conjugation and may be made to enter cells under special conditions, which are now widely exploited in genetic engineering and may well obtain sometimes in nature. In any case, there is no doubt that plasmids move around a great deal. They are fairly well behaved in the cells that they invade and generally replicate to only a few copies. Nevertheless, they can cause effects of very great importance. One reason is that a number of proteins that mediate resistance to antibiotics, as well as certain highly poisonous toxins produced by bacteria, are encoded by plasmid-associated genes. Thus plasmids may confer redoubtable properties on pathogenic microbes. Another property of plasmids, and of some viruses, is that they can serve as vectors for "passenger" or "hitchhiking" genes, which are removed from the genome of one cell and transferred to that of another by transposition phenomena (transduction).

With rare exceptions (an example is reovirus), RNA viruses are single stranded. They all have to code for a special polymerase, since cells do not possess enzymes capable of copying an RNA template. In the usual virulent viruses, this polymerase makes a complementary RNA strand, starting with a single nucleoside triphosphate as primer. It resembles transcriptases and primases, therefore, with the difference that it is instructed by an RNA strand instead of by a DNA strand. In one class, of

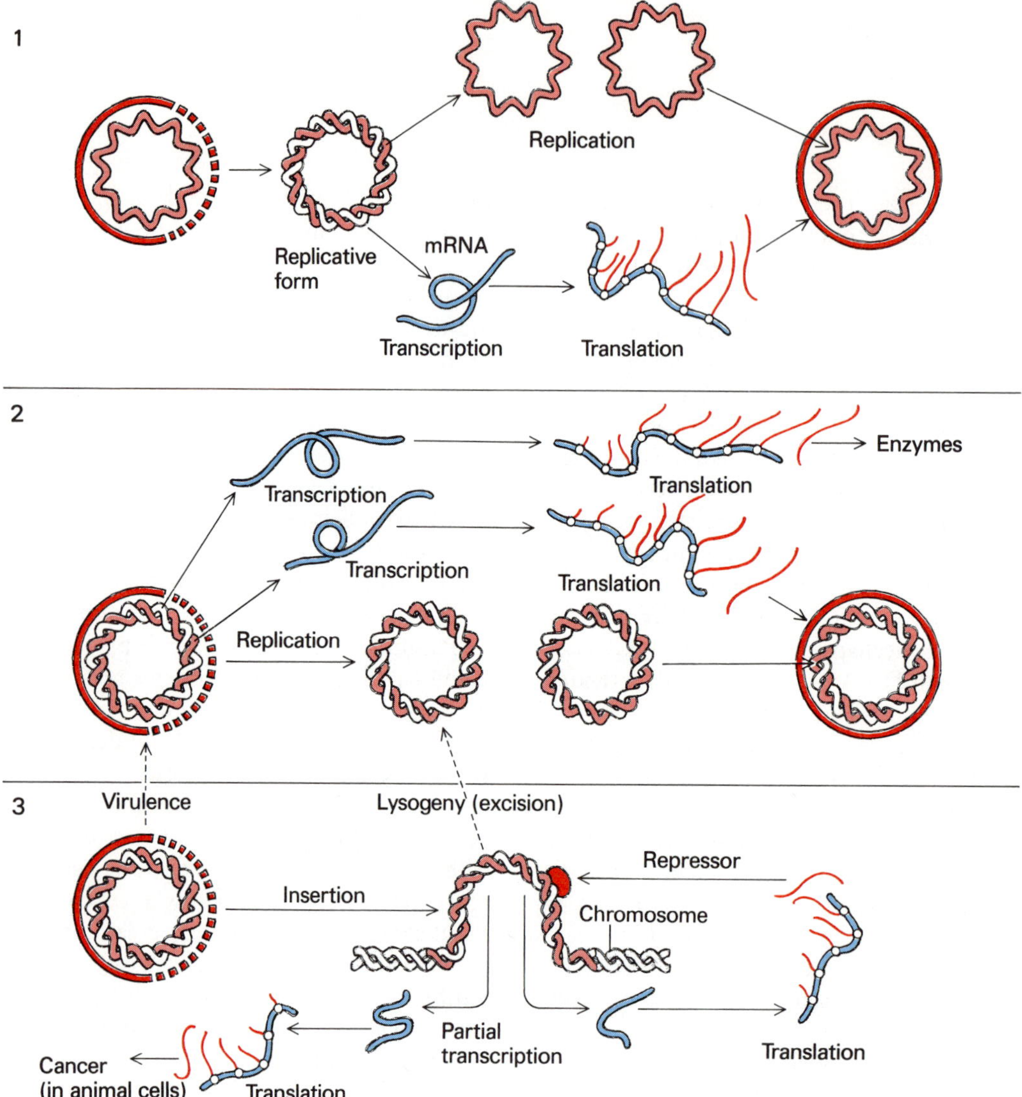

Examples of viral growth cycles: DNA viruses.

1. Single-stranded (e.g., phage ϕX174). The viral DNA is replicated to a double-stranded replicative form, which serves as template for the synthesis of new viral DNA and for transcription to mRNAs that are translated into viral proteins.

2. Double-stranded, virulent (e.g., phage T4, adenovirus). The viral DNA is replicated and transcribed. The mRNAs are translated into viral proteins, as well as enzymes used by the virus to establish dominance over the host cell.

3. Double-stranded, temperate (e.g., phage λ, simian virus SV40). The viral DNA is inserted into a chromosome. In this situation, transcription of the DNA is restricted by a repressor encoded by the virus, and replication of the viral DNA occurs only in concert with duplication of the chromosome. There is no proliferation of the virus. This kind of behavior is observed only in certain host cells or under certain conditions. In other circumstances, the virus is virulent (as in part 2). A temperate virus may be returned to the virulent state (lysogeny) by treatments that lead to excision of the viral DNA. With animal viruses of this type, insertion of the viral DNA into a chromosome may cause the development of cancer.

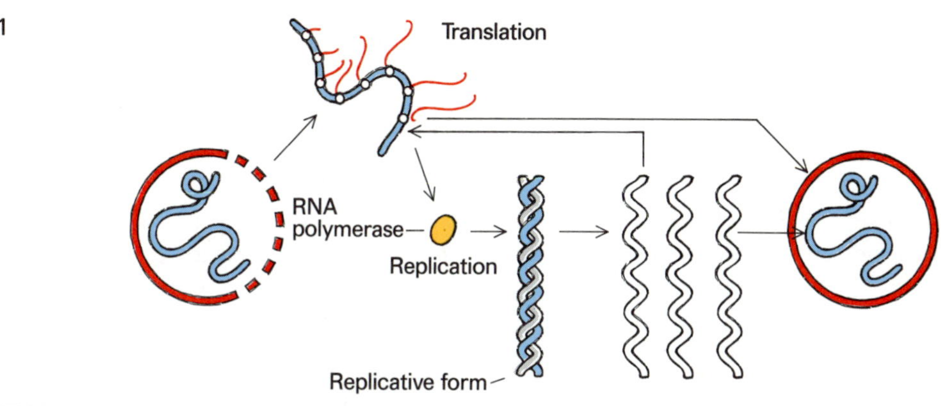

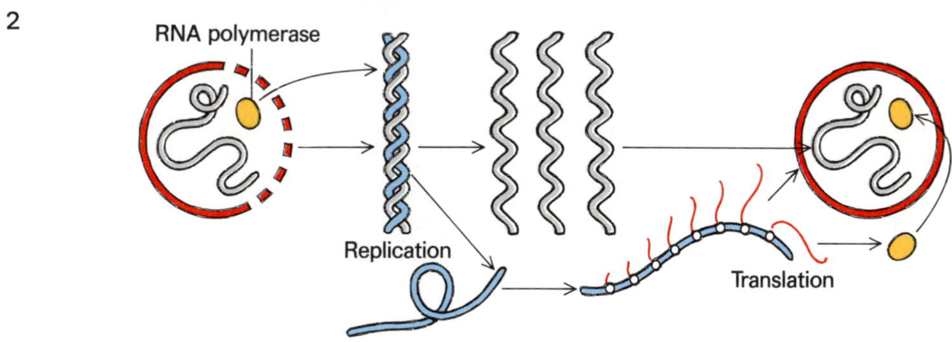

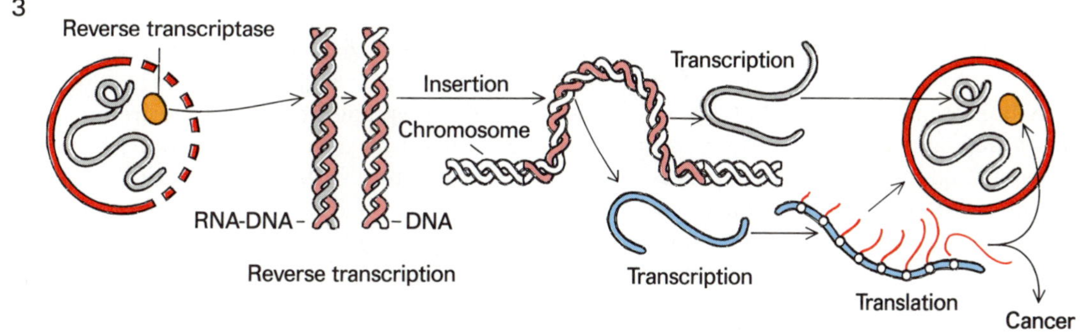

Examples of viral growth cycles: RNA viruses.

1. Single-stranded, plus (e.g., phage f_2, tobacco mosaic virus, poliomyelitis virus). The viral RNA is an mRNA (including cap and poly-A tail in eukaryotic viruses). Its translation produces viral proteins and the RNA polymerase needed for replication of the viral RNA by way of a double-stranded replicative form.

2. Single-stranded, minus (e.g., measles, influenza, rabies viruses). The viral RNA does not serve as mRNA and must first be replicated by a virus-associated polymerase. The plus strands that are generated are translated into viral proteins, and the double-stranded replicative form serves for the synthesis of new viral minus strands.

3. Single-stranded retroviruses (e.g., Rous sarcoma virus). The viral RNA is of the minus type and is transcribed by a virus-associated reverse transcriptase to give a hybrid RNA-DNA duplex, which is further replicated into double-stranded DNA, which in turn is inserted into a chromosome. New virus particles are then formed by transcription and translation of appropriate parts of the incorporated viral genome. If the virus carries an oncogene, its insertion causes a highly virulent cancer. If not, cancer may still ensue in rare cases as a result of the activation of an indigenous proto-oncogene by the inserted virus.

4. Double-stranded (e.g., reovirus). Not shown, this type of virus behaves essentially like a double-stranded replicative form (as in part 2), which is replicated and transcribed with the help of a virus-associated RNA polymerase.

which polio virus is a member, the viral RNA itself serves as mRNA. It is translated soon after infection to provide, among other proteins, the RNA polymerase needed for its replication. In the other class of single-stranded viruses, exemplified by rabies, the messages for the viral proteins, including the polymerase, are encoded in the strand complementary to the viral RNA. Such viruses need a preformed polymerase to make the complementary mRNA that will induce the cell to make more polymerase molecules. They carry this enzyme with them in their capsid luggage. Both virus types need a double-stranded intermediate (replicative form) for replication, as do the single-stranded DNA viruses. Double-stranded RNA viruses also carry their polymerase with them, to produce mRNA molecules from one of the two strands.

There is a special group of animal-cell RNA viruses, known as retroviruses or sometimes as oncorna viruses (from oncogenic RNA), in which the virus-coded polymerase is a DNA polymerase. It is called reverse transcriptase, because it makes complementary DNA (cDNA) from an RNA template, in contrast to transcriptase, which makes RNA from a DNA template. The single-stranded DNA transcribed from the viral RNA is replicated to form a duplex and is then integrated into the genome by some kind of transposition process. This process is essential for the multiplication of the virus, which takes place by normal transcription from the integrated cDNA. Occasionally, integration of the cDNA in the genome results in the cancerous transformation of the infected cell by a mechanism that is currently raising an enormous amount of interest and may be related to gene transposition (see pp. 364–367).

Playing Evolutionary Roulette

The game has been on for more than 4 billion years. Barring a cosmic cataclysm, it is due to continue for at least as long. It is being played simultaneously over the whole surface of our planet by myriads of living forms ceaselessly engaged in replicating, transcribing, and translating genes, as well as recombining them in various ways. There are so many twists to the game, so many possible moves, as to put it forever beyond the grasp of any conceivable brain, whether natural or artificial. Yet the rules are extremely simple and can be summed up in three words: fidelity, variability, selection.

Fidelity, we have seen, is a compelling condition, especially as regards DNA replication. No species can survive if its genome is not handed over essentially intact from generation to generation. Understandably, such species as do survive have available elaborate copying, proofreading, and other controlling devices that ensure the required level of fidelity. It is remarkable that these devices rely almost exclusively on the correct fitting of two pairs of small planar molecules joined at their edges by two or three weak electrostatic bonds. No chemist in his senses would trust such a flimsy arrangement. Proteins, however, are endowed with powers of discrimination undreamed of by any chemist. But even proteins can err and, furthermore, no safeguard exists that provides absolute protection against every possible sort of accident.

Mistakes and accidents are the sources of variability, which is just as important as fidelity, provided it is kept within acceptable bounds. There are two kinds of deaths, one caused by slavishness, the other by laxity. The game of life is played in the narrow strait that separates the two. It is there that have been generated all the changes whereby living beings have diversified, adjusted to different ecological niches, adapted to new conditions, evolved, and progressed.

Genes are the exclusive targets of these changes, with, as the only possible exception, certain multimolecular assemblies that act as templates for homomorphic growth. The pattern of implantation of cilia on the surface of certain protozoa belongs to this rare category. And so, perhaps, does the assembly of membranes, although this requires further validation. On the whole, however, hereditary changes are, with overwhelming frequency, alterations of genes: mutations. They have as a fundamental characteristic that they occur by chance.

Coming at the end of this tour, such an assertion cannot but sound totally incredible. How many times did we stop in breathless admiration before some piece of molecular machinery that could be described only as "exquisitely designed" or some sort of equivalent? And now we are asked to see in all those wonders nothing but the products of blind, fortuitous events. It is simply ludicrous. Yet the proof is in front of our eyes, both in the actual manner in which genes are seen to operate and in the historical record that this operation has left in their structure and in that of their products.

What converts randomness into order is natural selection: any genetic change that enhances the survival potential of the individual concerned—more specifically, its ability to produce progeny—will tend to be retained at the expense of those that do not. This statement is, you will notice, essentially tautological and requires no demonstration. And it does not seem to matter that favorable changes happen only very infrequently compared with the others. The game of life is played at high speed, with huge numbers, over indefinite time. It can afford to wait for the occasional stroke of luck. And so it is, to quote the beautiful words of Jacques Monod (see p. 357), that "out of a source of noise, selection has been able alone to extract all the music of the biosphere."

This noise is of two kinds, depending on whether a single base or a whole stretch of DNA is affected. Changes of the first kind are called point mutations. It is easy to see by looking at the genetic code (Chapter 15) what the possible consequences may be when one base is replaced by another in a genetic message. Four types may be distinguished:

1. The altered codon is replaced by a synonym, as CAU by CAC, which likewise calls for histidine. The mutation is silent. It will appear only in the DNA and in its transcript but not in its protein translation product.

2. The altered codon specifies a new amino acid, but the biological properties of the modified protein remain unaffected. For example, a change from CUU to AUU, which replaces leucine by the closely similar isoleucine, is most unlikely to have a significant effect on the structure or function of the protein concerned. Mutations of this sort are called neutral, or indifferent. We discovered in Chapter 15 that the structure of the genetic code is such as to favor mutational indifference, a strong argument in support of the theory that the code is itself a product of natural selection.

3. The altered codon specifies a new amino acid, but in this case the biological properties of the affected protein are significantly modified, usually for the worse. Mutations that change the electrical charge of a residue or greatly modify its hydrophilic or hydrophobic character are likely to have such effects. For instance, a change from GAG to GUG will substitute valine (uncharged, hydrophobic) for glutamic acid (negatively charged, hydrophilic). Such a change suffices to convert the normal human hemoglobin A into the severely pathological hemoglobin S, which is characteristic of sickle-cell anemia.

4. The altered codon is converted into a stop sign (e.g., UAC into UAG) resulting in abortive translation of the message, with the production of a usually inactive N-terminal fragment of the original polypeptide.

The consequences of mutations of the third and fourth kinds depend on how they affect the survival and reproductive potential of their victims. A priori, one would expect them to be lethal, or at least deleterious. But this is not necessarily so. It is remarkable that mammals lack many enzymes that are present in the lowly bacteria. Humans even lack some enzymes that are found in most of their mammalian relatives and, as a consequence, are subject to serious additional liabilities. We are, with other primates and, strangely enough, the Dalmatian dog, the only potential gout-sufferers among the mammals, because we lost at evolutionary roulette the capacity to break down uric acid. And we share only with our primate cousins and with guinea pigs the inability to manufacture vitamin C and the resulting distinction of being subject to scurvy. Clearly, these handicaps have not prevented us from getting on in the world. Perhaps—who knows?—they may even have helped, by conferring a selective premium on some genetically endowed piece of resourcefulness needed to overcome the handicaps.

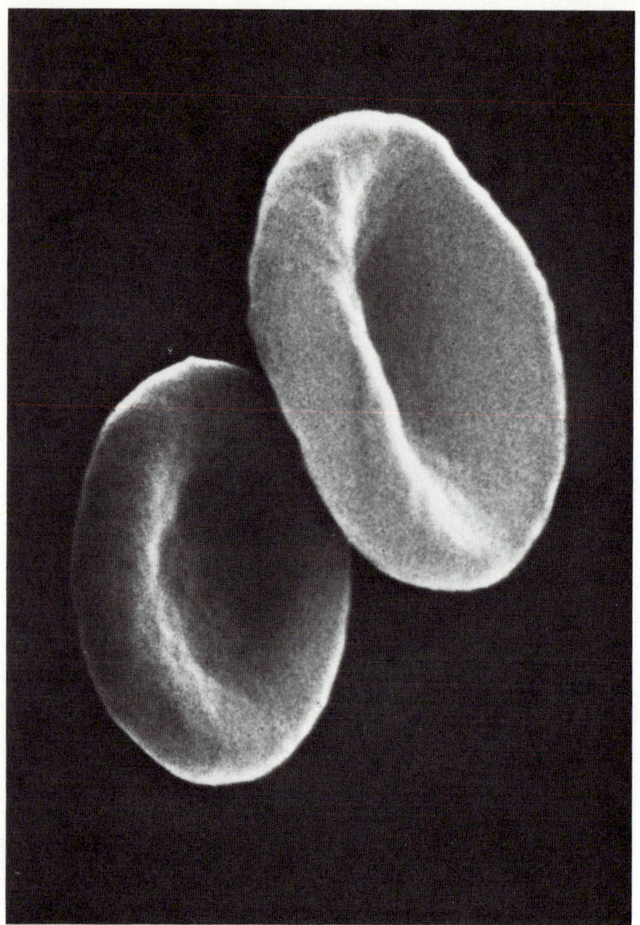

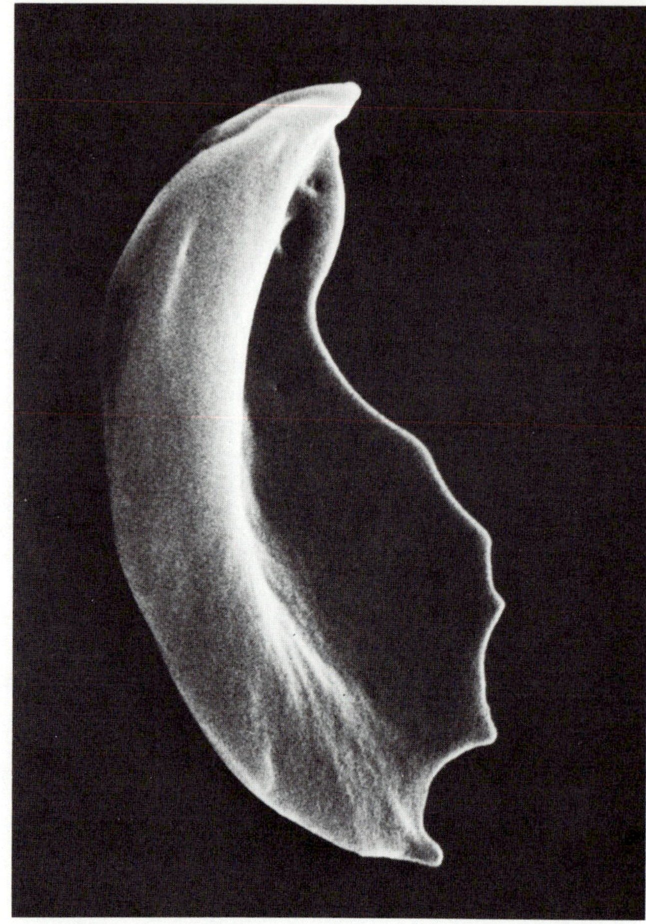

Sickle-cell anemia, the prototype of a "molecular disease." The scanning electron micrograph at the left shows the biconcave disk structure of normal red blood cells. That on the right illustrates "sickling," the structural deformation that the cells undergo at low oxygen tension in bearers of the disease. Such cells are abnormally fragile and suffer rapid destruction. Severe anemia and various vascular complications, leading to early death, are the final result. Remarkably, this dramatic unfolding of events is the consequence of a single amino-acid substitution (valine replaces glutamic acid in position 6 of the β chain) in hemoglobin. This abnormal protein can still function as oxygen carrier but tends to precipitate out of solution when it has released its oxygen. This precipitation is responsible for the alteration in cell shape. The sickling trait is widespread in the black population. Its spreading is believed to have been favored by the relative resistance of heterozygotes to malaria.

Another important point is sexual reproduction and the resulting diploidy of the offspring. Many defective genes are deleterious or lethal only in homozygotes, a fact that is not likely to oppose their spreading by phenotypically normal heterozygote carriers until these carriers start making up a significant proportion of an interbreeding population. Sometimes the defective gene may even confer an advantage on the heterozygote. It is believed that the high frequency of the sickle-cell gene in people of African descent is due to the protective effect that it exerts against malaria in the heterozygotes.

Point mutations are largely responsible for what is sometimes called micro- or molecular evolution—that is, the progressive replacement, in the course of evolutionary time, of bases in homologous genes and of amino acids in the corresponding proteins. But they probably had little to

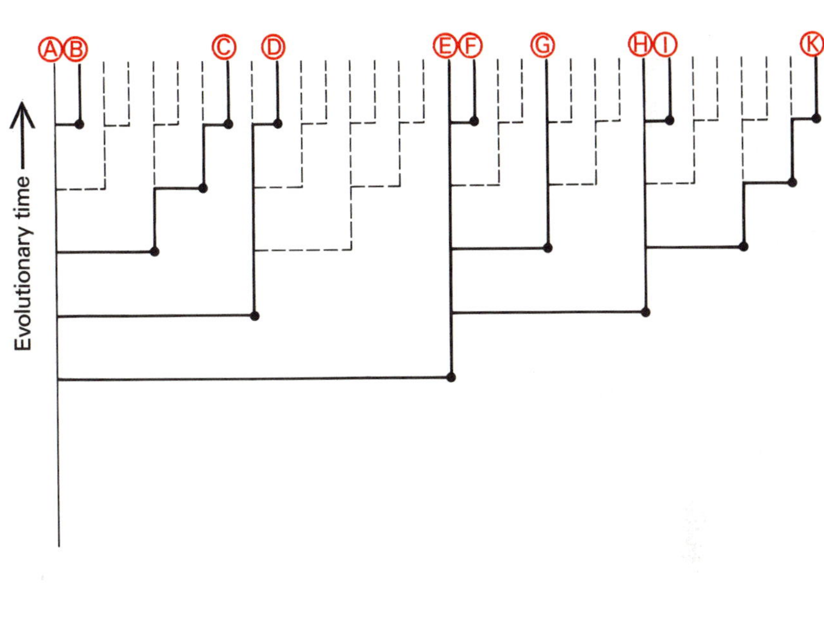

Number of amino-acid substitutions

	A	B	C	D	E	F	G	H	I	K
A	0									
B	1	0								
C	3	4	0							
D	2	3	5	0						
E	1	2	4	3	0					
F	2	3	5	4	1	0				
G	2	3	5	4	1	2	0			
H	2	3	5	4	1	2	2	0		
I	3	4	6	5	2	3	3	1	0	
K	5	6	8	7	4	5	5	3	4	0

Principle of molecular dating. The "tree" at the top illustrates a brief part in the evolutionary history of a hypothetical protein molecule. Each ramification point in the tree is the result of a mutation leading to the replacement of one amino acid by another. The mutated branch is indicated by a dot. Proteins A through K are evolutionary siblings differing from each other by the number of amino-acid substitutions indicated in the table. The dotted lines represent extinct proteins. The sequencing of homologous proteins in different species yields the kind of information given in the table. The most probable phylogenetic tree can then be constructed from the data. The method has many complications arising from the structure of the genetic code, the possibility of more than one mutation affecting the same site, the occurrence of mutations other than simple amino-acid substitutions (deletions, insertions, inversions), and the fact that different trees can accommodate the same data. But, as more and more proteins are being sequenced, the information assembled is becoming increasingly solid and secure. Such investigations are now being extended, with similar results, to genes and to RNAs. Soon, a highly detailed history of living organisms on earth will be available, as deciphered from the structure of their macromolecules.

do with the broader evolutionary phenomenon that has led to the development of increasingly complex forms of life. They seem to have happened with the same frequency throughout evolution and to bear no correlation to any of the events that have, at certain times, led rather abruptly to the appearance of new species.

This is so true that sequencing has become a favorite tool of evolutionary genealogists. Simply by counting the number of differences—bases in a nucleic acid or amino acids in a protein—between two homologous molecules belonging to different species and applying a suitable correction to account for silent mutations, consecutive replacements at the same site, and other possible complications, this new breed of historians will tell you how long ago the two species diverged from a common ancestor. The results of this elegant molecular-dating procedure

Comparative biochemistry of cytochromes. The table below gives the number of amino-acid substitutions in cytochrome *c* from twenty-five different species, including mammals and other vertebrates, invertebrates, plants, and fungi. The computer-generated phylogenetic tree of cytochrome *c* shown at the right is derived from sequencing data (most of which appear in the table below). In this calculation, amino-acid substitutions have been converted into the minimum number of nucleotide substitutions in DNA (according to the genetic code) needed to account for the observed amino-acid replacements. The branches are measured in terms of these substitutions. Agreement between molecular and more conventional methods of dating based on the fossil record is not perfect but remarkably close in view of the fact that only a single protein is considered.

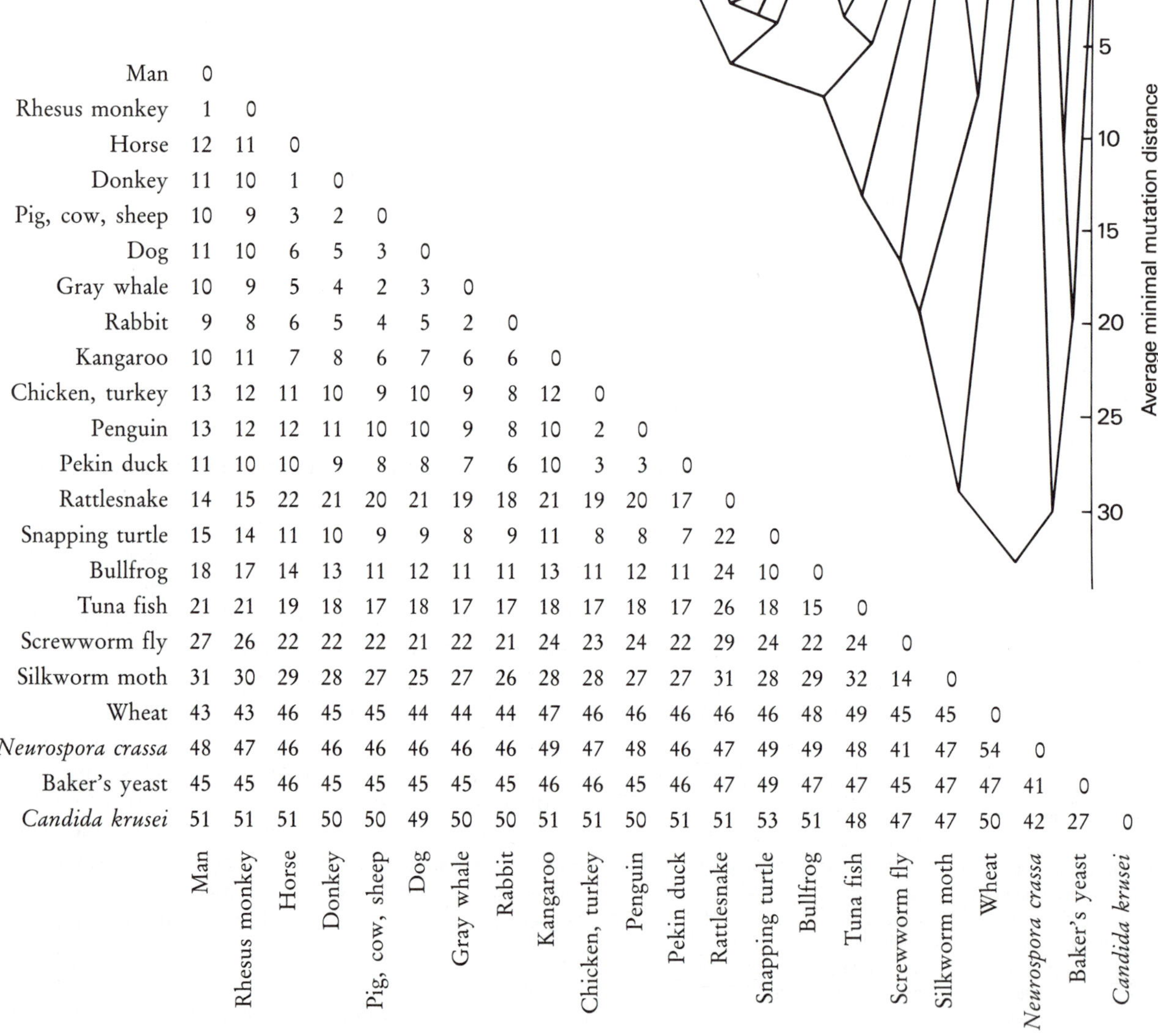

	Man	Rhesus monkey	Horse	Donkey	Pig, cow, sheep	Dog	Gray whale	Rabbit	Kangaroo	Chicken, turkey	Penguin	Pekin duck	Rattlesnake	Snapping turtle	Bullfrog	Tuna fish	Screwworm fly	Silkworm moth	Wheat	*Neurospora crassa*	Baker's yeast	*Candida krusei*
Man	0																					
Rhesus monkey	1	0																				
Horse	12	11	0																			
Donkey	11	10	1	0																		
Pig, cow, sheep	10	9	3	2	0																	
Dog	11	10	6	5	3	0																
Gray whale	10	9	5	4	2	3	0															
Rabbit	9	8	6	5	4	5	2	0														
Kangaroo	10	11	7	8	6	7	6	6	0													
Chicken, turkey	13	12	11	10	9	10	9	8	12	0												
Penguin	13	12	12	11	10	10	9	8	10	2	0											
Pekin duck	11	10	10	9	8	8	7	6	10	3	3	0										
Rattlesnake	14	15	22	21	20	21	19	18	21	19	20	17	0									
Snapping turtle	15	14	11	10	9	9	8	9	11	8	8	7	22	0								
Bullfrog	18	17	14	13	11	12	11	11	13	11	12	11	24	10	0							
Tuna fish	21	21	19	18	17	18	17	17	18	17	18	17	26	18	15	0						
Screwworm fly	27	26	22	22	22	21	22	21	24	23	24	22	29	24	22	24	0					
Silkworm moth	31	30	29	28	27	25	27	26	28	28	27	27	31	28	29	32	14	0				
Wheat	43	43	46	45	45	44	44	44	47	46	46	46	46	46	48	49	45	45	0			
Neurospora crassa	48	47	46	46	46	46	46	46	49	47	48	46	47	49	49	48	41	47	54	0		
Baker's yeast	45	45	46	45	45	45	45	45	46	46	45	46	47	49	47	47	45	47	47	41	0	
Candida krusei	51	51	51	50	50	49	50	50	51	51	50	51	51	53	51	48	47	47	50	42	27	0

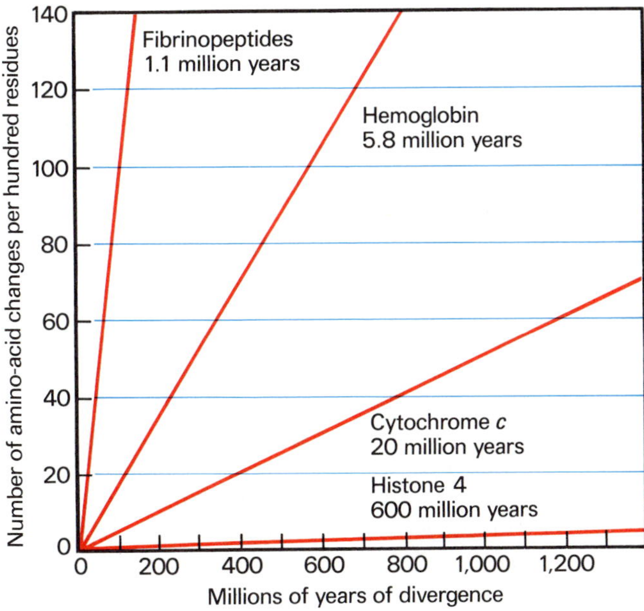

The linear relationship between molecular and fossil clocks of evolution suggests that, on an average and in first approximation, the mutation rate has remained constant throughout evolution and is the same in different species. More refined analysis indicates that this is not strictly true. What is striking in the graph, however, is the wide difference in the slopes of the different lines. This difference is not primarily an expression of differences in the mutability of the corresponding genes (although such differences may exist) but reveals mostly the degree to which changes in the amino-acid sequences of the proteins can be tolerated without loss of functional properties (nonviable mutants).

agree remarkably well with those of conventional paleontology. Not being restricted to the few traits that have been preserved in the fossil record, they have allowed us to extend, complete, and sometimes even revise the classical phylogenetic reconstructions. The relative time scale does, however, vary, sometimes considerably, from one molecular variety to another. Extremes among the proteins are the polypeptides (fibrinopeptides) that make up the meshwork of blood clots, which have suffered one amino-acid substitution in one hundred residues every 1.1 million years, and histone H4, which has needed about 600 million years for a single such replacement to occur. In between are hemoglobin, with 5.8 million years, and cytochrome *c*, with 20 million years.

These differences do not primarily reflect differences in the mutability of the genes concerned, although such differences may exist; rather, they reveal the stringency of the structural constraints on the functional properties of the corresponding proteins. Most likely, the genes coding for these proteins all mutated with about the same frequency, which sufficed, judging from the fibrinopeptides, to bring about at least one amino-acid substitution per hundred residues in their respective products every million years. However, for every hundred such substitutions that could be accommodated by the fibrinopeptides without loss of function (indifferent mutations), some eighty proved severely disabling to hemoglobin, ninety-five to cytochrome *c*, and essentially all to histone H4. They were consequently eliminated by natural selection, leaving no trace in contemporary living organisms.

Fascinating as they are, these examples illustrate only the negative, conservative aspect of natural selection—its strong tendency to keep things as they are and to allow only changes that make no difference. Clearly, such timorous toe-dipping into the waters of the unknown as is afforded by point mutations cannot account for the leaps that have propelled the upward surge of evolution. Something more drastic must have been at work.

That something is presumably represented by the deletions, the duplications, the recombinations, the transpositions, the splicings, the exchanges, the viral invasions, and all the other genetic migratory events that recent explorations have brought to light. Most of the time, such imprudent tamperings with orthodoxy must have been sanctioned without delay. But on certain rare occasions, perhaps coincidental with a sudden climatic or other environmental change, the freakish misfit proved to be better adapted than its "normal" congeners. It became a phagocytic hunter in a nutrient-depleted sea, or a fish out of water, or a tree dweller exiled to the savanna. Sometimes the resulting adaptations had far-reaching consequences.

All these events are buried in the depths of geological ages, and we will probably never get to know exactly how they happened, although we should not, perhaps, underestimate the power of molecular-biological analysis in reconstructing the past. Even more difficult to unearth are the events that preceded, and led to, the appearance of the first living cells. This difficulty has, however, not deterred biomolecular archaeologists from digging for clues. Already they have made the fundamental observation that many of the small organic molecules that are found in living matter, including amino acids, sugars, purines, and pyrimidines, can arise spontaneously under conditions mimicking those believed to have prevailed on the earth's surface in prebiotic times. Even primitive polymers have been generated in this way, among them oligonucleotides of sorts, as well as "proteinoids" endowed with catalytic activities resembling those of enzymes.

Mind you, this has nothing to do with creating life in the test tube, or even reproducing anything remotely resembling the mixture of molecules one gets from breaking apart a living organism. These artificial mixtures are really very "dirty." In addition to substances authentically found in living beings, they contain all kinds of other molecules, assembled in a very haphazard way. They could, however, approximate to some extent the "primeval soup" out of which the first living cells arose. How this emergence took place is a matter of conjecture, but it most likely involved, on a simpler chemical level, the same cardinal rules of fidelity, variability, and selection that governed biological evolution. Primitive self-maintaining and self-correcting systems must have formed and evolved progressively into dynamic structures of increasing complexity and stability. Other possibilities have been considered, including, as noted in Chapter 10, insemination of the earth by germs from outer space (panspermia). But this only pushes the problem back to the origin of those germs and stretches the available duration hardly more than three or four times, unless you reject the "big bang" theory and believe, with the iconoclastic British astronomer Fred Hoyle, in a steady-state universe of "enormous antiquity," whatever is meant by that expression.

In actual fact, the age of the universe is essentially irrelevant to the discussion. If you equate the probability of the birth of a bacterial cell to that of the chance assembly of its component atoms, even eternity will not suffice to produce one for you. So you might as well accept, as do most scientists, that the process was completed in no more than 1 billion years and that it took place entirely on the surface of our planet, to produce, as early as 3.3 billion years ago, the bacteriumlike organisms revealed by fossil traces. After that, bacteria ruled the world alone for more than 2 billion years, during which they developed such important attributes as photosynthesis, oxidative phosphorylation, and, presumably, the phagocytic way of life; at the same time they profoundly changed the properties of the atmosphere and of the earth's surface. And then, a little more than 1 billion years ago, what may have been the most important single event in the history of life took place: the formation of the first eukaryotic cells. (Some possible steps in this event have been alluded to in Chapters 6, 9, and 10.) Once the first eukaryotes were there, the evolutionary process started picking up at an ever-increasing pace. The first multicellular invertebrates arose some 800 million years ago, the first vertebrates about 200 million years later, and then, in accelerating succession, the higher fishes, the amphibians, the reptiles, and finally the mammals, which appeared approximately 300 million years ago. Primates are probably no more than 60 million years old, and their human offshoot branched out only 2 or 3 million years ago.

These are still very long time spans. Nevertheless, the speed at which evolution started moving once it discovered the right track, so to speak, and the apparently autocatalytic manner in which it accelerated are truly astonish-

This graph shows the rate of increase in the complexity of the genome as a function of evolutionary time. The shape of the curve suggests that complexity favored further complexity in some sort of autocatalytic fashion. (Certain species diverge greatly from the figures shown because, for some ununderstood reason, their genome contains unusually large amounts of "ballast DNA.")

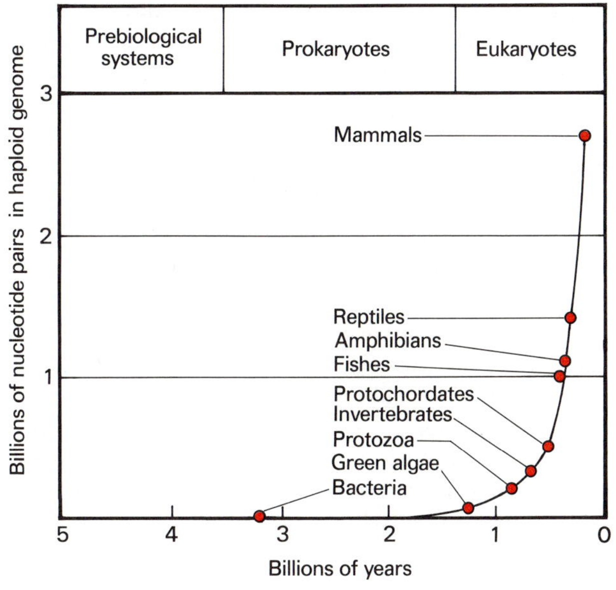

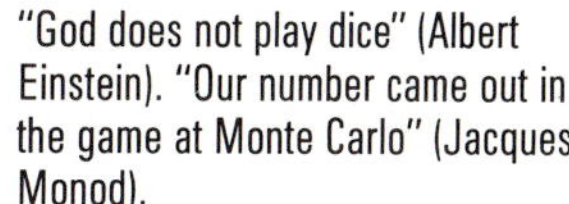
"God does not play dice" (Albert Einstein). "Our number came out in the game at Monte Carlo" (Jacques Monod).

ing, especially in regard to the increasing generation time of the organisms involved, which should have slowed down the process. It took no more than 150,000 generations for an ape to develop into the inventor of calculus, whereas some 30,000 billion bacterial generations may have been needed to produce an amoeba. Which brings us back to the question of chance and design: How much chance? How much design?

The answer of modern molecular biology to this much-debated question is categorical: chance, and chance alone, did it all, from the primeval soup to man, with only natural selection to sift its effects. This affirmation now rests on overwhelming factual evidence. But it is not, as some would have it, the whole answer, for chance did not operate in a vacuum. It operated in a universe governed by orderly laws and made of matter endowed with specific properties. These laws and properties are the constraints that shape the evolutionary roulette and restrict the numbers that it can turn up. Among these numbers are life and all its wonders, including the conscious mind. Faced with the enormous sum of lucky draws behind the success of the evolutionary game, one may legitimately wonder to what extent this success is actually written into the fabric of the universe. To Einstein, who once said: "God does not play dice," one could then answer: "Yes, He does, because He is sure to win." In other words, there may be a design. But it started with the "big bang."

Such a view is shared by some, not by others. The French scientist Jacques Monod, one of the founders of molecular biology and the author of the celebrated book *Chance and Necessity*, published in 1970, argued eloquently in favor of the opposite view. "Our number," he wrote, "came out in the game at Monte Carlo." And further: "The Universe was not pregnant with life, nor the biosphere with man." His final conclusion reflects the stoically (and romantically) despairing existentialism that greatly affected his generation of intellectuals in France: "Man now knows that he is alone in the indifferent vastness of the Universe from which he has emerged by chance."

This is nonsense, of course. Man knows nothing of the sort. Nor does he have any proof to the contrary, either. What he does know, however—or, at least, should know—is that, with the time and amount of matter available, anything resembling the simplest living cell, let alone a human being, could not possibly have arisen by blind chance were the universe not pregnant with them. Making such a statement does not in any way mean espousing a rigidly deterministic or, alternatively, a vitalistic view of the origin and evolution of life. It leaves full scope for essentially stochastic processes to operate behind these events, as conceived by modern Darwinian theory. But it does, by emphasizing the significance of the built-in constraints within which stochastic processes operate, provide a rational basis for a more optimistic philosophy than that of Monod.

There are many who have no need for such a philosophy and many more who cheerfully dispense with any sort of rational basis for the beliefs—or disbeliefs—they hold. Those, perhaps few in number, who are at the same time of a religious mind and respectful of scientific objectivity may derive some satisfaction from the view offered here. Surely they should find it more gratifying and uplifting than that of the so-called creationists, whose creed is really an insult to the Creator they believe in, making him into some sort of frivolously facetious, if not malicious, deity who has filled his work with innumerable molecular red herrings apparently designed for the sole purpose of leading scientists up some fantastically elaborate cosmic garden path.

Genetic Engineering

Certain genes, as we have just seen, are naturally endowed with a considerable degree of mobility. They can move from one place to another within a genome; they can jump in and out of a genome with the help of small satellite resting places, such as plasmids, bacteriophages, or viruses; finally, they can travel from one genome to another as passengers of these particles. Such migrations, which occur on a large scale in nature, illustrate the "selfish" behavior of genes already referred to. Indeed, if the mechanisms involved are considered in a Darwinian fashion, the *raison d'être* for their selection often seems to be little more than the perpetuation of the genes concerned. The cells involved may or may not benefit from them, depending on the extent to which the wandering genes need healthy cells for their own survival and multiplication.

Seen in this context, the whole world of viruses and related particles appears as an intricate fleet of gene carriers, which probably originated in the first place through the evolutionary emancipation of mobile genes from complex prokaryotic or eukaryotic genomes. Once these facts became appreciated and scientists began to understand the mechanisms involved, the idea of using the fleet for the transport of certain specifically chosen passenger genes naturally presented itself. Nature provided the tools, and so genetic engineering was born, perhaps to become one of the most powerful techniques ever developed by mankind.

Indeed, the prospects evoked by genetic engineering appeared so awesome that scientists themselves took fright. In July, 1974, in an unprecedented action, a group of eminent scientists in the field declared a voluntary moratorium on further experimentation until its possible effects could be assessed. Predictably, this commendable manifestation of prudence and social responsibility created a tremendous stir and sparked a public debate of highly charged emotional content. Fears of uncontrollable, worldwide epidemics and of massive cancer invasions were raised. Accusations of "playing God," "tampering with the laws of nature," "crossing the sacred species barrier," were leveled. Prometheus, Pandora, the Sorcerer's Apprentice, Frankenstein, and other mythical figures were conjured to re-express mankind's deeply rooted distrust of new knowledge and technological innovation. Dark allusions were made to *Herrenvolk* philosophy and the manipulation of human heredity by unscrupulous rulers. A town council was even called to vote the outright banning of all DNA recombinant research from one of the world's most famous campuses. In a more sober vein, but yet responsive to the powerful currents stirred up within and without the scientific community, health agencies all over the world took over the problem and drew up elaborate guidelines that almost imposed more safeguards on the handling of genetically altered microorganisms than on that of the deadliest bacteria and viruses.

Today the furor has died down, or rather it has been replaced by another, of the gold-rush type. Companies have been formed to exploit the new technology, patents have been applied for, scientists have turned into entrepreneurs, and universities have become partners in commercial ventures, not without lengthy, soul-searching debates on the freedom of scientific information and on the disinterestedness of academic research. What, then, is the cause of all this turmoil? And, first, what is genetic engineering, known technically as artificial DNA recombinant technology, and what are its main purposes?

In theory—and now also in practice—what you do is simple. You take a piece of DNA and introduce it into a foreign cell in such a way that it will be replicated and, in some applications, also transcribed and translated. If only replication is sought, the manipulation is called molecular cloning, and the host cell is then invariably a bacterial cell chosen to produce as many DNA copies as possible in the shortest possible time. Cloning has no practical applications (except as a first step toward subsequent expression) but is an extremely powerful tool in basic research in that it can provide in almost unlimited quantities any segment of genetic information one may wish to analyze. In combination with other techniques of molecular biology, it is destined to allow a complete mapping of the genome of a cell, as well as a decoding of all the instructional elements that govern gene transcription and translation.

When actual expression of the DNA is aimed at, the objective may still be purely scientific, such as the elucidation of exactly what counts in a promoter sequence or what makes a gene oncogenic. But, in addition, two important types of applications are to be considered. One is large-scale manufacturing of a gene product, such as human insulin, interferon, antigens to be used for making vaccines, or any other rare protein that may be of practical value. The other application is genetic transformation of the recipient cell. Endowing certain crop plants with the ability to utilize atmospheric nitrogen is one such aim that is currently attracting great interest. Correcting genetic diseases is another, though not yet within the realm of possibility.

A number of techniques are available for introducing the DNA into its recipient cells. It can be micro-injected, helped across the plasma membrane by special treatments of the cells, or enclosed within small membranous sacs or artificial phospholipid vesicles (liposomes, pp. 43–44) that fuse with the plasma membrane. As a rule, however, the yield of such procedures is low and their outcome uncertain. When applicable, the method of choice is to attach the DNA to the DNA of a vector, usually a plasmid or some viral particle, that happens to be naturally endowed with the appropriate means for introducing its DNA content into selected cells. This technique has the additional advantage that the vectors used often carry genetic markers, such as resistance to certain antibiotics, that render the recognition and isolation of the transformed cells very easy.

The tools for attaching the DNA to its vector are all borrowed from nature. They include restriction enzymes for cutting the DNAs at well-defined sites; nonspecific terminal deoxynucleotidyl transferases for fitting the vector and its passenger with sticky ends (e.g., poly-dG on one and poly-dC on the other), unless the restriction enzymes themselves are relied on to do this; DNA polymerase to fill in gaps; and DNA ligase to do the final stitching. The passenger DNA may be a simple piece of native DNA, a more complex mixture of such pieces, or the contents of a complete genome fragmented by some restriction enzyme. In the last case, the particular recombinant that one is interested in has to be fished out by special techniques from the highly motley population of transformed cells. Such experiments have been given the suggestive name of "shotgun experiments."

Quite often, the DNA is modified in various ways, either for the purpose of investigating the significance of some structural detail or in order to ensure its proper replication or transcription in the host cell. If the DNA comes from a eukaryote and the recipient cell is a prokaryote, the appropriate signals have to be given in

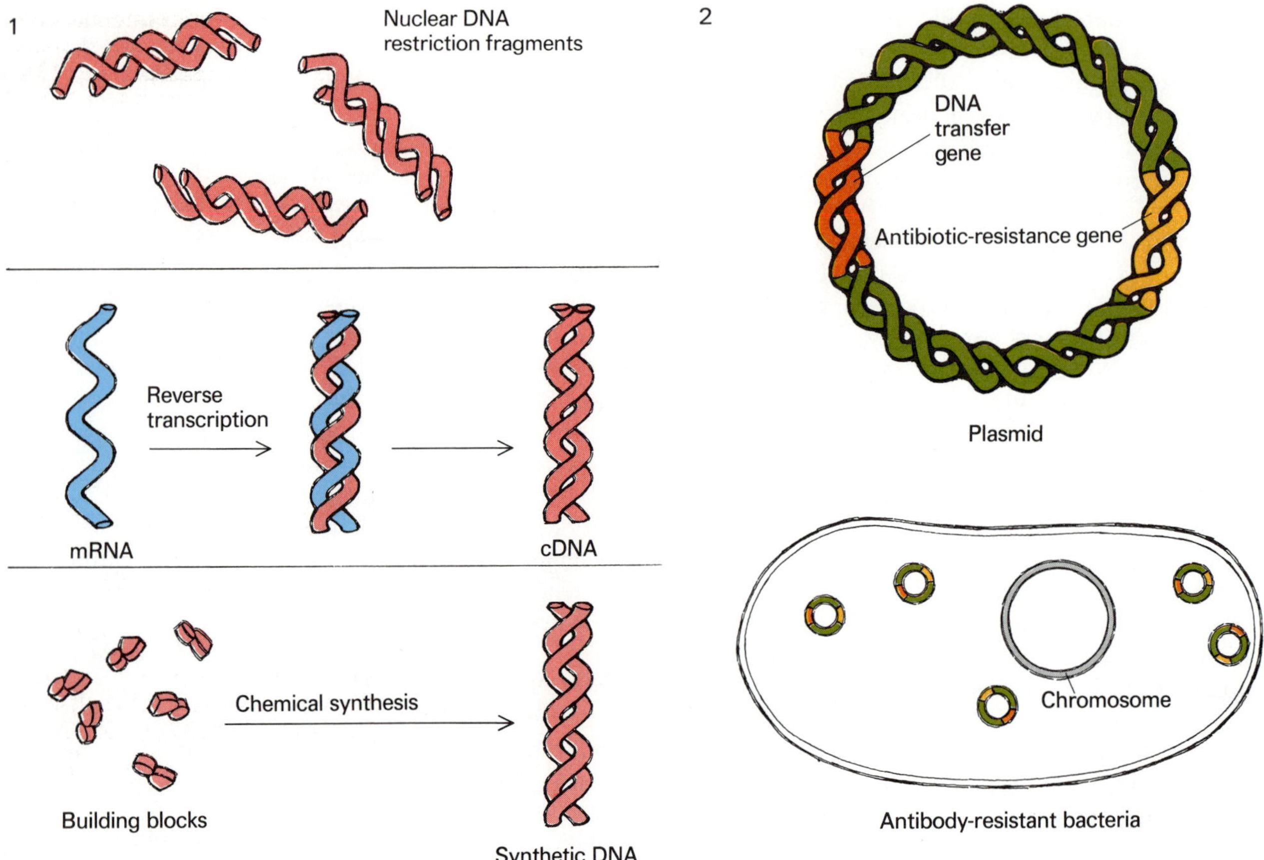

prokaryotic language. The difficulty of crossing the prokaryote-eukaryote line is particularly great when expression of the transplanted gene is desired and splicing of the mRNA is required, as it is for most eukaryotic genes. Bacteria do not splice their own messengers; they lack the necessary machinery. To overcome this difficulty, the engineers use either reverse transcripts (cDNAs) of mature mRNAs or completely synthetic DNAs made to order according to the amino-acid sequence of the protein, natural or artificial, they wish the transformed cells to make. As can be seen, the possibilities of these new techniques are endless, and the amount of molecular juggling that can be accomplished by means of the currently

DNA recombinant technique (genetic engineering).

1. The passenger DNA, to be cloned, may consist of fragments cut from the whole genome by means of restriction endonucleases ("shotgun" experiment), of cDNA transcribed from purified mRNA by means of reverse transcriptase, or of completely synthetic DNA.

2. The vehicle is usually a plasmid obtained from antibiotic-resistant bacteria and bearing one or more antibiotic-resistance genes, as well as genes that facilitate its transfer into its host. Sometimes a phage DNA is preferred and reinserted into the phage coat for transfer to the host (cosmid). With eukaryotic cells as hosts, rare plasmids or, occasionally, viruses may provide the vehicle DNA. Sometimes the DNA is introduced into the host cell without the help of a vehicle (transfection).

3

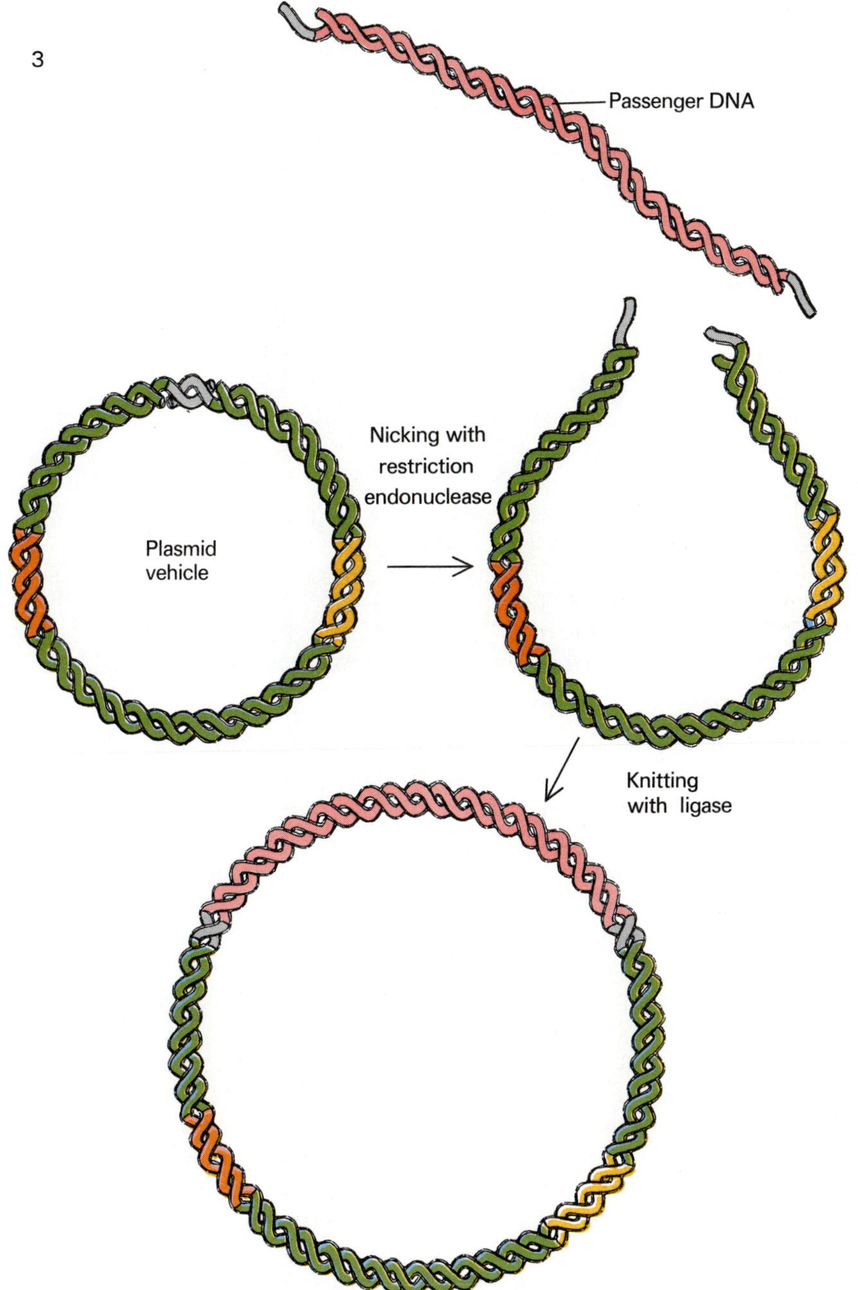

3. To attach the passenger DNA to the plasmid vehicle, the plasmid must be nicked by a suitable restriction endonuclease and both passenger and vehicle must be fitted with complementary sequences (sticky or cohesive ends). Such will be the case if the passenger DNA has been fragmented with the same restriction endonuclease as has been used to open the plasmid (see the accompanying illustration, as well as that on p. 335). When passenger and vehicle are mixed, some molecules join by their sticky ends as shown, and the bonds can be sealed by a ligase. The drawback of this method is that many other combinations are formed in addition to the desired one. A more specific procedure is to fit the passenger DNA with one kind of ends and the vehicle DNA with the corresponding complementary ends. Sequences of dA for one, and of dT for the other, may serve the purpose. Alternatively, dC and dG may be used. This fitting is done with terminal deoxynucleotidyl transferase—an enzyme acting like DNA polymerase, but nonspecifically—and the appropriate deoxynucleoside triphosphate as substrate (dATP or dCTP for one and, correspondingly, dTTP or dGTP for the other). Then, when passenger and vehicle are mixed, only the desired combination can form, but there will be gaps at the junctions owing to the unequal lengths of the added ends. DNA polymerase is used to fill the gaps and the final knitting is then done by ligase action, as in the first case.

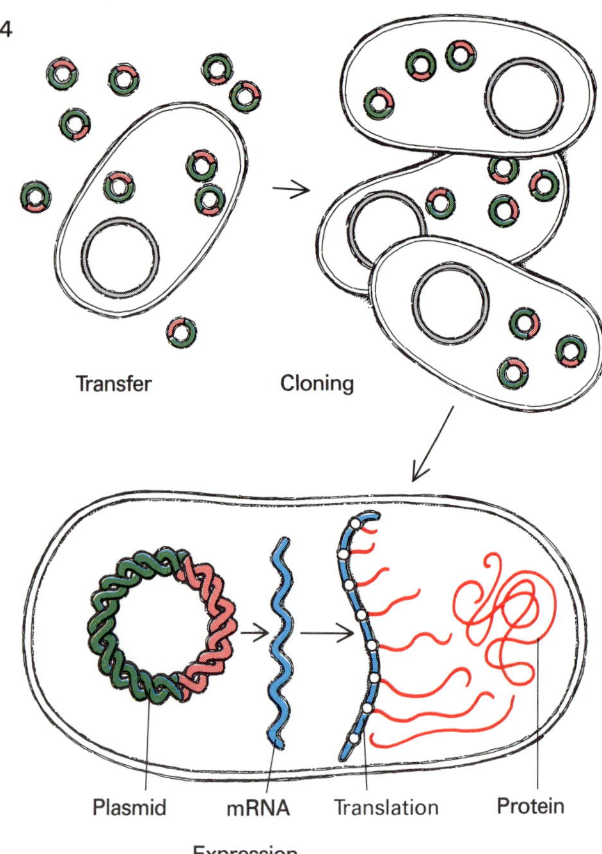

A genetically engineered giant. The mouse at the left is issued from an egg cell into which the human growth hormone gene, isolated by cloning and appropriately engineered for expression, was introduced by microinjection. The animal grew to twice the size of its control sibling.

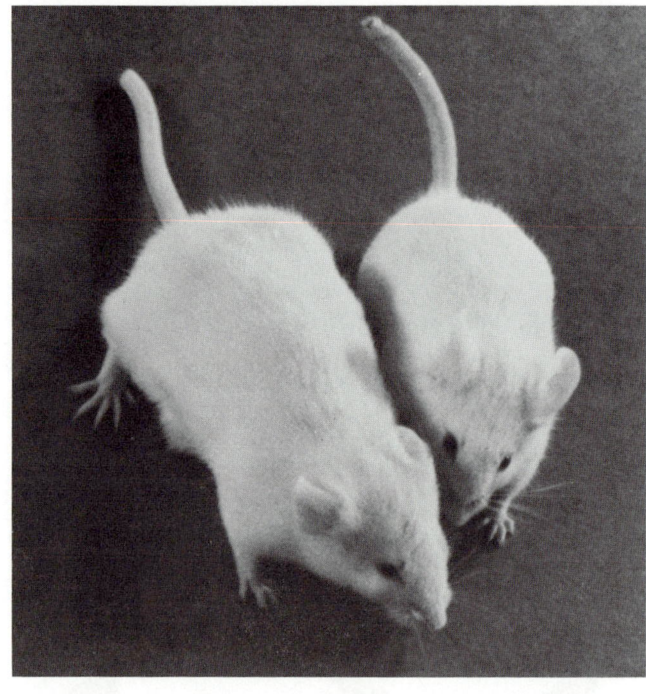

4. Exposure of the prospective bacterial (or eukaryotic) hosts to recombinant plasmids leads to the successful incorporation of the selected foreign gene into some of the host cells. If simple cloning of the gene (for the purpose of analysis, for instance) is desired, all that is needed is the correct replication of the plasmid and of its passenger in synchrony with cell multiplication. If the gene product is desired (industrial production of insulin or interferon, for example), engineering of the plasmid will have to include promoter and other regulatory sequences needed to induce the correct transcription and translation of the gene. A third application, not illustrated, involves the use of the modified cells themselves (cereals programmed to take up atmospheric nitrogen, correction of genetic defects, for instance). In all cases, the preliminary selection and cloning of the correctly changed host cells (usually a very small minority) is required. Advantage is taken of the antibiotic-resistance genes borne by the plasmid to select the cells that have taken up a plasmid. Various other "probes" are used to select the cells that contain the right gene in appropriate form. Such a selection is particularly stringent in experiments of the shotgun type, where thousands of different genes and gene fragments are handled simultaneously.

available recombinant kits is truly remarkable. There is no doubt that a new age has just started.

Bacteria are the obvious recipients for all cloning and manufacturing applications, in view of their rapid generation rate, ease of cultivation and selection, and large number of possible vectors. Some of the initial objections to genetic engineering arose largely from the fear that some bacteria, unwittingly transformed into highly pathogenic species, might escape into the environment. This risk appeared particularly hazardous because the most widely used bacterium is *E. coli*, a normal inhabitant of the human digestive tract. At first, strict physical containment of the facilities used was imposed as a safeguard against such conjectural accidents. Nowadays, reliance is put mostly on biological containment, itself a product of genetic manipulation (by conventional procedures). Mutant strains of *E. coli* have been produced that have strict temperature and nutritional requirements, absolutely incompatible with their survival in the human body or in any environment other than the highly specialized conditions provided in the culture vats.

The use of bacteria has, however, raised a number of difficulties, and eukaryotic hosts, especially the time-honored yeast cells, are also being developed. If transformation of a eukaryotic organism (for instance, a plant) is the final objective, then the host must, by definition, be of eukaryotic nature. Vectors for eukaryotes are less numerous and more difficult to manipulate than are bacterial vectors. Some of them have the additional disadvantage of being oncogenic, a fact that provided another reason for the original opposition to DNA recombinant techniques.

In certain important applications, vectors are dispensed with and the DNA is introduced into the cells by some other means. When the few cells that express the property that is sought are easily identified and cloned, this technique can be very powerful. It has played an important role in recent studies on cancerous transformation.

The Cancer Enigma

The Greeks saw it as a malevolent crab clawing deeply into its victim's flesh and gave it the crab's name of *karkinos* (*cancer* in Latin). They did not know how or whence it came. But they knew that once it has taken hold it never leaves off and just goes on eating away, spreading and growing to obscene proportions, until there is nothing left to consume. It attacks young and old indiscriminately but hits the old more frequently because of their longer exposure to its repeated assaults. And so it has become the main cause of death after cardiovascular disease, and certainly the most feared, in all areas of the world where removal of the primitive scourges of infection and starvation has left it enough scope to strike.

One reason why cancer is so frightening is its insidiousness. Not the slightest sign announces its onset, and it may remain silent and invisible for years while it establishes impregnable positions. All it needs is for a single cell to be transformed in such a way that it goes on dividing in undisciplined fashion (p. 320). This transformation appears to be irreversible and is in some way hereditary, in that the cell's progeny share its refractoriness to population control. What causes the transformation? And what does it consist of? Those are the two main questions investigators are trying to answer, in line with the two great strategies of medical research: the one that treats the cell or patient as a black box and simply looks for causes in the hope that their knowledge may serve as an empirical guide to effective prevention; the other that tries to open the black box and to elucidate the disease-provoking mechanisms as a means toward rational cure or protection.

In 1760, a British surgeon, Sir Percival Pott, noted that chimney sweeps were peculiarly subject to cancer of the scrotum and deduced that soot retained between skin folds could be responsible for causing the disease. Since then, a number of occupational cancers have been recognized in workers exposed to such materials as radioactive minerals, tar, aniline dyes, asbestos, and vinyl chloride, and appropriate protective measures have been enacted. In addition, other potential causes of cancer have been looked for by widespread epidemiological investigations. This search has produced a wealth of statistical information linking the incidence of certain cancers to geographical factors, diet, smoking or drinking habits, and other forms of exposure either imposed by the environment or self-administered. In practice, however, except for some occupational cancers of restricted incidence, little progress in prevention has been achieved through this approach, either because the nature of the link is not yet sufficiently established to support a clear-cut recommendation or because the recommended measure turns out to be socially, economically, or politically unacceptable. A flagrant example of this is cigarette smoking, which is the most important single cause of cancer in the world, accounting for at least one-third of all human tumors, including some of the nastiest ones, such as lung cancer. Yet it is there for everyone to indulge in, while massive rallies are organized against purely potential hazards and interminable discussions center on the banning of this or that trace chemical accused of possibly causing one additional case of cancer in 100,000 inhabitants.

A valuable offshoot of the occupational and epidemiological approach has been the search for means of inducing cancer in experimental animals. It has allowed the identification of many carcinogenic agents, physical or chemical, and has produced important tools for the screening of chemicals, drugs, food additives, dusts, fumes, and other products or by-products of human industry. In addition, and perhaps more importantly, it has provided investigators with experimental models for the detailed study of the carcinogenic process itself.

An important notion revealed by these studies is that most carcinogens are mutagens, and vice versa. This discovery is clearly related to the hereditary character of the cancerous transformation mentioned above. Yet there is an important and puzzling difference between carcinogenesis and mutagenesis. Whereas a mutation occurs as an immediate consequence of the chemically induced genetic lesion, the development of cancer seems to be a multistep process. It generally requires repeated exposure to the carcinogen, and it is influenced by numerous factors, including the presence of substances called promoters or cocarcinogens that are not carcinogenic in themselves but enhance considerably the potency of carcinogens. No doubt genetic changes are involved in the transformation of a normal cell into a cancer cell. But the changes could be more complex or more subtle than ordinary mutations.

A revealing clue to the nature of these changes was discovered in 1911 by a young American scientist, Peyton Rous, who found that he could transmit a chicken sarcoma (a kind of cancer) to other chickens by injecting them with an extract of the tumor containing no living cells. The properties of the carcinogenic agent in the extract suggested that it might be a virus. As a rule, the scientific community behaves with the objectivity and open-mindedness that is expected from it. But occasionally it will greet with scepticism or disbelief, or simply ignore, a report that conflicts with some sort of consensus view. This happened to Peyton Rous's discovery. He did, however, live long enough to see his work vindicated and even recognized by a Nobel Prize—55 years after his finding was first reported. Had he lived to be a centenarian—he was 91 when he died in 1970—he would have been given an inkling of the truly momentous significance of his discovery, not only for our understanding of cancer, but for the whole development of science and technology.

Indeed, the Rous sarcoma virus has turned out to be the prototype of retroviruses, which, as explained earlier in this chapter, are those peculiar RNA-containing viruses that rely for multiplication on transcription of their RNA into cDNA by reverse transcriptase, followed by insertion of the cDNA into the host cell's genome (pp. 349–350). Thus the Rous virus served as a vehicle for the discovery of reverse transcriptase, one of the landmarks in the development of molecular biology. From the theoretical point of view, the existence of this enzyme contradicted the "central dogma" according to which genetic information is always transferred unidirectionally from DNA to RNA and from RNA to protein. (Note that the second part of the statement remains true and is likely to stay that way, as it amounts to a denial of the widely discarded Lamarckian view of the heredity of acquired characters). Reverse transcriptase has also provided genetic engineering with one of its most valuable tools and may, in this respect, be credited indirectly with many recent advances both in basic knowledge and in practical applications.

Another major product of Rous's discovery sprang from the genetic analysis of the virus. The oncogenic agent has a simple genome, made of only four genes. Of these, three carry all the information that is needed for multiplication of the virus; two code for the two viral proteins and one for the reverse transcriptase. The fourth gene, called the *src* gene, is in a way gratuitous. It has no function in multiplication of the virus but exerts a powerful effect on the host cell, being solely responsible for the cell's cancerous transformation. It is an oncogene—more correctly an *onc* gene—literally a cancer-causing gene. Some twenty such *onc* genes are known, each carried by certain kinds of retroviruses. They bear names such as *ras*, *myc*, *sis*, *erb*, *yes*, or *ski*—acronyms referring to the names of their viral carriers and of the diseases they cause.

The greatest surprise—and probably the most far-reaching gift of the Rous sarcoma virus to science—came after the DNA wizards arrived on the scene and started probing normal cells with cloned transcripts of the viral *onc* genes. They found that these genes do not, in fact, rightfully belong to the viruses that carry them, but rather were purloined from an infected cell in some earlier episode in the history of the viruses. Technically, as already mentioned (pp. 338 and 347), the phenomenon is known as transduction: a virus picks up a gene from its host cell and transfers it to another cell in the course of a subsequent invasion. Because of their peculiar mode of multiplication, which requires insertion of the viral genome into the host cell's genome, retroviruses are particularly well suited to act as transducing agents. Even so, the event remains a rare one: many retroviruses exist that do not carry an *onc* gene. To distinguish between the viral and cellular forms of the genes, the nomenclature uses the terms v-*onc* and c-*onc*, or, alternatively, oncogene and proto-oncogene, respectively.

As a cause of cancer, gene transduction is probably exceptional, but its discovery has brought to light an all-important clue: normal cells harbor genes that can cause a cancerous transformation under certain conditions (or after undergoing some modification). This started the sleuths looking for evidence implicating the same genes in the production of human cancers. And they have already uncovered plenty of it. At present, the whole field is in ferment. There are investigators hot on the trail in all parts of the world, and almost no week goes by without some new finding of significance being announced. The present account is written just before going to press. Yet, it will be outdated by the time it is published.

Two questions dominate the searches: What are the functions subserved by proto-oncogenes in normal cells? How are these functions subverted, or, otherwise put, how does a proto-oncogene become an oncogene? To answer the first question, the researchers have focused their attention on the proteins encoded by oncogenes and on their counterparts in normal cells. They have identified several. Initiates refer to them as p21, p53, pp60, which, if you know the cipher, often tells you almost as much about these proteins as their discoverers know of them: namely, that their molecular mass is 21, 53, or 60 kilodaltons (pp means that the protein is phosphorylated).

Some oncogene products, however, are beginning to be characterized. Among them is pp60, the product of the *src* gene. This protein has all the hallmarks of a key regulator. It is a protein kinase, belonging to a group of enzymes that we encountered before as playing a central role in the control of various enzyme activities (Chapters 13 and 14). Within this group, pp60 occupies a special position in that it phosphorylates tyrosine residues in its protein substrates, not serine or threonine residues, as other known protein kinases do. Its home is on the cytosolic face of the plasma membrane, where all the major transducing devices connected to surface receptors are situated (Chapter 13). Indeed, there are strong indications that pp60 belongs to a normal mitogenic system that is subject to stimulation by such growth factors as EGF (epidermal growth factor) or PDGF (platelet-derived growth factor). Remarkably, two other oncogene products have been traced to this system. One, the product of the *erb* gene, is related to the EGF receptor. The other, coded for by the *sis* gene, is a close relative of PDGF. It is of interest that these relationships were uncovered by computers programmed to deduce the amino-acid sequences of the proteins from the nucleotide sequences of the oncogenes that code for them and to match these sequences with those of proteins of known structure.

Exciting as they are, these findings do not, of course, signal the end of the trail. From the enzyme, we must now move to its substrate, and hence no doubt to new complexities in the chain of events underlying the form of mitogenic stimulation controlled by the tyrosine-phosphorylating pp60. Vinculin, which participates in the anchoring of actin microfilaments to adhesion plaques (Chapter 12), is a pp60 substrate and appears as a possible link in the chain. Adhesion plaques are dismantled in cancer cells and may have something to do with the phenomenon of contact inhibition mentioned at the beginning of Chapter 17.

Several other oncogene products, apparently unrelated to pp60, share with this protein the ability to catalyze the phosphorylation of tyrosine residues in certain proteins, suggesting the existence of a whole family of tyrosine-phosphorylating protein kinases that participate in

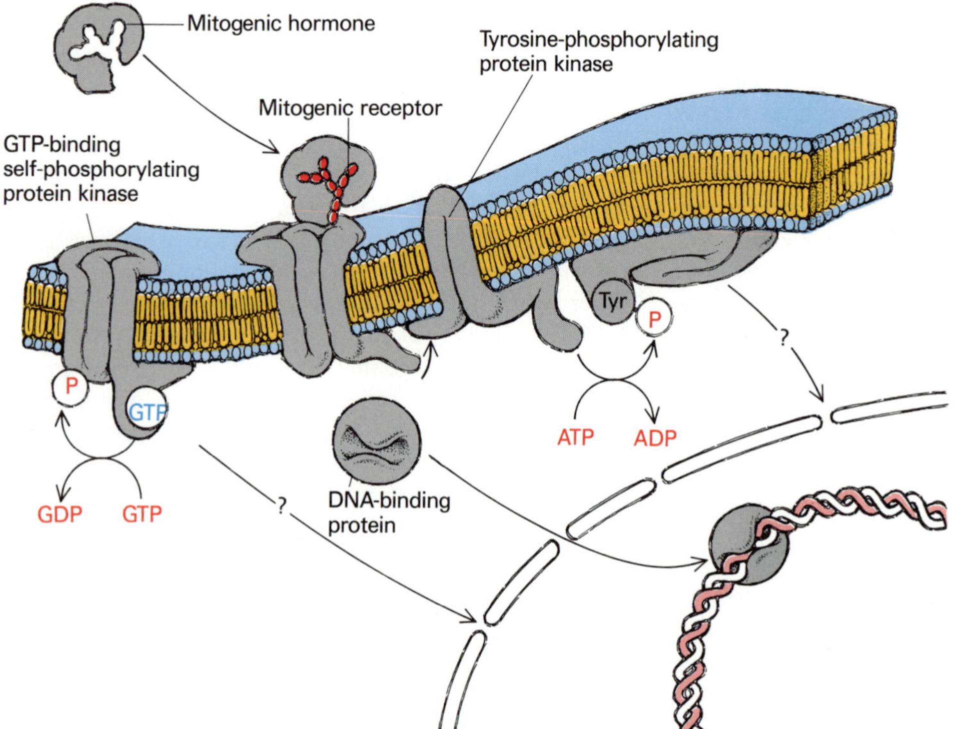

A schematic view of some proto-oncogene products and their presumptive function in normal mitogenesis.

growth control. A separate group of oncogene products, coded for by *ras* genes, comprises proteins that, like pp60 and its congeners, are bound to the plasma membrane and endowed with protein-kinase activity but that so far have been found to phosphorylate only themselves, and not on tyrosine residues at that; they have a strong affinity for GTP. A completely different family of oncogene products (e.g., of the *myc* gene) includes proteins that are situated in the nucleus, where they may bind to DNA.

The picture that emerges from all this is still very sketchy and hazy. But one important conclusion is already clear: many oncogenes—perhaps all—are derived from proto-oncogenes that code for a piece of some normal growth-controlling machinery. As to what causes a proto-oncogene to become a true oncogene, many different mechanisms seem to be involved. One is mutation, resulting in an altered gene product. How frequently this happens is not known, but at least one case has been recognized in which a single point mutation accounts for the difference between oncogene and proto-oncogene. It concerns an oncogene of the *ras* family that has been extracted from a human bladder cancer (designated T24 or EJ). It differs from its normal counterpart by a single base-pair substitution, which results in the replacement of a glycine by a valine residue in position 12 of p21, the gene product. Replacement in the same position by an arginine or by a serine residue has been found in two murine *ras* oncogenes. In a number of other cases, the differences between oncogene and proto-oncogene are much greater, indicating multiple mutations, massive deletions, translocations, or other major changes. For example, the *erb* oncogene mentioned earlier seems to code for a maimed EGF receptor lacking, among other parts, the actual binding site for EGF, though not, however, the active mitogenic domain. This part of the molecule has presumably become unleashed as a result of the alteration and stimulates cell multiplication in uninterrupted, uncontrolled fashion.

Sometimes the cancer-causing change is quantitative rather than qualitative and consists simply in excessive expression of the gene. This may happen in many different ways: by gene amplification; by insertion of a strong promoter upstream of the gene, or of what is known as an enhancer of transcription in its neighborhood; by mutations that have the same effect; by various forms of transposition or recombination—including chromosome breakages and exchanges, which are prominent characteristics of several human cancers—resulting in translocation of the gene to a region where it becomes subject to such influences. The mechanism may even be posttranslational. The product of an oncogene carried by the DNA-containing polyoma virus is a protein that binds to pp60, presumably stabilizing it and thereby enhancing its expression. Incidentally, the oncogenes of DNA viruses are not stolen from host cells. They are authentic viral genes and play an essential role in virus multiplication (even though they do not code for viral proteins proper). Only in those rare cases in which the viral genome becomes inserted into the host-cell genome do these genes behave as oncogenes, as already seen.

In the gathering of all this new information, a technique known as transfection has rendered great services. Cells in culture are exposed to DNA under conditions such that some of the DNA enters the cells and, in a small number of them, becomes incorporated in the genome and undergoes normal replication and transcription. In this way, transforming DNAs can be recognized by the characteristic changes that they induce in the transformed cells and thus be isolated for cloning and sequencing. Interestingly, such experiments succeed only with established cell lines—the NIH 3T3 cells are a favorite—that is, with cells that have been "immortalized" by in vitro culturing and no longer exhibit the phenomenon of programmed senescence typical of fully normal diploid cells. The resistance of normal cells to transformation by transfection can, however, be overcome if two distinct genes are used—for instance, one of the *myc* type, whose product goes to the nucleus and may be responsible for immortalization, and one of the *ras* type, which gives a plasma-membrane-bound product possibly responsible for the actual transformation. This finding provides an objective basis for the apparent multistep character of the cancerization process alluded to above.

In the intact animal, the successful establishment of a tumor and its ability to form metastatic colonies elsewhere in the body are subject to additional factors, some of which are genetic. In particular, mutations that block the expression of certain histocompatibility antigens on the surface of transformed cells (Chapter 3) may protect the cells against immune recognition and destruction by cytotoxic T cells and thereby precipitate the onset of a virulent neoplastic disease.

All in all, cancer researchers have reasons for satisfaction and optimism, especially considering that most of the new findings have been made in a span of less than five years. The cancer sphinx may not have met its Oedipus yet, but the day on which it does may not be too far away. What we will be able to do with the secret once we have it is another question. It is a tenet of the scientist's faith that understanding leads to control or, to quote Francis Bacon, that knowledge is power. But the example of genetic diseases, many of which we understand but do not yet control, shows that the road from understanding to control may not always be easy.

Interferon

Under the title "The IF Drug for Cancer," it made the cover of *Time* magazine before ever curing a patient. On the strength of very preliminary experimental evidence, millions of dollars have already been spent on clinical trials performed with the world's rarest, most expensive—$22 billion a pound until recently—and most unproved cancer drug. Future historians of science may well reflect on this strange phenomenon, which illustrates the increasing power that the public handling of scientific news is acquiring in our modern world over the directions that are given to scientific and medical investigation.

Let one thing be clear. Interferon is no quack remedy thrown on the market by some unscrupulous exploiter of human suffering. It is a natural substance of enormous interest, a prime agent in the defense of cells against viruses. The facts are as follows: When cells are attacked by a virus, the survivors, after a couple of days, become able to repel an attack not only by the same virus, but by many others. This phenomenon, called viral interference, is mediated by a protein that is specifically manufactured by the infected cells. That protein is interferon, of which there are several varieties, depending on the cells that make them.

As an antiviral agent, interferon may well have an important therapeutic future. The reason it is not in widespread use lies in its scarcity, itself the result of its specificity. To treat humans, you need human interferon. And to get human interferon—and in minute amounts, at that—you need human cells infected by a virus. But there is a bright side to the interferon coin. It happens to be one of the few proteins made from unspliced mRNA. This intriguing fact has allowed the successful transfer of the interferon gene in translatable form to bacteria by a shotgun type of experiment. If and when this laboratory finding is converted into a cheap industrial process, such irritating and often costly ills as the common cold, influenza, and herpes, as well as more serious conditions, such as hepatitis and viral pneumonia, could become things of the past, although the crucial tests still remain to be made.

In addition to its antiviral effect, interferon has a number of other biological properties, which include, at high enough dosage, the killing of certain tumor cells in culture and the stimulation of immune rejection mechanisms. These attributes, together with the possible involvement of viruses in the etiology of cancer, have sparked the first hopes that interferon may be an effective adjunct in the treatment of cancer. It is this cautiously raised possibility that somehow projected interferon into the limelight, to the point of draining considerable resources that many scientists felt might be used more profitably on less "iffy" and more cost-effective projects. Only time will tell whether this adventurous leap—an unusual event in the prudent and critical world of science—was justified.

19 Making Two out of One: Mitosis and Meiosis

During our various digressions in pursuit of wandering genes, our nucleus has continued to copy its genome. Now all is quiet. The replication fever has died down. Even transcription is grinding to a stop. But it is an oppressive quiet, pent up, suffocating. The nucleus is literally bursting under its double load of chromatin and cracking at its seams. Something is bound to happen. Indeed, the cell is going over into the M phase. It is about to enter mitosis (Greek *mitos*, thread).

Prophase: a Major Packing Job and the Erection of a Giant Crane

The first intimation of things to come is the reappearance of the chromosomes. Remember how we saw them dissolve into a tangle of chromatin fibers when the nucleus first closed around us. Now we are witnessing just the opposite. The dispersed strings of nucleosomes re-spiral into regularly coiled ropes, which themselves roll and fold back to re-form the massive bodies that are such conspicuous components of the nucleus in dividing cells.

As we observe this remarkable sight, we get a better view of the anatomy of a chromosome than we had before. Its main component, you may recall, is the chromatin fiber, which is itself made by the helical coiling of a nucleosome-beaded string of DNA. Chromatin fibers are

Model of metaphasic chromosome. Twin chromatids are linked by their centromeric regions to form a dyad. Each chromatid is supported by a framework of nonhistone proteins to which tightly bundled loops of chromatin (microconvules) are attached, possibly in a helical arrangement.

Electron micrograph of a metaphase chromatid dyad that has been depleted of histones. DNA loops attached to the protein framework are now completely unwound. Compare with the model at the left.

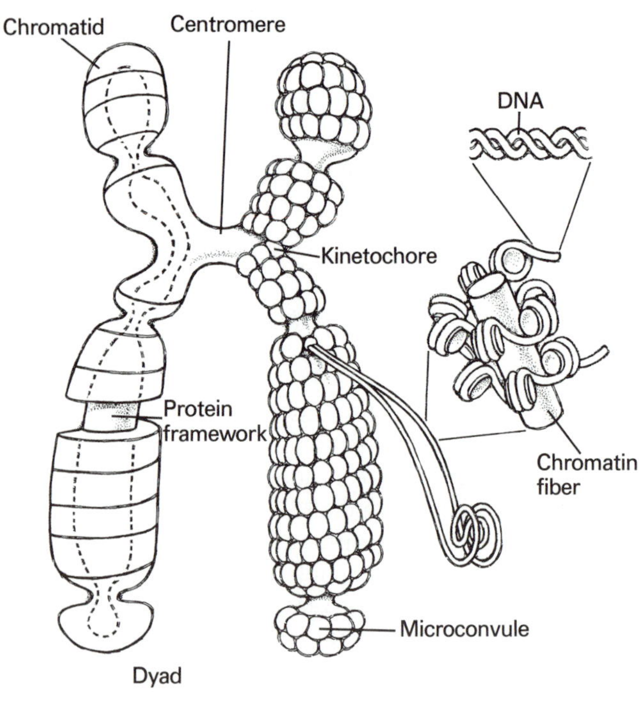

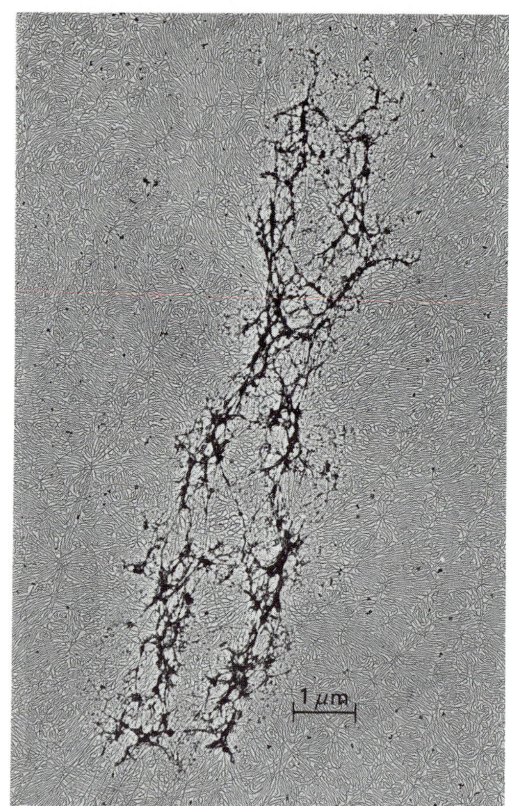

about 25 nm thick, they have a pitch of about 10 nm, and contain some 400 nm of double-stranded DNA per helical turn. To form a chromosome, each chromatin fiber folds into a series of loops of 400- to 800-nm total length, which further bundle up into small, oblong balls, sometimes called microconvules. The resulting string of microconvules is itself garlanded around a network of nonhistone proteins that provides a structural skeleton to the chromosome. As we have seen, it is likely that the main lineaments of this arrangement persist around the matrix in the interphase nucleus, where the structure simply unfolds as a sort of multidimensional accordion to allow selected parts of the DNA to be transcribed. Lampbrush chromosomes give us some idea of how this might occur. What we see now is simply a refolding of the loops into microconvules, and their regrouping around the protein framework. We would dearly like to know how such an extraordinary packing operation takes place. Unfortunately, nothing can be distinguished through this impenetrable tangle. We can only admire the final result: thousands of genes and intercalating sequences neatly folded and coiled to about one eight-thousandth of their length into 46 (for a human cell) separate bundles, or, rather, pairs of bundles.

The DNA, remember, has just been replicated and now exists as twin copies, identical in virtually every single one of their six billion base pairs. In synchrony with this operation, freshly made histone molecules have entered the nucleus and supplied the replicating DNA with supplementary nucleosome cushions. From what now goes on before our eyes, we may deduce that cohorts of nonhistone structural proteins also came in and formed a second set of chromosomal frames for the new DNA strands to hang their nucleosome garlands on. Conse-

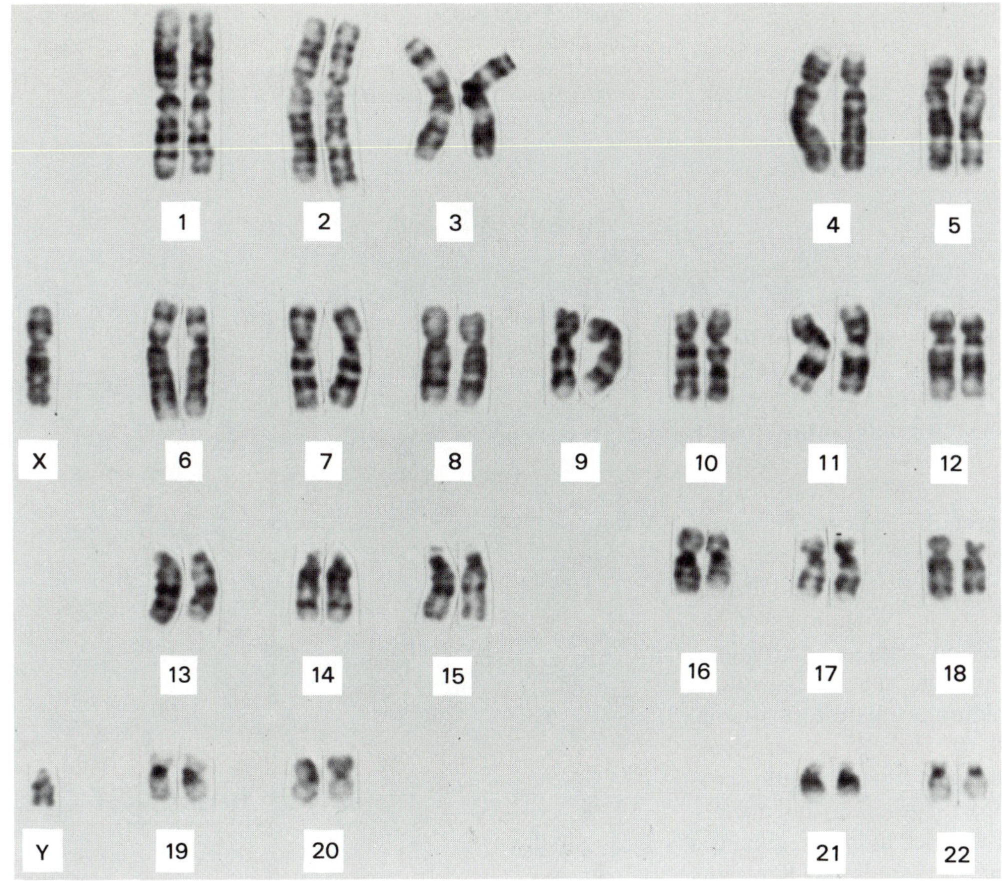

Karyotype of a normal human male. Metaphase chromosomes (note typical dyad structure) in a squashed cell preparation have been stained so as to reveal their banding patterns. The forty-six chromosomes in the photograph that was taken have been regrouped to construct this map.

quently, for each chromosome that we saw dispersing at the beginning of this visit, there are now two, in every way identical with each other.

The name chromatid is given to each member of such a pair. As they are seen to materialize during the first phase, or prophase, of mitotic division, the chromatids of each pair face each other like Siamese twins and are similarly interconnected by a bridge. Called the centromere, this junction exists between characteristically constricted parts of the chromatid bodies. It is a solid link, capable of resisting fairly harsh treatments, including the kind that need to be applied in order to squash cells and break open their nuclei. When dividing cells are squashed in this way on a glass slide and appropriately stained, the joined chromatid pairs, or dyads, appear under the microscope as typical X-shaped structures (sometimes Y- or V-shaped if the junction is eccentric), each of which can be identified by the length of its twin chromatids and by the position of the connection between them.

This technique, known as karyotyping, has become a valuable tool in genetic research and in the diagnosis of chromosome abnormalities. Thanks to special staining procedures, it has revealed an elaborate, longitudinal organization of the chromosomes into segments, or bands, which are distinguished by their degree of packing, sensitivity to thermic denaturation or proteolytic attack, ability to take up certain stains, or other physicochemical properties. Different bands are replicated at different times during the S phase and, as illustrated most vividly in the polytene chromosomes of diptera, correspond to different transcription domains (p. 317). More than a thousand bands have now been identified on the forty-six human chromosomes, and some five hundred specific genes have already been localized in them.

High-magnification electron micrograph of a centriole observed in a cultured mouse-embryo cell. Nine microtubule triplets are associated in a typical "waterwheel" pattern, as in the basal body or kinetosome of a cilium.

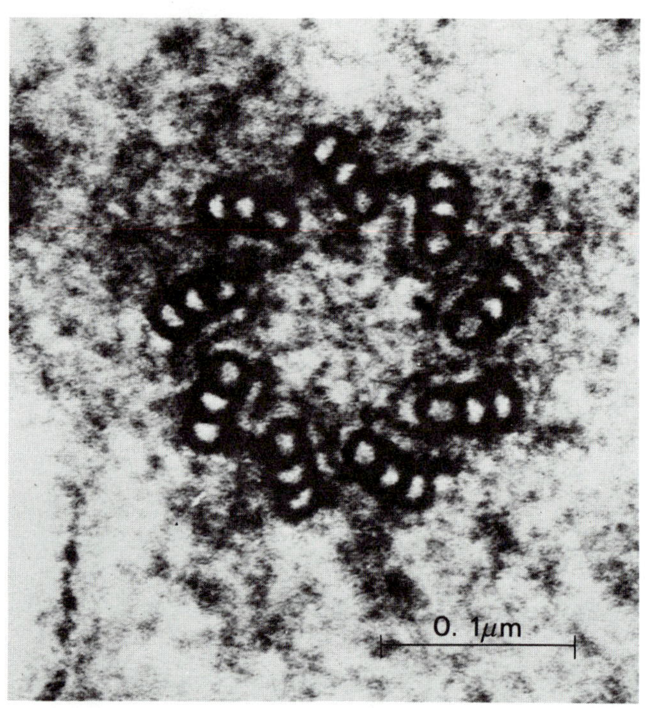

Chromosomes, we now clearly realize, are constructed in such a way that they can, without loss of their essential integrity, take up a protean variety of configurations. They can unwind and stretch out, severally or together, any of the thousands of individual segments of which they are composed, to the extent of adopting a completely dispersed state. Or, conversely, they can retract any of these segments, to the point of condensing fully into the compact packages that are needed to allow the genetic material to be moved around conveniently during mitosis.

Watching the enthralling spectacle of the mitotic metamorphosis of the chromosomes has been a rare privilege. But all the hustle and bustle it has entailed has hardly made our position more comfortable. Now, however, the pressure is beginning to ease, and feeble shafts of light come filtering in from the cytoplasm, announcing a new, remarkable event in the unfolding of the mitotic drama: the nuclear envelope is breaking open. With surprising rapidity, its continuous double sheet fragments into separate vesicles; the lamina falls apart; cytoplasm and nucleoplasm mingle their contents. But let us not rejoice too soon. We are not out of the nuclear woods yet. Replacing the padded wall of our prison is a huge, spindle-shaped cage made of hundreds of curved bars. Individual molecules and even small particles such as ribosomes readily pass between the bars, but not larger objects. Mitochondria remain outside, whereas chromosomes stay inside, as does our party. At least, we now have breathing space, and our view of the cell is almost unobstructed.

The cage that surrounds us was nowhere to be seen when we roamed through the cytoplasm, and it must have formed in direct connection with cell division. Indeed, it is none other than the mitotic spindle, so lovingly described by the first pioneers of cell exploration. They correctly saw it as playing an essential role in the forcible separation of the two chromosome sets during mitosis, but they could have had no inkling of how this extraordinary machine operates. Even today, our understanding of its mechanisms is fragmentary.

Construction of the spindle starts with one of the most mysterious phenomena of cell life: duplication of the centrosome, a small body situated near one of the poles of the nucleus. The centrosome consists of an amorphous matrix in which are embedded two short cylinders made of nine microtubule triplets and closely resembling the basal bodies, or kinetosomes, that form the roots of cilia (pp. 215–217). Called centrioles, these cylinders are characteristically oriented with their long axes at a right angle to each other. Such a pair of centrioles is called a diplosome. When the cell prepares to divide, the two centrioles move somewhat away from each other, and each grows a new centriole perpendicular to its flank to form two diplosomes, which organize further into centrosomes. How this occurs is not known. What follows is a little clearer, at least in certain cells. Nothing is more frustrating to the simple tourist in search of an "average" cell than the bewildering variety of mitotic events.

Starting from the two centrosomes, which apparently

act as microtubule organizing centers, microtubules start growing in all directions. Somehow, this growth process seems to push one of the centrosomes away from the other and to drive it around the nucleus down to the opposite pole. Eventually, long microtubules extend from one centrosome to the other, either continuously or in interdigitating fashion, all around the nuclear sphere and build the cage in which we found ourselves imprisoned when the nuclear envelope broke open. Shorter microtubules radiating from the centrosomes cap the two poles of the spindle with star-shaped crowns, the asters of classical cytology. And so the stage is set for the next phase, the metaphase.

Metaphase: Rigging the Chromosomes

After breakdown of the nuclear envelope, a second wave of microtubule assembly is set off inside the nuclear space, thanks to the influx of tubulin dimers from the cytoplasm. The organizing centers from which the microtubules grow are called kinetochores (Greek *kinein*, to move; *khôra*, place). They are situated on the main constriction that exists on each chromatid opposite the centromere. As a result of this disposition, each chromatid dyad sprouts, perpendicular to its main axis, two diametrically opposed sets of microtubules. At first, the direction of growth of the microtubules is variable because the chromatid pairs are more or less randomly oriented within the nuclear space. But as the microtubules grow longer, they align themselves parallel to the bars of the spindle cage, perhaps because of some constraint imposed by these bars.

The final outcome is the metaphase plate. The chromatid dyads are spread out (pushed by the growing microtubules?) in the equatorial plane of the spindle in such a way that this plane passes through their centromeres and the chromatids of each pair face opposing spindle poles (amphitelic disposition, see p. 381). The microtubules attached to them form two conical sheaves, each of which converges at one of the poles. The symmetry of this arrangement is remarkable and suggestive of some sort of balance of forces between the halves. The spindle itself, from being a simple, cagelike structure, is now filled with inner bars as well. These do not, like many of the cage bars, stretch all the way from one pole to another. They are always restricted to a single half-spindle and extend only from a kinetochore to the nearest spindle pole.

This elaborate rigging is completed by the addition of a number of poorly identifiable components, which may include short dynein arms connecting microtubules, as well as long actin fibers joined by myosin shafts. As we are about to see, the cell has just constructed a very special bidirectional chromosome hoist, all the more remarkable for being a purely temporary arrangement that will be dismantled immediately after having done its job, only to be reassembled at the next mitotic division.

Anaphase: Disjoining the Siamese Twins

We have come to the star turn of the mitotic show, one of the highlights of our tour: the anaphase, literally the ascension (Greek *ana*, upward). After what looks like a fair amount of pulling and wrenching, all the chromatid twins aligned on the metaphase plate become disjoined at about the same time and move away from each other as free chromosomes toward whatever spindle pole their kinetochore is attached to. It is a truly majestic sight, like watching 46 aerial gondolas lifting off in almost perfect synchrony over an unruffled lake in which their flight is mirrored in reverse. Their ascent is regally slow and may take from 5 to 10 minutes to cover the few microns (15–30 feet at our millionfold magnification) that separate them from a spindle pole. As the chromosomes rise, they are clearly lifted by their kinetosomes, while their hanging arms trail behind.

Many cell explorers have watched this display with awe and admiration, turning on it every instrument and technical device they could think of. Nevertheless, they are still debating about what does the pulling. That microtubules are involved is evident. In fact, the easiest way to "freeze" cells in midmitosis is to treat them with drugs, such as colchicine, vinblastine, or vincristine, that bind to tubulin dimers and inhibit their assembly. But where does the motive power come from? There are three possibilities, not necessarily exclusive of one another. One is "treadmilling" of microtubules—that is, their assembly at one end and disassembly at the other, such that any object attached sideways near the assembly end will be moved progressively toward the end where disassembly takes place. Another is "jacking" by means of dynein side arms attached to one microtubule and grappling onto an adjacent one, as in ciliary movement. Finally, there is the ubiquitous actin-myosin type of cytomuscle known to be involved in many forms of cell movement. The spindle contains parts of all three machineries, and all three may therefore be involved.

Another unresolved question is how the twin chromatids actually become separated. Are they wrenched away from each other mechanically? Or are their centromeres severed chemically by the action of some enzyme? Some of the movements we observe are suggestive of the mechanical explanation. But we may rightly wonder whether the tiny cytomuscles that do the pulling have the strength to break a connection capable, as we have seen, of resisting the brutal treatments that we inflict on the cells we prepare for karyotyping.

Amid all these uncertainties, some facts, at least, are beginning to emerge. It is clear that the kinetochore microtubules are directly engaged in the pulling mechanism, perhaps with the help of ATP-consuming cytomuscles of dynein type or actin-myosin type or both. As the microtubules pull their attached chromosomes, they shorten. Therefore, they are disassembled faster than they are assembled (if assembly takes place at all during this step, which is far from sure). This movement clears a widening space, essentially free of microtubules, between the two separating chromosome sets. For the first time since we entered a nucleus, we have space to relax. Except for the outer bars of the spindle cage, we could very well be in the cytoplasm.

In actual fact, we are not there yet. But the opportunity to get there is as good now as any we may encounter and we would be well advised to take advantage of it, even if it means bending or breaking a few microtubules. The spindle cage, you may have noticed, does not remain static while the chromosomes separate. It becomes both longer and narrower. The first change moves the poles further apart and thereby helps to widen the separation between the two sets of chromosomes. The second brings the microtubular bars of the spindle cage closer together and causes them to join progressively into bundles, called stembodies, in which they seem to be glued together by some amorphous material. As we shall see, these stembodies will be pressed even closer by the cleavage furrow. It is indeed time to leave. We may now watch the final phase, or telophase (Greek *telos*, end), from a safer vantage point in the cytoplasm.

Telophase: the Final Parting

By looking at either of the two sets of chromosomes after they have completed their ascent and become shorn of microtubules, we can follow from the outside the process of nuclear reconstruction that we saw earlier from the inside. Pieces of new lamina start forming around the clustered chromosomes while patches of endoplasmic reticulum converge upon them and fuse, leaving, however, the centrosome outside. Fenestrations form in this envelope and develop into pores. Inside, the chromosomes begin to disperse. Two typical interphase nuclei, each containing the same library of genetic texts, are the final outcome of these rearrangements, which obviously follow a complex script of instructions. Unfortunately, these are still totally undeciphered.

Occasionally, a cell may remain binucleated after mitosis. This occurs fairly often in the liver. As a rule, however, the cell undergoes division, or cytokinesis. In most animal cells, this happens by strangulation. A ring of plasma membrane, lined on its cytoplasmic face by hefty parallel bundles of actin-myosin fibers, is made to narrow

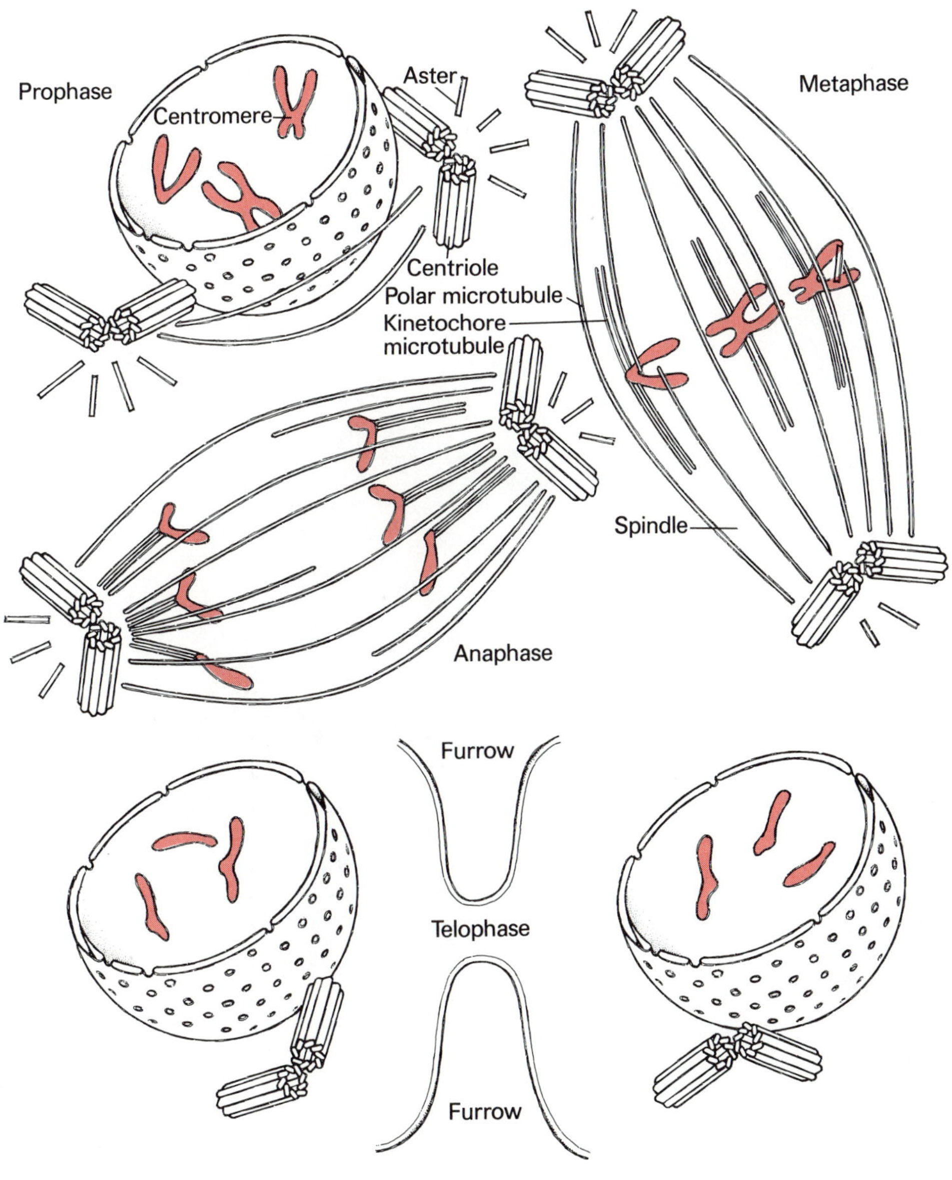

The different phases of mitosis

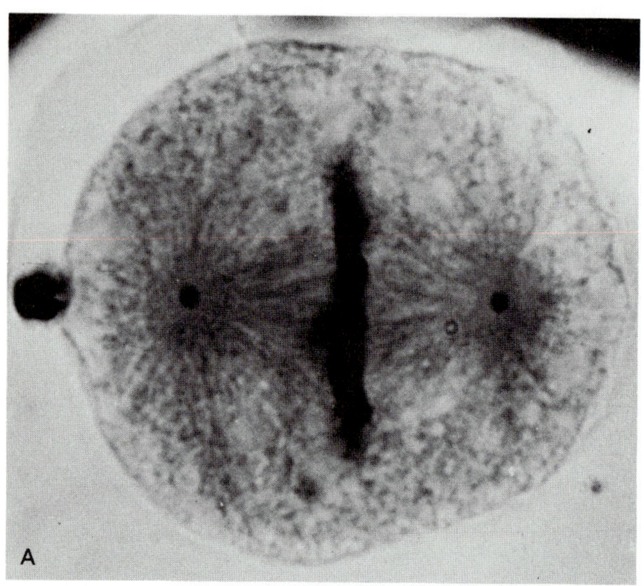

Mitosis in a fertilized egg of *Ascaris megalocephala* (a roundworm, parasite of the horse intestine). These preparations were made and stained in 1897 by J. B. Carnoy and photographed recently by L. Waterkeyn: (A) metaphase, transverse view; (B) metaphase, equatorial view; (C) anaphase.

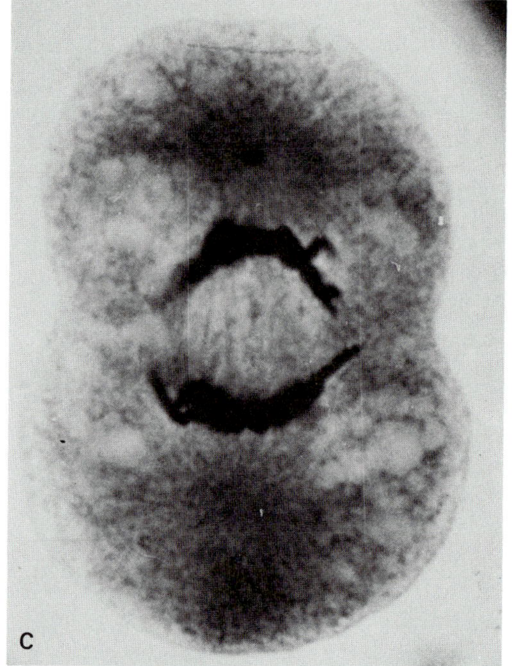

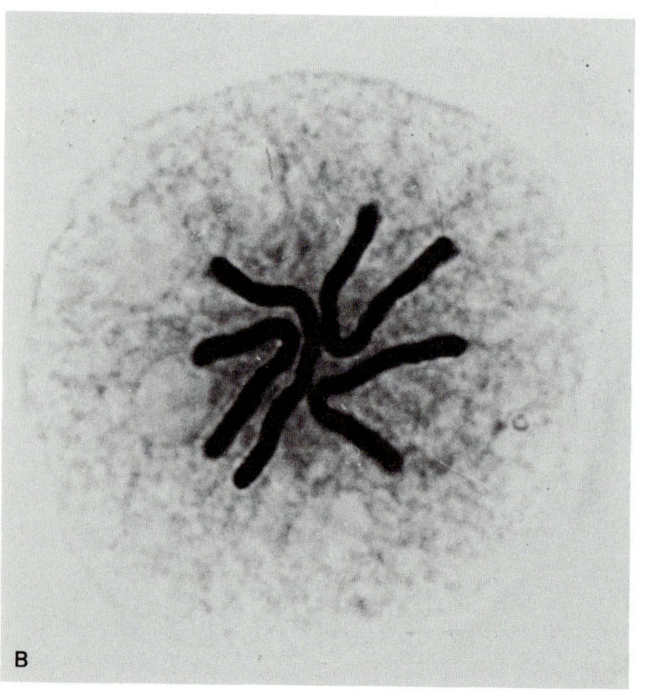

Cytokinesis.

A. Animal cell. A ring-shaped constriction (furrow) progressively narrows the cytoplasm between the two daughter nuclei, compressing the spindle microtubules into a tight bundle (stembody).

B. Plant cell. Secretion vesicles containing cell-wall material (phragmoplasts) align in the equatorial plane and fuse.

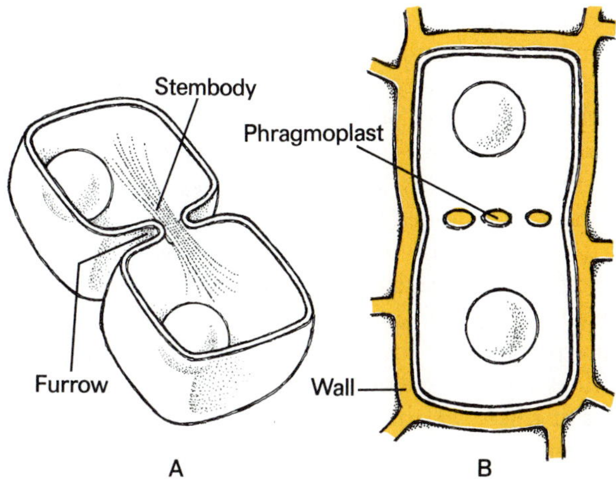

progressively into an ever-deepening furrow by the closing of this contractile noose. As the ring tightens, fluid cytosolic constituents are squeezed out on both sides, and the aggregated spindle microtubules (stembodies, p. 374) are pressed closer and closer together, eventually to form a single, dense shaft that finally becomes severed. When the lips of the constriction meet, they join by a typical *cis* type of fusion, and cell division is completed. The whole process has taken about an hour.

In plant cells, the mode of division is different. Only an incipient furrow is made by the plasma membrane, and the rest of the space between becomes filled with flattened vesicles originating from the Golgi apparatus. Called phragmoplasts (Greek *phragma*, partition), these vesicles fuse, also by *cis* fusion, to create a single flat cisterna that almost completely divides the cell into halves. Circular fusion with the surrounding plasma membrane furrow, by what may be viewed as a concerted annular exocytic process, cuts this cisterna into two sheets, each of which becomes part of the plasma membrane of one of the daughter cells. The term "exocytosis" is quite appropriate here, for phragmoplasts are authentic secretion granules and their contents contribute to the formation of a wall between the two membranes. Often, small cytoplasmic bridges, called plasmodesmata, are spared by the fusion of the phragmoplasts and remain as direct connections between the two daughter cells.

Meiosis, or the Art of Getting Ready for Sex

A hen, it has been said, is only an egg's way of making another egg. Or, in the words of the nineteenth-century scientist August Weismann, *soma*, the body, serves to perpetuate the *germen*. But bodies do not arise out of single germ cells by parthenogenesis (Greek *parthenos*, virgin), but rather from two such cells by fertilization. Here is where sex comes in and, with it, meiosis (Greek *meioein*, to reduce).

When male and female germ cells, or gametes (Greek for spouses), unite upon fertilization, their nuclei combine into a single nucleus that contains the sum of the chromosomes contributed by each cell (p. 313). A diploid egg cell thus arises from the joining of two haploid germ cells. All the somatic cells that originate from this egg cell by successive mitotic divisions are likewise diploid. Should this be true also of the germ cells, the next generation would be tetraploid, the next one octoploid, and so on. This obviously would not do. There can be no sexual reproduction without some device for reducing the number of chromosomes in the germ-cell line. This device is meiosis.

To understand the mechanism of meiosis, we must remember that a diploid cell contains two complete haploid sets of chromosomes, one derived from the male, the other from the female germ cell. We will use the terms paternal and maternal to indicate the origin of any chromosome or gene, and we will call homologous the two chromosomes that bear the same genes or alleles in the paternal and maternal sets. Meiosis is a special kind of division that occurs uniquely in the maturation of both spermatozoa and ova. It results in the random segregation of one chromosome out of each homologous pair. As has been pointed out, even with intact chromosomes such a lottery would suffice to ensure a great deal of genetic diversity: 2^{23}, or more than eight million, different combinations for any human individual. But, in actual fact, diversification is immeasurably greater, thanks to crossing-over and the resulting recombination events that are characteristic of meiosis. Now that we are briefed, let us make a short excursion down the germ-cell maturation line to see how things actually happen.

In its early stages, meiosis is not very different from mitosis. DNA is replicated during interphase. And when prophase starts, the chromosomes recondense in the form of twin chromatids. From now on, however, things become very different. In meiosis, this prophase stage drags on for a very long time, sometimes many months or years—in the human female, oocytes go into meiotic prophase during the fifth month of fetal life and remain blocked at the diplotene stage (see p. 379) until the onset of puberty. The nuclear envelope remains intact during all that time, and the chromosomes uncoil into long, thin threads, which go through a great variety of contortions and changes in shape, whose main function is pairing, or synapsis, of homologous chromatid dyads into tetrads, followed by reciprocal exchanges of segments between paternal and maternal chromatids by recombination.

At the leptotene stage (Greek *leptos*, thin; *tainia*, tape), the chromatids are at their thinnest and longest, and they become attached by their extremities to specific points on the inner face of the nuclear envelope. Somehow, the matching ends of homologous chromatid pairs are brought close together by this process, so that the homologues become clearly distinguishable as coupled loops of various lengths. It is the zygotene stage. Then, starting from the attachment points, a synaptonemal complex builds up between the homologous chromatid pairs and "zippers" them intimately together into single thickened loops in which the two components are no longer distinguishable. During this pachytene stage (Greek *pakhys*, thick), homologous stretches of paternal and maternal DNA bearing almost identical base sequences are brought into register with each other. The sole apparent function of this elaborate matching machinery is to allow the aligned partners to exchange homologous segments with each other (p. 337). The result is multiple recombination—of the "legitimate" type, of course, as befits the perfectly licit, and even obligatory, character of this kind of molecular intercourse. After a while, the tetrads are released from their attachment points on the nuclear envelope, the synaptonemal complexes disintegrate, and the homologous chromatid twins again separate, except where recombination has joined the DNA of a paternal chromatid to that of a maternal one. These junctions generate the characteristic chiasmata, or crosses, which revealed crossing-over to the early cytologists and opened the way to the interpretation of the phenomenon of genetic linkage. This final stage of meiotic prophase is called diplotene.

It is at this stage that meiosis often comes to rest, as already mentioned, especially in the female sex. Oocytes may, in the meantime, pursue their development and differentiation with the help of active transcription of their DNA. Their chromosomes then adopt the typical "lampbrush" shape, which results from the symmetrical unfolding and transcription of identical stretches on closely joined chromatid twins (Chapter 16). If you follow a lampbrush chromosome over a long enough distance, you are likely to come to a chiasma, from which you will find out that the dyad is really part of a tetrad.

Another phenomenon that takes place during the first meiotic prophase, actually at the pachytene stage, is the tremendous amplification of rRNA-coding genes in certain amphibian oocytes. Mention has been made of this selective replication process, which leads to the formation of hundreds of free nucleoli in a single nucleus and enables the cell to step up manifold its ribosome-manufacturing capacity (p. 302).

When this prolonged prophase finally ends, the subsequent events recall the similar stages of mitosis. A spindle is constructed around the nucleus in the cytoplasm; the

Prophase I

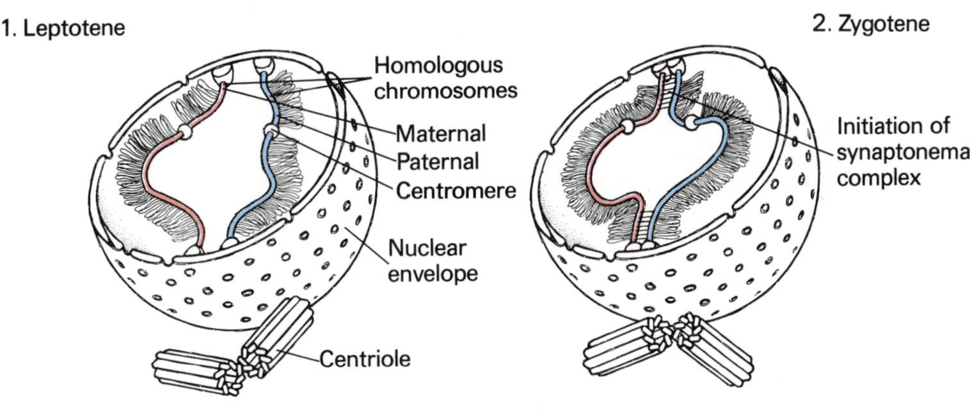

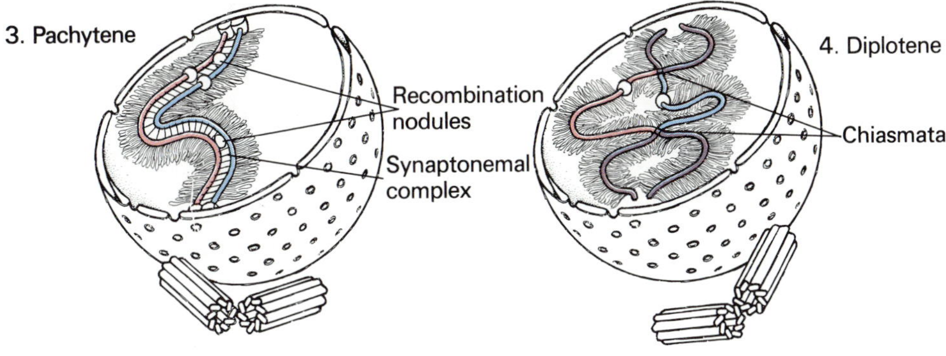

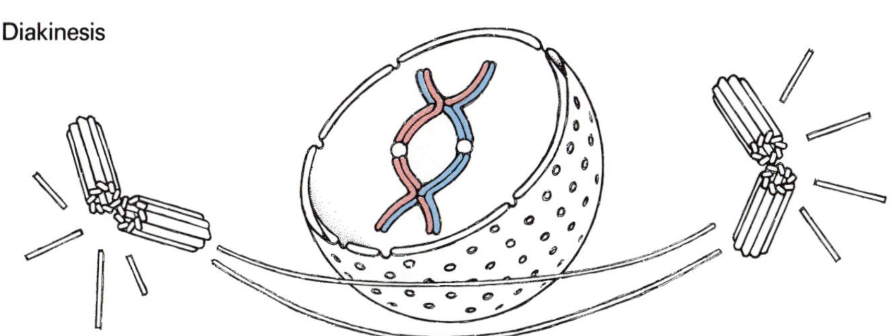

Details of prophase in the first division of meiosis.

1. Leptotene stage. Elongated dyads of closely apposed sister chromatids (not individually distinguishable at this stage) become attached to the nuclear envelope.

2. Zygotene stage. Homologous chromatid dyads of maternal and paternal origin are brought together in closely matched register by a proteinaceous synaptonemal complex linking the protein frameworks of the chromatids.

3. Pachytene stage. Synapsis is completed. Recombination (of the legitimate kind, Chapter 18) between homologous DNA loops takes place within recombination nodules.

4. Diplotene stage. Desynapsis has taken place. Dyads remain joined by one or more chiasmata at sites of recombination. Chromosomes have a typical lampbrush shape (Chapter 16).

5. Diakinesis. Chromosomes detach from the nuclear envelope and become progressively shorter and thicker. Their tetrad structure becomes visible.

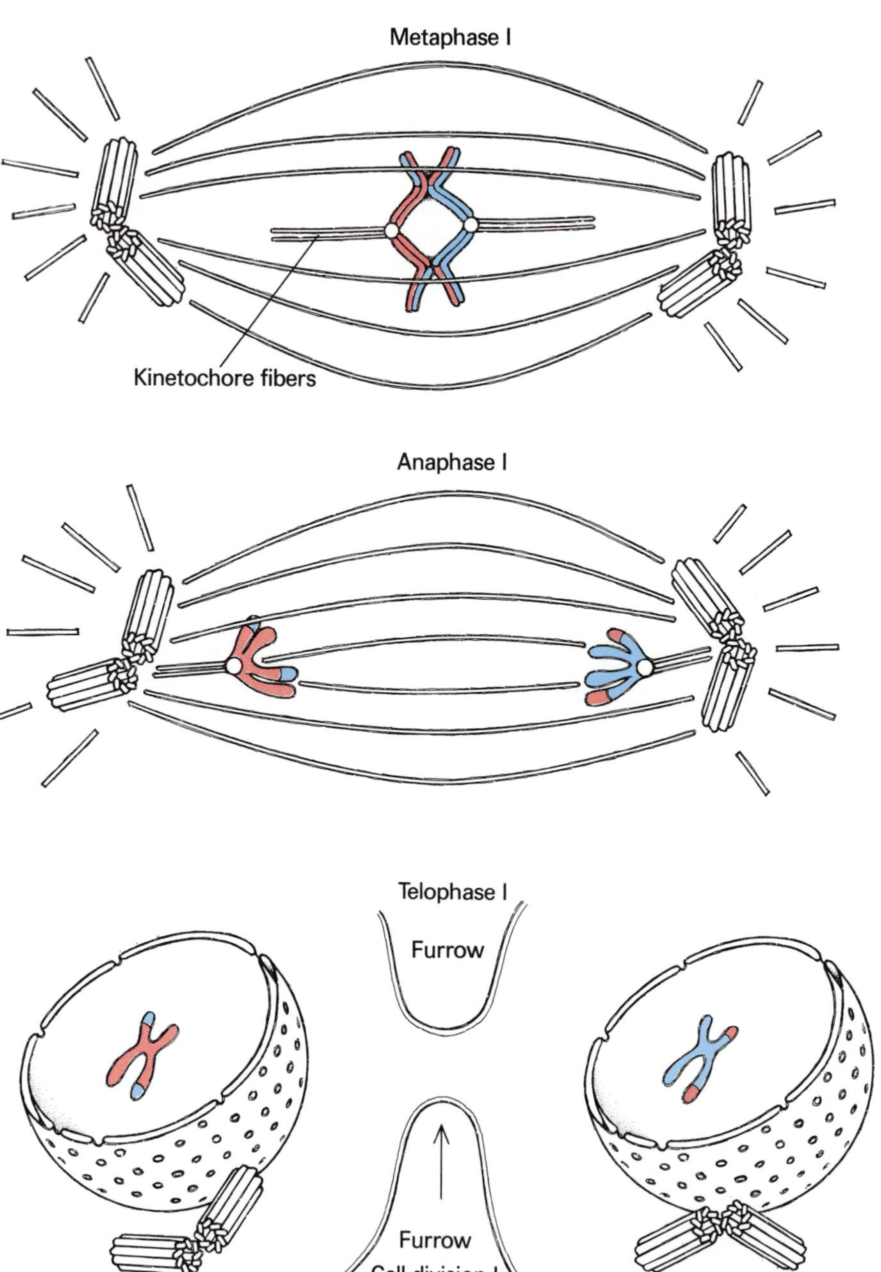

Details of metaphase, anaphase, and telophase in the first meiotic division.

Metaphase. The nuclear envelope disintegrates and a spindle forms. The tetrads line up in the equatorial plane in syntelic disposition (with the equatorial plane passing through the chiasmata).

Anaphase. Homologous dyads are pulled away from each other, with breakage of the chiasmata. Segments that have been exchanged by recombination accompany their new chromosomal owners.

Telophase. Two nuclei form and the cell divides. At this stage, each daughter nucleus contains a single set of dyads, some of which are of paternal, the other of maternal, origin (except for the segments exchanged by recombination).

nuclear envelope disintegrates; and microtubules assemble on the kinetochores. But there is one fundamental difference. Not forty-six dyads, but twenty-three tetrads come to be spread out on the metaphase plate. Furthermore, in their disposition, the equatorial plane does not cut through the centromeres and separate identical dyad twins (amphitelic disposition). Rather, the equatorial plane passes through chiasmata and separates paternal from maternal dyads (syntelic disposition). There are no rules for this separation, so that some paternal and some maternal dyads end up on each side of the plane. Furthermore, in many dyads, the twin chromatids are no longer in their original state or identical with each other, owing to the exchanges of homologous segments caused by crossing-over.

Because of the syntelic disposition, the subsequent anaphase severs chiasmata, not centromeres, and it separates two nonidentical sets of dyads, themselves made up of nonidentical twin chromatids. Genetic diversification has been achieved, but not yet the reduction in chromosome number. The two cells that form through this first meiotic division are still diploid. They will, however, go through a second division without an intervening S phase of DNA replication. In this second division, the chromosomes adopt the amphitelic disposition at metaphase, and the subsequent anaphase severs centromeres and separates twin chromatids, as in mitosis. But the number of dyads involved is only half that obtaining at mitosis, and their separation produces two haploid nuclei.

The mechanisms, the control, and the evolutionary origin of this extraordinary process are all beyond our present understanding. Its significance is of paramount importance. Meiosis is indissociably linked to sexual reproduction, which is itself considered the single most powerful generator of genetic diversity and therefore the most important agent of evolutionary experimentation and innovation. It is likely that, without sexual reproduction, evolution would have produced only very simple organisms. Even bacteria are known to indulge in some form of sexual activity. They differentiate into two distinct forms, which then conjugate and exchange genetic material, particularly plasmids.

Details of the second division of meiosis.

The two daughter cells of the first division divide again by a mechanism similar to normal mitotic division (amphitelic disposition and splitting of the centromeric bridge between sister chromatids at anaphase), except that division is not preceded by duplication of the DNA. Four haploid cells, each with a different and virtually unique complement of maternal and paternal genes, are produced.

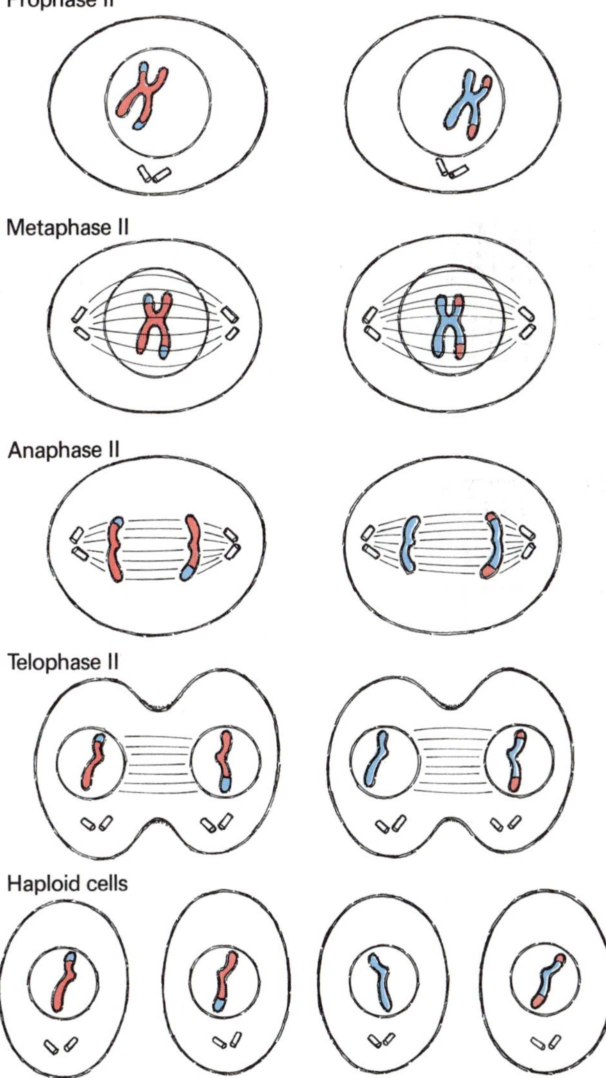

Boy or Girl, Who Decides?

The answer to this question is: the father. (At least, if you are human; there are species in which it is the opposite.) But don't draw any sexist conclusions. The father has not the slightest control over his decision. It so happens that males, but not females, have one asymmetric pair of chromosomes. It is designated XY, as opposed to the XX combination that is characteristic of females. After meiotic reduction, one half of the mature spermatozoa carry an X chromosome, and the other half a Y chromosome. All ova, on the other hand, carry an X chromosome. It is easy to see that the sex of a conceptus will depend on whether an X-carrying or a Y-carrying spermatozoon fertilizes the ovum. Both are equally capable of doing so, and all the father can claim is the privilege of tossing a coin. Chance can, however, be influenced to the extent that X or Y spermatozoa can be separated and used selectively for artificial insemination. Animal breeders have attempted to do this, with some success.

Whatever vanity males may derive from their role in sex determination, it is more than offset by the price they pay for this distinction, which is the possession of only single copies of all the genes associated with the sex chromosomes. A woman carrying an abnormal gene on one of her X chromosomes may be rescued by a normal allele on the other X chromosome. But half her sons, on an average, will have only the bad gene. Hemophilia, a defect of blood coagulation, is the best known of these X-linked hereditary diseases that affect males and are transmitted by females. It has affected a number of members of the royal House of Hanover (Queen Victoria is the most famous carrier of the X-linked hemophilia defect).

Rather remarkably, females have chosen not to take full advantage of their superiority. At some stage in embryological development, their cells inactivate irreversibly one of their X chromosomes, which henceforth is seen as a highly condensed dot (Barr body). After this happens, both X chromosomes still undergo replication at each cell division, but only one is transcribed. Which one remains functional is determined by chance. The female organism thus consists of two types of cellular colonies, or clones, differing by the origin, paternal or maternal, of the X chromosome they express, and therefore by any X-linked character represented by different alleles on the two X chromosomes. There are many fascinating aspects to this genetic "mosaicism."

Making One out of Two

Cell fusion is the opposite of cell division. The most important such event is fertilization, the fusion of two gametes. It also occurs in a number of other processes, most characteristically in the development of muscle fibers. These arise from hundreds of embryonic muscle cells (myoblasts), which fuse to form a multinucleated syncytium. The fusion mechanism depends on a *trans* type of merger between the plasma membranes of the two partners. How it is triggered is not known.

However, ways have been found to make cells fuse in vitro. Such devices work not only with cells of the same type, as in the natural fusion of myoblasts, but also with cells of different types, even with cells from different species. This technique has become a powerful tool in the analysis of nucleocytoplasmic interactions and in the genetic mapping of chromosomes. It has also given rise to an extremely valuable procedure for manufacturing monospecific (monoclonal) antibodies.

Very early on our tour, we encountered the antibody-producing B lymphocytes, and we noted the important fact that each cell makes only a single kind of antibody. Any large-scale production of a given antibody is elicited by mitogenic stimulation of the few competent cells. These multiply into a clone of cells that all make the same antibody. There are numerous reasons why we would like to raise and maintain such clones in vitro and, thereby, to have available permanent antibody-producing factories of whatever specificity we need. We can isolate the B lymphocytes, but, unfortunately, we are unable to culture them. Those we get are postmitotic cells that do not divide.

The problem has now been solved by the artificial fusion of B lymphocytes with cancerous lymphoid cells

Monoclonal antibodies. Fusion of a nondividing antibody-producing cell with a dividing tumor cell gives rise to a hybridoma that combines the ability to produce antibodies with the capacity to divide. Descendants of a single hybridoma form a clone of cells that manufacture the same kind of antibody molecules (monoclonal antibodies). Such clones can be used for the industrial production of antibodies.

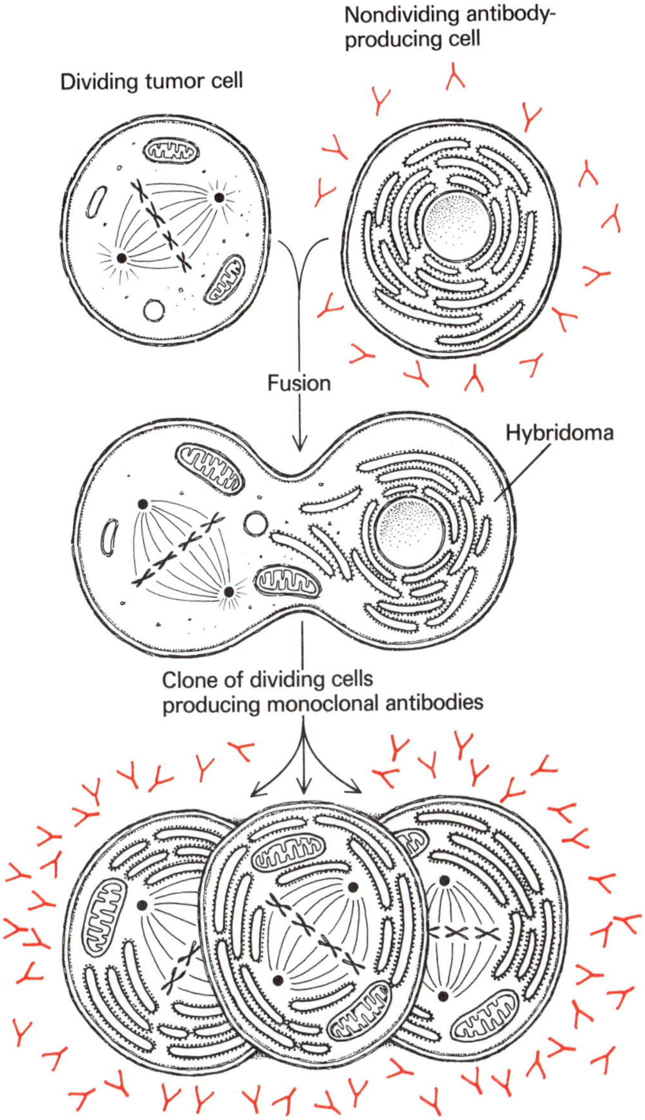

(lymphoma cells), which possess the capacity of indefinite multiplication characteristic of the cancerous transformation. A small but significant fraction of the hybridomas obtained in this manner combine the antibody-producing capacity of the B lymphocytes with the multiplication potential of the lymphoma cells. By appropriate procedures, the desired clone can be isolated and used to set up a self-perpetuating antibody factory.

Final Exit

As promised, we have escaped from the nucleus, thanks to the disruption of the nuclear envelope during mitosis. It remains for us to find an exit from the cytoplasm. If we wish to respect the integrity of the cell, the only way we can get out of it is by budding.

Many viruses use this method. They are the membrane-wrapped viruses that provided us with our means of entry into the cytosol (p. 103). Such viruses actually acquire their membranous envelope when they bud from their original cellular birthplace. They do this in a very sophisticated fashion and carefully prepare their exit by instructing the cell to manufacture virus-specific membrane glycoproteins. Thanks to this specificity, the biogenetic pathway of the viral glycoproteins has been accurately traced. It begins in the rough-surfaced ER to which the nascent proteins are directed by a typical signal sequence and in which their glycosylation starts cotranslationally. After completion, the glycosylated proteins move on to the smooth-surfaced ER, probably by lateral diffusion, and are then transported to the Golgi apparatus, where they undergo further processing and glycosylation before being finally transported to the plasma membrane. In both transport steps, the viral glycoproteins move as parts of clathrin-coated vesicles. This information has provided much of the support for our belief that normal plasma membrane proteins follow the same biogenetic itinerary (Chapter 6).

While these events go on in the export department, viral nucleic acids and capsid proteins are made elsewhere in the cell, assemble into a complete nucleocapsid, and move near the cell surface, where they attract the virus-specific glycoproteins that have become inserted in the plasma membrane. Here, again, the virus has helped to throw light on important cellular mechanisms by showing how an object in the cytosol can induce the clustering of membrane-associated proteins. Presumably, these are mobile transmembrane proteins, capable of moving freely in the plane of the membrane by lateral diffusion and extending projections on both the outer surface of the plasma membrane—the carbohydrate "hairs," or "spikes," of the viral envelope—and its cytoplasmic face. Association between these inner projections and the nucleocapsid is what induces the clustering of the virus-specific glycoproteins.

Thanks to this mechanism, a patch of plasma membrane fitted with special viral components assembles in direct contact with the nucleocapsid and progressively surrounds it until the stem of the resulting bud snaps off by a *cis* type of membrane fusion. Thanks to this mechanism, which is illustrated on page 103, the virus is free to go and infect another cell using its membrane coat as means of ingress. Incidentally, some of these viruses, in particular a Japanese virus known as Sendai, can be made to try to invade two cells at the same time by merging simultaneously with the plasma membranes of the two cells. In doing this, they cause the two cells to fuse with each other. They provide one of the means used to create cellular hybrids.

It remains for us to copy the membrane-wrapped viruses to effect our final exit. By necessity, our technique will be cruder, since we cannot ask the cell to prepare tourist-specific membrane glycoproteins for us. In true human fashion, we will substitute ingenuity and expediency for the lack of built-in automatisms. Keeping in mind the fluidity, plasticity, and self-sealing properties of biomembranes, we will simply push gently but firmly against the plasma membrane. As expected, it yields and bulges under our pressure, flows and reshapes around us, until, as with the virus, a *cis* type of fusion lets us loose. This process seals the plasma membrane behind us and leaves the cell essentially intact, except for the hardly noticeable loss of a small patch of outer membrane.

This patch has ceased to be of use to its erstwhile owner. As for us, it could only help us to return into the cell. This, no doubt, we will want to do many times in the future, for much that could not be encompassed in this first visit remains to be seen, admired, and enjoyed in the living cell. Indeed, we will return. But, for now, the tour is over. You may take off your coats.

APPENDIX I The Building Blocks of Living Cells

Little more than fifty different micromolecular building blocks account for the bulk of the organic matter found in any living organism. Diversity arises from the manner in which the building blocks are assembled into macromolecules, of which innumerable varieties exist, and, secondarily, from the activities of these macromolecules, in particular the enzymic proteins, which catalyze the formation of the thousands of small molecules that participate in metabolism and other specialized processes. The compendium that follows gives the structures of all major biological building blocks, together with those of most of the molecules mentioned explicitly in the book.

Carbohydrates

Monosaccharides

Glycoses. Carbohydrates are made of sugars, or monosaccharides. The simplest sugars, or glycoses, have the gross formula $(CH_2O)_n$. (Hence the name carbohydrate, which is actually a misnomer: carbohydrates are by no means hydrates of carbon.) Of the n oxygen atoms of a glycose molecule, $n - 1$ belong to alcohol groups, OH, and one to a carbonyl group, CO, which may be either aldehydic (in aldoses) or ketonic (in ketoses):

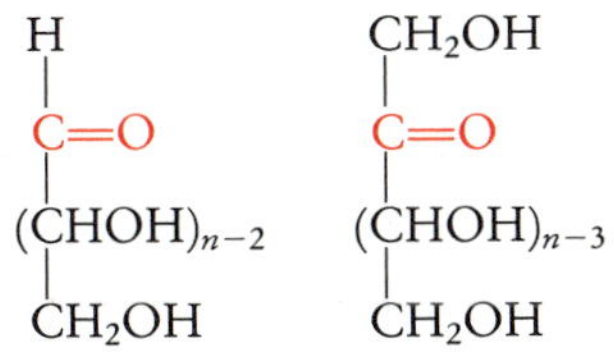

Aldose	Ketose
H	CH_2OH
|	|
C=O	C=O
|	|
$(CHOH)_{n-2}$	$(CHOH)_{n-3}$
|	|
CH_2OH	CH_2OH

According to the number, n, of carbon atoms in their molecules, the glycoses are called trioses, tetroses, pentoses, hexoses, etcetera.

In glycoses, all the carbons bearing a secondary alcohol group (CHOH) are asymmetric—that is, such a carbon atom bears four different chemical groupings. These can be arranged around the carbon in two distinct, nonsuperimposable configurations, resulting in the existence of stereoisomers (Greek *stereos*, solid). Thus, glyceraldehyde (Chapter 7) can exist in two nonsuperimposable forms:

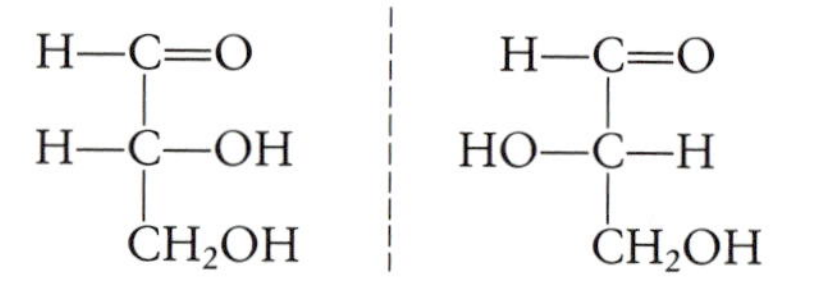

D-Glyceraldehyde	L-Glyceraldehyde
H—C=O	H—C=O
H—C—OH	HO—C—H
CH_2OH	CH_2OH

If there is more than one asymmetric carbon, the number of stereoisomers increases in exponential fashion. If the number of asymmetric carbons is represented by a, the number of stereoisomers is 2^a, forming $2^{(a-1)}$ pairs of enantiomorphs (Greek *enantios*, opposite; *morphê*, form), which are mirror images of each other. Each such pair is designated by a single name composed of a specific prefix followed by the suffix *ose*. The enantiomorphs are distinguished from each other by the letters D and L, depending on whether, in the planar projection of the molecular structure, the hydroxyl group on the penultimate carbon is situated at the right (as in D-glyceraldehyde) or at the left. For example:

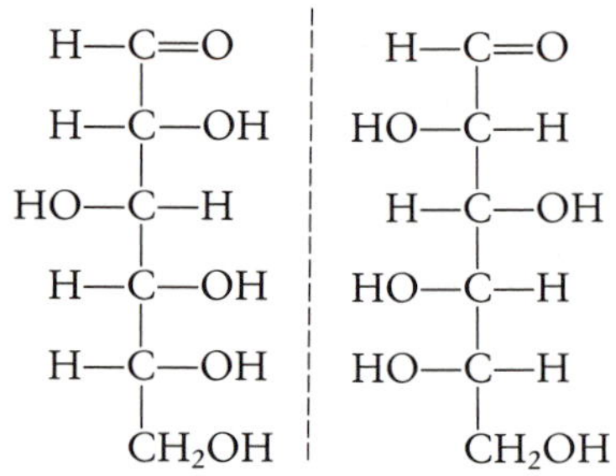

D-Glucose	L-Glucose
H—C=O	H—C=O
H—C—OH	HO—C—H
HO—C—H	H—C—OH
H—C—OH	HO—C—H
H—C—OH	HO—C—H
CH_2OH	CH_2OH

Sugars that have a chain of three or more carbons attached to the carbonylic carbon can undergo reversible cyclization by the addition of an alcohol group onto the double bond of the carbonyl group (hemiacetal bond). Two such cyclic structures can form without undue strain of the bond angles: the pentagonal furanose ring (4 carbons plus 1 oxygen) and the hexagonal pyranose ring (5 carbons plus 1 oxygen):

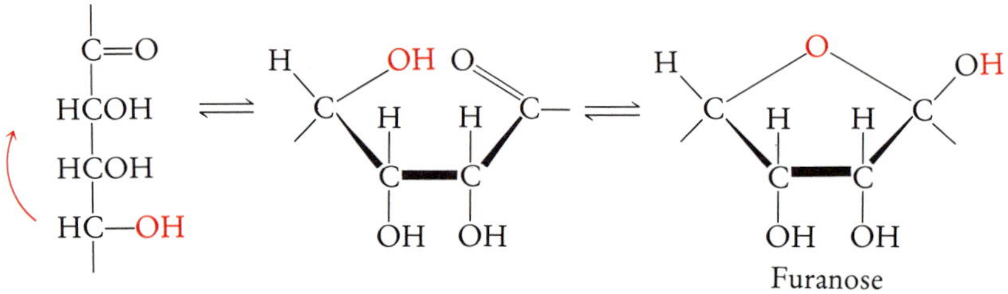

C=O / HCOH / HCOH / HC—OH ⇌ [open-chain form, OH approaching C=O] ⇌ [cyclic form]

Furanose
(β form)

C=O / HCOH / HCOH / HCOH / HC—OH ⇌ [open-chain form, OH approaching C=O] ⇌ [cyclic form]

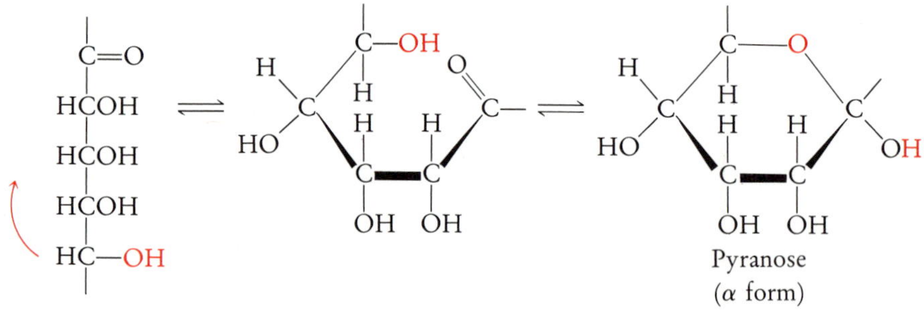

Pyranose
(α form)

As a result of this cyclization, the carbonylic carbon (carbon atom number 1 of aldoses, most often number 2 of ketoses) becomes itself asymmetric, so that two distinct cyclic stereoisomers exist of each linear variety. These are called anomers and are designated α and β, depending on the position of the anomeric hydroxyl with respect to the plane of the ring.

The cyclic forms of the sugars are of cardinal importance, as they are the real building blocks in the formation of all complex carbohydrates, thanks to their ability to combine by their anomeric carbon with a variety of molecules of carbohydrate or noncarbohydrate nature. Such combinations are called R-glycosides, or glycosyl-Rs. They are sometimes called hologlycosides if R is a carbohydrate, heteroglycosides if it is not. They can be of the α or β configuration:

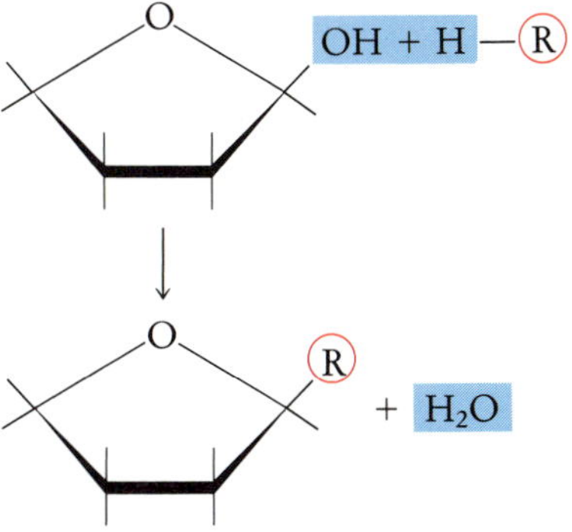

R-β-Furanoside, or β-Furanosyl-R

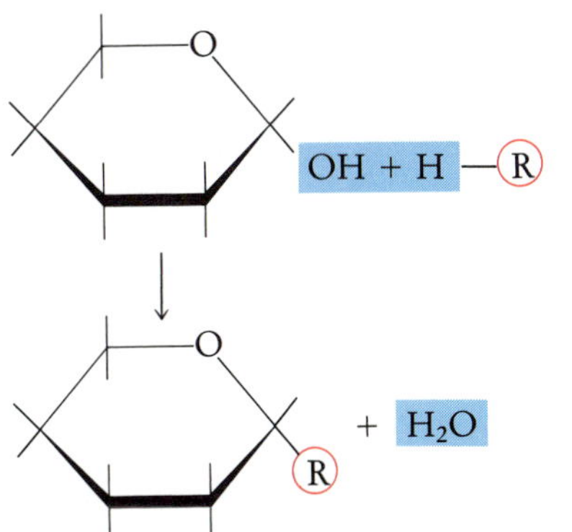

R-α-Pyranoside, or α-Pyranosyl-R

The structures of some of the most important glycoses follow, shown in their dominant cyclic configuration. Strictly speaking, deoxyribose, which lacks an oxygen atom in position 2, does not conform to the general formula of glycoses. Its structure is given here next to that of ribose because of the comparable functions of the two sugars in the formation of the two classes of nucleic acids.

PENTOSES

D-Ribose (β)

D-Deoxyribose (β)

(Color shows difference between the two pentoses.)

HEXOSES

D-Glucose (α)

D-Fructose (β)

D-Galactose (β)

D-Mannose (α)

(Color shows where configuration differs from that of glucose.)

Glycose derivatives. In addition to the glycoses, the monosaccharides include a number of related substances derived from the glycoses by removal of oxygen (deoxyglycoses); by oxidation, either of the aldehyde group in aldoses (aldonic acids) or of the terminal hydroxyl (glycuronic acids); by replacement of a hydroxyl by an amino group (glycosamines), which is generally acetylated (*N*-acetylglycosamines); by reduction (glycitols); or by some other modification. Except for the aldonic acids and glycitols, which have lost the capacity to cyclize by hemiacetal formation, these derivatives can form the same types of glycosides as do the glycoses from which they derive.

We have already encountered an example of a deoxyglycose in deoxyribose. Another is L-fucose, the 6-deoxy derivative of L-galactose, which is found at the end of some oligosaccharide side chains of glycoproteins. Among the aldonic acids, mention deserves to be made of ascorbic acid (vitamin C), which is a derivative of L-gulonic acid. The most important glycuronic acids are those derived from the three main aldohexoses—glucose, galactose, and mannose—by oxidation of the primary alcohol group in position 6. The most common glycosamines are derived from the same hexoses by substitution of a primary amino group for the hydroxyl in position 2. Representative examples are shown below.

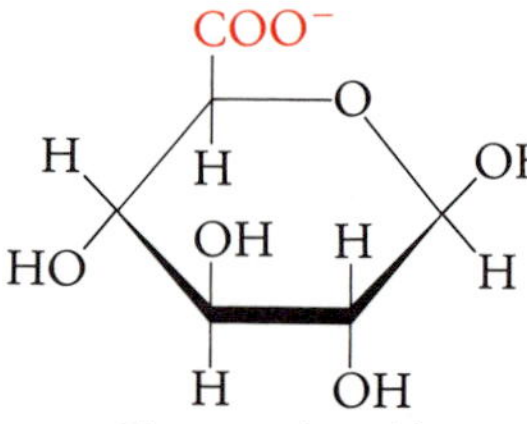

D-Glucuronic acid (β)

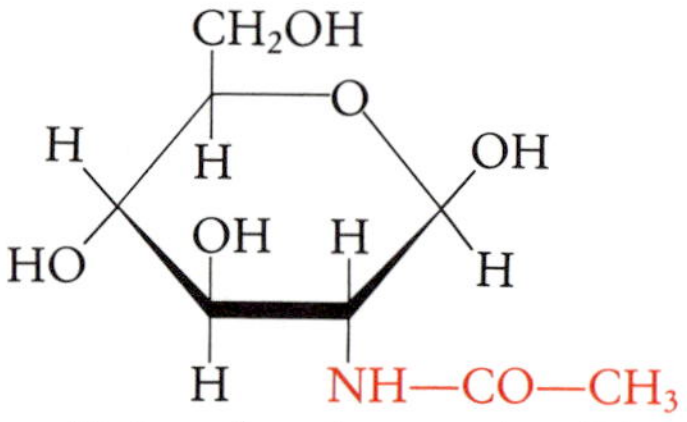

N-Acetyl-D-glucosamine (β)

Among the more common glycitols are glycerol (reduced glyceraldehyde), a major component of lipids, and ribitol (reduced ribose), a constituent of flavin coenzymes (see p. 399). Related to the glycitols, except that they have only secondary alcohol groups, are the cyclitols, cyclic polyols obtained by isomerization of hexoses. *meso*-Inositol, found in certain phospholipids, forms in this manner from D-glucose:

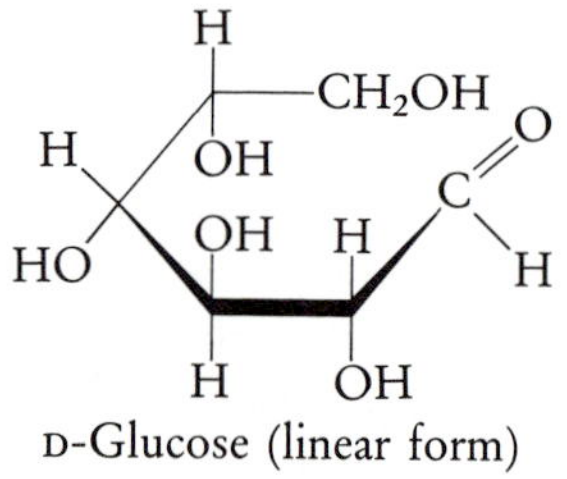

D-Glucose (linear form)

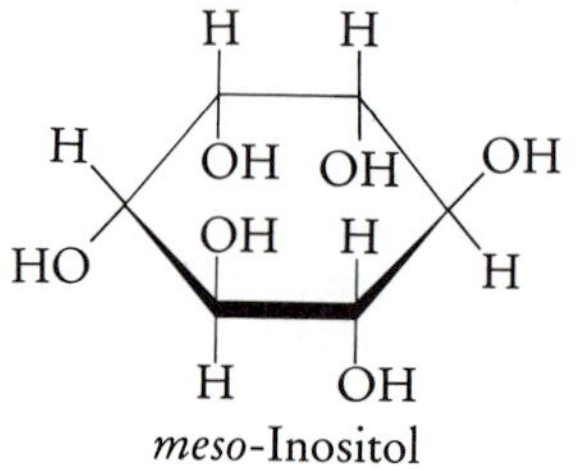

meso-Inositol

In this case, cyclization occurs by the addition of carbon-6 (instead of oxygen-5, as in pyranose formation, see above) to the carbonyl double bond.

Mention must finally be made of *N*-acetylneuraminic acid (NANA), or sialic acid, a molecule derived from the association of *N*-acetylmannosamine with pyruvic acid. Like fucose, it often occupies the end of oligosaccharide side chains.

Disaccharides

Sucrose, also named saccharose, is the ordinary sugar that we use for sweetening. It is made of α-D-glucopyranose and β-D-fructofuranose, linked by a rare form of mutual glycosylation.

Lactose, or milk sugar, consists of D-galactopyranose attached by a β-glycosidic bond to the oxygen in position 4 of D-glucopyranose.

Oligosaccharides

This group of substances includes a variety of linear and branched assemblies of up to fifteen sugar molecules, constructed mostly with glycoses and *N*-acetylglycosamines, often ending, as already mentioned, with either a fucosyl or a sialyl group. They form the side chains of glycoproteins and of glycolipids and thereby play a key role in the function of receptors (Chapters 3 and 13).

Polysaccharides

Nature, especially in the plant world, abounds in variously structured polymers of all the major glycoses. Starch, glycogen, and cellulose are all polyglucosides, or glucans. The first two have α-glucosidic linkages, whereas in cellulose the linkages are of the β-configuration. This apparently trivial difference accounts for the remarkable resistance of cellulose to degradation, commented upon in Chapter 2. Other homogeneous polysaccharides include mannans, galactans, and fructans, as well as polymers of pentoses, of glycuronic acids (pectins), and of *N*-acetylglycosamines (the chitin of crustacean shells).

A particularly important class of animal polysaccharides, sometimes referred to as mucopolysaccharides, is made of repeating disaccharide units usually consisting of an *N*-acetylglycosamine and either a glycose or a glycuronic acid. The hyaluronic acid of connective tissue (Chapter 2) is an example of such a molecule. It is made of *N*-acetyl-D-glucosamine and D-glucuronic acid. In many of these substances, as in most of the proteoglycans of connective tissue and in the anticoagulant heparin, some of the hydroxyls (occasionally, also, an amino group) bear a sulfuryl ester group. Such glycan sulfates are polyanions with a highly acidic character.

Heteroglycosides

The nucleosides (see p. 395) are the most important glycosides in which the R group is not a carbohydrate.

Proteins

Amino Acids

Proteins are made of L-α-amino acids, which, with one minor variant (proline, in which the R and NH_2 groups are linked), have the following structure:

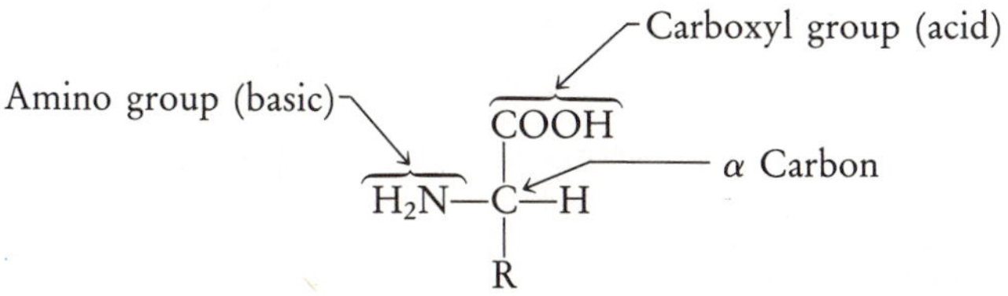

Carboxyl group (acid)
Amino group (basic)
α Carbon

$$\begin{array}{c} COOH \\ | \\ H_2N—C—H \\ | \\ R \end{array}$$

Under physiological conditions (pH 7), amino groups are protonated for the most part, and carboxyl groups dissociated. Therefore, the most abundant natural structure of the amino acids is the amphoteric (Greek *amphô*, both), or zwitterionic (German *zwitter*, hermaphrodite), form:

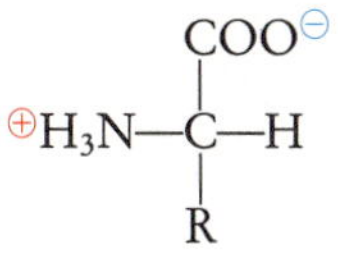

$$\begin{array}{c} COO^{\ominus} \\ | \\ {}^{\oplus}H_3N—C—H \\ | \\ R \end{array}$$

Except in glycine, the α carbon is asymmetric, leading to the existence of two stereoisomers. They are designated D and L by analogy with the glyceraldehydes (p. 386):

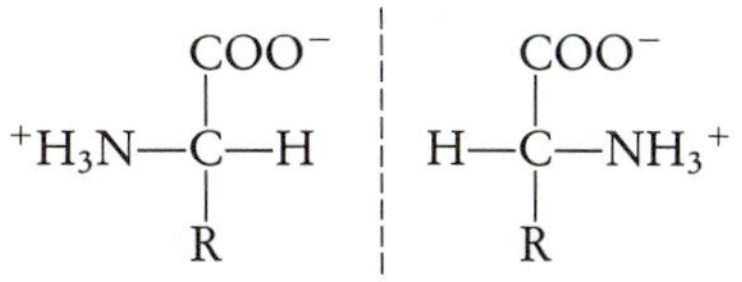

$$\begin{array}{c} COO^- \\ | \\ {}^+H_3N—C—H \\ | \\ R \end{array} \qquad \begin{array}{c} COO^- \\ | \\ H—C—NH_3{}^+ \\ | \\ R \end{array}$$

L-Amino acid D-Amino acid

Only L-amino acids are found in proteins, but their D stereoisomers are components of the bacterial cell wall and other bacterial products, including some antibiotics.

As explained in Chapters 2 and 15, twenty distinct amino acids are genetically coded and serve as building blocks for the synthesis of proteins. A few more—the hydroxyproline of collagen is an example—arise by posttranslational modification. The structural formulas of the genetically coded amino acids are given on the next page, arranged approximately in order of increasing hydrophilicity, a character that plays a dominant role in the folding and associative properties of proteins (Chapters 2 and 3). The critical importance of the physical properties of the R groups for the conformation and biological activity of proteins is illustrated by the structure of the genetic code. As pointed out in Chapters 15 and 18, the code seems such as to minimize the proportion of point mutations that lead to the replacement of amino-acid residues by residues of significantly different physical character. Most likely, the genetic code acquired this particular structure in the course of evolution by the operation of natural selection.

Peptides and Proteins

Amino acids join by amide bonds (called peptide bonds in this particular case) to form linear assemblies, or peptides. Except for a number of special oligopeptides, mostly hormones, the vast majority of natural peptides are macromolecules containing up to several hundred amino-acid residues. When fully extended, these polypeptide chains have a zigzag shape, with the R groups alternating on each side of the backbone:

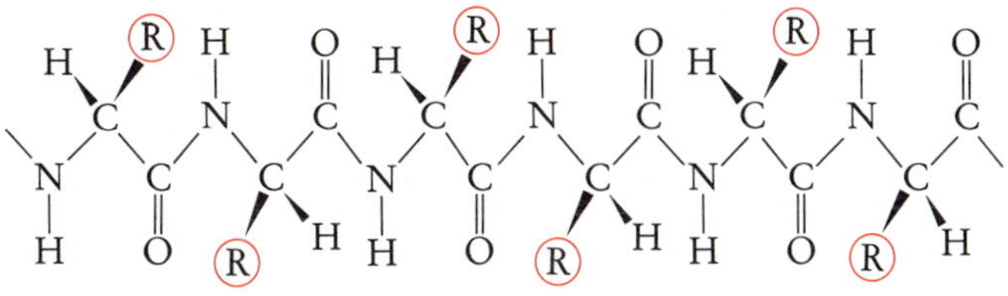

H R H O H R H O H R H O
C N C C N C C N C
N C C N C C N C C
H O R H H O R H H O R H

This basic structure may then undergo further coiling and folding, to create the convoluted three-dimensional forms adopted by most proteins. Examples of such structures have been considered in Chapters 2, 3, 12, and 15, among others.

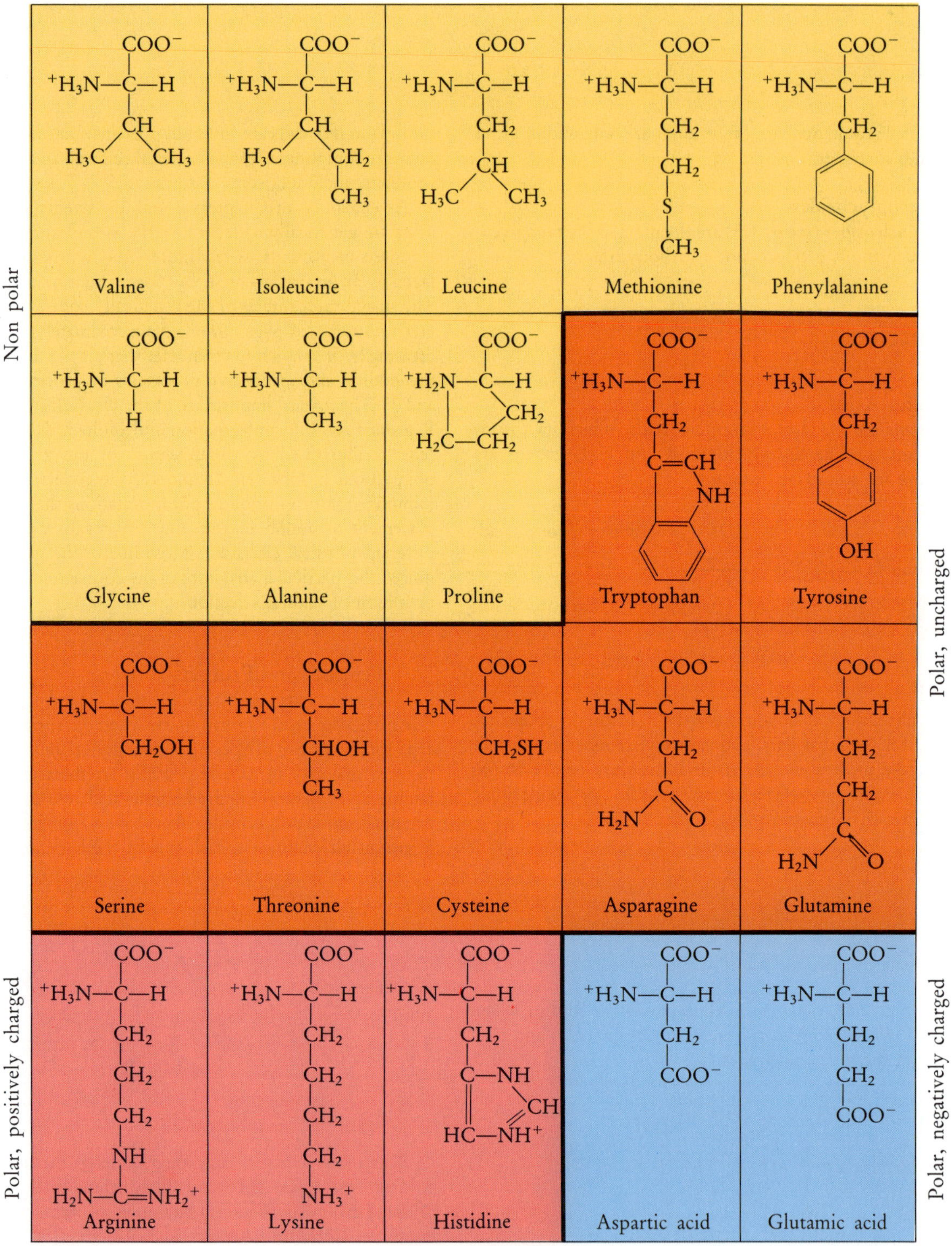

Non polar

COO^- $^+H_3N-C-H$ CH H_3C CH_3 Valine	COO^- $^+H_3N-C-H$ CH H_3C CH_2 CH_3 Isoleucine	COO^- $^+H_3N-C-H$ CH_2 CH H_3C CH_3 Leucine	COO^- $^+H_3N-C-H$ CH_2 CH_2 S CH_3 Methionine	COO^- $^+H_3N-C-H$ CH_2 (benzene ring) Phenylalanine
COO^- $^+H_3N-C-H$ H Glycine	COO^- $^+H_3N-C-H$ CH_3 Alanine	COO^- $^+H_2N-C-H$ CH_2 H_2C-CH_2 Proline	COO^- $^+H_3N-C-H$ CH_2 $C=CH$ NH (benzene ring) Tryptophan	COO^- $^+H_3N-C-H$ CH_2 (benzene ring) OH Tyrosine
COO^- $^+H_3N-C-H$ CH_2OH Serine	COO^- $^+H_3N-C-H$ CHOH CH_3 Threonine	COO^- $^+H_3N-C-H$ CH_2SH Cysteine	COO^- $^+H_3N-C-H$ CH_2 C H_2N O Asparagine	COO^- $^+H_3N-C-H$ CH_2 CH_2 C H_2N O Glutamine
COO^- $^+H_3N-C-H$ CH_2 CH_2 CH_2 NH $H_2N-C=NH_2^+$ Arginine	COO^- $^+H_3N-C-H$ CH_2 CH_2 CH_2 CH_2 NH_3^+ Lysine	COO^- $^+H_3N-C-H$ CH_2 $C-NH$ CH $HC-NH^+$ Histidine	COO^- $^+H_3N-C-H$ CH_2 COO^- Aspartic acid	COO^- $^+H_3N-C-H$ CH_2 CH_2 COO^- Glutamic acid

Polar, uncharged

Polar, positively charged

Polar, negatively charged

In the table on the facing page, the structures of the twenty genetically coded amino acids are arranged in the approximate order of increasing hydrophilicity. The first horizontal row, which includes valine, isoleucine, leucine, methionine, and phenylalanine, comprises the most hydrophobic amino acids; they tend to be buried in regions from which water is excluded. The next row contains amino acids that, in spite of essentially hydrophobic R groups, are less strictly segregated from water—glycine and alanine, thanks to the small size of their R groups; proline because of the rigidity imposed on its molecule by the cyclization of the R group with the primary amino group; and tryptophan and tyrosine, owing to their possession of a polar group (NH or OH) that can participate in hydrogen bonding. In the third row, serine and threonine (with their OH group), cysteine (with its SH group), asparagine and glutamine (with their CO and NH_2 groups) all have a clearly hydrophilic character. Their R groups are, however, electrically neutral and can be accommodated also in hydrophobic regions. Cysteine, in addition, can join to another cysteine by a disulfide bridge, S—S, which has an essentially nonpolar character. The last row contains the five amino acids that have an electrically charged R group. Strongly hydrophilic, they are confined to hydrated areas. Arginine, lysine, and histidine are basic and have a positively charged R group. (Only about half the histidines, on average, are so charged at any given moment.) Aspartic and glutamic acids are negatively charged.

Lipids

Fatty Acids

The main building blocks of lipids are long-chain fatty acids with an even number of carbon atoms, mostly sixteen and eighteen. Some are fully saturated, as is the C_{16} palmitic acid:

$$CH_3—(CH_2)_{14}—COOH$$

Others are unsaturated. In the C_{18} series, for example, oleic acid has one, linoleic acid two, and linolenic acid three double bonds. A specially important polyunsaturated fatty acid is arachidonic acid, with twenty carbon atoms and four double bonds. It gives rise by oxidation to the prostaglandins, a family of molecules of high biological activity.

The double bonds in natural fatty acids are mostly of the *cis* type, with the result that the carbon chain is bent at an angle where they occur. Such kinks are of obvious importance for the organization of lipid bilayers (Chapter 3). The alteration in molecular shape caused by double bonds can be appreciable, as shown by the following comparison between oleic acid and its saturated C_{18} congener, stearic acid:

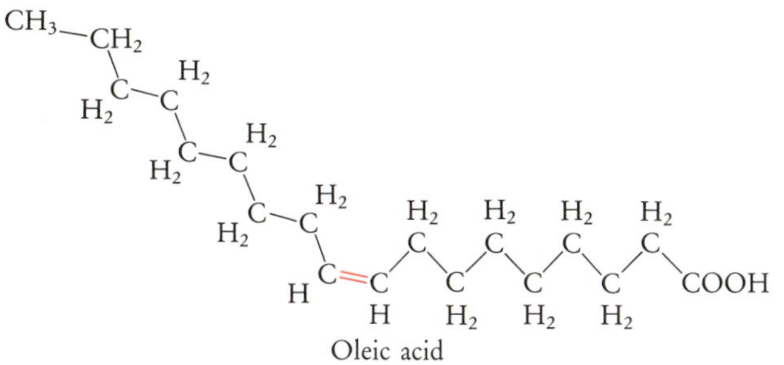

$CH_3—CH_2—C(H_2)—C(H_2)—C(H_2)—C(H_2)—C(H_2)—C(H_2)—C(H)=C(H)—C(H_2)—C(H_2)—C(H_2)—C(H_2)—C(H_2)—C(H_2)—C(H_2)—COOH$

Oleic acid

$CH_3—C(H_2)—C(H_2)—C(H_2)—C(H_2)—C(H_2)—C(H_2)—C(H_2)—C(H_2)—C(H_2)—C(H_2)—C(H_2)—C(H_2)—C(H_2)—C(H_2)—C(H_2)—C(H_2)—COOH$

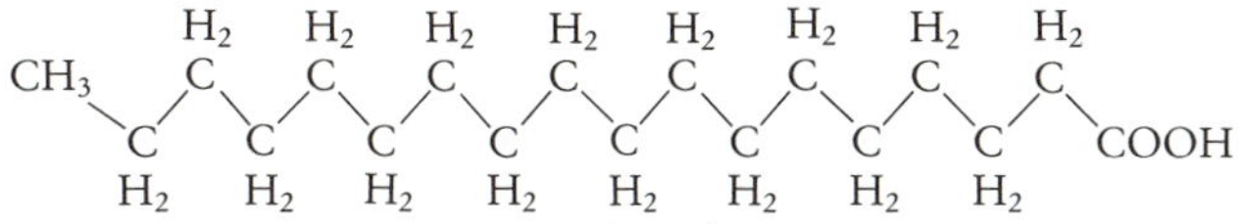

Stearic acid

As indicated in Chapter 3, the fatty acids provide many of the long hydrophobic tails characteristic of most lipids. In free form they are fully dissociated under physiological conditions and thereby acquire the hydrophilic, negatively charged head to which soaps owe their amphipathic character. Mostly, however, fatty acids occur as compounds of lipids, in which their carboxyl group is engaged in an ester or amide linkage.

Neutral Lipids

Triglycerides. Triglycerides are fatty acyl esters of the trialcohol glycerol, the glycitol derived from glyceraldehyde (see pp. 386 and 388):

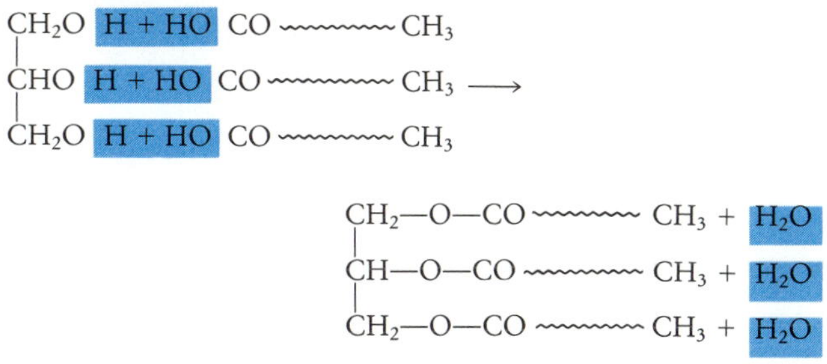

CH_2O H + HO CO ~~~~~ CH_3
CHO H + HO CO ~~~~~ CH_3 ⟶
CH_2O H + HO CO ~~~~~ CH_3

$CH_2—O—CO$ ~~~~~ CH_3 + H_2O
$CH—O—CO$ ~~~~~ CH_3 + H_2O
$CH_2—O—CO$ ~~~~~ CH_3 + H_2O

Such molecules have almost no hydrophilic character and are completely immiscible with water. They constitute the bulk of animal and vegetable fats and oils. They fulfill essentially no structural function, but have the advantage of providing the highest calorie count (energy of oxidation) per unit of weight of all biological constituents. Mostly, they are stored to serve as fuel.

Waxes. These are fatty acyl esters of long-chain alcohols, sometimes of very great length—the myricylic alcohol of beeswax contains as many as thirty carbon atoms. Except for the ester group, waxes are structurally very close to the high-molecular-weight hydrocarbons that make up petroleum jelly. Like such alkanes, they are extremely hydrophobic.

Phospholipids

Except for sphingomyelin (see p. 393), the phospholipids, which make up the lipid bilayers essential to the structure of all biomembranes (Chapter 3), are derivatives of phosphatidic acids. These may be viewed as triglycerides in which an external fatty acyl group has been replaced by a phosphoryl group:

$$
\begin{array}{l}
{}^{-}O{-}P(=O)(O^{-}){-}\,OH + H\,O{-}CH_2{-}CH(O{-}CO{\sim\sim}CH_3){-}CH_2{-}O{-}CO{\sim\sim}CH_3 \\
\text{Diglyceride} \\
\downarrow \\
{}^{-}O{-}P(=O)(O^{-}){-}O{-}CH_2{-}CH(O{-}CO{\sim\sim}CH_3){-}CH_2{-}O{-}CO{\sim\sim}CH_3 + H_2O \\
\text{Phosphatidic acid}
\end{array}
$$

Mostly, this phosphoryl group forms a second ester bond with some alcohol (represented by ROH):

$$
\begin{array}{l}
Ⓡ{-}O\,H + {}^{-}O\,{-}P(=O)(O^{-}){-}O{-}CH_2{-}CH(O{-}CO{\sim\sim}CH_3){-}CH_2{-}O{-}CO{\sim\sim}CH_3 \\
\text{Phosphatidic acid} \\
\downarrow \\
Ⓡ{-}O{-}P(=O)(O^{-}){-}O{-}CH_2{-}CH(O{-}CO{\sim\sim}CH_3){-}CH_2{-}O{-}CO{\sim\sim}CH_3 + OH^{-} \\
\text{Phosphatidyl-R}
\end{array}
$$

Among the alcohols participating in such associations are glycerol itself and some of its combinations, the cyclic hexitol inositol (p. 388), and especially the aminoalcohol ethanolamine, its methylated derivative, choline (also a constituent of the neurotransmitter acetylcholine, Chapter 13), and its carboxylated derivative, the amino acid serine:

$$
\begin{array}{ccc}
CH_2OH{-}CH_2{-}NH_3^{+} & CH_2OH{-}CH_2{-}\overset{+}{N}(CH_3)_3 & CH_2OH{-}CH(NH_3^{+}){-}COO^{-} \\
\text{Ethanolamine} & \text{Choline} & \text{Serine}
\end{array}
$$

These are all highly polar molecules, which, together with the phosphoryl group, give the phospholipids their hydrophilic heads. Here, for example, is the structure of phosphatidylcholine, known earlier as lecithin (Greek *lekithos*, egg yolk), a major constituent of all biomembranes:

$$
\underbrace{(CH_3)_3\overset{\oplus}{N}{-}CH_2{-}CH_2{-}O{-}P(=O)(O^{\ominus}){-}O}_{\text{Hydrophilic head}}{-}CH_2{-}\underbrace{CH(O{-}CO{\sim\sim}CH_3){-}CH_2{-}O{-}CO{\sim\sim}CH_3}_{\text{Hydrophobic tails}}
$$

Sphingolipids

This diverse group of complex lipids is composed of derivatives of sphingosine, a long-chain aminoalcohol with the following structure:

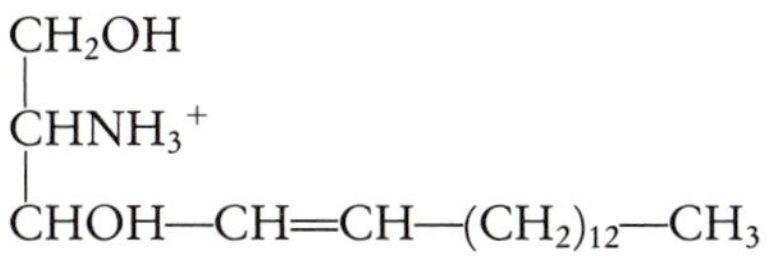

$$\begin{array}{l} CH_2OH \\ | \\ CHNH_3^+ \\ | \\ CHOH{-}CH{=}CH{-}(CH_2)_{12}{-}CH_3 \end{array}$$

As a rule, sphingosine has a long-chain fatty acid attached to its amino group by an amide linkage. Called ceramide, this combination resembles diglycerides in having a two-pronged hydrophobic tail and a free hydroxyl to which various groups can be linked. The diverse family of sphingolipids arises in this manner. In association with phosphorylcholine, for example, ceramide forms sphingomyelin, a phospholipid very similar in shape and in physical properties to phosphatidylcholine:

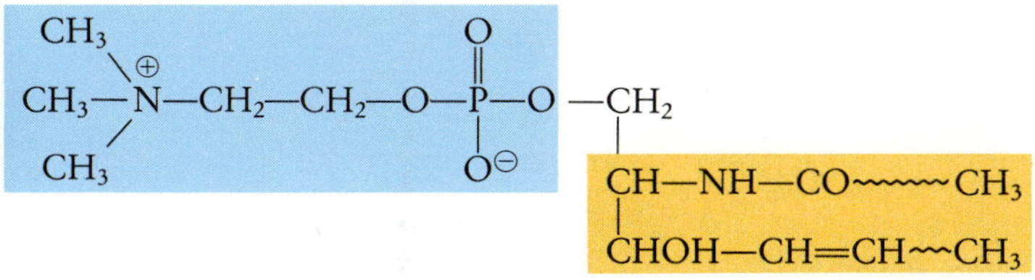

$$(CH_3)_3N^{\oplus}{-}CH_2{-}CH_2{-}O{-}P(=O)(O^{\ominus}){-}O{-}CH_2{-}CH(NH{-}CO{\sim\sim}CH_3){-}CHOH{-}CH{=}CH{\sim}CH_3$$

Replacement of the phosphorylcholine group in sphingomyelin by a β-D-galactosyl group gives a cerebroside:

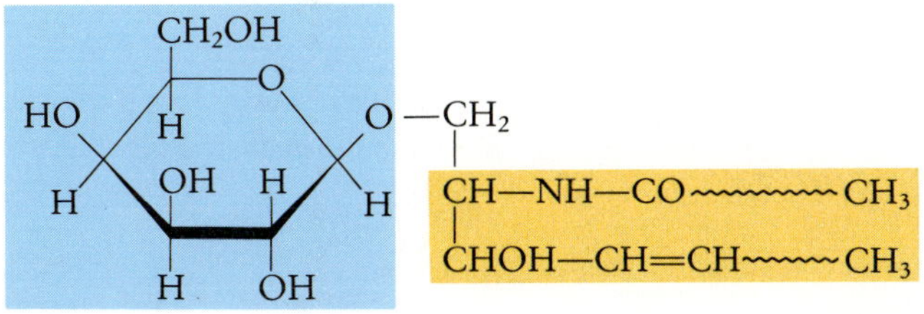

$$\text{Galactosyl}{-}O{-}CH_2{-}CH(NH{-}CO{\sim\sim}CH_3){-}CHOH{-}CH{=}CH{\sim\sim}CH_3$$

With glucose instead of galactose, a glucocerebroside is obtained. These substances, in turn, give rise to a variety of derivatives, including the sulfocerebrosides, in which some of the hydroxyl groups of the sugars are esterified by sulfuric acid, and the gangliosides and other complex glycolipids, in which oligosaccharide assemblies are attached to the ceramide, usually by a terminal β-D-glucosyl group.

Terpenoids

This vast family of natural substances, which derives its name from the same Greek root as turpentine, is also known as the isoprene group, from the name of its basic structural unit, which is itself related to its main biosynthetic precursor, isopentenyl pyrophosphate (Chapter 8):

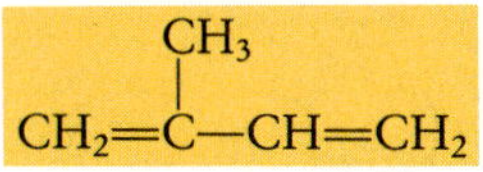

$$CH_2{=}C(CH_3){-}CH{=}CH_2$$

Isoprene

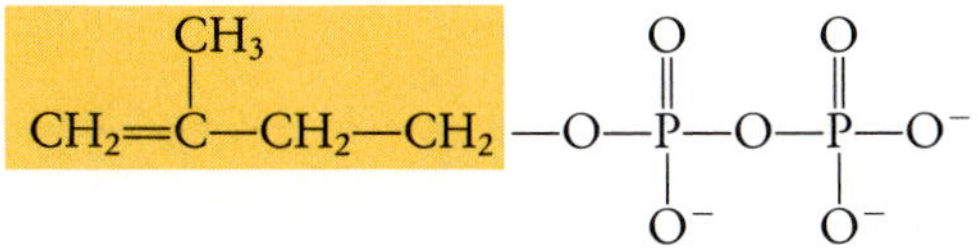

$$CH_2{=}C(CH_3){-}CH_2{-}CH_2{-}O{-}P(=O)(O^-){-}O{-}P(=O)(O^-){-}O^-$$

Isopentenyl pyrophosphate

An important terpene derivative is phytol, a long-chain alcohol made of four such units, which provides one of the building blocks in the formation of chlorophyll and several fat-soluble vitamins (A, E, K):

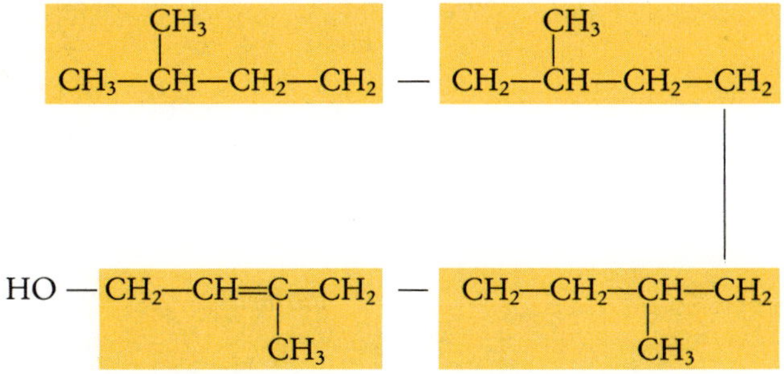

$$CH_3{-}CH(CH_3){-}CH_2{-}CH_2{-}CH_2{-}CH(CH_3){-}CH_2{-}CH_2{-}CH_2{-}CH(CH_3){-}CH_2{-}CH_2{-}CH_2{-}C(CH_3){=}CH{-}CH_2{-}OH$$

Another is squalene, a symmetrical, six-unit molecule first discovered in shark oil. Squalene has turned out to be a major biosynthetic intermediate as precursor of the steroid nucleus. The quinone electron carriers (see p. 402) and dolichol (see p. 405) are other examples of terpene derivatives, as are latex, camphor, resins, essential oils, and countless other similar plant products.

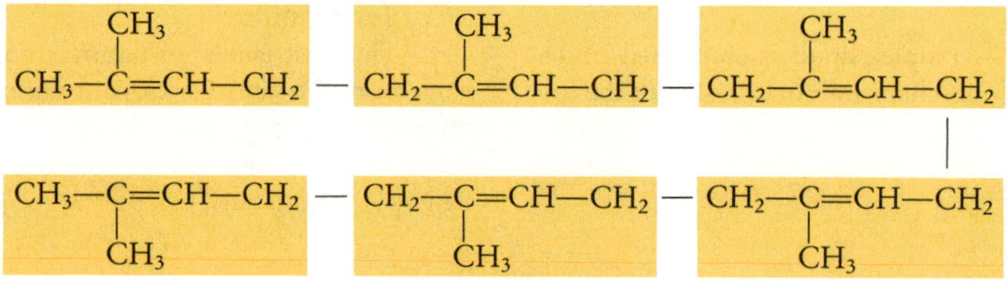

$CH_3-C(CH_3)=CH-CH_2-CH_2-C(CH_3)=CH-CH_2-CH_2-C(CH_3)=CH-CH_2-CH_2-CH=C(CH_3)-CH_2-CH_2-CH=C(CH_3)-CH_2-CH_2-CH=C(CH_3)-CH_3$

Squalene

Sterols and steroids

The mother substance of this group is cholesterol, a key constituent of the plasma membrane (Chapter 3). It is a polycyclic compound that arises from squalene by a complex process of cyclization and trimming:

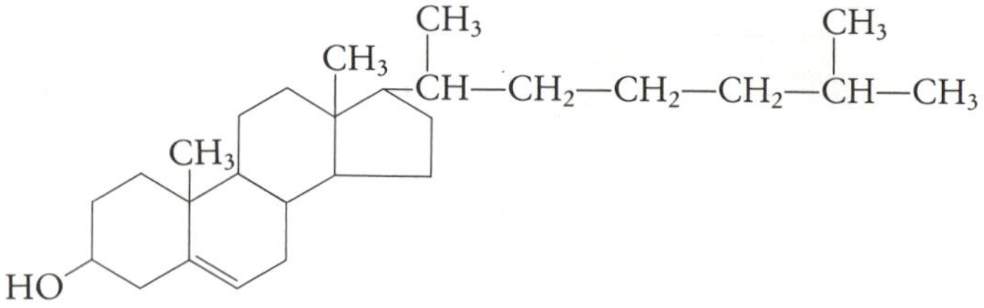

CH_3, CH_3, CH_3, $-CH(CH_3)-CH_2-CH_2-CH_2-CH(CH_3)-CH_3$, HO

Esterification of the hydroxyl group of cholesterol by fatty acids gives cholesteryl esters, which are an essentially inert storage form. Other modifications of the cholesterol molecule lead to the bile acids, the male and female sex hormones, and cortisone and the other hormones manufactured by the adrenal cortex. Ultraviolet irradiation of cholesterol and related substances produces vitamin D.

Pyrimidines:

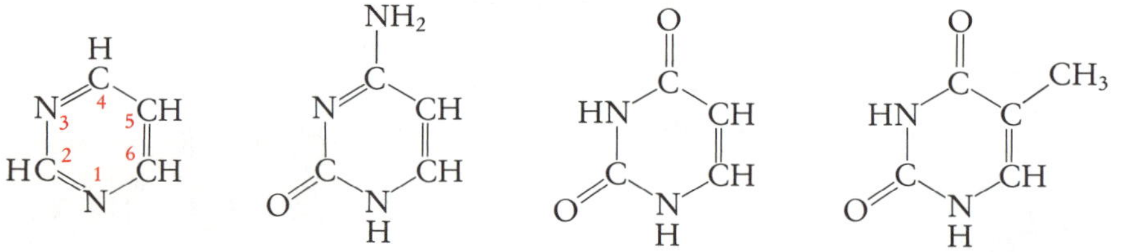

Pyrimidine ring (Py)

Cytosine (2-keto-4-amino-Py)

Uracil (2,4-diketo-Py)

Thymine (2,4-diketo-5-methyl-Py)

Purines:

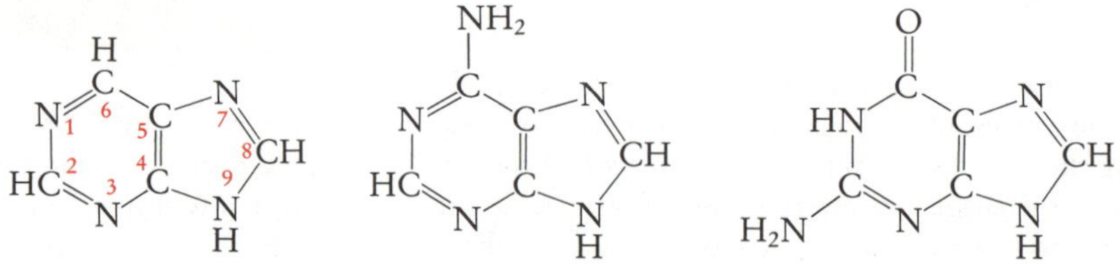

Purine ring (Pu)

Adenine (6-amino-Pu)

Guanine (2-amino-6-keto-Pu)

Nucleic Acids

Bases

As explained in Chapters 8, 15, and 16, nucleic acids are made of three kinds of building blocks: bases, pentoses (already seen on p. 387), and phosphoric acid. The bases are derived either from the pyrimidine or from the purine ring, as shown on the facing page.

Nucleosides

Heteroglycosides in which D-ribose or D-deoxyribose is linked to a purine or pyrimidine base are called nucleosides. The linkage is of the β-glycosidic type, with nitrogen 1 of the pyrimidine or nitrogen 9 of the purine. To avoid confusion, the atom numbers of the pentose are marked with a prime:

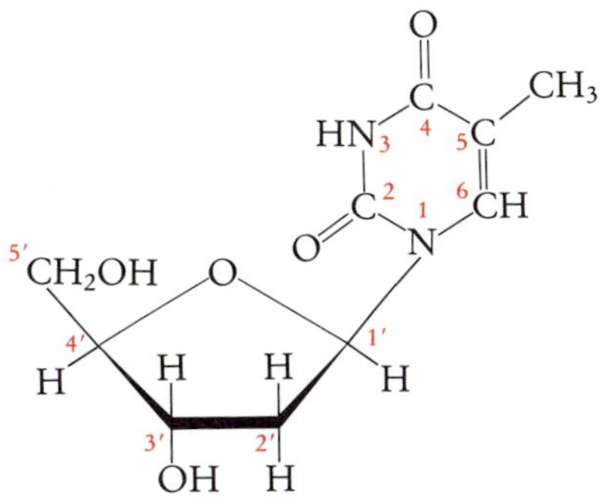

A pyrimidine deoxynucleoside
Deoxythymidine

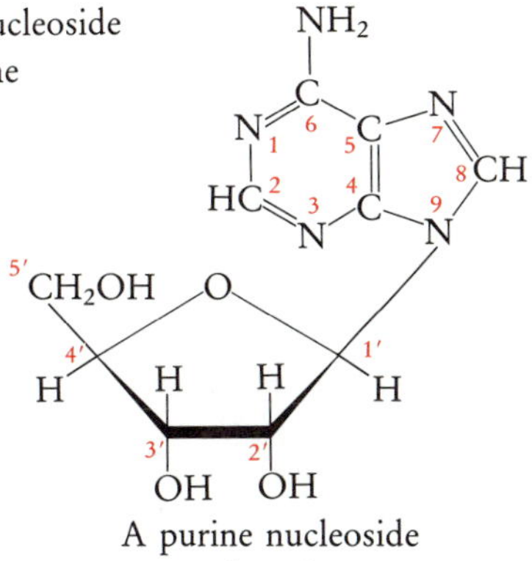

A purine nucleoside
Adenosine

The nomenclature and symbols used to denote the natural nucleosides and their derivatives are given in Chapter 8 (p. 127).

Nucleotides

Mononucleotides arise from nucleosides by phosphorylation of one of the hydroxyl groups of the pentoses, most often that in position 5′:

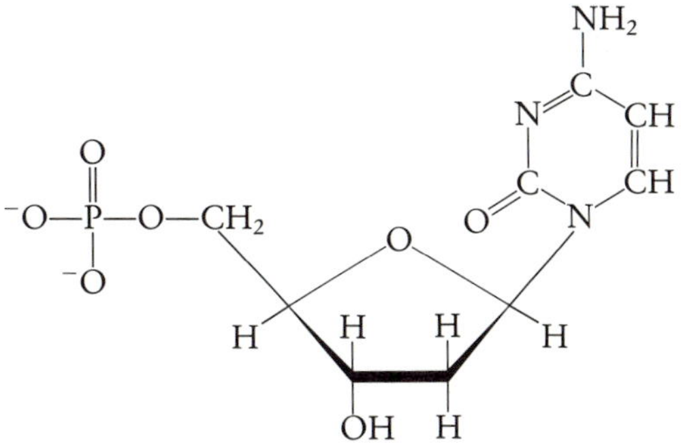

A pyrimidine 5′-mononucleotide
Deoxycytidine monophosphate (dCMP)
or deoxycytidylic acid

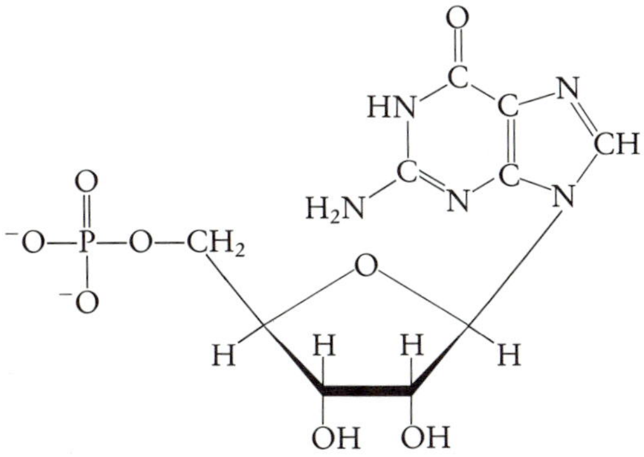

A purine 5′-mononucleotide
Guanosine monophosphate (GMP)
or guanylic acid

The phosphoryl group of 5′-mononucleotides can form several types of combinations of major biological importance. As seen in Chapter 8, it can bind one or two additional phosphoryl groups to form the various nucleoside diphosphates and triphosphates. It can also join by a second phosphoester bond with another hydroxyl of the pentose to which it is attached. Cyclic nucleotides arise in this way—for instance, cyclic AMP (Chapters 8 and 13):

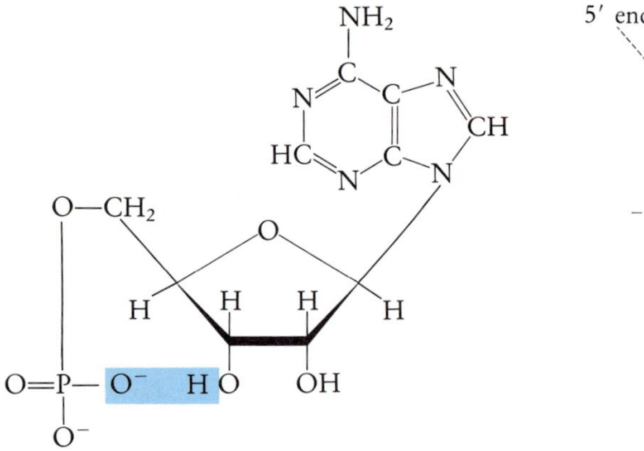

Adenosine monophosphate (AMP)

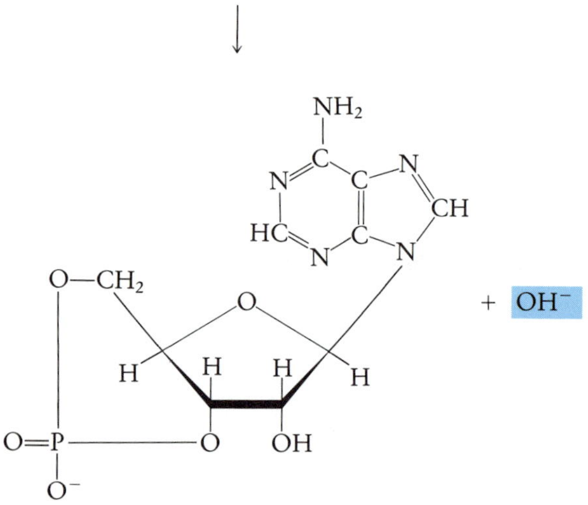

↓

$+ \; OH^-$

3′,5′-Cyclic AMP

In a third type of linkage, the phosphoryl group of a 5′-mononucleotide binds by a pyrophosphate bond to the phosphoryl group of another mononucleotide or analogous substance. Several important coenzymes arise in this manner (see pp. 397–399 and p. 402).

Finally, the phosphoryl group of 5′-mononucleotides can join with the 3′ hydroxyl group of another mononucleotide. Such a dinucleotide can lengthen by the similar attachment of another mononucleotide unit to its free 3′-hydroxyl group to give a trinucleotide. Repetition of the process leads to oligonucleotides and polynucleotides.

Nucleic Acids

The polynucleotide chains of nucleic acids have the following structure, in which the free valences in position 2′ are occupied either by OH (RNAs) or by H (DNAs):

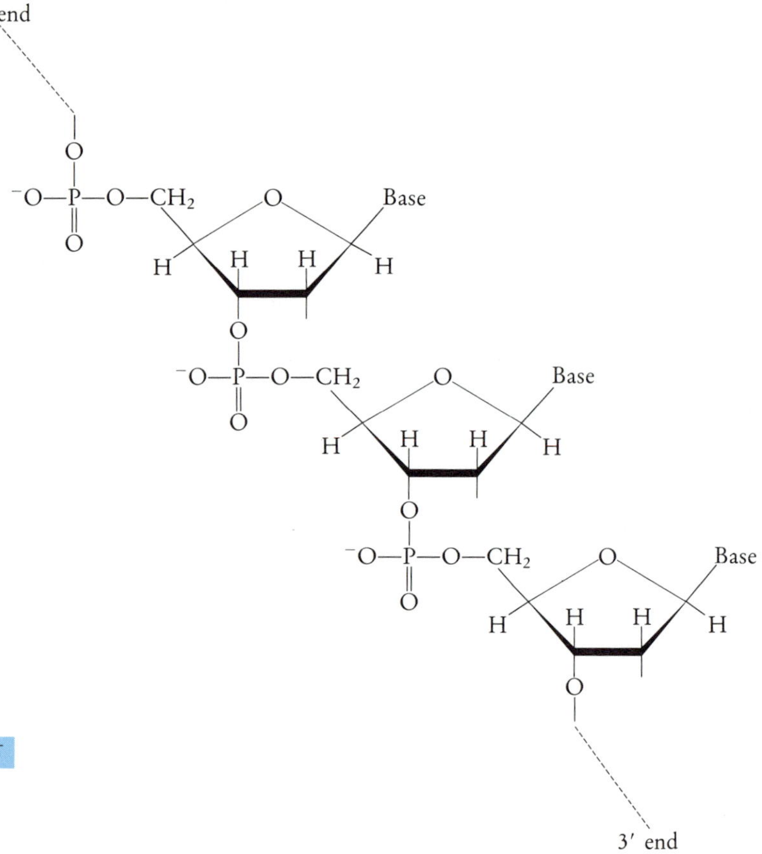

5′ end

3′ end

Or, in simplified form (N stands for nucleoside, p for phosphoryl):

5′ end . . . . pNpNpNpN . . . . 3′ end

The 5′ end of such chains usually bears a phosphoryl group. It may also, as in newly made RNAs (Chapter 16), bear a triphosphoryl group or, as in most eukaryotic mRNAs, a 7-methyl-GTP cap (see p. 397). In a number of cases, the terminal 5′-phosphoryl and 3′-hydroxyl groups of a chain join to produce a cyclic molecule with no free ends.

Modified Nucleic Acids

After their formation, nucleic acids may undergo a variety of chemical modifications, mostly of the bases—for example:

methylation (of adenine, guanine, cytosine, and, in some RNAs, the 2′ hydroxyl group of ribose); hydroxymethylation (of cytosine in certain phage DNAs); glycosylation of such a hydroxymethyl group (also in phages); deamination (of adenine, to give 6-ketopurine, named hypoxanthine, whose nucleoside is inosine, designated I); reduction (of the double bond between carbons 5 and 6 of uridine, to give dihydrouridine, or DHU); and transglycosylation (in uridine from nitrogen-1 to carbon-5, to give pseudouridine, designated ψ). Transfer RNAs (Chapter 15) are particularly rich in chemically modified bases. Typical examples of modified structures are:

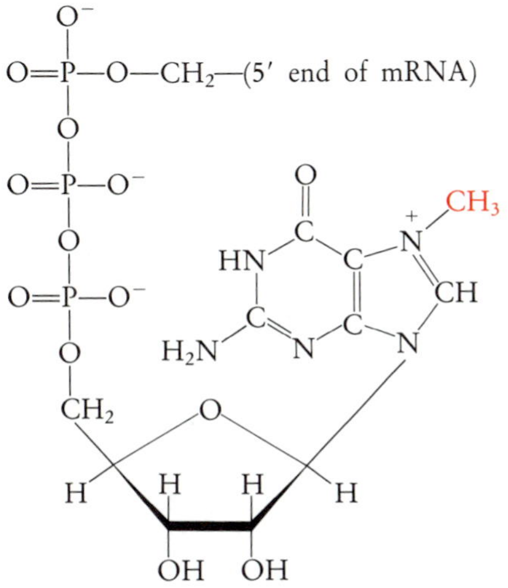

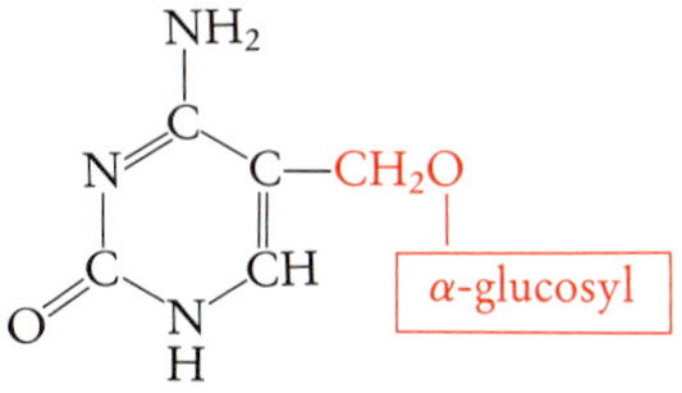

α-Glucosyl-hydroxymethylcytosine

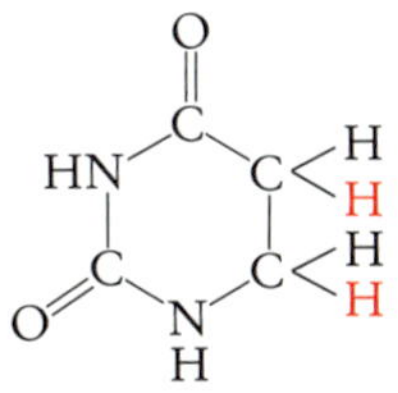

Dihydrouracil

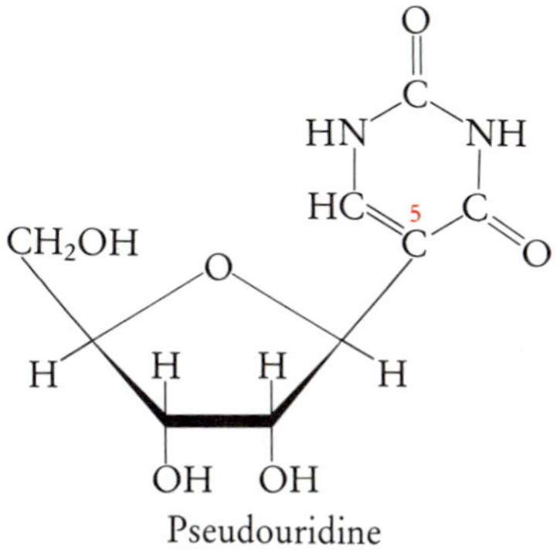

7-Methyl-GTP cap of mRNA

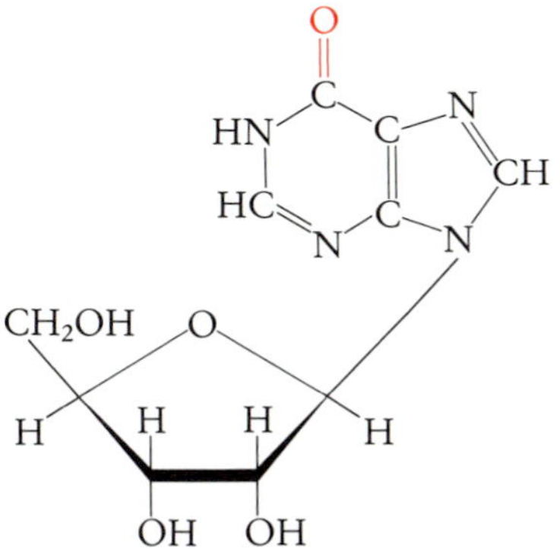

Pseudouridine
ψ

Inosine

Electron Carriers

Nicotinamide Coenzymes

Nicotinamide adenine dinucleotide (NAD), the most important coenzyme in electron-transfer reactions (Chapters 7, 9, 10, and 14) is a dinucleotide in which a molecule of AMP is linked by a pyrophosphate bond to a molecule of nicotinamide mononucleotide (NMN), a ribonucleotide in which the position of the base is occupied by nicotinamide (vitamin PP):

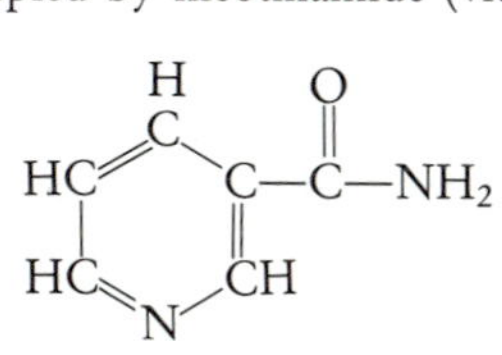

This molecule provides the electron-acceptor part of the coenzyme. Nicotinamide adenine dinucleotide phosphate (NADP), which plays a special role in photochemical and other biosynthetic reductions (Chapter 10), is NAD with an additional phosphoryl group in position 2′ of the AMP moiety:

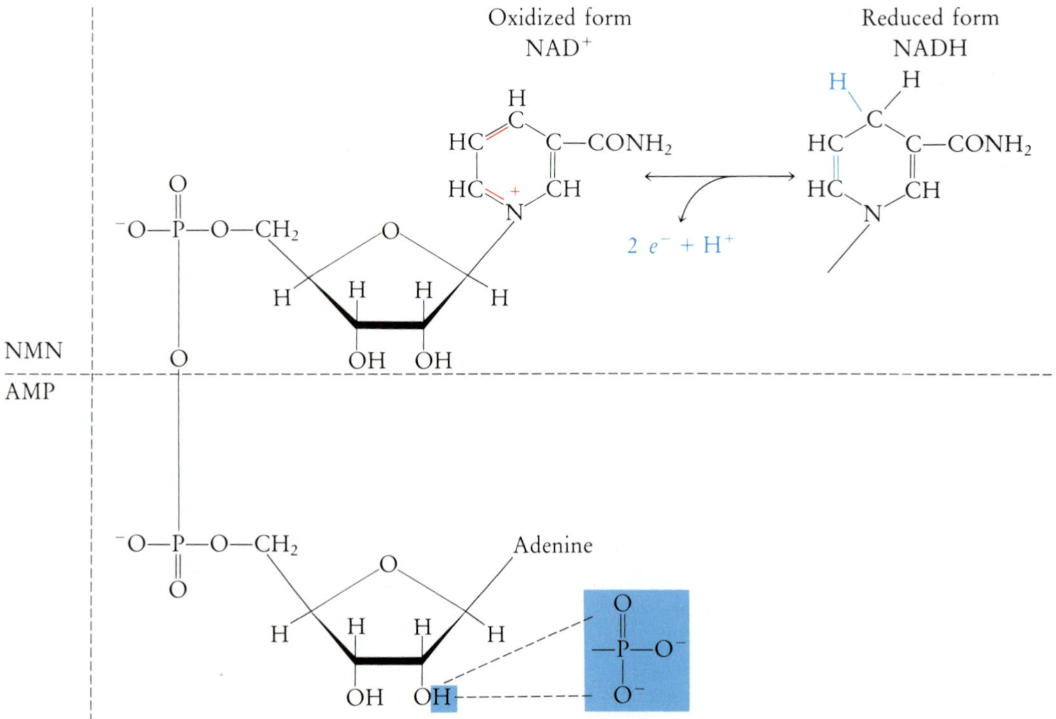

Oxidized form
NAD^+

Reduced form
NADH

$2\,e^- + H^+$

NMN

AMP

Adenine

in NADP

Nicotinamide adenine dinucleotide (phosphate), NAD(P)

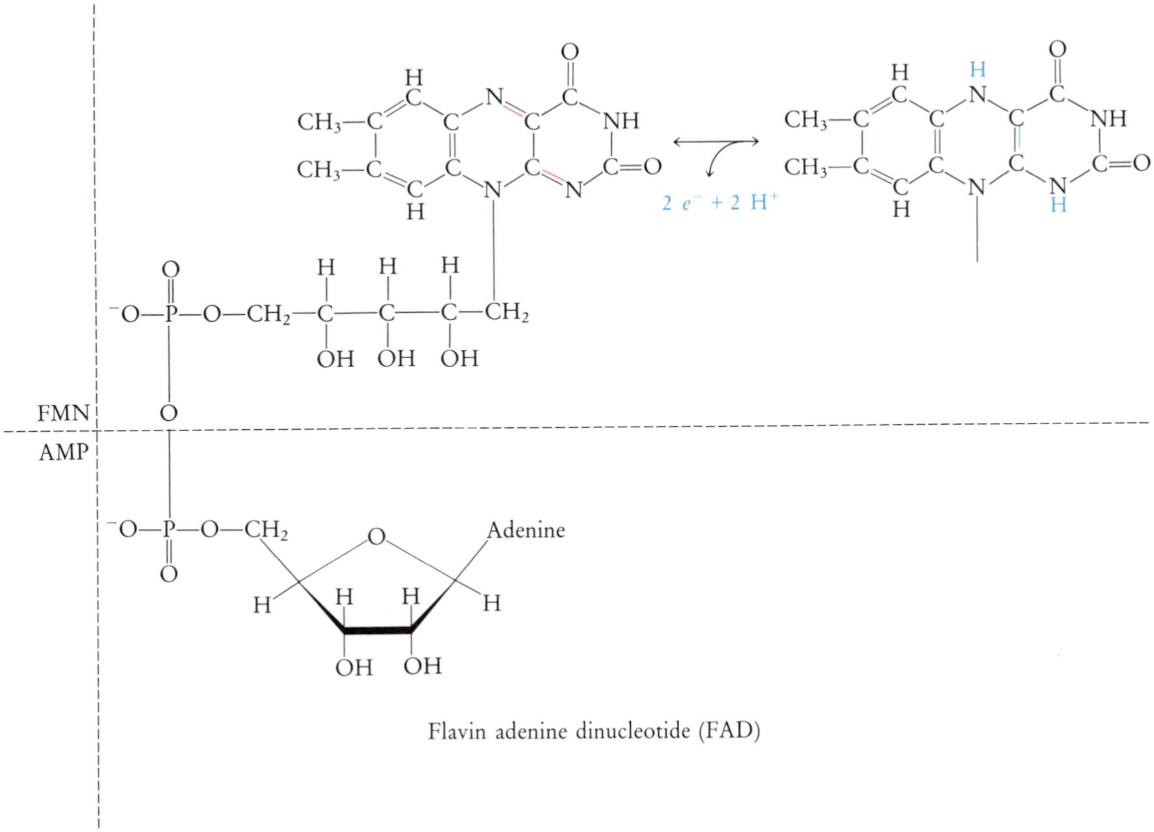

Flavin adenine dinucleotide (FAD)

Flavin Coenzymes

The flavin coenzymes also participate in a number of electron-transfer reactions (Chapters 9 and 10). They include flavin mononucleotide (FMN) and flavin adenine dinucleotide (FAD), which consists of FMN linked to AMP by a pyrophosphate bond, as are the two mononucleotide parts of NAD. The nucleoside part of FMN is really a pseudonucleoside. Known as riboflavin (vitamin B_2), it consists of the pentitol ribitol, the reduction product of ribose, and of a derivative of the polycyclic isoalloxazine ring, which is the reactive part of the coenzymes.

Porphyrin Cofactors

The heme groups of hemoglobins, cytochromes, and other hemoproteins and the chlorophylls of photosynthetic organisms all derive from the porphin, or tetrapyrrole, ring—a planar, polycyclic molecule made of four pyrrole rings linked by methene (—CH═) bridges. Porphyrins are derived from the porphin ring by various substitutions on carbons 1 through 8 and sometimes on some of the methene bridges.

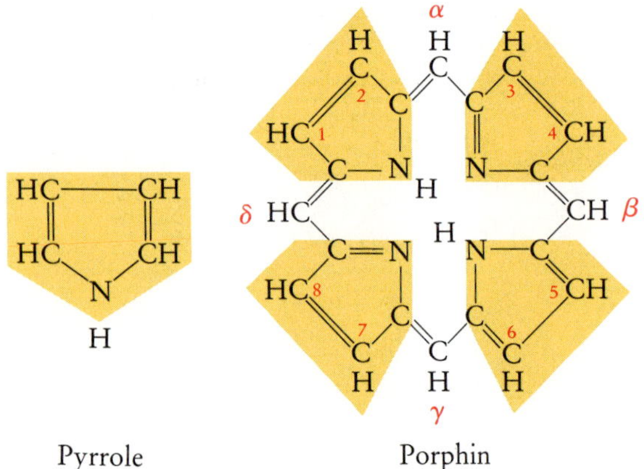

Pyrrole   Porphin

Hemes are iron-porphyrin complexes in which the two central hydrogens are replaced by an iron atom. The most important such molecule is protoheme IX, derived from protoporphyrin IX, or 1,3,5,8-tetramethyl-2,4-divinyl-6,7-dipropionic acid porphin:

This molecule associates with a variety of proteins to give rise to hemoproteins. In hemoglobins, the central iron is in the ferrous (Fe^{2+}) form and serves as binding site for oxygen. It is in the ferric (Fe^{3+}) form in peroxidases and catalase and binds hydrogen peroxide, the electron acceptor used by these enzymes. In cytochromes, the iron alternates between the two forms to serve in electron transport. These three possibilities are shown schematically as follows with the porphyrin ring cut perpendicularly to its plane through two of the four central nitrogens:

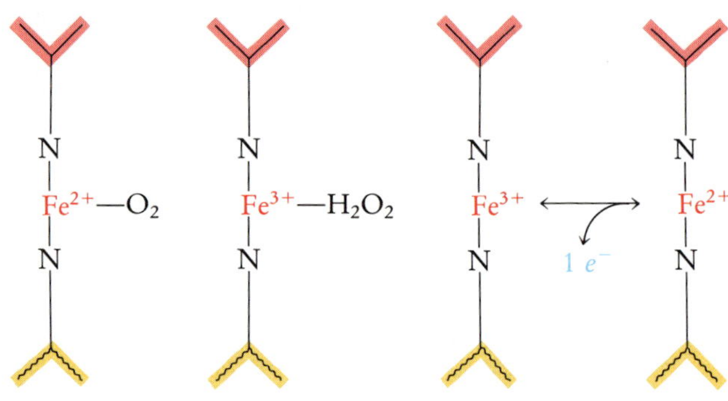

O_2 complex of hemoglobins H_2O_2 complex of peroxidases and catalase   Electron transport by cytochromes

Cytochrome oxidase is a complex molecule made of two distinct cytochromes (their heme component is different from protoheme IX) and containing copper. It catalyzes a concerted transfer of four electrons to a bound molecule of oxygen:

$$O_2 + 4\ e^- + 4\ H^+ \longrightarrow 2\ H_2O$$

In chlorophylls, the central position of the porphyrin is occupied by magnesium, and the substituents include a molecule of phytol (p. 393).

Metalloproteins

Iron also participates in electron transport in nonheme form as protein-bound iron-sulfur complexes. These consist of a single iron atom, of two irons and two sulfides (Fe_2S_2), or of four irons and four sulfides (Fe_4S_4), in each case cradled by a cluster of four cysteine residues belonging to the protein. The schematic structure of an Fe_2S_2 complex is shown on the facing page.

The iron in iron-sulfur proteins oscillates between the ferrous and ferric states, as in hemoproteins. Some of the iron-sulfur proteins, called ferredoxins, accept electrons at a particularly high level of energy (low redox potential) and participate in this capacity in the photoreduction of NADP (Chapter 10) and the formation of molecular hydrogen (Chapter 11). The respiratory chain of mitochondria also includes iron-sulfur proteins.

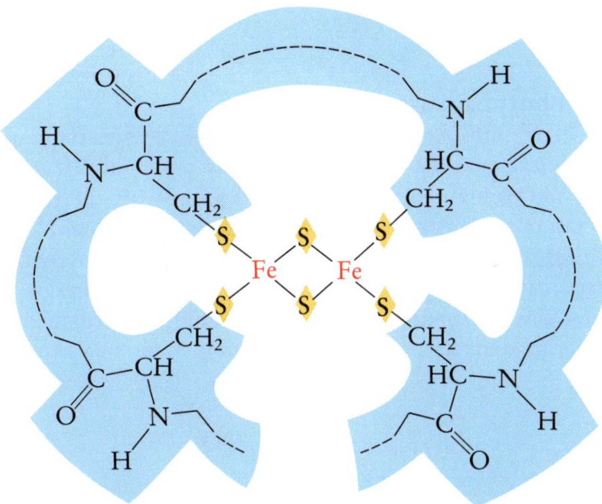

Another important thiol is lipoic acid, which bears two sulfhydryl groups on an eight-carbon fatty-acid skeleton. Its carboxyl group is normally attached by an amide bond to the free amino group of a lysine residue belonging to the enzyme for which it serves as coenzyme. Upon oxidation, lipoic acid forms an intramolecular disulfide bridge:

$$\underset{\textstyle |\atop \textstyle \mathrm{SH}}{\mathrm{CH_2}}\mathrm{-CH_2-}\underset{\textstyle |\atop \textstyle \mathrm{SH}}{\mathrm{CH}}\mathrm{-CH_2-CH_2-CH_2-CH_2-}\overset{\textstyle \mathrm{O}\atop \textstyle \|}{\mathrm{C}}\mathrm{-NH}\ (\text{Enzyme})$$

$$\updownarrow\ \to 2\,e^- + 2\,\mathrm{H^+}$$

$$\underset{\textstyle |\atop \textstyle \mathrm{S}}{\mathrm{CH_2}}\mathrm{-CH_2-}\underset{\textstyle |\atop \textstyle \mathrm{S}}{\mathrm{CH}}\mathrm{-CH_2-CH_2-CH_2-CH_2-}\overset{\textstyle \mathrm{O}\atop \textstyle \|}{\mathrm{C}}\mathrm{-NH}\ (\text{Enzyme}) \quad (\mathrm{S}\text{—}\mathrm{S}\ \text{bridge})$$

Other metals also play a role in some electron-transfer reactions. The most important ones are copper and manganese, which are found in a number of oxidizing enzymes.

Thiols

The reaction whereby two molecules of cysteine combine oxidatively by a disulfide bridge can play a role in electron transport:

$$2\ \mathrm{R{-}SH} \rightleftharpoons \mathrm{R{-}S{-}S{-}R} + 2\ e^- + 2\ \mathrm{H^+}$$

Particularly important in this respect is the tripeptide glutathione (GSH), which dimerizes oxidatively to GSSG (GSH is γ-glutamylcysteinylglycine):

In its reduced form, lipoic acid also functions as an acyl carrier, like coenzyme A (see pp. 402–403). Thanks to this dual ability to participate in both electron and group transfer, lipoic acid plays a key role in several important substrate-level phosphorylation reactions (Chapters 7 and 8). Because of the covalent attachment of lipoic acid to the enzyme, its transporting capacity is limited to the distance that can be covered by its flexible arm. We will encounter a similar situation with biotin. (see p. 403).

$$2\ \begin{array}{c} \mathrm{COO^-} \\ | \\ \mathrm{^+H_3N{-}C{-}H} \\ | \\ \mathrm{CH_2} \\ | \\ \mathrm{CH_2} \\ | \\ \mathrm{O{=}C{-}NH{-}} \end{array}\begin{array}{c} \mathrm{COO^-} \\ | \\ \mathrm{CH_2} \\ | \\ \mathrm{NH} \\ | \\ \mathrm{C{=}O} \\ | \\ \mathrm{C{-}H} \\ | \\ \mathrm{CH_2{-}SH} \end{array} \ \underset{2\,e^- + 2\,\mathrm{H^+}}{\longleftrightarrow}\ \begin{array}{c} \mathrm{COO^-} \\ | \\ \mathrm{^+H_3N{-}C{-}H} \\ | \\ \mathrm{CH_2} \\ | \\ \mathrm{CH_2} \\ | \\ \mathrm{O{=}C{-}NH{-}} \end{array}\begin{array}{c} \mathrm{COO^-} \\ | \\ \mathrm{CH_2} \\ | \\ \mathrm{NH} \\ | \\ \mathrm{C{=}O} \\ | \\ \mathrm{C{-}H} \\ | \\ \mathrm{CH_2{-}S} \end{array}\begin{array}{c} \mathrm{COO^-} \\ | \\ \mathrm{CH_2} \\ | \\ \mathrm{NH} \\ | \\ \mathrm{C{=}O} \\ | \\ \mathrm{H{-}C} \\ | \\ \mathrm{{-}S{-}CH_2} \end{array}\begin{array}{c} \mathrm{COO^-} \\ | \\ \mathrm{H{-}C{-}NH_3^+} \\ | \\ \mathrm{CH_2} \\ | \\ \mathrm{CH_2} \\ | \\ \mathrm{{-}NH{-}C{=}O} \end{array}$$

$$2\ \mathrm{GSH} \rightleftharpoons \mathrm{GSSG} + 2\ e^- + 2\ \mathrm{H^+}$$

Quinones

Quinones, which can be reduced to the corresponding diphenols, act in electron transfer in the respiratory chains of both mitochondria and chloroplasts. They are generally derived from *p*-benzoquinone and bear various substituents, including a long terpenoid chain. The mitochondrial coenzyme Q_{10}, or ubiquinone 50, has the following structure:

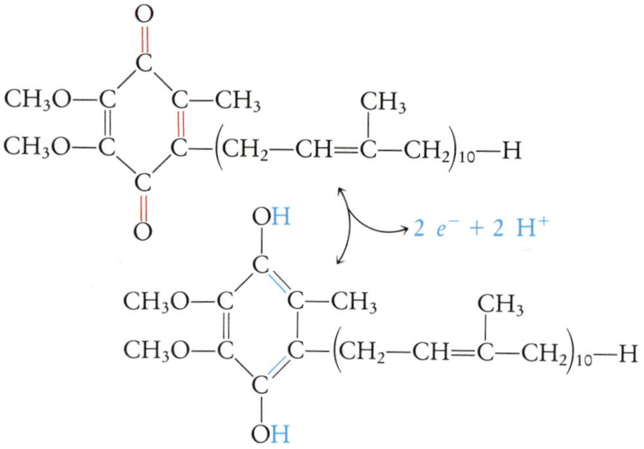

$$CH_3O-C,\ CH_3O-C,\ C-CH_3,\ C-(CH_2-CH=C(CH_3)-CH_2)_{10}-H$$

$$\rightleftharpoons 2\,e^- + 2\,H^+$$

The number 10 in Q_{10} identifies the number of isoprene units in the side chain, whereas the number 50 affixed to ubiquinone refers to the number of carbon atoms of the side chain.

Group Carriers

Phosphagens

The phosphagens—literally phosphate generators—are phosphorylated substances of high group potential that serve as reservoirs of high-energy phosphoryl groups for the rapid regeneration of ATP from ADP in muscle tissue (Chapter 8). They have the distinction that the phosphoryl group is attached to a nitrogen atom belonging to a guanidine group. In many invertebrates, the phosphoryl carrier is the amino acid arginine. It is creatine, a derivative of the amino acid glycine, in vertebrates.

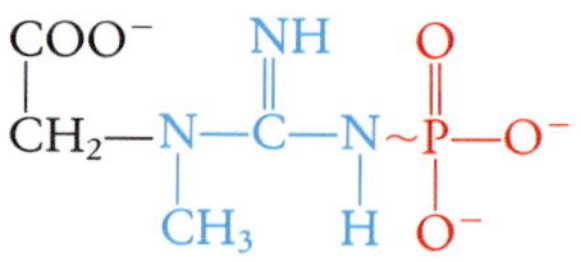

$$^{-}OOC-CH_2-N(CH_3)-C(=NH)-NH\sim PO_3^{2-}$$

Creatine phosphate

$$^{-}OOC-CH(^{+}H_3N)-CH_2-CH_2-CH_2-NH-C(=NH)-NH\sim PO_3^{2-}$$

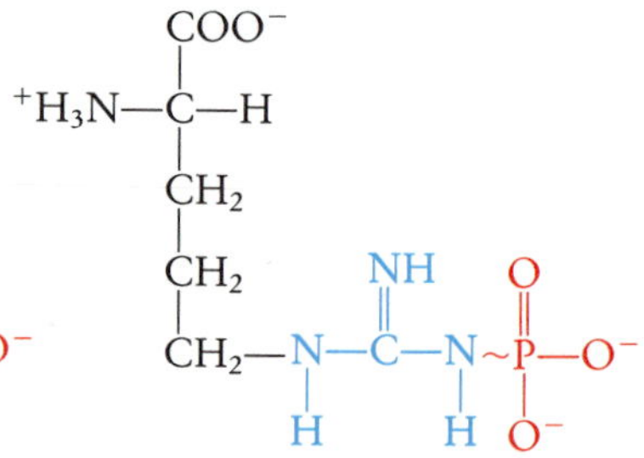

Arginine phosphate

Phosphopantetheine

Pantetheine

Pantothenic acid | Cysteamine

$$^{-}O-P(=O)(-O-)-O-CH_2-C(CH_3)_2-C(H)(OH)-C(=O)-NH-CH_2-CH_2-C(=O)-NH-CH_2-CH_2-SH$$

AMP

$$^{-}O-P(=O)-O-CH_2\ (\text{ribose, Adenine; 3'-}O-P(=O)(O^-)-O^-)$$

$$-CH_2-S-C(=O)-R$$ Acylation

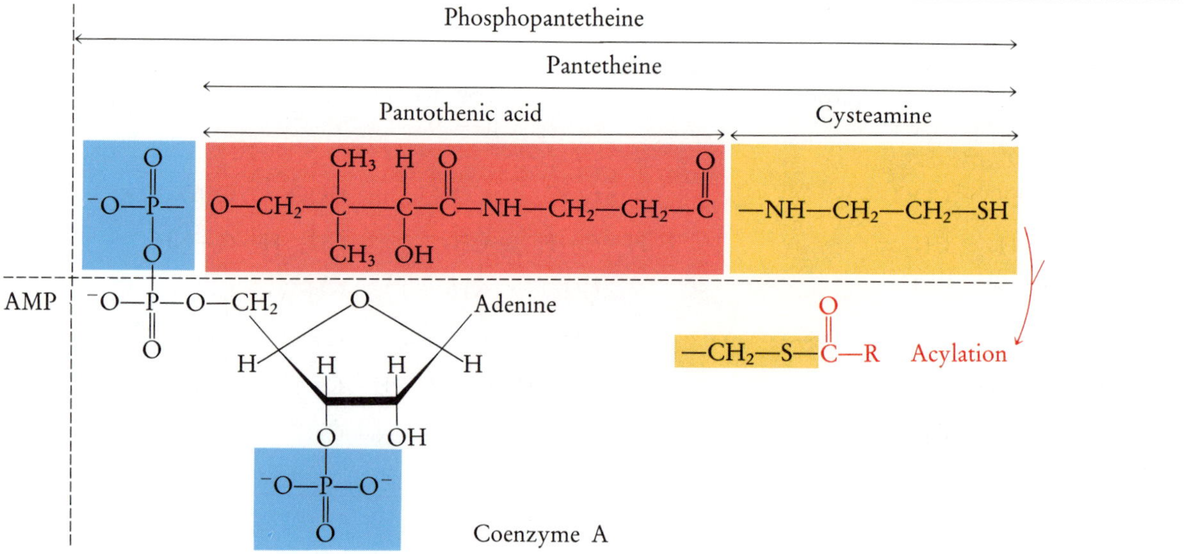

Coenzyme A

Coenzyme A

This central carrier of acyl groups (Chapter 8) consists of a molecule of 3′-phospho-AMP linked by a pyrophosphate linkage to phosphopantetheine, which itself is formed through the association, by an amide bond, of cysteamine (the product of decarboxylation of the amino acid cysteine) with pantothenic acid, or vitamin F. The thiol group of cysteamine serves as carrier for the acyl group. The structure of coenzyme A is shown on the facing page.

In acyl-carrier protein (ACP, Chapter 8), the phosphopantetheine part of coenzyme A is attached to the protein without the participation of AMP.

Carnitine

Carnitine (Chapter 13) is a fatty acyl carrier that serves in the transfer of coenzyme A-linked fatty acyl groups across the inner mitochondrial membrane:

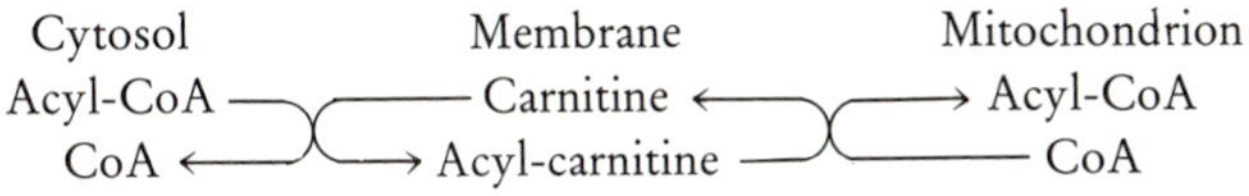

Cytosol	Membrane	Mitochondrion
Acyl-CoA →	Carnitine ⇄	Acyl-CoA
CoA ←	Acyl-carnitine	CoA

Carnitine is a derivative of choline, already encountered as a constituent of phospholipids, and of acetic acid. These two moities are attached by a carbon–carbon bond, not by an ester bond as in acetylcholine. The hydroxyl group of the choline part serves as acyl carrier:

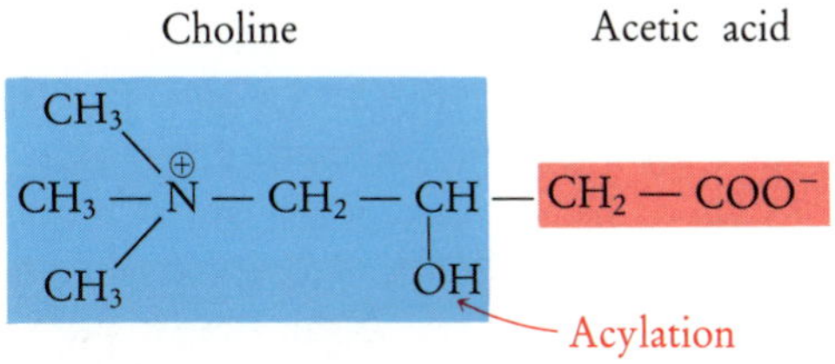

Choline — Acetic acid

$$(CH_3)_3\overset{\oplus}{N}-CH_2-\underset{\underset{OH}{|}}{CH}-CH_2-COO^-$$

Acylation (at OH)

Biotin

Biotin (vitamin H), the carboxyl carrier in carboxylation reactions (Chapter 8), is a saturated sulfur-containing molecule fitted with a five-carbon acid side chain, which is covalently attached to the transferase molecule and serves as a flexible arm in the transport of the activated carboxyl group to its final acceptor. (Lipoic acid, considered on p. 401, presented us with a similar situation.) The carboxyl group is attached to one of the nitrogens of the ring:

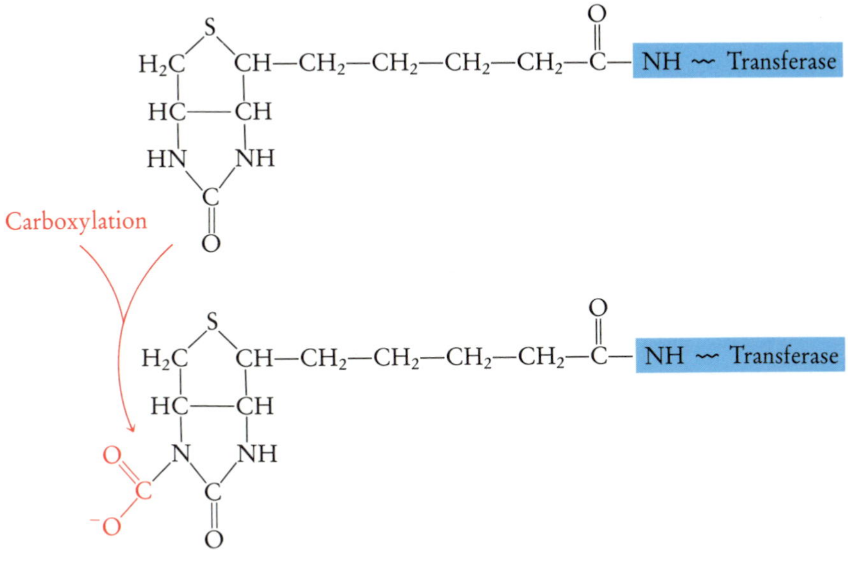

Biotin ring (S, H_2C, HC—CH, HN, NH, C=O) $-CH-CH_2-CH_2-CH_2-CH_2-\overset{O}{\overset{\|}{C}}-$ NH ~ Transferase

Carboxylation ↓

Carboxybiotin ring ($^-O-\overset{O}{\overset{\|}{C}}-N$, NH, C=O) $-CH-CH_2-CH_2-CH_2-CH_2-\overset{O}{\overset{\|}{C}}-$ NH ~ Transferase

Folic acid

Three components are assembled in the formation of this vitamin of the B group: a derivative of the pterine ring, *p*-aminobenzoic acid, and glutamic acid:

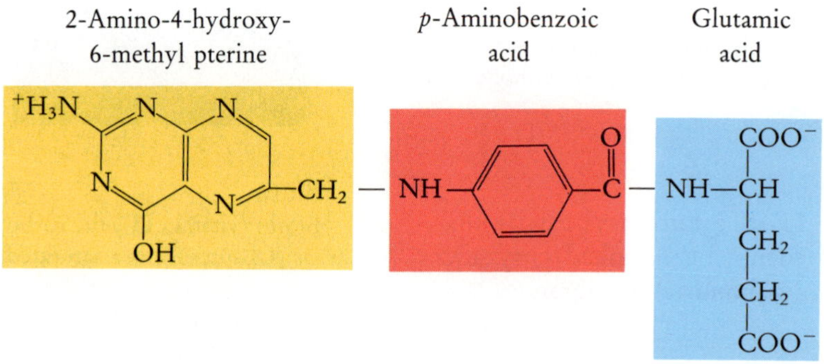

2-Amino-4-hydroxy-6-methyl pterine

p-Aminobenzoic acid

Glutamic acid

As pointed out in Chapter 8, tetrahydrofolic acid (THF), the product of reduction of folic acid, serves in the transport of the formyl group and of a number of its derivatives. Following is the structure of formyl-THF:

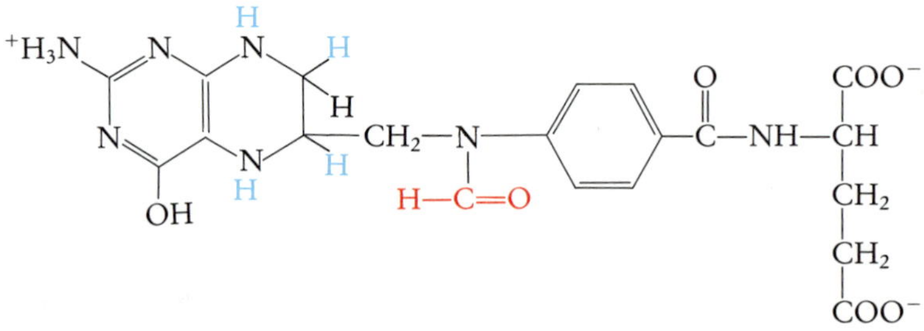

Dolichyl phosphate

As mentioned in Chapters 6, 8, and 13, dolichyl phosphate plays an important role as glycosyl carrier in the assembly of oligosaccharide side chains of glycoproteins and glycolipids in the ER. It is a very long terpenoid alcohol containing about twenty isoprene units. This highly hydrophobic chain serves to anchor the molecule firmly in the lipid bilayer of the ER membrane. Phosphorylation of the alcohol group provides the molecule with a hydrophilic head on which the transported glycosyl groups are attached by a glycosyl phosphate bond:

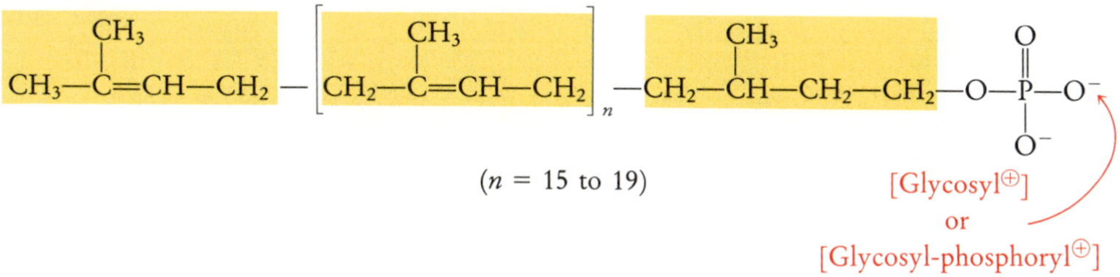

$$CH_3{-}\overset{\overset{CH_3}{|}}{C}{=}CH{-}CH_2 - \left[CH_2{-}\overset{\overset{CH_3}{|}}{C}{=}CH{-}CH_2\right]_n {-}CH_2{-}\overset{\overset{CH_3}{|}}{C}H{-}CH_2{-}CH_2{-}O{-}\overset{\overset{O}{\|}}{\underset{\underset{O^-}{|}}{P}}{-}O^-$$

$(n = 15 \text{ to } 19)$

[Glycosyl$^{\oplus}$]
or
[Glycosyl-phosphoryl$^{\oplus}$]

In some reactions, dolichyl phosphate carries a glycosyl phosphate molecule linked by a pyrophosphate bond.

Thiamine

First of the vitamins to be discovered, thiamine, or vitamin B_1, was singled out in Chapter 8 as one of the rare recipients of the pyrophosphoryl group. The product of this reaction, thiamine pyrophosphate (TPP), serves as coenzyme in a number of decarboxylation reactions, such as the decarboxylation of pyruvic acid to acetaldehyde in alcoholic fermentation (Chapter 7). TPP is highly unusual among group carriers in that it serves to transport a negatively charged group, the carbanion group R—CO$^-$. Thiamine is formed by the association of a pyrimidine derivative with a thiazol derivative:

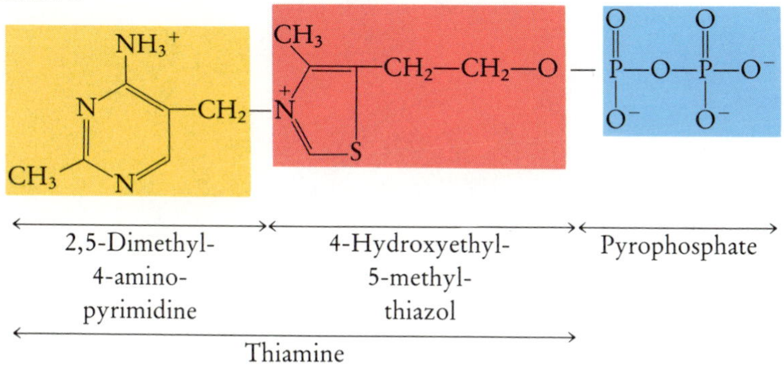

NH$_3^+$, N, CH$_3$, N, —CH$_2$—$\overset{+}{N}$, CH$_3$, S, —CH$_2$—CH$_2$—O — $\overset{\overset{O}{\|}}{\underset{\underset{O^-}{|}}{P}}{-}O{-}\overset{\overset{O}{\|}}{\underset{\underset{O^-}{|}}{P}}{-}O^-$

2,5-Dimethyl-4-amino-pyrimidine	4-Hydroxyethyl-5-methyl-thiazol	Pyrophosphate
Thiamine		

Pyridoxal phosphate

Derived from pyridoxine, or vitamin B_6, pyridoxal phosphate is the coenzyme of transamination reactions. These group-transfer reactions do not conform to the general scheme of group transfer given in Chapter 8. Instead of being exchanged for a proton, the transferred amino group is exchanged for a ketonic oxygen. As a rule, the partners of the reaction are an amino acid and an α-keto acid:

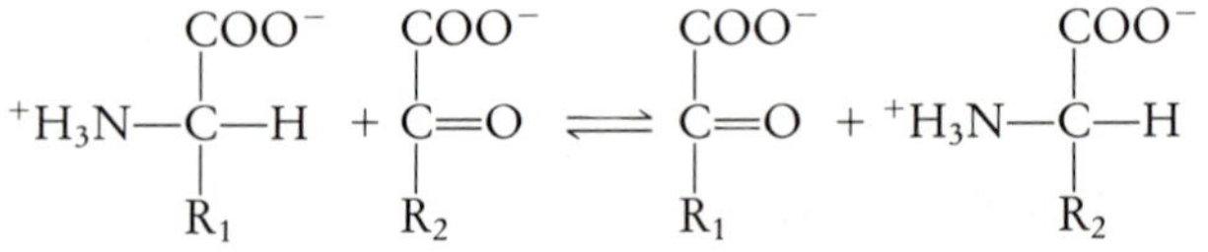

$$^{+}H_3N{-}\underset{R_1}{\overset{COO^-}{C}}{-}H + \underset{R_2}{\overset{COO^-}{C}}{=}O \rightleftharpoons \underset{R_1}{\overset{COO^-}{C}}{=}O + {}^{+}H_3N{-}\underset{R_2}{\overset{COO^-}{C}}{-}H$$

Pyridoxine is a pyridine derivative (like nicotinamide). It is 2-methyl-3-hydroxy-4,5-dihydroxymethylpyridine. In its functional form, it is phosphorylated in position 5 and either oxidized (pyridoxal phosphate) or aminated (pyridoxamine phosphate) in position 4:

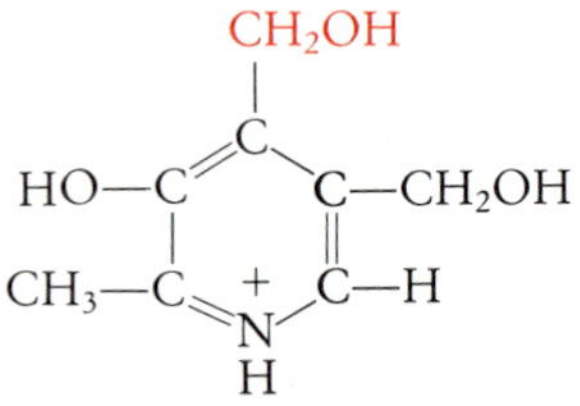

Pyridoxine: ring with 4-CH_2OH, 3-HO, 5-CH_2OH, 2-CH_3, N^+–H

Pyridoxine

Pyridoxal phosphate: ring with 4-(H–C=O), 3-HO, 5-$CH_2{-}O{-}PO_3^{2-}$ ($C{-}CH_2{-}O{-}P(=O)(O^-){-}O^-$), 2-$CH_3$, N^+–H

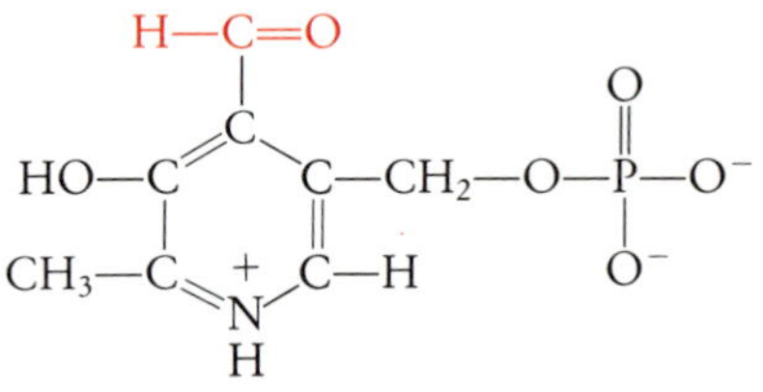

Pyridoxal phosphate

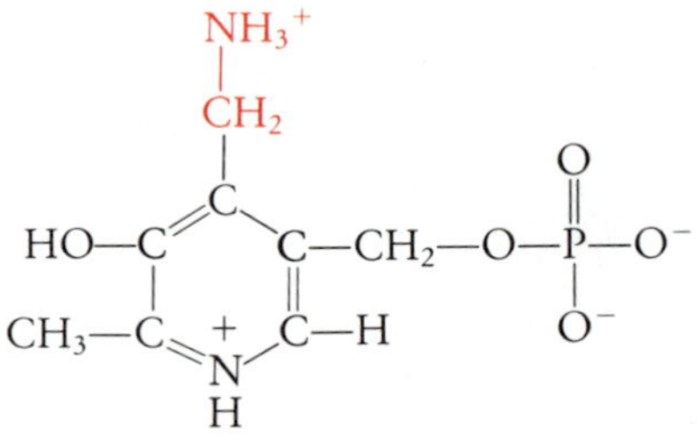

Pyridoxamine phosphate: ring with 4-($CH_2{-}NH_3^+$), 3-HO, 5-$CH_2{-}O{-}P(=O)(O^-){-}O^-$, 2-$CH_3$, N^+–H

Pyridoxamine phosphate

The involvement of the coenzyme in the transaction is due to its ability to form Schiff bases reversibly with amino acids (as pyridoxal) and with α-keto acids (as pyridoxamine). Isomerization of the Schiff base allows the exchange of groups between coenzyme and substrate in the manner illustrated below. As indicated, an amino group is exchanged for a ketonic oxygen at each turn of the cycle, while the coenzyme oscillates between its pyridoxal and pyridoxamine forms. Every step in this sequence is, of course, reversible. One-way arrows have been used to show a complete reaction cycle:

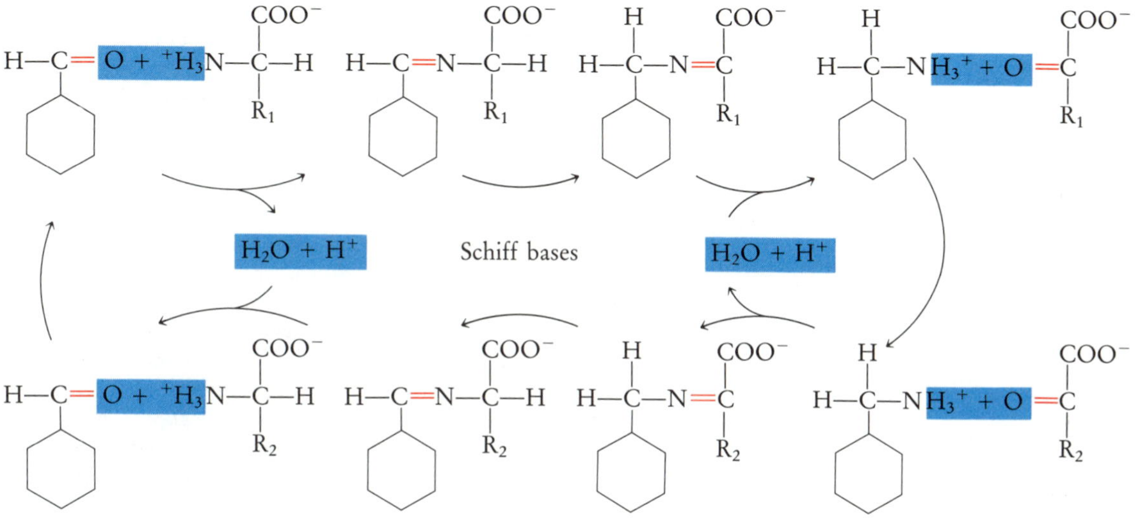

$H_2O + H^+$ Schiff bases $H_2O + H^+$

APPENDIX 2 Principles of Bioenergetics

Free Energy, the Source of Work

In terms of everyday experience, energetics is very simple: when we perform work, we do so at the expense of something that enables us to do work. That something we call energy. Scientists, when they want to be rigorous, call it *free energy,* to take care of the fact that the conversion of certain forms of energy—heat, for example—into work is subject to constraints.

As we all know from having felt exhausted after physical exertion, the more work we perform, the more (free) energy we spend. We also know, although these concepts are less intuitive, that there are different kinds of work—mechanical, electric, chemical—and that they are interconvertible under certain conditions; that is, one type of work can supply free energy for another with the help of an appropriate *transducer.* The electric generator in our automobile is an example of a mechanoelectric transducer. The main practical function of *energetics* (originally named *thermodynamics* because it was developed to explain and assess the performance of steam or heat engines) is to put these familiar notions into quantitative terms to allow accurate bookkeeping of the free-energy gains and losses for every kind of work and, thus, for every kind of conversion. *Bioenergetics* does this for living organisms, which, it should be stressed, obey exactly the same laws as do inanimate objects. But living organisms carry out a remarkable set of energy conversions with the help of unique transducers.

All of Nature's Streets Are One Way

Consider the following statements: "Newton's apple could not, under its own power, jump back to the branch from which it fell" or "Nobody has ever seen, or expects

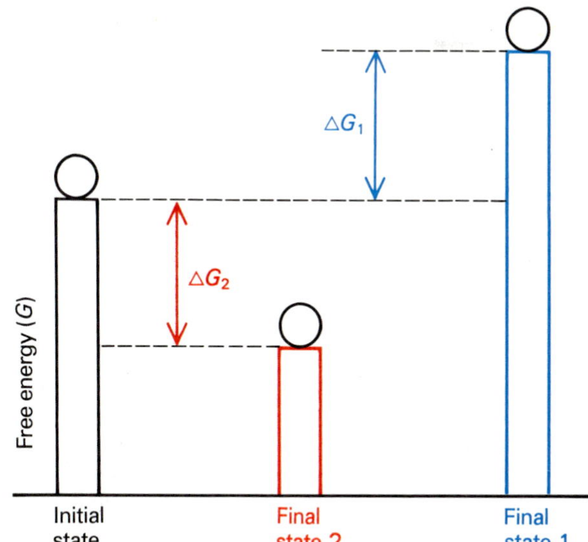

The difference in ability to perform work between final state and initial state is indicated by ΔG.

1. Endergonic transformation: G increases, $\Delta G_1 > 0$.

2. Exergonic transformation: G decreases, $\Delta G_2 < 0$.

Only exergonic transformations can take place spontaneously. However, an endergonic transformation can be made to occur by coupling to an exergonic transformation, provided $\Delta G_1 + \Delta G_2 < 0$.

to see, sugar gathering back into a lump in a cup of coffee." "Unexceptionable," you will say, or perhaps even "Self-evident! Do you take me for a moron?" Yet, if I write:

$$\Delta G \leqslant 0 \qquad (1)$$

or state that "The free energy of an isothermic and isobaric closed system that can exchange only heat with its surroundings cannot increase spontaneously," you may well ask with equal furor whether I take you for Einstein.

Actually, both types of statements say the same thing: Some events are possible; others are not. The more obscure statement simply uses the formalism that scientists had to devise to put the obvious into rigorous, quantitative terms of universal significance. It expresses what is known as the second law of thermodynamics. The first law is the law of *conservation of energy* (now combined with the law of *conservation of matter*, since Einstein's demonstration that matter and energy are interconvertible); it is easy to grasp. The second law, sometimes called the law of *degradation of energy*, is more abstruse. Much of energetics is concerned with its applications.

Free energy has already been defined as the ability to perform work. The italic letter G is the symbol chosen to represent this quantity. The capital Greek delta, Δ, stands for *difference*—between a final and an initial state. Therefore, ΔG represents the *change in free energy*—positive if there is a gain, negative if there is a loss—that is associated with a given event or transformation. When we write, as in equation 1, that ΔG is smaller than, or equal to, zero—that is, either negative or null—we actually state an interdiction: ΔG cannot be positive; G cannot increase spontaneously. Which, when you come to think of it, is the reason why Newton's apple does not levitate: it would thereby increase its G content, its ability to perform work by falling; and Newton would get something for nothing. Similarly, although this may be less obvious, dissolved sugar does not spontaneously reform into a lump because this would, at no cost, reinstate its initial ability to perform work, of an osmotic type in this case, by dissolving.

This, one can readily appreciate once it is pointed out, is an absolutely general observation. For every natural phenomenon, there is an authorized and a forbidden direction. Stones roll downhill, not uphill; heat flows from hot to cold, not in the reverse direction; molecules diffuse from regions where they are more abundant to regions where they are less; Nature abhors a vacuum; the list is endless. Energetics gives us the universal key to unidirectionality by telling us that the one-way sign always points in the direction of decreasing G. This direction is defined by the term *exergonic*, as opposed to *endergonic*, when G increases. The special case in which ΔG equals zero characterizes the state in which no change can occur, whether in one direction or in the other, because G is at a minimum. This state of authentic standstill is known as *stable* or *thermodynamic equilibrium*.

So far, so good. But we must watch out. Falling apples can be lifted back to their starting position; sugar can be crystallized; stones can be pushed uphill; heat can be forced out of a refrigerator; vacuum can be created. In other words, prohibited phenomena can be made to take place. The telling word, here, is "made." Endergonic events never occur spontaneously, but they can be made to occur, which, of course, is what work is all about. When we work, or have a machine work for us, it is always to accomplish something that would not happen on its own, to back up one of Nature's one-way streets. This is not against the law, provided we spend at least as many G units in doing the work as we gain from it. Let ΔG_1 be the free-energy change (positive) of the endergonic event that has been accomplished and ΔG_2 the free-energy change (negative) of the exergonic event that drives the former. Then, if ΔG_2 is equal to, or larger than, ΔG_1 in absolute value, we may write:

$$\Delta G_1 + \Delta G_2 \leqslant 0 \qquad (2)$$

Life is supported by the sun; it does not violate the second law because $\Delta G_B + \Delta G_S < 0$.

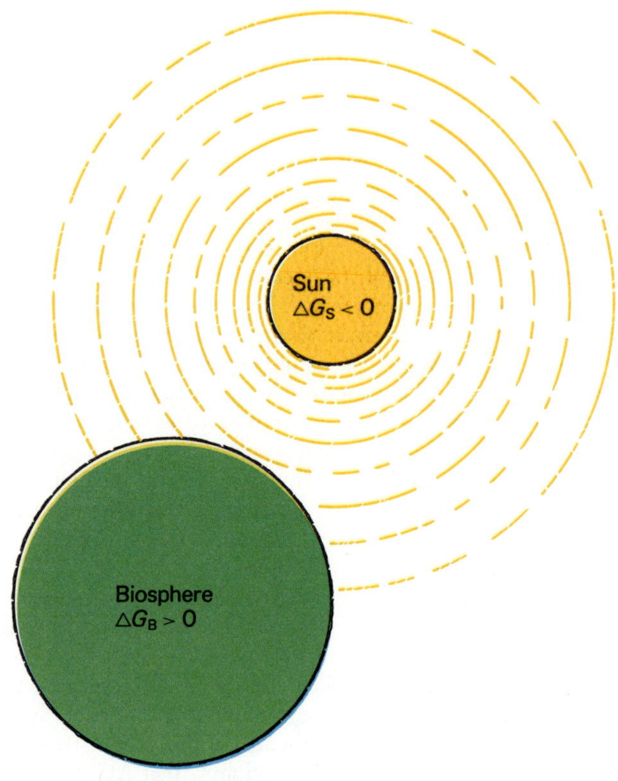

For the combined events, the second law is obeyed; Nature's ways have not been violated. There is a net loss of free energy—or, at least, no net gain—although part of the system has gained free energy. Thus, free-energy gains can take place, provided they are associated with free-energy losses of at least the same magnitude. An essential condition of success in such an operation is *coupling* between the two events. Machines, including the "bio-engines" that we see at work in the living cell, are no more than coupling devices of this sort. The *efficiency* of a machine is given by the ratio of free energy gained to free energy lost. Expressed in percentage:

$$\% \text{ efficiency} = \frac{\Delta G_1}{-\Delta G_2} \times 100 \qquad (3)$$

As a rule, the efficiency is lower than its maximal permissible value of 100 per cent, and a fraction of the free energy expended is lost as heat. We will come back to this point. Note that the loss can be very small. Especially in the living cell, efficiency values approaching 100 per cent are not uncommon, which means that the coupled systems concerned operate near thermodynamic equilibrium.

Of Systems and Their Boundaries

With energetics, we can find out what allows any part of the world to run as it does. But, as illustrated by the examples that we have just seen, we must make sure of exactly what part of the world we are looking at. Should we focus on the biosphere, for example, without including the sun, we would be left with the impression of a tremendous build-up of free energy—as fuel, to say the least—characterized by an enormous positive ΔG; we might even conclude, as many did in the past, that life does not obey the second law of thermodynamics. But, if we add the even more enormous negative ΔG associated with the provision of light to photosynthetic organisms, the need to believe that life is outside the law vanishes. In order to avoid such traps and to respect the requirements of perfect rigor, the scientists who worked out the energetic code have, like lawyers, had to resort to a forbiddingly formal language, stuffed with precautionary clauses and hair-splitting definitions that have dampened many enthusiasms. The sentence that made you exclaim that you are not an Einstein belongs to that pedantic-sounding, but indispensable, phraseology.

First the word *system.* In energetics, it designates any kind of defined assembly of matter. It can be you or me, a living cell, a mitochondrion, a reaction vessel in a chemical plant, a locomotive, the earth, even the universe—anything material that can be defined by appropriate limits, even imaginary ones. A system may be open, closed, or isolated. An *open system* is one that can exchange both matter and energy with the outside. A *closed system* can exchange only energy, not matter. An *isolated system* can exchange nothing. Living organisms are open systems, traversed by fluxes of matter and energy. As such, they belong to the domain of *nonequilibrium thermodynamics,* which uses a fairly sophisticated kind of formalism. For practical purposes, it is often possible to simplify a bioenergetic problem by artificially closing the system under consideration and applying to it the laws of *equilibrium* or *classical thermodynamics*. This is what we have done.

Several additional assumptions must be made about the closed systems that we are looking at if we are to apply the second law in the simple form enunciated above. First we assume, unless explicitly stated otherwise and taken into account, that heat is the only form of energy that the systems can exchange with the outside. We assume further that this exchange is carried out at a perfectly constant temperature. Finally, we assume that the systems are kept at constant pressure (or at constant volume, but the assumption of constant pressure is clearly more appropriate for living organisms).

The assumption of *isothermicity* implies the existence around the system of a reversible heat exchanger, connected with a constant-temperature calorie reservoir of infinite capacity; so any change inside the system that would tend to raise or lower its temperature is compensated immediately, continually, and perfectly by an appropriate flow of heat out of or into the system. Heat exchangers of this sort do not exist, of course; heat flows only if there is a difference in temperature between two spots. But this does not matter. We can just think of the temperature difference as being "infinitesimally small," "negligible," or whatever. Physics—remember the "perfect gas"—needs such idealized "limit" situations to formulate manageable "laws," which can then be applied, and adapted if need be, to the real world.

The assumption of *isobaricity* (constant pressure; Greek *baros*, weight) means that, if the change affecting the system is such as to raise or lower its internal pressure, the system is free to expand or contract at constant pressure. This is an idealized situation; in practice, a little extra pressure is needed to set off such a volume change. The corresponding work performed or received by the system is taken to be automatically compensated by an equivalent flow of heat into or out of the system through the heat exchanger just mentioned. This heat is automatically included in the bookkeeping.

The Two Sources of Free Energy

Systems that perform work generally draw mostly on their own supply of *internal energy.* We can readily find out how much energy they have available for doing the work by uncoupling the transformation that supplies the energy from the work process that utilizes it—the equivalent of disengaging a clutch and letting a motor idle. The energy then appears as heat. In chemistry, it is called the *heat of reaction* and is represented by ΔH.

In agreement with the general convention followed in thermodynamics, ΔH is defined from the point of view of the system; it represents the change in internal energy—more precisely, but less intuitively, *enthalpy* (Greek *thalpein,* to heat) for a system at constant pressure—suffered by the system in association with its transformation. If ΔH is positive, this change is a gain: energy must be supplied from the outside for the transformation to occur at constant temperature; the transformation is *endothermic.* If ΔH is negative, the system loses energy; the transformation is *exothermic.*

Should a system's capacity to work depend simply on an adequate supply of internal energy, as our personal experience of physical toil might make us believe, there would be no need to introduce the concept of free energy and to distinguish between ΔG and ΔH. But a second factor, much more elusive, may allow the system to draw additional free energy from external heat or, on the contrary, may force it to supply heat to the outside at the expense of its own internal energy. This factor confronts us with the more subtle notions of thermodynamics—namely, the nature of heat as a *degraded* form of energy and, especially, the concept of *entropy* (Greek *tropê*, turn).

Heat, or caloric energy, is the sum of the individual motion energies of all the atoms and molecules that compose a system. It is, thus, essentially kinetic in nature, as is the energy of a moving projectile. But it is random instead of directed. This is an absolutely fundamental difference, which explains why heat so often lies at the end of Nature's one-way streets. It can always be generated, but it is not readily harnessed. A moving projectile can dissipate its entire kinetic energy and convert it into heat—upon hitting an armor-plated surface, for example; nobody has ever seen a bullet picking up speed by cooling. Yet the steam engine works, and its offspring, the science of thermodynamics, was not misnamed.

Indeed, the founders of thermodynamics have solved the problem for us: it is a question of degree of randomness. There is something in the very fabric of the universe that favors randomness over regularity, chaos over order, disorganization over structure. This is so obvious, so much a matter of everyday experience, that it hardly needs demonstrating. Actually, it is very easy to demonstrate, at least in principle, by simple statistical reasoning. If you consider the various ways in which the components of a system—a collection of bricks, for example—can be arranged, you will find invariably that the possible random configurations are more numerous than the orderly ones. Thus disorder is favored, not because of some inherent sloppiness of the universe, but simply because disorder is more probable—that is, has more ways of being realized—than order.

But probability is a quantitative concept that can be evaluated. And so, therefore, must randomness be: some things are more random than others—bricks can be scattered, heaped, stacked, or assembled into a house. Thanks to the occurrence of such differences, random agitation can be converted into directed motion and made to work, as it is in a steam engine, provided more randomness is generated than is consumed in the process. Substitute the word entropy for randomness and you have the condition that limits the conversion of heat into kinetic energy or into any other kind of energy. Such a conversion is possible to the extent that the overall entropy increases or, at least, does not decrease. The italic letter S being the accepted symbol for entropy, this rule is expressed by the following equation, which is an alternative formulation of the second law:

$$\Delta S \geqslant 0 \tag{4}$$

This equation tells us that entropy cannot decrease spontaneously. Order cannot arise out of chaos without help. The latter qualification is important. As with the ban against the increase of G, we must clearly define the limits of the system.

The relationship of entropy to heat is given by the following formula:

$$\Delta S = \frac{Q}{T} \tag{5}$$

in which Q is a quantity of heat gained by a system, ΔS is the corresponding increase in entropy, and T the absolute temperature.

Without going into details, you can see from equation 5 that more entropy is gained by transferring a given quantity of heat at low than at high temperature. The trick, therefore, to making a heat engine is to remove heat from a hot source and transfer just enough of it to a cold collector to satisfy equation 4. The remainder can now be converted into work, provided you have arranged to make the transfer process contingent on such a conversion. A heat engine is a device of this sort; it is a machine in which the conversion of heat into work is harnessed to heat transfer. If Q is the heat given out by the hot source, Q' the heat gathered by the cold collector ($Q - Q'$ being the work obtained), and T and T' the absolute temperatures of the source and collector, respectively, you can easily find, by applying equations 4 and 5, that the fraction of the total heat produced that is converted into work is given by:

$$\text{Yield} = \frac{\text{work}}{\text{heat}} = \frac{Q - Q'}{Q} \leqslant \frac{T - T'}{T} \tag{6}$$

Principle of heat engine. Of the Q calories delivered from the combustion chamber to the transducer at a temperature of T K, Q' are transferred to the cooler at T' K, and the difference $Q - Q'$ is converted into work. The maximum work obtainable is determined by the condition that

$$\Delta S = -\frac{Q}{T} + \frac{Q'}{T'} \geq 0$$

or

$$Q' \geq \frac{T'}{T} \times Q$$

For example, if T is 120°C (393 K) and T' is 20°C (293 K),

$$Q' \geq \frac{293}{393} \times Q = 0.75\,Q$$

At least 75 per cent of the heat delivered to the transducer must be returned to the cooler. As much as 25 per cent of this heat may be converted into work. If T is 300°C (573 K), almost 50 per cent of the heat can be converted into work.

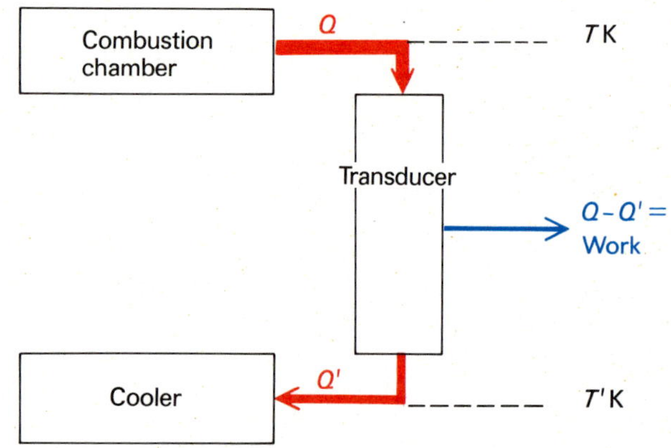

Thus, the yield, or efficiency, of a heat engine depends on the difference between the two temperatures. With T' limited to ambient temperature, you can gain only by increasing T. The internal combustion engine is superior to the steam engine in this respect. And Mr. Diesel achieved world fame by inventing an engine capable of running at a higher temperature than the ordinary internal combustion engine. On the other hand, without a temperature difference, there can be no conversion of heat into other forms of energy. A monothermic heat engine is impossible.

This statement disqualifies living systems as possible heat engines but not, however, as entropy engines. Randomness, it must be stressed, can be generated by many ways other than molecular agitation. Sugar is more random in a cup of coffee than in a lump; amino acids are more random in a mixture than

Two types of engine. In the internal energy engine, the system loses internal energy ($\triangle H < 0$) in the transformation from the initial state to the final state, and the energy lost is converted more or less completely into work. In the entropy engine, the internal energy content of the system does not change, but its entropy (degree of randomness) increases in the passage from the initial state to the final state. This transformation is harnessed to the conversion of external heat into work. The decrease ($\triangle S'$) in the entropy of the environment associated with this conversion is compensated by the increase ($\triangle S$) in the entropy of the system. Thus the maximum work obtainable from external heat in this manner is $T\triangle S$ (total entropy change $\triangle S + \triangle S' = 0$).

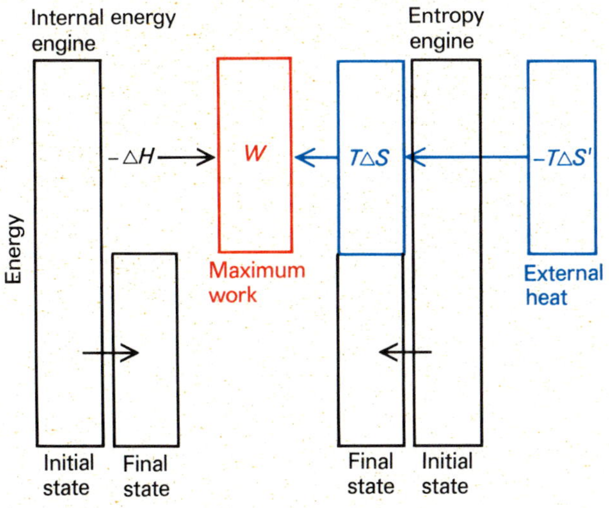

in a protein. An entropy engine is a device whereby a change that increases randomness—that is, has a positive ΔS—is coupled to the conversion of heat into work. If the system is isolated, it will cool, as does ether when it evaporates or a stretched rubber band that is allowed to contract. If the system is isothermic, as in our case, it will gain the corresponding amount of heat from the outside. Therefore, an isothermic entropy engine converts external heat into work, thanks to the increase of its internal entropy. These relationships work both ways. If the change undergone by the system has a negative ΔS, the entropy debt must be paid for by the conversion of internal energy into heat and its transfer to the outside world.

In connection with these heat exchanges, the entropy of the outside world suffers a change, $\Delta S'$, the amount of heat transferred being $T\Delta S'$ (equation 5). Thus, the total entropy change of system plus outside world is $\Delta S + \Delta S'$. According to equation 4, this total change in entropy cannot be negative. Therefore:

$$\Delta S + \Delta S' \geq 0 \tag{7}$$

If ΔS is positive and $\Delta S'$ negative (as with the entropy engine), $T\Delta S$ is the maximum amount of outside heat that can be used for work. If ΔS is negative and $\Delta S'$ positive, $T\Delta S$ is the minimum amount of internal energy that the system must convert into heat and transfer to the outside.

If we combine the contributions of internal energy (enthalpy) and entropy to the capacity of a system to carry out work by undergoing a given transformation, we arrive at the definition of the free-energy change associated with the transformation (free energy of reaction):

$$\Delta G = \Delta H - T\Delta S \tag{8}$$

As long as the entropy component is null or small, the free-energy change is about the same as the internal-energy change (internal-energy engine). But, if ΔS becomes appreciable, the relation between ΔG and ΔH changes to the point that a reaction can be exergonic (ΔG negative) in spite of being endothermic (ΔH positive), or endergonic in spite of being exothermic.

If the system undergoes the change without carrying out any work, ΔH is made entirely of heat given out to (ΔH negative) or received from (ΔH positive) the outside world. Then, $\Delta H = -T\Delta S'$, in which $\Delta S'$ is the entropy change of the outside world. Therefore, in the absence of work:

$$\Delta G = -T(\Delta S + \Delta S')$$

which, by virtue of equation 7, reduces to equation 1:

$$\Delta G \leq 0$$

As you can see, equations 1 and 4 indeed express the same principle. They are different formulations of the second law of thermodynamics. But beware of confusion. In equation 4, ΔS refers to the entropy change of a completely isolated system (or of the universe). In equation 8, on the other hand, ΔS represents the entropy change of a closed system capable of exchanging heat with the outside. To find out what happens to the entropy of the universe, we must include, as we have done, the change $\Delta S'$ in the entropy of the outside world.

A last remark. As already pointed out, the equality sign in every formula derived from the second law (see equations 1, 2, 4, 6, and 7) corresponds to an ideal limit situation. In the real world, entropy always increases when something happens; the work we accomplish is always less than the energy we spend; some energy is always degraded to heat. There is an "entropy tax" levied on every energy transaction. This explains why time machines are impossible: the direction of entropy increase defines the arrow of time; there is no going back. Sometimes it is also said that the universe runs on its reserve of negative entropy, or *neguentropy.* When this runs out, nothing will happen any more, except statistical fluctuations. This gloomy prediction need hardly worry us, however. The sun alone has, we are told, another 5 billion years to go.

The Many Currencies of Energy and Work

Energy and work take many forms—most of them encountered in the living world—that are all interconvertible within the framework of the first law of thermodynamics, subject to the constraints imposed by the second law. Each form of work, or kind of energy, is defined in terms of its own internal parameters and evaluated in corresponding units.

Mechanical work is accomplished by a force that displaces its point of application a certain distance in the direction of action of the force. It is the product of the force times distance and is expressed in ergs (dynes × centimeters), joules (10^7 ergs), kilogram-meters (9.81 joules), foot-pounds (1.36 joules), or any other useful combination of the variables.

A related concept is *power,* which is the work performed, or energy consumed, per unit of time. The unit of power is the watt (one joule per second). The horsepower (550 foot-pounds per second, or 748 watts) and its French equivalent, the "cheval-vapeur" (75 kilogram-meters per second, or 736 watts, or 0.98 horsepower) illustrate the bewildering parochialism that still dominates the world's energy currency system.

Electric work depends on an electric charge moving across a potential difference and is given in terms of these two variables. It is expressed in joules (coulombs × volts) or in electron-volts (1.6×10^{-19} joules), which, multiplied by Avogadro's number ($N = 6.023 \times 10^{23}$), gives Faraday-volts (96,500 joules), useful in electrochemistry. Electric energy is often also expressed as power multiplied by time. The most commonly used such unit is the kilowatt-hour (3.6×10^6 joules).

Electromagnetic energy, including that of visible *light,* is defined by the number of quanta or photons and by their energy, which is a function of the frequency (ν) or wavelength (λ) of the light, according to the well-known equation of Planck:

$$\text{Energy of one photon} = h\nu = h\frac{c}{\lambda} \tag{9}$$

in which h is Planck's constant (6.624×10^{-27} erg · sec), and c is the velocity of light (3×10^{10} centimeters per second). A number of photons equal to Avogadro's number is called an einstein. If wavelength is expressed in centimeters, the energy of one einstein is $12/\lambda$ joules.

The unit of heat, or *caloric energy,* is the calorie. It is the amount of heat required to raise the temperature of 1 gram of water from 15 to 16 degrees centigrade and is worth 4.186 joules. Its insular counterpart, the British thermal unit (BTU), raises the temperature of 1 pound of water at maximum density by 1 degree Fahrenheit. It equals 250 calories, or 1,046 joules.

Chemical energy in its various forms is traditionally expressed in calories, a heritage of thermochemistry and calorimetry. So is biological energy. Recently, however, the rulers of energy currency exchanges have decreed that henceforth the joule (symbol, J) is to be the official energy unit. But habits die hard, and few biochemists have as yet converted to the new system. This book is no exception. All energy values quoted in it are expressed in kilocalories (kcal). They can be readily converted into kilojoules (kJ) by multiplying by 4.186.

The Work of Molecular Transport

Translocation—by diffusion, for example—is the simplest molecular event that can occur, and it can serve as guide to more complex transformations. The associated free-energy change is evaluated by the change in the *chemical potential* of the substance undergoing translocation. Represented by μ, and expressed in kilocalories per gram-molecule (6.023×10^{23} molecules), the chemical potential is defined, in the legalistic jargon that must be used sometimes if rigor is to be respected, as the increase in free energy associated with the addition of an infinitesimally small amount of the substance (negligible with respect to the amount present) to an isothermic and isobaric system in which the substance is present at a defined partial pressure (if in gaseous phase) or concentration (if in solution).

Alternatively, and perhaps more intuitively, we may define the chemical potential as the increase in free energy associated with the addition of 1 gram-molecule of a given substance to a system—infinitely large, in fact—such that this addition does not appreciably alter the partial pressure, or concentration, of the substance in the system.

Both definitions are equivalent. They include the essential stipulation that the partial pressure or concentration of the substance must remain constant, because the chemical potential is a *function of the state of the system.* This relationship is given by an equation of absolutely fundamental importance, which dominates the whole field of chemical energetics:

$$\mu = \mu^\circ + RT \ln a \tag{10}$$

In this equation, μ° is a constant—characteristic for each substance—called the *standard chemical potential,* R is the gas constant (1.987×10^{-3} kcal per gram-molecule per degree), T is the absolute temperature in kelvins (K), ln is the natural logarithm (2.3 × the decadic logarithm), and a is the activity of the substance. At mammalian temperature (37°C = 310K), $RT \ln a = 1.417 \log a$, or, in first approximation, $1.4 \log a$.

Activity is another of those ideal limit concepts energetics is fond of. It is a measure of the effective abundance of the substance concerned, and corresponds exactly to its *partial pressure* (p) if it is a perfect gas, to its *concentration* (C) if it is an ideal solute. For real systems, it approaches partial pressure or concentration the more closely, the more rarefied the gas or dilute the solution. We will adopt these approximations for simplicity's sake and write:

$$\mu = \mu^\circ + RT \ln p \tag{11}$$

for a gas and:

$$\mu = \mu^\circ + RT \ln C \tag{12}$$

for a substance in solution.

The key message of these equations may be stated succinctly: *the more there is, the harder it is to add more.* Intuitively, we are not surprised. Still, there is something of a contradiction, considering that what we add (a certain number of molecules) is rigorously the same, whatever the partial pressure or the concentration. Therefore, the increase in the internal energy of the system associated with the addition should be—and is—the same under all circumstances. What is not the same, however, is the entropy change. The more molecules there are segregated in a given space, the greater the odds against an even greater degree of segregation, and the greater, therefore, in absolute value the (negative) entropy change associated with addition. This fact explains the activity-dependence of the chemical potential, expressed by equation 10. The chemical potential, it must be remembered, is a free-energy change, a function, therefore, of both internal energy and entropy, according to equation 8.

The logarithmic form of the second term of equations 10, 11, and 12 makes the increment in chemical potential additive for a multiplicative increase in abundance. A simple calculation shows that, at body temperature (37°C, 310 K), μ increases by about 0.42 kcal per gram-molecule for every doubling of partial pressure or concentration and by 1.4 kcal per gram-molecule for every tenfold increase in partial pressure or concentration. This increment is independent of the nature of the substance concerned—at least in the ideal world of theory, where molecules are taken to be either inert or too far apart to interact significantly with each other. In the real world things are different.

The nature of the substance does, however, enter into consideration with the standard chemical potential, μ°, which is a characteristic of each substance. It is defined as the value of μ when the substance is in the *standard state,* which is itself defined by the state in which the activity, partial pressure, or concentration equals unity; that is, its logarithm is equal to zero. In this event, indeed, equations 10, 11, or 12 reduce to $\mu = \mu^\circ$. Defining the standard state, therefore depends on the units we choose for expressing partial pressures or concentrations.

It is customary to express pressures in *atmospheres* and concentrations in *gram-molecules per liter* of solution (molarity). Therefore, μ° is the chemical potential of a gas at a partial pressure of 1 atmosphere, or of a solute at a concentration of 1 gram-molecule per liter (1 molar). We are perfectly free to choose other units if we so wish. This will simply mean that we give ourselves a different standard state and have to adjust the μ° values accordingly, so that the equation always gives us the same correct value of μ for a given partial pressure or concentration.

With this lengthy but necessary introduction we can now address our translocation problem. Imagine that a small amount of a substance passes from a compartment where its concentration is C_1 to one where its concentration is C_2. By definition of the chemical potential, the free energy of compartment 1 decreases by μ_1 kcal per gram-molecule, while that of compartment 2 increases by μ_2 kcal per gram-molecule. The overall free-energy change, expressed in the same units, is:

$$\Delta G = \mu_2 - \mu_1$$

or, by virtue of equation 12:

$$\Delta G = RT\,(\ln C_2 - \ln C_1) = RT \ln \frac{C_2}{C_1} \simeq 1.4 \log \frac{C_2}{C_1} \text{(at 37°C)} \tag{13}$$

It is the basic equation of molecular transport. Particularly simple because $\mu°$ cancels out in the operation, it tells in quantitive terms what we all know qualitatively from everyday experience—namely, that substances always move down concentration gradients. If C_1 is greater than C_2, ΔG is negative; transport can take place spontaneously. If C_1 is equal to C_2, $\Delta G = 0$; the system is at equilibrium. If C_1 is less than C_2, ΔG is positive; transport can occur only with the help of an external supply of energy equal at least to ΔG. It is the story of the sugar and the coffee and of countless other familiar phenomena. Both kinds of transport occur in living cells. They are called passive and active, depending on whether they can take place spontaneously or need to be driven. (Chapters 3, 13, and 14).

The Work of Ionic Transport

Equation 13 applies identically to electrically neutral and to electrically charged substances, or *ions* (within the limits of our ideal world; in the real world, ions of the same charge are less companionable than are neutral molecules because of the Coulomb repulsions between them). If an electric potential is interposed between the two compartments (membrane potential), a second term must be added to the free-energy equation to account for the electric work.

If z is the electric charge of the transported ion, then the total charge transferred with 1 gram-molecule of the ion is $z \times 96{,}500$ coulombs. If the membrane potential is represented by V (in volts), taken positively if compartment 2 is positive, then the electric work, in kilocalories per gram-molecule transported, is:

$$\text{Electric work} = \frac{96{,}500}{4{,}186} \cdot z \cdot V = 23\ z \cdot V \qquad (14)$$

and (at 37°C):

$$\Delta G = 1.4 \log \frac{C_2}{C_1} + 23\ z \cdot V \qquad (15)$$

The electric term may oppose transport (e.g., a positive ion moving against a positive potential) or it may help transport. Biology abounds in examples of both. The sodium-potassium pump, for instance (Chapter 13), forces Na^+ ions out of cells against a fifteen-fold difference in concentration (+1.65 kcal per equivalent) and against a membrane potential of the order of 0.07 volts (+1.61 kcal per equivalent), with a total cost of 3.26 kcal per equivalent. In contrast, the reverse transport of K^+ ions

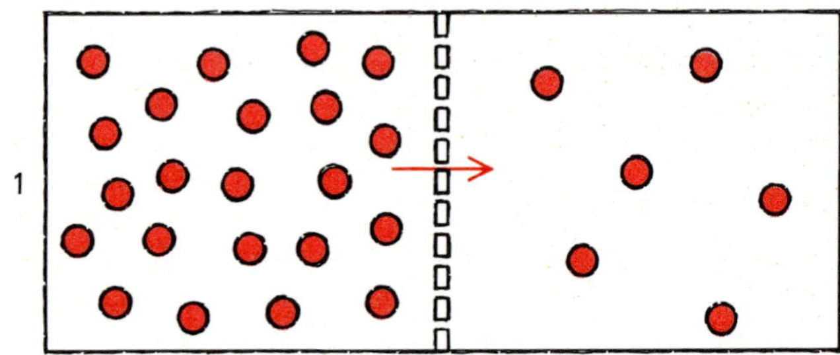

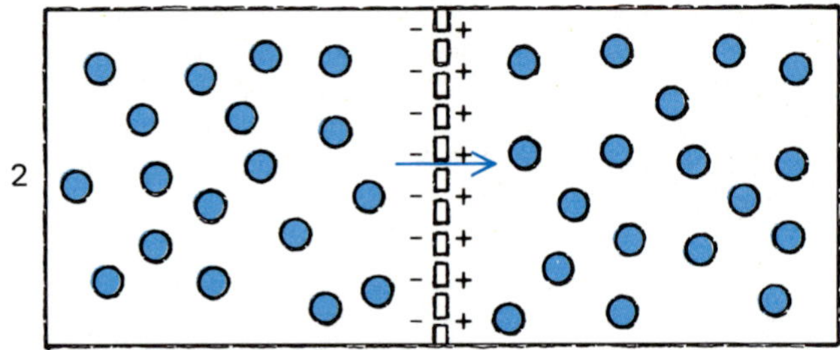

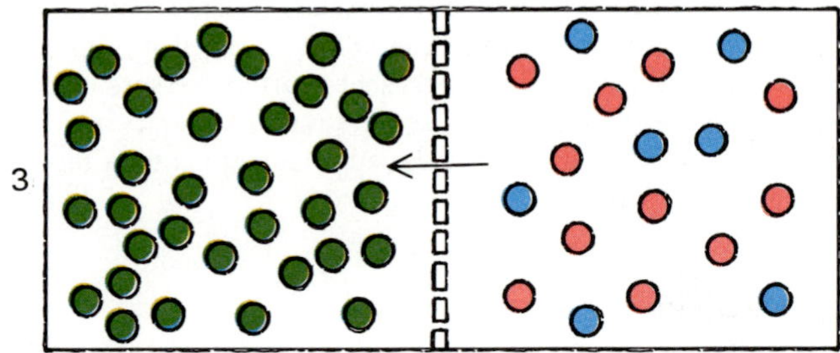

Some forms of chemical transport. Arrows show spontaneous direction ($\Delta G < 0$).

1. Simple molecular transport. A permeable membrane separates two regions of unequal chemical potential (partial pressure for gas or concentration for solute).

2. Ionic transport. In the example chosen, the chemical potential of the ion is the same in the two compartments. The sign of the membrane electric potential (V positive) determines the direction of movement of the ions, which are taken to be negatively charged (z negative).

3. Water transport through a semipermeable membrane impermeable to solutes. Total osmolalities on each side of the membrane determine the direction of water transport.

into the cells against a comparable gradient is helped by the membrane potential and costs essentially nothing. Or, more correctly stated, the active outward transport of sodium pays for itself and for the inward transport of potassium.

The Work of Water Transport

Theoretically, and also in practice, we can increase the concentration of a solution either by adding solute or by removing solvent. We tend to speak of *osmotic work* when water is transported rather than solute, as occurs, for instance, in the convoluted tubules of the kidneys.

The simplest form of osmotic work is accomplished when water is transported across a membrane that allows only water molecules, but none of the solute molecules, to pass through (semipermeable membrane). By the same reasoning as before, but applying equation 10 rather than equation 12, we find:

$$\Delta G = RT \ln \frac{a_2}{a_1} \tag{16}$$

in which a_1 and a_2 are the activities of water in the two compartments. It is customary in the thermodynamic analysis of solutions to express the activity of water by the *mole fraction*—that is, the number of molecules of water divided by the total number of molecules ($a = 1$ for pure water), which is related to the *osmolal* concentration of solutes, C', by:

$$a = \frac{55.56}{55.56 + C'} \tag{17}$$

In this formula, C' represents the sum of the molal concentrations of all the solutes present. Molal concentrations are expressed in gram-molecules per 1000 grams (55.56 gram-molecules) of water, and not per liter of solution (molarity), as we do elsewhere. For dilute solutions, molality and molarity are not very different. The prefix *os* (from osmotic) indicates that all concentrations have been added up.

Introducing equation 17 into equation 16, we find:

$$\Delta G = RT \ln \frac{55.56 + C'_1}{55.56 + C'_2} \tag{18}$$

which represents the free-energy change, in kilocalories per gram-molecule of water transported, associated with the transport of water from a compartment of osmolality C'_1 to one of osmolality C'_2.

It is easy to see that water transport is not a costly business in biological systems. For example, returning 1 liter of water from concentrated urine (1.3 osmolal) to blood (0.3 osmolal) requires no more than 0.6 kcal, which, therefore, also represents the amount of free energy that would be released by 1 liter of water flowing freely from urine to blood through a semipermeable membrane. Surprisingly, the pressure that would have to be applied on the blood side to oppose water transfer (*osmotic pressure*) is very large, of the order of 25 atmospheres. The paradox is only apparent, of course: 25 liter-atmospheres are, in fact, equal to 0.6 kcal.

In conclusion, large differences in osmotic pressure, such as those that keep the vacuoles of plant cells (tonoplasts) under tension and maintain turgidity, are not expensive to generate. The fundamental reason for this is that, because of the great abundance of water with respect to solutes in most biological systems, its thermodynamic activity suffers only small variations and remains close to that of pure water. In most bioenergetic calculations, the activity of water is, in fact, assumed to be constant and equal to unity (see p. 422).

Chemical Work

Formally, a chemical transformation can be treated in terms of translocation. Take the simple reaction:

$$\mathrm{A} \rightleftharpoons \mathrm{B}$$

From left to right, it is equivalent to adding molecules of B to the system and removing (adding negatively) the same number of molecules of A from the system. Therefore, by virtue of the definition of the chemical potential, the free-energy change associated with the transformation (*free energy of reaction*), expressed in kilocalories per gram-molecule of A consumed or of B formed, may be formulated as follows:

$$\Delta G = \mu_B - \mu_A$$

or, according to equation 12 (assuming the reaction takes place in solution):

$$\Delta G = \mu^\circ_B - \mu^\circ_A + RT (\ln [B] - \ln [A])$$

in which brackets are used to express concentrations. This relationship can be rewritten:

$$\Delta G = \Delta G^\circ + RT \ln \frac{[B]}{[A]} \tag{19}$$

Chemical reaction. Formally, a chemical reaction may be treated as chemical transport from and into outside reservoirs. In the example shown, B enters the system and A leaves it in identical amounts. Therefore, for the system itself (central compartment):

$$\Delta G = \mu_B - \mu_A$$

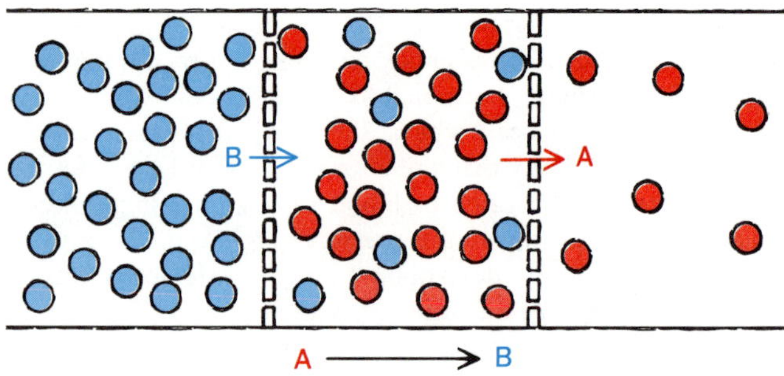

in which $\Delta G° = \mu°_B - \mu°_A$ = *the standard free energy of the reaction.*

We could consider the reaction from right to left in the same manner and would find exactly the same value of ΔG, but with the opposite sign. Therefore, there is necessarily a direction for which ΔG is negative; it is the exergonic, spontaneous direction. The reverse direction is endergonic and cannot be taken without an external supply of free energy. Note, however, that these directions are not immutable. Even if $\Delta G°$ is positive, ΔG can still be negative, and the spontaneous direction can be from left to right. All that is needed is for the logarithmic term in equation 19 to be negative ($[B] < [A]$) and greater in absolute value than $\Delta G°$. In contrast (assuming again that $\Delta G°$ is positive), the reaction will proceed spontaneously from right to left if the logarithmic term is positive ($[B] > [A]$) or if it is negative and smaller in absolute value than $\Delta G°$. If the two terms cancel each other out, $\Delta G = 0$, the system is at equilibrium.

We encounter the same entropy factor as we did in simple transport. The internal energy bill of making B from A is the same, whatever the state of the system. Not so the entropy bill. The more of B that is already present, the harder it is to make more, and vice versa.

By definition, $\Delta G°$ ($=\mu°_B - \mu°_A$) is the free energy of the reaction when both A and B are in the standard state. According to equation 19, it is also the free energy of the reaction in all situations in which the logarithmic term is equal to zero—that is, whenever $[A] = [B]$.

Note, finally, that at equilibrium ($\Delta G = 0$), we have:

$$\Delta G° = -RT \ln\left(\frac{[B]}{[A]}\right)_{eq}$$

But we know from chemical kinetics (law of mass action) that the *equilibrium constant*, K, of the reaction is none other than:

$$K = \left(\frac{[B]}{[A]}\right)_{eq}$$

Therefore:

$$\Delta G° = -RT \ln K \qquad (20)$$

This is an important relationship, which underlies a major method of determining free energies of reaction through the equilibrium constant. In our example, all we need to know are the concentrations of A and B at equilibrium.

The preceding notions can be generalized to every possible chemical transformation. Consider the following, entirely general, formulation of a chemical reaction:

$$m_A\ A + m_B\ B + m_C\ C + \cdots \rightleftarrows m_M\ M + m_N\ N + m_Q\ Q + \cdots$$

Extending our earlier reasoning, we readily find:

$$\Delta G = m_M\ \mu_M + m_N\ \mu_N + m_Q\ \mu_Q + \cdots - m_A\ \mu_A - m_B\ \mu_B - m_C\ \mu_C - \cdots$$

or, using the capital Greek sigma, Σ, for sum and representing the products (M, N, Q, . . .) of the reaction by P and the reactants (A, B, C, . . .) by R:

$$\Delta G = \Sigma m_P\ \mu_P - \Sigma m_R\ \mu_R \qquad (21)$$

Replacing the chemical potentials μ by their values (equation 13) and rearranging the terms of the equation according to the rules of logarithmic calculation, we obtain:

$$\Delta G = \Delta G° + RT \ln \frac{\Pi[P]^{m_P}}{\Pi[R]^{m_R}} \qquad (22)$$

in which:

$$\Delta G° = \Sigma m_P\ \mu°_P - \Sigma m_R\ \mu°_R \qquad (23)$$

and the capital Greek pi, Π, stands for product.

Equation 22 is the generalized form of equation 19. It reduces identically to equation 20 at equilibrium. In it, ΔG and $\Delta G°$ are expressed in kilocalories per m_A gram-molecules of A, or m_B of B, m_C of C, etcetera, consumed, or per m_M gram-molecules of M, or m_N of N, m_Q of Q, etcetera, formed—that is, in kilocalories per *stoichiometric equivalent* (Greek *stoikheion*, element) of reactant consumed or of product formed.

Living cells are chemical machines; they carry out thousands of reactions, many of which play key roles in providing free energy, or using it, for the performance of work through coupled systems or transducers. If we wish to understand these mechanisms, which is the object of bioenergetics, we must know the ΔGs of the reactions. For this purpose, biochemists have determined a large number of ΔG° values from equilibrium studies (equation 20) or by other methods. Lists of these values can be found in most textbooks, usually corrected for hydrogen-ion concentration, which, in living cells, is of the order of 10^{-7} equivalents per liter (pH 7.0), or 7 orders of magnitude away from the standard state, corresponding to a decrease in chemical potential by $1.4 \times 7 = 9.8$ kcal per equivalent.

Even so corrected, these $\Delta G^{\circ\prime}$ values are rarely directly applicable to physiological situations because the concentrations of metabolites and cofactors in living cells are all much lower than standard. They range from about 0.01 gram-molecule per liter, for the most abundant substances, to concentrations that may be as low as 10^{-6} gram-molecule per liter or lower. Furthermore, these concentrations vary according to circumstances; in addition, they may require fairly extensive corrections before they can be substituted for the activities of the substances because the situation in living cells is very different from that of the very dilute solution that must be assumed if we are to use equation 12 instead of the rigorously correct equation 10.

For these reasons it is extremely difficult to know any physiological ΔG value accurately. But in many cases an estimate of the normal range is available. The ΔG values quoted in this book, referred to as "physiological," are based on such estimates, as is explained at the end of Chapter 7 and elsewhere.

The Work of Electron Transfer

The most striking revelation of bioenergetics has been the universal importance of electron transfer as almost exclusive purveyor of energy throughout the living world. Such reactions also occur in the nonliving world. Their analysis, following the development of *electric cells*, provided classical thermodynamics with a new important chapter, which turned out to be particularly useful for the understanding of biological mechanisms.

Consider what happens when metallic zinc is exposed to sulfuric acid:

$$\mathrm{Zn} + \mathrm{H_2SO_4} \longrightarrow \mathrm{ZnSO_4} + \mathrm{H_2}\nearrow$$

The zinc dissolves as zinc sulfate and hydrogen gas is evolved. Actually, the sulfate ion does not participate in the reaction, which is really a transfer of electrons from zinc atoms to hydrogen ions:

$$\mathrm{Zn} + 2\,\mathrm{H^+} \longrightarrow \mathrm{Zn^{2+}} + \mathrm{H_2}$$

This becomes particularly clear when the preceding reaction is broken down formally into two half-reactions:

$$\mathrm{Zn} \longrightarrow \mathrm{Zn^{2+}} + 2\,e^-$$
$$2\,\mathrm{H^+} + 2\,e^- \longrightarrow \mathrm{H_2}$$

By definition—a historical consequence of the powerful electron-stripping ability of oxygen—removal of electrons, as undergone by zinc, is called an *oxidation*; acquisition of electrons, as by the hydrogen ions, is a *reduction*. Electron transfer is an *oxidation-reduction*.

The reaction between zinc and protons is exothermic. Under standard conditions, it produces some 35 kcal per gram-atom of zinc oxidized (65 grams) or per pair of proton-equivalents reduced (2 grams). Most of this energy is free energy ($\Delta H \simeq \Delta G$); it need not appear as heat and could be used for work. This is achieved by the physical separation of the two half-reactions in two distinct half-cells in such a manner that the electrons delivered by the zinc can reach the hydrogen ions only through an outside conductor. Any electric transducer can then be interposed in the pathway of the electrons, thereby coupling electron transfer to the performance of work. All electric cells function on this principle.

The maximum amount of work, W, that can be accomplished is readily computed from the difference in electric potential, E (volts), existing between the zinc terminal and the hydrogen terminal. Expressed in kilocalories (4,186 J) per pair of electron-equivalents ($2\,\mathcal{F} = 2$ Faradays $= 2 \times 96{,}500$ coulombs):

$$W = -\frac{2\,\mathcal{F}}{4{,}186}\,E = -46\,E \qquad (24)$$

By definition, $W = -\Delta G$. In consideration of equation 22, we find:

$$-W = \Delta G = \Delta G^\circ + RT\ln\frac{[\mathrm{Zn^{2+}}]\,p_{\mathrm{H_2}}}{[\mathrm{Zn}]\,[\mathrm{H^+}]^2} \qquad (25)$$

In the particular case where the components of the hydrogen half-cell are in the *standard state* ($p_{\mathrm{H_2}} = 1$ atmosphere; $[\mathrm{H^+}] = 1$ equivalent per liter), we obtain:

$$-W = \Delta G = \Delta G^\circ + RT\ln\frac{[\mathrm{Zn^{2+}}]}{[\mathrm{Zn}]} \qquad (26)$$

Schematic representation of electron transfer in an electrochemical cell. The transfer of an electron from A^- to B can take place only by way of the electrodes and of the outside conductor. The free energy of the transfer can be harnessed to do work. It can be estimated from the electric potential difference of the cell.

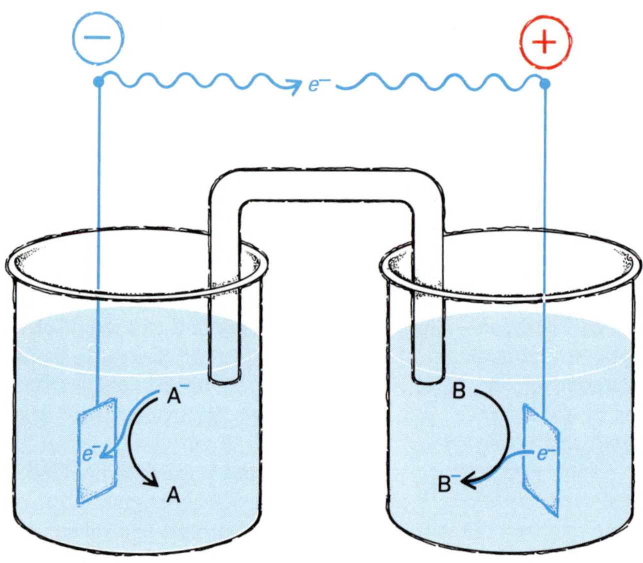

which, combined with equation 24, gives:

$$E = \frac{4{,}186\ \Delta G^\circ}{2\ \mathcal{F}} + \frac{4{,}186\ RT}{2\ \mathcal{F}} \ln \frac{[Zn^{2+}]}{[Zn]} \tag{27}$$

or:

$$E = E^\circ + \frac{R'T}{2\ \mathcal{F}} \ln \frac{[Zn^{2+}]}{[Zn]} \tag{28}$$

in which R' is the gas constant expressed in joules per degree per gram-molecule. (Earlier, the gas constant R was expressed in kilocalories per gram-molecule per degree; $R' = 4{,}186\ R = 8.317$.) Note that [Zn] does not have to figure explicitly in this equation because the thermodynamic activity of a solid metal is equal to unity under all conditions.

By definition (i.e., when measured against a standard hydrogen half-cell), E, as given by equation 28, is the *oxidation-reduction potential* of the Zn/Zn^{2+} couple, and E° is its standard oxidation-reduction potential. In general, for any *redox couple*:

$$\text{Red} \rightleftharpoons \text{Ox} + n\ e^-$$

Equation 28 becomes:

$$E = E^\circ + \frac{R'T}{n\ \mathcal{F}} \ln \frac{[\text{Ox}]}{[\text{Red}]} \tag{29}$$

It is the Nernst equation.

One of the main advantages of oxidation-reduction potentials is that they allow a very easy determination of the ΔG for any electron-transfer reaction between redox couples whose standard oxidation-reduction potentials are known. Consider the following general reaction:

$$A_{red} + B_{ox} \rightleftharpoons A_{ox} + B_{red}$$

which decomposes to:

$$A_{red} \rightleftharpoons A_{ox} + n\ e^-$$
$$B_{ox} + n\ e^- \rightleftharpoons B_{red}$$

From left to right, in consideration of equations 24 and 25:

$$\Delta G = \frac{n\ \mathcal{F}}{4{,}186} E_{(A-B)} \tag{30}$$

in which $E_{(A-B)}$ is the potential difference that would exist between the A and B half-cells should we build an electric cell with these two couples. Actually, we need not do that if we know the oxidation-reduction potentials, E_A and E_B, of A and B. Then, for obvious reasons (both E_A and E_B have been measured against the same standard hydrogen half-cell):

$$E_{(A-B)} = E_A - E_B$$

or, by virtue of equation 29:

$$E_{(A-B)} = E^\circ_A - E^\circ_B + \frac{R'T}{n\ \mathcal{F}} \ln \frac{[A_{ox}]\ [B_{red}]}{[A_{red}]\ [B_{ox}]}$$

which, combined with equation 30, gives:

$$\Delta G = \frac{n\ \mathcal{F}}{4{,}186} (E^\circ_A - E^\circ_B) + RT \ln \frac{[A_{ox}]\ [B_{red}]}{[A_{red}]\ [B_{ox}]}$$

It follows, in view of equation 22, that:

$$\Delta G^\circ = \frac{n\ \mathcal{F}}{4{,}186} (E^\circ_A - E^\circ_B) = n \cdot 23 \cdot (E^\circ_A - E^\circ_B) \tag{31}$$

Standard oxidation-reduction potentials are known for most important biological redox couples. They are listed in textbooks, generally as $E^{\circ\prime}$ (i.e., corrected for the hydrogen-ion concentration, as are $\Delta G^{\circ\prime}$ values, p. 419).

In bioenergetics, it is often useful to know the total amount of free energy that can be made available by the oxidation (defined as the removal of one pair of electrons) of a given substance, with atmospheric oxygen as electron acceptor. This is estimated from the oxidation-reduction potential of oxygen, which, under "physiological" conditions, is of the order of +0.805 volts. Therefore:

$$\Delta G_{ox} = \frac{2\,\mathcal{F}}{4{,}186}(E - 0.805) = 46\,E - 37 \qquad (32)$$

in which E is the "physiological" oxidation-reduction potential of the redox couple under consideration. In Chapter 7, ΔG_{ox} is referred to as the *electron potential*, not to be confused with oxidation-reduction potential (it is measured in kilocalories per pair of electron-equivalents, not in volts).

The Work of Generating Electricity

Electricity is always created in biological systems through an unequal distribution of electric charges on the two sides of a membrane. The maintenance of this unequal charge distribution is ensured by the lipid bilayer, and inserted proteins make up the necessary channels, gates, and pumps (Chapters 3 and 13). One can imagine charge separation to be caused either by electron transport or by ion transport. In practice, it seems that only ion transport is directly involved, although the two processes may be intimately linked, as in the chemiosmotic transducers of mitochondria (Chapter 9) and chloroplasts (Chapter 10).

By definition, an *electrogenic* ion pump is an energy-driven transport system that translocates ions across a membrane impermeable to these ions (otherwise the pump is short-circuited) without built-in charge compensation (such as the obligatory accompaniment by a counter-ion of opposite sign, or exchange for an ion of the same sign). An electrogenic pump may, however, drive the transport of a compensating ion by means of the membrane potential it generates. For this to happen, the membrane must contain a gate or a carrier (ionophore) for the ion transported in this way (Chapters 9 and 13).

The main electrogenic pumps transport sodium ions or protons; but there may be others. The potentials they generate vary from a range of 50 to 70 mV (plasma membrane) up to as much as 200 mV or more (mitochondria, chloroplasts). *Currents* are produced as a result of a local disturbance that renders the membrane temporarily permeable to the electrogenic ion. Depolarization is transmitted laterally along the membrane—of an axon, for instance—by some sort of domino effect. In the electric organs of certain fishes, such as *Torpedo*, or the electric eel, *Electrophorus*, electricity-generating cells (electroplaxes) are organized so as to link their potentials in series in the form of batteries capable of producing discharges of several hundred volts.

The energetic cost of these various activities is paid in the currency of ion transport (see p. 416).

The Work of Biosynthesis

As explained in Chapter 8, most of the innumerable endergonic assembly reactions involved in the construction of the complex constituents of living cells are driven by exergonic processes to which they are coupled by *group-transfer* reactions. The energy bookkeeping of such processes is done by means of the *group potentials*—that is, the free energies of hydrolysis, ΔG_{hy}, of the substances involved.

Consider the following B-transfer reaction:

$$\text{A—B} + \text{CH} \rightleftharpoons \text{AH} + \text{B—C}$$

We decompose it formally into the hydrolysis of A—B and the dehydrating condensation (reversal of hydrolysis) of B—C:

$$\text{A—B} + H_2O \rightleftharpoons \text{AH} + \text{BOH}$$
$$\text{B—OH} + \text{CH} \rightleftharpoons \text{B—C} + H_2O$$

and write:

$$\Delta G_{(\text{transfer})} = \Delta G_{\text{hy (A—B)}} - \Delta G_{\text{hy (B—C)}}$$

This elementary accounting system can be extended to sequential group transfers and coupled processes of any sort of complexity. It also serves to weigh group transfer against electron transfer and to quantitate oxidative phosphorylation. Most biologically important group-potential values have been determined. They are usually listed as $\Delta G^{\circ\prime}_{hy}$, or standard free energies of hydrolysis corrected to physiological pH.

Even so corrected, such values can be misleading. The value of $\Delta G^{\circ\prime}_{hy}$ for the hydrolysis of ATP to ADP and inorganic phosphate is of the order of −7.0 kcal per gram-molecule. The real value, in the mitochondria of living cells, may reach twice this amount (Chapter 9). This is because the [ATP] to [ADP] ratio may exceed 100, and the concentration of inorganic phosphate, $[P_i]$, may fall as low as 0.001 gram-molecule per liter. Introduced into equation 22, these values give:

$$\Delta G = \Delta G^{\circ\prime} + RT \ln \frac{[\text{ADP}]\,[P_i]}{[\text{ATP}]}$$

$$= -7.0 + 1.4 \log \frac{0.001}{100} = -14.0$$

Note that the concentration (activity) of water is not taken into account explicitly in the computation of group potentials, even though water is the key reactant in all hydrolysis reactions. As discussed earlier (p. 417), the activity of water varies very little in biological systems. In view of this fact, it is assumed to be invariable, at a value close to the activity of pure water, and is included in the $\Delta G^{\circ\prime}$ term. This is legitimate as long as we are dealing with aqueous systems. Things are different in nonaqueous media. For hydrolysis reactions taking place in an oil droplet or in the lipid bilayer of a membrane, water may become a critical reactant in determining the direction of the reaction.

Photochemical Work

The conversion of light energy into work always requires as first step the *absorption* of a quantum of light, or photon, by a molecule, which is said to become *excited* in the process. Molecules can store energy in four different forms: linear motion (translation), rotation, vibration, and electronic excitation. The last three can vary only by discontinuous jumps (quanta) and are involved in the absorption of light. The rule is that a photon can be absorbed only if its energy content, $h\nu$, is such as to allow the absorbing molecule to jump to an authorized energy level. These energy levels are readily identified from the *absorption spectrum*. Each absorption band corresponds to an energy jump equal to $28{,}600/\lambda$ kcal per gram-molecule (equation 9, but with λ expressed in nanometers).

Compared with the ΔG range of biochemical reactions, the amounts of energy that molecules equipped with the appropriate light-absorbing chemical groupings (*chromophores*) can collect from sunlight are considerable. For visible light (λ = 800–400 nm), they vary between 36 and 72 kcal per gram-molecule. They exceed 100 kcal per gram-molecule in the ultraviolet region, where proteins (λ = 280 nm) or nucleic acids (λ = 260 nm) have absorption bands. There is, indeed, great power in sunlight, provided you can catch it and channel it through an effective transducer.

If the transducer is photochemical, it has to act fast because molecules do not remain excited very long. Usually, they lose their excitation energy at the first collision with another molecule and convert it into translation energy. All we are left with is useless heat. Sometimes—this happens with special chromophores called fluorophores—part of the absorbed light is re-emitted in the form of light of higher wavelength. This phenomenon, called *fluorescence*, is used on a large scale to convert "cold" ultraviolet light into visible light. It is also widely used in the study of cells (Chapter 12). In certain favorable arrangements, as in the chloroplast thylakoids (Chapter 10), the excited state may be transferred to a neighboring molecule with little loss of utilizable free energy.

In a relatively small number of cases, absorbed light may serve to drive a photochemical reaction. Most often this reaction is exergonic, and the role of light is simply to help molecules jump over the energy barrier (activation energy) that prevents the reaction from taking place (*photocatalysis*). Some photocatalytic reactions are chain reactions and have an explosive character. Sometimes the photochemical reaction is actually endergonic, and light provides the necessary ΔG. The fundamental reaction of photosynthesis is of this kind (Chapter 10). With photons of sufficiently high energy, molecules may break apart (excessive vibration) or lose electrons (excessive electronic excitation). The *free radicals* generated in this manner are highly reactive and may cause many transformations, including mutations.

Some photochemical transducers operate in the reverse direction. The molecules are excited chemically with the help of an exergonic reaction and fall back to ground level by emitting photons. Molecules of this kind are involved in biological *luminescence*.

The Work of Information

The Scottish physicist James Clerk Maxwell, the founder of the electromagnetic theory of light, is also remembered for throwing an embarrassing pebble in the entropy pond. "Imagine," he said, "two adjacent compartments filled with the same gas at the same temperature, communicating with each other by a small opening, where stands a 'demon' who allows only fast-moving molecules to pass in one direction, and slow-moving ones in the other. Soon, one compartment will be warmer, the other will be colder. Entropy will have decreased, order arisen out of disorder, with no other help than *information* at the molecular level."

It took almost a century before the paradox was solved: the demon, in order to obtain the information he needs, consumes neguentropy. There is a relationship between information and entropy, which is readily grasped when both concepts are expressed in terms of probability. Entropy, as we have seen, is a measure of randomness, which itself depends on the probability of the particular configuration considered. According to statistical mechanics, the relationship is a simple logarithmic one and is given by:

$$\Delta S = k \ln \mathcal{P} \tag{33}$$

in which ΔS represents the entropy change associated with the creation of a configuration of probability $\mathcal{P}$, and k is the Boltzmann constant ($R/N = 3.3 \times 10^{-27}$ kcal per degree).

Information, on the other hand, is measured in binary units, or *bits*, the number of +/− choices needed for its expression. If we represent information content by I, we may write:

$$\mathcal{P} = 2^{-I} \tag{34}$$

or:

$$I = -\log_2 \mathcal{P} = -1.45 \ln \mathcal{P} \tag{35}$$

Hence, according to equation 33:

$$\Delta S = -0.69\, kI \tag{36}$$

As an example, take the information content of the human haploid genome (Chapter 16). With 3×10^9 base pairs, and 2 bits per base pair, $I = 6 \times 10^9$ bits. Hence, $\Delta S = -13.7 \times 10^{-18}$ kcal per degree, and the minimum work ($-T\Delta S$) in instructing the synthesis of one molecule at body temperature is 4.24×10^{-15} kcal, or 2.55×10^9 kcal per gram-molecule. This can be compared with the minimum chemical work involved in assembling 3×10^9 pairs of nucleotides into a double-stranded DNA ($\Delta G_{hy} \simeq -12$ kcal per gram-molecule for the phosphodiester bonds of DNA), or $3 \times 10^9 \times 2 \times 12 = 72 \times 10^9$ kcal per gram-molecule. The cost of information is small with respect to that of assembly, but not negligible.

The Work of Not Changing

If a Ping-Pong ball is at rest on a table, no work is required to keep it there. But, if it is floating in air, we must perform work in a continuous fashion—by means of a water jet, for example—to prevent it from falling. Living organisms are, like the floating ball, away from equilibrium and require a continuous expenditure of energy to be maintained in this unstable *steady state*. This kind of work obviously depends on the intrinsic instability of the systems considered. The sodium-potassium pump must work harder to maintain a sodium-ion gradient across a leaky membrane than across a tight membrane. A cell with a high degree of autophagy must carry out more biosynthetic repair than a cell that engages in little self-destruction.

When pure maintenance work is performed, nothing actually changes in the system concerned. Therefore, the work consists simply in the conversion of a given amount of free energy into heat. It is given per unit of time (t) and is expressed as the *flux* of free energy supplied to the system, $\Delta G/\Delta t$, or as the flux of entropy, $\Delta S/\Delta t$, generated in the system, which is equal to the free-energy flux divided by the absolute temperature.

A new branch of thermodynamics, called nonequilibrium thermodynamics, or thermodynamics of irreversible processes, has been developed for the formal analysis of systems traversed by such fluxes. In practice, the fluxes themselves can be measured by determining the rate of heat production under conditions in which the systems may be assumed to perform no work other than maintenance work. Physicians try to approximate such conditions when they measure the *basal metabolic rate*. For an adult human being, it amounts to about 2,000 kcal per day, which is equivalent to the consumption of a 100-watt bulb. That is what each of us has to use up simply to keep alive.

Sources of Illustrations

page 7 (left)
Scanning electron micrograph courtesy of Dr. Brian J. Ford, Cardiff University.

page 7 (right)
Replica made by Carl Zeiss Inc., Thornwood, New York.

page 10
Reproduction of a portrait by Jacques Louis David, *Lavoisier and His Wife*. All rights reserved, The Metropolitan Museum of Art.

page 11
Reproduced from *The Journal of Experimental Medicine*, 1945, vol. 81, p. 233, fig. 2, copyright permission of The Rockefeller University Press.

page 13
Electrophoretic patterns courtesy of Dr. Yukio Fujiki.

page 14 (upper right)
Electrophoretic patterns courtesy of Dr. Yukio Fujiki.

page 14 (bottom)
Autoradiograph courtesy of Dr. James D. Jamieson and Dr. George E. Palade.

page 25
Electron micrograph courtesy of Dr. Eldon H. Newcomb.

page 27
Electron micrograph courtesy of Dr. David Chase.

page 29 (top)
Photography by Etienne Bertrand Weill. Arp, Hans. *Serpent* © by ADAGP, Paris 1984.

page 29 (bottom)
Photography by Edward Leigh/courtesy of Sir John Kendrew.

page 33
Adapted from R.E. Dickerson, in volume II, of *The Proteins* (H. Neurath, editor). Academic Press Inc. Copyright © 1964.

page 34
Scanning electron micrograph courtesy of Dr. Etienne de Harven and Nina Lampen. Sloan Kettering Institute, New York.

page 37 (top)
Illustration from *The Naturalist on the River Amazons*, by H.W. Bates (1864 by John Murray, London, reprinted 1962 by the University of California Press).

page 38
Electron micrograph courtesy of Dr. Jerome Gross.

page 42
From *Tissues and Organs: A Text-Atlas of Scanning Electron Microscopy*, by Richard G. Kessel and Randy H. Kardon. W.H. Freeman and Company. Copyright © 1979.

page 44
Electron micrograph courtesy of Dr. Pierre Baudhuin and Dr. Pierre Guiot.

page 48 (left)
Picture courtesy of *Proceedings of the Royal Society of London*, vol. 66, plate 6.

page 54
Electron micrographs courtesy of Dr. Samuel C. Silverstein.

page 63
Electron micrograph courtesy of Helen Shio.

page 69
Electron micrograph courtesy of Dr. Pierre Baudhuin.

page 70
Electron micrographs courtesy of Dr. Marilyn Gist Farquhar.

page 73
Electron micrographs courtesy of Dr. Pierre Baudhuin.

page 76
Phase-contrast micrograph courtesy of Helen Shio.

page 83
Electron micrograph courtesy of Dr. George E. Palade.

page 88
Electron micrograph courtesy of Dr. George E. Palade.

page 91
Electron micrograph courtesy of Dr. George E. Palade.

page 92
Electron micrograph courtesy of Dr. Marilyn Gist Farquhar.

page 94
de Chirico, Giorgio. *The Anxious Journey.* 1913. Oil on canvas, $29\frac{1}{4} \times 42$ inches. Collection, the Museum of Modern Art, New York. Acquired through the Lillie P. Bliss Bequest.

page 105
Electron micrograph courtesy of Dr. George E. Palade.

page 107
Drawing by Louis Pasteur reproduced by René Dubos in his chapter "Pasteur and Modern Science," pp. 17–32, in *The Pasteur Fermentation Centennial 1857–1957,* copyright 1958 by Chas. Pfizer & Co., Inc.

page 150
Electron micrograph courtesy of Dr. Robert A. Bloodgood, University of Virginia.

page 151 (top)
Photograph of model courtesy of Dr. Hans-Peter Hoffmann and Dr. Charlotte Avers.

page 151 (bottom)
Reclining Figure (1), 1945, courtesy of Henry Moore.

page 152 (left)
Electron micrograph courtesy of Dr. George E. Palade.

page 157
Electron micrograph courtesy of Dr. Donald F. Parsons.

page 174 (bottom)
Electron micrograph courtesy of Dr. Leo P. Vernon.

page 175 (top)
Electron micrograph taken by W.P. Wergin/courtesy of Dr. Eldon H. Newcomb.

page 181 (top left)
Electron micrograph courtesy of Dr. Arvid B. Maunsbach.

page 181 (top right)
Electron micrograph courtesy of Dr. Pierre Baudhuin.

page 181 (bottom)
Electron micrograph courtesy of Dr. George E. Palade.

page 183
Electron micrograph courtesy of Dr. Eugene Vigil.

page 184
Electron micrograph courtesy of Dr. Phyllis Novikoff and Dr. Alex Novikoff.

page 185
Electron micrograph taken by Sue Ellen Frederick/courtesy of Dr. Eldon H. Newcomb. Published in E.H. Newcomb, "Ultrastructure and Cytochemistry of Plant Peroxisomes and Glyoxysomes," *Annals of The New York Academy of Science,* vol. 386, p. 230 (1982).

page 186
Electron micrographs courtesy of Dr. Marten Veenhuis. Published in J.P. Van Dijken, M. Veenhuis, and W. Harder, "Peroxisomes of Methanol-Grown Yeast," *Annals of The New York Academy of Science,* vol. 386, p. 201 (1982).

page 187
Electron micrograph courtesy of Dr. Marten Veenhuis.

page 189
Electron micrograph courtesy of Helen Shio.

page 190
Electron micrograph taken by Isabelle Coffens/courtesy of Dr. Pierre Baudhuin.

page 192
Fluorescence micrograph courtesy of Dr. Werner W. Franke.

page 199
Electron micrograph and model courtesy of Dr. Hugh Huxley.

page 201
Electron micrograph courtesy of Dr. John Heuser.

page 203
Electron micrograph courtesy of Dr. Hugh Huxley.

page 204
Electron micrograph courtesy of Dr. Hugh Huxley.

page 209
Electron micrograph courtesy of Dr. Keith R. Porter.

page 210
Light micrograph courtesy of Dr. L.E. Roth.

page 211
Electron micrograph courtesy of Dr. L.E. Roth and Dr. Y. Shigenaka, University of Tennessee, Knoxville, and Hiroshima University, respectively.

page 212
Light micrograph courtesy of Dr. Keith R. Porter.

page 213
Electron micrographs courtesy of Dr. Keith R. Porter.

page 214
Scanning electron micrograph courtesy of Dr. Sidney L. Tamm.

page 215
Electron micrographs courtesy of Dr. Ian R. Gibbons.

page 217
Electron micrograph courtesy of Dr. John Heuser.

page 218
Photograph courtesy of Buckminster Fuller Archives.

page 219
Electron micrographs courtesy of Dr. John Heuser.

page 221
Electron micrographs courtesy of Dr. John Heuser.

page 224
Electron micrograph courtesy of Dr. George E. Palade.

page 235
Electron micrograph courtesy of Dr. Morris Karnovsky. From M.J. Karnovsky, E.R. Unanue, and M. Leventhal. 1972. *Journal of Experimental Medicine* 136:907.

page 256
Photograph courtesy of the Vatican Museum.

page 260
Structure from Dr. Walter Fiers.

page 267
Photographs of model courtesy of Dr. Heinz-Gunter Wittmann.

page 281
Electron micrograph courtesy of Dr. Daniel Branton.

page 282
After Dr. Werner W. Franke.

page 283
Electron micrograph courtesy of Dr. Werner W. Franke.

page 287 (left)
Drawing by Frans Janssens originally published in *La Cellule* in 1909.

page 290
Diagrammatic illustration of DNA from J.D. Watson and F.H.C. Crick. 1953. Molecular Structure of Nucleic Acids, *Nature* 171:737.

page 291
Photograph of model courtesy of Dr. Maruice H.F. Wilkins, Biophysics Department, King's College, London.

page 294
Electron micrograph courtesy of Dr. Françoise Haguenau.

page 295
Electron micrograph courtesy of Dr. Oscar L. Miller.

page 296
Electron micrograph courtesy of Dr. Oscar L. Miller.

page 303
Phase-contrast micrograph courtesy of Dr. Joseph G. Gall.

page 304
Electron micrograph courtesy of Dr. Oscar L. Miller.

page 306
Electron micrograph courtesy of Dr. Oscar L. Miller. From O.L. Miller, B.A. Hamkalo, and C.A. Thomas. 1970. *Science* 169:392.

page 309
Electron micrograph courtesy of Dr. Pierre Chambon.

page 317
Light micrograph courtesy of Dr. Wolfgang Beermann.

page 328
Autoradiograph courtesy of Dr. John Cairns.

page 329
Electron micrograph courtesy of Dr. David S. Hogness.

page 333
After Dr. Ronald W. Hart.

page 334
Photographs courtesy of Dr. Karl F. Koopman, American Museum of Natural History.

page 338
Electron micrograph courtesy of Dr. David Dressler and Dr. Huntington Potter, Harvard University.

page 352
Scanning electron micrographs courtesy of Dr. Marcel Bessis.

page 354
Phylogenetic tree courtesy of Dr. Emanuel Margoliash.

page 362 (right)
Photograph courtesy of Dr. Ralph L. Brinster.

page 370 (right)
Electron micrograph courtesy of Dr. Ulrich K. Laemmli.

page 371
Karyotype courtesy of Dr. Herman Van den Berghe.

page 372
Electron micrograph courtesy of Dr. Etienne de Harven.

page 376
Light micrograph courtesy of Dr. Luc Waterkeyn.

Index

Page numbers in boldface type refer to illustrations; those in italic type refer to definitions, etymologies, and structural diagrams.